GEOLOGIC TIME

Millions of years before the present	Era	Period	Epoch	Duration in millions of years	The biological record	
65	Cenozoic	Quaternary	Recent	0.01		
225			Pleistocene	2	Rise of man; large mammals abundant	
600		Tertiary	Pliocene	5	Flowering plants abundant	Age of Mammals
	Mesozoic		Miocene	19	Grasses abundant; rapid spread of grazing mammals	
			Oligocene	12	Apes and elephants appear	
			Eocene	16	Primitive horses, camels, rhinoceroses	
			Paleocene	11	First primates	
		Cretaceous		71	First flowering plants; dinosaurs die out	Age of Reptiles
	Paleozoic	Jurassic		44	First birds; dinosaurs at their peak	
		Triassic		45	Dinosaurs and first mammals appear	
		Permian		45	Rise of reptiles; large insects abundant	
		Pennsylvanian		35	Large nonflowering plants in enormous swamps	
		Mississippian		45	Large amphibians; extensive forests; sharks abundant	
		Devonian		50	First forests and amphibians; fish abundant	
		Silurian		40	First land plants and air-breathing animals (scorpions)	
3,500		Ordovician		60	First vertebrates (fish) appear	
		Cambrian		100	Marine shelled invertebrates (earliest abundant fossils)	
	Precambrian time	Proterozoic		1,900?	Marine invertebrates, mainly without shells	
4,500		Archeozoic		2,000?	Marine algae (primitive one-celled plants)	

THE SOLAR SYSTEM

Body	Mean distance from sun, millions of miles	Mean diameter, thousands of miles	Relative mass, $m_{earth} = 1$	Period of revolution[a]	Period of rotation[b]	Mean density, g/cm³[c]	Acceleration of gravity, g[d]	Escape velocity, mi/sec[e]	Known satellites
Sun		865	333,000		25–30 days	1.4	28	383	(9 planets)
Moon	[f]	2.16	0.01	27⅓ days	27⅓ days	3.4	0.2	1.2	—
Mercury	36	3	0.05	88 days	59 days	5	0.4	2.7	0
Venus	67	7.8	0.82	225 days	243 days[g]	5.0	0.9	6.5	0
Earth	93	7.9	1.00	365 days	24 hr	5.5	1.0	7.0	1
Mars	142	4.2	0.11	687 days	24.6 hr	4.0	0.4	3.2	2
Jupiter	483	89	318	12 years	9.8 hr	1.3	2.7	38	12
Saturn	887	75	95	29 years	10.3 hr	0.7	1.2	23	10
Uranus	1,784	31	15	84 years	10.8 hr[h]	1.5	1.0	14	5
Neptune	2,795	28	17	165 years	15.7 hr	2.0	1.4	16	2
Pluto	3,672	4?	0.2?	248 years	6 days?	6?	?	?	0

[a] Orbital period.
[b] Spin period.
[c] Density of water = 1.0 g/cm³.
[d] 1 g = 32 ft/sec² = 9.8 m/sec².
[e] Speed necessary for permanent escape from gravitational field of the body.
[f] The mean distance of the moon from the earth is 238,000 miles.
[g] The direction of rotation of Venus is opposite to that of the other planets.
[h] The axis of rotation of Uranus is only 8° from the plane of its orbit.

FUNDAMENTALS OF PHYSICAL SCIENCE

FUNDAMENTALS OF PHYSICAL SCIENCE

Konrad B. Krauskopf
Professor of Geochemistry
Stanford University

Arthur Beiser
Formerly Associate Professor of Physics
New York University

Sixth Edition

McGraw-Hill Book Company

New York   St. Louis   San Francisco   Düsseldorf   Johannesburg   Kuala Lumpur
London   Mexico   Montreal   New Delhi   Panama   Rio de Janeiro
Singapore   Sydney   Toronto

FUNDAMENTALS OF PHYSICAL SCIENCE

Copyright © 1959, 1966, 1971 by McGraw-Hill, Inc. All rights reserved. Copyright 1941, 1948, 1953 by McGraw-Hill, Inc. All rights reserved. Printed in the United States of America. No part of this publication may be reproduced, stored in a retrieval system, or transmitted, in any form or by any means, electronic, mechanical, photocopying, recording, or otherwise, without the prior written permission of the publisher.

Library of Congress Catalog Card Number 76-152006

07-035440-5

5 6 7 8 9 10   VHVH   82 81 80

This book was set in Bodoni Book by York Graphic Services, Inc.; printed on permanent paper by Newman-Rudolph Lithographing Company, and bound by Von Hoffman Press, Inc. The designer was Richard Paul Kluga; picture editor, Gabriele Wunderlich; the drawings were done by John Cordes, J & R Technical Services, Inc.; cover photo by J. Paul Kirouac. The editors were Bradford Bayne, Hiag Akmakjian, and Andrea Stryker-Rodda. Peter D. Guilmette supervised production.

Contents

preface　　xi

Part One: MATTER IN MOTION

1　Motion　　2
　　how motion is described　　4
　　the acceleration of gravity　　11
　　analyzing motion　　15
　　glossary　　17
　　exercises　　18

2　Force and Motion　　21
　　inertia　　22
　　the role of force　　25
　　action and reaction　　33
　　vectors　　34
　　glossary　　39
　　exercises　　40

3　Gravitation　　43
　　circular motion　　44
　　universal gravitation　　47
　　the "scientific method"　　60
　　glossary　　63
　　exercises　　63

4　Energy and Momentum　　67
　　work　　68
　　energy　　71
　　momentum　　80
　　glossary　　85
　　exercises　　86

5　Relativity　　89
　　special relativity　　89
　　mass and energy　　99
　　general relativity　　101
　　glossary　　104
　　exercises　　104

Part Two: COLLECTIVE MATTER

6　Heat　　109
　　temperature　　110
　　heat　　113
　　the nature of heat　　118
　　glossary　　121
　　exercises　　122

7　Solids, Liquids, and Gases　　125
　　matter in bulk　　125
　　pressure and density　　130
　　the gas laws　　136
　　glossary　　140
　　exercises　　141

8　Kinetic Theory of Matter　　145
　　kinetic theory of matter　　146
　　heat　　149
　　second law of thermodynamics　　156
　　glossary　　160
　　exercises　　161

9 Waves   163
     wave motion   164
     wave behavior   168
     sound   171
     glossary   180
     exercises   181

Part Three: ELECTRICITY AND MAGNETISM

10 Electricity   185
     electric charge   186
     coulomb's law   189
     fields of force   193
     glossary   196
     exercises   196

11 Electricity and Matter   199
     the electron   200
     electrical properties of matter   202
     electric current   206
     glossary   215
     exercises   215

12 Magnetism   219
     permanent magnets   219
     the magnetic field   222
     magnetic forces   226
     electromagnetic induction   229
     glossary   234
     exercises   235

13 Light   237
     electromagnetic waves   237
     light   245
     wave nature of light   248
     glossary   253
     exercises   253

Part Four: THE ATOM

14 Waves and Particles   259
     quantum theory of light   260
     matter waves   268
     the uncertainty principle   271
     glossary   274
     exercises   275

15 Atoms and Molecules   277
     elements and compounds   277
     atoms and molecules   287
     atomic structure   292
     glossary   295
     exercises   296

16 The Nucleus   299
     nuclear structure   299
     nuclear energy   308
     glossary   316
     exercises   317

17 Elementary Particles   319
     radioactivity   320
     nuclear stability   325
     elementary particles   333
     glossary   338
     exercises   338

18 Theory of the Atom   341
     atomic spectra   341
     bohr model of the atom   345
     quantum theory of the atom   350
     glossary   356
     exercises   357

Part Five: ATOMS IN COMBINATION

19 The Periodic Law   361
     the periodic law   361

atomic structure 371
glossary 377
exercises 377
20 The Chemical Bond 379
molecular formation 379
the ionic bond 381
the covalent bond 388
glossary 394
exercises 395
21 The Solid State 397
crystal structure 397
bonding in solids 404
band theory of solids 411
glossary 414
exercises 415

Part Six: BASIC CHEMISTRY

22 Chemical Calculations 419
valence 419
chemical equations 424
stoichiometry 426
glossary 432
exercises 433
23 Eight Elements 435
eight elements 435
the language of chemistry 449
glossary 451
exercises 452
24 Ions and Solutions 455
electrolysis 459
solubility 462
ions in solution 468
glossary 469
exercises 469

25 Acids, Bases, and Salts 471
acids 471
bases 474
neutralization 477
glossary 486
exercises 486

Part Seven: CHEMISTRY IN ACTION

26 Chemical Reactions 491
chemical energy 492
reaction rates 500
chemical equilibrium 504
glossary 509
exercises 510
27 Oxidation and Reduction 513
oxidation-reduction reactions 514
oxidation number 519
metallurgy 526
glossary 530
exercises 531
28 Organic Chemistry 535
introduction 536
structures of organic molecules 538
carbon bonds 544
hydrocarbon derivatives 549
industrial organic chemistry 554
glossary 557
exercises 558
29 The Chemistry of Life 561
carbohydrates and lipids 561
proteins 565
nucleic acids 568
glossary 573
exercises 573

Part Eight: MATERIALS OF THE EARTH

30 Earth Materials   577
    silicates   578
    minerals   584
    glossary   591
    exercises   592
31 Rocks   595
    rocks   595
    soil   602
    glossary   605
    exercises   606
32 The Atmosphere   609
    the atmosphere   610
    weather   616
    climate   627
    glossary   631
    exercises   633
33 The Oceans   635
    properties of the oceans   635
    movements of the oceans   640
    glossary   648
    exercises   649

Part Nine: BASIC GEOLOGY

34 Erosion and Sedimentation   653
    erosion   654
    sedimentation   663
    groundwater   668
    glossary   670
    exercises   671
35 Vulcanism and Diastrophism   673
    vulcanism   674
    diastrophism   681
    glossary   688
    exercises   689
36 Within the Earth   691
    properties of the earth   692
    earthquake waves   697
    interior structure   699
    terrestrial magnetism   704
    glossary   709
    exercises   710

Part Ten: THE EVOLVING EARTH

37 Interpreting the Rock Record   715
    the principle of uniform change   716
    methods of historical geology   720
    geologic time   727
    glossary   731
    exercises   732
38 Continental Drift   735
    the floating crust   735
    continental drift   740
    sea-floor spreading   744
    glossary   752
    exercises   753
39 Earth History   755
    precambrian time   756
    the paleozoic era   758
    the mesozoic era   761
    the cenozoic era   765
    glossary   770
    exercises   770

CONTENTS

Part Eleven: THE SUN AND ITS FAMILY

40 The Solar System   775
- the family of the sun   776
- the planets   784
- glossary   792
- exercises   792

41 The Moon   795
- motions of the moon   796
- properties of the moon   800
- origin of the moon   813
- glossary   815
- exercises   815

42 The Sun   817
- tools of astronomy   817
- the sun   824
- glossary   832
- exercises   833

43 The Stars   839
- the stars in space   840
- stellar properties   845
- stellar evolution   849
- glossary   856
- exercises   857

44 Structure of the Universe   859
- our galaxy   859
- island universes   865
- cosmic rays   869
- glossary   876
- exercises   877

45 Evolution of the Universe   879
- the expanding universe   880
- evolution of the universe   885
- glossary   891
- exercises   892

MATHEMATICS REFRESHER

- algebra   893
- positive and negative quantities   895
- exponents   896
- equations   899
- powers of ten   901

glossary   905
answers to odd-numbered exercises   915
index   937

Preface

The impact of science on modern culture has been so strong and so pervasive that some degree of scientific literacy is required of a well-informed citizen of any of today's worlds, whether intellectual, political, economic, or social. Although different routes are possible to such literacy for a student whose primary interests lie elsewhere, a synthesis of several of the major scientific disciplines has much in its favor. In this way a broad perspective of the knowledge and insights that science has afforded about the universe can be presented, while the point of view of the scientist and his approach to nature are demonstrated in a variety of situations.

Physics, chemistry, geology, and astronomy form the subject matter of "Fundamentals of Physical Science." How can one book treat four such extensive sciences and still communicate something of their distinctive flavors as well as a real understanding of their component facts and theories? Our approach is a selective one that concentrates on those topics that are fundamental to each discipline and that shows where they fit into the fabric of physical science. Since we are not trying to prepare the reader for a career in science, we have been able to pass lightly over the more formal aspects of many subjects, in most cases without sacrificing an appreciation of the ideas involved. There is enough physics, chemistry, geology, and astronomy in the pages that follow, we believe, to provide more than a superficial survey.

It is not easy to decide how extensively to use mathematics in a book of this kind. On the one hand, the essence of all the physical sciences is their quantitativeness, their continual search for precision in measurement and theory, and nobody can claim acquaintance with science without some conception of how experiment and hypothesis are linked mathematically. On the other hand, not all the students who use this book have the training needed to follow extensive algebraic and trigonometric arguments, not to mention the need for calculus in many cases if a proper analysis is to be given. It is debatable whether the time required to familiarize the less-prepared student with mathematic methods might not better be spent on topics accessible without such familiarity. Our approach has been to employ mathematical reasoning only where there is no alternative, as in mechanics, or to underline notable concepts, such as the kinetic-molecular theory of matter. An instructor who feels his class is reasonably at home with basic mathematics can expand with profit upon the modest amount of formal material here, but another instructor faced with a less-qualified audience might find it more sensible to abbreviate much of this material. A Mathematics Refresher designed for self-study is given at the end of the book.

In preparing this edition of "Fundamentals of Physical Science" we have tried to strengthen its treatment of basic concepts and to expand its coverage of recent developments. For example, the material on atomic and nuclear structure and the chemical bond has been rearranged and rewritten, and there are new chapters on The Chemistry of Life, The Oceans, Continental Drift, and The Moon. The sequence of topics has been altered so that physics and chemistry are treated separately, with the theory of the chemical bond forming the transition between them; relativity is introduced early in the text; chemical calculations are grouped together in a single chapter; and so on. Aids to the student besides the Mathematics Refresher now include a glossary of new terms in each chapter as well as a complete glossary at the end of the book, and outline solutions are given for the odd-numbered exercises, not just numerical answers. Although British units have not been eliminated, metric units are emphasized to a greater extent than before.

We are grateful to those users of previous editions of "Fundamentals of Physical Science" who have been kind enough to send us comments on their experience with the book. Feedback of this kind is invaluable in meeting the needs of successive generations of students.

Konrad B. Krauskopf
Arthur Beiser

part one
MATTER IN MOTION

A body whose position changes with time is said to be moving.

We all know—or think we know—what motion is, but how many of us have thought about the different kinds of motion there are or just how bodies are set into motion and how their states of motion may be changed? We shall find that introducing the notions of inertia, force, mass, and weight helps organize our thinking and makes it possible to formulate three quite general laws of motion. In turn, these are the laws of motion that permitted Newton to interpret data on the solar system in terms of a single universal theory of gravitation.

Our study of mechanics continues with a look at three of the most important "conservation principles" in science, those concerned with energy, momentum, and angular momentum. We shall learn that it is possible to regard all motions, changes, and events of any kind as involving interchanges of energy from one form to another, a concept that is illustrated in almost every chapter of the book.

Finally we come to the theory of relativity. In relativity, our intuitive feelings about space and time, mass and energy are discarded in favor of a careful analysis of just what these quantities are in terms of how they are measured. The result is a series of relationships among space, time, mass, and energy, which had formerly been thought to be independent of one another.

chapter 1
Motion

HOW MOTION IS DESCRIBED

    speed
    converting units
    constant speed
    instantaneous speed
    velocity
    acceleration
    constant acceleration

THE ACCELERATION OF GRAVITY

    free fall
    air resistance
    galileo
    the meaning of equations

ANALYZING MOTION

    speed and acceleration
    distance and acceleration

We shall begin our study of physical science with an analysis of motion: what it is, how things move, and why they do so. For several reasons this is the logical point of departure from the casual realm of our daily lives to the precise realm of science. Everything in the universe is in ceaseless movement, from the most minute constituents of atoms to the immense galaxies of stars that populate space, and this movement is an essential factor in the structure and evolution of the physical world. Another consideration is the simplicity of the phenomena involved in motion, which permits us to become familiar with the methods of science before applying these methods to more complicated situations. And, as a historical note, modern science began with the study of motion. The classic conflict between the ptolemaic (earth-centered) and copernican (sun-centered) concepts of the solar system brought into clear focus the relative nature of motion, while Galileo's observations on moving bodies, which contradicted ancient but unfounded beliefs, showed the power and primacy of measurement in the exploration of nature.

HOW MOTION IS DESCRIBED

When a body changes its position with respect to something else as time goes on, we say that it *moves*.

The above statement is straightforward enough, but, like many other apparently simple statements in science, it contains subtleties of great importance. Here the significant idea is that all motion is relative to some reference object; so-called "absolute motion," without regard to an external reference, has no meaning. A body may go from one place to another—a train from New York to Boston—or it may repeat the same path indefinitely—a child on a merry-go-round. In either case the body travels through a certain distance *relative to the earth's surface* in each specific time interval during its motion. Under different circumstances the appropriate reference object may be something other than the earth: an atomic nucleus, the deck of a ship, the sun, the center of our galaxy.

All reference objects are equally "correct" provided they are properly identified, but usually a certain unique choice permits us to perceive the essential elements in a particular phenomenon. To somebody on the earth, the sun and stars appear to circle the earth daily, while to somebody outside the earth, it is equally obvious that the earth is merely rotating on its axis. Both persons are correct in their observations, but the latter point of view presents a clearer picture of what is happening, a picture that can be interpreted more simply and straightforwardly than the former. We might say that regarding the earth as rotating on its axis is the more *fruitful* notion, since it is the key to understanding other phenomena

motion implies a frame of reference

All motion is relative to a chosen frame of reference. Here the photographer has turned his camera to keep pace with the first car, and relative to this car the other cars and the building are moving. There is no fixed frame of reference in nature, and therefore no such thing as "absolute motion": all motion is relative.

MOTION

as well as astronomical ones. As we proceed to examine the quantitative language used to describe motion, we must continue to bear in mind the relative character of motion and the importance of a proper choice of reference object (or *frame of reference*, as a physicist would say).

speed

Now that we have explored the relative nature of motion and why the proper selection of a frame of reference is so important, we shall examine the quantitative language which is used to describe motion. A point worth noticing is the very specific definitions given such words as *speed, velocity,* and *acceleration,* which everybody uses in conversation although seldom with an awareness of just what they really mean. This insistence on precision of thought is characteristic of the sciences and is in large part responsible for their success.

As we said above, a moving body may go from one place to another or it may repeat the same path indefinitely. In either case the body travels through a certain distance in each interval of time while it is actually moving. For instance, a train going from New York to Boston may cover 68 miles during a certain hour of its trip, and a child on a merry-go-round may complete a circle 30 feet (ft) in circumference every 6 seconds (sec). **The *speed* of a moving body is the rate at which it covers distance,** that is, the distance it covers per unit time. This definition is conveniently expressed in the equation

$$\text{Speed} = \frac{\text{distance traveled}}{\text{time required}}$$

speed is rate of change of position

In symbolic form,

$$v = \frac{d}{t}$$

where v = speed of body
d = distance traveled in a period of time
t = time required

Thus a train that covers 68 miles in an hour has a speed of

$$v_{\text{train}} = \frac{68 \text{ miles}}{1 \text{ hr}} = 68 \text{ mi/hr}$$

and a child on a merry-go-round who covers 30 ft in 6 sec has a speed of

$$v_{\text{child}} = \frac{30 \text{ ft}}{6 \text{ sec}} = 5 \text{ ft/sec}$$

constant speed

In each case the speed is relative to the earth's surface.

When a body moves at constant speed, v is the same at all times. If this is the case, we can rewrite the definition of speed in the useful forms

$$d = vt$$

and

$$t = \frac{d}{v}$$

The first of these permits us to determine how far the body goes in any period of time t. Thus a train whose speed is 68 mi/hr travels

$$d = 68\frac{\text{mi}}{\text{hr}} \times \frac{1}{2}\text{hr} = 34 \text{ miles}$$

in ½ hr and

$$d = 68\frac{\text{mi}}{\text{hr}} \times 3 \text{ hr} = 204 \text{ miles}$$

in 3 hr. The second formula tells us how to find the time needed by a body whose speed is v to cover a distance d. Thus the train needs

$$t = \frac{100 \text{ miles}}{68 \text{ mi/hr}} = 1.47 \text{ hr}$$

to travel 100 miles.

converting units

It is sometimes convenient to express quantities given in one set of units in terms of a different set. To perform the conversion we note first that units are treated in an equation just like ordinary algebraic quantities and may be multiplied or divided by one another. Second, we recall that multiplying or dividing a quantity by 1 does not affect its value. Let us see how these rules enable us to convert a speed of 5 ft/sec into other units. Suppose we require the speed expressed in inches per second. We note that

units are treated as algebraic quantities

12 in. = 1 ft

so that

12 in./ft = 1

Hence, multiplying or dividing anything by 12 in./ft merely changes its units, but leaves it otherwise unaltered. If we multiply 5 ft/sec by

MOTION

12 in./ft, we have

$$5\frac{\text{ft}}{\text{sec}} \times 12\frac{\text{in.}}{\text{ft}} = 60\frac{\text{in.}}{\text{sec}}$$

since ft/ft = 1 and so cancels out. Similarly, we can express 5 ft/sec in feet per minute by multiplying it by 60 sec/min:

$$5\frac{\text{ft}}{\text{sec}} \times 60\frac{\text{sec}}{\text{min}} = 300\frac{\text{ft}}{\text{min}}$$

To perform the reverse conversion, we divide by 60 sec/min and so obtain

$$\frac{300 \text{ ft/min}}{60 \text{ sec/min}} = 5\frac{\text{ft}}{\text{sec}}$$

since min/min = 1.

Time, min	Distance, miles
0	0
1	0.5
2	1.0
3	1.5
4	2.0
5	2.5
6	3.0

constant speed

A graph is often helpful in analyzing motion. Suppose we measure at intervals of 1 min how far a car has gone from a certain starting point, and obtain the data shown in the table. To plot these data, we mark off a time scale along the bottom of a sheet of graph paper, letting each large division represent 1 min. On the left-hand side of the paper we mark off a distance scale, letting each large division represent 1 mile. Now we plot each measurement of time and distance as a dot on the graph. This procedure is shown in Fig. 1.1 for the dot that represents $t = 4$ min and $d = 2$ miles. Finally we draw a line through the various points. The line is a straight one, which always signifies that the quantities being plotted are *directly proportional* to each other. The car goes twice as far in 2 min as it does in 1 min, three times as far in 3 min, four times as far in 4 min, and so on, which is what is meant by saying that d is proportional to t.

Because d and t are proportional, their ratio has the same value at all times. This ratio is

$$v = \frac{\text{distance}}{\text{time}} = \frac{d}{t} = 0.5\frac{\text{mi}}{\text{min}}$$

Since there are 60 min in an hour,

$$v = 0.5\frac{\text{mi}}{\text{min}} \times 60\frac{\text{min}}{\text{hr}} = 30\frac{\text{mi}}{\text{hr}}$$

The ratio v is, of course, the car's speed, and the fact that d is proportional to t evidently means that this speed is constant. When a graph of d versus t results in a straight line, then, the speed of the moving body is constant.

a graph is a picture of an equation

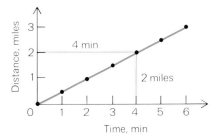

1.1 *A straight-line graph signifies a direct proportionality, here between the distance a car has traveled and the time taken.* When d is proportional to t, the speed is constant.

instantaneous speed

Sometimes one quantity is related to another quantity without being directly proportional to it. The table shows data taken from observations made on another car. In this case d is *not* directly proportional to t because the ratio $v = d/t$ has a different value for every pair of measurements. The data are given and plotted in Fig. 1.2, and the curved line that results confirms that d is not proportional to t. A moving body whose speed is not constant is, by definition, accelerated, and a graph of d versus t for such a body always yields a curved line.

Although the speed of the second car is not constant, it has a certain value at every moment called its *instantaneous speed*. The instantaneous speed can be found for any particular time, say $t = 4$ sec, by the method shown in Fig. 1.3. What we do is draw a straight line tangent to the distance-time curve at the point where $t = 4$ sec, and then find the speed that corresponds to this line. By applying this procedure to the curve at $t = 1, 2, 3, 4,$ and 5 sec, we obtain the instantaneous speeds shown.

velocity

Velocity is a term that is often confused with speed. The difference between them is that speed refers only to the distance covered by an object as it moves; velocity also takes into account the *direction* in which the object is moving. There are many situations in which it is important to know direction as well as speed. To a sailor an east wind at 32 mi/hr

Time, min	Distance, miles
0	0
1	0.25
2	1.00
3	2.25
4	4.00
5	6.25
6	9.00

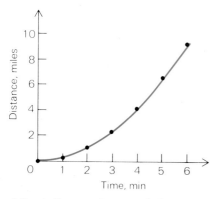

1.2 *A distance-time graph for an accelerated body is always a curved line.* The speed of the body is not constant.

Time, min	Instantaneous speed, mi/min
1	0.5
2	1.0
3	1.5
4	2.0
5	2.5

1.3 *The instantaneous speed of an accelerated body may be obtained from its distance-time graph for any specific time by drawing a tangent to the curve at that time.*

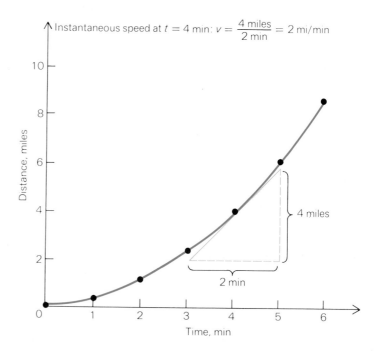

may be vastly different from a west wind at 32 mi/hr. An automobile traveling at 50 mi/hr to the northeast is going to find itself at the end of its trip in an entirely different place from an automobile whose velocity is 50 mi/hr to the south. When we speak of the *velocity* of a moving body, then, we refer both to its *speed* (how fast it is going) and to its *direction* (where it is headed).

velocity refers to both speed and direction

The distinction between speed and velocity may seem, at first, to be a trivial matter, but we shall see that its implications are extremely important in understanding the effects of forces on motion.

acceleration

When a body is moving with constant velocity, neither its direction nor its speed changes. A child on a merry-go-round may have a constant speed, but because he is traveling along a curved path, his velocity is not constant. A train may be proceeding on a straight track so that its direction is always the same, but if it goes faster or slower, its velocity is not constant. A body whose velocity changes is said to be *accelerated* (Fig. 1.4).

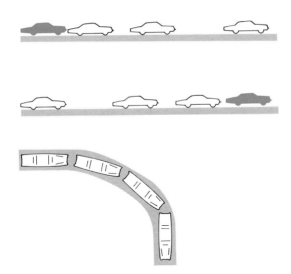

1.4 Three cases of accelerated motion, showing successive positions of a body after equal periods of time. *At the top the intervals between the positions of the body increase in length because the body is traveling faster and faster. Below it the intervals decrease in length because the body is slowing down. At the bottom the intervals are the same in length because the speed is constant, but the direction of motion is constantly changing.*

The rate of change of the velocity of a body is called its *acceleration*. In this chapter and the next we shall restrict ourselves to motion along a straight line, and in Chap. 3 we shall take up the more complicated problem of accelerations that involve a change in direction. Only accelerations whose magnitudes are constant will be considered.

The acceleration of a body restricted to straight-line motion can be written

$$\text{Acceleration} = \frac{\text{change in speed}}{\text{time required}}$$

acceleration is rate of change of velocity

In symbols,

$$a = \frac{v_2 - v_1}{t}$$

where a = body's acceleration
t = time interval
v_1 = its speed at start of t
v_2 = its speed at end of t

As an example, let us consider a man traveling at 50 ft/sec in a car who presses down on the gas pedal until he is going at 90 ft/sec. His speed has therefore increased by

$$v_2 - v_1 = 90\frac{\text{ft}}{\text{sec}} - 50\frac{\text{ft}}{\text{sec}} = 40\frac{\text{ft}}{\text{sec}}$$

Time, min	Instantaneous speed, mi/min
1	0.5
2	1.0
3	1.5
4	2.0
5	2.5

Suppose that 80 sec is required for his car to reach the higher speed. Then his acceleration is

$$a = \frac{v_2 - v_1}{t} = \frac{40 \text{ ft/sec}}{80 \text{ sec}} = 0.5\frac{\text{ft/sec}}{\text{sec}} = 0.5\frac{\text{ft}}{\text{sec}^2}$$

This result means that the speed of the car increases by 0.5 ft/sec during each second the acceleration continues.

constant acceleration

The table shows the instantaneous speed in miles per minute at several moments for the car whose motion was described in Figs. 1.2 and 1.3. We can see from these values that its instantaneous speed is proportional to the elapsed time. This proportionality suggests that, since

$$\text{Acceleration} = \frac{\text{change in instantaneous speed}}{\text{time interval}}$$

the car has a constant acceleration. Figure 1.5 is a graph of instantaneous speed versus time for the second car, and the straight line verifies that the acceleration is constant. The value of the acceleration is

$$a = \frac{v_2 - v_1}{t} = 0.5\frac{\text{mi/min}}{\text{min}}$$

$$= 0.5 \text{ mi/min}^2$$

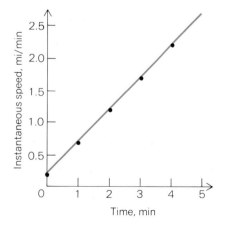

1.5 A straight-line graph of the instantaneous speed of a body versus the time that has elapsed since it started moving signifies that the body's acceleration is constant.

If we like, we can express the latter result in other units, for instance as

MOTION

$$a = 0.5\frac{\text{mi/min}}{\text{min}} \times 60\frac{\text{min}}{\text{hr}} = 30\frac{\text{mi/hr}}{\text{min}}$$

An acceleration of 30 (mi/hr)/min means that the car's speed increases by 30 mi/hr in each minute of its travel. Clearly the car cannot maintain this acceleration indefinitely, but over the time interval spanned by the measurements given, Fig. 1.5 indicates that the acceleration is constant.

When a body is slowing down, it is also accelerated, but in this event a graph of distance versus time yields a curve that gets flatter, as in Fig. 1.6, instead of steeper, as in Fig. 1.2. The instantaneous speed versus time graph that corresponds to the motion of Fig. 1.6 is shown in Fig. 1.7; the straight line indicates a constant acceleration, and its downward slope indicates that this acceleration is negative, as we would expect for a decreasing speed.

THE ACCELERATION OF GRAVITY

The most familiar acceleration is that of gravity. When something is dropped, it does not fall with uniform speed (neglecting air resistance). A stone released from the top of a cliff strikes the ground with a much higher speed than a stone released at shoulder level (Fig. 1.8). If we jump off a table, we strike the floor with greater impact than if we jump off a chair.

free fall

The significant aspect of the acceleration of freely falling bodies is that it is always the same under ideal conditions. Large objects and small, heavy ones and light, all descend with an acceleration of 32 ft/sec² [9.8 meters per second squared (m/sec²) in metric units]. This acceleration is denoted by the symbol g. Because of it, a body falling from rest has a speed of 32 ft/sec at the end of the first second, 64 ft/sec at the end of the next second, and so on.

By using the formula

Speed = acceleration × time

or, in symbols,

$v = gt$

we can calculate the speed v of a falling body t sec after it has been dropped. When we let $g = 32$ ft/sec², v comes out in feet per second; if $g = 9.8$ m/sec², v comes out in meters per second.

It does not matter whether a falling body starts out from a stationary position or is initially moving. If a ball is simply held in the air and dropped, its velocity increases steadily until it strikes the ground. If it

a positive acceleration means an increasing speed; a negative acceleration means a decreasing speed

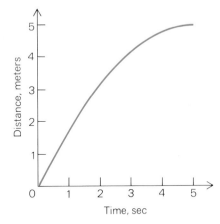

1.6 Distance versus time graph for a ball rolling in a straight line on a level floor. *The ball comes to a stop after it has gone 5 m.*

the acceleration of gravity near the earth's surface is the same everywhere

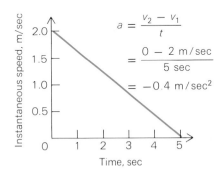

1.7 Instantaneous speed versus time for the ball of Fig. 1.6. *The initial instantaneous speed of the ball is 2 m/sec, and its final speed is 0. Because the ball is slowing down, its acceleration is negative.*

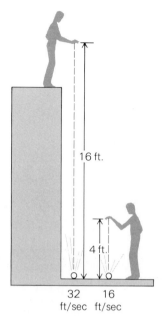

1.8 Falling bodies are accelerated. *A stone released from the top of a cliff strikes the ground with a much higher speed than a stone released at shoulder level.*

1.9 The acceleration of gravity does not depend upon horizontal motion. *When one ball is thrown horizontally from a building at the same time that a second ball is dropped vertically, the two reach the ground simultaneously because both have the same downward acceleration.*

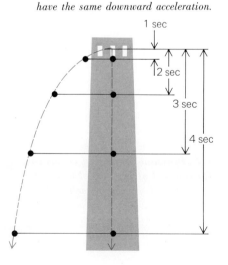

is thrown in a horizontal direction, its motion is determined both by its tendency to keep on moving in the direction it was thrown and by the pull of gravity. The pull of gravity accelerates its motion downward, so that the ball moves in a curved path which grows steeper as it nears the ground (Fig. 1.9). If the ball is tossed vertically upward (Fig. 1.10), the acceleration, acting downward, at first is in a direction opposite to the ball's motion. Hence the ball's velocity steadily diminishes, becoming zero at the top of its climb, and then increases steadily downward, the acceleration remaining constant throughout.

air resistance

For simplicity, in the preceding paragraph we have neglected air resistance. In the motion of small, light objects like raindrops, air resistance exerts a retarding influence opposing the accelerating tendency of gravity; otherwise raindrops would attain bulletlike velocities, and we could not safely venture out in the lightest shower. But for moderately heavy objects dropping at moderate speeds, air resistance is only a minor influence.

The retarding action of air often gives an erroneous idea of the relative rates with which different objects fall. A feather and a lead bullet indisputably fall with very different speeds. But this difference is entirely due to the buoyancy and large surface of the feather; if the two objects or any other two objects are enclosed in an evacuated cylinder, they fall at precisely the same rate (Fig. 1.11). Since the acceleration of gravity is the same for everything, the speeds of any two objects dropped simultaneously will increase at the same rate in a vacuum, and the objects will reach the ground together.

The faster a certain object falls, the greater the retarding force on it due to air resistance. Eventually the retarding force becomes so great that the object cannot fall any faster, and it then continues its descent at a constant speed called its *terminal velocity*. The terminal velocity of a man in free fall is about 120 mi/hr, but if he uses a parachute, his terminal velocity is only about 14 mi/hr.

galileo

The conclusion that objects near the earth's surface fall with the same acceleration regardless of their weight was first clearly stated by Galileo Galilei (1564–1642), as a result of experiments on the speeds of balls rolling down inclined planes. This directly contradicted part of the teachings of Aristotle, whose ideas were widely accepted in the early seventeenth century. Aristotle's reasoning depended on a general theory of the universe in which each different kind of material had a "natural" place appropriate to it and toward which it tried to move. Thus fire has a "natural motion" upward toward the sun and stars, whereas rocks or sticks of wood had an "earthy" quality that caused them to move downward toward their natural home in the earth. Furthermore, objects might contain different proportions of earthiness or fieriness, and their motions upward or downward would be accordingly fast or slow. It followed that a big stone, possessing more earthiness than a small one, ought to move downward more rapidly.

This fragment of Aristotle's reasoning, taken out of context, sounds rather absurd in the light of present knowledge. Actually this is not fair to the great philosopher, for the conclusion about falling bodies is only a minor detail in a grand scheme that sought to explain the operation of the universe and man's relation to it. The central ideas of Aristotle's universe—a stationary earth at the center of things, with mankind as the most important part of creation—fitted well into Christian theology and were in fact made a part of Church dogma in the writings of St. Thomas Aquinas and his followers. With this powerful backing, it is small wonder that Aristotle's ideas survived almost unaltered for nearly 2,000 years and that to challenge them in the early 1600s took more than ordinary courage.

Yet challenge them Galileo did. The particular question of falling bodies he worked on as a young professor at the University of Pisa, and, when he found that his experiments contradicted prevailing ideas, he attacked the conclusions of Aristotle with all the brashness and heedlessness of youth. A story has come down to us that he sought to dramatize his ideas by climbing to the top of Pisa's famous leaning tower, dropping from there a bullet and a cannonball and so demonstrating to a skeptical crowd that two objects would indeed reach the ground together. The story is not confirmed by historical records, but history does tell us that his outspoken championing of experimental results in opposition to the authority of Aristotle caused him to lose his professorship at Pisa. This was only a momentary setback in a brilliant career. Galileo continued his experimental work in other Italian cities, his learning bringing him fame as a scholar and his personal charm winning him friends among the ruling families of Italy. Despite fame and friends, his continued criticism of accepted opinions brought recurrent trouble with Church authorities. Especially serious was his advocacy of the copernican system. For many years he managed to escape serious reprisals, but finally, as an old man, he was hailed before the Inquisition at Rome and forced, under threat of torture, to recant his heretical beliefs.

The most radical new idea that Galileo introduced, which no amount of recanting could obliterate, was not a specific scientific discovery but the general concept that useful information about the world can be gained from accurate observation and experiment. Aristotle had sought to explain the world largely by reasoning alone—reasoning based on supposedly self-evident principles, such as the principle that all materials seek to move toward their natural levels. He would have regarded Galileo's experiments as trivial, as a wrongheaded way of getting information, because they did not relate motion to general principles. Galileo, on the other hand, ignored hypothetical principles in favor of finding out by actual trial just how falling bodies do behave. From his experiments he could derive mathematical rules or laws describing how fast a falling body will move and how far it will go in a given time, and he could check the rules by further experiment; but these rules are entirely different from the broad principles of Aristotle.

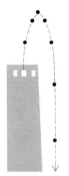

1.10 *When a ball is thrown upward, its downward acceleration eventually causes its velocity to decrease to zero.* This point is the top of its path, and it then begins to fall as if it had been dropped from there.

1.11 *In a vacuum all bodies fall with the same acceleration since air resistance and buoyancy are absent.*

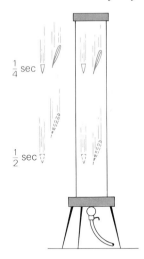

Galileo emphasized observation and experiment as the key elements in the scientific study of nature

The emphasis on conclusions drawn from observation and experiment rather than from "self-evident" principles is the characteristic that distinguishes modern science from all previous methods of trying to gain understanding and control of the natural world.

the meaning of equations

An equation is a summary of a relationship, not merely a device that helps in obtaining a numerical answer to a problem. In this book our concern is with how scientists attack problems in nature and not with how, for instance, engineers attack problems in structural design, and so the connection between equations and the phenomena they describe is of primary interest here.

equations summarize relationships among quantities

The falling-body experiment is a good example of the way in which a scientist tries to express the result of an experiment by an equation. He begins by selecting two variables to study, here distance and time, and arranges things so that other possible variables (friction, electric forces, air currents, and so on) have the least possible effect. Then he permits the two variables to change by letting the body fall, and measures the values of one variable that correspond to different values of the other. With these data in hand he now tries to find an algebraic relation between the two variables. This is often pretty much a trial-and-error process. First he might see if the variables are directly proportional, either by computing quotients for several pairs of values or by drawing a graph. If this is not successful (as it would not be in this example), he may try squaring the values of one variable, or cubing them, or taking the square root, and then again testing for proportionality. Unless the relationship is very complicated (as it may well be) he will eventually hit upon an equation that seems to fit the data approximately (in this example, $d = Kt^2$), using a constant to change the proportionality to an equality. With further work he may find that his K has a definite physical meaning, like velocity or acceleration or force, but in setting up the equation initially he is concerned only in showing that it is constant.

Usually, of course, things are not quite so simple. In some experiments it is not possible to pick out two variables and control all the others. The measurement itself may be very difficult. The relationship between the variables may be too complicated to express in a simple formula with one proportionality constant. But despite these possible complications, the method as just outlined is the general procedure by which the results of many scientific experiments are expressed in algebraic form.

We can reverse the above process by using an equation to visualize the results of an experiment. For example, we shall find in Chap. 3 the equation

gravitational force

$$F = G\frac{m_1 m_2}{r^2}$$

It looks formidable: a force is expressed as the product of the constant G and two masses divided by the square of a distance. Let us see what

MOTION

the equation means if it is interpreted simply as the result of a series of experiments.

The force, we note at once, is directly proportional to each of the masses, provided that the distance is held constant; if either mass is doubled, the force also is doubled, and so on. If the masses are held constant while the distance is changed, the force is inversely proportional to the square of the distance; if the distance is doubled, the force must be reduced to one-fourth its value. Thus without knowing anything about what the particular forces, masses, and distances may be, we can read from the equation the sort of experimental results from which the equation was derived. When used in this way, interpreted as a combination of proportionalities, an equation shows much more clearly its relation to experiment than when it is used simply as a tool to derive numerical results.

interpreting an equation

In future chapters we shall use equations both to obtain numbers and to show relationships between variables. Numerical results are important in making an equation real, in showing just how it works when applied to practical situations. But the true significance of an equation lies in what it is that is thus summarized, in what is revealed about the natural world in so compact a manner.

ANALYZING MOTION

speed and acceleration

The defining formula of acceleration,

$$a = \frac{v_2 - v_1}{t}$$

may be rewritten in the form

$$v_2 = v_1 + at$$

speed and acceleration

This equation states that the final speed of an object undergoing the acceleration a is equal to its speed at the start of the acceleration v_1 plus the product of the acceleration a and the time t during which it acts. Thus a ball rolling across the floor may be retarded by friction so that it slows down at the rate of 2 centimeters per second squared (cm/sec^2). If its initial speed is 10 cm/sec, at the end of 1 sec its speed is

$$v_2 = v_1 + at$$
$$v_2 = 10 \text{ cm/sec} - 2 \text{ cm/sec}^2 \times 1 \text{ sec}$$
$$= 8 \text{ cm/sec}$$

At the end of 2 sec its speed is

$$v_2 = 10 \text{ cm/sec} - 2 \text{ cm/sec}^2 \times 2 \text{ sec}$$
$$= 6 \text{ cm/sec}$$

and so on. The acceleration is negative in this case since its effect is to reduce the ball's speed.

distance and acceleration

Let us suppose we have a body of initial speed v_1 which experiences an acceleration a. We have seen that its speed after a time t is $v_2 = v_1 + at$. How far does it go during this time? To answer this basic question we note that, during the time interval t, the body's *average* speed $\bar{v}$ is

$$\bar{v} = \frac{v_1 + v_2}{2}$$

Since its speed is increasing uniformly, the body travels as far as if it had the constant speed v equal to its average speed $\bar{v}$, and we have just

$$d = \bar{v}t$$

Substituting for $\bar{v}$,

$$d = \tfrac{1}{2}(v_1 + v_2)t$$

and inserting $v_2 = v_1 + at$, we finally find that

$$d = v_1 t + \tfrac{1}{2}at^2$$

This formula gives the distance traveled by a body of initial speed v_1 during the time t in which it undergoes the acceleration a.

If there is no acceleration, the speed is constant and

$$d = v_1 t$$

which we already know. On the other hand, if its initial speed v_1 is 0, and it has the acceleration a, then

$$d = \tfrac{1}{2}at^2$$

What the latter formula does is give a prescription for calculating the distance a body starting from rest travels in terms of its acceleration a and the time t. The formula is not restricted to any particular object nor only to motion in a horizontal line. To adapt these results to the case of a body falling freely from rest, we note that $v_1 = 0$ and $a = g$, the acceleration of gravity. Hence the speed of a falling body a time t after it has been dropped is

The Leaning Tower of Pisa. While legend has it that Galileo dropped a bullet and a cannonball from the Leaning Tower to show that all bodies fall with the same acceleration, his conclusions about the behavior of moving bodies actually came from experiments with inclined planes to permit accurate measurements of time and distance.

$v_2 = gt$

During this period of time the body will have fallen

$d = \tfrac{1}{2}gt^2$ *falling bodies*

To see how these results can be applied to particular problems, let us consider an imaginary stone dropped from the top of the Pan Am Building in New York City, 830 ft above street level. Neglecting air resistance, how long does the stone take to reach the ground, and what is its final speed? We begin by substituting $d = 830$ ft and $g = 32$ ft/sec^2 in the formula $d = \tfrac{1}{2}gt^2$, and obtain

$$d = \frac{1}{2}gt^2$$

$$830 \text{ ft} = \frac{1}{2} \times 32 \frac{\text{ft}}{\text{sec}^2} \times t^2$$

$$t^2 = \frac{2 \times 830}{32} \text{ sec}^2 = 51.9 \text{ sec}^2$$

and so, taking the square root of both sides of the equation,

$t = 7.2$ sec

The stone requires 7.2 sec to reach the ground, and its final speed is therefore

$v = gt$

$= 32 \dfrac{\text{ft}}{\text{sec}^2} \times 7.2 \text{ sec}$

$= 230$ ft/sec

the speed with which an object must be thrown upward to reach a certain height is the same as the speed the object will have when it reaches the ground if dropped from that height

This is also the speed with which the stone would have to be thrown upward in order to reach the top of the building from the ground.

GLOSSARY

The *speed* of a body is the rate at which it covers distance; more precisely, it is the distance that the body moves in a period of time divided by that period of time. Speed is expressed in such units as *feet per second* (*ft/sec*), *meters per second* (*m/sec*), *miles per hour* (*mi/hr*), and so on.

A *graph* is a pictorial representation of the relationship between two quantities.

The *velocity* of a body refers both to its speed and to the direction of its motion. A car has a speed of 30 mi/hr; its velocity is 30 mi/hr to the northwest.

The *acceleration* of a body is the rate of change of its velocity. A body is accelerated when its speed changes, when its direction of motion changes, or when both change. Acceleration is expressed in such units as *feet per second squared* (ft/sec^2), *meters per second squared* (m/sec^2), and so on.

An *equation* is a mathematical statement of the relationship between a certain quantity and others upon which it depends.

The *acceleration of gravity* is the acceleration with which bodies fall in the absence of friction. The acceleration of gravity is the same for all bodies near the earth's surface, and its value is 32 ft/sec^2, or 9.8 m/sec^2.

EXERCISES

1 A certain pendulum consists of a bob at the end of a string. What are the directions of the ball's velocity and acceleration (*a*) at the left-hand end of its path, (*b*) at the lowest point of its path as it swings to the right, and (*c*) at the right-hand end of its path?

2 A rifle is aimed directly at a squirrel in a tree. Should the squirrel drop from the tree at the instant the rifle is fired or should it remain where it is? Why?

3 Suppose that you are in a barrel going over Niagara Falls and that, during the fall of the barrel, you drop an apple inside the barrel. Would it appear to move toward the top of the barrel or toward its bottom, or would it remain stationary within the barrel?

4 Is it necessary that an accelerated body be moving at all times? If not, give some examples of a body at rest that is accelerated.

5 Light from the sun requires about $8\frac{1}{3}$ min to reach the earth, a distance of about 150 million km or 93 million mi. What is the speed of light, in kilometers per second and in miles per second?

6 If an airplane has a cruising speed of 300 mi/hr and is bucking a 60-mi/hr head wind, what distance can it cover in 6 hr? If the same airplane has a tail wind of 60 mi/hr, what distance can it cover in 6 hr?

7 The speed of sound in air is about 1,100 ft/sec. If a gun is fired 1 mi from you, how long will it take before you hear the sound?

8 It is approximately 2,300 miles by car from Chicago to San Francisco. If a motorist averages 40 mi/hr and drives for 9 hr each day, how many days will he require for the trip?

MOTION

9 On a graph of speed versus time, how would the motion appear of a car moving along a straight road with the constant acceleration of 3 ft/sec^2? With the constant acceleration of -3 ft/sec^2?

10 Construct a graph of y against x from the following data:

x	y	x	y
0	0	4	28
1	7	5	35
2	14	6	42
3	21		

What is the relationship between x and y?

11 Construct a graph of y against x from the following data:

x	y	x	y
0	1.5	4	3.5
1	2	5	4
2	2.5	6	4.5
3	3		

What is the nature of the relationship between x and y?

12 Construct a graph of y against x from the following data:

x	y	x	y
0	0	8	320
2	20	10	500
4	80	12	720
6	180		

What is the nature of the relationship between x and y?

13 A car moving at 50 mi/hr is brought to a stop in 3 sec. What is its average acceleration in this period?

14 A pitcher requires about 0.1 sec to throw a baseball. If the ball leaves his hand with a speed of 120 ft/sec, what is its acceleration?

15 A car is moving with a speed of 20 ft/sec. Four seconds later its speed is 14 ft/sec. What is its acceleration?

16 A stone dropped from a cliff falls, as we know, with an acceleration of 9.8 m/sec^2. How fast will the stone be moving 1 sec after it is dropped? 3 sec after it is dropped?

17 A ball is thrown vertically downward with an initial speed of 10 m/sec. What is its speed after 1 sec? After 5 sec?

18 A ball is thrown vertically upward with an initial speed of 10 m/sec. What is its speed after 1 sec? After 5 sec?

19 A ball is thrown vertically upward with an initial velocity of 96 ft/sec. How long will it take the ball to reach the highest point in its path? How long will it take the ball to return to its starting place? What will the ball's speed be there?

20 An object whose initial speed is 90 m/sec is acted upon by an acceleration of -20 m/sec^2. How far will the object go while its speed decreases to 30 m/sec?

21 A car whose acceleration is constant attains a speed of 50 mi/hr in 20 sec starting from rest. How much additional time is required for it to attain a speed of 80 mi/hr?

22 A man in a descending elevator drops a stone from a height of 6 ft above the elevator floor. How long does the stone take to reach the floor when the elevator is falling (*a*) at a constant speed of 10 ft/sec and (*b*) at a constant downward acceleration of 2 ft/sec^2?

23 A man in an ascending elevator drops a stone from a height of 6 ft above the elevator floor. How long does the stone take to reach the floor when the elevator is rising (*a*) at a constant speed of 10 ft/sec and (*b*) at a constant acceleration of 2 ft/sec^2?

chapter 2

Force and Motion

INERTIA
- the first law of motion
- inertia
- mass
- the relativity of mass

THE ROLE OF FORCE
- force
- acceleration and force
- the second law of motion
- friction
- mass and weight
- british units

ACTION AND REACTION
- the third law of motion

VECTORS
- vector and scalar quantities
- vectors
- vector addition
- vector components

In the previous chapter we learned how to analyze the motion of a body in terms of velocity and acceleration, time and distance, and saw how these quantities are related. Now it is appropriate to inquire into the origin of motion. Why does anything move? Why do some bodies maintain constant velocities while others go faster or slower or change direction? These are fundamental questions, and the answers to them constitute the equally fundamental laws of motion formulated nearly three centuries ago by Isaac Newton.

INERTIA

Imagine a ball at rest on a level table. Given a gentle push, the ball rolls a short distance and gradually comes to a stop. That is common experience. The harder and smoother the ball and table top are made, the farther the ball rolls before stopping. Suppose that the ball could be made perfectly round and the table flawlessly smooth and perfectly level. In other words, suppose there were no friction between them; suppose further that all air (which resists motion through it) could be removed from the path of the ball and the table were infinitely long. If the ball were now set in motion, would it ever stop rolling?

the first law of motion

The conditions of this experiment cannot, of course, be realized in our laboratories. But they can be closely approximated, and the result is that, as resistance to its motion becomes less and less, the ball shows less and less inclination to stop. It is reasonable to conclude that under ideal conditions it would keep on rolling forever.

This conclusion was first expressed in the writings of Galileo. Later it was stated by Isaac Newton in a general form that has come to be known as Newton's first law of motion: **Every object continues in its state of rest or of uniform motion in a straight line if no net force acts upon it.**

first law of motion

In other words, objects about us do not start moving spontaneously; once set in motion, they continue with constant speed in a straight line until some resistance (for instance, friction) makes them stop. In our daily life such influences as friction and air resistance cannot be eliminated, and, consequently, all moving bodies in our experience tend to stop. To keep them moving at constant speed it is necessary that something push them continually, the push being used to overcome friction and air resistance. But here and throughout this chapter we are considering ideal conditions, under which friction and air resistance are absent. When we speak of balls we shall mean perfectly smooth, hard spheres; our tables and floors and roads will be smooth, level surfaces of indefinite extent. Granted these conditions, a moving ball has no reason to stop. A push is required to set it moving, but, once started, it continues of its own accord. Ideally, motion at a constant speed in a straight line is a condition quite as natural as a state of rest.

inertia

When we say that a motionless body tends to remain at rest, we imply that it offers a definite resistance to any attempt to make it begin moving. This fact is strikingly evident when we try to start a stalled car moving by pushing it. Newton's first law also states that a body, once in motion, resists attempts at stopping it or changing its motion. Again, if we have managed to push the car hard enough to get it rolling, we find great difficulty in stopping it.

The resistance that a material object offers to any change in its motion is an important property of matter called *inertia*. Inertia is significant because it gives us a means of measuring the quantity of matter present in an object.

inertia is resistance to change in state of motion

The expression *quantity of matter* is another of those phrases whose meaning seems to be clear, yet which science insists upon providing with a very specific definition that may seem at first to be rather peculiar. The trouble is that a number of different concepts are mixed up in most of our minds when we speak of matter. There is, first, the notion of bulk, the feeling that one of the essential properties of matter is that it occupies space. This is a very difficult idea to put into the precise terms of science, because, as we shall see later, it is impossible to ascribe a definite volume to the fundamental building blocks of which matter is composed.

Another concept involved in our mental picture of matter is *weight*. In everyday life we measure the amount of matter present in an object by simply weighing it. Something weighing 2 pounds (lb) contains twice as much matter as something else weighing 1 lb. The reason that weight is unsatisfactory for an adequate definition of quantity of matter is that **weight is the gravitational pull of the earth on an object.** This pull is not the same everywhere on the earth; it is less at high elevations (on mountaintops, for example) than at sea level, less near the equator than near the North and South Poles because the earth bulges slightly at the equator. These differences are not large, of course, but they do exist. For example, a body weighing exactly 1 ton (2,000 lb) in Lima, Peru, would weigh about 8 lb more if taken to Oslo, Norway. Also, the idea of weight becomes confusing if we leave the vicinity of the earth altogether. A 150-lb man "weighs" 25 lb on the moon (that is, the moon exerts a pull of 25 lb on him), but he would weigh 2 tons on the sun and nothing at all in empty space far from the solar system and other stars. It is hard to believe that it is the amount of matter in an object that varies so greatly when we take it from place to place, and so weight is not a suitable quantity to use in describing matter from a scientific point of view.

weight is gravitational pull of earth

The property of matter that gives rise to inertia, though, is much more promising. Imagine two balls of equal size, one made of lead, the other of wood. Here on earth we would say immediately that the lead ball contains more matter because it weighs more. If we were blindfolded and forbidden to weigh them, we could still tell the balls apart by kicking them along a level floor, because the inertia of the lead ball would resist our kick more than that of the wooden ball.

Now suppose that we and the two balls are transferred to some point in the depths of space. Can we find any method of establishing the greater quantity of matter in the lead ball under these new conditions? Weighing them would be quite useless, for neither ball would exert any downward push on our hand or on the pan of a balance. But inertia would still give the answer. The lead ball would again appear to contain more matter, because our toe would again hurt more after kicking it. The resistance that the two balls offer to any attempt to make them move is a property quite independent of their weights—and much more fundamental than their weights, since it does not depend upon their position with respect to the earth (Fig. 2.1).

2.1 Inertia is an inherent property of matter. *The inertia of a lead ball is greater than that of a wood ball everywhere in the universe. When both are kicked with the same force, the wood ball receives the greater acceleration because its mass is less.*

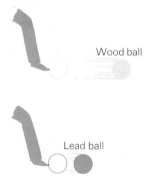

mass

the mass of a body is a measure of its inertia at rest

The name *mass* is given to the property of matter that manifests itself as the inertia of a body at rest. Since this property is an intrinsic characteristic of matter and does not depend upon the location, shape, size, and so forth of a particular object, mass may properly be thought of as quantity of matter; the more mass something has, the more matter it contains. (In Chap. 5 we shall see how relativity provides a more sophisticated analysis of mass.)

In the metric system mass is measured in terms of *grams*. The gram was originally defined as the mass of a cubic centimeter of water (a cubic centimeter is the volume of a box in the form of a cube each of whose edges measures 1 cm). Because the gram (g) is such a small unit—1 g is equivalent to 0.0353 ounce (oz)—the *kilogram* (kg), equal to 1,000 g, has come into common use. Today, masses in the metric system are all referred to the mass of the *standard kilogram,* a platinum-iridium cylinder kept at the International Bureau of Weights and Measures in France. The mass of the standard kilogram is, by definition, exactly 1 kg.

A rotating turntable at an amusement park. The participants slide outward because of inertia, the tendency of a body in motion to continue in motion along a straight line.

The unit of mass in the British system, the *slug*, is unfamiliar to most people, because we normally deal with the weights of things rather than with their masses. We shall postpone a discussion of slugs until later in this chapter in order not to be sidetracked from our present purpose, which is to understand force and motion and the relationship between them.

the relativity of mass

As we know, velocity is a relative quantity that depends upon the frame of reference of the observer. Relative to somebody at the side of a road, a certain car is moving north, while relative to the driver of the car it is stationary. Less familiar is the fact that *mass*, too, is a relative quantity: The mass of a body as measured by an observer depends upon the body's speed with respect to the observer. The mass is least when the body is at rest relative to the observer, and it increases as the relative speed increases.

<small>the mass of a body increases with its relative speed</small>

As we shall learn when the theory of relativity is discussed in Chap. 5, the mass increase is extremely small until the relative speed v is in the vicinity of the speed of light, which is 3×10^8 m/sec (186,000 mi/sec). For example, let us consider a body whose mass is 1,000 g when measured at rest. If v were 10 percent of the speed of light, the mass of the body would be 1,005 g, while if v were 90 percent of the speed of light, the mass of the body would be 2,294 g—more than twice its rest mass! At a relative speed equal to the speed of light, the mass of the body would be infinite, which is why the speed of light represents the absolute limit to the speed anything can have. The relativity of mass cannot be detected in everyday life because ordinary objects cannot be given sufficiently high speeds, but in the microscopic world of such atomic particles as electrons, protons, and neutrons, which often have speeds near that of light, the variation of mass with speed is a significant effect.

THE ROLE OF FORCE

force

Let us now return to the last clause of Newton's first law of motion, which states that the motion of a body can be altered by the action of a force. Most of us have a hazy notion of what force means; we think of a horse pulling a wagon, a man pushing a wheelbarrow or lifting a flour sack. Other familiar examples we have met in the preceding discussion: the force of gravity, which pulls us and objects about us to the earth's surface; the force of friction, which retards the motion of any object moving in contact with another. The pull of a magnet and the force of water pushing against the vanes of a turbine are further illustrations.

In all these examples the central idea is one of pushing or pulling, lifting or throwing—a process either involving muscular effort or producing the same results as the exertion of muscular effort. We shall speak of forces immensely greater than any muscle could produce and forces immeasurably smaller than the most delicate touch could detect, but we call them *forces* because they produce results on a larger or smaller scale similar to results accomplished by muscular effort. Force is thus a concept based on the direct evidence of our senses.

a force is any influence that can change the motion of a body

In physics, the definition of force is a restatement of the first law of motion, which specifies the result of a force's action: **A force is any influence capable of producing a change in the motion of a body.** Actual change of motion need not result from the application of a force. We may push with all our strength against a stone wall without affecting its state of rest in the slightest degree, yet we still call our muscular exertion the application of a force, since the same exertion would be capable of producing motion if applied to a more suitable object.

acceleration and force

What we have called simply *change of motion* in the above paragraph is, of course, *acceleration*. A moving body with no forces acting upon it proceeds with constant velocity. If a force should act on the body, the velocity of the body changes—either in magnitude or direction or both—and the body is therefore accelerated. A force, then, is something capable of causing an acceleration, and conversely, every acceleration is the result of an applied force.

every acceleration is the result of a force

It is important to remember that a body continues to be accelerated as long as a force acts upon it, and no longer. A force does not merely make a body move faster, for example, than it moved before; it causes the speed to increase steadily until the force is removed.

Here again results that would be obtained under ideal conditions seem to contradict everyday experience. In principle a small force acting on a ball rolling without friction would cause its speed to increase steadily, and the ball could be made to move with any conceivable speed if only the force acted for a long enough time. Again in principle, the gas pedal in an ideal automobile would not be needed at all to maintain a constant speed of 30 or 60 mi/hr; once set in motion at either of these speeds, an ideal car would keep on moving without any further force being applied to its wheels. To change from 30 to 60 mi/hr would require use of the gas pedal, that is, the application of a force. We might say, then, that the force would have produced a change from one constant speed to another; but the direct effect of the force, the effect while the force was acting, would have been to make the speed increase continually, and, unless the force was removed when 60 mi/hr was attained, the speed would go right on increasing. An *ideal* car could be made to go at any desired speed, however fast, merely by a slight momentary pressure on the gas pedal to bring it to that speed. *Ordinary* cars cannot, because at high speeds friction and air resistance increase so that all of the force that the engine applies to the rear wheels is used up in counteracting these opposing forces.

From similar reasoning we conclude that no force at all, or an infinitely small force, is required to move an ideal ball from one position of rest

FORCE AND MOTION

a

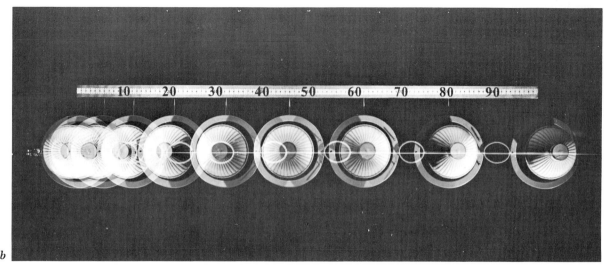

b

on a level table to another. Application of any finite force would set the ball moving forever—with constant velocity if the force were removed, with steadily increasing velocity if the force were continuous. In practice we have to supply a measurable force: we strike the ball and let it roll to a stop in its new position. During the fraction of a second while we are striking it, the force we exert changes the ball's velocity from zero to some small value; when the striking force is removed friction begins to alter this velocity, reducing it to zero when the ball reaches its new position of rest. We apply a force to make the ball change position, but the force is used entirely in overcoming the opposing force of friction. Ideally, the initial force needed would be infinitely small.

the second law of motion

Newton's second law of motion is a quantitative expression of the ideas we have discussed in the above paragraphs. It gives us a relation among

The disks shown moved from left to right on a horizontal sheet of glass while a light was flashed at regular intervals. The scale shown is in centimeters. Each disk consists of a piece of metal containing pieces of Dry Ice (solid carbon dioxide) which gradually vaporize, so that there is a layer of carbon dioxide gas between disk and glass that virtually eliminates friction. In (a) no external force is applied to the disk, and the disk travels equal distances in equal times, which means it has a constant velocity. In (b) the disk is pulled to the right by a constant force, and the distance traveled in each time interval increases with time: the disk is accelerated. The light flashed every 10/24 sec in each picture.

force, mass, and acceleration that can be treated mathematically: **The acceleration that a net force gives an object is directly proportional to the magnitude of the force and inversely proportional to the mass of the object; the acceleration is in the direction of the applied force.**

This means that, if we measure the accelerations produced by different forces on the same mass, doubling the force will double the acceleration; if we let the same force act on different masses, doubling the mass will cut the resulting acceleration in half (Fig. 2.2). Provided that proper units are employed, the law can be stated mathematically in the form

$$a = \frac{F}{m}$$

or, equivalently, as

$$F = ma$$

where F = force
m = mass
a = acceleration

2.2 Newton's second law of motion. *When different forces act upon the same mass, the greater force produces the greater acceleration. When the same force acts upon different masses, the greater mass receives the smaller acceleration.*

force equals mass times acceleration

In words, the second expression says that force is equal to the product of mass and acceleration. This is a precise statement of the definition of force given above.

It is impossible to exaggerate the importance of Newton's second law of motion. Replacing vague, intuitive feelings about the effects of forces with a definite, quantitative statement, this straightforward law provides the foundation for much of the science of physics.

An important part of Newton's second law concerns direction. The direction of the acceleration is always the same as the direction of the net force. An automobile begins to go faster and faster—therefore, the force exerted on it is in the same direction as the one in which it is moving. The same automobile slows down—therefore, the force exerted on it is in the direction *opposite* to that in which it is moving. This conclusion follows from the fact that the velocity of the automobile is decreasing, implying an acceleration (rate of change of velocity) that is negative in terms of its forward movement. A forward acceleration produces an increase in speed, whereas a backward acceleration produces a decrease in speed (see Fig. 2.3). When the automobile rounds a curve

acceleration is in same direction as force

2.3 The direction of a force is significant. *A force applied in the direction in which a body is moving produces a positive acceleration (increase in speed). A force applied opposite to the direction of motion produces a negative acceleration (decrease in speed).*

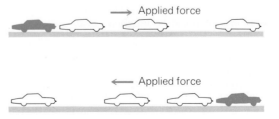

FORCE AND MOTION

at a constant speed, its direction changes, and there is an acceleration present. This acceleration is toward the inside of the curve; thus, the force that produced the turning must have been pointing toward the inside of the curve. We shall have more to say about force and acceleration with respect to bodies moving along curved paths, and the essential thing to keep in mind is that every change in velocity, whether a change in speed or direction of motion, or both, is caused by a force acting in the same direction as the change.

The second law permits us to define a unit for force in the metric system. If we express mass m in kilograms and acceleration a in meters per second per second (m/sec^2), the force F is given directly in terms of a unit called the *newton*. Thus in the metric system we can summarize the second law of motion in the form

the newton is the metric unit of force

$$F(\text{newtons}) = m(\text{kg}) \times a(\text{m/sec}^2)$$

As an example, suppose we apply a force of 15 newtons (about $3\frac{1}{2}$ lb) to a ball of mass 0.3 kg for 0.1 sec. The ball's acceleration while the force is acting upon it is

$$a = \frac{F}{m} = \frac{15 \text{ newtons}}{0.3 \text{ kg}} = 50 \frac{\text{m}}{\text{sec}^2}$$

If the ball started from rest, its speed after 0.1 sec (when the force is taken away) is

$$v = at = 50 \frac{\text{m}}{\text{sec}^2} \times 0.1 \text{ sec} = 5 \frac{\text{m}}{\text{sec}}$$

On the other hand, if we measure the acceleration of an object of known mass, we can compute what force must have been applied to it. We might observe a 2-kg ball rolling on the ground whose initial speed is $v_1 = 4$ m/sec and which comes to a stop 5 sec later because of friction (Fig. 2.4). Since the final speed of the ball is $v_2 = 0$, its acceleration is

$$a = \frac{v_2 - v_1}{t} = \frac{0 - 4 \text{ m/sec}}{5 \text{ sec}} = -0.8 \frac{\text{m}}{\text{sec}^2}$$

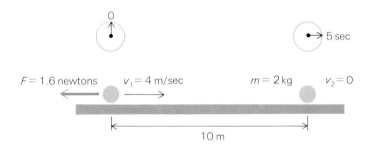

2.4 *A 2-kg ball whose speed is 4 m/sec comes to a stop in 5 sec when a retarding force of 1.6 newtons is applied to it.*

The minus sign means that the ball is slowing down. The force on the ball is (ignoring the minus sign)

$$F = ma = 2 \text{ kg} \times 0.8 \frac{\text{m}}{\text{sec}^2} = 1.6 \text{ newtons}$$

Since $v_1 = 0$, we can use the formula $d = \frac{1}{2}at^2$ to find out how far the ball rolled while coming to a stop. We have

$$d = \frac{1}{2} at^2 = \frac{1}{2} \times 0.8 \frac{\text{m}}{\text{sec}^2} \times (5 \text{ sec})^2 = 10 \text{ m}$$

Evidently the second law of motion, used with the formulas of elementary mechanics, is a powerful tool for understanding the behavior of moving bodies.

friction

frictional forces oppose motion

Friction refers to the forces that occur between two bodies in contact that impede the motion of one of them relative to the other. It is friction that prevents our feet from sliding on a rug, and it is lack of friction that permits our feet to slip on a highly polished floor. Lubricants such as oil and grease reduce friction by separating the surfaces in contact with a film of a substance that flows readily.

Frictional forces are real ones that always tend to oppose motion. Inertia is quite different, since there is no such thing as an "inertial force"; the term inertia is used to describe the fact that a body maintains its initial state of rest or of motion at constant speed in a straight line if no net force acts upon it. A very small force can cause a body to be accelerated regardless of its inertia, while if friction is present, a very large force may have to be applied to move it at all.

inertia is not a force

mass and weight

The force with which the earth attracts a body is called the *weight* of the body. If your weight is 160 lb, that means that the earth is pulling you down with a force of 160 lb. Weight is different from *mass*, which is a measure of the quantity of matter an object contains. There is a very intimate relationship between weight and mass, though, and it is extremely important for us to understand it.

Let us look at the problem in the following way. Whenever a net force F is applied to a mass m, Newton's second law of motion tells us that the mass will be so accelerated that its acceleration a will be in accordance with the formula

$$F = ma$$

In the case of a body on the earth, the force exerted upon it by gravity is its weight W. This force causes the body to fall with the constant

FORCE AND MOTION

acceleration $g = 32$ ft/sec^2. We may accordingly substitute W for F and g for a in the above formula, so that

$$W = mg$$

relation between mass and weight

Since g is a constant, the weight W of a body and its mass m are always directly proportional to each other; a large mass is heavier than a small one.

In the metric system, as we know, mass rather than weight is normally specified; a customer in a French grocery might ask for a kilogram of bread or 5 kg of potatoes. To find the weight in newtons (the metric unit of force) of something whose mass in kilograms is known, we simply turn to $W = mg$ and set $g = 9.8$ m/sec^2, its value in the metric system. Then we have

$$W \text{ (newtons)} = m \text{ (kg)} \times 9.8 \text{ m/sec}^2$$

mass and weight in metric units

Thus, the weight of 5 kg of potatoes is

$$W = 5 \text{ kg} \times 9.8 \text{ m/sec}^2 = 49 \text{ newtons}$$

This is the force with which the earth attracts a mass of 5 kg.

british units

In the British system of units we use pounds (lb) for weight and feet per second per second (ft/sec^2) for acceleration. The corresponding unit of mass is the *slug* (Table 2.1). In order to find the mass in slugs of some object whose weight W we know, we need only rearrange the above equation in the form

the slug is the british unit of mass

$$m \text{ (slugs)} = \frac{W \text{ (lb)}}{g \text{ (ft/sec}^2)}$$

Since g is 32 ft/sec^2,

TABLE 2.1 Mass and weight.

System of units	Unit of mass	Unit of weight	Acceleration of gravity g	To find mass m given weight W	To find weight W given mass m
Metric	Kilogram (kg)	Newton (nt)	9.8 m/sec^2	$m \text{ (kg)} = \dfrac{W \text{ (nt)}}{9.8 \text{ m/sec}^2}$	$W \text{ (nt)} = m \text{ (kg)} \times 9.8 \text{ m/sec}^2$
British	Slug	Pound (lb)	32 ft/sec^2	$m \text{ (slugs)} = \dfrac{W \text{ (lb)}}{32 \text{ ft/sec}^2}$	$W \text{ (lb)} = m \text{ (slugs)} \times 32 \text{ ft/sec}^2$
Conversion of units:	1 slug = 14.6 kg 1 kg = 0.0685 slug			1 lb = 4.45 newtons 1 newton = 0.225 lb	1 ft = 0.305 m 1 m = 3.28 ft

mass and weight in british units

$$m \text{ (slugs)} = \frac{W \text{ (lb)}}{32 \text{ ft/sec}^2}$$

In other words, dividing by 32 ft/sec² the weight of something measured in pounds gives its mass in slugs.

It is legitimate to ask at this point why we should ever want to know the mass of a body in terms of so peculiar-sounding a unit as the slug, when, after all, weights rather than masses are what we normally deal with. We go to the store for 10 lb of onions, not for $\frac{1}{3}$ slug of onions. The usefulness of knowing the mass of some object comes in whenever we wish to apply Newton's second law of motion to its behavior. Suppose that we push a car weighing 3,200 lb with a force of 50 lb. To find its acceleration we must first find its mass; since

$$m \text{ (slugs)} = \frac{W \text{ (lb)}}{32 \text{ ft/sec}^2}$$

we have

$$m = \frac{3{,}200 \text{ lb}}{32 \text{ ft/sec}^2} = 100 \text{ slugs}$$

Now we can use this mass value in the second law of motion,

$$F = ma$$

to find the value of the acceleration a. With $F = 50$ lb and $m = 100$ slugs,

$$a = \frac{F}{m} = \frac{50 \text{ lb}}{100 \text{ slugs}}$$
$$= \tfrac{1}{2} \text{ ft/sec}^2$$

If there were no friction present, the car's acceleration would be $\frac{1}{2}$ ft/sec². At the end of 1 sec its speed would be $\frac{1}{2}$ ft/sec, at the end of 10 sec its speed would be 5 ft/sec, and so forth.

As we know, the second law of motion can be used in reverse, so to speak, to determine the force that is producing a known acceleration. The same car we spoke of above might increase its speed, say, from 20 to 50 mi/hr in 20 sec. Again neglecting friction, what force did the car's motor have to supply in order to cause this acceleration? We start by calculating the acceleration. The car increased its speed by (50 − 20) mi/hr = 30 mi/hr in 20 sec. Since 30 mi/hr is equal to 44 ft/sec,

$$\text{Acceleration} = \frac{\text{change in speed}}{\text{time interval}}$$

$$a = \frac{44 \text{ ft/sec}}{20 \text{ sec}} = 2.2 \text{ ft/sec}^2$$

Thus, the force acting on the car was

$$F = ma = 100 \text{ slugs} \times 2.2 \text{ ft/sec}^2$$
$$= 220 \text{ lb}$$

ACTION AND REACTION

Forces in the world about us always turn out, upon close examination, to consist of combinations of forces. When we think we have an example of just one force at work, further inspection reveals others—less obvious perhaps, but nevertheless present. Let us look into the question of whether it is ever possible for a single force to exist.

We push downward on a table. As far as we are concerned, that seems to be a single force—an elemental push, giving us the sort of sensation which we must use in any intuitive definition of force. But immediately we are conscious of a second force: the resistance of the table, pushing upward against our hand as we push downward on its top. The harder we press down, the more stubbornly the table resists. Seemingly, we cannot exert a force on the table without its exerting a force on us.

Suppose that we transfer ourselves and the table to the frozen surface of a lake—to a sheet of ice, which we shall imagine to be smooth and slippery, so that it can offer no resistance to the table's motion. Again we push on the table, horizontally now instead of vertically, and watch it accelerate, as it should under the influence of a single constant force (Fig. 2.5). But again we meet difficulties: We can stick to the ice no better than the table can, and as we push it away from us, we find ourselves starting to move in the opposite direction. Even here we cannot

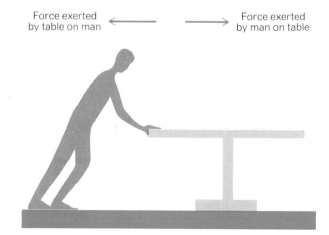

2.5 Action and reaction forces act on different bodies. *Pushing a table on a frozen lake results in man and table moving apart in opposite directions.*

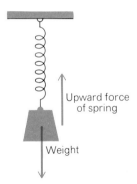

2.6 *Newton's third law. For every force there is an equal and opposite force: the spring pulls upward on the weight with the same force as that with which the weight pulls downward on the spring.*

action and reaction forces act on different bodies

seem to exert a force on the table without its pushing back on us, as revealed by our motion.

the third law of motion

Considerations of the above sort led Newton to his third law of motion: For every force there is an equal and opposite force. More precisely, **Whenever one object exerts a force on a second object, the second exerts an equal force in the opposite direction on the first.**

No force ever occurs singly. A weight hangs on a spring balance; the weight pulls downward on the spring, and the spring pulls upward on the weight (Fig. 2.6). A chair pushes downward on the floor; the floor presses upward on the chair. The firing of a rifle exerts force on a bullet; the firing simultaneously exerts a backward push (recoil) on the rifle.

Sometimes the reality of the opposite force is difficult to appreciate. A book resting on a table exerts the downward force of its own weight; but just how can an apparently rigid object like the table exert a real upward force on the book? If the table top were made of rubber, the book would depress it, and the upward force would obviously result from the elasticity of the rubber. A similar explanation actually holds for table tops of wood or metal, although the depressions are extremely small ones; nothing can be perfectly rigid. Again, a falling apple experiences a downward pull from the earth and, by the third law, must itself pull upward on the earth with an equal force. We cannot observe the effects of this force, simply because the earth is so very much larger than the apple, but we have no reason to doubt that the force exists.

Newton's third law always applies to *two different forces on two different objects*—the force that one exerts on the second and the opposite force that the second exerts on the first.

It is not always entirely clear how the third law of motion operates in various situations. A boy pulls a cart with a force of 10 lb. According to the third law, the cart pulls right back on the boy with the same force, 10 lb, but in the opposite direction. How can the cart move? Seemingly, the boy's force is matched by the reverse force of the cart, leaving a total of zero force. What we have overlooked in making this statement is the fact that the two forces, that of the boy and that of the cart, act on different things. The boy pulls the cart; the cart experiences this force; the cart begins to move. The cart pulls backward on the boy; the boy must exert himself to overcome this force—which is what he is doing when he pulls on the cart. The third law of motion is sometimes called the law of action and reaction in order to make clear the distinction between the natures of the forces involved.

VECTORS

vector and scalar quantities

Physical quantities that are completely specified by a number and a unit are called *scalar quantities*. The number answers the question "How

FORCE AND MOTION

many?" and the unit answers the question "Of what?" Volume is an example. Since it is sufficient to describe the volume of a barrel by calling it 55 gallons (gal), or to describe the volume of a room by calling it 3,000 ft³, or to describe the volume of a bottle by calling it 12 oz, volume is a scalar quantity. Other scalar quantities are mass (the mass of a certain bag of potatoes is 2 kg), frequency (the frequency of ordinary alternating current is 60 Hz), and power [the engine of a certain car develops 280 horsepower (hp)].

a scalar quantity has magnitude only

Some physical quantities require more information if they are to be completely specified. It is not enough to say that an airplane has flown for 200 miles; after all, an airplane leaving Boston that flies southwest for 200 miles will end up in the neighborhood of New York, while if the same airplane heads southeast for 200 miles, it will end up over the Atlantic Ocean. A *vector quantity* is one whose direction is significant. Thus a vector quantity has both a magnitude and a direction.

a vector quantity has both magnitude and direction

The displacement of a body that moves from one place to another is an example of a vector quantity. When we say that an airplane has gone 200 miles to the southwest, we have completely described its displacement from wherever it started. Other vector quantities are velocity (the velocity of a certain car is 50 mi/hr *to the north*), acceleration (the acceleration of gravity is 32 ft/sec² *directed downward*), and force (a man applies an *upward* force of 20 lb to a box). The arithmetic of vector quantities is more complicated than is the case with scalar quantities because of the need to take directions into account. If an airplane flying at 120 mi/hr is in a 40-mi/hr wind stream, its velocity relative to the ground may be anything from 80 to 160 mi/hr depending upon the relative directions of the velocity of the airplane and the velocity of the wind.

vectors

Both the magnitude and the direction of a vector quantity can be represented by a straight line with an arrowhead at one end. The length of the line is proportional to the magnitude of the quantity, and the line is drawn so that its direction indicates the direction of the quantity. Such a line is called a *vector*. Figure 2.7 shows how a displacement of 200 miles to the southwest is represented by a vector. The scale used here is ⅝ in. = 100 miles, and a compass rose is included in the diagram for orientation.

All vector quantities can be represented by suitable vectors. Figure 2.8 shows a force vector on a scale of 1 cm = 15 lb and Fig. 2.9 shows

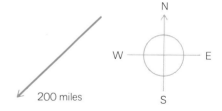

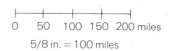

2.7 *Vector representation of a 200-mile displacement to the southwest.*

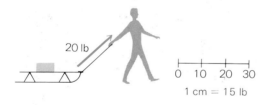

2.8 *A force vector.*

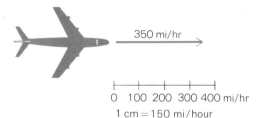

2.9 *A velocity vector.*

symbols for vector and scalar quantities

a velocity vector on a scale of 1 cm = 150 mi/hr. Vector quantities are usually printed in boldface type (**F** for force, **v** for velocity) while italic type is used for scalar quantities (*m* for mass, *V* for volume) and also for the magnitudes of vector quantities (*F* is the magnitude of the force **F**, *v* is the magnitude of the velocity **v**). For instance, the magnitude of a displacement **D** of 200 miles to the southwest is $D = 200$ miles. A vector quantity is usually expressed in handwriting by an arrow over its symbol, so that $\vec{F}$ means the same thing as **F**.

vector addition

If there are 5 gal of gasoline in the tank of a car and we add 10 gal more, the total in the tank is 15 gal. Scalar quantities of the same kind are added by using ordinary arithmetic. Similarly, if we walk eastward for 5 miles and then a further 10 miles in the same direction, we have traveled a total of 15 miles to the east. However, if we walk eastward for 5 miles and then walk northward for 10 miles, we do *not* find ourselves 15 miles from our starting point, even though we have walked for 15 miles.

To determine the actual displacement in the latter situation, we make use of the *vector diagram* of Fig. 2.10. The vector **A** represents the first

A Civil War cannon being fired. Note the furrow plowed by the cannon's rear wheel as it recoils: the forward force on the cannonball is matched by a backward force on the cannon itself.

walk of 5 miles to the east and the vector **B** represents the second walk of 10 miles to the north. Then we connect the starting point and the finishing point with a single vector **R**, which represents the net displacement. The magnitude of **R** is 11.2 miles, which we find by comparing its length with the distance scale, and its direction is 27° east of north, which we find with the help of a protractor. If we had headed 27° east of north at the beginning, we would have had to walk only 11.2 miles to reach the same place as the two walks, one of 5 miles to the east and the other of 10 miles to the north, finally got us.

The above procedure is a quite general one. To add a vector **B** to another vector **A**, we draw **B** so that its tail is at the head of **A**. The vector sum **A** + **B** is the vector **R** that joins the tail of **A** and the head of **B**, as in Fig. 2.11. **R** is usually called the *resultant* of **A** and **B**. In equation form,

A + **B** = **R**

where it is understood that vector addition requires the taking into account of directions as well as magnitudes. The order in which the vectors **A** and **B** are added is unimportant:

A + **B** = **B** + **A**

Here is another example of vector addition, this time involving forces. Two tugboats are towing a ship, as in Fig. 2.12. Each tugboat exerts a force of 6 tons on the ship, the angle between the tow ropes is 60°, and we are to find the magnitude of the resultant force on the ship. The procedure we follow is shown in the diagram; the answer is that the magnitude of **R** is 10.4 tons. If the angle between the tow ropes is smaller than 60°, *R* is greater; if the angle is greater than 60°, *R* is smaller.

Exactly the same procedure is followed when more than two vectors of the same kind are to be added. The vectors are strung together head to tail (being careful to preserve their correct lengths and directions), and the resultant **R** is the vector drawn from the tail of the first vector to the head of the last (Fig. 2.13). The order in which the vectors are added does not matter.

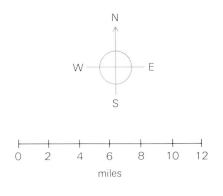

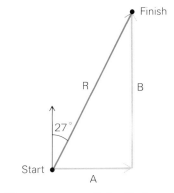

2.10 *The vector sum of the displacements A and B is the resultant displacement R.*

adding more than two vectors

2.11 *The general rule for vector addition: To add the vectors A and B, shift B parallel to itself so that its tail is at the head of A; the sum of A and B is a vector drawn from the tail of A to the head of B.* The order in which *A* and *B* are added is unimportant.

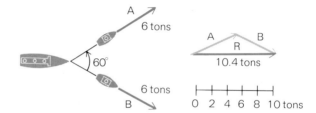

2.12 *The net force exerted on the ship by the two tugboats is 10.4 tons.*

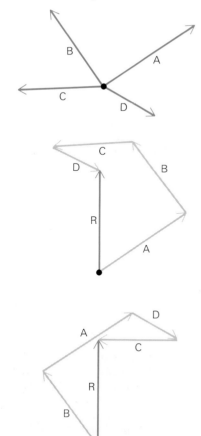

2.13 *To add a number of vectors together, string them together head to tail and draw a vector from the tail of the first to the head of the last; this vector is the sum of all the others.* The order in which the vectors are added together does not matter.

vector components

Just as two or more vectors can be added together to yield a single resultant vector, so it is possible to break up a single vector into two or more other vectors. If the vectors **A** and **B** are together equivalent to the vector **C**, then the vector **C** is equivalent to the two vectors **A** and **B**, as illustrated in Fig. 2.14. When we replace a vector by two or more other vectors, the process is called *resolving* the vector, and the new vectors are known as the *components* of the initial vector.

The resolution of a vector into components is a valuable technique for solving many problems, and by dealing with their components rather than with the vectors themselves, the addition of more than two vectors is greatly simplified.

The components into which a vector is resolved are nearly always chosen to be perpendicular to one another, which makes matters more convenient. Figure 2.15 shows a wagon being pulled by a boy with the force **F**. Because the wagon moves horizontally, the entire force is not effective in producing its motion. The force **F** may be resolved into two component vectors, $\mathbf{F}_x$ and $\mathbf{F}_y$, where

$\mathbf{F}_x$ = horizontal component of **F**

$\mathbf{F}_y$ = vertical component of **F**

The magnitudes of these component vectors are

$F_x = F \cos \theta$

$F_y = F \sin \theta$

$\mathbf{F}_x$ is the projection of **F** in the horizontal direction, and $\mathbf{F}_y$ is its projection in the vertical direction. Evidently the component $\mathbf{F}_x$ is responsible for the wagon's motion, and if we were interested in working out the details of this motion, we would need to consider only $\mathbf{F}_x$ and not **F**.

Let us suppose the magnitude F of the force the boy exerts on the wagon is 10.0 lb and the angle θ is 30°. In this case

$F_x = F \cos \theta = 10.0 \text{ lb} \times \cos 30°$

$ = 8.66 \text{ lb}$

FORCE AND MOTION

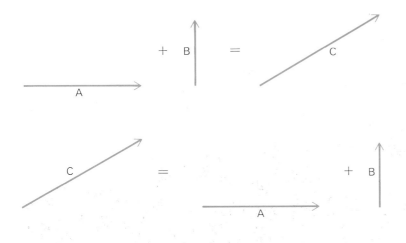

2.14 *If two vectors A and B can be added to yield the single vector C, then the vector C can be resolved into the two vectors A and B.*

GLOSSARY

A *force* is any influence capable of producing a change in the motion of a body, in other words, capable of accelerating it. In the British system the unit of force is the *pound* (*lb*); in the metric system it is the *newton*.

Newton's *first law of motion* states that every body continues in its state of rest or of uniform motion in a straight line if no force acts upon it.

Newton's *second law of motion* states that the acceleration of a body is directly proportional to the magnitude of the force acting upon it and inversely proportional to its mass; the acceleration is in the direction of the applied force.

Newton's *third law of motion* states that for every force there is an equal and opposite force acting on the body that exerts the first force.

Inertia is the apparent resistance a material body offers to any change in its state of motion.

The *mass* of a body is the property of matter that manifests itself as inertia; it may be thought of as the quantity of matter in the body.

The *weight* of a body is the force with which gravity pulls it toward the earth.

A *vector* is an arrow whose length is proportional to the magnitude of some quantity and whose direction is that of the quantity. A *vector quantity* has both magnitude and direction, whereas a *scalar quantity* has only magnitude.

A vector can be *resolved* into two or more other vectors whose sum is equal to the original vector. The new vectors are the *components* of the original vector, and are usually chosen to be perpendicular to one another.

2.15 *Horizontal and vertical components of a force.*

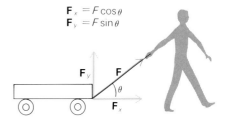

$$\mathbf{F}_x = F \cos\theta$$
$$\mathbf{F}_y = F \sin\theta$$

The design of almost every structure involves vector analysis. Here fertilizer from the plant at left is loaded into a barge by means of a conveyor belt supported by a girder which in turn is supported by cables from the top of the tower.

EXERCISES

1 Is the moon's motion around the earth accelerated? Is a force acting on the moon? If so, what is the direction of the force?

2 When you whirl a stone at the end of a string, the stone seems to be pulling outward against your hand. When you release the string, however, the stone moves along a straight path. Explain each of these effects.

3 Why is it less dangerous to jump from a high wall onto loose earth than onto a concrete pavement?

4 A book rests on a table. What is the reaction force equal and opposite to the force the book exerts on the table? To the force that gravity exerts on the book?

5 Since the opposite forces demanded by Newton's third law of motion are equal in magnitude, how can anything ever be accelerated?

6 If you were set down in the center of a frozen lake whose surface was so smooth and slippery that it offered no frictional resistance to any kind of motion, how could you get off the lake?

7 Why is it more sensible to pull a sled with an 8-ft rope than it is to push it with an 8-ft pole? Both rope and pole are assumed to be held at the same height above the ground.

8 A force of 50 lb and another of 75 lb act on the same object. What is the maximum total force they can exert on the object? What is the minimum total force?

9 Three forces, each of 5 lb, act on the same object. What is the maximum total force they can exert on the object? What is the minimum total force?

10 An airplane is coming in for a landing along a straight path. Explain how a passenger can tell by watching the curtains of his window whether the airplane is traveling with a constant speed, an increasing speed, or a decreasing speed.

11 A rifle bullet is fired horizontally. Does it begin to fall only when it has lost most of its speed? Does its rate of fall depend on its horizontal speed? On its size and shape? Would the answers to these questions be different if the rifle were fired in a vacuum instead of the earth's atmosphere?

12 A force of 4,000 newtons is applied to a 1,400-kg car. What is its acceleration? What will its velocity be after 8 sec if it started from rest?

13 What force is required to impart a 5-m/sec^2 acceleration to a 6-kg body?

14 A loaded elevator weighs 1,000 kg. The cable supporting it can safely withstand a tension of 12,000 newtons. What is the maximum upward acceleration that the elevator can have if the tension in the cable is not to exceed this figure?

15 A man whose mass is 80 kg is riding in an elevator whose upward acceleration is 2 m/sec^2. What force does he exert on the floor of the elevator?

16 What is the mass of a 160-lb man? With how much force is he attracted to the earth? If he falls from a cliff, what will his downward acceleration be?

The three chief forces that affect the motion of a skier are (1) the component of his weight parallel to the slope, (2) the friction between his skis and the snow, which is proportional to the force pressing the skis against the snow (and hence to the component of his weight perpendicular to the slope), and (3) air resistance. The steeper the slope, the greater the component of the skier's weight parallel to the slope and the smaller the component perpendicular to the slope, so a relatively small change in the steepness of the slope can have a large effect on the acceleration of the skier.

17 How much force is required to give a 5-slug mass an acceleration of 12 ft/sec^2?

18 An automobile weighing 3,500 lb goes from 30 mi/hr (44 ft/sec) to 60 mi/hr (88 ft/sec) in 5 sec. What is the average force acting upon it? The brakes of the automobile are able to exert a force of 500 lb. How long will it take for them to slow the automobile to a stop from an initial speed of 60 mi/hr?

exercises 19–28 may be done graphically with a ruler and protractor.

19 In going from one town to another a car travels 40 miles north, 40 miles west, and then 10 miles south. What is the actual straight-line distance between the towns?

20 A body weighing 75 lb is acted upon by a horizontal force of 75 lb. What are the magnitude and direction of the resultant force on the body?

21 An ocean liner is steaming due west at a speed of 25 mi/hr. There is a wind from the northwest at 10 mi/hr. What is the speed of the smoke from the liner's funnel relative to the boat? What is the direction of the smoke?

22 A motorboat whose speed through the water is 10 mi/hr heads directly across a river 1 mile wide. The current in the river is 2 mi/hr. When the boat reaches the other side of the river, how far downstream will it be?

23 A bullet is fired horizontally from a great height at a speed of 100 m/sec. As soon as it leaves the gun it starts falling with an acceleration of 9.8 m/sec^2. Find its approximate speed after 1, 5, and 10 sec, respectively. (Neglect air resistance and winds.)

24 A horizontal and a vertical force combine to produce a force of 10 newtons that acts in a direction 40° above the horizontal. Find the magnitudes of the horizontal and vertical forces.

25 Two forces, one of 12 newtons and the other unknown, act at right angles on an object. If the net force on the object is 30 newtons, what is the magnitude of the unknown force?

26 Find the vertical and horizontal components of a 200-lb force that is directed 50° from the vertical.

27 A man pushes a lawnmower with a force of 20 lb. If the handle of the lawnmower is 40° above the horizontal, how much downward force is being exerted on the ground?

28 An airplane is heading northeast at a speed of 550 mi/hr. What is the northward component of its velocity? The eastward component?

chapter 3

Gravitation

CIRCULAR MOTION

    centripetal force

    derivation of centripetal force formula

UNIVERSAL GRAVITATION

    ptolemy and copernicus

    kepler's laws

    the law of gravity

    center of mass

    newton's proof

    experimental proof

    artificial satellites

    the discovery of neptune and pluto

THE "SCIENTIFIC METHOD"

One of the truly great triumphs of the human mind is the law of universal gravitation, which has lost none of its luster in the three centuries since its discovery. At one stroke this seemingly simple relationship showed that the motions of the planets about the sun, the motion of the moon about the earth, and the behavior of falling bodies on the earth are all manifestations of the same basic phenomenon of nature, gravitation. The man who first perceived the law of gravitation together with all its implications was Isaac Newton (1642–1727), the same genius to whom we owe the laws of motion that formed the subject of the last chapter. Newton proved that the same laws of nature apply in the universe as are valid on the earth's surface, so that the gravitational force that attracts apples to the ground is identical with the gravitational force that keeps the planets in their orbits. Newton accomplished more than merely explaining the various regularities in the motions of the planets that had been discovered; he showed that these features of the solar system, so puzzling at first glance, are absolutely inevitable consequences of natural laws, thereby establishing the existence of a much more profound kind of order in the universe than his predecessors had conceived.

The facts of Newton's life are simple and undramatic. He was the son of an obscure farmer who died before Isaac was born. At first an undistinguished student, young Newton soon revealed his scientific aptitude and was sent to Cambridge to complete his studies. In his twenties Newton was appointed professor of mathematics there, and he remained at Cambridge, living quietly and never marrying, for 30 years. Then, at fifty-four, Newton was appointed an official of the British mint, where he remained for the rest of his life. Honors came to Newton in profusion, and he was buried in Westminster Abbey with the noblest of England's dead.

In contrast to this uneventful life are the adventures of his far-ranging mind, adventures which cannot but amaze all who read of his work. In the law of gravitation Newton solved the problem of planetary motion and gave science a powerful tool for understanding natural phenomena that remained the final word on the subject for 250 years until Einstein, in his general theory of relativity, showed how gravitation fits into a broader picture of the structure of time and space. Newton's formulation of the three laws of motion placed the science of mathematics on a solid foundation. By inventing the calculus, Newton initiated a new and powerful kind of mathematics indispensable in modern science and engineering. Finally, his work in optics was among the earliest systematic investigations of the properties of light. Newton's great work, the *Principia*, was published in 1687. This event is one of the most important landmarks in the whole history of science.

Of all the resounding tributes that have been paid to Newton's greatness, perhaps the most elegant is that by the mathematician Joseph Louis Lagrange: "Newton was the greatest genius who ever lived, and the most fortunate, for there cannot be more than once a system of the world to establish."

CIRCULAR MOTION

We shall begin our study of gravitation by considering the general problem of motion along a curved-path, the kind of motion we must be able to understand before we tackle the forces at work on the planets.

centripetal force

A simple example of motion along a curved path is the whirling of a ball tied to the end of a string (Fig. 3.1). Even though the speed of the ball may be the same at all times, its direction constantly changes and therefore the ball must be accelerated. We find that to hold the ball in its circular path requires the continual exertion of a force by our hand on the string, which is the force responsible for the continual acceleration of the ball. The force that we exert on the ball through the string is called *centripetal* ("center-seeking") force. It is the external force that must be applied in order to cause an object to move in a curved path. The acceleration produced by a centripetal force is called centripetal acceleration, and, like the force that gives rise to it, is directed toward the center of curvature of the object's path.

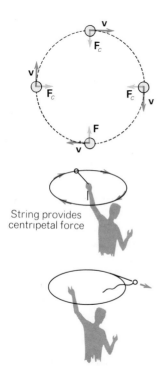

String provides centripetal force

3.1 A centripetal force is necessary for circular motion. *An inward centripetal force F_c acts upon every object that moves in a curved path. If the force is removed, the object continues moving in a straight line tangent to its original path.*

centripetal force produces motion in a curved path

GRAVITATION

Let us consider the ball's motion between any two points fairly close together on the circle it describes, such as A and B in Fig. 3.2. At A the ball moves with a velocity represented by the tangent vector $\mathbf{v}_A$ and, if unrestrained, would continue to move with this velocity along a straight line. At B the ball's velocity is $\mathbf{v}_B$, and, as shown, the total change in its velocity between positions A and B is given by the vector $\Delta\mathbf{v}$. The ball's velocity at B is the vector sum of $\mathbf{v}_A$, its velocity at A, and $\Delta\mathbf{v}$, its change in velocity, as shown in the right-hand diagram in Fig. 3.2. If the centripetal force acting from O is of the proper magnitude to keep the ball moving with uniform speed in a circle, then $\mathbf{v}_B$ will have the same length as $\mathbf{v}_A$ and will be tangent to the circle at B.

As discussed in the next section, the velocity change $\Delta\mathbf{v}$ is directed toward the center of the circle, which means that the centripetal acceleration of the ball and (by Newton's second law of motion) the centripetal force on it are both also directed toward the center.

A detailed calculation shows that the centripetal force $\mathbf{F}_c$ that must be provided to cause an object of mass m and speed v to travel in a circular path of radius r has the magnitude

$$F_c = \frac{mv^2}{r}$$

To whirl a given body in a circle requires a greater force if it moves fast or if the circle is small, and the more massive the object, the more force is needed. The direction of $\mathbf{F}_c$, which is a vector quantity, is toward the center of the circle as mentioned above.

From the formula for F_c we can see, for example, why cars rounding a curve are so difficult to steer when the curve is sharp (small r) or the speed is high (a large value for v means a very large value for v^2). A heavy car (large m) is evidently going to be generally harder to maneuver than a light car (small m). In the case of a car, the centripetal force is supplied by friction between the car's tires and the road; if the force needed to make a particular turn at a particular speed is too great, the car skids.

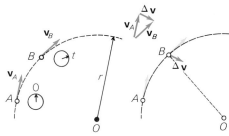

3.2 *A and B are two successive positions t sec apart of a ball undergoing uniform circular motion at the speed v in a circle of radius r. The difference between the ball's velocity $\mathbf{v}_A$ at A and its velocity $\mathbf{v}_B$ is $\Delta\mathbf{v}$. The magnitude of the centripetal acceleration is $\Delta v/t$, which detailed analysis shows is equal to v^2/r.*

centripetal force

On a level curve, friction between its tires and the road provides a slow-moving car with the required centripetal force. If the car's speed is high, the curve must be banked so that the reaction force of the road (which arises in response to the car's weight) can furnish enough centripetal force to prevent an outward skid.

Let us consider a 1,000-kg car which is traveling at 5 m/sec (about 11 mi/hr) around a turn 30 m in radius. The centripetal force needed to make the turn is

$$F_c = \frac{mv^2}{r} = \frac{1{,}000 \text{ kg} \times (5 \text{ m/sec})^2}{30 \text{ m}} = 833 \text{ newtons}$$

This force (which is about 190 lb) is readily transmitted by the road to the car if the pavement is dry and in good condition. However, if the car's speed were 20 m/sec, the required force would be sixteen times greater, and the car would be unlikely to be able to make the turn.

derivation of centripetal force formula

From Fig. 3.3 it is evident that the vector triangle whose sides are $\mathbf{v}_A$, $\mathbf{v}_B$, and $\Delta \mathbf{v}$ is similar to the space triangle whose sides are OA, OB, and s, respectively. Since corresponding sides of similar triangles are proportional in length,

$$\frac{\Delta v}{v} = \frac{s}{r}$$

where Δv = length of $\Delta \mathbf{v}$
v = length of both $\mathbf{v}_A$ and $\mathbf{v}_B$
r = radius of circle
= OA and OB

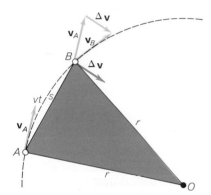

3.3 How centripetal acceleration is calculated. *The chord joining A and B is s, while the actual distance the particle traverses is vt. In calculating the instantaneous acceleration of the particle we are restricted to having A and B an infinitesimal distance apart, in which case the chord and arc have the same length and v points radially inward to the center of the circle at O. Here A and B are shown relatively far apart for clarity, and so v is not quite radially inward.*

The actual distance the ball travels in going from A to B is vt, where t is the time required for it to cover this distance. In the diagram vt, an arc of the circle, is clearly longer than s, the chord that joins A and B. However, the closer together the points A and B are, the more nearly equal vt and s are. When A and B are infinitely close together, $vt = s$. This limiting situation is what we are concerned with, since the quantity of interest here is the *instantaneous acceleration* of the ball (that is, its acceleration at any particular instant). Hence we can substitute vt for s in the above formula and obtain

$$\frac{\Delta v}{v} = \frac{s}{r} = \frac{vt}{r}$$

so that

$$\frac{\Delta v}{t} = \frac{v^2}{r}$$

In Chap. 1 we learned that the acceleration of a body is given by the change in its velocity during a certain time interval divided by that time interval. Here the ball changes its velocity by Δv during the time t, so its acceleration is

$$a_c = \frac{\Delta v}{t} = \frac{v^2}{r}$$

centripetal acceleration

The quantity a_c is the centripetal acceleration of a body of speed v which moves along a circular path of radius r.

Newton's second law of motion permits us to relate the centripetal force F_c on an object in circular motion to its centripetal acceleration a_c. Because, in general,

$$F = ma$$

here we have

$$F_c = ma_c$$

The centripetal force on a body is the product of its mass m and its centripetal acceleration. The direction of the force is the same as that of the acceleration, so $\mathbf{F}_c$, as we know from experience, points toward the center of the circular path. By substituting for a_c we obtain

$$F_c = \frac{mv^2}{r}$$

UNIVERSAL GRAVITATION

ptolemy and copernicus

If we forget our modern knowledge, there are two possibilities around which a satisfactory hypothesis of the astronomical universe may be constructed: either the earth is stationary, as it appears to be, with the celestial objects revolving about it, or the earth is moving, its motion being then responsible for a part of the apparent motion of other objects. For instance, the apparent daily rotation of the sky may represent actual motions of sun, moon, planets, and stars, or it may be explained by a rotation of the earth on its axis. These alternatives were clear to the philosophers of ancient Greece, most of whom advocated a stationary earth while a few argued for a moving earth. In their day scientific knowledge was not sufficiently advanced to settle the matter. It is hardly surprising, therefore, that the majority of the Greeks favored the common-sense view that the earth is stationary.

The hypothesis most widely accepted by the later Greek and Roman scholars was originally devised by Hipparchus. Ptolemy of Alexandria subsequently incorporated Hipparchus's ideas into his *Almagest*, a compendium of astronomy which was to remain the standard reference on the subject for over a thousand years, and this picture of the universe became known as the ptolemaic system.

It was an intricate and ingenious system. Our earth stands at the center, motionless, with all other objects in the universe revolving about it in paths that are either circles or combinations of circular motions—since

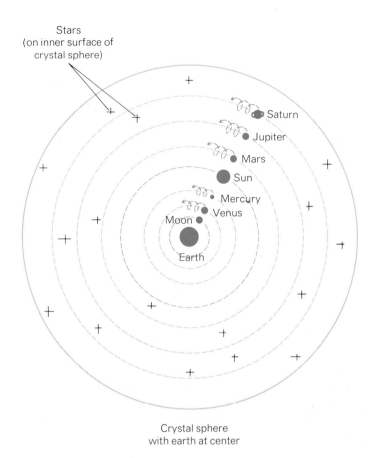

3.4 The ptolemaic system, showing the assumed arrangement of the members of the solar system within the celestial sphere. Each planet is supposed to travel around the earth in a series of loops, while the orbits of the sun and moon are circular. Only the planets known in Ptolemy's time are shown.

in the ptolemaic system, the earth is stationary at the center of the universe

to the Greeks the circle was the only "perfect" curve, hence the only conceivable path for a celestial object. Enclosing all is a gigantic crystal sphere studded with the fixed stars, making approximately one revolution each day. Somewhere inside is the sun, moving around the earth exactly once a day. The difference in motion between sun and stars is just enough so that the sun appears to move among the constellations, completing its circuit once a year. Near the earth in a small orbit is the moon, revolving more slowly than the sun. The planets Venus and Mercury come between moon and sun, the planets Mars, Jupiter, and Saturn between sun and stars. To account for the observed peculiarities of planetary motion, Ptolemy imagined each planet to move in a small circle about a point which in turn described a large circle about the earth (Fig. 3.4). By a combination of these circular motions each planet travels in a series of loops; since we observe these loops edgewise, it appears to us as if the planets moved with variable speeds and sometimes reversed their motions.

From observations made by himself and his predecessors Ptolemy calculated the relative speed with which each celestial object moved in its orbit. Using these speeds he could then compute the location of an object in the sky for any date in the past or future. These computed positions checked fairly well, though not perfectly, with observed positions that had been recorded for several centuries before his time, and his predictions also agreed fairly well with observations made in succeeding years. So Ptolemy's system fulfilled all the requirements of a scientific hypothesis: it was based solidly on observational facts, it explained adequately all the facts about celestial motions that were known in his time, and it made possible the prediction of facts that could be verified in the future.

By the sixteenth century it had become obvious that something was not quite right in the ptolemaic system. Observed positions of the planets simply did not agree with the positions calculated from Ptolemy's complicated orbits. Discrepancies were not large but could be detected even by inexpert observers. There were two possible ways of removing the discrepancies: either slight changes could be introduced into the ptolemaic orbits, making the system still more complicated, or the ptolemaic hy-

A 1598 portrait of Tycho Brahe in his laboratory. The man at the right is determining the position of a celestial body by shifting a sighting vane along a giant protractor until the body is visible through the aperture at upper left.

in the copernican system, the earth turns on its axis and revolves around the sun

pothesis could be discarded in favor of a completely new hypothesis based on different assumptions.

The first to defy tradition by setting up a new explanation for the universe was Nicolaus Copernicus, a versatile and energetic Pole of the early sixteenth century. Copernicus lived in the years following Columbus's great discovery, years when mental as well as geographical horizons were receding before eager explorers. Let us consider the earth, said Copernicus, as one of the planets, a sphere rotating once a day on its axis. Let us further imagine that the planets, including the earth, revolve in circular orbits about the sun (Fig. 3.5), that the moon is relatively close to the earth and revolves about it, and that the stars are situated at great distances beyond. In this picture rotation of the earth on its axis explains the daily rising and setting of celestial objects. The apparent motion of the sun among the stars is due to the earth's motion in its orbit; as we swing around the sun, it appears to us as if the sun were constantly shifting its position against the background of fixed stars. The moon's eastward drift relative to the stars is due in large part to its actual orbital motion. Apparent movements of the planets are explained as combinations of their actual motions around the sun and our shift of position as the earth moves.

The idea behind Copernicus's hypothesis was not new, for some of the Greeks, notably Aristarchus, had been well aware that apparent celestial motions could be the result of motions of the earth. But Copernicus went beyond these earlier speculations in working out the planetary motions mathematically. From observations of the positions of the planets he

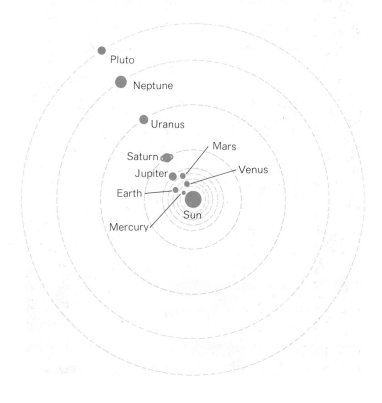

3.5 The copernican system, showing the assumed arrangement of the members of the solar system. The planets, including the earth, are supposed to travel around the sun in circular orbits. The earth rotates daily on its axis, and the moon revolves around the earth. All planets are shown here. The actual orbits are ellipses and are not spaced as shown here, though they do lie in approximately the same plane.

calculated how big each orbit must be in comparison with that of the earth and how fast each planet must be moving. With these figures he could compute the apparent positions for any time in the past or future, just as Ptolemy had done with figures based on a different hypothesis. Copernicus found that these calculated positions agreed with observation fairly well but hardly better than those calculated from the ptolemaic system.

Despite lack of complete check with observation, Copernicus felt that his simple circular orbits gave a truer picture of the universe than the complex orbits of the ptolemaic system. With the publication of Copernicus's manuscript there began a long and bitter argument. To us, growing up in the calm assurance that the earth moves, it seems odd that this straightforward idea was so long and so violently opposed. But in the sixteenth century, before the invention of the telescope, a decision between the ptolemaic and copernican systems was by no means easy. Predictions from both hypotheses agreed only moderately well with observations. Good scientific arguments could be brought forward on both sides. To settle the debate, precise observations were necessary both of celestial objects and of the earth itself, and instruments to make these observations simply were not available.

the copernican system is much simpler than the ptolemaic system

kepler's laws

Fortunately, significant improvements in astronomical measurements—the first since the time of the Greeks—were not long in coming. Tycho Brahe, a Dane of noble descent, built an observatory on the island of Hven near Copenhagen in which the instruments were as rigid and precise as possible. With the help of these instruments, Tycho, blessed with exceptional eyesight and patience, made thousands of measurements, a labor that occupied much of his life. Even without the telescope, which had not yet been invented, Tycho's observatory was able to determine celestial angles to better than one-hundredth of a degree.

At his death in 1601, Tycho left behind his own somewhat peculiar theory of the solar system, a body of superb data extending over many years, and an assistant named Johannes Kepler. Kepler regarded the copernican scheme "with incredible and ravishing delight" and fully expected that Tycho's improved figures would prove Copernicus correct once and for all. But this was not the case; after 4 years of work on the orbit of Mars alone, Kepler could not reconcile the observational data with any of the models of the solar system that had by then been proposed. If the facts do not agree with the hypothesis, then the hypothesis, no matter how attractive, must be discarded. Kepler began a search for a new cosmic design that would be in better accord with Tycho's observations.

After considering every possibility, which meant years of drudgery in performing calculations by hand, Kepler found that circular orbits for the planets were out of the question even when modified in various plausible ways. He abandoned circular orbits only reluctantly, for he was something of a mystic and believed, like Copernicus and the Greeks, that circles were the only fitting type of path for celestial orbits. Discarding the circle led to consideration of other geometrical figures, and here Kepler found the key to the puzzle: **The paths of the planets around the sun are ellipses** (Fig. 3.6). This is called Kepler's first law.

kepler's first law

Even this epochal discovery was not sufficient, as Kepler realized, for the virtue of the older hypotheses was that they could be used to *predict*—with adequate accuracy for most purposes—the course of the planets through the sky. What was needed now was a law that could relate the speeds of the planets to their positions in their elliptical orbits. Kepler could not be sure that such a law even existed, and he was overjoyed when he came upon the answer: **The planets move so that their radius vectors** (which are imaginary lines drawn from the sun to the planets) **sweep out equal areas in equal times.** Thus, in Fig. 3.7, each of the shaded areas is covered in the same period of time, which means that the planets travel more rapidly when they are near the sun than when they are far from it. The earth, for instance, has a speed of 18.8 miles per second (mi/sec) when it is nearest the sun and 18.2 mi/sec when it is farthest away, a difference of over 3 percent. This rule of equal areas is known as Kepler's second law.

kepler's second law

A great achievement, but Kepler was not satisfied. He was obsessed with the idea of order and regularity in the universe, and the notion that the courses of the planets were not manifestations of some general pattern seemed incredible to him. So Kepler persevered and, after 10 more years of labor, could announce his third law of planetary motion, the so-called harmonic law: **The ratio between the square of the time required by a planet to make a complete revolution around the sun and the cube of its average distance from the sun is a constant for all the planets.**

kepler's third law

To satisfy ourselves that the harmonic law is correct, we can refer to the planetary data of Table 3.1. The average radius of the earth's orbit,

TABLE 3.1 The solar system.

Body	Mean distance from sun, millions of miles	Mean diameter, thousands of miles	Relative mass, $m_{earth} = 1$	Period of revolution[a]	Period of rotation[b]	Mean density, g/cm^3[c]	Acceleration of gravity, g[d]	Escape velocity, mi/sec[e]	Known satellites
Sun		865	333,000		25–30 days	1.4	28	383	(9 planets)
Moon	[f]	2.16	0.01	27⅓ days	27⅓ days	3.4	0.2	1.2	—
Mercury	36	3	0.05	88 days	59 days	5	0.4	2.7	0
Venus	67	7.8	0.82	225 days	243 days[g]	5.0	0.9	6.5	0
Earth	93	7.9	1.00	365 days	24 hr	5.5	1.0	7.0	1
Mars	142	4.2	0.11	687 days	24.6 hr	4.0	0.4	3.2	2
Jupiter	483	89	318	12 years	9.8 hr	1.3	2.7	38	12
Saturn	887	75	95	29 years	10.3 hr	0.7	1.2	23	10
Uranus	1,784	31	15	84 years	10.8 hr[h]	1.5	1.0	14	5
Neptune	2,795	28	17	165 years	15.7 hr	2.0	1.4	16	2
Pluto	3,672	4?	0.2?	248 years	6 days?	6?	?	?	0

[a] Orbital period.
[b] Spin period.
[c] Density of water = 1.0 g/cm³.
[d] 1 g = 32 ft/sec² = 9.8 m/sec².
[e] Speed necessary for permanent escape from gravitational field of the body.
[f] The mean distance of the moon from the earth is 238,000 miles.
[g] The direction of rotation of Venus is opposite to that of the other planets.
[h] The axis of rotation of Uranus is only 8° from the plane of its orbit.

GRAVITATION

for instance, is 93 million miles (9.3×10^7 miles) and its period of revolution about the sun is 365 days. For the earth, then,

$$\frac{T^2}{R^3} = \frac{(365 \text{ days})^2}{(9.3 \times 10^7 \text{ miles})^3} = \frac{1.33 \times 10^5 \text{ days}^2}{8.04 \times 10^{23} \text{ miles}^3}$$
$$= 1.65 \times 10^{-19} \text{ day}^2/\text{mile}^3$$

A similar calculation of T^2/R^3 for each of the other planets yields the same value of 1.65×10^{-19} day^2/mile^3.

Thus, finally, the solar system was reduced to a set of regular motions. Planetary positions computed from Kepler's ellipses agree not only with Tycho's data but with observations made thousands of years earlier. Predictions could be made regarding positions of the planets in the future—accurate predictions this time, no longer mere approximations. Finally, Kepler's laws showed that the speed of a planet in different parts of its orbit was governed by a simple rule and that the speed was related to the size of the orbit.

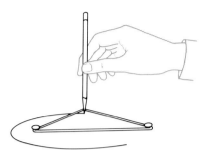

3.6 To draw an ellipse, place a loop of string over two tacks a short distance apart. *Then move the pencil as shown, keeping the string taut. By varying the length of the string, ellipses of different shapes can be drawn. The points in an ellipse corresponding to the positions of the tacks are called focuses; the orbits of the planets are ellipses with the sun at one focus, which is Kepler's first law.*

the law of gravity

That some force is necessary to hold the planets in their elliptical orbits had been recognized before Newton, but the nature of the force had remained a matter of speculation. It was Newton's great inspiration that this force might be the same as the familiar force that pulls objects to the surface of the earth. Perhaps, thought Newton, the moon revolves around the earth much as a stone on the end of a string revolves around a finger, with gravity taking the place of the pull on the string. In other words, perhaps the moon is a falling object, pulled toward the earth just as we are pulled but moving so fast that the pull is just sufficient to keep it from flying off in a straight line away from the earth (Fig. 3.8). Further, the earth and its sister planets might well be held in their orbits by a greater gravitational attraction from the sun.

It was a remarkable idea, but would it work? Galileo had discovered certain exact laws expressing the relations among distance, velocity, and time for ordinary objects falling freely toward the earth; could it be shown that the moon falls according to the same laws? Kepler had been able to obtain three precise statements concerning planetary motion; could a force between planets and sun explain those three laws?

Newton's problem was to express in mathematical terms the gravitational force between different bodies. His method of attack had essentially three steps. He began with Kepler's second law, which states that the line joining any planet with the sun sweeps out equal areas in equal time intervals (Fig. 3.7). Newton showed, with the help of his newly invented calculus, that the force must act directly along the line between sun and planet—just as gravity on the earth pulls objects directly toward the earth's surface.

Kepler's third law gave a more specific hint about the nature of the force. This law states that the square of the time T required for a planet's revolution is proportional to the cube of the radius R of its orbit, so that

gravity holds the moon to the earth and the planets to the sun

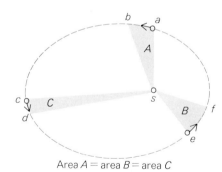

Area $A =$ area $B =$ area C

3.7 Kepler's second law. *As a planet goes from a to b in its orbit, its radius vector (an imaginary line joining it with the sun) sweeps out the area A. In the same amount of time the planet can go from c to d, with its radius vector sweeping out the area C, or from e to f, with its radius vector sweeping out the area B; the three areas A, B, and C are equal.*

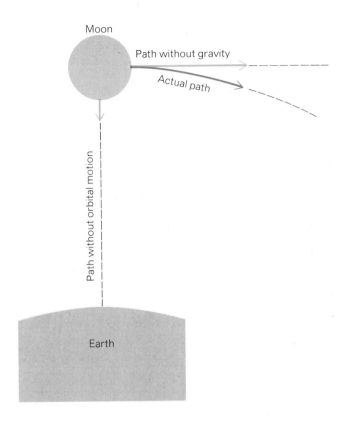

3.8 *The motion of the moon around the earth represents a balance between the downward pull of gravity and the tendency of the moon to travel in a straight line; gravity provides the centripetal force required to keep the moon in its orbit.*

T^2/R^3 has the same value throughout the solar system. Together with the fact that the gravitational attractive force on a planet must provide for the exact amount of centripetal force involved in holding it in its orbit, this law was all that Newton needed to determine the way in which gravitational force varies with distance. His result states that the force of gravitation exerted by the sun on a planet is inversely proportional to the square of the average distance R between them. That is, the force varies as $1/R^2$. All other things staying the same, a planet twice as far from the sun as it normally is would feel only $1/2^2$, or one-fourth as much attractive force; if it were half as far from the sun, the force on it would be $1/(1/2)^2$, or four times greater; and so on. This dependence upon distance is shown in Fig. 3.9.

gravity is an inverse-square force

3.9 *The gravitational force between two bodies depends upon the square of the distance between them. The gravitational force on a planet would drop to one-fourth its usual amount if the distance of the planet from the sun were to be doubled. If the distance is halved the force would increase to four times its usual amount.*

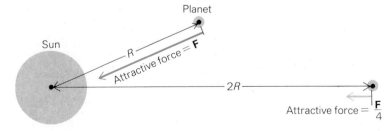

The important conclusion that Newton reached regarding the variation of gravitational force with distance was bolstered by Kepler's first law: each planet moves in an ellipse with the sun at one focus. Again making use of the calculus, Newton showed that a planet attracted to the sun with a force inversely proportional to the square of its distance away must travel in an ellipse.

Galileo's work on falling bodies supplied the final clue. All objects in free fall at the earth's surface have the same acceleration g, and

Weight of object = mass of object × value of g

Therefore the weight of an object, which is the force of gravity upon it, is always proportional to its mass m. Newton's third law of motion (action-reaction) requires that, if the earth attracts an object, the object also attracts the earth. If the earth's attraction for a stone depends upon the stone's mass, then the reaction force exerted by the stone on the earth depends upon the earth's mass. Hence the gravitational force between two bodies is proportional to *both* of their masses.

We can summarize the above conclusions in a single statement: **Every particle in the universe attracts every other particle with a force that is directly proportional to the product of their masses and inversely proportional to the square of the distance between them.** In equation form, Newton's law of gravitation states that the force F present between two bodies whose masses are m_1 and m_2 is

$$F = \frac{Gm_1m_2}{R^2}$$

law of gravitation

where R is the distance between them and G is a constant of nature, the same number everywhere in the universe. The value of G is 6.670×10^{-11} newton-m²/kg² in metric units and 3.44×10^{-8} lb-ft²/slug² in British units.

center of mass

Newton was still confronted with one difficult problem: From what points in two objects should R be measured? The force between an apple and the earth is inversely proportional to the square of their distance apart; but what distance? From the apple to the earth's surface, to the earth's center, or to some other point in the earth's interior? Again using the calculus, Newton succeeded in showing that, for two uniform spherical objects, R is the distance between their centers; in other words, spheres behave as if their masses were concentrated at their centers (Fig. 3.10). Solving the problem is more difficult for objects of other shapes, but in general, for any body, a *center of mass* can be found from which R is to be measured.

The inverse-square relationship between gravitational force and distance of separation R is a most important one. Let us see what this law means

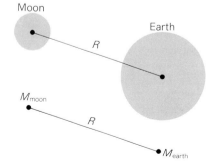

3.10 *For computing gravitational effects, spherical bodies (such as the earth and moon) may be regarded as though their masses are located at their geometrical centers, provided that they are uniform spheres or consist of concentric uniform spherical shells.*

in terms of the earth's attraction for a ball whose weight at sea level is 1 lb. At sea level the force of gravity on the ball is, of course, 1 lb, and the ball is roughly 4,000 miles from the earth's center. If this distance is multiplied by 2, so that the ball is 8,000 miles from the earth's center (in other words, the ball is at an altitude of 4,000 miles above sea level), the force of gravity decreases by a factor of 2^2, or 4; thus its weight would be only $\frac{1}{4}$ lb. If we take the ball much farther out, so that it is 40,000 miles from the earth's center, or ten times farther out than it is at sea level, its weight drops by a factor of 10^2, or 100, to only 0.01 lb. As the distance increases, the gravitational force decreases very rapidly (Fig. 3.11).

newton's proof

Gravitational forces between objects on the earth's surface are so exceedingly small that their measurement is very difficult. In Newton's day such measurement was impossible, so that the law of gravitation could not be established by direct laboratory experiment. Agreement of the law with Kepler's generalizations was of course good evidence in its favor but provided no convincing connection between planetary forces and the familiar force of gravity at the earth's surface. There was one astronomical body, however, which could give Newton this necessary connection and which at the same time could furnish a proof of the gravitational law that did not depend on Kepler's laws. This body was the moon.

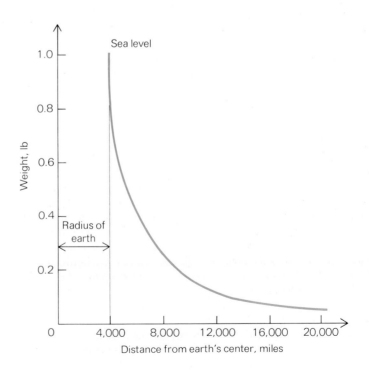

3.11 A body whose mass is $\frac{1}{32}$ slug weighs 1 lb at sea level. *This graph shows how the weight of such a body varies with distance from the center of the earth. The mass of the body is the same everywhere in the universe.*

GRAVITATION

Freely falling objects at the earth's surface move downward with an acceleration of about 9.8 m/sec². At the moon's distance, 240,000 miles, the pull of the earth is reduced to $\frac{1}{3,600}$ of its value here on the surface, so that objects should move toward the earth with an acceleration of $(\frac{1}{3,600}) \times 9.8$ m/sec², which is 0.00272 m/sec². If the law of gravitation is correct, this means that the moon should be falling toward the earth with a velocity increasing by 0.00272 m/sec each second; since $d = \frac{1}{2}at^2$, it should fall 0.00136 m [1.36 millimeter (mm)] in 1 sec, 4.90 m in 1 min, and 10,200 km (about 6,300 miles) in 1 day.

What does it mean to say that the moon "falls" toward the earth 6,300 miles in a day? Obviously our satellite is not directly approaching us at this rate. The "falling" of the moon is simply its deflection away from the straight-line path which it would follow if the earth's pull were absent. In Fig. 3.12, the arc *TV* represents part of the moon's orbit *M* around the earth *E*. At the instant when the moon passes *M*, its velocity is given by the vector **A** along the tangent *MR*. This straight line would represent the moon's path away from *M* if the earth did not attract it, *MS* indicating the distance it would cover in one day. The earth's attraction pulls the moon away from this path, so that in one day it moves to *P* rather than *S*. The line *SP* represents, in effect, the distance that the moon "falls" in a day's time.

To establish the correctness of his gravitational law, Newton's problem was to show that *calculated* values for the moon's fall, like those given above, corresponded with *observed* values of its deflection from a straight-line path. These observed values are found very simply from observations of the moon's distance and rate of motion. We know that the moon completes its nearly circular orbit once in 27.3 days, hence that it moves in one day through an arc of 360°/27.3, or 13.2°. We know further that its average distance from the earth is 240,000 miles. In Fig. 3.12 this 240,000 miles is represented by 6 cm; the angle *MEP* is 13.2°. By direct measurement on the diagram, the amount of the moon's deflection from a straight line *SP* is found to be approximately 1.5 mm. Since every centimeter represents 40,000 miles, 0.15 cm is equivalent to 6,000 miles. This is not far from the calculated value, 6,300 miles, for the moon's fall in one day.

Obviously so small a distance as *SP* cannot be measured with any great accuracy. To make the comparison with the precision demanded in scientific work, we should need to use more accurate values for the earth's radius, for the moon's distance and time of revolution, for the acceleration of gravity; we should have to remember that the moon's orbit is actually an ellipse rather than a circle; and we should carry out the calculation mathematically rather than by measuring distances in a diagram. When Newton performed this computation, he found close agreement between observed values and those calculated from his gravitational law. The force that holds the moon in its orbit was proved identical with the pull of the earth on objects at its surface.

experimental proof

Nearly a century after Newton's time another great English scientist, Henry Cavendish, succeeded in measuring directly the force of gravitation.

the moon's orbit is the result of its falling toward the earth owing to gravity

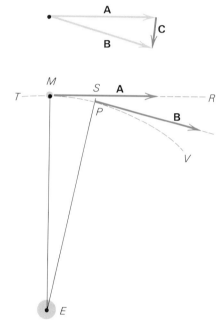

3.12 The moon as a falling body. *In one day the moon "falls" a distance SP away from a straight-line path. The small triangle is a vector diagram of velocities. Scales: Distances, 1 cm = 40,000 miles. Velocities, 1 cm = 0.2 mi/sec.*

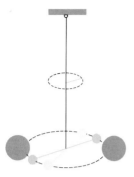

3.13 Scheme employed by Cavendish in demonstrating the law of gravitation. *The gravitational force between each pair of spheres can be determined from the amount of twist in the wire when the two large spheres are brought near the apparatus.*

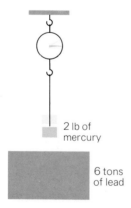

3.14 Another method for measuring gravitational forces. *The increase in the downward force on the mercury when the lead is placed underneath it is a measure of the gravitational attraction between them.*

His method was simple: two small metal spheres were attached to the ends of a light rod, which was suspended at its center from a slender wire; two heavy lead spheres were brought near the small ones, and the twist in the wire was observed (Fig. 3.13). Simple as it sounds, the completion of the experiment was a real feat, for the forces involved are incredibly small. By these delicate measurements Cavendish obtained experimental proof of the law of gravitation.

An even more direct experiment was performed later by a German named Phillip von Jolly, who weighed a few pounds of mercury carefully and then measured the increase in the weight of the mercury when 6 tons of lead was placed directly below it (Fig. 3.14). The Cavendish method, however, is superior in practice and figured in the most recent determination of the constant G at the National Bureau of Standards in Washington. In the latter experiment $1\frac{1}{2}$-oz gold spheres were placed at the ends of an 8-in. rod, and the external masses were in the form of steel cylinders weighing 145 lb each. In the absence of the cylinders, the rod required about 30 min to swing back and forth; when the cylinders were put in position, the period of vibration was reduced to about 25 min owing to the gravitational attraction between them and the gold spheres at the ends of the rod, and from this difference G could be calculated with great accuracy.

It is hard to realize just how small gravitational forces are, so that enormous masses—such as that of the sun or that of the earth—are needed to exert an appreciable pull on nearby objects. Two 40,000-ton ocean liners, for instance, if they are $\frac{1}{2}$ mile apart, attract each other with a force of only $\frac{1}{2}$ oz!

artificial satellites

The first artificial earth satellite, Sputnik I, was launched by the Soviet Union in 1957, and since then hundreds of others have been put in orbits about the earth by both the Soviet Union and the United States. What keeps them from falling down? The answer is that a satellite *is* actually falling down, but, like the moon (which is a natural satellite), at exactly such a rate as to circle the earth in a stable orbit. "Stable" is a relative term, to be sure, since friction due to the extremely thin atmosphere present at the altitudes of actual satellites will eventually bring them down. Satellite lifetimes in orbit range from a matter of days to hundreds of years.

It is not hard to calculate the speed v an artificial satellite the distance r from the center of the earth must have. What we do is set equal the centripetal force mv^2/r needed to keep the satellite in its orbit and the gravitational force mg on the satellite, whose mass is some amount m. Thus

$$F_{\text{centripetal}} = F_{\text{gravitational}}$$
$$\frac{mv^2}{r} = mg$$

GRAVITATION

$$v^2 = rg$$
$$v = \sqrt{rg}$$

speed required for stable satellite orbit

The mass of the satellite turns out to be irrelevant.

For an approximate value of v, we can use the radius of the earth, 6.4×10^6 m, and the sea-level acceleration of gravity g, 9.8 m/sec^2, which yields

$$v = \sqrt{6.4 \times 10^6 \text{ m} \times 9.8 \text{ m/sec}^2} = 7.9 \times 10^3 \text{ m/sec}$$

This speed is about 18,000 mi/hr. At a lower speed than this, a projected vehicle would simply fall to the earth, while at a higher speed it could have an elliptical rather than a circular orbit.

If the speed is high enough, at least 25,000 mi/hr, the vehicle would be able to escape from the earth entirely (Fig. 3.15). The speed required for an object to leave the gravitational field of an astronomical body permanently is called the *escape velocity*. Escape velocities for the sun, moon, and planets are listed in Table 3.1.

escape velocity

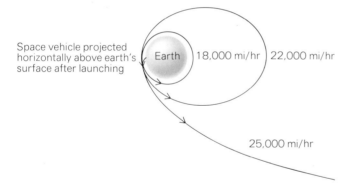

3.15 The minimum speed an earth satellite can have is 18,000 mi/hr, while the escape velocity from the earth is 25,000 mi/hr.

the discovery of neptune and pluto

We have seen something of the immense usefulness of the law of gravitation in astronomical explanations and calculations. But historically the most spectacular application of this law was the prediction of the existence of the two outermost members of the sun's family.

The planet Uranus was found by accident in 1781. Observations during the next few years enabled astronomers to work out details of the new planet's orbit and to predict its future positions in the sky. To make these predictions, it was necessary to consider not only the sun's attraction but the minor attractions of the neighboring planets Jupiter and Saturn as well. The calculations were long and tedious, but their results were accepted unquestioningly. For 40 years, about half the time required for Uranus to make one complete revolution, calculated positions of the planet agreed accurately with observed positions. Then an error seemed to creep in. Little by little the planet moved away from its predicted path among the stars. The calculations were checked and rechecked, but no mistake could be found; the attractions of all known bodies had been correctly allowed for. One of two conclusions seemed necessary: either the law of gravitation, on which the calculations were based, was not strictly accurate, or else some unknown body was attracting Uranus away from its predicted path.

So firmly established was the law of gravitation that two young men, Urbain Leverrier in France and John Couch Adams in England, set themselves the prodigious task of calculating the position of an unknown body that might be responsible for the discrepancies in Uranus's position. Adams, completing his computations first, sent them to England's Astronomer Royal. Busy with other matters, the Astronomer Royal put the calculations away for future checking. Meanwhile, Leverrier sent his paper to a young German astronomer, Johann Gottfried Galle, who lost no time in turning his telescope to the part of the sky where the new planet should appear. Very close to the position predicted by Leverrier, Galle found a faint object, which proved to be the eighth member of the sun's family. A little later the Astronomer Royal showed that Neptune's position had been correctly given in Adams's work also.

When Neptune had been observed for several decades, very slight discrepancies between the observed and calculated positions of this planet led the American astronomer Percival Lowell to predict the existence of yet another planet. The discovery of Pluto was hardly as dramatic as that of Neptune; the discrepancies in Neptune's orbit were so small that Pluto's position could not be predicted with accuracy, and the search dragged on for 25 years before the ninth planet was finally located in 1930. Curiously enough, though Pluto's position when found and its orbit are both close to Lowell's predictions, these predictions turned out to be based upon faulty data. Pluto's discovery therefore seems to have been more a lucky coincidence than anything else.

THE "SCIENTIFIC METHOD"

We have already made considerable progress in our attempt to explore the universe about us. Ours has been a passive role, in the sense that we have confined ourselves to learning facts and interpretations based on centuries of scientific work. It is appropriate at this point to pause and look into the procedures by which scientists acquired this knowledge in the first place. Diverse as these procedures are, we find underlying them in all scientific work a basic pattern of inquiry, a general scheme of attacking problems, which has become known as the *scientific method*.

the scientific method

In grossly simplified terms, we may describe the scientific method as comprising four steps: (1) formulating a problem, (2) observation, (3) generalization from the observed facts, and (4) checking the generalization by further observations. Observations of the material world constitute the heart of the scientific method: they serve both as the foundation on which a scientist builds his theories and as the ultimate proof of the correctness of these theories.

1. *Formulating a problem* may involve no more than selecting a certain field to study, but more often a scientist has some specific notion in mind that he wishes to investigate further. In many cases there is a good deal of overlap between formulating a problem and making a generalization: the scientist has a speculation, a hunch, about some aspect of nature, but he cannot come to any definite conclusion without further work. *the problem*

2. *Scientific observation* is carried out painstakingly, to be sure that all pertinent facts are collected and that each fact is checked for accuracy. If the number of facts is large, a considerable amount of classification and analysis of the observational data may be necessary before generalization is possible. *observation*

3. *Generalization* may be merely the statement of a rule or pattern to which the observed facts seem to conform. Or it may be a more ambitious attempt to explain the observations in terms of simpler rules and processes. In any case it involves the extension of results from one series of observations to similar observations in new and untried circumstances, in other words, the *prediction* of results in other experiments. *generalization*

4. *Checking a generalization* implies the setting up of new experiments whose results can be predicted from the generalization. If the new observations agree with the predictions, the generalization has proved its usefulness. The new observations may lead to further generalizations or to refinements of the old one, which in turn must be checked by further experiment, and so on indefinitely. *checking*

As put forward originally, a scientific generalization is commonly called a *hypothesis*. When checked and rechecked in a variety of ways, so that there is no longer any doubt of its correctness, the hypothesis becomes a *law*. The word *theory* is usually reserved for a larger logical structure, built on two or three fundamental generalizations, designed to explain a wide variety of phenomena. But there is no uniformity in scientific circles regarding the use of these three terms. *hypothesis, law, and theory*

A breakdown of the scientific method into four steps does not mean that a working scientist always consciously follows these steps in order. Sometimes all four processes go on nearly simultaneously and quite unconsciously; while he is collecting data a possible hypothesis may suggest itself as a hunch, and almost immediately he may be aware that the hunch has already been tested by previous experiments designed for some other purpose. The four steps are simply a formalized expression of a process that a scientist, consciously or unconsciously, uses over and over again as he works.

Nor does this simple description of the scientific method mean that any neophyte going into a laboratory can make startling discoveries merely by telling himself to observe, generalize, and check. He would not know, first and most important, what questions to ask of nature; nor would he know the techniques of observation or how to devise experiments to serve as valid checks; and finally, he would not have the background or the intuition or the trained imagination to arrive at reasonable generalizations.

Making the scientific method sound simple as a logical process does not make the individual steps simple, nor does it detract from the greatness of scientists who are unusually skillful in formulating problems, in securing data, or in originating hypotheses.

An appropriate example of the scientific method is the history of attempts to explain planetary motions. The original observations are records of the positions of the planets with respect to the earth and the fixed stars. From these data the Greeks made several attempts to piece together a reasonable generalization, the most successful being the ptolemaic hypothesis. From this hypothesis future positions of the planets could be predicted, and observations checked the predictions satisfactorily. As centuries passed and observational methods were improved, it appeared that the check was not quite accurate. Some modification of the original hypothesis was necessary, but modification seemed impossible without introducing unreasonable complexities.

Then appeared a rival generalization, the hypothesis of Copernicus. Predictions from this hypothesis gave at first no better checks with observation than Ptolemy's hypothesis. Kepler refined Copernicus's idea, basing his more accurate generalizations on the observations of Tycho Brahe. With these modifications the copernican system now made possible predictions of planetary positions that agreed well with the data. Because of these accurate checks and because of its greater simplicity, the copernican hypothesis soon replaced the ptolemaic hypothesis. Its correctness was established even more firmly by the telescopic observations of Galileo and his successors.

Behind Kepler's laws and Galileo's law of motion Newton discerned an even broader generalization, the law of gravitation. This generalization he checked by observations of the moon's motion. Newton and the physicists who followed him used his law to make one prediction after another, which have been checked by observation.

It is safe to say that the scientific method has made possible the technological civilization of today. Without it science and engineering would still be in a primitive trial-and-error stage, far removed from the present-day advanced state of these disciplines. It is hard to see, for instance, how the law of gravitation, in its glittering combination of simplicity and universality, could have been developed in the framework of the ptolemaic system.

limits of scientific method

But we must remember that the scientific method has strict limitations. It is a means of discovering and organizing facts about the physical world, and no more than that. Observed facts are the foundation of its structure and the ultimate proof of its results. Usually the facts must be based on observations that can be repeated at will; always they must be facts that would be clear to anyone with normal senses and sufficient training to understand them. Scientists may disagree about how observations should be interpreted, but about the observations themselves there should be no dispute. This insistence on accurate observational data is what sets the various natural sciences apart from other modes of intellectual endeavor. But it is likewise a severe limitation, a limitation that must be clearly recognized, especially if the scientific method is to serve any useful purpose in fields of thought into which human values enter—because human values are by nature subjective, the products of an individual's emotions, and therefore impossible to deal with in an objective, quantitative fashion.

GLOSSARY

The *centripetal force* on a body moving in a circle is the inward force that must be exerted to produce this motion. It always acts toward the center of the circle in which the body is moving.

Kepler's laws of planetary motion state that (1) the paths of the planets around the sun are ellipses, (2) the planets move so that their radius vectors sweep out equal areas in equal times, and (3) the ratio between the square of the time required by a planet to make a complete revolution around the sun and the cube of its average distance from the sun is a constant for all the planets.

Newton's *law of gravitation* states that every body in the universe attracts every other body with a force that is directly proportional to the product of their masses and inversely proportional to the square of the distance between them.

The *scientific method* is a general scheme for attacking scientific problems that may be thought of as consisting of four steps: (1) *formulating a problem* concerning some aspect of the physical world; (2) *observation*, that is, the collection of facts that bear upon the problem; (3) *generalization*, that is, the statement of the pattern to which the observed facts seem to conform, or an explanation of the observations in terms of simpler patterns or processes; (4) *checking the generalization* by performing new experiments or making new observations to verify predictions made on the basis of the generalization.

A *hypothesis* is a scientific generalization as originally presented; a *law* is a scientific generalization whose correctness has been clearly demonstrated; a *theory* is logical structure built on several fundamental generalizations that explains a wide variety of phenomena. (These terms are often used in senses slightly different from the ones indicated, but the definitions stated here indicate their usual meanings.)

EXERCISES

1 A track team on the moon unencumbered by space suits could set new records for the high jump or pole vault because of the smaller gravitational force. Could sprinters also improve their times for the 60-yd dash?

2 Why would a stone dropped down a deep well land a short distance east of a point vertically below the point from which it is dropped?

3 Where on the earth's surface is the acceleration due to its rotation greatest? Where is it least? Why?

4 Compare the weight and mass of an object at the earth's surface with what they would be at the earth's center.

5 Is the sun's gravitational pull on the earth the same at all seasons of the year? Explain.

6 According to the basic premise of gravitation, the earth must be continually "falling" toward the sun. If this is true, why does the average distance between earth and sun not grow smaller?

7 According to Kepler's second law, the earth travels fastest when it is closest to the sun. Is this consistent with the law of gravitation? Explain.

8 An artificial satellite is placed in orbit half as far from the earth as the moon is. Would its time of revolution around the earth be longer or shorter than the moon's if its orbit is to be stable?

9 What centripetal force is required to keep a 1-kg mass whose speed is 5 m/sec moving in a circle of radius 2 m?

10 What centripetal force is required to keep a 1-lb weight whose speed is 10 ft/sec moving in a circle of radius 4 ft?

11 The maximum force a road can exert on the tires of a certain 3,500-lb automobile is 800 lb. What is the greatest speed with which the automobile can round a turn of radius 1,200 ft?

12 A string 3 ft long is used to whirl a 2-lb stone in a vertical circle at a speed of 12 ft/sec. What is the tension in the string when the stone is at the top of the circle? At the bottom? Halfway between?

13 The minute hand of a clock is 0.1 m long. Find the centripetal acceleration of its tip.

14 A string 1 m long is used to whirl a $\frac{1}{2}$-kg stone in a vertical circle at a speed of 5 m/sec. Find the tension in the string when the stone is at the top and at the bottom of its path.

15 A man swings a pail of water in a vertical circle 4 ft in radius. If the water is not to spill, what is the maximum time that each revolution of the pail should take?

16 What is the acceleration of a meteor when it is 8,000 miles from the earth's surface?

GRAVITATION

17 A man whose weight on the earth is 160 lb is in a rocket 20,000 miles from the earth's surface. With what force is he attracted to the earth?

18 Venus has a diameter of 7,800 miles and a mass 0.8 of the earth's mass. How much will a girl whose weight on the earth is 100 lb weigh on the surface of Venus?

19 The sun's mass is 2×10^{30} kg and the average radius of Jupiter's orbit is 7.8×10^{11} m. Find the gravitational force the sun exerts on Jupiter and the orbital speed Jupiter must have in order not to be drawn into the sun nor to fly off into space.

20 Find the acceleration of gravity at the surface of an imaginary planet whose mass and radius are both half those of the earth.

21 A 2-kg mass and a 5-kg mass are 1 m apart. Find the gravitational force each one exerts on the other, and find their respective accelerations if they are free to move.

22 With the help of the data in Table 3.1, find the minimum speed artificial satellites must have to pursue stable orbits about the moon and about Jupiter. (Note that 1 mile = 5,280 ft.)

chapter 4

Energy and Momentum

WORK

- definition of work
- the joule
- power
- british units of work and power

ENERGY

- kinetic energy
- derivation of kinetic-energy formula
- potential energy
- energy transformations
- reference level
- energy in other forms

MOMENTUM

- linear momentum
- rocket propulsion
- angular momentum
- conservation principles and symmetry

We use the word *energy* so often and in such a variety of ways that the very specific definition of this term given in physics may be something of a surprise. Usually we associate energy with activity or motion: a falling stone possesses energy, an energetic person is constantly doing things. Sometimes, though, we speak of certain foods as being rich in energy or of the earth as receiving radiant energy from the sun. What is it that a piece of pie and a falling stone have in common? To the physicist the answer is obvious: the ability to do *work*. Let us begin our discussion of energy, then, by considering work.

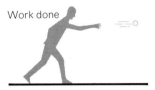

4.1 *The work done by a force on a body is the product of the force and the distance through which the body moves while the force acts upon it.* For a force to do work on a body, the body must undergo a displacement while the force acts on it. No work is done by pushing against a rigid wall.

WORK

Changes that take place in the physical world are invariably the result of forces. Forces are needed to pick things up, to move things from one place to another, to squeeze things, to stretch things, and so on. However, not all forces produce changes, and it is the distinction between forces that accomplish change and forces that do not that is central to the idea of work.

definition of work

If we push against a stone wall, nothing happens. We have applied a force, but the wall has not yielded in any way and shows no effects. However, if we apply the same force to one of the stones that make up the wall, the stone flies through the air for some distance (Fig. 4.1). Now something has been accomplished as the result of our push. The basic difference between the two situations is that, in the first case, our hand did not move while it pushed against the wall. The force was a stationary one. In the second case, when we threw the stone, our hand did move while the force was being applied and before the stone actually left our grasp. It is the motion of the body while the force acts upon it that constitutes the difference between the two cases and that is responsible for the difference in the two results.

When we analyze the problem of forces and the actions that they produce, we find that all forces that produce effects such as motion or distortions in an object undergo displacements in so doing; that is, a moving force accomplishes something, while the stationary force does not. The physicist makes this concept definite by defining a quantity called *work*. To a physicist, **the work done by a force acting on a body is equal to the magnitude of the force multiplied by the distance through which the force acts.** If the distance is zero, no work is done by the force, no matter how great it is. And even if something moves through a distance, work is not done on it unless a force was acting.

The above formal definition is in agreement with our observations. However, it is clear that a person can become tired without doing work, so how tired he becomes is no valid index of the amount of work he has performed. Surely, pushing against a wall for an afternoon in the hot sun is more exhausting than simply throwing a stone, but in the latter case work is being done while in the former nothing is being accomplished at all.

In algebraic form,

$$W = Fd$$

where W = work
F = applied force
d = distance through which force has acted

work is force times displacement

there is no work without motion

In this equation **F** is assumed to be in the same direction as **d**. If it is not, for example in the case of a body pulling a wagon with a rope

ENERGY AND MOMENTUM

not parallel to the ground, we must use for F the component of the applied force that acts in the direction of the motion (Fig. 4.2).

The component of a force in the direction of a displacement **d** is

$$F \cos \theta$$

where θ is the angle between **F** and **d**. Hence the most general equation for work is

$$W = Fd \cos \theta$$

general formula for work

When **F** and **d** are parallel, $\theta = 0$ and $\cos \theta = 1$, so that $Fd \cos \theta$ reduces to just Fd. When **F** and **d** are perpendicular, $\theta = 90°$ and $\cos \theta = 0$, so that no work is done. A force that is perpendicular to the motion of an object can do no work upon it. Thus gravity, which results in a downward force on everything near the earth, does no work on objects moving horizontally along the earth's surface. However, if we drop an object, as it falls to the ground work is definitely done upon it.

no work is done by a force acting perpendicular to a displacement

the joule

In the metric system, the unit of work is the *joule*, where one joule is the amount of work done by a force of one newton when it acts through a distance of one meter. That is,

$$1 \text{ joule} = 1 \text{ newton-m}$$

the joule is the metric unit of work

Hence we can write the definition of work in the form

$$W \text{ (joules)} = F \text{ (newtons)} \times d \text{ (meters)}$$

If we push a box for 8 m across a floor with a force of 100 newtons, the work we perform is

$$W = Fd = 100 \text{ newtons} \times 8 \text{ m} = 800 \text{ joules}$$

How much work is done in raising a 500-kg elevator cab from the ground floor of a building to its tenth floor, 30 m higher? We note that the force needed is equal to the weight of the cab, which is mg, and so

$$W = Fd = mgd = 500 \text{ kg} \times 9.8 \text{ m/sec}^2 \times 30 \text{ m}$$
$$= 147{,}000 \text{ joules} = 1.47 \times 10^5 \text{ joules}$$

power

It is often of interest to know the rate at which work is being done by some force. This quantity is called *power*, P, and, by definition, is equal to the work W performed divided by the period of time t involved:

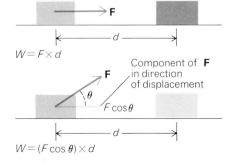

4.2 *When a force and the distance through which it acts are parallel, the work done is equal to the product of F and d. When they are not in the same direction, the work done is equal to the product of d and the component of **F** in the direction of d, namely $(F \cos \theta) \times d$.*

power is rate of doing work

$$P = \frac{W}{t}$$

In the metric system the unit of power is the *watt*, where

1 watt = 1 joule/sec

the watt is the metric unit of power

Thus a motor with a power output of 5,000 watts is capable of doing 5,000 joules of work per second.

A *kilowatt* (kw) is equal to 1,000 watts. Hence the above motor has a power output of 5 kw.

How much time does the elevator cab of the previous example need to ascend 30 m if it is being lifted by a 5-kw motor? We rewrite $P = W/t$ in the form

$$t = \frac{W}{P}$$

and then substitute $W = 1.47 \times 10^5$ joules and $P = 5 \times 10^3$ watts to find that

$$t = \frac{W}{P} = \frac{1.47 \times 10^5 \text{ joules}}{5 \times 10^3 \text{ watts}} = 29.4 \text{ sec}$$

See Fig. 4.3.

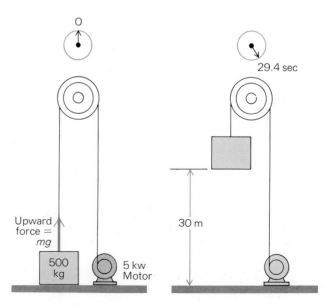

4.3

british units of work and power

In the British engineering system of units, the unit in which work is measured is called the *foot-pound*. One foot-pound is the amount of work done by a force of one pound that acts through a distance of one foot. As an example, lifting a 50-lb crate to a height of 4 ft means that a force of 50 lb acts through a distance of 4 ft. Hence the work done is

the foot-pound is the british unit of work

$$W = Fd = 50 \text{ lb} \times 4 \text{ ft} = 200 \text{ ft-lb}$$

To convert from one system of units to another, we note that

1 joule = 0.738 ft-lb

1 ft-lb = 1.36 joules

In the British system the proper unit of power is the foot-pound per second (ft-lb/sec). However, the horsepower (hp) is the customary unit of power in much engineering work. The origin of this unit is interesting. In order to sell the steam engines he had invented, James Watt was usually compelled to compare their power output with that of a horse. After various tests he found that a typical horse could perform work at the rate of 22,000 ft-lb/min for as much as 10 hr/day. To avoid possible disputes about full measure, Watt increased this figure by one-half in establishing the unit that he called the horsepower. Thus Watt's horsepower represents a rate of doing work of 33,000 ft-lb/min, or 550 ft-lb/sec. In metric units 1 hp is equal to 746 watts. (The early steam engines ranged from 4 to 100 hp, with the 20-hp model the most popular.)

the horsepower

ENERGY

What makes it possible for a force to do work? The answer is *energy*. Energy may be thought of as that property of something which enables it to do work. When we say that something has energy, we suggest that it is capable of exerting a force on something else and performing work on it. When work is done on something, on the other hand, energy has been added to it. Energy is measured in the same units as those of work, the foot-pound and the joule.

energy is the capacity to do work

kinetic energy

Energy may take many forms. A familiar example is the energy a moving body possesses by virtue of its motion. Every moving object has the capacity to do work. By striking another object that is free to move, the moving object can exert a force and cause the second object to shift its position. It is not necessary that the moving object actually do work; it may keep on moving, or friction may slowly bring it to a stop. But while it is moving, it has the *capacity* for doing work. It is this specific property

kinetic energy is energy of motion

that defines energy, since energy means the ability to do work, and so all moving things have energy by virtue of their motion. This type of energy is called *kinetic energy*.

A good example of the relation between work and kinetic energy is furnished by the pile driver, a simple machine that lifts a heavy weight and allows it to fall on the head of a pile, thereby driving the pile into the ground. Just before the pile is struck, the hammer possesses considerable kinetic energy, which it loses by exerting a force on the pile. During the fraction of a second in which the force acts, the pile moves a short distance against the frictional resistance of the ground; hence, work is accomplished by the energy of the falling weight (Fig. 4.4).

The energy that an object possesses because it is moving, its kinetic energy KE, is described formally by the expression

kinetic energy
$$KE = \tfrac{1}{2}mv^2$$

in which m is the mass of the object and v its speed. That mass and speed should determine kinetic energy seems reasonable enough; a train going 60 mi/hr should have more energy than a hummingbird traveling the same rate and more energy than a similar train going 10 mi/hr. But it is not obvious why the m and v should be combined in this peculiar fashion.

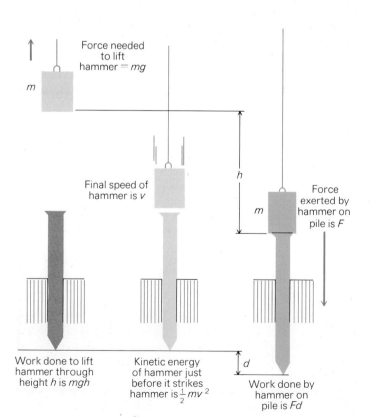

4.4 Work and energy. *In the operation of a pile driver, work mgh done in raising the hammer is converted into kinetic energy $\tfrac{1}{2}mv^2$ when the hammer is released. The kinetic energy is converted into work Fd when the hammer strikes the pile.*

ENERGY AND MOMENTUM

Without going into detail, we can see how the v^2 factor arises with a simple illustration. If two balls are thrown vertically upward, with the starting speed of one twice that of the other, the former will rise not twice as high but *four times* as high as the latter (this conclusion follows simply from the laws of falling bodies developed in Chap. 1). Thus there is some property of the motion that depends on the square of the speed—the property we call kinetic energy.

The squared v term means that kinetic energy increases very rapidly with increasing speed. At 90 mi/hr a car has nine times as much kinetic energy as at 30 mi/hr—and requires nine times as much force to bring it to a stop in the same distance. A meteor entering the earth's atmosphere at 50 km/sec and slowed by friction to 5 km/sec thereby has its kinetic energy reduced a hundredfold. The variation with mass is, of course, less spectacular: a 2-ton car going 50 mph has just twice the kinetic energy of a 1-ton car at the same speed.

kinetic energy depends upon the square of the speed

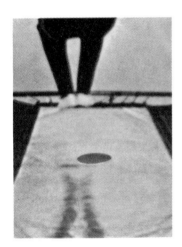
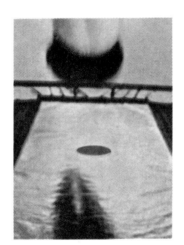
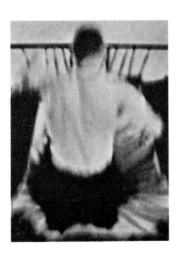

A man bouncing on a steel-foil trampoline. At the top of his path, he has only gravitational potential energy. As he falls, the PE changes into kinetic energy. When he strikes the trampoline, the KE is converted into the potential energy of the stretched springs that hold the foil to its supports. The PE of the springs then becomes KE as the man rises into the air again to begin another cycle.

the metric energy unit is the joule

The unit of energy in the metric system is the same as that of work: the joule. Thus the kinetic energy of a 1,000-kg car whose speed is 20 m/sec (about 45 mi/hr) is

$$KE = \tfrac{1}{2}mv^2 = \tfrac{1}{2} \times 1{,}000 \text{ kg} \times (20 \text{ m/sec})^2$$
$$= 200{,}000 \text{ joules} = 2 \times 10^5 \text{ joules}$$

the british energy unit is the foot-pound

In the British system the unit of energy is the foot-pound. A numerical example may be helpful to illustrate the calculation of kinetic energy in the British system. What is the kinetic energy of a 2-ton car going 50 mi/hr? The mass of this car is

$$m = \frac{w}{g} = \frac{4{,}000 \text{ lb}}{32 \text{ ft/sec}^2} = 125 \text{ slugs}$$

and its speed is

$$v = 50 \frac{\text{mi}}{\text{hr}} \times 1.47 \frac{\text{ft/sec}}{\text{mi/hr}} = 73.5 \text{ ft/sec}$$

The kinetic energy of the car is therefore

$$KE = \tfrac{1}{2}mv^2 = \tfrac{1}{2} \times 125 \text{ slugs} \times (73.5 \text{ ft/sec})^2 = 3.38 \times 10^5 \text{ ft-lb}$$

derivation of kinetic-energy formula

The formula $KE = \tfrac{1}{2}mv^2$ for the kinetic energy of a moving body can be derived from the relationship between work and energy. The force required to lift a pile-driver hammer of mass m is its weight of mg (Fig. 4.4). If the hammer is raised to a height h, the work done is

$$W = Fd = mgh$$

Now we drop the hammer, and when it reaches the ground it must have a kinetic energy equal to W; that is,

$$KE = W = mgh$$

The vertical distance h is related to the time of fall t of the hammer by

$$h = \tfrac{1}{2}gt^2$$

as we saw in Chap. 1, and so

$$KE = \tfrac{1}{2}m(gt)^2$$

The hammer's speed v at the bottom of its fall is its acceleration a multiplied by the time t,

$v = gt$

and we therefore obtain

$KE = \tfrac{1}{2}mv^2$

Even though we have obtained this formula for the specific case of a falling body, it is a perfectly general conclusion that holds for all moving things.

potential energy

The statement that energy is the capacity to do work is not restricted to kinetic energy but is a perfectly general definition. Many objects possess energy because of their position. Consider the pile driver of the preceding section: when the hammer has been lifted to the top, it has only to be released to fall and do work on the pile. The capacity for doing work is present in the hammer as soon as it has been lifted, simply because of its position several feet above the ground. The actual work on the pile is done at the expense of kinetic energy gained during the hammer's fall, but the capacity for working is present before the fall starts. Energy of this sort, depending merely on the position of an object, is called *potential energy*.

potential energy is energy of position

Examples of potential energy are everywhere. A book on a table has potential energy, since it can fall to the floor; a skier poised at the top of a slope, water at the brink of a cataract, a car at the top of a hill, anything capable of moving toward the earth under the influence of gravity has energy because of its position. Nor is the earth's gravity necessary: a planet has potential energy with respect to the sun, since it can do work in falling toward the sun; a nail placed near a magnet has potential energy, because it can do work in moving to the magnet.

It is easy to obtain a formula for the gravitational potential energy PE that an object has at or near the earth's surface. The work W required to raise a body of mass m to a height h above its original position, as in Fig. 4.5, is

$W = Fd = mgh$

Hence the potential energy of the body at the height h is

$PE = mgh$

When the weight w of the body is specified instead of its mass m, this formula becomes

$PE = wh$

since mass and weight are related by

$w = mg$

Weight = mass × acceleration of gravity

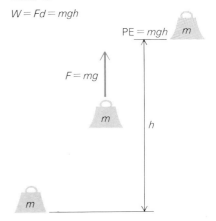

4.5 *The potential energy of a raised object is equal to the work used to raise it.*

$W = Fd = mgh$

$PE = mgh$

$F = mg$

4.6 Conservation of energy. *In the absence of friction, a car can coast from the top of one hill into a valley and then up to the top of another hill of the same height as the first. In doing this the initial potential energy of the car is converted into kinetic energy as it goes downhill, and this kinetic energy then turns into potential energy as it climbs the next hill.*

energy transformations occur in many processes

4.7 Energy transformations in planetary motion. *Near the sun the potential energy of a planet is a minimum and its kinetic energy a maximum, while far from the sun its potential energy is a maximum and its kinetic energy a minimum. The total energy of the planet is the same at all points in its orbit.*

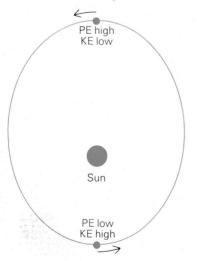

These results for PE are in accord with our experience. For example, from the formula $PE = wh$ we would expect the effectiveness of the hammer of a pile driver to depend upon its weight w and the height h to which it is lifted, which of course is borne out in practice.

energy transformations

Nearly all familiar mechanical processes actually consist of interchanges of energy among its kinetic and potential forms and work. An example is the case of a pile driver. The first step in the operation of a pile driver is the raising of its hammer above the ground by an engine, which performs an amount of work equal to the weight of the hammer multiplied by the height to which it is raised. The hammer in its upper position now has potential energy. When the hammer is released, it falls faster and faster, its potential energy becoming converted to kinetic energy. When the pile is struck by the hammer, the hammer's kinetic energy becomes work as the pile is driven into the ground, and the hammer comes to a halt.

We can obtain a great deal of insight into mechanical processes by thinking in terms of such transformations of energy. Thus, when the car of Fig. 4.6 drives to the top of a hill, its engine must do work in order to lift the car up. At the top, the car has an amount of potential energy equal to the work done in getting it up there (neglecting friction). If the engine is turned off, the car can still coast down the hill, and its kinetic energy at the bottom of the hill will be the same as its potential energy at the top.

Changes of a similar nature, from kinetic energy to potential and back, are exhibited in the motion of a planet in its orbit around the sun (Fig. 4.7) and in the motion of a pendulum (Fig. 4.8). The orbits of the planets are ellipses with the sun at one focus, and each planet is therefore at a constantly varying distance from the sun. At all times the total of its potential and kinetic energies remains the same. When the planet is close to the sun, its potential energy is low and its kinetic energy high. The additional speed due to the increased kinetic energy keeps the planet from being pulled into the sun by the greater gravitational force it experiences at this point in its path. When the planet is far from the sun, its potential energy is high and its kinetic energy correspondingly lower, the reduced speed exactly keeping pace with the reduced gravitational force.

A typical pendulum, as in Fig. 4.8, consists of a ball suspended by a string whose mass is negligible compared with that of the ball. When the ball is pulled to one side with its string taut and then released, it swings back and forth indefinitely. When it is released, the ball has a potential energy relative to the bottom of its path of mgh. At its lowest point all this potential energy has become kinetic energy, and the speed of the ball is found by setting PE and KE equal:

$$PE_{top} = KE_{bottom}$$
$$mgh = \tfrac{1}{2}mv^2$$
$$v = \sqrt{2gh}$$

ENERGY AND MOMENTUM

This is the same speed the ball would have if dropped from a height h, but here the ball is moving horizontally because of the constraining influence of the string. However, energy is not a vector quantity, and the direction of the ball's motion has no bearing on its kinetic energy. Since h is the vertical distance through which an object falls in acquiring the speed $\sqrt{2gh}$, the fact that the ball's path is an arc of a circle, and not merely straight down, is also irrelevant. After reaching the bottom, the ball continues in its motion until it rises to a height h on the opposite side from its initial position. Then, momentarily at rest since all its kinetic energy is now potential energy, the ball begins to retrace its path back through the bottom to its initial position.

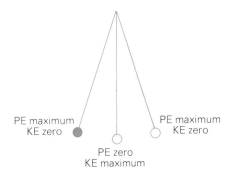

4.8 Energy transformations in pendulum motion. *The constant total energy of the ball is continuously exchanged between kinetic and potential forms.*

reference level

The gravitational potential energy of an object depends upon the level from which it is reckoned. Often the earth's surface is convenient, but sometimes other references are more appropriate. Suppose we hold a 10-g (0.01-kg) pencil 10 cm (0.1 m) above a table whose top is 1 m above the floor (Fig. 4.9). The pencil has a potential energy of

$$PE = mgh = 0.01 \text{ kg} \times 9.8 \text{ m/sec}^2 \times 0.1 \text{ m} = 0.0098 \text{ joule}$$

The collisions of billiard balls illustrate the conservation principles of energy and momentum.

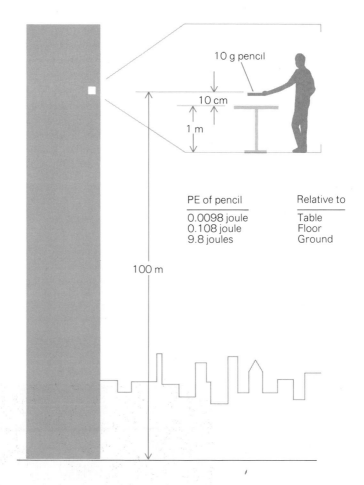

4.9 *The gravitational potential energy of a body depends upon the choice of reference level. Potential energy is a relative quantity, and only the differences between the potential energies of a body at points in its motion are significant.*

potential energy is a relative quantity

relative to the table, and a potential energy of

$$PE = mgh = 0.01 \text{ kg} \times 9.8 \text{ m/sec}^2 \times 1.1 \text{ m} = 0.108 \text{ joule}$$

relative to the floor. The room in which we are might well be, say, 100 m above the ground, so that the potential energy of the pencil is

$$PE = mgh = 0.01 \text{ kg} \times 9.8 \text{ m/sec}^2 \times 100 \text{ m} = 9.8 \text{ joules}$$

relative to the ground.

What, then, is its true potential energy? The answer is that there is no such thing as "true" potential energy; potential energy is intrinsically a *relative* quantity, meaningful only in terms of a specific reference location. However, the *difference* between the potential energies of a body at two points in its motion *is* significant. If we drop the pencil and want to compute its speed upon striking the floor, we equate its kinetic energy of

$$KE = \tfrac{1}{2}mv^2$$

with its loss of potential energy in falling through 1.1 m. Hence

$$PE = KE$$
$$mgh = \tfrac{1}{2}mv^2$$
$$v = \sqrt{2gh}$$
$$= \sqrt{2 \times 9.8 \text{ m/sec}^2 \times 1.1 \text{ m}}$$
$$= 4.6 \text{ m/sec}$$

The fact that the floor of the room in which the pencil falls may be at ground level or 100 m above it is irrelevant; only the actual distance of fall is involved.

It is important to note that h in the formulas $PE = mgh$ and $PE = wh$ is always the vertical distance (not necessarily the total distance the body may travel) between the starting point and the lowest point of fall.

In the British system of units, where weights rather than masses are customarily given, the potential energy of a body is just its weight (in pounds) multiplied by its height (in feet) above the selected reference level. Thus a 2-ton car at the top of a hill 100 ft high has a potential energy of

$$PE = wh$$
$$= 4{,}000 \text{ lb} \times 100 \text{ ft}$$
$$= 4 \times 10^5 \text{ ft-lb}$$

relative to the foot of the hill, which is only a trifle more than its kinetic energy when moving at 50 mi/hr. If a 50 mi/hr car crashes into some fixed object, nearly as much work (that is, damage) will be done on car and object as if the car drops through 100 ft of free fall.

energy in other forms

The two kinds of energy—kinetic and potential—we have spoken of are not the only kinds that occur in nature. Energy in other forms can also perform work. The *chemical energy* of gasoline is used to drive our automobiles; the chemical energy of food enables our bodies and the bodies of domestic animals to perform work. *Heat energy* from burning coal or oil is used to form the steam that drives ships. *Electric energy* and *magnetic energy* turn motors in home and factory. *Radiant energy* from the sun, though man has yet to learn how to harness it efficiently, performs very necessary work in lifting water from the earth's surface into clouds, in producing inequalities in atmospheric temperatures that cause winds, in making possible chemical reactions in plants that produce foods.

Just as kinetic energy can be converted into potential energy and potential into kinetic, so other forms of energy can readily be transformed. In the cylinders of an automobile engine, for example, chemical energy stored in gasoline and air is changed first to heat energy when the mixture is ignited, then to mechanical energy as the expanding gases push down on the pistons. This mechanical energy is in large part transmitted to the wheels, but some is used to turn the generator and thus produce electrical energy for charging the battery, and some is changed to heat by friction in bearings. Energy transformations go on constantly, all about us. In every case, the energy involved is *conserved*—its total amount remains the same. Conservation of energy is considered in detail in Chap. 6.

conservation of energy

Almost all the energy available to us on the earth today has come ultimately from a single source—the sun. Light and heat reach us directly from the sun; food and wood owe their chemical energy to sunlight falling on plants; water power exists because the sun's heat evaporates water constantly from the oceans. Coal and petroleum were formed from plants and animals that lived and stored energy derived from sunlight millions of years ago.

the sun is the source of most of the earth's energy

Modern civilization owes its spectacular development in large measure to the discovery of vast sources of energy and to the development of new methods for storing and transforming it. Within less than 200 years man has learned to use efficiently the chemical energy of coal and petroleum, to change heat into useful mechanical energy, to store chemical energy in explosives, to get electric energy from moving water, and to use electrical energy for heating, lighting, mechanical work, and communication. In the development of atomic bombs and nuclear reactors a new energy source has been tapped—the energy stored in the interior of atoms. Other possible sources, still being explored by industry, are the energy of tides and radiant energy direct from the sun.

MOMENTUM

Another quantity that often must be taken into account in situations that involve moving bodies is *momentum*. Momentum may be thought of as a quantitative expression of the inertia of a moving body. There are two kinds of momentum: *linear momentum*, which is a measure of the tendency of a moving body to continue in motion along a straight line, and *angular momentum*, which is a measure of the tendency of a rotating body to continue to spin about an axis that does not change its orientation.

linear momentum

The linear momentum of a body of mass m and velocity $\mathbf{v}$ is defined as the product of m and $\mathbf{v}$:

linear momentum

$$\text{Momentum} = m\mathbf{v}$$

We note that the product of a scalar and a vector quantity, here m and $\mathbf{v}$ respectively, is a vector quantity with the same direction as $\mathbf{v}$ and with a magnitude equal to mv.

ENERGY AND MOMENTUM

The important thing about momentum is that it is a *vector quantity*, meaning that it has a direction associated with it, namely, the direction of motion of the body. Kinetic energy, given by the somewhat similar formula $\frac{1}{2}mv^2$, has a completely different significance, since it is a quantity having only magnitude. In the early days of physics, even after the time of Newton, the distinction between momentum and kinetic energy was as troublesome to scientists as it frequently is to students today.

Momentum considerations are most useful in situations that involve (in general terms) explosions and collisions. When external forces do not act upon the bodies involved, their momentum is *conserved:* **The total momentum of all the bodies before they interact is exactly the same as their total momentum afterward.** The total kinetic energy of the bodies need not be the same before and after, however; the kinetic energy involved in an explosion is initially zero, since it all comes from the chemical energy stored in the explosive material, and the kinetic energy of bodies that collide may disappear, in whole or in part, into heat and sound energy. In both cases the total momentum does not change.

momentum is a vector quantity, energy is a scalar quantity

momentum is conserved in a system when no outside forces act

Apollo 11, a multistage rocket, lifts off its pad to begin man's first lunar voyage. Conservation of linear momentum underlies rocket propulsion.

When two objects collide and stick together, as in Fig. 4.10, the vector sum of the initial momenta of the objects (which means taking into account their directions of motion), which we might write $m_1\mathbf{v}_1$ and $m_2\mathbf{v}_2$, must be equal in magnitude and direction to the momentum $m\mathbf{v}$ of the composite body they form when they stick together.

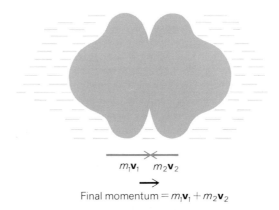

4.10 Momentum is conserved in collisions. *When two objects collide, the sum of their initial momenta is equal to the momentum of the composite body they form when they stick together. Momentum is a vector quantity, and the directions of motion of the objects must be taken into account when their momenta are added together.*

Let us examine what happens when a pistol is fired (Fig. 4.11). Before the trigger is pulled, both pistol and bullet are stationary, so the initial momentum is zero. When the trigger is pulled, however, the bullet flies out of the barrel with the momentum $m\mathbf{v}$, where m is its mass and $\mathbf{v}$ its velocity. Because no outside forces are acting on the bullet and pistol at the moment of firing, the total momentum after firing must be the same as it was initially. This means that the pistol itself must move *backward* if its momentum is to balance out the forward momentum of the bullet. If the mass of the pistol is M and its backward velocity is $\mathbf{V}$,

$$M\mathbf{V} = -m\mathbf{v}$$

The minus sign indicates that $M\mathbf{V}$ is opposite in direction to $m\mathbf{v}$.

Suppose that the bullet has a mass of 10 g (0.01 kg) and a muzzle velocity of 800 m/sec. This gives it a momentum of magnitude

$$mv = 0.01 \text{ kg} \times 800 \text{ m/sec} = 8 \text{ kg·m/sec}$$

If the pistol's mass is 2 kg, its velocity can be found from the fact that

$$MV = -mv = -8 \text{ kg·m/sec}$$

to be

$$V = \frac{-8 \text{ kg·m/sec}}{2 \text{ kg}} = -4 \text{ m/sec}$$

It is this momentum that is felt as the recoil of a pistol, rifle, or shotgun; the heavier the bullet and the greater its muzzle velocity, the more "kick"

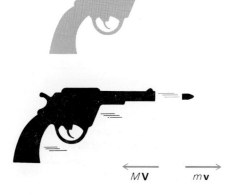

4.11 Momentum is conserved in explosions. *When a pistol is fired, the backward momentum of the pistol is equal to the forward momentum of the bullet. The speed of the bullet is greater than that of the pistol because it is lighter. The initial momentum of the system of pistol plus bullet is zero, and the final momentum of the system is also zero.*

will be felt by the shooter. Notice that tracing the various energies involved through their transitions from one form to another gives us no indication that there will be any recoil.

rocket propulsion

The operation of a rocket illustrates conservation of momentum. When the rocket stands stationary on its launching platform, its momentum is zero. When it is fired, the momentum of the gases rushing out downward is balanced by the momentum in the other direction of the rocket body moving upward; the total momentum of all the constituents of the rocket, gases, and body, remains zero, because momentum is a vector quantity and the upward and downward momenta cancel out (Fig. 4.12). Thus a rocket does not operate by "pushing" against anything and functions best in the near vacuum of space, where friction is virtually absent. Energy also is conserved in a rocket, the kinetic energy of the rocket and of the exhaust gases after firing being equal to the chemical energy expended in producing the gases at high velocity.

rocket propulsion is based upon conservation of momentum

The ultimate speed a single rocket can attain is determined by the amount of fuel it can carry and by the speed of its exhaust gases. Because both these quantities are limited, *multistage rockets* are used in the exploration of space. The first stage is a large rocket which has a smaller one mounted in front of it. When the fuel of the first stage has burnt up, its motor and empty fuel tanks are cast off. Then the second stage is fired. Since the second stage is already moving rapidly and does not have to carry the motor and fuel tanks of the first stage, it can reach a much higher final speed than would otherwise be possible.

multistage rockets

Depending upon the final speed needed for a given mission, three or even four stages may be required. The Saturn V launch vehicle that propelled the Apollo 11 spacecraft to the moon in July 1969 had three stages, as shown in Fig. 4.13. Just before takeoff the entire assembly was 363 ft long and weighed 3,240 tons.

angular momentum

We have all observed the tendency of rotating bodies to continue to spin unless they are slowed down by an external agency. A simple example is a top, which would spin indefinitely were it not for friction between its tip and the ground. Another example is the earth, which has been rotating for billions of years and is likely to continue doing so for many more to come. The earth, too, has a retarding influence acting on its rotation—the friction of the tides, which are held in place by the moon's gravitational pull while the earth rotates underneath—but the resulting deceleration is barely detectable: each rotation of the earth takes about 0.000,000,025 sec longer than the one before.

The rotational quantity analogous to ordinary momentum is called *angular momentum,* and *conservation of angular momentum* is a formal way to describe the tendency of spinning bodies to keep spinning.

The precise definition of the angular momentum of a body is complicated because it depends not only upon the mass of the body and upon

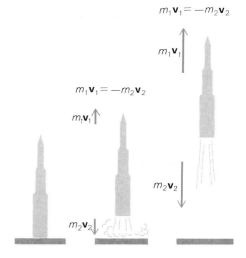

4.12 Rocket propulsion is based upon conservation of momentum. At all times the downward momentum of the exhaust gases is equal in magnitude and opposite in direction to the upward momentum of the rocket itself.

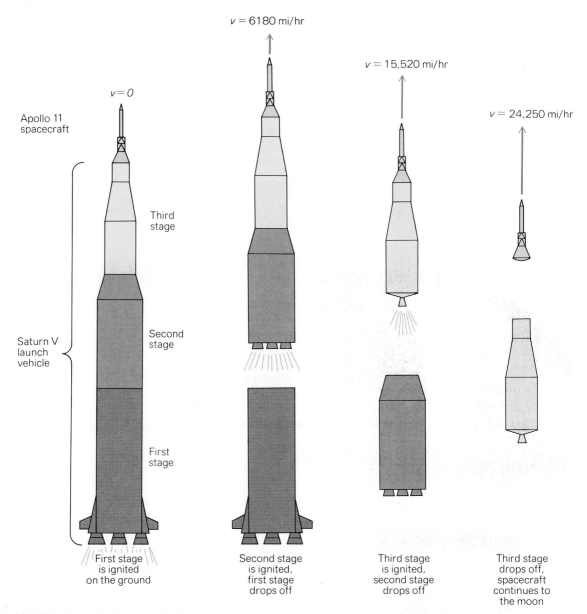

4.13 *The Saturn V launch vehicle that propelled the Apollo 11 spacecraft to the moon for the first manned landing had three rocket stages.*

how fast it is turning, but also upon how the mass is distributed in the body. As we might expect, the greater the mass of a body and the more rapidly it rotates, the more angular momentum it has and the more pronounced its tendency to continue to spin. Furthermore, the farther away from the axis of rotation the mass is distributed, the more the angular momentum. An illustration of both the latter peculiarity and the conservation of angular momentum is a skater doing a spin (Fig. 4.14). When the skater starts the spin, he pushes against the ice with one skate to start turning. Initially both arms and one leg are extended, so that his mass is spread as far as possible from the axis of rotation. Then he brings his arms and the outstretched leg in tightly against his body, so that now all his mass is as close as possible to the axis of rotation. As a result he spins faster: to make up for the change in the mass distribution, the speed must change as well if angular momentum is to be conserved.

Like ordinary momentum, angular momentum is a vector quantity with direction as well as magnitude. A spinning body therefore tends to maintain the *direction* of its spin axis besides the amount of angular momentum it has. A stationary top falls over at once, but a rapidly spinning top stays upright because its tendency to keep its axis in the same position by virtue of its angular momentum is greater than its tendency to fall over. Footballs and rifle bullets are sent off spinning to prevent them from tumbling during flight, which would increase air resistance and hence shorten their range.

Axis
Slow spin

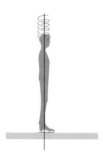
Fast spin

4.14 Conservation of angular momentum. *Angular momentum depends upon both the speed of turning and the distribution of mass; when the skater pulls in his arms and extended leg, he spins faster to compensate for the change.*

conservation principles and symmetry

The conservation principles of energy, linear momentum, and angular momentum are especially interesting because of what they reveal about the underlying symmetrical character of the physical universe. Using advanced mathematics it is possible to show the following:

1 If the laws of nature are the same at all times, past, present, and future, then energy must be conserved.
2 If the laws of nature are the same everywhere in the universe, then linear momentum must be conserved.
3 If the laws of nature are independent of direction, then angular momentum must be conserved.

Hence these conservation principles, which are observed to be obeyed in all processes, indicate that there is a profound order in the physical universe, despite the random appearance of many aspects of it.

A number of other conservation principles have been found—for instance, the conservation of electric charge—most of which can be traced to symmetries of some kind. A few conservation principles that apply to transformations of elementary particles into one another and to their appearance and disappearance in certain processes have not as yet been associated with symmetries, and it is hoped that the discovery of the latter will further illuminate the basic structure of the universe.

GLOSSARY

Work is the product of a force and the distance through which it acts. If the force is not parallel to its displacement, the component of the force parallel to the displacement must be used in calculating the work it does. The unit of work in the British system is the *foot-pound* (*ft-lb*); in the metric system, the *joule*.

Power is the rate at which work is being done. The unit of power in the metric system is the *watt*. The *horsepower* is a unit of work equal to 746 watts or 550 ft-lb/sec.

Energy is the property something has that enables it to do work. Energy has the same units as work. Some forms of energy are kinetic energy, potential energy, heat energy, chemical energy, electrical energy, magnetic energy, radiant energy, and mass energy. Energy is *conserved* in every process.

Kinetic energy is the energy a body has by virtue of its motion. The kinetic energy of a moving body is equal to $\frac{1}{2}mv^2$, one-half the product of its mass and the square of its speed.

Potential energy is the energy a body has by virtue of its position. The gravitational potential energy of a body is equal to Wh, the product of its weight and its height above some reference level. (In the metric system gravitational potential energy is given as mgh, the product of the mass of the body, the acceleration of gravity, and the height.)

The *linear momentum* of a body is the product of its mass and its velocity. Linear momentum is a vector quantity, possessing both a magnitude and a direction; the direction is that of the body's motion. It is a measure of the tendency of a moving body to continue moving in a straight line.

Conservation of linear momentum states that, when several bodies interact with one another (for instance, in an explosion or a collision), if outside forces do not act upon the bodies involved, the total momentum of all the bodies before they interact is exactly the same as their total momentum afterward.

The *angular momentum* of a rotating body is a measure of its tendency to continue to spin at the same rate about an axis that does not change in orientation. Angular momentum is a vector quantity. The angular momentum of an isolated body or system of bodies is *conserved*.

EXERCISES

1 Is it correct to say that all changes in the physical world involve energy transformations of some sort? Why?

2 Can you suggest any kinds of energy found on the earth that do not have their ultimate source in the sun's radiation?

3 In what part of its orbit is the earth's potential energy greatest with respect to the sun? In what part of its orbit is its kinetic energy greatest? Explain your answers.

4 Is it possible for a body to have simultaneously more kinetic energy but less momentum than another body? Is the converse possible?

5 If the polar ice caps melt, the length of the day will increase. Why?

6 The orbital motion of the earth around the sun possesses angular momentum. Use this fact to explain why the earth's orbital speed is greatest when it is closest to the sun and least when it is farthest away.

7 Two objects, one with a mass of 1 kg and the other with a weight of 1 newton, both have potential energies of 1 joule relative to the ground. What are their respective heights above the ground?

8 Two objects, one with a mass of 1 slug and the other with a weight of 1 lb, both have potential energies of 1 ft-lb relative to the ground. What are their respective heights above the ground?

9 A total of 490 joules of work is needed to lift a body of unknown mass through a height of 10 m. What is its mass?

10 A force of 20 lb is required to move a 50-lb crate across a floor. How much work must be done to move the crate 30 ft?

11 Ten thousand joules of work is expended in raising a 90-kg box. How high was it raised? What is its potential energy?

12 A 3,200-lb car, initially at rest, coasts down a hill 50 ft high. If there were no frictional forces present, what would its speed be at the foot of the hill?

13 A pile-driver hammer weighing 1,000 lb is dropped from a height of 19 ft above the head of a pile. If the pile is driven 6 in. into the ground with each impact of the hammer, what is the average force on the pile when it is struck?

14 A child on a swing is 2 m above the ground at his highest point and 1 m above the ground at his lowest point. What is the child's maximum speed?

15 A 160-lb weight is raised 5 ft from the ground by a man who uses a rope and a system of pulleys. He exerts a force of 45 lb on the rope and pulls a total of 20 ft of rope through the pulleys while lifting the weight. How much work does the man do? What is the change in the potential energy of the weight? If the answers to these questions are different, explain why.

16 A centripetal force of 50 lb is used to keep a 0.2-slug ball moving in a circle with uniform speed at the end of a string 3 ft long. How much work is done by the force in each revolution of the ball?

17 An automobile weighing 3,500 lb and traveling at 30 mi/hr (44 ft/sec) strikes a stationary automobile of the same weight. The two cars stick together after the collision. What is their final velocity? How much kinetic energy has been lost?

18 A 180-lb man dives horizontally from a 300-lb boat with a speed of 5 ft/sec. What is the recoil speed of the boat?

19 A bullet weighing 0.08 lb is fired from a 10-lb rifle at a muzzle velocity of 2,000 ft/sec. What is the recoil speed of the rifle?

20 Two boys, one weighing 75 lb and the other 100 lb, are facing each other on frictionless roller skates. The smaller boy pushes the larger one, so that the latter moves away at a speed of 2 mi/hr. With what speed does the smaller boy move, and in what direction?

chapter 5
Relativity

SPECIAL RELATIVITY
    frames of reference
    postulates of special relativity
    simultaneity
    the fourth dimension
    time and space
    muon decay

MASS AND ENERGY
    the relativity of mass
    mass and energy

GENERAL RELATIVITY
    gravitation
    the twin paradox

All physical science is ultimately concerned with measurement, and the theory of relativity is in essence an analysis of how measurements depend upon the observer as well as upon what is observed. From relativity emerges a more general mechanics in which there are intimate relationships between space and time, mass and energy. Without these relationships it would be impossible to understand the microscopic world within the atom, the elucidation of which is the central problem of modern physics.

SPECIAL RELATIVITY

frames of reference

Let us begin with a simple experiment. Suppose that you are on a moving train and that I am observing you from the station platform (Fig. 5.1). You proceed to roll two balls along a level table, ball 1 toward the front end of the car and ball 2 toward the rear, and both of us measure the

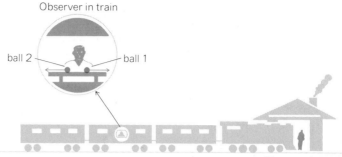

5.1 *All motion is relative to a frame of reference, which must be specified. Observers in motion relative to one another will see the same event differently.*

speed of each ball. Your results will indicate that the balls are moving with the same speed, say 20 mi/hr. But to me their speeds will seem very different. The ball that you rolled forward would have not only the speed given it by your arm but, in addition, the speed of the train; if the train's speed is 20 mi/hr, I should see the ball moving $20 + 20 = 40$ mi/hr. The second ball would appear to me motionless, for you have given it just enough speed backward to overcome the train's motion forward. The speed of each ball is evidently *relative* to the position of the observer, which we can summarize as follows.

Relative to train: $v_{\text{ball 1}} = +20$ mi/hr

$v_{\text{ball 2}} = -20$ mi/hr

Relative to ground: $v_{\text{ball 1}} = +40$ mi/hr

$v_{\text{ball 2}} = 0$

All this is obvious enough. A moving observer always views the motions of other objects differently from a stationary observer, since his own

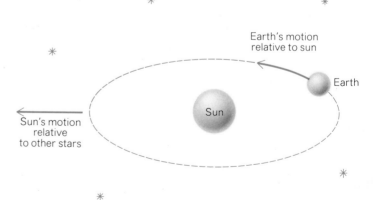

5.2 *There is no "fixed" frame of reference in the universe.*

motion is added to other motions. But now we face an interesting question: Which is correct, your measurement of the ball's speed or mine? You might well concede that my measurements show the "real" motion, since I am standing still while you are obviously moving. But why should measurements from a station platform be any more valid than measurements from a moving train? Perhaps the "real" motion of the balls could only be seen by an observer on the sun, who would add to your observations not only the speed of the train but also the speed of the earth as it rotates and moves in its orbit. Yet the sun itself is moving with respect to the stars (Fig. 5.2), so why should an observer there be any more correct than one of us? We might refer the problem to an observer at the center of our galaxy, but even he would admit that he is moving with respect to other galaxies. We are driven to conclude that there is no "real" motion of the balls—or, better, that the motion seen by any observer is just as "real" as that seen by any other. Motion has no meaning unless it is referred to an observer.

all motion is relative

Let us look at the motion of an observer himself. How can he tell if he is stationary or moving? Again the answer is, he cannot tell anything about any motion, including his own, except with respect to a stated frame of reference. If we are in a free balloon above a uniform cloud bank and see another free balloon change its position relative to us, we have no way of knowing which balloon is "really" moving (Fig. 5.3). Should we be isolated in the universe, there would be no way in which we could determine whether we are in motion or not, because without a frame of reference the concept of motion has no meaning.

The theory of relativity resulted from an analysis of the physical consequences implied by the absence of a universal frame of reference. The special theory of relativity, developed by Albert Einstein in 1905 when he was 26 years old, treats problems involving *inertial frames of reference*, which are frames of reference moving at constant velocity with respect to one another. The general theory of relativity, proposed by Einstein a decade later, treats problems involving frames of reference accelerated with respect to one another. An observer in an isolated laboratory *can* detect accelerations, as anybody who has ever been in an elevator or on a merry-go-round can verify from his own experience. The special theory has had a profound influence on all of physics, and we shall look into its chief results before a brief glance at the general theory.

special and general relativity

postulates of special relativity

The special theory of relativity is based upon two postulates. The nature of the experiments whose results are summarized in these postulates may be understood with the help of the railroad car mentioned above.

If you were to try a variety of simple experiments in your moving railroad car while I looked on, we should find that your conclusions and mine were in good agreement. In other words, the ordinary laws of physics hold as well on the moving car as on the station platform. No experiment gives us any basis for believing that your measurements are any more correct than mine, nor mine than yours. Einstein phrased this conclusion

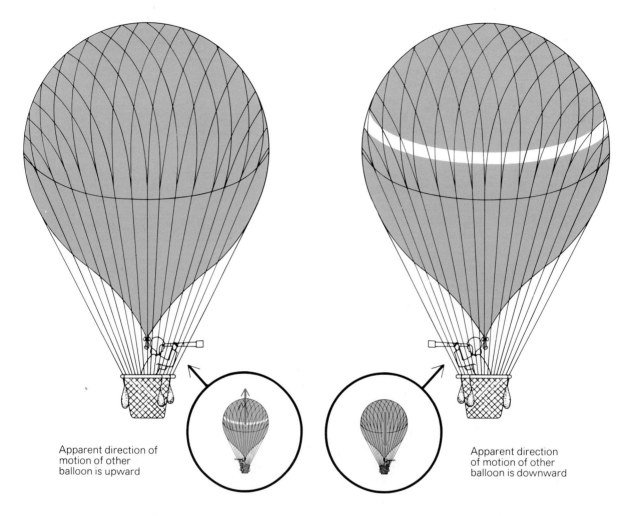

5.3 *All motion is relative to the observer.*

in general terms: **All laws of nature are the same in all frames of reference moving relative to one another at uniform velocity.**

This statement leads to no difficulty until we try experiments that involve the speed of light. Let us first try an experiment with another form of wave motion, sound. Suppose that you stand in the middle of the moving car and say something. If we both had means of determining when the sound of your voice reached the two ends of the car, you would find that the sound moved in each direction with its normal speed, about 1,100 ft/sec, and that sound waves reached the two ends at the same instant. From my position on the station platform, I should find, of course, that the sound waves moving forward were traveling somewhat faster than normal because of the train's motion and that those moving backward were somewhat slower. I should agree with you, however, that sound waves reach the two ends simultaneously, since the front end of the car is moving away from your mouth and the rear end toward your mouth just fast enough to make up for the differences in speed. There is nothing new here, since the motion of sound waves is exactly analogous to the motion of the two balls.

Now suppose that you switch on a light source in the middle of the car. If we undertake the difficult measurement of the speed of light toward the front and rear of the car, would you again expect to find equal speeds in the two opposite directions? Should I expect to find one speed increased and the other decreased by the train's motion? These are subtle questions, impossible to answer without trying the experiment. Our results for sound waves depended on the fact that the air molecules that transmitted the sound were being carried along by the train. Now light is not transmitted by any material particles but consists of changes in an electromagnetic field in empty space (the nature of light is discussed in Chap. 13). When we try the experiment, we find, contrary to the case of sound, that both of us get the *same* figure for speed of light in the two directions. Neither motion of the light source nor motion of the observer seems to affect the speed of light at all. The speed of light in free space is 3×10^8 m/sec (186,000 mi/sec) and is denoted by the symbol c.

Actual experiments the same in principle as the one just described were of enormous significance in the history of physics. During most of the nineteenth century light was thought to consist of wave motion in a hypothetical all-pervading medium called the *ether*, just as sound is wave motion in air or in other materials. Belief in an ether became untenable when measurements showed that the speed of light was not influenced by the direction in which the light moved or by the speed of its source.

Einstein generalized this unexpected property of light in the statement that **the speed of light in vacuum is the same in all systems of** second postulate of special relativity

A bowling ball sent down an alley in a jet airplane flying horizontally at 600 mi/hr behaves exactly as it would if the airplane were stationary on the ground. The laws of physics are the same in all frames of reference moving at constant velocity relative to one another; this is one of the two principles that underlie Einstein's special theory of relativity.

reference moving uniformly relative to one another. This statement and the one above are the foundation for the special theory of relativity.

simultaneity

Let us examine some of the remarkable consequences of the two postulates of special relativity. From inside the moving railroad car you can see light traveling with equal speed from a source in the middle toward the two ends, and you find that light waves reach the two ends simultaneously. From outside the car, I also see light traveling in opposite directions with the same speed. But, while the light is traveling, I see the rear end of the car *approaching* the backward-moving light waves and the front end of the car *receding* from the forward-moving waves. Hence I report that light reaches the rear end before it reaches the front end. Two events that are simultaneous from your point of view seem from my point of view to be separated by a short interval of time. At the instant when you find that each light ray is striking an end of the car, I find that one ray has already struck and the other has not yet arrived.

Again we ask, which observation is correct? And again the answer must be, *both* are correct. Like velocity, the concept of *simultaneity* is relative to the motion of the observer. This means a revision in our notions of past, present, and future; an event which seems to be happening now from my point of view may be an event of the past to a second observer and may still be in the future of a third. Our different conclusions about the same event depend simply on our motions relative to each other.

Although time is a relative quantity, not all of the notions of time formed by everyday experience are incorrect. Time does not run backwards to *any* observer, for instance: a sequence of events that occur somewhere in a certain order will appear in the same order to all observers everywhere, though not necessarily with the same time intervals between each pair of events. Similarly no observer, regardless of his state of motion, can see an event before it happens since the speed of light c is finite and signals require the minimum period of time L/c to travel a distance L. There is no way to peer into the future, although the appearance in time and space of past events may be different to different observers.

the fourth dimension

In the special theory of relativity, *events* play an important role—for instance, the arrival of a light ray at a certain place at a certain time. An event is fully described only when both the place of occurrence and the time of occurrence are specified. Thus we say, "The Declaration of Independence was signed in Philadelphia on July 4, 1776." If we said only that the signing took place in Philadelphia or if we merely named the date of signing, the description of the event would be incomplete. To designate the place where an event occurs requires three numbers: thus, to locate Philadelphia with respect to the earth's center, we might give its latitude, its longitude, and its distance from the center. Three numbers are necessary simply because we regard space as three-dimensional—as having length, breadth, and thickness, so to speak. Time, on the other hand, may be designated by a single number, like the number of hours since noon or the number of days since the birth of Christ. To describe an event, then, requires a minimum of four numbers, three referring to space dimensions and one referring to time.

In relativity the numbers referring to time have the same status as the numbers referring to space. In effect, time is a fourth dimension: events are located in a unified *space-time* rather than separately in space and in time. We considered one instance of the intimate connection between space and time in the section headed "Simultaneity," where we found that concepts of present, past, and future depend on motion through space. There is nothing particularly strange or difficult in the idea: it is merely a change in point of view which makes certain natural phenomena possible to describe.

time and space

Let us try another experiment. Suppose that I point a flashlight upward and switch it on momentarily. We may imagine its light, a short train of electromagnetic waves, traveling rapidly outward into space. Suppose that, just as the light is flashed, you set out in the same direction in a rocket, at a speed, say, of 180,000 mi/sec, while I remain on the ground. If we both measure the speed of the retreating waves, we both get 186,000 mi/sec, since motion of the observer does not affect the speed of light (Fig. 5.4). But how is this possible? We are both using the same kind of clocks to measure time and the same sort of meter sticks to measure distance. How can you possibly maintain that the light flash is moving away from you at 186,000 mi/sec, when to me it is perfectly obvious that your rocket is moving nearly as fast?

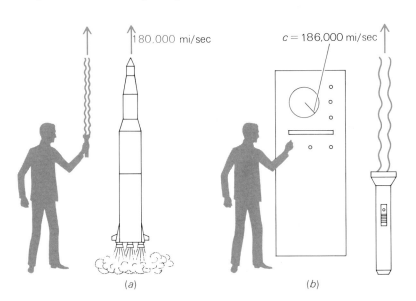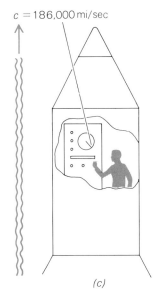

5.4 (a) A pulse of light is sent off from the ground at the same time as a spacecraft. (b) An observer on the ground finds the speed of the pulse of light to be c = 186,000 mi/sec. (c) An observer in the spacecraft also finds this speed for the pulse of light.

lengths and time intervals are relative quantities

Our measurements can both be right only if your motion has somehow changed the characteristics of your measuring instruments. The changes required are a decrease in the length of your meter stick and a slowing down of your clock: if your meter stick is shortened, then the distances you measure will seem longer, and if your clock runs slow, the times you measure will seem shorter. If the changes are large enough, your measurement of 186,000 mi/sec for a speed that to me seems only 6,000 mi/sec becomes understandable. You, of course, would not be aware of the changes; your meter stick and your clock would seem perfectly normal. You would conclude, in fact, that my instruments had been altered in the same way, so that my measurements were peculiar rather than yours.

The shortening of a meter stick at high velocities means that *length* also must be relative to the motion of the observer. The change in length is only in the direction of the relative motion; the meter stick on your rocket would appear to me to have its normal length if you turned it sidewise to your motion, but to shrink when you turned it into the line of motion.

The equations governing the variation of time intervals and length with relative velocity were found by Einstein to be

time dilation
$$t = \frac{t_0}{\sqrt{1 - v^2/c^2}}$$

and

lorentz contraction
$$L = L_0 \sqrt{1 - \frac{v^2}{c^2}}$$

where t_0 and L_0 = time interval and length measured when clock or object is at rest

t and L = time interval and length measured by an observer who is stationary with respect to moving clock or object

v and c = relative speed and speed of light, respectively

The quantity $\sqrt{1 - v^2/c^2}$ is always less than 1 for a moving object, and so a clock moving relative to an observer ticks more slowly than a clock stationary relative to the same observer. The slowing down of a moving clock is called *time dilation*. ("Dilate" means to become larger.)

Suppose we are watching from the ground a clock on a spacecraft whose speed is 18,600 mi/sec, which is $0.1c$. The minute hand will make a complete revolution, signifying the passage of $t_0 = 1$ hour *to the crew of the spacecraft*, in a time *to us on the ground* of

$$t = \frac{t_0}{\sqrt{1 - v^2/c^2}} = \frac{1 \text{ hr}}{\sqrt{1 - (0.1c)^2/c^2}} = \frac{1 \text{ hr}}{\sqrt{1 - 0.01}}$$

$$= 1.005 \text{ hr} = 1 \text{ hr } 0 \text{ min } 18 \text{ sec}$$

In this case the time dilation amounts to 18 sec/hr. The greater v is, the greater the time dilation.

The shortening of the length of an object in motion relative to an observer is called the *Lorentz contraction*. Only lengths in the direction of motion are affected: the width of a spacecraft is unchanged, though its length is reduced. If the above spacecraft is 100 ft long on the ground, in flight as measured *by an observer on the ground* its length is reduced to

$$L = L_0 \sqrt{1 - v^2/c^2} = \sqrt{1 - (0.1c)^2/c^2} \times 100 \text{ ft}$$
$$= \sqrt{1 - 0.01} \times 100 \text{ ft} = 99.5 \text{ ft}$$

The length contraction here is 0.5 percent. The greater v is, the greater the Lorentz contraction.

muon decay

A striking illustration of both the slowing down of a moving clock and the contraction in length of a moving object occurs in the decay of unstable elementary particles called *muons* (or *mu mesons*). These particles may have positive or negative electric charges and have masses 207 times greater than the mass of the electron. For the moment our interest lies in the fact that a muon "decays" into an electron an average of 2×10^{-6} sec after it comes into being. Now muons are created high in the atmosphere by fast cosmic-ray particles arriving at the earth from space, and they reach sea level in profusion—about eight muons pass through each square inch of the earth's surface per minute. (Chapter 44 contains further information on cosmic rays.) Such muons have a typical speed of 2.994×10^8 m/sec, which is 0.998 of the velocity of light c. But in $t_0 = 2 \times 10^{-6}$ sec, the muon lifetime, they can travel a distance of only

(a) Bubble-chamber photograph of meson decay. (b) A positive pion (π^+) enters at lower left and decays into a positive muon (μ^+) at top. The muon's lifetime is so brief that it travels only a short distance before decaying into a positive electron (β^+). There is a magnetic field perpendicular to the plane of the paper which is responsible for the curved paths of the particles. The details of meson decay and the operation of the bubble chamber are described in Chap. 17.

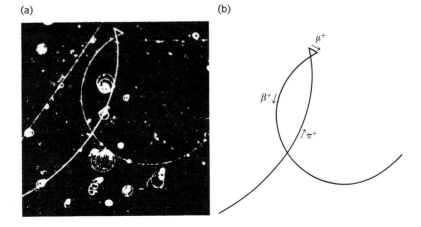

$$L = vt_0$$
$$= 2.994 \times 10^8 \text{ m/sec} \times 2 \times 10^{-6} \text{ sec}$$
$$= 600 \text{ m}$$

whereas they are actually created at altitudes more than ten times greater than this.

the distance a muon can travel before decaying is short in its reference frame, long in our reference frame.

We can resolve the muon paradox by using the results of the special theory of relativity. Let us examine the problem from the frame of reference of the muon, in which its lifetime is 2×10^{-6} sec. While the muon lifetime is unaffected by the motion, its distance to the ground appears shortened by the factor

$$\frac{L}{L_0} = \sqrt{1 - \frac{v^2}{c^2}}$$

That is, while we, on the ground, measure the altitude at which the muon is produced as L_0, the muon "sees" it as L. If we let L be 600 m, the maximum distance the muon can go *in its own frame of reference* at the speed $0.998c$ before decaying, we find that the corresponding distance L_0 *in our reference frame* is

$$L_0 = \frac{L}{\sqrt{1 - v^2/c^2}}$$
$$= \frac{600}{\sqrt{1 - (0.998c)^2/c^2}} \text{ m}$$
$$= \frac{600}{\sqrt{1 - 0.996}} \text{ m}$$
$$= \frac{600}{0.063} \text{ m}$$
$$= 9{,}500 \text{ m}$$

Hence, despite its brief lifespan, it is possible for the muon to reach the ground from quite respectable altitudes.

Now let us examine the problem from the frame of reference of an observer on the ground. From the ground the altitude at which the muon is produced is L_0, but its lifetime *in our reference frame* has been extended because of the relative motion to the value

$$t = \frac{t_0}{\sqrt{1 - v^2/c^2}}$$
$$= \frac{2 \times 10^{-6}}{\sqrt{1 - (0.998c)^2/c^2}} \text{ sec}$$
$$= \frac{2 \times 10^{-6}}{0.063} \text{ sec}$$

$$= 31.7 \times 10^{-6} \text{ sec}$$

almost sixteen times greater than when it is at rest with respect to us. In 31.7×10^{-6} sec a muon whose speed is $0.998c$ can travel a distance

$$L_0 = vt$$
$$= 2.994 \times 10^8 \text{ m/sec} \times 31.7 \times 10^{-6} \text{ sec}$$
$$= 9{,}500 \text{ m}$$

the same distance we found before. *The two points of view give identical results.*

MASS AND ENERGY

Evidently such fundamental physical quantities as length and time have meaning only when the particular reference frame in which they are measured is specified. Given this frame, we may compute what the values of these quantities would be if measured in other reference frames in motion relative to the specified frame by applying the proper transformation equations. An event in time and space—a collision between two bodies, for instance—will have a different appearance in different frames of reference. However, according to the postulates of special

The Cosmotron at Brookhaven National Laboratory accelerates protons to 3-Bev energies. A proton with this energy moves at over 97 percent of the speed of light and has a mass over four times its rest mass.

relativity, the laws of motion arrived at by observing events of this kind must have the same form in all frames of reference. It is this requirement that led to the prediction of the relativity of mass, which was subsequently confirmed by experiment.

the relativity of mass

If m_0 is the mass of an object as measured when it is at rest with respect to an observer, its mass m as measured when it is moving at the speed v with respect to an observer is

relativistic mass increase
$$m = \frac{m_0}{\sqrt{1 - v^2/c^2}}$$

The mass of a body moving at the speed v relative to an observer is larger than its mass when at rest relative to the observer by the factor $1/\sqrt{1 - v^2/c^2}$. This mass increase is reciprocal, just like the relative length contraction. As seen from the earth, a rocket ship in flight is shorter than its twin still on the ground and its mass is greater. To somebody on the rocket ship in flight, the ship on the ground also appears shorter and to have a greater mass.

Relativistic mass increases are significant only at speeds approaching that of light (Fig. 5.5). At a speed one-tenth that of light the mass increases amount to only 0.5 percent, but at a speed nine-tenths that of light the increase is over 100 percent. Only atomic particles such as electrons, protons, and mesons have sufficiently high speeds for relativistic effects to be measurable, and in dealing with these particles the "ordinary" laws of physics cannot be used. The above equation, like the others of special relativity, has been verified by so many experiments that it is recognized as one of the basic formulas of physics.

As an example, let us calculate the mass of an electron whose speed is 99 percent of the speed of light. The rest mass of an electron is

$$m_0 = 9.1 \times 10^{-31} \text{ kg}$$

When $v = 0.99c$, the electron mass m is

$$m = \frac{m_0}{\sqrt{1 - v^2/c^2}} = \frac{9.1 \times 10^{-31}}{\sqrt{1 - (0.99c)^2/c^2}} \text{ kg} = \frac{9.1 \times 10^{-31}}{\sqrt{1 - 0.98}} \text{ kg}$$
$$= \frac{9.1 \times 10^{-31}}{0.14} \text{ kg} = 65 \times 10^{-31} \text{ kg}$$

which is over seven times its rest mass.

mass and energy

The most famous relationship Einstein obtained from the postulates of special relativity concerns mass and energy. This relationship can be

5.5 The relativity of mass. *The quantity m_0 is the mass of an object when it is at rest with respect to an observer and m is its mass as measured when it is moving at the speed v with respect to the observer. The velocity of light is c.*

derived directly from the definition of the kinetic energy KE of a moving body as the work done in bringing it from rest to its state of motion by taking into account the variation of mass with speed. When this is done, the result is

$$KE = mc^2 - m_0c^2$$
$$= (m - m_0)c^2$$

which means that the kinetic energy of a body is equal to the increase in its mass consequent upon its relative motion multiplied by the square of the speed of light.

The above formula may be rewritten

$$mc^2 = KE + m_0c^2$$

If we interpret mc^2 as the *total energy* E of the body, it follows that, when the body is at rest and $KE = 0$, it nevertheless possesses the energy m_0c^2. Accordingly m_0c^2 is called the *rest energy* E_0 of a body whose mass at rest is m_0. Thus

rest energy equals m_0c^2

$$E = m_0c^2 + KE$$

Total energy = rest energy + kinetic energy

the total energy of a body is its rest energy plus its kinetic energy

In addition to its kinetic, potential, electromagnetic, thermal, and other familiar guises, then, energy can manifest itself as mass. The conversion factor between the unit of mass (kilogram) and the unit of energy (joule) is c^2, so 1 kg of matter has an energy content of 9×10^{16} joules. Even a minute bit of matter represents a vast amount of energy, and, in fact, the conversion of matter into energy is the source of the energy liberated in all the power-producing reactions of physics and chemistry.

Since mass and energy are not independent entities, the separate conservation principles of energy and mass are properly a single one, the principle of conservation of mass energy. Mass *can* be created or destroyed, but only if an equivalent amount of energy simultaneously vanishes or comes into being, and vice versa.

conservation of mass and energy

GENERAL RELATIVITY

The special theory of relativity shows us how to formulate the equations describing natural laws so that they apply to the measurements of observers moving uniformly with respect to one another, regardless of their speed. A more difficult undertaking is embodied in the *general theory of relativity*, which Einstein published in 1915. Here his problem was to express the equations for natural laws so that they apply to any part of space, no matter how that part of space might be moving with respect to an observer. The special theory refers only to uniform motion whereas the general theory includes accelerated motion as well.

general relativity applies to accelerated as well as to uniform relative motion

gravitation

A simple example will show something of the additional considerations that accelerated motion introduces. Suppose that you had lived all your life in a large box falling freely toward the earth from somewhere far out in space. The box and everything within it would be moving with uniform acceleration, the speed increasing steadily as it approached the earth. Since you would be moving with the box, you would not be conscious of any gravitational pull toward the earth; your feet would not press down on the floor, and, if you tried to let a ball drop, it would hang motionless in mid-air.

From my position on solid earth I would see things in your box very differently from you. If I watched you drop a ball, I would say that the ball approached the earth with a steadily accelerated motion; you would say that the ball remained motionless with respect to your box. I would say that the ball was impelled downward by a force; you would say that you find no evidence of any force whatever. One of us sees accelerated motion and the action of a force; the other sees no motion and no force. We are both right, as usual. Not only motion but force as well are relative to our points of view.

principle of equivalence

The above conclusion can be expressed formally by saying that **there is no way for an observer in a closed laboratory to distinguish between the effects produced by a gravitational field and those produced by an acceleration of the laboratory.** This statement is known as the principle of equivalence, and it is verified by the experimental observation that the *inertial mass* of a body is always exactly the same as its *gravitational mass*. (The inertial mass of a body determines the acceleration **a** it is given by an applied force **F** according to Newton's second law of motion, $\mathbf{F} = m\mathbf{a}$. The gravitational mass of the body determines the force it experiences due to the gravitational attraction of another body.)

light is deflected in a gravitational field

The principle of equivalence implies something very significant about the behavior of light. Light waves carry energy from one place to another, which means that, since $E = mc^2$, we can think of a certain amount of inertial mass as being in some way associated with a given bundle of light waves. From the principle of equivalence, a gravitational mass of the same amount must also be associated with the bundle, so that in general light ought to be subject to gravity! This is a startling prediction, and must be verified by experiment. The amount of mass involved is extremely small, so for a perceptible effect a very large body is needed. Such a very large body is the sun, and, if Einstein was right, light rays that pass near the sun should be deflected toward it, just as the paths of the planets are deflected toward the sun to produce their orbits. Einstein calculated that light rays that just miss the sun should be deviated by $0.0005°$, and measurements made on light from stars near the sun in the sky during solar eclipses indeed show almost exactly the same deviation.

Since we normally think of light rays in vacuum as traveling in "straight lines"—in fact, a straight line is commonly defined as the path a light ray would take—the finding that light is bent in a gravitational field might lead us to reexamine our notions of the structure of space. It has proved a fruitful notion to regard gravity as an alteration in the configuration of space near a body of matter, so that moving bodies and light rays naturally pursue curved paths there instead of straight ones. What we think of as the force of gravity can thus be attributed to a warping of space. It may seem as though one abstract concept is merely being substituted for another, but the fact is that the new point of view has led to a variety of unexpected discoveries that could not have come to light in the older way of thinking.

gravity can be thought of as a warping of space

the twin paradox

One of the predictions of general relativity is that a stationary clock runs more slowly in a strong gravitational field than in a weak one. A clock that has an interval of t_0 between ticks when it is infinitely far from all other matter will have an interval between ticks of

$$t = \frac{t_0}{\sqrt{1 - 2GM/R}}$$

slowing of a clock in a gravitational field

when it is the distance R from a mass M; G is the constant of gravitation found in Newton's law of gravitation.

We are now in a position to understand the famous relativistic phenomenon known as the twin paradox. This paradox involves two identical clocks, one of which remains on the earth while the other goes on a voyage into space at the speed v and returns a time t later. It is customary to replace actual clocks with a pair of identical male twins named A and B; this substitution is perfectly acceptable, because the processes of life—heart beats, respiration, and so on—constitute biological clocks on the average as regular as any other.

Twin A takes off when he is 20 years old and travels at a speed of $0.99c$. To his brother B on the earth, A seems to be living more slowly, in fact at a rate only

$$\sqrt{1 - \frac{v^2}{c^2}} = \sqrt{1 - \frac{(0.99c)^2}{c^2}} = 0.14 = 14 \text{ percent}$$

as fast as B does, according to special relativity. For every breath that A takes, B takes 7; for every meal that A eats, B eats 7; for every thought that A thinks, B thinks 7. Finally, after 70 years have elapsed by B's reckoning, A returns home, a man of only 30 while B is then 90 years old.

Where is the paradox? If we examine the situation from the point of view of twin A in the space ship, B on the earth is in motion at $0.99c$. Therefore we might expect B to be 30 years old upon the return of the space ship while A is 90 at this time—the precise opposite of what was concluded in the preceding paragraph.

The resolution of the paradox depends upon the fact that the spaceship is accelerated at various times in its journey: when it takes off, when it turns around, and when it finally comes to a stop. Therefore the formulas of special relativity, which hold only for frames of reference in relative motion at constant velocity, cannot be applied to the situation at all, and we must turn to general relativity. By the principle of equivalence, a large acceleration produces effects indistinguishable from those produced by a strong gravitational field—and clocks tick more slowly in strong gravitational fields. (The earth's gravitational field is small compared with the gravitational field needed to cause accelerations of the magnitude a spaceship of final speed $0.99c$ must undergo.) Hence A, the space traveler, *is indeed younger* on his return than his twin brother. Of course, A's lifespan has not been extended *to him*, since however long his 10 years on the spaceship may have seemed to his brother B, it has only been 10 years as far as he is concerned.

GLOSSARY

The *special theory of relativity* deals with problems that arise when one frame of reference moves at a constant velocity with respect to another. Its two postulates are that all laws of nature are the same in all frames of reference moving relative to one another at constant velocity and that the speed of light in vacuum is the same in all such frames of reference.

The *relativistic time dilation* is the increase in the length of time intervals when measured by an observer in motion relative to a clock compared with their length when measured by an observer at rest relative to the clock.

The *Lorentz contraction* is the decrease in the measured length of an object that is moving relative to an observer compared with its length when measured by an observer at rest relative to the object.

The *relativity of mass* refers to the greater mass of an object when measured by an observer in relative motion as compared with the mass measured by an observer at rest relative to the object.

The *total energy* of an object is the product of its mass and the square of the speed of light, namely mc^2. The *rest energy* of the object is the product of its mass measured when it is at rest and the square of the speed of light, namely $m_0 c^2$. The difference between these energies is equal to the kinetic energy of the object.

The *general theory of relativity* deals with problems that arise when one frame of reference is accelerated with respect to another.

EXERCISES

1 State the two postulates of the special theory of relativity. Were experiments necessary to verify these postulates, or do we accept them because they were stated by Einstein?

2 An observer is situated in a windowless laboratory. Can he determine whether the earth
   a. is traveling through space with a constant velocity?

b. is traveling through space with a constant acceleration?
c. is rotating?

3 The length of a rod is measured in several frames of reference, in one of which the rod is at rest. How can the latter frame be identified?

4 It is possible for the electron beam in a television picture tube to move across the screen at a speed faster than the speed of light. Why does this not contradict special relativity?

5 An astronaut 6 ft tall on earth is lying along the axis of a spaceship whose speed is 1.5×10^8 m/sec relative to the earth. An observer on the earth measures his height. What value does the observer find?*

6 What would the speed of a spaceship have to be in order that each week on the ship corresponds to 8 days on the earth?

7 A rocket ship in flight is observed to be 99 percent of its length when it was on the earth. What is its speed?

8 Calculate the mass of a 1-g object when it is traveling at 10 percent of the speed of light. Make the same calculation for speeds of 90 and 99 percent of the speed of light.

9 A rocket ship leaves the earth at 1.5×10^8 m/sec. How much time (as measured on the earth) must elapse before a clock in the ship and on the earth differ by 1 min?

10 With what speed must a body travel if its mass is to double?

11 The energies associated with chemical reactions are usually several "electron volts" per molecular change, where 1 ev = 1.6×10^{-19} joule. What is the mass equivalent of 1 ev?

12 Approximately 5.4×10^6 joules of chemical energy is released when 1 kg of dynamite explodes. What fraction of the total energy of the dynamite is this?

13 Approximately 4×10^9 kg of matter is converted into energy in the sun per second. Express the power output of the sun in watts.

* When v is $0.5c$ or less, the following approximations are accurate to a few percent or better and can be used to simplify relativistic calculations:

$$\sqrt{1 - \frac{v^2}{c^2}} \approx 1 - \frac{1}{2}\left(\frac{v^2}{c^2}\right)$$

$$\frac{1}{\sqrt{1 - \frac{v^2}{c^2}}} \approx 1 + \frac{1}{2}\left(\frac{v^2}{c^2}\right)$$

part two
COLLECTIVE MATTER

The problem of the ultimate structure of matter has intrigued man for thousands of years.

In recent times it has become clear that all matter in bulk is composed of one or more of the 103 known elements, which are substances that cannot be decomposed or transformed into one another by ordinary chemical or physical means.

Today we know that every element consists of tiny, almost identical particles called *atoms*; when two or more elements combine to form a compound, their atoms join together to form molecules or crystals of the compound. The kinetic energy of the constantly moving molecules is what we recognize as heat; the internal energy of a sample of matter is the total molecular kinetic energy it possesses, while its temperature is a measure of the average kinetic energy of the individual molecules. The kinetic-molecular theory of matter is able to account for a wide variety of everyday phenomena involving solids, liquids, and gases as well as for the more formal data of the laboratory.

chapter 6

Heat

TEMPERATURE

thermometers

temperature scales

HEAT

the kilocalorie

specific heat

changes of state

THE NATURE OF HEAT

mechanical equivalent of heat

conservation of energy

quantization of energy

Not always does the disappearance of kinetic or potential energy result in the performance of work or a change in rest mass. A skier slides down a high slope and comes to rest at the bottom. What has become of the potential energy he originally had? The engine of an automobile is shut off, while the automobile itself is permitted to coast along a level road from some initial speed; eventually the automobile slows down and comes to a halt. What has become of its original kinetic energy? All of us can give many similar illustrations that exhibit the apparent disappearance of kinetic or potential energy. What these examples have in common, as careful examination would reveal, is that *heat* is always produced. As a general observation, whenever mechanical energy disappears from a system of objects that does not interact with its surroundings, all this energy is converted either into mechanical work or into heat. This is one way of stating the law of conservation of energy, one of the most fundamental of all physical principles. In this chapter we shall explore the related topics of heat and temperature, together with the conservation of energy.

The color of an object heated to incandescence permits its temperature to be accurately determined. Here a 182-ton steel billet is heated prior to forging.

TEMPERATURE

Heat and temperature are often confused. We say commonly, and usually correctly, that the higher its temperature, the more heat an object possesses. But we cannot say that one object possesses more heat than *another* object because its temperature is higher. The filament of a light bulb, for instance, has a much higher temperature than does a piece of burning wood, yet we should hardly choose to warm ourselves on a cold day by a light bulb in preference to a wood fire. A cup of boiling water is hotter than is a pailful of warm water, but the warm water would melt a larger quantity of ice. The cupful of boiling water evidently contains a smaller amount of heat despite its higher temperature (Fig. 6.1).

Temperature, like force, is a physical quantity primarily meaningful to us in terms of the responses of our sense organs. We can touch an object and estimate its temperature, but just how we do this is not easy to explain in words. Some objective means of specifying the notion of temperature is required. We shall postpone the proper interpretation of temperature in terms of molecular motion until Chap. 8 and for the

moment invoke a familiar experimental observation. When a hot and a cold body are placed in contact, they eventually reach the same intermediate temperature. Hot coffee poured into a cold cup becomes cooler, while the cup simultaneously becomes warmer. A working definition of temperature can be based upon this phenomenon: **When heat flows from one body to another, the former is at a higher temperature than the latter.**

heat flows from hot bodies to cold ones

thermometers

A *thermometer* is a device that measures temperature. Most substances expand when heated and shrink when cooled, and the thermometers we use in everyday life are designed around this property of matter. More precisely, they are based upon the fact that different materials react to a given temperature change to different extents. The familiar mercury-in-glass thermometer (Fig. 6.2) functions because mercury expands more than glass when heated and contracts more than glass when cooled.

a thermometer measures temperature

6.1 The heat content of a given substance depends upon both its mass and its temperature. *A pail of cool water contains more heat than a cup of boiling water.*

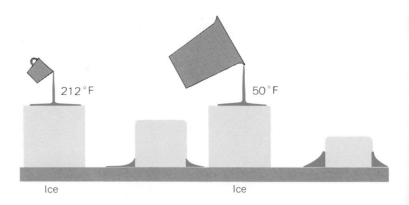

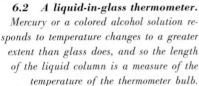

6.2 A liquid-in-glass thermometer. *Mercury or a colored alcohol solution responds to temperature changes to a greater extent than glass does, and so the length of the liquid column is a measure of the temperature of the thermometer bulb.*

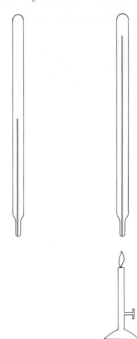

Thus the length of the mercury column in the glass tube provides a measure of the surrounding temperature.

Another common thermometer makes use of the different rates of expansion of different kinds of metals. Two straight strips of dissimilar metals are joined together at a particular temperature (Fig. 6.3). At higher temperatures the bimetallic strip bends so that the metal with the greater expansion is on the outside of the curve, and at lower temperatures it bends in the opposite direction. In each case the exact amount of bending depends upon the temperature. Thermometers of this kind are often employed in measuring fairly high temperatures, such as in ovens and furnaces.

Thermal expansion is not the only property of matter that can be used to construct a thermometer. As another example, the color and amount of light emitted by a very hot body vary with temperature. A poker thrust in a fire first glows dull red, then successively bright red, orange, yellow, and finally, if the poker achieves a high enough temperature, it becomes "white hot." The precise color of the light given off by an incandescent object is thus a measure of its temperature.

temperature scales

Given a physical property that varies with temperature, the next step is to establish a scale of temperature to make possible quantitative measurements. Unfortunately there are two such scales in common use, one devised by Fahrenheit in 1714 and the other by Celsius in 1742. Fahrenheit chose for the zero point of his scale the lowest temperature he could obtain with an ice-salt mixture and arbitrarily picked 96° as the temperature of the human body. (His measurement was inaccurate; today we regard 98.6° as normal body temperature.) Such temperatures are difficult to reproduce accurately, and so today the Fahrenheit scale is defined by calling 32° the freezing point of water and 212° its boiling point.

More practically, Celsius chose the easily determined temperatures of freezing and boiling water as the 0° and 100° marks on his scale. The Fahrenheit scale is an awkward makeshift and remains in use only because England and America cling to it with the same obstinacy that preserves the British system of weights and measures. Most other civilized countries,

HEAT

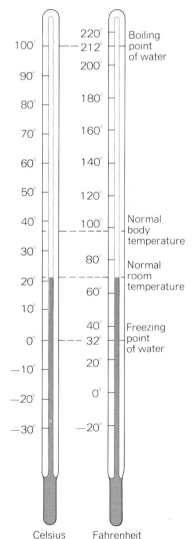

6.3 A bimetallic strip thermometer. *No matter on which side the heat is applied, the bend is away from the more expansive metal. The higher the temperature, the greater the deflection. At low temperatures the deflection is in the opposite direction.*

6.4 Comparison of Celsius and Fahrenheit temperature scales. *A Celsius temperature T_C may be changed to its equivalent Fahrenheit temperature T_F by means of the formula $T_F = \frac{9}{5}T_C + 32$; to change from a Fahrenheit temperature to its equivalent Celsius temperature, the formula $T_C = \frac{5}{9}(T_F - 32)$ may be used.*

and scientists throughout the world, use the more convenient Celsius (or centigrade) scale. A comparison of the two is shown in Fig. 6.4. To go from a Fahrenheit temperature T_F to a Celsius temperature T_C, and vice versa, we note that Fahrenheit degrees are five-ninths as large as Celsius degrees. Hence, also taking into account the difference between the zero point on the two scales, we see that

$$T_F = \tfrac{9}{5} T_C + 32°$$
$$T_C = \tfrac{5}{9}(T_F - 32°)$$

Thus the Celsius equivalent of the normal body temperature of 98.6°F is

$$T_C = \tfrac{5}{9}(98.6° - 32.0°)$$
$$= \tfrac{5}{9} \times 66.6°$$
$$= 37.0°C$$

HEAT

the kilocalorie

The *heat* possessed by an object depends not only on its temperature, but also on the amount and kind of material in it. A cupful of water at 100°C has less heat than a pailful at 30°C simply because less water is present. To raise the temperature of the cupful on a stove from 0 to 100°C would require less time than to raise the temperature of the pailful from 0 to 30°C. In order to include quantity of material as well as temperature in the idea of heat, we define the unit of heat, the *kilocalorie*, as follows: **The *kilocalorie* is the amount of heat required to raise the temperature of one kilogram of water by one degree Celsius.** Raising the temperature of 1 kg of water from 20 to 25°C would require 5 kilocalories (kcal); raising the temperature of 5 kg of water through the same range would require 25 kcal; and so on. The kilocalorie is an energy unit, like the joule and the foot-pound.

If we wish to lower the temperature of something, heat must be removed from it. The amount of heat involved is the same as would be needed to raise the temperature through the same temperature interval; thus, to cool 1 kg of water through 1°C, exactly 1 kcal has to be extracted.

The "calorie" used to express the energy contents of foods is actually the kilocalorie; it is sometimes written Calorie to distinguish it from the ordinary calorie, which is equal to 0.001 kilocalorie. Here is a list of the typical Calorie contents of some common foods:

Food	Calories	Food	Calories
1 raw onion	5	1 broiled hamburger patty	150
1 dill pickle	15	1 glass milk	165
6 asparagus	20	1 cup bean soup	190
1 gum drop	35	½ cup tuna salad	220
1 poached egg	75	1 ice cream soda	325
8 raw oysters	100	½ broiled chicken	350
1 banana	120	1 lamb chop	420
1 cupcake	130		

specific heat

The temperatures of most other substances can be changed by the transfer of less heat than that required by an equal mass of water. For instance, to increase or decrease the temperature of 1 kg of ice by 1°C we must add or remove only 0.5 kcal, and to do the same for 1 kg of silver a mere 0.056 kcal need be added or removed. The *specific heat* of a substance is defined as the heat required to change the temperature of one kilogram of the substance by one degree Celsius. A list of the specific heats of various substances is given in Table 6.1. For a given mass, the smaller the specific heat, the less the amount of heat needed to effect a given temperature change.

From the definition of specific heat (symbol c) we can obtain a simple formula for the heat (symbol Q) involved in changing the temperature

TABLE 6.1 Specific heats of various substances. The values given are averages, since the actual specific heats vary to some extent with temperature.

Substance	Specific heat, kcal/kg-°C	Substance	Specific heat, kcal/kg-°C
Alcohol (ethyl)	0.58	Lead	0.030
Aluminum	0.22	Marble	0.21
Copper	0.093	Mercury	0.033
Glass	0.20	Silver	0.056
Granite	0.19	Steam	0.48
Ice	0.50	Water	1.00
Iron	0.11	Wood	0.42

of a mass m of a substance by ΔT. This formula is

$$Q = mc\,\Delta T$$

quantity of heat

and it holds whether the heat is added or removed provided that the substance does not change state (from solid to liquid, for example) during the process. Thus, in order to raise the temperature of 10 kg of iron from 20 to 50°C, we must supply

$$Q = 10 \text{ kg} \times 0.11 \text{ kcal/kg·°C} \times 30°\text{C}$$
$$= 33 \text{ kcal}$$

since ΔT is 30°C.

changes of state

An important effect brought about by an appropriate change in temperature is the change of one form of matter to another. All gases can be liquefied and all liquids frozen if the temperature is decreased sufficiently. Nearly all solids can be liquefied and vaporized, except for those like paper and coal which decompose below their melting points. Since almost every substance can exist as a solid, liquid, and gas, these are considered as different *states* of matter rather than different kinds, and changes from one to another are called *changes of state*.

the states of matter are solid, liquid, and gas

Let us consider the temperature changes accompanying the melting of ice and the vaporization of water under ordinary conditions (normal atmospheric pressure). If we start with a dish of chipped ice below its melting point, the first effect of supplying heat is to raise the temperature. The rise continues steadily up to 0°C, then stops abruptly as the ice begins to melt. During the melting the temperature remains at 0°C, no matter how much heat we supply, provided the ice-water mixture is kept well mixed. If heating continues after the last of the ice disappears, the temperature again starts upward and rises continuously to 100°C. During this rise we find vapor given off in increasing quantities. Water vapor itself is colorless and therefore invisible, but above the heated dish it may cool and condense into visible white clouds of tiny liquid droplets. If we had had the proper means of detecting it, we should have found that vapor was given off during the entire heating process, from the cold ice as well as the heated water; the rising temperature simply increases the rate of vaporization. At 100°C the liberation of vapor becomes so rapid that it no longer takes place from the surface alone: bubbles of vapor form all through the liquid, and we say that the water is boiling. At the boiling point, as at the melting point, the temperature remains constant. We may use as hot a flame as we like, or as many burners, but the thermometer will remain at 100°C until the water disappears. Thereafter, if we arrange to heat the water vapor (usually called *steam* above the boiling point of water) in a closed container, its temperature will rise indefinitely as more heat is supplied. This sequence of changes is shown graphically in Fig. 6.5.

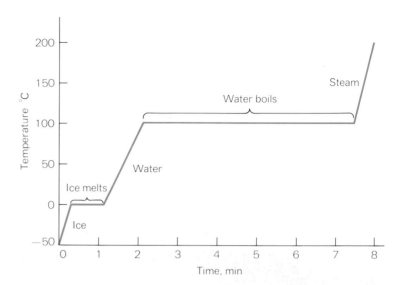

6.5 *Graph of temperature versus time for 1 kg of water, initially ice at* $-50°C$, *to which heat is added at the constant rate of 100 kcal/min. The process is assumed to take place with no loss of heat to the surroundings of the water.*

boiling and condensation occur at the same temperature

melting and freezing occur at the same temperature

the heat needed to melt 1 kg of a solid at its melting point is its heat of fusion

the heat needed to vaporize 1 kg of a liquid at its boiling point is its heat of vaporization

When steam at sea-level atmospheric pressure is cooled to 100°C, it condenses to liquid water, and the liquid freezes to ice at 0°C; the sequence of changes shown in Fig. 6.5 is exactly reversed by decreasing the temperature. The condensation point of steam and the freezing point of water are identical, respectively, with the boiling point of water and the melting point of ice. Because these temperatures are the same from whichever side approached, and because they remain constant while heat is added or removed, they are eminently suitable for fixing the points on a thermometer scale, as Celsius noted two centuries ago.

The only factors that influence the temperatures of freezing and boiling are the surrounding pressures and the purity of the water. The boiling point is particularly susceptible to pressure changes: The decrease in atmospheric pressure in going from sea level to a high mountain, for instance, lowers the boiling point so markedly that eggs and vegetables, to be thoroughly cooked, must be boiled considerably longer. The advantage of a pressure cooker lies in the fact that increased pressure raises the boiling point and hence decreases the time necessary for cooking.

Note that in the melting of ice and the vaporization of water, the material *absorbs heat without becoming hotter*. Just to melt 1 kg of ice, with the temperature constantly at 0°C, requires the addition of 80 kcal of heat—enough to raise the temperature of 1 kg of liquid water from room temperature to the boiling point. To vaporize 1 kg of water at its boiling point, without any change in temperature, requires the addition of considerably more heat—nearly 540 kcal. These two figures are called, respectively, the *heat of fusion* and the *heat of vaporization* of water (Fig. 6.6).

The reverse changes of state give off heat instead of absorbing it. The condensation of 1 kg of steam liberates 540 kcal, and the freezing of 1 kg of water liberates 80 kcal. That steam indeed gives off heat when it condenses is attested by the severe burns which live steam can produce. The liberation of heat when lakes freeze in winter is sufficient to keep

HEAT

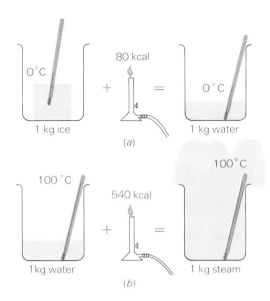

6.6 (a) *The heat of fusion of ice is 80 kcal/kg.* (b) *The heat of vaporization of water is 540 kcal/kg.*

their shores temporarily at a somewhat higher temperature than that of surrounding regions.

Changes of state in other substances are similar to those we have just discussed for water. Most pure substances have definite melting points and boiling points and characteristic heats of fusion and vaporization (Table 6.2). A few materials, such as glass, have no sharp melting point, but gradually soften when heated. Some solids vaporize readily on heating and ordinarily pass directly into the gaseous state without liquefying; iodine, Dry Ice (solid carbon dioxide), and camphor are familiar examples. The direct formation of vapor from a solid, called *sublimation*, is not

sublimation is direct change from solid to vapor

TABLE 6.2 Melting and boiling points of various substances together with their heats of fusion and vaporization.

Substance	Melting point, °C	Heat of fusion, kcal/kg	Boiling point, °C	Heat of vaporization, kcal/kg
Alcohol (ethyl)	−114	25	78	204
Bismuth	271	12.5	1560	190
Bromine	−7	16	59	43
Lead	327	5.9	1744	175
Lithium	179	160	1317	511
Mercury	−39	2.8	357	71
Nitrogen	−210	6.1	−196	48
Oxygen	−218	3.3	−183	51
Sulfuric acid	11	39	326	122
Water	0	80	100	540
Zinc	419	24	907	475

a peculiar property of these few substances—a light snowfall, for example, gradually disappears by sublimation even on a very cold day—but for most solids the amount of vaporization is so slight that we do not ordinarily observe it.

THE NATURE OF HEAT

The work done in raising an object through a height h is transformed into potential energy, and this potential energy is available for further conversion into kinetic energy or back into work. When work is done against friction, however, it ceases to be directly available for such conversion. To give a specific example, we might push a crate 50 ft across a warehouse floor by exerting a horizontal force of 30 lb on it, and in so doing we perform

$$W = Fd = 30 \text{ lb} \times 50 \text{ ft} = 1{,}500 \text{ ft-lb}$$

of work. But at the end of its path the crate is stationary, so it has acquired no kinetic energy, and it is at the same level as when it started, so it has gained no potential energy.

At first glance the work we did seems to have vanished, but a closer look reveals a new phenomenon: The crate and floor are now at a higher temperature than they were before. This is hardly a novel observation; even primitive man must have found that rubbing two sticks together both involves effort on his part and causes the sticks to grow warm. But a more basic question is less easy to answer: Does the dissipation of a given amount of energy always liberate *exactly* the same amount of heat? Only if this is true can we be sure that heat is really a form of energy.

mechanical equivalent of heat

Although it comes as no surprise to us today to learn that heat is a form of energy, our predecessors were not so clear on this point. Not much more than a century ago most scientists regarded heat as an actual substance, which they called *caloric*. The absorption of caloric caused an object to rise in temperature, while the escape of caloric from something caused its temperature to fall. Because the weight of an object does not change when it is heated, caloric was considered to be a weightless or imponderable substance, one of several imponderables whose convenient behavior was invoked by learned men in the latter part of the eighteenth century to account for mysteries in nature for which no other explanation was known. Caloric was invisible, odorless, and tasteless, as well as weightless, attributes which, of course, were responsible for its failure to be perceived directly.

Actually, the idea of heat as a substance worked fairly satisfactorily for materials heated over a flame, but it did not furnish an explanation for the heat generated by friction. One of the first to appreciate this difficulty was an adventurous American, born Benjamin Thompson in 1753, who fled this country during the Revolution and was made Count

The heat produced by friction at high speeds forms the basis for the friction-welding process. Here a rod of tungsten carbide and another of steel are rotated in opposite directions while in contact. When the ends of the rods become sufficiently hot, the rotation is stopped and they fuse together.

Rumford during a spectacular career in Europe. One of Rumford's many occupations was supervising the boring of cannon for a German prince, and he was impressed by the large amount of heat evolved by friction in the boring process. He showed that the heat could be used to boil water and that heat could be produced again and again from the same piece of metal. If heat was a fluid, it was not unreasonable that boring a hole in a piece of metal should liberate it. However, even a dull drill which cut no metal liberated a large quantity of heat. It was hard to conceive of a piece of metal as containing an infinite amount of caloric, and Rumford preferred to regard heat as a form of energy. Actually, in his own terminology, Rumford considered heat to be "motion" rather than a substance; but his inference is clear.

James Prescott Joule was an English brewer after whom the metric unit of energy is named in recognition of his classic experiment which settled the nature of heat. Joule's experiment employed a small paddle wheel within a container of water (Fig. 6.7). Work was done in turning the paddle wheel against the resistance of the water, and Joule determined exactly how much heat was supplied to the water by friction during this process. He found that a given amount of mechanical energy invariably produced the same amount of heat. Not only was heat intimately associated with motion, as Rumford had found, but the *amount of work* performed against friction precisely determined the *amount of heat* produced. This was a clear demonstration that heat is energy, and not matter.

The amount of mechanical energy that produces 1 kcal of heat, according to modern experiments, is 4,185 joules. This figure is often called the *mechanical equivalent of heat*.

the mechanical equivalent of heat is equal to 4,185 joules/kcal

conservation of energy

Joule's work led to one of the basic laws of physical science. His experiments showed clearly that, in the transformation of mechanical energy

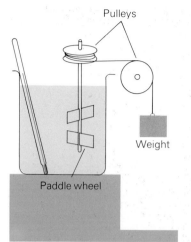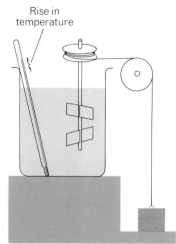

6.7 Joule's experimental demonstration that energy is conserved. *As the weight falls, it turns the paddle wheel, which heats the water by friction. The potential energy of the weight is converted first into the kinetic energy of the paddle wheel and then into heat energy (4,185 joules of mechanical energy is equivalent to 1 kcal of heat).*

into heat, no energy is lost, nor is any new energy created. One kind of energy is simply converted into another, and every bit of mechanical energy expended reappears as heat energy. In further experiments on other changes of mechanical energy into heat and in conversions of electric, magnetic, and radiant energy into heat, the amount of heat produced was always shown to be precisely equal to the amount of some other kind of energy that had vanished.

In other types of energy transformation, from potential to kinetic, for instance, some energy at first glance does seem to disappear; thus the hammer of a pile driver never has quite so much kinetic energy when it strikes the pile as it had potential energy before it fell, because some energy is lost in friction. But, if the heat energy produced by the friction is added in with the kinetic energy, the sum of the two is precisely equal to the original potential energy. For all ordinary transformations of energy that have been studied in detail a similar statement holds: if all the different sorts of energy that go into a transformation are added together and if all the energy produced is accounted for, the two sums are precisely equal. In other words, so far as we know, **energy cannot be created or destroyed.** This sweeping generalization is the law of conservation of energy.

law of conservation of energy

In a number of experiments there are what seem, at first, to be violations of the law of conservation of energy. In the theory of relativity, as we learned in Chap. 5, Einstein proved that mass must be thought of as a form of energy, and this equivalence has maintained conservation of energy in its place as one of the basic principles of physics. We shall learn later of the various ways in which mass is converted into energy and energy into mass in nature.

Just how far-reaching the law of conservation of energy is can perhaps be appreciated only by a physicist, who has applied it to a great variety of situations and seen it time and again bring order into tangled mazes of observational data. It is probably the broadest exact principle in all science, applying with equal force to remote stars and to the intricate biological processes in living cells. In the practical world, it enables technicians to calculate accurately the amount of energy obtainable from a machine, and it has shown clearly the fallacy in the age-old dream of perpetual motion.

quantization of energy

In all stable systems of particles, energy is not only conserved but is also *quantized*, which means that it can have only certain specific values. These specific energy values are very close to each other, and as a result energy quantization is conspicuous only in the behavior of such tiny objects as electrons, atoms, and molecules.

something whose energy is quantized can have only certain specific amounts of energy

For example, an electron confined to a hypothetical box 10^{-10} m wide (the order of magnitude of atomic dimensions) can only have energies of $6 \times 10^{-18} \, n^2$ joule, where $n = 1, 2, 3, \ldots$; no other energies, including zero, are possible. An electron in such a box with the minimum kinetic energy of 6×10^{-18} joule would have a speed of 3.6×10^6 m/sec, which would certainly be detectable. Thus energy quantization is an important factor in the atomic world, and many examples of it will appear later in this book.

In the macroscopic world, on the other hand, energy quantization is insignificant. The possible energies of a 10-g marble in a box 10 cm wide are $5.5 \times 10^{-64} \, n^2$ joule (the marble's vastly greater mass leads to a closer spacing between permitted energy levels). The minimum speed of such a marble is 3.3×10^{-31} m/sec, which is so close to zero as to be impossible to measure.

energy quantization is very important for atomic particles, but not for objects in the everyday world

Every object trapped in some region of space is subject to quantization of energy—the electrons in an atom, the atoms in a molecule or crystal, the planets in the solar system. A particle moving freely, however, can have any energy at all, including zero.

GLOSSARY

The *temperature* of a body is its degree of hotness or coldness; when two bodies are in contact, heat flows from the body at the higher temperature to the one at the lower temperature.

A *thermometer* is a device for measuring temperature. In the *Fahrenheit* temperature scale, the freezing point of water is defined as $32°F$ and the boiling point of water as $212°F$; in the *Celsius* temperature scale, the freezing point of water is defined as $0°C$ and the boiling point of water as $100°C$.

Heat is a form of energy. The heat that a body possesses depends upon its temperature, its mass, and the kind of material of which it is composed. The unit of heat called the *kilocalorie* is defined as the heat necessary to raise the temperature of 1 kg of water by $1°C$.

The *specific heat* of a substance is the heat required to change the temperature of 1 kg of the substance by $1°C$.

The *heat of vaporization* of a substance is the amount of heat required to change 1 kg of the substance from liquid to vapor at its boiling point; it is also the heat that 1 kg of the substance liberates when it changes from vapor to liquid at its boiling point.

The *heat of fusion* of a substance is the amount of heat required to change 1 kg of the substance from solid to liquid at its freezing point; it is also the heat that 1 kg of the substance liberates when it changes from liquid to solid at its freezing point.

The *mechanical equivalent of heat* is the amount of energy equivalent to one unit of heat; it is 4,185 joules/kcal.

EXERCISES

1 If an egg is dropped into boiling water over a gas flame, the length of time necessary to cook the egg is not changed by turning the gas higher, although more heat is being supplied to the water. Explain. What becomes of the extra heat?

2 How does perspiration give the body a means of cooling itself?

3 A car whose cooling system contains an alcohol antifreeze is more likely to overheat in summer than one whose cooling system contains pure water. Explain this observation with the help of Table 6.1.

4 Why does a nail become hot when it is hammered into a piece of wood?

5 Why is a piece of ice at 0°C more effective in cooling a drink than the same weight of cold water at 0°C?

6 In order to cool a room during the summer, a man turns on an electric fan and leaves. If the room is completely insulated from the outside, what will happen to its temperature and why?

7 The boiling point of oxygen is -183°C. What is the temperature on the Fahrenheit scale?

8 What is the Fahrenheit equivalent of a temperature of 25°C?

9 What is the Celsius equivalent of a temperature of 100°F?

10 How many kilocalories are required to raise the temperature of 20 kg of water from the freezing point to the boiling point?

11 How much heat is required to raise the temperature of 1 kg of copper from 100 to 140°C? From 140 to 180°C?

HEAT

12 How much steam at $120°C$ is required to melt 2 kg of ice at $0°C$?

13 Two kilograms of punch at $0°C$ is poured into a 3-kg silver punchbowl at $20°C$. The specific heat of the punch is 0.7 kcal/kg-°C. Find the final temperature of the punch.

14 How much heat is required to change 50 g of ice at $0°C$ into water at $20°C$?

15 How much heat is given off when 1 kg of steam at $100°C$ condenses and cools to water at $20°C$?

16 How much energy does an 80-kg man use in climbing a staircase 8 m high? To how many calories of heat does this energy correspond?

17 A 10-kg stone is dropped into a pool of water from a height of 100 m. How much energy in joules does the stone have when it strikes the water? If all this energy goes into heat, how many calories of heat are added to the water? If the pool contains 10 m^3 of water, by how much is its temperature raised? (The mass of 1 m^3 of water is 10^3 kg.)

18 How high is a waterfall if the water at its base is $1°C$ higher in temperature than the water at the top?

19 How many kilocalories of heat are evolved per hour by a 600-watt electric heater?

20 A lead bullet whose initial temperature is $100°C$ strikes a hardened steel plate and melts. What was the minimum speed at which it was moving?

21 If all the heat given off by a certain quantity of water at $0°C$ when it turns to ice at $0°C$ could be used to lift the ice vertically upward, how high would it be raised?

22 One kilogram of water at $0°C$ contains 80 kcal of energy more than 1 kg of ice at $0°C$. What is the mass equivalent of this amount of energy?

chapter 7

Solids, Liquids, and Gases

MATTER IN BULK

    solids

    hooke's law

    liquids

    surface tension

    gases

PRESSURE AND DENSITY

    pressure

    pressure in a fluid

    atmospheric pressure

    density

    buoyancy

THE GAS LAWS

    boyle's law

    charles' law

We now turn to matter itself and the characteristics of its three states, solid, liquid, and gas. In everyday life matter shows no direct sign that it really consists of myriad tiny particles in incessant motion. However, there are plenty of indirect signs that such is the case, and we shall find a number of them as we examine some of the mechanical properties of solids, liquids, and gases in this chapter. Then, in Chap. 8, we shall see exactly how the physics of matter in bulk follows from the physics of individual particles.

MATTER IN BULK

Almost everyone has a fairly clear idea of the distinctions among solids, liquids, and gases. Solids maintain their sizes and shapes no matter where they are placed. Liquids, however, spread out and assume the shapes of

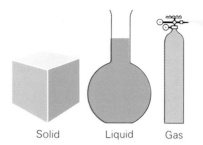

7.1 Solids, liquids, and gases. *A solid maintains its shape and volume no matter where it is placed; a liquid assumes the shape of its container while maintaining its volume; a gas expands indefinitely unless stopped by the walls of a container.*

Solid   Liquid   Gas

their containers while maintaining their volumes constant. Gases maintain neither shape nor volume but expand until they fill completely whatever container they are placed in (Fig. 7.1).

A related distinction involves the ability of gases and liquids to flow readily, which is why these phases of matter together are called fluids. Solids resist deformation, although, of course, if enough force is applied they too will change shape. Thus the dividing line between solids and liquids is not always perfectly clear-cut. Pitch, for instance, is hard and brittle like a solid; yet, if it is allowed to stand long enough, it will spread out to fill its container in the manner of a liquid.

solids

Solids surround us in enormous variety. To such naturally occurring substances as rocks, minerals, and products of plant and animal life, man has added an endless list of artificial materials. Iron, wood, feathers, and salt are all solids, since all will retain their shapes indefinitely regardless of the shapes of their containers. The constituents of most solids occur in the form of regularly shaped *crystals*. The formation of a crystalline solid is a wonderful sight if the time is taken to watch it carefully. When water freezes to ice, for example, slender ice needles radiate out over the liquid surface, branching and interlocking, with each new-formed bit of solid appearing just where it should to preserve the shapes and pattern of the interlaced needles. We can hardly miss the conclusion that the shape of an ice needle or the shape of any other growing crystal reflects something about the inner structure of the solid and that perhaps the secret of this inner structure might be revealed by a painstaking study of the crystal, a notion we shall return to in later chapters.

most solids are crystalline

When a solid is bent, stretched, or squeezed, it does not flow as a fluid would. If the deforming force is small, the solid changes shape only slightly and then springs back to its original form when the force is removed. This behavior is called *elastic*; some materials, of course, are much more elastic than others. Subjected to a great force, a solid may be permanently deformed. Thus an iron wire, if bent slightly, springs back to its original shape, but if subjected to a greater force, is permanently bent. Brittle solids, like ceramics, can undergo almost no permanent deformation without breaking; other solids, like some metals, can be deformed almost indefinitely—an ounce of gold, for instance, can be

solids are elastic

SOLIDS, LIQUIDS, AND GASES

hammered into 300 ft² of thin, translucent foil, and platinum can be drawn into wires so fine that they are scarcely visible.

hooke's law

An important property of most solids is that an applied stress produces a change in size whose extent is proportional to the magnitude of the stress, provided that the latter is not too great. If a certain wire stretches by 0.1 in. when it is used to support a weight of 50 lb, a 100-lb weight would cause it to stretch by 0.2 in., a 150-lb weight would cause it to stretch by 0.3 in., and so on (Fig. 7.2). When the weight is removed, the wire contracts to its initial length. Because of this proportionality between the force F and the resulting elongation s, we can write

$$F = ks$$

where k, the force constant, varies with the size, shape, and nature of the material under stress. A thick wire stretches less than a thin one of the same kind under the same tension, so it has a higher force constant, as does a steel wire compared with an identical strand of rubber. In all cases, though, the ratio between s and the original length of the wire stays the same during the elongation, so that a long wire stretches more than a short one. This relationship is known as *Hooke's law*.

Figure 7.3 is a graph that shows the response of a particular iron rod to an applied force. With increasing tension the curve is at first a straight line, reflecting Hooke's law. Eventually a point called the *elastic limit* is reached, where further elongation leads to a permanent deformation. When the force exceeds that corresponding to the elastic limit, the rod stretches at a more rapid rate, until the breaking point is reached. Some metals, such as copper, can be stretched well beyond their elastic limits without rupture, while others, such as cast iron, are so brittle that they cannot go far beyond their elastic limits without breaking.

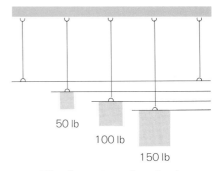

7.2 The elongation of a wire is proportional to the force applied to it.
As long as the elongation is not too great, when the force is removed, the wire returns to its original length.

the elastic limit

7.3 Elongation of an iron rod as a function of the force applied to it.
The approximate proportionality between the force and the elongation it produces is called Hooke's law; this proportionality only holds over a limited range. Permanent deformation occurs beyond the elastic limit.

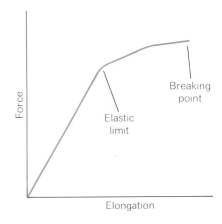

liquids

Water is the only common naturally occurring liquid. Petroleum, molten lava, and liquids produced by plants and animals occur in small amounts. A great variety of liquids has been prepared artificially, but no other liquid is remotely comparable with water in abundance or in multiplicity of uses.

In contrast with gases, liquids are virtually incompressible. The highest pressures obtainable with modern laboratory equipment are able to squeeze water only into about three-fourths of its original volume. Some liquids, water and alcohol, for instance, are *miscible* ("mixable") with each other in all proportions, but others, like water and oil, are immiscible. Two immiscible liquids shaken together give a cloudy mixture with tiny globules of one liquid suspended in the other; when the shaking stops, the globules coalesce and the liquids separate into distinct layers. If two miscible liquids are placed in contact without stirring, one will diffuse into the other, but

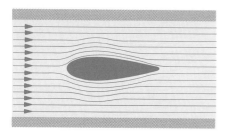

7.4 Flow lines in fluid motion around an obstacle. *Each line represents paths taken by successive fluid particles. These lines are called streamlines.*

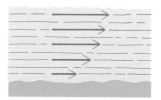

7.5 Relative motion at different levels in a slowly moving liquid. *The liquid moves by the sliding of one layer over another; at the bottom the motion is least, and it is fastest at the surface.*

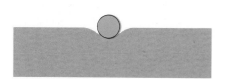

7.6 Enlarged cross section of needle on water surface. *The elastic character of a liquid surface is the result of surface tension.*

the process is very much slower than diffusion in gases. If a heavy gas (like carbon dioxide) is placed in the bottom of a container and a light gas (hydrogen) in the top, after a few hours the two will be completely mixed; but if the lighter of two miscible liquids (for instance, alcohol) is poured carefully on top of the heavier (for instance, water), months may pass before diffusion makes the mixing complete.

Like gases, liquids flow readily, but their rate of flow is much smaller. A thick liquid like honey, particularly when full of tiny bubbles or solid specks, shows well the details of liquid flow. As honey creeps slowly down an inclined surface, the bubbles gather in lines along the direction of flow, the distinctness of the lines showing roughly the amount of liquid movement in their vicinity. Lines are best formed at the top of the flow, scarcely detectable at its bottom. If the motion of the honey is confined to channels of different shapes, the bubble lines form graceful curves around obstacles and irregularities in the channel walls (Fig. 7.4). The most distinct lines are always at the surface of the liquid far from the confining walls. These observations suggest that the liquid moves by the sliding of one layer over another, layers near the bottom and sides of the channel clinging to the solid surfaces and scarcely moving at all (Fig. 7.5). How fast the liquid flows depends on how freely layers within the liquid can slide past one another. In honey there is evidently considerable resistance to the sliding motion, in water or alcohol much less. This inner resistance of liquids to flowing movement is called *viscosity*. Honey is a viscous liquid; water and alcohol are less viscous. Gases also show some resistance to fluid motion, but their viscosities are extremely low compared with those of liquids.

In rapidly flowing liquids and gases the mechanism of flow is not as simple as that just described. Flow lines are not straight or in smooth curves, but follow complex, ever-changing loops and eddies. Turbulent flow of this sort is well shown in a swift mountain brook.

surface tension

Some curious properties of liquids are associated with their free surfaces. For instance, steel is denser than water, but a steel needle can be made to float if lowered very carefully to a water surface. The needle rests in a slight depression of the surface, the water showing no tendency to "wet" the metal (Fig. 7.6). When we disturb the needle so that any part cuts through the surface, it sinks immediately to the bottom. If we touch the water surface very carefully with the end of a clean glass rod, then at first contact the glass, like the needle, simply depresses the surface without being "wet" by the liquid. Further lowering causes the surface to break, and a little ridge of water climbs up around the rod (Fig. 7.7). When we lift the rod out, its end is now "wet," that is, covered with a thin film of water. The water surface, after the momentary disturbance, becomes smooth and flat as before.

In experiments like these the surface layer of water seems to behave differently from the rest of the liquid; water will "wet" both glass and steel, but the surface layer must be broken before liquid and solid can make intimate contact. It is almost as if the water were covered with a thin, stretched sheet of rubber, which must be ruptured before the water can make contact with metal or glass. Unlike rubber, the surface layer has the capacity for repairing itself: No matter how many times the surface is broken, it shows the same behavior to each new disturbance. The apparent stretched, elastic character of liquid surfaces is shown also by the spherical shape of raindrops and by the spherical shape of globules formed when immiscible liquids are shaken together. Here each liquid fragment has a free surface on all sides, and the surface contracts to give the drop a shape with the least possible surface area—the sphere.

The force tending to decrease the area of a liquid surface is a property of the liquid called *surface tension* and is due to the attraction for one another of the individual molecules of which the liquid is composed.

surface tension is due to the mutual attraction of liquid molecules

gases

Gases have become familiar articles of commerce: the light gases hydrogen and helium, the heavy gas carbon dioxide, the pungent, yellow-green, poisonous gas chlorine. Surrounding us always, necessary to our existence, is the mixture of colorless and odorless gases that we call air. The odors of flowers, foods, and perfumes are all due to gaseous materials. Yet the ways of gases remain mysterious to the layman, because without special techniques he cannot isolate and handle them as he can liquids and solids. He often finds it startling even to learn that gases have weight (the air in a good-sized room, for instance, weighs perhaps a hundred pounds), since they spread so easily outward and upward in apparent defiance of gravity.

To the scientist, on the other hand, gases are by far the best understood of the three forms of matter. He knows that all gases have certain important characteristics that are much alike, and he has learned simple laws to describe their properties. We shall discuss some of these important laws later, but for the moment let us simply note a few of the more obvious facts about gas behavior.

The defining property of gases is their ability to expand indefinitely. Coupled with this is their extreme compressibility, with even a small increase in pressure causing a marked reduction in the volume of a gas. A gas will expand either into a vacuum or into another gas, the second gas serving merely to slow down the rate of expansion. The spreading, or diffusion, of one gas into another can be well shown by opening a bottle of ammonia in one corner of a room: the irritating odor of ammonia gas spreads quickly through the air into all parts of the room. Nor is there any limit to the amount of one gas that can diffuse into another: gases are miscible with one another in all proportions. Finally, gases, unless highly compressed, are characterized by extreme lightness in comparison with liquids and solids.

Matter can also exist in the unfamiliar but important *plasma* state. A plasma is essentially a gas whose constituent particles are electrically charged, and because of this their behavior depends strongly upon electric and magnetic forces. Though plasmas are relatively rare on the earth's

7.7 Glass rod (a) before breaking water surface and (b) after breaking surface. The surface layer of water must be broken for the rod to be "wet" by the water.

plasma

surface, most of the astronomical universe is in the plasma state, and indeed the earth's outer atmosphere is a plasma.

PRESSURE AND DENSITY

pressure

We have already noted the great compressibility of gases and the near-incompressibility of liquids and solids. Let us examine a little more closely the influence of pressure on various materials.

First of all, what do we mean by pressure? By definition, pressure is the perpendicular force per unit area acting on a surface. In equation form,

pressure is force per unit area

$$\text{Pressure} = \frac{\text{force}}{\text{area}}$$

or, symbolically,

$$P = \frac{F}{A}$$

Pressure may be expressed in such units as pounds per square inch ($lb/in.^2$, sometimes written "psi"), pounds per square foot (lb/ft^2), newtons per square meter ($newtons/m^2$), and so on.

As an example, we can consider the cylinder of Fig. 7.8 which has a fluid of some kind within it. The cylinder has a cross-sectional area of 0.05 m^2 and a force of 200 newtons is applied to the piston. The pressure on the fluid is therefore

$$P = \frac{F}{A} = \frac{200 \text{ newtons}}{0.05 \text{ m}^2} = 4{,}000 \text{ newtons/m}^2$$

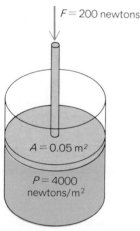

7.8 *Pressure is force per unit area. A force of 200 newtons applied to a piston of area 0.05 m^2 results in a pressure of 4000 newtons/m^2.*

pressure in a fluid

pressure increases with depth

In a fluid, since gravity pulls downward on all parts of it, the bottom layers are under pressure from the weight of overlying fluid, whether external pressure is acting or not. At the bottom of the cylinder just described, for instance, the pressure must be slightly greater than 4,000 newtons/m^2, since the bottom layer must support not only the piston's weight but a considerable weight of fluid as well. In ordinary laboratory experiments with gases, pressures due simply to the weights of the gases involved are usually negligible, but in experiments with liquids such pressures are important.

Except for the steady downward increase due to gravity, **pressures in all parts of a fluid are the same.** It would be unthinkable, for instance, to have the fluid confined in the cylinder of Fig. 7.8 under a pressure of 4,000 newtons/m^2 in the upper part of the

SOLIDS, LIQUIDS, AND GASES

The weight of a car is transferred to the road through the air in its tires. The compressibility of air enables the tires to absorb much of the shock of going over irregularities in the road.

cylinder and under a pressure of 100 newtons/m^2 in the lower part. Since the fluid is capable of flowing, it would move from the region of high pressure to the region of low pressure until the pressure was equalized. If the air pressure in an automobile tire is 30 lb/in.^{2}, then the air exerts an outward force of 30 lb on every square inch of the tire's inner surface. Different parts of a solid, on the other hand, can be under very different pressures, since the solid cannot flow. One end of a timber can be placed under immense pressure in a vise, while the other end is under no pressure except that of the surrounding atmosphere.

Also because a fluid can flow, **the pressure at any one point within it must be the same in all directions.** It is this fact that makes pressure so useful a quantity in physics. Suppose that a piece of paper can be supported horizontally in the middle of the cylinder of Fig. 7.8. If the downward pressure on its upper surface is 4,000 newtons/m^2, then the upward pressure on its lower surface must be 4,000 newtons/m^2. If the two pressures were different, the paper would move toward the direction of lower pressure—in other words, the fluid would flow until the pressure at each point became the same in all directions. This characteristic of fluid pressure is well shown by the device of Fig. 7.9, which consists of a funnel capped with a thin rubber sheet and connected with an instrument for measuring pressure. If the funnel is held with the center of the membrane at any one level in a liquid or gas, turning the funnel in different directions has no effect on the indicated pressure.

fluid pressure is independent of direction

atmospheric pressure

Since we live at the base of the atmosphere, we are under pressure from the weight of gas above us—a pressure amounting to nearly 15 lb/in.^{2} at sea level, sufficient to crush a container from which air has been removed unless its walls are fairly stout. We are not conscious of the 15-lb force pushing inward on every square inch of our bodies, simply because our bodies are sufficiently permeable to air so that pressures inside are maintained equal to those without.

7.9 Pressure is the same in all directions at the same level in a fluid.

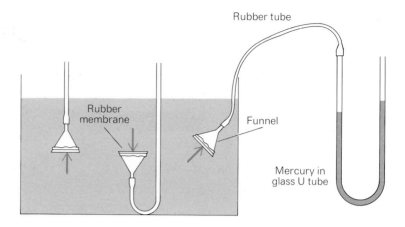

a barometer measures atmospheric pressure

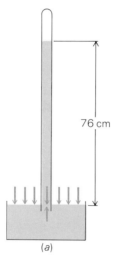

7.10a *A mercury barometer. The weight of the mercury column is balanced by atmospheric pressure.*

Atmospheric pressures are measured with instruments called *barometers*, of which a common type is a closed tube of mercury about 1 m long inverted in a dish of mercury (Fig. 7.10a). The upper part of the tube is evacuated, either directly with a vacuum pump or by taking care that no air enters the tube when it is inverted in the dish. At the bottom of the tube, the downward force is equal to the weight of the mercury column. This must be balanced by an upward force, else the mercury would flow out of the tube. The upward force is maintained by the downward push of the atmosphere on the mercury surface in the dish. Thus the pressure of 100 miles or so of atmosphere is balanced by the pressure of 76 cm of mercury. The pressure of the atmosphere at any one time is measured by the height of the mercury column, which fluctuates with varying weather conditions over a range of about 12 cm. The familiar *aneroid* barometer is illustrated in Fig. 7.10b.

Normal atmospheric pressure at sea level corresponds to a barometric height of 76 cm (about 30 in.) of mercury. This pressure is frequently used as a pressure unit called an *atmosphere*. Thus a pressure of 2 atmospheres (atm) would correspond to 152 cm of mercury. Smaller pressures, since they are often measured with mercury columns, are commonly described in terms of mercury heights; thus a pressure of 5 cm of mercury is equal to about 0.066 atm, or roughly 1 lb/in.²

Atmospheric pressure is often expressed in *millibars* (mb) by meteorologists, where 1 mb = 100 newtons/m². Sea-level atmospheric pressure averages 1,013 mb, which is 14.7 lb/in.²

density

density is mass per unit volume

An important physical property of a substance is its *density*, symbol d, which is its mass per unit volume. Table 7.1 is a list of the densities of various common substances. In metric units, density is properly expressed in kilograms per cubic meter; the density of ice, for example, is 920 kg/m³. Often densities in the metric system are instead given in grams per cubic centimeter. Since there are 1,000 g in a kilogram and 100 cm in a meter,

$1 \text{ g/cm}^3 = 10^3 \text{ kg/m}^3 = 1,000 \text{ kg/m}^3$

$1 \text{ kg/m}^3 = 10^{-3} \text{ g/cm}^3 = 0.001 \text{ g/cm}^3$

Thus the density of ice can also be expressed as 0.92 g/cm^3.

In the British system of units, density should properly be expressed in slugs per cubic foot, since the slug is the unit of mass in this system. In these units the density of ice is 1.8 slugs/ft^3. However, because weights rather than masses are usually specified in the British system, a quantity called *weight density* is commonly used. As the name suggests, weight density is weight per unit volume, and its units are pounds per cubic foot. Because the weight of something of mass m is mg,

Weight density = mass density $\times g = dg$

The last column of Table 7.1 lists weight densities.

Pressures in liquids rapidly become greater with increasing depth because of the weight of the overlying material. The stoutest submarine must be wary of venturing more than a few hundred feet down, for fear of collapsing. At a depth of 6 miles in the ocean the pressure is about 8 tons/in.², or roughly 1,000 atm—sufficient to compress water by some 3 percent of its volume. Fish that inhabit these depths can withstand such enormous pressures for the same reason that we can endure the pressures at the bottom of our ocean of air: pressures inside their bodies are kept

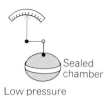

(b)

7.10b *An aneroid barometer.* The flexible ends of a sealed metal cylinder are forced in by a high atmospheric pressure, out by a low atmospheric pressure.

TABLE 7.1 Densities of various substances at atmospheric pressure and room temperature.

Substance	Mass density			Weight density
	kg/m^3	g/cm^3	$slugs/ft^3$	lb/ft^3
Air	1.3	1.3×10^{-3}	2.5×10^{-3}	8×10^{-2}
Alcohol (ethyl)	7.9×10^2	0.79	1.5	48
Aluminum	2.7×10^3	2.7	5.3	1.7×10^{-2}
Balsa wood	1.3×10^2	0.13	0.25	8
Bromine	3.2×10^3	3.2	6.2	2×10^2
Carbon dioxide	2.0	2.0×10^{-3}	3.8×10^{-3}	0.12
Concrete	2.3×10^3	2.3	4.5	1.4×10^2
Gasoline	6.8×10^2	0.68	1.3	42
Gold	1.9×10^4	19	38	1.2×10^3
Helium	0.18	1.8×10^{-4}	3.5×10^{-4}	1.1×10^{-2}
Hydrogen	0.09	9×10^{-5}	1.7×10^{-3}	5.4×10^{-2}
Ice	9.2×10^2	0.92	1.8	58
Iron	7.8×10^3	7.8	15	4.8×10^2
Lead	1.1×10^4	11	22	7×10^2
Mercury	1.4×10^4	14	26	8.3×10^2
Nickel	8.9×10^3	8.9	17	5.5×10^2
Nitrogen	1.3	1.3×10^{-3}	2.4×10^{-3}	7.7×10^{-2}
Oak	7.2×10^2	0.72	1.4	45
Oxygen	1.4	1.4×10^{-3}	2.8×10^{-3}	9×10^{-2}
Water, pure	1.00×10^3	1.00	1.94	62
Water, sea	1.03×10^3	1.03	2.00	64

constantly equal to pressures outside. When brought quickly to the surface, deep-ocean fish often explode because of their high internal pressures.

The fluid pressure at a depth h below the surface of a fluid is proportional to both h and the weight density dg of the fluid. In addition, of course, the atmosphere also exerts the pressure P_{atm} on the fluid surface, which in turn is passed on to the base of the tank. Hence the total pressure is

pressure at depth h in a fluid

$$P = P_{atm} + P_{fluid}$$
$$= P_{atm} + dgh$$

This equation holds for any depth h in a fluid of density d, since the fluid below this depth does not affect the weight of the fluid *above* it, and the actual dimensions of the container are irrelevant, since only the vertically downward force of the overlying fluid affects the pressure (Fig. 7.11).

As an example, let us find the pressure a skin diver experiences at a depth of 30 ft in sea water. Since $P_{atm} = 14.7$ lb/in.^{2} and dg for sea water is 64 lb/ft^3,

$$P = 14.7 \frac{\text{lb}}{\text{in.}^2} + 64 \frac{\text{lb}}{\text{ft}^3} \times 30 \text{ ft} \times \frac{1 \text{ ft}^2}{144 \text{ in.}^2}$$
$$= (14.7 + 13.3) \text{ lb/in.}^2 = 28 \text{ lb/in.}^2$$

which is nearly twice atmospheric pressure.

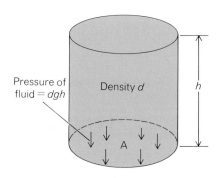

7.11 Density is mass per unit volume. *The volume of a tank of height h and cross-sectional area A is Ah. When it is filled with a fluid of density d, the mass of the fluid is dAh and its weight is $dgAh$. Hence the pressure on the bottom due to the fluid is $dgAh/A$, which is simply dgh.*

buoyancy

When something is immersed in water (or other fluid), it seems to weigh less than it does in air. This effect is called *buoyancy*, and it is responsible for such diverse phenomena as balloons floating in the atmosphere, ships floating in the sea, and the continents floating in the plastic rock that constitutes most of the earth's interior. If the upward buoyant force on a submerged body is greater than its weight, the body floats; if the force is less than its weight, the body sinks.

buoyancy refers to the upward force exerted by a fluid on an object placed in it

Let us consider a solid body of any shape and any material whose volume is V that is immersed in a tank of water. An identically shaped body of water in the tank is supported by a buoyant force equal to its weight of Vdg, since the water in the tank is at rest. The buoyant force is the vector sum of all of the forces that the rest of the water in the tank exerts on this particular volume, and it is always upward because the pressure underneath the volume is greater than the pressure above it; the pressures on the sides cancel one another out. If the body of water is replaced by the solid body, the forces on it are the same (Fig. 7.12), and the buoyant force remains Vdg.

This conclusion can be generalized to state that

archimedes' principle

Buoyant force on a body in a fluid
$$= \text{weight of fluid displaced by the body}$$

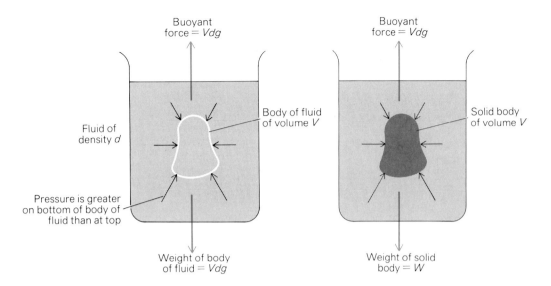

since Vdg is the weight of the fluid equivalent to the volume occupied by the solid body. The above statement is known as *Archimedes' principle*, and it holds whether the body floats or sinks. If the body floats, the volume V refers only to the submerged part.

As an illustration of Archimedes' principle, we can compute the proportion of an iceberg that is below the surface of the sea. In order for the iceberg to float in equilibrium, its weight must be exactly balanced by the buoyant force of the sea water. If we think of the iceberg as a cylinder and let its area be A and its height be L, as in Fig. 7.13, and call its submerged depth L_{sub}, we have

7.12 Archimedes' principle. *The buoyant force on a body in a fluid is equal to the weight of fluid displaced by the body. If the body's weight W is greater than Vdg, the body sinks; otherwise it floats at such a level that the weight of the displaced fluid equals W.*

Weight of iceberg = weight of displaced water

$$(Vdg)_{ice} = (Vdg)_{water}$$
$$LAd_{ice}g = L_{sub}Ad_{water}g$$
$$\frac{L_{sub}}{L} = \frac{d_{ice}}{d_{water}} = \frac{9.2 \times 10^2 \text{ kg/m}^3}{1.03 \times 10^3 \text{ kg/m}^3} = 0.9$$

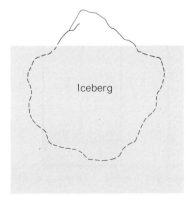
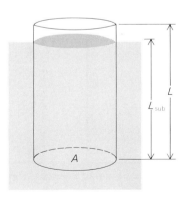

7.13 *Ninety percent of the volume of an iceberg is submerged because the density of ice is 90 percent that of water.*

Ninety percent of the volume of an iceberg is below the surface of the water.

THE GAS LAWS

In many ways the gaseous state is the one whose behavior is the easiest to describe and account for. As an important example, the pressures, volumes, and temperatures of gas samples are related by simple formulas that have no counterpart in the cases of liquids and solids. The existence of these formulas led to a search for their explanation in terms of the microscopic structure of gases, a search rewarded by the discovery of the kinetic theory of matter.

boyle's law

Suppose that a sample of some gas is placed in the cylinder of Fig. 7.14 and a pressure of 1×10^5 newtons/m² is applied. The final volume of the sample is 1 m³. If we double the pressure to 2×10^5 newtons/m², the piston will move down until the gas volume is 0.5 m³, half its original amount, if the gas temperature is kept unchanged. If the pressure is made ten times greater, the piston will move down farther, until the gas occupies a volume of 0.1 m³, again if the gas temperature is kept unchanged.

The above findings illustrate a general observation: The volume of a given quantity of a gas at constant temperature is inversely proportional

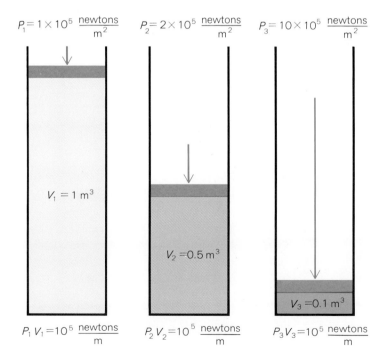

7.14 Boyle's law: At constant temperature, the volume of a given quantity of any gas is inversely proportional to the pressure applied to it. Here $P_1V_1 = P_2V_2 = P_3V_3$.

SOLIDS, LIQUIDS, AND GASES

to the pressure applied to it. (By "inversely proportional" is meant that as the pressure increases, the volume decreases by the same proportion, and vice versa.) If the volume of the gas is V_1 when the pressure is P_1 and the volume changes to V_2 when the pressure is changed to P_2, the relationship among the various quantities is

$$\frac{P_1}{P_2} = \frac{V_2}{V_1} \qquad \text{(at constant temperature)}$$

boyle's law

This relationship is called *Boyle's law*, in honor of the English physicist who discovered it.

charles' law

Just as changes in gas volume are simply related to pressure changes, so they are simply related to temperature changes. If a gas is cooled steadily, starting at 0°C, while its pressure is maintained constant, its volume decreases by $\frac{1}{273}$ of its volume at 0°C for every degree the temperature falls. If the gas is heated, its volume increases by the same fraction (Fig. 7.15). If volume rather than pressure is kept fixed, the

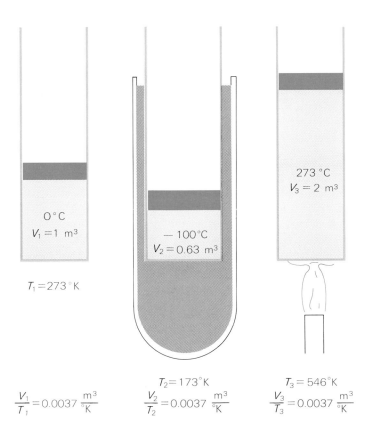

7.15 Charles' law: At constant pressure, the volume of a given quantity of a gas is directly proportional to its absolute temperature. (To change a Celsius temperature to its equivalent absolute temperature, add 273°.) Here $V_1/T_1 = V_2/T_2 = V_3/T_3$.

The Santa Mariana weighs 20,000 tons and so, by Archimedes' principle, she displaces 20,000 tons of water when afloat.

pressure increases with rising temperature and decreases with falling temperature, again by the fraction $\frac{1}{273}$ of its 0°C value for every degree change.

These figures suggest an obvious question: What would happen to a gas if we could lower its temperature to -273°C? If we should try to maintain constant volume, the pressure at this temperature would seem to fall to zero; if the pressure remained constant, the volume should fall to zero. It is hardly probable, however, that our experiment would have so startling a result. In the first place, we should find it impossible to attain quite so low a temperature, and in the second place, all known gases liquefy before that temperature is reached. Nevertheless, this temperature, -273°C, has a very special significance, a significance that will become clearer shortly. It is called *absolute zero*.

absolute zero

For many scientific purposes it is more convenient to reckon temperatures from absolute zero than from the freezing point of water. Temperatures on such a scale, given as degrees Celsius above absolute zero, are called *absolute temperatures*. Thus the freezing point of water is 273° absolute, written 273°K in honor of the English physicist Lord Kelvin, and the boiling point of water is 373°K. Any Celsius temperature can be changed to its equivalent absolute temperature by adding 273°:

absolute temperature scale

$$°K = °C + 273°$$

The density of hot air is less than that of cool air at the same pressure, and this difference provides the buoyancy that keeps this balloon afloat. A propane burner in the balloon's base keeps its 30,000 ft^3 of air about 130°F above the temperature of the outside air.

Using the absolute scale, we may express the above relationship between gas volumes and temperatures very simply: The volume of a gas is directly proportional to its absolute temperature (Fig. 7.16). This relation may be expressed mathematically as

charles' law
$$\frac{V_1}{T_1} = \frac{V_2}{T_2} \quad \text{(at constant pressure)}$$

where the V's are volumes and the T's are absolute temperatures. Discovered by two eighteenth-century French physicists, Jacques Alexandre Charles and Joseph Gay-Lussac, this relation is commonly known as *Charles' law*. Like Boyle's law, it holds to a fair approximation for all gases at ordinary pressures, but becomes inaccurate at high pressures.

These laws can be combined in the single formula

$$\frac{P_1 V_1}{T_1} = \frac{P_2 V_2}{T_2}$$

At constant temperature, $T_1 = T_2$ and we have Boyle's law, while at constant pressure, $P_1 = P_2$ and we have Charles' law. A common way of writing this formula is

ideal-gas law
$$\frac{PV}{T} = \text{constant}$$

since it reflects the fact that this particular combination of variables does not change in value even though the individual variables P, V, and T may vary. In Chap. 22 we shall see how the value of the constant is calculated for a given gas sample.

It is very significant that $PV/T = $ constant holds fairly well for *all* gases, and it therefore furnishes a specific goal for theories which attempt to explain the gaseous state of matter. This equation is known as the *ideal-gas law*.

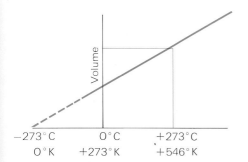

7.16 Graphic representation of Charles' law, showing the proportionality between volume and absolute temperature for gases. If the temperature of a gas could be reduced to absolute zero, its volume would fall to zero. Actual gases liquefy at temperatures above absolute zero.

GLOSSARY

A *solid* tends to maintain its volume and shape no matter where it is placed; that is, it resists deformation.

A *liquid* tends to maintain its volume but assumes the shape of whatever container it is placed in.

A *gas* maintains neither its volume nor its shape and expands to fill whatever container it is placed in; if allowed to escape, a gas expands indefinitely.

A *fluid* is a substance able to flow readily; liquids and gases are fluids.

Viscosity is the resistance of fluids to flowing motion; liquids are more viscous than gases.

The *pressure* on a surface is the force acting perpendicular to the surface divided by the area of the surface. A pressure is a force per unit area. Pressure units are pounds per square inch ($lb/in.^2$), pounds per square foot (lb/ft^2), and newtons per square meter ($newtons/m^2$).

A *barometer* is a device for measuring atmospheric pressure. *Atmospheric pressure* is the force with which the atmosphere presses down upon each unit area at the earth's surface. Atmospheric pressure at sea level is normally about 14.7 $lb/in.^2$, corresponding to the pressure of 76 cm (about 30 in.) of mercury.

Archimedes' principle states that the buoyant force on an object immersed in a fluid is equal to the weight of the fluid displaced by the object.

Boyle's law states that the volume of a gas is inversely proportional to its pressure provided that the temperature is held constant.

Absolute zero is that temperature at which the volume of an ideal-gas sample would shrink to zero if its pressure were held constant. This temperature is $-273°C$. The *absolute temperature scale* gives temperatures in degrees Celsius above absolute zero, denoted °K; thus the freezing point of water is $273°K$.

Charles' law states that the volume of a gas is proportional to its absolute temperature provided that the pressure is held constant.

EXERCISES

1 A wooden block is submerged in a tank of water and pressed down against the bottom of the tank so that there is no water underneath it. The block is released. Will it rise to the surface or stay where it is?

2 Why do helium-filled balloons rise to a certain altitude and float there instead of rising indefinitely?

3 A kilogram of lead and a kilogram of feathers are placed on opposite pans of a sensitive balance. What happens and why?

4 A small amount of water is boiled for a few minutes in a large tin can, and the can is then covered tightly while it is still hot. As the can cools, it collapses. Why does this happen?

5 Would a barometer in the upper atmosphere show a higher or a lower reading than on earth? Would water boil at a lower or a higher temperature?

6 When the volume of a gas sample is held constant, what is the relationship between its absolute temperature and pressure?

7 A certain spring has a force constant of 2 lb/ft. By how much will it stretch when it is used to support an object whose mass is 0.1 slug?

8 A certain steel wire stretches by 0.01 in. when it is used to support a weight of 50 lb. What is its force constant in pounds per foot?

9 A 180-lb man wears shoes whose area is 20 in.^{2} each. How much pressure does he exert on the ground?

10 A 50-kg woman balances on the heel of her right shoe, which is 1 cm in diameter. How much pressure (in atmospheres) does she exert on the ground?

11 Mercury is 13.6 times as dense as water. If water were used in a barometer instead of mercury, how high would the column of liquid be?

12 A certain room is 20 ft long, 15 ft wide, and 10 ft high. How many pounds does the air in the room weigh?

13 (a) How many cubic feet does a ton of water occupy? (b) How many cubic feet does a ton of iron occupy?

14 (a) How many cubic meters does a metric ton (1 metric ton = 1,000 kg) of pure water occupy? (b) How many cubic meters does a metric ton of air occupy?

15 A man whose mass is 60 kg is standing on a rectangular swimming raft 3 m long and 2 m wide which is floating in fresh water. By how much does the raft rise after he dives off?

16 Find the buoyant force on a 160-lb man in air under the assumption that his density is the same as that of fresh water.

17 A 50-g bracelet is suspected of being gold-plated lead instead of pure gold. It is dropped into a full beaker of water and 4.0 cm^3 of water overflows. Is the bracelet pure gold? If not, what percentage by weight of it is gold?

18 What is the pressure on a gas confined in a cylinder 4 in. in diameter if the force on the piston is 100 lb? (Do not consider atmospheric pressure.)

19 A certain quantity of hydrogen occupies a volume of 1,000 cm^3 at 0°C (273°K) and ordinary atmospheric pressure. If the pressure is

tripled but the temperature is held constant, what will the volume of the hydrogen be? If the temperature is increased to 273°C but the pressure is held constant, what will the volume of the hydrogen be?

20 An air bubble at the bottom of a certain lake has a volume of 1 cm^3. When it reaches the surface, the bubble has expanded to a volume of 10 cm^3. If the water temperature is assumed constant, how deep is the lake?

21 An automobile tire gauge measures the amount by which the air pressure in a tire exceeds the atmospheric pressure of 14.7 lb/in.^{2} If the "gauge" pressure of a certain tire is 24 lb/in.^{2} at 0°C, what will it be at 27°C if the tire volume remains the same?

22 The tires of a 3,000-lb car are inflated to a gauge pressure of 25 lb/in.^{2} What area of each tire is in contact with the ground?

chapter 8
Kinetic Theory of Matter

KINETIC THEORY OF MATTER
- brownian movement
- kinetic theory of gases
- absolute temperature

HEAT
- molecular energy
- molecular speeds
- liquids and solids
- evaporation
- melting

SECOND LAW OF THERMODYNAMICS
- heat engines
- heat as disordered energy
- laws of thermodynamics

Suppose there were no limit to the power of our microscopes, so that we could examine a drop of water under stronger and stronger lenses indefinitely. What sort of microscopic world would we discover when the drop was enlarged, say, a million times? Would we still see structureless, transparent, liquid water? Or would we perhaps see distinct particles, the building blocks, as it were, of the substance that to our gross senses is structureless, transparent, and liquid? These are questions as old as civilization, posed whenever men have speculated deeply on the nature of things. Our first record of a serious approach to the problem dates from the fifth century B.C. in Greece, where Democritus championed our modern view that matter ultimately consists of individual particles. Four centuries afterward, in the Rome of Caesar and Cicero, Democritus' views were elaborated by the great scholar-poet Lucretius.

Lucretius and Democritus used their hypothesis that matter is made up of tiny particles chiefly for philosophical speculation and tried only superficially to connect it with actual observations. Physicists of the nineteenth century, less interested in philosophy than in factual knowledge, found in the 2,000-year-old idea a powerful tool for explaining and correlating a great variety of observations and simple experiments. Their extension of the theory of Democritus is known as the *kinetic theory of matter*. The word kinetic suggests the nature of the extension: Matter consists of myriad particles in constant motion, not a static assembly of them.

KINETIC THEORY OF MATTER

The kinetic theory is an attempt to account for a wide variety of physical and chemical properties of matter in terms of a simple model of its structure. According to this model,

all matter consists of molecules in motion

1 All matter is composed of tiny, discrete particles called *molecules*.
2 These molecules are in constant motion.

These assumptions were originally made as guesses (intelligent ones, to be sure), and physicists and chemists tried to show that the familiar characteristics of matter follow from the application of ordinary physical laws to molecules. The attempt was highly successful; the behavior of matter could be described accurately, or *explained*, in terms of this model. Furthermore, the model enabled scientists to predict unsuspected aspects of the behavior of matter, and these predictions could be checked by experiment. As the original guesses proved themselves capable of explaining more and more observations and of predicting the results of new experiments, they were accepted ever more widely as facts rather than as assumptions. Today the kinetic theory of matter is among the dozen or so theories that constitute the foundation of all modern science.

brownian movement

Later on we shall discuss molecules in some detail, but for the time being we can simply define them as the individual primitive particles characteristic of a substance.

Today we have much information about the actual sizes, speeds, even shapes of the molecules in various kinds of matter, which, of course, was not available when the kinetic theory was formulated. Our information is obtained indirectly, since even today we cannot directly measure the dimensions of most molecules, but it is confirmed in so many different ways that we have every reason to believe it to be accurate. For example, a molecule of nitrogen, the chief constituent of air, has a diameter of about 0.18 billionths of a meter (1.8×10^{-10} m) and weighs 4.7×10^{-26} kg. It travels (at 0°C) with an average speed of about 500 m/sec (1,600 ft/sec, a typical speed for a rifle bullet), and in each second collides with more than a billion other molecules. Of similar dimensions and moving with similar speeds in each cubic centimeter of air are nearly 3×10^{19} other molecules.

The extreme smallness of ordinary molecules will probably make it impossible for us ever to see them directly because the resolving power of our microscopes is limited by the nature of light (Chap. 13). But particles not many times larger than molecules are visible in the highest-powered microscopes, particles small enough to move in response to the blows of swiftly traveling molecules against their sides. Molecular motions are highly erratic, and so the visible particles are buffeted about in irregular, zigzag paths (Fig. 8.1). The smallest visible particles move rapidly under molecular bombardment, darting this way and that, now brought to a stop, now starting out in a new direction. Larger particles show only a slight jiggling motion or do not move at all, since so many molecules strike them that at any one instant the forces on all sides are nearly the same. This motion of microscopic particles, called the *brownian movement*, is well shown by particles of certain dyes and by tiny oil droplets suspended in water, and by smoke particles suspended in air. Brownian movement is among the most direct and convincing evidence we have of the reality of molecules and their motions.

brownian motion

kinetic theory of gases

We go on now to discuss the specific assumptions of the kinetic theory regarding gas molecules. We begin with gases, because the uniformity in behavior of different gases and the simple mathematical rules that describe this behavior suggest that the structure of gases is simpler than that of liquids and solids.

The three basic assumptions of the kinetic theory for gas molecules are these:

1 Gas molecules are small compared with the average distance between them.
2 Gas molecules collide without loss of kinetic energy.
3 Gas molecules exert practically no forces on one another, except when they collide.

assumptions of kinetic theory of gases

A gas, then, is assumed to be mostly empty space, its isolated molecules moving helter-skelter like a swarm of angry bees in a closed room. Each molecule collides with others several billion times a second, changing its speed and direction at each collision but uninfluenced by its neighbors between collisions. If a series of collisions brings it momentarily to a stop, new collisions will set it in motion; if its speed becomes greater than the average, successive collisions will slow it down. There is no order in the motion, no uniformity of speed or direction; we can say merely that the molecules maintain a certain average speed and that at any instant as many molecules are moving in one direction as in another.

This animated picture suggests immediately some general explanations for the more obvious properties of gases. The ability of a gas to expand and to leak through small openings is a consequence of the rapid motion of its molecules and their lack of attraction for one another. Gases are easily compressed, because the distances between molecules can be shortened with scarcely any deformation of the molecules themselves. One gas mixes readily with another, because the wide spaces between molecules leave ample room for others. Gases are light in weight, because their volume consists so largely of empty space.

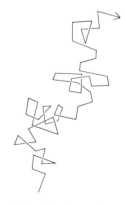

8.1 The irregular path of a microscopic particle bombarded by molecules. *The line joins the positions of a single particle observed at 10-sec intervals. This phenomenon is called brownian movement and is direct evidence of the reality of molecules and their random motions.*

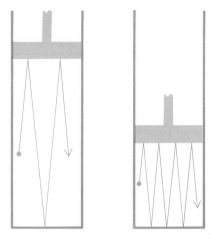

8.2 Gas pressure is the result of molecular bombardment. *When a gas is compressed into a smaller volume its molecules strike the walls of the container more often than before, leading to an increase in pressure. For simplicity, only vertical molecular motions are shown.*

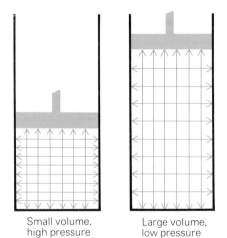

Small volume, high pressure   Large volume, low pressure

8.3 Origin of Boyle's law according to the kinetic theory of gases. *Expanding a gas sample means that its molecules must travel farther between successive impacts on the container wall and that their blows are spread over a larger area, so the gas pressure drops.*

Gas pressure is the effect of bombardment of millions and millions of molecules, the same sort of bombardment that causes suspended smoke particles to show the brownian movement; the myriad tiny, separate blows affect our crude senses and measuring instruments as a continuous force. When a gas is squeezed into a smaller volume, more molecules strike a square centimeter of surface each second, and the pressure is increased. When the gas is expanded to a larger volume, each square centimeter is struck less often, and the pressure decreases (Fig. 8.2). This is the general relationship summarized in Boyle's law.

The kinetic theory is able to show that the pressure and the volume of a gas at constant temperature should be exactly inversely proportional to each other. Suppose that the molecules of a gas confined in a cylinder (Fig. 8.3) are thought of as moving in a completely regular manner, some of them vertically between the piston and the base of the cylinder and the remainder horizontally between the cylinder walls. If the piston is raised so that the gas volume is doubled, the vertically moving molecules have twice as far to go between collisions with top and bottom and hence will strike only half as often; the horizontally moving molecules must spread their blows over twice as great an area, and hence the number of impacts per unit area will be cut in half. Thus the pressure in all parts of the cylinder is halved, as Boyle's law would predict. It is not hard to extend this reasoning to a real gas whose molecules move at random.

The third assumption mentioned above is not strictly accurate: Gas molecules do exert slight attractive forces on one another. These forces become conspicuous when a gas is greatly compressed and account (in part) for the fact that Boyle's law does not hold at high pressures.

absolute temperature

To explain the effects of temperature, one further assumption is required:

4. The absolute temperature of a gas is directly proportional to the average kinetic energy of its molecules.

Just why this assumption is stated in this particular manner is not obvious without a mathematical discussion. As a matter of fact, it is more like a conclusion than an assumption, because it is the key to agreement between the kinetic theory and the experimental data. Assumption 4 is not unreasonable: The fact that temperature should be closely related to molecular speeds, and hence to molecular energies, follows from the simple observation that the pressure of a confined gas increases as its temperature rises. Increases in pressure must mean that the molecules are striking their confining walls more forcefully and so must be moving faster.

In Chap. 7 we learned that the pressure of a gas approaches zero as its temperature falls toward $-273°C$. For the pressure to become zero, molecular bombardment must cease. Thus absolute zero finds a logical explanation, in terms of the kinetic theory, as the temperature at which gas molecules would lose their kinetic energies completely. There can be no lower temperature, simply because there can be no smaller amount

KINETIC THEORY OF MATTER

of energy than none at all. The regular increase of gas pressure with absolute temperature if the volume is constant and the similar increase of volume if the pressure is constant (Charles' law) are understandable from this definition of absolute zero, although the formal mathematical demonstration of direct proportionality is a bit complicated.

at absolute zero gas molecules would cease to move

As a consequence of our temperature assumption, we might guess that compressing a gas in a cylinder will cause its temperature to rise. For, while the piston is moving down, molecules rebound from it with increased energy (Fig. 8.4), just as a baseball rebounds with increased energy from a moving bat. Hence the average kinetic energy of the gas molecules should be raised, and the temperature of the gas should increase. To verify this prediction, one need only pump up a bicycle tire and observe how hot the pump becomes after the air in it has been compressed a few times. If, on the other hand, a gas is expanded by pushing a piston outward, its temperature should fall, since each molecule that strikes the retreating piston gives up some of its kinetic energy. In a steam engine, for instance, compressed steam at a temperature well above the boiling point of water is cooled nearly to its condensation point as it expands by pushing against the piston of the engine. The cooling effect of gas expansion underlies the operation of most refrigerators.

The fourth assumption is the basis of a general definition of temperature: **The temperature of a gas is a measure of its average molecular kinetic energy.** It follows that all gases at the same temperature have the same average molecular kinetic energies.

From the latter statement we can draw an interesting conclusion: Heavy molecules should move more slowly than light molecules at the same temperature. The average kinetic energy $\overline{KE}$ of a molecule is

heavy molecules move more slowly than light ones at the same temperature

$$\overline{KE} = \tfrac{1}{2} m \overline{v^2}$$

where m is the molecule's mass and $\overline{v^2}$ is the average value of its squared speed; a light molecule (small m) must have greater speeds on the average than a heavy molecule if $\tfrac{1}{2} m \overline{v^2}$ is to be the same for both. (A bar over a quantity indicates that its average value is being considered.) One easy way to test this prediction is to measure the rates at which two gases leak out through small openings. The gas carbon dioxide, for instance, has molecules 22 times heavier than molecules of hydrogen; through small openings hydrogen escapes much more rapidly, with the difference in rates (about 4.7 : 1) being just sufficient to make the average kinetic energies of the two kinds of molecules the same.

HEAT

As we have mentioned, heat is a form of energy. What meaning can we attach now to the heat content of a gas in the language of the molecular theory? Since supplying heat to a gas raises its temperature, thereby increasing the kinetic energy of its molecules, we might reasonably guess that the heat content of a gas is precisely this energy of molecular motion. But such a guess must be examined carefully.

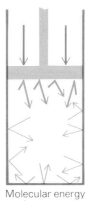
Molecular energy increases

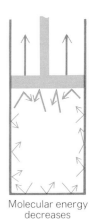
Molecular energy decreases

8.4 *Compressing a gas causes its temperature to rise because molecules rebound from the piston with increased energy; expanding a gas causes its temperature to drop because molecules rebound from the piston with decreased energy.*

A snow-making machine in operation. A mixture of compressed air and water is blown through a nozzle, and the expansion of the air cools the mixture sufficiently to freeze the water into the ice crystals of snow.

molecular energy

Up until now, energy of molecular motion has implied energy associated with the movement of molecules from place to place, which we might call translational movement. If heat energy is merely the kinetic energy of translational motion, then a given amount of heat ought to affect the temperature of all gases alike. But this is emphatically not true; the heat absorbed by a gas is not all used in making its molecules move around more quickly.

Studies of the effects of heat on different gases, coupled with results of other research, have shown that complex molecules may absorb heat energy in three principal ways: (1) their translational motion may be increased; (2) one part of a molecule may be set to vibrating with respect to another part, as if the parts were connected by a spring; and (3) each molecule may be set in rotation (Fig. 8.5). When heat is supplied to a gas, it becomes molecular energy in one or more of these different forms, but only the energy that goes into one particular form, kinetic energy

the temperature of a gas is a measure of the average energy of translational motion of its molecules

of translation, affects the temperature of the gas. Thus it makes more sense to speak of the *internal energy* of a gas than of its "heat content."

Count Rumford and several other scientists before him had surmised that heat was a form of motion. The kinetic theory finally presented, for gases at least, a definite picture of just what kind of motion heat is—a complex, disordered motion of tiny particles, spinning around, vibrating back and forth, flying about in all directions.

This picture of heat answers what seems at first glance to be an obvious question: What keeps the molecules moving? We picture them as something like tiny billiard balls, moving rapidly, colliding with one another many times a second. Real billiard balls, after a few seconds of such motion, would lose their energy in the friction of collisions and come to rest at the bottom of their container. All other motions of our experience, save perhaps the motion of stars and planets, are similarly brought to a halt by friction unless some outside force maintains them. Molecular motions are maintained by no outside force, yet continue indefinitely with no sign of diminishing speed. Why is it that friction does not bring these tiny particles to rest, as it does other moving objects?

The answer is that friction involves a transformation of mechanical energy into heat energy, which is the same as molecular energy. Friction between molecules would mean simply a transformation of molecular energy into molecular energy—which is not a transformation at all. Molecules keep moving because there is nothing to make them stop. The question of the last paragraph is, in fact, quite meaningless; it appears to make sense only because we are so accustomed to thinking of heat and motion as distinct concepts. Motion in the molecular world cannot produce heat, as it does in the larger world of everyday life, because molecular motion *is* heat.

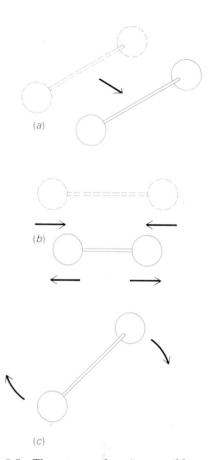

8.5 *Three types of motion possible for diatomic molecules (like those of hydrogen, oxygen, and chlorine): (a) translation, (b) vibration, (c) rotation. The bond between the atoms is shown as a rod for convenience in visualizing the motions.*

molecular speeds

The precise relationship between average molecular kinetic energy $\overline{KE}$ and absolute temperature T is usually written

$$\overline{KE} = \tfrac{3}{2}kT$$

The quantity k is known as *Boltzmann's constant* and its value of 1.38×10^{-23} joule/°K is the same for *all* molecules. Since $\overline{KE} = \tfrac{1}{2}m\overline{v^2}$, we have

$$\tfrac{1}{2}m\overline{v^2} = \tfrac{3}{2}kT$$

$$\sqrt{\overline{v^2}} = \sqrt{\frac{3kT}{m}}$$

For gas molecules, the average molecular speed $\overline{v}$ is a few percent greater than $\sqrt{\overline{v^2}}$, so we can write

$$\overline{v} \approx \sqrt{\frac{3kT}{m}}$$

where the symbol $\approx$ means "approximately equal to."

Let us compute the average speed of a nitrogen molecule at 0°C. Since 0°C = 273°K and the mass of a nitrogen molecule is 4.7×10^{-26} kg, we have

$$\bar{v} \approx \sqrt{\frac{3 \times 1.38 \times 10^{-23} \text{ joule/°K} \times 273\text{°K}}{4.7 \times 10^{-26} \text{ kg}}}$$

$$\approx 490 \text{ m/sec}$$

This is how the figure of "about 500 m/sec" quoted at the beginning of this chapter for such a molecule was obtained.

liquids and solids

The kinetic theory grew from the efforts of several nineteenth-century physicists to find an explanation for the behavior of gases. Its greatest triumphs have come from work on gases, but in recent years the theory has been applied with much success to the vastly more complicated inner structures of liquids and solids.

The incompressibility of liquids and solids suggests the assumption that their molecules are closer together, probably as close as possible. The brownian movement in liquids is ample basis for assuming that liquid molecules, like gas molecules, are in constant, random motion. The assumption that particles in solids are held more or less rigidly in fixed positions follows from the definite shapes that solids maintain and from their inability to diffuse except under unusual circumstances.

Strong attractive forces are assumed to be holding together the particles of liquids and solids. Some of these intermolecular attractions are the same as the very slight attractions between gas molecules that are responsible for deviations from Boyle's law at high pressures. In liquids and solids the forces are more conspicuous because the molecules are so much closer together. The nature of the attractive forces, which are of electrical origin, is discussed in Chaps. 20 and 21.

To explain the effects of temperature on liquids and solids, we may keep the assumption that was so useful for gases: absolute temperature is proportional to average kinetic energies of moving particles. (This is not quite exact for liquids and solids however.) In liquids a rise in temperature means an increase in average speed of translational motion; in solids a rise in temperature means an increase in vibrational motion of the particles about their fixed positions. As for gases, we further assume that liquid and solid particles can take up heat energy in other ways, so that absorption of 1 kcal of heat by 1 kg of different materials may produce different increases in temperature.

the molecules of a liquid are able to move past one another

If a gas resembles a swarm of angry bees, the particles of a liquid may fairly be likened to bees in a hive, crawling over one another incessantly, each one continually in contact with its neighbors. The slowness of liquid diffusion is explained by the difficulties of motion when each particle is closely surrounded by others. Liquids can flow because their molecules slide easily over one another, but they are more viscous than gases because intermolecular attractions impede the motion. The viscosity of a liquid is a rough measure of how strongly its particles attract one another.

Attractive forces between the particles of solids are stronger than in liquids, so strong that the particles are no longer free to move about. They are far from motionless, however: held in position as if by springs attached to its neighbors, each particle oscillates back and forth rapidly and continuously. A solid is elastic because its particles retreat back into their normal positions after being stretched apart or pushed together. A solid breaks or is permanently deformed when it is subjected to a force larger than the forces of attraction between its particles. In a brittle solid, rupture takes place suddenly along a single surface, and the particles are pulled so far apart that healing is impossible. In a solid that can yield by slow deformation to excessive forces, tiny fractures develop all through the solid, with each fracture healing itself as the particles slide to new positions and find new partners for their attractive forces.

the particles of a solid vibrate about fixed positions

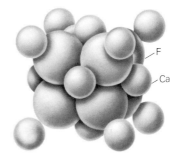

8.6 The structure of a fluorite (CaF_2) crystal, showing the regular arrangement of the calcium (Ca) and fluorine (F) atoms.

Crystal form and crystal growth suggest arrangements of particles in patterns of equally spaced rows and planes. Each particle, so to speak, faces the same way, and the distances to its neighbors on all sides are determined by its attractive forces in various directions. The smooth faces and sharp angles of the crystal are, then, the outer expression of this inner regularity. The microscope and X-rays prove convincingly that many solids that do not show crystal faces nevertheless have a regular pattern in their inner structures (Fig. 8.6). Solids of this sort, whether or not they occur in well-shaped crystals, are called *crystalline solids;* salt, diamond, quartz, and most metals are familiar examples. Solids whose particles have no regularity of arrangement (for instance, glass and rubber), and which, of course, never show crystal forms, are called *amorphous solids.*

8.7 Evaporation. Ether evaporates more rapidly than water because the attractive forces between its molecules are smaller.

evaporation

As a further example of the relationship between molecules in motion and the properties of matter, let us consider the characteristic temperature-energy relationships involved in changes of state, that is, changes from solid to liquid, liquid to solid, solid to gas, gas to solid, liquid to gas, or gas to liquid.

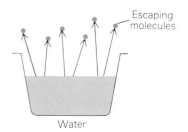

Suppose that two liquids, water and ether, are placed in open dishes. Particles in each are moving in all directions, with a variety of speeds. At any instant some particles are moving fast enough upward to break through the liquid surface and escape into the air, in spite of the attractions of their slower neighbors. By this loss of its faster molecules each liquid gradually evaporates; since the molecules remaining behind are the slower ones, evaporation leaves cool liquids behind. The ether evaporates more quickly (or is more *volatile*) and cools itself more noticeably because the attraction of its molecules for one another is smaller and a greater number can escape (Fig. 8.7).

Now suppose that a tight cover is placed on each dish. Molecules of vapor can no longer escape completely; by collisions with one another and with the cover, they are sooner or later knocked back into the liquid. So the fast particles that leave each liquid return to it, and it no longer is cooled. Evaporation goes on as before but is balanced now by the reverse process of condensation. How many molecules are present in the space above each liquid at any given instant is determined by how rapidly they leave its surface; over the ether considerably more vapor is present than over the water, since ether evaporates more easily. The amount of vapor is conventionally expressed in terms of its pressure; thus we say that the *vapor pressure* of the ether is higher than that of water. The vapor pressure of any liquid at a given temperature is defined as the pressure that its vapor exerts when confined above the liquid. The vapor pressure of a substance is a measure of how readily its particles escape from its surface.

vapor pressure

Let us now remove the covers from our dishes of water and ether and heat each liquid slowly. We aid the process of evaporation by supplying heat, giving more and more particles the energy necessary to escape. Vapor rises from each dish in steadily increasing quantities; if we stop the heating at intervals, cover each dish, and measure the vapor pressures, we find that these pressures are growing rapidly larger. At a temperature of 35°C even the particles of average speed in the ether dish apparently gain sufficient energy to vaporize, for bubbles of vapor begin to form all through the liquid. We say now that the liquid is *boiling*. Its vapor pressure has become equal to the pressure of the surrounding atmosphere; particles within the liquid form bubbles because now the pressure that they exert as vapor is sufficient to overcome the downward pressure of the atmosphere on the liquid surface. At this temperature the vapor pressure of water is still only a small fraction of an atmosphere (4.2 cm of mercury); only when the temperature approaches 100°C does its vapor pressure also become equal to atmospheric pressure, so that the liquid can boil. We may define the boiling point of a liquid as the temperature at which its vapor pressure becomes equal to the surrounding pressure (Fig. 8.8). Standard boiling points listed in tables are the temperatures at which vapor pressures become equal to normal atmospheric pressure.

boiling occurs when the vapor pressure of a liquid equals atmospheric pressure

Whether evaporation takes place spontaneously from an open dish or is aided by heating, the formation of vapor from a liquid requires energy. In the first case energy comes from the internal energy of the liquid

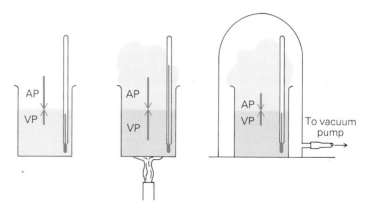

8.8 Boiling. *A liquid boils when its temperature is high enough for its vapor pressure VP to equal the atmospheric pressure AP. A liquid may be made to boil at a lower temperature than usual by reducing the external pressure.*

itself (since the liquid grows cooler), in the other case from the external source of heat. For water at its boiling point, 540 kcal (the heat of vaporization) are used up in changing each kilogram of liquid into vapor. Here there is no difference in temperature between liquid and vapor, hence no difference in their average molecular kinetic energies. If not into kinetic energy, into what form of molecular energy do the 540 kcal of heat energy go?

Intermolecular forces provide the answer. In the liquid these forces are strong, because the molecules are close together. To tear the molecules apart, to separate them by the wide distances that exist in the vapor, requires that these strong forces be overcome. Each molecule must be moved against the attraction of its neighbors, moved to a new position in which their attraction for it is very small. Just as a stone thrown upward against the earth's attraction acquires potential energy, so molecules moved apart in this fashion acquire potential energy—potential energy with reference to intermolecular forces. So heat supplied in evaporating a liquid goes into another form of molecular energy besides the translational, vibrational, and rotational energies we have considered: molecular potential energy. When a vapor condenses to a liquid, its molecules "fall" toward one another under the influence of their mutual attractions, and their potential energy is then given off as heat energy to the surroundings.

origin of heat of vaporization

melting

Temperature changes accompanying the melting of crystalline and amorphous solids are quite different and afford one easy experimental method for distinguishing the two types of material. A crystalline solid like ice melts sharply, at one definite temperature, and requires the addition of a certain quantity of energy at this temperature simply to effect the change from solid to liquid. An amorphous solid like glass softens gradually on heating, so that no one temperature can be given as its melting point.

crystalline solids have definite melting points, whereas amorphous solids soften gradually

In terms of the kinetic theory, this difference in behavior is a consequence of differences in inner structure. Particles of a crystalline solid are arranged in a definite pattern, each one oriented so that the forces binding it to its neighbors on all sides are as large as possible. To overcome these forces and give the particles the disorderly arrangement of a liquid structure requires that they gain potential energy, just as liquid particles must gain potential energy during evaporation. This potential energy is the heat of fusion (80 kcal/kg for water), which must be supplied to melt any crystalline solid and which is given out when the liquid crystallizes again (Fig. 8.9). Particles of amorphous solids, on the other hand, have

origin of heat of fusion

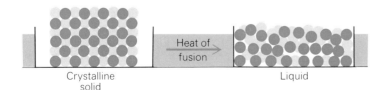

8.9 The orderly arrangement of particles in a crystalline solid is converted to the random arrangement of particles in a liquid when enough energy is supplied to the solid to overcome the binding forces within it.

no definite pattern but are already in the random, disorderly arrangements characteristic of liquids. Melting involves merely a gradual loosening of the ties between adjacent particles, without any marked increase in potential energy at a certain temperature.

Thus, without any new assumptions, the kinetic theory offers a rational interpretation of vaporization and melting in terms of the motions, the potential energies, and the arrangements of tiny particles.

SECOND LAW OF THERMODYNAMICS

Different kinds of energy can be transformed from one into another. But heat energy is unusual in that it cannot be transformed into other kinds *efficiently*. From the heat energy of burning coal we may obtain mechanical, chemical, or electric energy, but always in these transformations a large fraction of the heat energy is wasted. There is no escape from this waste; the transformations will not take place without it. This distinctive characteristic of heat energy was discovered early in the nineteenth century, at the beginning of the Industrial Revolution, as a result of attempts to improve the recently invented steam engine. Attacked both by engineers trying to get as much mechanical work as possible out of a ton of coal and by scientists more interested in the properties of heat as a form of energy, the problem of why transformations involving heat should be so wasteful was finally solved by the kinetic theory's picture of heat as random, disorderly motion of molecules.

heat engines

The only practical method of obtaining mechanical energy directly from heat, the method used in both steam and internal-combustion engines, is to supply heat to a compressed gas and let it expand against a piston or the vanes of a turbine.

Without concerning ourselves with the details of a real engine, we may picture the fundamental process involved as follows. Suppose that a gas under pressure in a cylinder (Fig. 8.10) is heated and allowed to expand by pushing the piston outward. Heat energy is thereby converted into mechanical energy of the piston, which may be used for turning a dynamo, running a pump, or for any purpose we wish. Now when expansion has reduced the pressure of the heated gas to atmospheric pressure, or when the piston reaches the end of the cylinder, further heating accomplishes little. To give the piston more mechanical energy, we must somehow recompress the gas and let it expand again. Recompressing the gas, however, requires the application of mechanical energy from outside. If the gas is compressed while hot, just as much energy will be needed as the gas has given out by expansion, so that the net gain will be zero. But if we cool the gas first, we find that less energy is required to compress it than we gain from the expansion. To make the engine do useful work, we must arrange to compress the gas while it is cold, then heat it and allow it to expand while hot, then cool it again for the next compression.

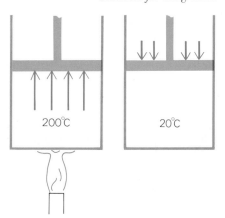

8.10 An idealized heat engine. A gas at 200°C gives out more energy in expanding than is required to compress the gas at 20°C, and this excess energy is available for doing work.

KINETIC THEORY OF MATTER

Note carefully what happens to the heat supplied to the gas: a part of it is used to drive the piston, but some must be allowed to escape when the gas is cooled before compression. For the engine to run, we need both a source of heat and something to which the gas can give part of its heat, usually the surrounding atmosphere. In effect, heat flows through the engine from the heat source to the atmosphere, and during the flow we manage to change some of the heat into mechanical energy. All heat engines work on this principle, taking advantage of the flow of heat from a hot reservoir to a cold reservoir in order to recover some of it as mechanical energy (Fig. 8.11). A *difference of temperature* between two reservoirs is essential.

heat engines extract energy from the flow of heat through them

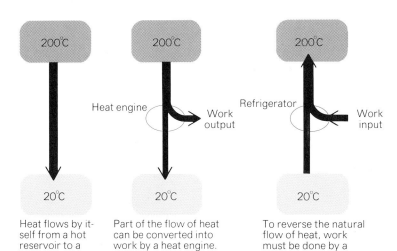

8.11 A heat engine is a device that converts part of the heat flowing from a hot reservoir to a cold one into work. A refrigerator is a device that extracts heat from a cold reservoir and delivers it to a hot one by performing work that is converted into heat.

Contained in the molecular motions of the atmosphere is a vast quantity of heat energy, but we cannot recover it because no cold object is available to which the heat can flow. Of course, we might set up a refrigerating machine to maintain a low temperature, but we should find that more energy is required to run the refrigerator than we could get by using it as the cold side of a heat engine. A refrigerator, in fact, is the reverse of a heat engine: it employs mechanical energy to force heat from a cold object to a warmer object, whereas a heat engine uses the natural flow of heat from hot objects to cold as a means of obtaining mechanical energy. Setting up a refrigerator to maintain a low temperature for running a heat engine is just the old perpetual-motion dream in a new guise.

We are strictly limited, then, in our ability to turn heat energy into mechanical energy. We can obtain mechanical energy only by letting heat flow from a region of high temperature to a region of low temperature, and in the best of circumstances the conversion is incomplete. This limitation is not due to friction, which saps the energy of all kinds of mechanical devices, but is a limitation imposed by the fundamental nature of heat. Why should heat energy in this respect be so different from other forms of energy?

no heat engine can be completely efficient even in principle

The operating cycle of a four-stroke gasoline engine. In the intake stroke, a gasoline vapor-air mixture from the carburetor is drawn into the cylinder as the piston moves downward. In the compression stroke the fuel-air mixture is compressed to perhaps 200 lb/in.² At the end of the compression stroke the spark plug is fired, which ignites the fuel-air mixture. The expanding gases force the piston downward with pressures of 600–700 lb/in.² in the power stroke. Finally the piston moves upward again to force the spent gases out through the exhaust valve.

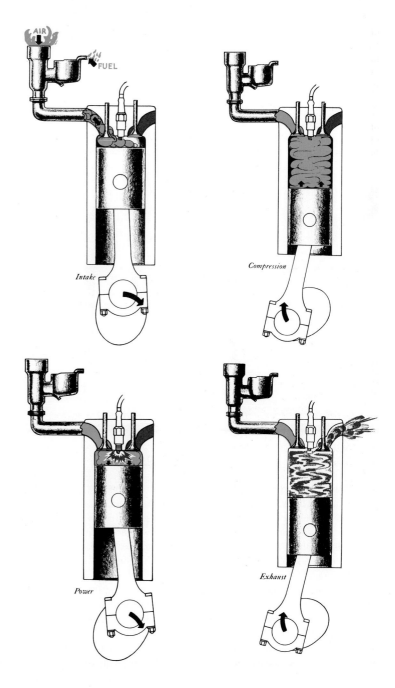

heat as disordered energy

The kinetic theory states that heat is energy of random, disorganized motion. When heat is supplied to a gas, its molecules increase their speeds in all directions. We use only the increase in speed in a particular direction, the direction in which the piston moves; but to obtain an increase in this direction, we must speed up motions in all other directions as well. If we could line up the molecules and fire them all, like tiny bullets, straight at the piston, then all the energy we give them would go to make the piston move. In a real gas most of our bullets go astray and only a few of them give their excess energy to the piston. From the energy of random, disorderly motion we can extract only a fraction as the energy of ordered motion in definite directions which we need to run the world's machinery.

In contrast to the difficulty of obtaining mechanical energy efficiently from heat is the ease with which other forms of energy may be converted into heat energy. When coal burns, its chemical energy changes directly to heat. When a pendulum swings, kinetic energy is transformed to potential and back again, but with each swing, friction removes some energy as heat. Electric energy in a light filament is changed partly to heat, partly to light; the light, falling on walls and furniture, is absorbed and converted into heat energy. In all the energy transformations of our acquaintance some, and more often all, of the original energy becomes eventually energy of disordered molecular motion.

The jet engines of the DC-9 in the foreground are more efficient than the piston engines of the Super Constellation in the background, but no heat engine can be 100% efficient even if friction were absent because of the necessity of rejecting a certain proportion of the input heat in order that mechanical energy be produced.

Once in the form of heat, energy cannot easily be restored to its original form. Some may be recovered, as we have seen, by making the heat flow to a cooler object, but to recover even this amount is possible only for a brief time after the heat is produced. Heat energy spreads quickly into its surroundings, and the temperature difference necessary to make a heat engine operate soon disappears. A stone falls to the pavement, and kinetic energy turns into heat; for a few moments particles in the stone and concrete move a little faster, then the faster movement affects neighboring molecules, and the heat energy spreads in ever-widening circles. Wood burning in a fireplace warms the air of a room; when the fire goes out, the warmth persists for a little while, but spreads gradually to the walls, to adjoining rooms, to the outside air, and presently is distributed so widely through the surroundings that we can no longer detect it.

From any hot object heat flows to cooler objects about it, spreading indefinitely until it becomes a part of the general molecular motion of earth and atmosphere. From this reservoir of heat we cannot recover even a fraction of the original energy, for a perceptible temperature difference no longer exists. The energy has not disappeared, but it is no longer available for conversion into other forms.

laws of thermodynamics

the second law of thermodynamics refers to the continual dissipation of other forms of energy into heat throughout the universe

We may summarize these observations on heat by the statement: **In every energy transformation, some of the original energy is always changed into heat energy not available for further transformations.** This statement is one way of stating the *second law of thermodynamics*. It is merely a formal expression of the everyday observations that other forms of energy eventually become heat energy and that heat is dissipated into its surroundings.

the first law of thermodynamics is the law of conservation of energy

So far as we know, this law of the "wastage" of energy applies quite as universally as the law of conservation of energy, which is the *first* law of thermodynamics. The radiant energy of stars, the mechanical energy of planetary motions, the chemical energy of food, all are being steadily changed into the energy of disordered molecular motion. The law implies that the universe in the past had more energy in forms capable of doing work than it has at present. It seems to imply also that the distant future will bring a time when there is no energy but heat energy, heat energy evenly distributed so that no part of the universe is warmer than another, the so-called "heat death" of the universe.

GLOSSARY

The *kinetic theory of matter* states that matter is composed of tiny discrete particles called *molecules* and that these molecules are in constant motion.

According to the *kinetic theory of gases*, the absolute temperature of a gas is directly proportional to the average kinetic energy of its molecules.

A *crystalline solid* is one whose molecules are arranged in a definite pattern; an *amorphous solid* is one whose molecules exhibit no regularity of arrangement.

The *vapor pressure* of a substance at a given temperature is the pressure its vapor exerts when confined above a sample of the substance. It is a measure of how readily the molecules of the substance escape from its surface.

The *first law of thermodynamics* is the law of conservation of energy.

The *second law of thermodynamics* states that, in every energy transformation, some of the original energy is always changed into heat energy not available for further transformations.

KINETIC THEORY OF MATTER

EXERCISES

1. Why does bombardment by air molecules not produce brownian movement in large objects such as chairs and tables?

2. Can you account for the ability of gases to leak rapidly through very small openings?

3. How can the conclusion of kinetic theory that molecular motion ceases only at absolute zero be reconciled with the observation that solids have definite shapes and volumes at temperatures much higher than absolute zero?

4. Why does moving air feel cooler than still air?

5. A sample of hydrogen is compressed to half its original volume, while its temperature is held constant. What happens to the average speed of the hydrogen molecules?

6. The oceans contain an immense amount of heat energy. Why can a submarine not make use of this energy for propulsion?

7. How could you tell experimentally whether a fragment of a clear, colorless material is glass or a crystalline solid?

8. When they are close together, molecules attract one another slightly. As a result of this attraction, are gas pressures higher or lower than expected from the ideal-gas law?

9. A tank contains 5 kg of carbon dioxide gas at 20°C and a pressure of 1 atm. What happens to the pressure when another 5 kg of carbon dioxide is added to the tank at the same temperature? Explain this result in terms of the kinetic theory of gases.

10. Each molecule of nitrogen has a mass very close to 14 times as great as that of a hydrogen molecule. Find the temperature of a hydrogen sample in which the average molecular speed is equal to that in a sample of nitrogen at a temperature of 300°K.

11. The average speed of a hydrogen molecule at room temperature and atmospheric pressure is about 1 mi/sec. What is the average speed of a nitrogen molecule, 14 times as heavy, under the same conditions?

12. Each molecule of the gas sulfur dioxide has a mass almost exactly double that of an oxygen molecule. If both gases are at the same temperature, which will have molecules with the higher average speed? If the speed of oxygen molecules at a particular temperature averages 4×10^2 m/sec, what is the average speed of the sulfur dioxide molecules?

13 What is the average kinetic energy of gas molecules at 0°C? At 100°C?

14 What is the average speed of silver atoms in silver vapor at 1500°K? The mass of a silver atom is 1.8×10^{-25} kg.

15 The table at left shows the vapor pressures of alcohol and water for various temperatures.
   a Which is the more volatile liquid?
   b What is the approximate boiling point of alcohol at normal atmospheric pressure?
   c At what temperature would water boil on a mountain top where atmospheric pressure is 60 cm of mercury?
   d How low must the pressure be reduced to make water boil at 0°C?
   e How could you make alcohol boil at 100°C?

16 The first assumption of the kinetic theory of gases is that gas molecules are small compared with the average distance between them. The oxygen and nitrogen molecules in air are roughly 2×10^{-10} m in diameter, and there are 2.7×10^{25} molecules in a cubic meter of air at sea level and room temperature. How many molecular diameters is the average molecular separation? (*Hint:* First find the average volume per molecule. Consider this volume as a cube, and calculate how long such a cube is on each edge; this figure represents the average molecular separation.)

Temperature, °C	Vapor pressure of water, cm of mercury	Vapor pressure of alcohol, cm of mercury
0	0.46	1.27
20	1.74	4.45
40	5.49	13.37
60	14.89	35.02
70	23.33	54.11
75	28.88	66.55
80	35.49	81.29
85	43.32	98.64
90	52.55	118.93
95	63.37	142.51
100	76.00	169.95

chapter 9

Waves

WAVE MOTION
- water waves
- transverse and longitudinal waves
- sound waves
- describing waves
- waveform

WAVE BEHAVIOR
- refraction
- reflection
- interference
- diffraction

SOUND
- speed
- frequency and wavelength
- reflection and refraction
- other properties
- the doppler effect
- musical sounds

If we fasten one end of a rope securely, stretch it, and then jerk the other end back and forth, a wave travels along the rope away from our hand. If we sharply strike one end of a steel rod, a vibration can be felt at the other end almost immediately afterward. These are examples of a general phenomenon: If the particles in one part of an object are disturbed, the disturbance is transmitted to adjacent particles and may travel long distances through the object. The only requirement is that the material be reasonably elastic, which means that a disturbance sets up forces within the object that tend to restore it to its undisturbed condition. Soft dirt or loose cotton has very little elasticity, so that the effect of a sharp blow is damped out and does not travel beyond the point of disturbance. But most materials, both fluids and solids, are sufficiently elastic so that the effect of a sudden disturbance can be transmitted from particle to particle. The phenomenon is particularly striking if the original disturbance is periodic—repeated over and over, with a constant rhythm. In this case impulses are transmitted from particle to particle at regular intervals, and we say that *waves* are traveling through the material.

wave motion consists of the passage of a periodic disturbance through a medium

WAVE MOTION

water waves

Let us consider exactly what is meant by a wave before going on to explore some of the properties of wave motion. If we stand on an ocean beach, watching the waves roll in and break one after the other, we are impressed with their ceaseless progress toward the shore. At first we might guess that masses of water are moving bodily shoreward, carrying along pebbles and shells and bits of driftwood. A few minutes' observation, however, convinces us that this cannot be true, for between the breakers water rushes out to sea, and there is no accumulation of water on the shore. The overall motion seems to be merely an endless back-and-forth movement. We can see the details of the motion better if we move out beyond the breakers, say at the end of a pier or in a boat; now if we fix our attention on a floating cork or piece of seaweed, we find that its actual position changes very little. As the crest of each wave passes, the cork rises and appears to move shoreward; in the following trough it drops and moves an equal distance backward. On the whole its path is approximately circular, and we can guess that adjacent water particles must follow similar circular paths.

The illusion of an overall movement toward the shore is explained by the fact that each particle of water performs its circular motion a trifle later than the last, as if the motion of one particle caused the next to move, and this one imparted its motion to another, so that the impulse was transmitted shoreward. In other words, each particle has a periodic motion in a circle, and a regular succession of these periodic motions, each a little later than the last, gives the illusion of bodily movement toward shore (Fig. 9.1).

waves transport energy

What does move shoreward is *energy*. Most water waves at sea are produced by wind, and it is energy from the wind far out at sea that is transmitted by means of wave motion to the shore. All familiar wave motions have this same characteristic: by a succession of periodic motions of individual particles, they accomplish a transfer of energy from place to place, but produce no bodily movement of matter in the direction of the waves.

transverse and longitudinal waves

Surface waves on water are the most familiar, but among the most complex, of the wave motions that a physicist must unravel. Much simpler

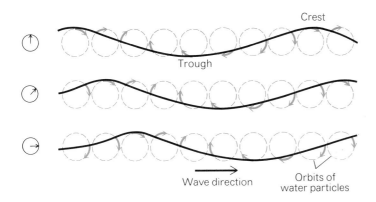

9.1 Nature of a water wave in deep water. Each particle performs a periodic motion in a small circle, and because successive particles reach the tops of their circles at slightly later times, their combination appears as a series of crests and troughs moving along the surface of the water. There is no net transfer of water by the wave.

are the waves set up in a taut rope by a succession of jerks at one end (Fig. 9.2). Obviously, the rope as a whole does not change position, but waves in the rope carry energy from the hand to the point of attachment. If we fix our attention on one small segment of the rope, we will note that it moves up, then down, then up again as each wave passes, and that this motion is transmitted from one segment to the next down the length of the rope. In other words, the motion of particles here is mostly perpendicular to the motion of the wave itself. Such waves are called *transverse waves*.

in a transverse wave, the particles of the medium move perpendicular to the wave direction

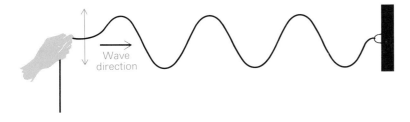

9.2 Transverse waves. The waves travel along the rope in the direction of the colored arrow. The individual particles of the rope move back and forth (black arrows) perpendicular to the direction of the waves.

Another type of wave can be illustrated by a long horizontal coil spring (Fig. 9.3). If the left-hand end of the spring is moved back and forth, a series of *compressions* and *rarefactions* moves along the spring, compressions being places where the loops of the spring are pressed together and rarefactions places where they are stretched apart. Any one loop simply moves back and forth, transmitting its motion to the next in line, and the regular succession of back-and-forth movements gives rise to the compressions and rarefactions. A similar motion would be produced in

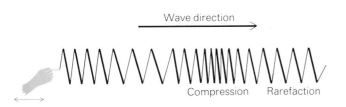

9.3 Longitudinal waves. Successive regions of compression and rarefaction move along the spring (colored arrow). The particles of the spring move back and forth parallel to the spring (black arrows).

in a longitudinal wave, the particles of the medium move parallel to the wave direction

a line of people, each with his hands on his neighbor's shoulders, if someone were to push and pull alternately at one end of the line; the back-and-forth motion would be transmitted from one person to the next down the line. Waves of this kind, in which the motion of individual units is along the same line that the wave itself travels, are called *longitudinal waves*.

Water waves, on the basis of these definitions, could be described as having both transverse and longitudinal characteristics inasmuch as each particle moves partly horizontally, in the direction of the waves, and partly vertically, at right angles to the waves.

only longitudinal waves can travel through a fluid

One reason the distinction between the two kinds of waves is important is that transverse waves are restricted to solids, whereas longitudinal waves can travel in any medium, solid or fluid. Why this is true is not hard to see. Transverse motion requires that each particle, as it moves, drag with it adjacent particles to which it is tightly bound; this is impossible in a fluid, where molecules have no rigid attachment to their neighbors. Longitudinal motion, on the other hand, requires simply that each particle exert a push on its neighbors, which can happen as easily in a gas or liquid as in a solid. (Surface waves on water—in fact any waves at the boundary between two fluids—are an exception to the rule, for in part they involve transverse motion.) The fact that longitudinal waves that originate in earthquakes are able to pass through the center of the earth while transverse earthquake waves cannot is one of the reasons the earth is believed to have a liquid core.

sound waves

The longitudinal nature of the wave motion in sound can be made clear by a study of the tuning fork (Fig. 9.4). As the prongs move outward, air molecules are pushed together; these push adjacent ones, and a region of compression starts away from the fork. When the prongs move inward the molecules on the outside are spread apart; adjacent molecules move toward the vacancy, others move to take their place, and so a region of rarefaction follows the region of compression—much like the loops of the long spring in Fig. 9.3. The continued vibration of the fork sends regularly spaced layers of compression and rarefaction into the air, and these moving layers constitute the sound waves. Note that the actual movement of each molecule in the sound wave is very slight, only a tiny fraction of an inch. The movement is superimposed on the normal random movements of the molecules, so that for any one molecule the actual motion is a combination of the motion due to thermal energy and the motion impressed on it by the tuning fork.

sound waves are longitudinal

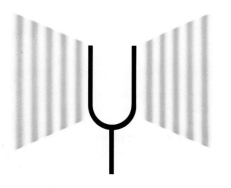

9.4 *Sound waves produced by a tuning fork. Alternate regions of compression and rarefaction move outward from the vibrating tines of the fork.*

The great majority of sounds consist of waves of this type, but a few—the crack of a rifle, the first sharp sound of a thunderclap, the noise of an explosion—are essentially single, sudden compressions of the air rather than periodic phenomena.

describing waves

All waves can be represented by a graph like that in Fig. 9.5. The resemblance to transverse wave motion is easiest to see; in fact the curve is an idealized picture of continuous waves in a rope like that of Fig. 9.2. As the wave moves to the right, each point on the curve can be thought of as moving up or down along a path whose extremes are the heights of the high point and low point of the curve, just as any small segment of the rope would move. The resemblance to longitudinal waves is not so obvious. Here we should say simply that high points of the curve represent maximum displacements of individual particles in one direction, low points maximum displacements in the other direction.

A representation like Fig. 9.5 permits us to assign numbers to certain characteristics of a wave, so that we can compare different waves quantitatively. The distance from crest to crest (or from trough to trough) is called the *wavelength*, usually symbolized by the Greek letter λ (lambda). The *speed* of the waves v is the rate at which each crest appears to move, and the *frequency* f is the number of crests that pass a given point each second. Frequency is measured in cycles/sec, a unit that is nowadays often called the *Hertz*, abbreviated Hz, after Heinrich Hertz, a pioneer in the study of electromagnetic waves.

wavelength and frequency

the hertz is equal to 1 cycle/sec

The number of waves that pass per second multiplied by the length of each wave equals the speed with which the waves travel (Fig. 9.6); if 10 waves, each 2 ft long, pass in a second, then each wave must travel 20 ft during that second. In other words, frequency f times wavelength λ gives speed v:

wave speed equals wavelength times frequency

$$v = f\lambda$$

This is a general relationship that applies to waves of all kinds. For example, large water waves on the open sea sometimes have wavelengths as great as 1,000 ft; they travel at roughly 70 ft/sec, so

$$f = \frac{v}{\lambda} = \frac{70 \text{ ft/sec}}{1,000 \text{ ft}} = 0.07 \text{ Hz}$$

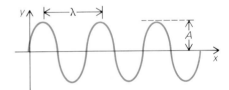

9.5 *A wave moving in the x direction whose displacements are in the y direction.* The wavelength is λ, and the amplitude is A.

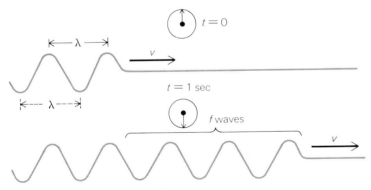

Wave speed = distanced traveled per sec
= (number of waves per sec) × (length per wave)
= frequency × wavelength
$v = f\lambda$

9.6 *Wave speed equals frequency times wavelength.*

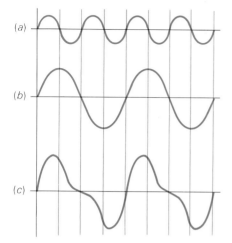

9.7 *Wave addition. Complex wave at c is equivalent to a combination of two simple waves a and b traveling over the same path.*

As another example, sound in air travels at 1,100 ft/sec; the musical note middle C has a frequency of 261 Hz and hence a wavelength of

$$\lambda = \frac{v}{f} = \frac{1{,}100 \text{ ft/sec}}{261 \text{ Hz}} = 4.2 \text{ ft}$$

The *amplitude* of a wave refers to the heights of the crests and troughs. It is represented by A in Fig. 9.5, and is defined in general as half the length of the path over which each particle moves. The amount of energy carried by waves evidently depends on their amplitude and their frequency, in other words, on the violence of the waves and the number of waves per second. It turns out that the energy is proportional to the square of each of these quantities.

waveform

Another characteristic of waves is called *waveform*. So far we have considered only waves produced by simple back-and-forth or up-and-down motions of particles, but the motions are often much more complicated. Such complications are represented in a diagram like Fig. 9.5 by irregularities in the smooth waveform, irregularities like those shown in the lower diagram of Fig. 9.7. This kind of complication often results when two (or more) kinds of vibrational motion are imposed on the same material simultaneously, say when two people try to shake the same stretched rope at different rates. Two simple wave motions may add together to give more complex motion, as Fig. 9.7 shows, and likewise a complex waveform can be analyzed into a combination of simple motions. In acoustics the waveform of a sound is what accounts for its *quality* or *timbre*, while the frequency of a sound is called its *pitch*.

the pitch of a sound wave refers to its frequency

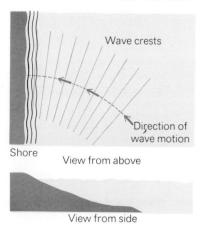

9.8 *Refraction of water waves. Waves approaching shore obliquely are turned because they move more slowly in shallow water near shore.*

WAVE BEHAVIOR

refraction

Water waves approaching a sloping beach provide a conspicuous illustration of the phenomenon of *refraction*. No matter what direction the wind may be blowing from, the direction of motion of waves near shore is practically at right angles to the shore (Fig. 9.8). Farther out in open water the direction of motion may be oblique, but the waves turn as they move in so that their crests become roughly parallel with the shoreline. The explanation is straightforward. As a wave moves obliquely shoreward, its near-shore end encounters shallow water where friction with the bottom slows it down. More and more of the wave is slowed as it continues to move, and the slowing becomes more pronounced as the water gets shallower. As a result the whole wavefront swings around until it is moving almost directly shoreward. The wave has turned because part of it is forced to move more slowly than the remainder. Thus refraction is produced by differences in speed along the wavefronts.

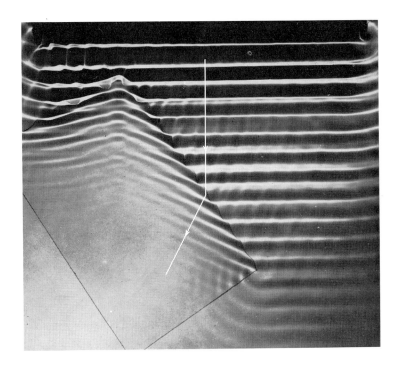

The refraction of water waves in a shallow tank. In the left side of the tank is a glass plate over which the water is shallower than elsewhere. Waves move more slowly in shallow water than in deeper water, and hence refraction occurs at the edge of the plate. The arrows show the direction of movement of the waves.

The bending may be at a sharp angle if waves move obliquely across a boundary between regions in which they move at different speeds. The photograph on this page shows ripples on a water tank moving obliquely from an area of deep water to an area of shallow water. The refraction of light waves when they move from water into air is responsible for the foreshortened appearance of poles in water, and a similar refraction of light waves when they move from air into glass and out again accounts for the focusing of light by lenses. The phenomenon of refraction is a perfectly general property of wave motion, occurring wherever waves move from one medium into another where the speed is different.

9.9 Reflection of water waves. *Waves approaching an obstacle obliquely appear to re-form and move away in a different direction.*

reflection

Another familiar property of waves is their ability to be *reflected* when they meet an obstacle. The stretched rope of preceding paragraphs illustrates reflection beautifully: if a single wave is sent down the rope by a jerk on the free end, the wave meets the attached end, re-forms itself, and travels back along the rope. The reflection occurs because the point of attachment of the rope is pulled and slightly displaced when the original wave arrives, and then springs back (because it is elastic) and in so doing gives the rope a jerk in the opposite direction. Waves on water are similarly reflected, as is shown by waves striking a breakwater obliquely and re-forming themselves as waves moving outward at a similar angle in the other direction (Fig. 9.9).

reflection is the change in direction of a wave when it strikes an obstacle

If a succession of waves is sent along a stretched rope, the reflected waves must meet the forward-moving waves head on. Each tiny segment of the rope, accordingly, must respond to two different impulses at the same time. The two impulses add together vectorially: if the segment is being pushed in the same direction by both wave impulses, it will move in that direction with an amplitude equal to the sum of the two motions; if the two wave impulses are in opposite directions, the segment will have an amplitude equal to the difference of the two. If the timing is properly adjusted, the two motions may cancel out completely for some points of the rope while points between these move with twice the normal amplitude. In this situation the waves appear not to travel at all; some parts of the rope are in violent back-and-forth motion and other parts remain at rest (Fig. 9.10). Waves of this sort are called *standing waves*. Vibrating strings in musical instruments are the most familiar examples. Longitudinal waves traveling in opposite directions over the same path may also set up standing waves, as may be illustrated with the long spring of Fig. 9.3 or by the vibrating air columns of whistles, organ pipes, and wind instruments.

standing waves

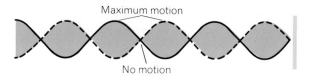

9.10 *Standing waves in a stretched rope or vibrating string.*

interference

Interference refers to the adding together of two or more waves of the same kind that pass by the same point in space at the same time. A simple demonstration of interference is shown in Fig. 9.11. If waves are started along the stretched strings AC and BC by shaking them at the ends A and B, the single string CD will be affected by both. Each portion of CD must respond to two different impulses at the same time, and its motion will therefore be determined by the sum of the effects of the two original waves. When A and B are shaken in unison, the waves add together, so that in CD the crests are twice as high and the troughs twice as deep as in AC and BC. This situation is called *constructive* interference.

On the other hand, if A and B are shaken just out of step with each other, wave crests in AC will arrive at C just when troughs get there from BC. As a result, crest matches trough, the wave impulses cancel each other out, and CD remains stationary. This situation is called *destructive* interference. As another possibility, if B is shaken twice as rapidly as A, the two waves add together to give the complex waveform pictured in Fig. 9.11d. The variations are endless, and the resulting waveforms depend upon the amplitudes, wavelengths, and timing of the incoming waves.

The interference of water waves is shown by ripples in Fig. 9.12. Ripples spreading out from the vibrating rods affect the same water particles: in some directions crests from one source arrive at the same

in constructive interference, the original waves are in step and combine to give a wave of greater amplitude

in destructive interference, the original waves are out of step and combine to give a wave of smaller amplitude

time as crests from the other source and the ripples are reinforced while between these regions of prominent motion are narrow lanes where the water is quiet, representing directions in which crests from one source arrive together with troughs from the other, so that the wave motions cancel.

diffraction

An important property of waves is their ability to bend around the edge of an obstacle in their path. This property is called *diffraction*.

A simple example of diffraction occurs whenever we hear a street noise, such as that of an automobile horn, around the corner of a building. The noise could not have reached us through the building, and refraction is not involved since the speed of sound does not change between the source of noise and our ears. What has happened is that the sound waves have spread out from the corner of the building into the "shadow" as though they originated at the corner, which is how diffraction occurs. The diffracted waves are not as loud as those that have proceeded directly to a listener, but they have gone around the corner in a way that a stream of particles, for example, cannot. Hence diffraction, as well as interference, is a wave property that provides a definitive criterion for establishing whether an unknown radiation consists of waves or of streams of particles, and in later chapters we shall see how this criterion figured in the development of modern physical concepts. Figure 9.13 illustrates the diffraction of water waves.

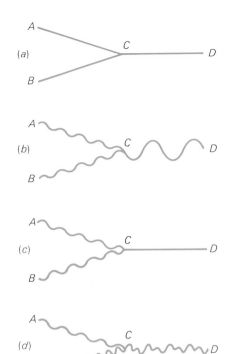

9.11 Interference. (a) *Waves started along stretched strings AC and BC will interfere at C.* (b) *Constructive interference.* (c) *Destructive interference.* (d) *A mixture of constructive and destructive interference.*

SOUND

speed

It is a common observation that the speed of sound must be less than the speed of light. A man chopping wood on a distant hillside is seen to strike the wood before the sound of the stroke is heard; steam emitted from a whistle at a distance is seen before the sound of the whistle is heard. Jet planes are a particularly good example: the sound of their motors

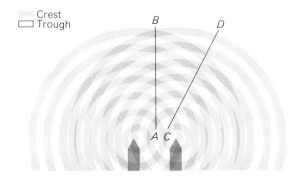

9.12 Interference of water waves. *A diagrammatic representation of the interference of water waves shown in the photograph on p. 172.*

The interference of water waves. Ripples are spreading out across the surface of a shallow tank of water from the two sources at the top. In some directions (for instance AB) the ripples reinforce each other and the waves are more prominent. In other directions (for instances CD) the ripples are out of step and cancel each other, so that the waves are small or absent.

does not appear to come from the planes at all, but from a point far behind them, simply because they travel so fast that they move a long distance in the time it takes the sound to reach our ears. The most direct way to measure the speed of sound is simply a refinement of such observations; one need only employ some method of accurately timing the production and reception of sound over a known distance, capitalizing upon the fact that the speed of light is so great that its time of travel over the same distance is negligible.

Measurements of this sort give a speed of about 1,100 ft/sec (340 m/sec) or 750 mi/hr in ordinary air. The speed varies somewhat with atmospheric conditions, being faster in warm air than in cold air. In liquids

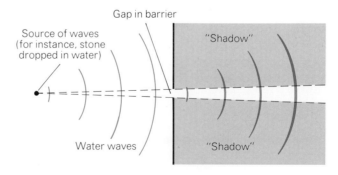

9.13 Water waves diffracted at a gap in a barrier. *The waves spread into the geometrical shadow region as though they had originated at the gap.*

and solids the speed is greater: about 4,700 ft/sec in water and 17,000 ft/sec in iron.

frequency and wavelength

The frequency of sound waves is the quantity we ordinarily call pitch. One way to demonstrate this is to hold a small card or reed against a notched wheel that can be turned rapidly. The card is made to vibrate as the notches strike it, and when the vibration becomes fairly rapid we hear it as a musical tone. When the wheel is turned faster the pitch of the tone rises. Even with this crude apparatus we could determine roughly what frequencies of vibration correspond to particular pitches.

More refined measurements indicate that the normal ear responds to sound vibrations in the range from about 16 to 20,000 Hz. Vibrations slower than 16 per second, if felt at all, are perceived as separate pulses; waves beyond the upper limit, called *ultrasonic* waves, are audible to some animals and may be detected by mechanical or electrical devices.

ultrasonic waves have frequencies too high to be audible to the human ear

Frequencies and wavelengths of sound waves are related by

$$v = f\lambda$$

where v is the wave speed. A frequency of 20 Hz, for example, corresponds to a wavelength in air of

$$\lambda = \frac{v}{f} = \frac{1{,}100 \text{ ft/sec}}{20 \text{ Hz}} = 55 \text{ ft}$$

while a frequency of 20,000 Hz, a thousand times higher, means the correspondingly shorter wavelength of 0.055 ft, a little over $\frac{1}{2}$ in.

Sound wavelengths can be measured directly with an apparatus like that shown in Fig. 9.14. The metal disk is set in vibration by stroking the rod with a cloth; sound waves travel through the air column in the glass tube and are reflected at the far end. The plug can be moved so as to change the length of the air column, and for certain values of the length the reflected waves combine with the forward-moving waves to produce standing waves, as explained in a previous section. If the glass tube contains a light powder scattered along its bottom, the particles of powder are stirred up by the vibrating air molecules and tend to collect in parts of the tube where the vibratory motion is least intense. Thus the standing waves, so to speak, are outlined by the distribution of powder, and the wavelength is twice the distance from one accumulation of powder to the next.

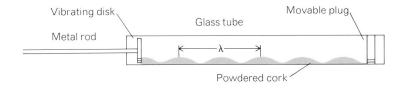

9.14 Kundt's tube is one type of apparatus for measuring the wavelength of sound. *Sound waves produced by the disk set up standing waves in the air column in the glass tube, and the powdered cork (or any very light material) collects at the points of minimum vibration.*

Diffraction of water waves around the end of a breakwater at Morro Bay, California.

The waveforms of sounds can be analyzed electronically with the help of an oscilloscope, a device that displays electric signals on the screen of a tube like the picture tube of a television set. A microphone is used to convert acoustical waves into electric waves, and these in turn can be displayed on the oscilloscope screen. "Pure" tones, like those produced by a tuning fork, have simple waveforms, while musical instruments and human vocal cords produce complex waveforms (Fig. 9.15). Ordinary nonmusical noises consist of waves with complex and rapidly changing forms.

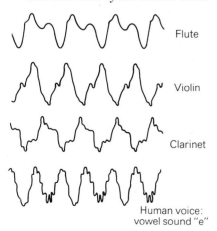

9.15 *Oscilloscope traces of sound waves produced by three musical instruments and by the human voice.*

reflection and refraction

The reflection of sound waves is beautifully illustrated by echoes. For the production of distinct echoes a reasonably smooth surface, such as a cliff or the wall of a cave, at some distance from the source of sound is necessary. If no smooth surface is present the sound is partly absorbed and partly scattered, so that a coherent reflected wave cannot form. If the surface is too close to the sound source, the reflected wave follows so soon after the original wave that the ear does not sense two separate sounds; the original sound is merely slightly prolonged, and we speak of it as *reverberating*.

The difference in the sound of voices in the open air and in a large room is due to reverberations from the walls of the room. Obviously, in a room or auditorium the reverberations can become complex and long-

In the ultrasonic testing of steel, a beam of high-frequency sound waves is sent into a steel object. Flaws in the interior of the object affect the passage of the sound waves and thereby reveal their presence.

continued as the sound waves bounce from walls to ceiling to floor and back again. A certain amount of reverberation is considered desirable, because without it sound becomes "flat" or "dead." On the other hand, too much reverberation must be avoided in order to keep one sound from overlapping and obscuring the following sounds. In the design of buildings, especially large auditoriums, such problems of *acoustics* become very important. The amount of reverberation can be controlled to a large extent by adjusting the shape of a room and by using materials on the walls that have the necessary ability to absorb or reflect sound.

acoustics refers to the science and engineering of sound

The refraction of sound waves is less familiar and less easy to demonstrate convincingly. When successive layers of air have different temperatures, the ability of sound to travel faster in warm air than in cold sometimes causes unusual deflections and distortions of sound. One example is the often-noted fact that sounds can be heard from abnormally long distances over bodies of water on quiet days. Air next to the water is cooler than air above, so that sound waves are slightly retarded near the water. Instead of being scattered and quickly dissipated near the

9.16 Refraction of sound. *Over open water a layer of cold air often underlies warm air. Sound travels faster in the warm air, hence is bent toward the water. The changing direction of motion of the sound is shown by the arrows.*

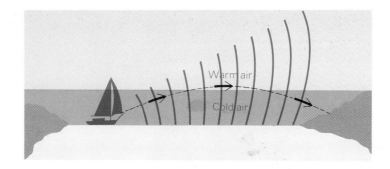

source, sound waves are therefore bent down toward the water and may follow it for long distances (Fig. 9.16).

other properties

The interference of sound waves is responsible for the formation of standing waves in the tube in Fig. 9.14, and for similar standing waves in organ pipes and other wind instruments. Another striking illustration is the phenomenon of *beats:* if two tuning forks differ in frequency by only a few vibrations per second and if the two are struck simultaneously, the two sound waves combine to produce an unpleasant throbbing sensation. The throbbing is due to interference, which alternately augments and reduces the amplitude of the sound waves (Fig. 9.17). If the frequency

beats

9.17 Interference of sound. *Two tones with frequencies of 30 and 40 Hz, respectively, interfere so as to give increased amplitudes every 0.1 sec and to cancel each other out in between. This produces a "beat tone" with a frequency of 10 Hz. In the lowest graph, the solid colored line shows the equivalent smoothed-out beat tone.*

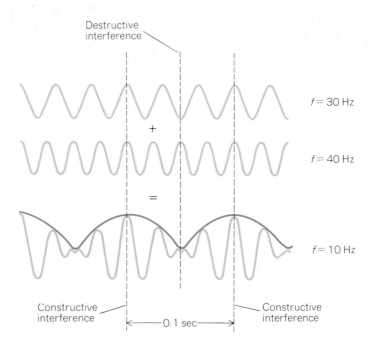

difference between the two forks is greater than a few vibrations per second, the beats may be so rapid as to produce a third sound, or beat tone, with a frequency equal to the difference of the two original frequencies.

The loudness of a sound is related to the *energy* of the waves, and this in turn to the energy of the vibrating source. The relation is a complex one, the human ear responding to much smaller differences in energy among faint sounds than among loud ones, and in addition perceiving sounds of different frequencies but similar energies as having different loudness. The range of loudness to which the ear can respond is enormous: the energy of the faintest audible sound differs from the energy of sounds loud enough to be painful by a factor of roughly 10^{13}.

the doppler effect

We all know that sounds produced by vehicles moving toward us seem higher-pitched than usual, whereas sounds produced by vehicles receding from us seem lower-pitched than usual. Anybody who has listened to the whistle of a train as it approaches and then leaves a station, or to the siren of a fire engine as it passes by at high speed, is aware of these apparent changes in frequency, called the *Doppler effect*. A similar effect occurs in light waves, and is one of the techniques by which astronomers detect and measure stellar motions. Stars emit light which contains only certain characteristic wavelengths; when a star moves either toward or away from the earth, these wavelengths appear, respectively, shorter or longer than usual. From the amount of the shift, it is possible to calculate the speed with which the star is approaching or receding.

the doppler effect is the change in frequency of a wave due to motion of source or observer

Let us consider a train that is stationary at some distance from an observer, as in Fig. 9.18a. Successive waves travel outward in all directions from the train's whistle, and their frequency is simply the number of these waves that strike the observer's ear each second. Now the train starts moving toward the observer, with the whistle still blowing (Fig. 9.18b). The number of waves that reach the observer each second is increased, because the whistle moves closer between any two successive waves; hence the apparent frequency of the waves is greater than when the train is stationary. When the train has moved past the observer (Fig. 9.18c), each wave starts from a point farther away than the preceding one, so that fewer waves can reach the observer per second and the frequency is decreased. The faster the train moves, the greater will be the fall in pitch.

The same sort of phenomenon occurs if we are moving while the source of sound stands still. When we are on a train, for example, moving past a crossing while the warning bell is ringing, we notice that the sound of the bell falls in pitch just as we pass it. When we are moving *toward* the bell, our ear intercepts more waves per second than it would if we were standing still, so we hear a sound of greater than normal frequency; moving *away from* the bell, our ear picks up fewer waves per second and we infer that the sound has dropped in frequency. Again the amount by which the pitch appears to fall depends on the speed of motion.

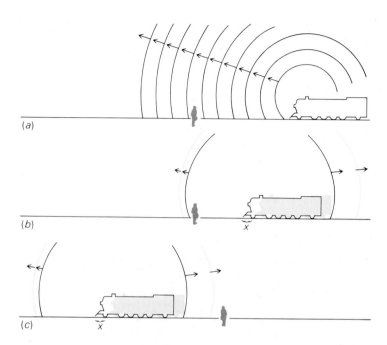

9.18 The Doppler effect. At (a), the train is standing still, and sound waves from the whistle reach the observer at their normal frequency. At (b), the train approaches the observer, moving a distance x between two successive waves; to the observer, the wavelength seems shorter and the frequency higher. At (c), the train recedes from the observer, again moving a distance x between successive sound waves; here the observer finds that the wavelength is longer and the frequency lower.

Thus the Doppler effect depends simply on *relative* motion of the observer and the source of sound. Either or both may be in motion. When the motion lessens their distance apart, sound frequencies are increased; when they are moving away from each other, sound frequencies are decreased.

A similar effect is readily observed with water waves. If we travel in a small boat against the wind, waves strike the boat with greater than normal frequency and the ride may be very choppy. On the other hand, if we travel with the wind, waves catch up with us only slowly, and the frequency with which they strike the boat is much less than normal.

musical sounds

Musical sounds consist of waves with a sustained frequency and waveform, as opposed to nonmusical noise in which the frequency and waveform change rapidly. Musical sounds are produced by vibrating objects —stretched gut or wire in stringed instruments, vocal cords in the larynx, stretched membranes in drums, air columns in wind instruments set in motion by vibrations of reeds, lips, or air currents directed against a sharp edge. The frequency, or pitch, of a musical tone can in most instruments be varied at will by the player; the waveform, or quality, of the tone is determined largely by the shape and kind of instrument but in part is also controlled by the player.

Some of the characteristics of musical sounds can be well illustrated with a stretched string. If the string is plucked, it may be set into various kinds of vibration. The simplest is a vibration in which a single standing

wave makes up the whole length of the string (Fig. 9.19a); the ends remain fixed, and the maximum vibration is in the middle. The frequency of the note emitted may be varied by changing the tension on the string, as a violinist does when he tunes his instrument; for a given tension the frequency may be varied by changing the length of the string, as the same violinist demonstrates when with the pressure of his fingers he controls the length of the vibrating part of the string. Depending on where the string is plucked or bowed, more complex vibrations may be set up: standing waves may form with two, three, or even more maxima (Fig. 9.19). Sound waves set up by these shorter standing waves will have higher frequencies, and the frequencies will clearly be related to the frequency of the single long wave by simple numerical ratios—$2:1$, $3:1$, and so on. The tone produced by the string vibrating as a whole is called the *fundamental*, and the higher frequencies produced when it vibrates in segments are called *overtones*.

In practice, the strings of a musical instrument are set in motion so that they vibrate not merely to give the fundamental or a single overtone but to produce a combination of the fundamental plus several overtones. The motion of the string, and correspondingly the form of the sound wave, may be very complex. To the ear the fundamental tone by itself seems flat and uninteresting; as overtones are added the tone becomes "richer," and the quality, or timbre, of the tone depends on which particular overtones are emphasized. The emphasis is secured in part by the shape of the instrument, which enables it to *resonate* with some frequencies

Sound waves can be generated by a variety of means. Shown here are vibrating strings (string bass), vibrating membranes (drums), and vibrating air columns (saxophone and horn).

9.19 *A few of the possible kinds of vibration in a stretched string.*

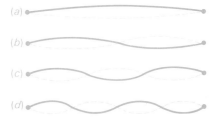

more than others. This means that the sounding part of the instrument—the belly of the violin or the soundboard of the piano—has certain natural frequencies of vibration, and that it is more readily set into vibration with these frequencies than with others. The resulting sounds may include a large number of overtones, but the greater emphasis on certain overtones provides the musical quality for which the instrument is especially prized.

In wind instruments the sound is produced by a vibrating air column, whose length determines the fundamental frequency. The entire air column may be set in motion as a unit, or overtones may be produced by standing waves whose length is any simple fraction of the total length. The performer can control the pitch of a note, in part by controlling the overtones with pressure from his lip and in part by using stops to vary the length of the vibrating column. The human voice resembles a wind instrument in that the vocal cords produce vibrations in a column of air extending from the throat into the mouth, nose, and sinuses. The shape of this column, consciously adjusted during speech or singing by manipulating the mouth and tongue, determines the different vowel sounds by emphasizing some overtones and suppressing others. The general shape of the mouth and sinuses is also responsible for the subtle differences in quality which enable us to distinguish one person's voice from another's.

Some combinations of musical sounds make pleasing or consonant chords; others impress us as harsh discords. There is no universal agreement as to which combinations are pleasant and which unpleasant—witness the great differences in Oriental and Western harmonies—but a few generalizations are possible. The combination of a fundamental tone and its first overtone (the interval between them is called an "octave"), with frequencies in the ratio $1:2$, is commonly regarded as pleasant; so also is the interval called the "fifth," with a frequency ratio $2:3$ (e.g., the notes C and G), and the interval called the "third," with a frequency ratio $4:5$ (e.g., the notes C and E). In general, to Western ears, only combinations with frequency ratios expressed by small numbers seem consonant; the combination C and D (ratio $8:9$) appears discordant, and the combination E and F (ratio $15:16$) is more discordant still.

GLOSSARY

In *wave motion* a change or distortion in a medium is propagated. A wave carries energy from one place to another, but there is no net transport of matter.

In a *transverse wave* the oscillations occur at right angles to the direction of motion of the wave. Waves in a stretched string are transverse waves.

In a *longitudinal wave* the oscillations occur back and forth in the same direction as that in which the wave is moving. Sound waves and compressional waves in a spring are longitudinal waves. Water waves are a combination of longitudinal and transverse waves.

The *wavelength* of a wave is the distance between adjacent crests, in the case of transverse waves, or between adjacent compressions, in the case of longitudinal waves.

The *frequency* of a wave is the number of waves that pass a given point per second. The frequency of a sound wave is its *pitch*.

The *amplitude* of a wave refers to the maximum value attained by whatever quantity is periodically varying.

Reflection occurs when a wave bounces off a surface; light is reflected by a mirror, sound by a wall.

Refraction occurs when a wave changes direction on passing from one medium to another in which its speed is different.

Interference occurs when two or more waves of the same kind pass the same point in space at the same time. If the waves are "in step" with each other, their amplitudes add together to produce a stronger wave; this situation is called *constructive interference*. If the waves are "out of step" with each other, their impulses tend to cancel out and the resulting wave is weaker; this situation is called *destructive interference*.

The *Doppler effect* is the change in frequency of a sound due to relative motion between its source and a listener.

EXERCISES

1 We have described ordinary sound waves as vibratory motions of air molecules. Is the assumption that air consists of molecules *essential* for an understanding of sound phenomena—at least for the phenomena we have considered in this chapter? Explain.

2 What happens ultimately to the energy of sound waves?

3 How can the phenomena of constructive and destructive interference be reconciled with the principle of conservation of energy?

4 It is observed that the speed of sound in a gas is related to the mean speed of the molecules in the gas. Why is such a relationship reasonable?

5 If you walk past a bell while it is ringing, you notice no change in pitch, but if you ride past the same bell in a rapidly moving train or car the pitch seems to decrease markedly. Explain.

6 In the experiment shown in Fig. 9.14 why is λ the distance from one pile to the second pile in sequence, rather than to the adjacent pile?

7 What is the frequency of ocean waves 500 ft long whose speed is 60 ft/sec?

8 What is the frequency of sound waves in air whose wavelength is 1 ft?

9 A violin string vibrates 1,044 times per second. How many vibrations does it make while its sound travels 50 ft?

10 The speed of sound waves in water is approximately 4,700 ft/sec. If an echo-sounding device on a ship measures an interval of 3 sec between the emission of a sound pulse and the reception of its echo from the sea bottom, how deep is the water?

11 Suppose that you time the interval between seeing a lightning flash and hearing the thunderclap. If the interval is 4 sec, how far away was the lightning?

12 A certain groove in a phonograph record travels past the needle at a speed of 40 cm/sec. The frequency of the sound that is produced is 3,000 Hz. Find the wavelength of the wavy indentations of the groove.

13 A man is watching as spikes are being driven to hold a steel rail in place. The sound of each sledgehammer blow reaches him 0.14 sec through the rail and 2 sec through the air after he sees the hammer strike the spike. Find the speed of sound in the rail.

14 The engineer of a train moving at 60 mi/hr (88 ft/sec) blows its whistle. How much later does a man in the caboose 1,000 ft behind the engine hear the sound?

part three
ELECTRICITY AND MAGNETISM

All matter is electrical in nature, and electromagnetic interactions are involved in all aspects of the universe, from the smallest atom to the largest galaxy of stars. Indeed, the electromagnetic interaction is one of the four basic interactions that are responsible for all physical processes. Another is gravitation, which we have met already. The other two only act over very short distances and their effects are confined to atomic nuclei; they are discussed in Chap. 17. Most of the familiar properties of matter—the existence of atoms, molecules, solids, and liquids, for instance—can be traced to electromagnetic forces.

The great triumph of electromagnetic theory was the discovery of how closely electricity and magnetism are related, so closely indeed that coupled electric and magnetic oscillations can travel as waves through empty space. The reality of these electromagnetic waves, first predicted a century ago, was demonstrated when it was shown that light, X-rays, and radio waves all consist of them.

chapter 10

Electricity

ELECTRIC CHARGE

    positive and negative charge
    conductors and insulators
    the nature of charge

COULOMB'S LAW

    coulomb's law
    the coulomb
    matter in bulk
    charge invariance

FIELDS OF FORCE

    what is a force field?
    lines of force

The kinetic theory asserts that all matter is made up of tiny moving particles. In gases, the particles are far apart and attract one another only to a negligible extent; in liquids, the attraction between them is great enough to keep the particles close together but not to prevent their moving about; in solids, the particles are held so firmly that their motion is restricted to vibrations about fixed positions. An increase in temperature represents faster motion of the particles and so a tendency for them to disperse in spite of their mutual attractions. In terms of these particles and the simple forces between them, such diverse phenomena as boiling, freezing, the expansion of gases, and the flow of liquids find reasonable explanation. Thus the kinetic theory introduces order and simplicity into the apparently complex behavior of ordinary materials.

  But for all the success of the kinetic theory, it can only account for a certain few aspects of the behavior of matter. For an example of a basic phenomenon upon which the kinetic theory cannot shed any light, all that is necessary is to comb one's hair with a hard rubber comb on a day when the air is dry: the hair seems to crackle, and its ends are attracted by the comb. If this is done before a mirror in a dark room, tiny sparks will be seen to jump between comb and hair.

ELECTRIC CHARGE

The first recorded investigator of such behavior was Thales of Miletus, a Greek philosopher who lived about 600 B.C. Lacking rubber, Thales experimented with amber, in his language *electron*. We immortalize his work by saying that amber (or hard rubber) rubbed with fur possesses an *electric charge*—by which we mean simply that it is capable of producing a spark and attracting small light objects.

positive and negative charge

Let us examine the behavior of electric charges more carefully. We begin by suspending a small pith ball from a silk thread, to serve as an indicator of charges in its vicinity. If touched with a rubber rod that has been stroked with fur, the pith ball jerks violently away and, thereafter, is strongly repelled whenever the rod is brought near (Fig. 10.1). We assume that the pith ball had no electric charge at the beginning of the experiment; at the instant of contact with the rod the ball acquired some of the charge on the rod and in this charged condition is repelled by the rod.

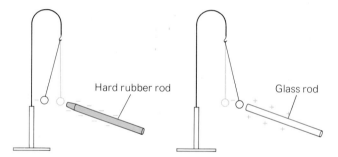

10.1 Like charges repel, unlike charges attract. *A pith ball touched by a rubber rod that has been stroked with fur is repelled by the rod. When a glass rod that has been stroked with silk cloth is brought near, however, the pith ball is attracted to it. Performing the experiment in the reverse order has the same effect; hence the conclusion that like charges repel each other, unlike charges attract each other.*

Now we bring near the same pith ball a glass rod that has been rubbed with silk. The ball is no longer repelled but strongly attracted. With the ball in this condition, therefore, the charge of the rubber rod repels it, and the glass rod attracts it. Next we try the experiment in reverse and charge a second pith ball by touching it with the glass rod. It flies away, evidently repelled. But the charged rubber rod attracts this second ball strongly.

We can draw only one conclusion: The charges on the two rods are somehow different. Furthermore, the kind of charge on one rod attracts the kind on the other, but each rod repels an object that has some of its own kind of charge. More simply, **like charges repel each other, unlike charges attract each other.**

like charges repel, unlike charges attract

Comprehensive experiments show that all electric charges that can be produced fall into one or the other of the two types described above; that is, they behave as though they originated on a rubber rod rubbed with fur or else on a glass rod rubbed with silk, regardless of their actual origin. Benjamin Franklin suggested names for these two basic kinds of electricity. He called the charge produced on a rubber rod rubbed with

ELECTRICITY

fur *negative charge*, the charge produced on a glass rod rubbed with silk *positive charge*. These definitions are the ones we follow today.

We have concentrated our attention on the positive charge of the glass, the negative charge of the rubber. However, we do not produce a positive charge alone by rubbing glass with silk, or a negative charge alone by stroking rubber with fur. If the fur used with the rubber is brought near a positively charged pith ball, the ball is repelled; if the fur is brought near a negatively charged ball, the ball is attracted. Thus the fur must have a positive charge. Similarly the silk used with the glass has a negative charge. In general, when electricity is produced by contact between two dissimilar objects, one acquires a positive charge and the other a negative charge, the distribution of charge depending on the nature of the two substances used.

positive and negative charge

conductors and insulators

Although pith balls are very convenient for simple experiments in electricity, they are hardly quantitative tools for the physicist. Somewhat better is an *electroscope*, a device consisting of two leaves of thin metal foil suspended from a metal support inside a glass-walled box (Fig. 10.2). A charge of either kind applied to the metal support spreads itself over the leaves; charged with the same kind of electricity, they fly apart, with the distance apart depending roughly on the amount of the charge.

Let us use the electroscope to study the motion of electric charges from one place to another. Our experiments have already shown that charges can move. For example, when a pith ball is touched with an electrified rubber rod, part of the charge on the rod went over to the ball. Suppose that we place an electroscope a foot or so from a charged object, say a large metal sphere, and that we connect the sphere and the metal of the electroscope with rods or wires of different materials (Fig. 10.3). If a copper wire is laid between them, the leaves of the electroscope fly apart at once. With about equal facility a charge is transferred from sphere to electroscope by wires of other metals. But if the connection is made with a dry glass rod or a silk thread or a piece of dry wood, the electroscope is scarcely affected. The ability of electric charge to pass from sphere to electroscope depends on the nature of the material making the connection. Materials like iron or copper, which carry charge readily, are called *conductors*. Materials like glass or dry wood or hard rubber, through which charges move with difficulty, are *insulators*.

A perfect conductor would be a substance along which charges could pass with no resistance to their motion, and a perfect insulator would be a substance through which no electric charge could be forced. No substance of either type is known to exist at ordinary temperatures, though near absolute zero certain metals and alloys lose virtually all their resistance. The latter phenomenon is called *superconductivity*. At room temperature there are good conductors like copper, good insulators like rubber and silk, but these fall far short of perfection. Between good conductors and good insulators can be listed a great variety of substances along which charges move with more or less difficulty. A few of these intermediate

the electroscope

A woman standing on a stool to insulate her from the ground is given an electric charge when she touches the terminal of a static electricity generator. Because all her hairs have the same sign of charge, they repel one another.

10.2 An electroscope. When the electroscope is uncharged its thin metal leaves hang freely, but when it is charged they fly apart because of their mutual repulsion.

substances may be considered either insulators or conductors, depending on circumstances; there is no sharp line between the two.

Ordinary air is a very poor conductor. If two strong opposite charges are brought close together, however, the air between them may momentarily become a good conductor. The air molecules become disrupted, in a manner we shall describe later, and the charges are able to leap across the gap. This type of sudden discharge is an *electric spark*. After the spark, unless the two charges are continuously renewed, the air reverts quickly to its normal nonconducting state.

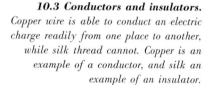 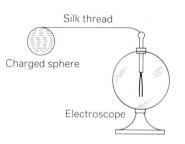 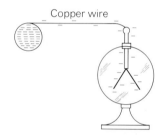

10.3 Conductors and insulators. Copper wire is able to conduct an electric charge readily from one place to another, while silk thread cannot. Copper is an example of a conductor, and silk an example of an insulator.

The conductivity of pure water is extremely small. But traces of dissolved impurities increase the conductivity enormously; since most water we use in daily life is somewhat impure, we usually think of it as a fair conductor. On humid days solids exposed to the atmosphere become coated with an invisible film of water, which makes even good insulators capable of conducting charge appreciably. Experiments with electric charge are difficult in humid weather, since the water films on insulators enable charges to leak away from pith balls and electroscopes; even in dry weather, experiments are improved if insulating materials are heated or rubbed with alcohol to remove adhering water molecules.

grounding

The earth as a whole, at least that part of it beneath the outer dry soil, is a fairly good conductor. Hence if a charged object is connected with the earth by a piece of metal, the charge is conducted away from the object to the earth. This convenient method of removing the charge from an object is called *grounding* the object. The shell of an electrical appliance is grounded by attaching it with a wire to a gas or water pipe (which is connected to other pipes below the surface), thereby giving electric charges in the shell a path to the ground. Ground connections for appliances or lightning rods must be carefully made, using metal all the way, but for rough experiments with small electric charges the connection is sufficiently good if a person simply touches a charged object with his finger. The charge travels through his body, through the floor or the walls of the building, to water pipes or directly to the ground.

The readiness of charges to move to the ground, either through the air or through solid supports, makes exact experimentation difficult. Small charges tend to leak off slowly; even a pith ball, hung from a silk thread in dry air, will not maintain a charge indefinitely. Larger charges may find their way to the ground by sparking unless sufficient insulation is provided.

The human body is not a good conductor—a fortunate circumstance, for otherwise we should be shocked more frequently and more severely. But charges can easily be produced in the laboratory that can be very painful if allowed to move through the body to the ground. When properly insulated from the ground, however, say by standing on a plate of glass or on a platform supported by glass or rubber insulators, a person can be given a considerable charge, large enough to make separate hairs stand out away from each other and to make tiny sparks jump from his finger tips, without any unpleasant sensation. The charge is dangerous only when it can actually move through the body, where it affects the nervous system.

the nature of charge

Thus far nothing has been said about the actual nature of electric charges. As we shall learn in a later chapter, electric charge is one of the various inherent properties *elementary particles* may have; these are the particles of which all matter is constituted, and their interactions are responsible for the properties of atoms and molecules and hence give rise to all natural phenomena. The most familiar kind of electricity in motion, the kind that flows in wires and operates our lights and heaters and household appliances, consists of tiny negative particles called *electrons*. The electricity that moves through gases and liquids and the electric charges that accumulates on pith balls and electroscopes may involve particles with both positive and negative signs. Movement of the tiny particles may still be likened to the movement of a fluid, and for many purposes this picture remains a useful one; practical electricians by custom regard electric current as a movement of positively charged "juice" from one place to another.

electric currents in wires consist of flows of electrons, which are negatively charged

COULOMB'S LAW

It is possible to study the behavior of electric charges quantitatively in an experiment that requires no knowledge of their properties except that they attract and repel one another. The repulsive force between a rubber rod stroked with fur and a negatively charged pith ball depends on two things: how close the pith ball is to the rod, and how much charge each possesses. The influence of these two factors can be shown by noting (1) that the pith ball is only slightly affected by the rod when it is some distance away but is increasingly repelled the closer the rod is brought, and (2) that an increased charged produced by more rapid stroking of

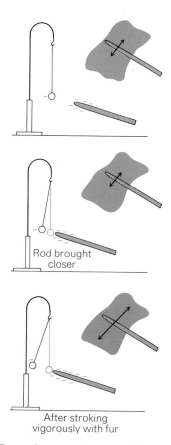

10.4 Forces between charges. *When a rubber rod that has been stroked with fur is brought near a negatively charged pith ball, the force on the ball is greater when the rod is held close to it and also greater when the rod has been vigorously stroked.*

the rod with fur makes the repulsion stronger (Fig. 10.4). A third factor is involved which we shall neglect—the influence of the air between the two charges. If the air were replaced by some other gas or a liquid or were removed altogether, the repulsive force would be different.

coulomb's law

The repulsive force between two objects of like sign is greater the smaller the distance between them. If the two objects are 2 cm apart, the force between them is one-fourth as great as if they are 1 cm apart and four times as great as when they are 4 cm apart (these relationships are exact only when the objects are very small compared with the distance between them). In other words, doubling the distance between the objects quarters the force between them, tripling the distance decreases the force ninefold, and so forth. Between force and distance there is evidently an inverse relationship—**as the distance increases, the force decreases as its square.**

We have not yet specified any way of determining the amount of charge on an object. One way of doing this requires two identical insulated metal spheres. We give one of them a charge, any charge we please; the amount does not matter. Then we bring them in contact and make the reasonable assumption that the charge distributes itself equally over the two. By separating them we provide ourselves with two charges known to be equal. Now we bring one sphere near to a third charged object and measure the force between them. Next we put the two spheres together at the same distance from the third object, and again measure the force. We find that the force is just twice as great (Fig. 10.5). It looks as if the force is directly proportional to amount of charge; we can verify this by obtaining other spheres identical with the first two, giving them all equal charges by letting them come in contact, and testing their effect in various combinations on the third object. Also, we could duplicate the third object and see how doubling its charge affects the force. We would find that **the force is proportional to each of the charges.**

Now we are in a position to state in mathematical terms exactly how the force between two charges depends upon the magnitudes of the charges and upon the distance between them. We have already used the symbols F for force and d for distance; electric charges are usually denoted by the letter q, and, if we have more than one, we distinguish them by calling one of them q_1, another q_2, and so on. On the basis of the experiments described in the preceding paragraphs, the force between two charges of magnitudes q_1 and q_2 that are separated by the distance r is

coulomb's law
$$F = K \frac{q_1 q_2}{r^2}$$

where K is a constant of proportionality whose numerical value depends upon the units in which we measure the charges. The above equation, called *Coulomb's law,* is a shorthand way of expressing the fact that the force between the two charges q_1 and q_2 depends directly upon the magnitudes of q_1 and q_2 and inversely on the square of the distance r between them. The larger the charge values, the greater the force, and the larger the separation, the smaller the force. When the charges have the same sign—both positive or both negative—the force is repulsive, tending to push the charges apart, and, when the charges have opposite signs—one positive and the other negative—the force is attractive, tending to pull the charges together.

the coulomb

The modern unit of charge is the *coulomb* (coul), named in honor of Charles Coulomb (1736–1806), who developed the above equation. With q_1 and q_2 expressed in coulombs, the constant K is almost exactly 9×10^9 newton-m²/coul². Thus we may rewrite Coulomb's law

the coulomb is the unit of electric charge

$$F = 9 \times 10^9 \frac{q_1 q_2}{r^2}$$

where the force F is expressed in newtons and the distance r in meters, the metric units for these quantities.

It is a significant fact that all charges in nature, both positive and negative, occur only in multiples of 1.6×10^{-19} coul. This quantity of charge is denoted e. Of the three elementary particles that all matter can ultimately be broken down into, the *electron* has a charge of $-e$, the *proton* has a charge of $+e$, and the neutron has no charge. Hence if we could somehow assemble 6.25×10^{18} electrons in one place, we would have a charge of exactly -1 coul, and if we could somehow assemble 6.25×10^{18} protons in one place, we would have a charge of exactly $+1$ coul.

all electric charges are multiples of $\pm e$, where $e = 1.6 \times 10^{-19}$ coul

It is easy to show that the electric forces between the elementary particles that make up an atom are vastly stronger than the gravitational forces between them. An example is the hydrogen atom, the simplest atom of all. A hydrogen atom consists of a proton (mass 1.7×10^{-27} kg, charge $+1.6 \times 10^{-19}$ coul) and an electron (mass 9.1×10^{-31} kg, charge -1.6×10^{-19} coul), whose mean separation is 5.3×10^{-11} m. The gravitational force of attraction between the proton and electron is

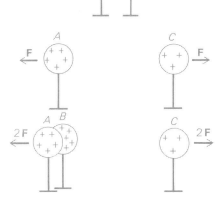

10.5 The force between charged bodies is proportional to the product of the charges.

$$F_{\text{grav}} = G \frac{m_p m_e}{r^2}$$

$$= \frac{(6.7 \times 10^{-11} \text{ newton-m}^2/\text{kg}^2) \times 1.7 \times 10^{-27} \text{ kg} \times 9.1 \times 10^{-31} \text{ kg}}{(5.3 \times 10^{-11} \text{ m})^2}$$

$$= 3.7 \times 10^{-47} \text{ newton}$$

The electric force of attraction is

$$F_{\text{elec}} = K \frac{q_p q_e}{r^2}$$

$$= \frac{(9 \times 10^9 \text{ newton-m}^2/\text{coul}^2) \times 1.6 \times 10^{-19} \text{ coul} \times 1.6 \times 10^{-19} \text{ coul}}{(5.3 \times 10^{-11} \text{ m})^2}$$

$$= 8.2 \times 10^{-8} \text{ newton}$$

which is more than 10^{39} times greater! Clearly, gravitational effects within atoms are negligible compared with electrical effects.

matter in bulk

Coulomb's law for the force between charges is one of the fundamental laws of physics, in the same category as Newton's law of gravitation. The latter, as we know, is written in mathematical form as

$$F = G\frac{m_1 m_2}{r^2}$$

Coulomb's law and the law of gravitation are remarkably similar, but they have one important difference: Gravitational forces are always attractive, tending to draw the objects involved together, but electric forces may be either attractive or repulsive.

Let us elaborate a significant consequence of this last fact. Because one lump of matter always attracts another lump gravitationally, matter in the universe as a general rule tends to assemble into large masses. Even though dispersive forces of various kinds exist, they must contend with this steady attraction; the stars and planets, which modern theories indicate to have condensed from originally diffuse matter, bear witness to this cosmic herd instinct. And, on a terrestrial scale, we can pile up as large a pyramid as we like with rather primitive means.

To collect a significant electric charge of either sign, however, is far more of a feat. Charges of opposite sign attract each other strongly, so it is hard to separate neutral matter into differently charged portions. (For example, the force between two 1-coul charges that are 1 m apart is 9×10^9 newtons, which is a little over a million tons.) And charges of the same sign repel each other, so even if the charge separation is performed a little at a time, the assembly of charges of one sign is difficult to manage.

gravitational forces are dominant on a cosmic scale, whereas electric forces are dominant on an atomic scale

To sum up the above argument, we might say that a system of neutral particles is most stable (that is, has a minimum potential energy) when the particles make up a single solid body, while a system of electric charges is most stable when charges of opposite sign pair off to cancel each other out. Hence, on a cosmic scale, gravitational forces are significant and electric ones are not; no single charge can be very great before it attracts others of opposite sign and starts to be torn apart by internal repulsion. On an atomic scale, however, the reverse is true. The masses of subatomic particles are too small for them to interact gravitationally to any appreciable extent, while their electric charges, though minute, are still enough for electric forces to exert marked effects.

charge invariance

The theory of relativity has shown, and experiments confirm, that measurements of length, mass, and time depend upon the frame of reference of the observer. If the object under study is moving relative to the observer, he will find that it is shorter and more massive than when it is at rest relative to him, and that characteristic time intervals are longer in duration.

What about electric charge? Unlike length, mass, and time, electric charge is relativistically invariant: if one observer finds the magnitude of a certain charge to be q, all other observers, regardless of their state of motion, will also find the magnitude of the charge to be q. A significant consequence of this is the electrical neutrality of matter. The atoms and molecules of matter contain equal numbers of protons and electrons, but the electrons are in rapid motion whereas the protons move only slowly. If electric charge varied with relative speed, as mass does for instance, then atoms and molecules would not be electrically neutral. In fact, the electrical neutrality of matter is what provides convincing proof of the relativistic invariance of charge. Careful experiments indicate that the electrons and protons in the hydrogen molecule (there are two of each in this molecule) have charges whose magnitudes are equal to at least one part in 10^{10}, despite the much higher speeds of the electrons.

electric charge is relativistically invariant

FIELDS OF FORCE

We are so familiar with gravitational, electric, and magnetic forces that they do not excite astonishment, yet if we really examine their natures it is hard to repress a feeling of wonder. Seemingly unlike other common forces, they act upon other objects without actual contact being made across the intervening space. We cannot move a book from a table by merely waving our hand at it; a golf ball will not move from its tee, however violent our exertions, until the golf club actually strikes it. But a charged pith ball does not wait until a charged glass rod touches it, but reacts to the presence of the rod while it is still some distance away. Remove the air, place rod and ball in the most perfect vacuum attainable, and the force is not diminished. The two react on each other without benefit of anything that our senses can detect. For the pith ball the region near the rod is somehow different from other space, since in this space it is impelled to move. Near any electric charge is such a region of altered space. A magnet also, though in a different way, changes the space around it, so that material of the proper sort is acted on by a magnetic force, and, as we know, every mass causes around it an alteration in space that results in an attractive gravitational force on any other masses in the vicinity.

what is a force field?

What precisely is meant by an "alteration in space," and how do the various kinds of such alterations differ from one another? Is there any relationship among these various kinds of alteration: that is, are gravitational, electric, and magnetic forces in some way connected? These questions, which occur very naturally, have intrigued scientists for many centuries. The chief source of difficulty has been the inability of most people's imaginations to conceive of what we call space—which we regard as emptiness, the complete absence of matter—as being capable of being altered. After all, how can *nothing* be altered?

But the facts of the matter are that as long as we are able to express mathematically just how forces act without the participating bodies actually touching one another (as we have already done in Newton's law of gravitation and Coulomb's law of electric forces and as we shall do in a later chapter for magnetic forces), we have no real need for a mental image of what is happening. To be sure, it is always helpful to be able to visualize what is going on in a physical process, but, if we are honest, we must admit that most of our mental pictures are surely not accurate. Consider a bat striking a ball: both bat and ball are made up of individual molecules rather far apart from one another—just how do these molecules interact when the ball is struck by the bat? Why do not the molecules of the ball and bat, which have so much space in between, simply mesh together into a single object? When this problem is analyzed in detail,

In a thunderstorm, the electric discharge we call lightning occurs when the electric field intensity between a cloud and the earth exceeds the insulating ability of air.

we find that it is the action of electric and magnetic forces on the molecular level that produces the observed transfer of energy and momentum from the bat to the ball. As long as we can describe phenomena quantitatively in terms of mathematical equations that agree with the results of experiment, we have as much information about these phenomena as we need. Mental pictures, however convenient as aids to the imagination, are not really required by the scientist.

The region of altered space around a mass, an electric charge, or a magnet is called a *field of force*. Strictly speaking, a field of force extends to infinite distances in all directions, since a mass, a charge, or a magnet presumably exerts a force everywhere in the universe. But in practice the force becomes negligible at some distance from the exciting object, and by its field of force we ordinarily mean the space immediately adjacent to the object, the region in which its influence is perceptible.

a force field is an alteration in the properties of space in a certain region that causes a force to be exerted on an appropriate object in that region

The *electric field intensity* **E** at any point in space is so defined as to equal numerically the force that it would exert on a positive charge of 1 coul located there. If a charge q is at that point, the force **F** it experiences is given by

$$\mathbf{F} = q\mathbf{E}$$

Electric field intensity is a vector quantity with both magnitude and direction; the direction of **E**, as we can see from the above equation, is that of the force on a positive charge. The units of electric field intensity are newtons per coulomb.

electric field intensity

lines of force

Fields of force are recognized by the tendency of objects in them to move; accordingly, we describe fields in terms of motions. Suppose we are to describe the electric field about a positively charged metal sphere: How would a small charge move in this field? Evidently, if the small charge is negative, it would move in a straight line toward the sphere; if it is positive, it would move directly away from the sphere. (We neglect gravity, air resistance, friction, the support of the sphere, and so on.) The paths of motion in the field are straight lines radiating out from the sphere's center (Fig. 10.6). By convention we consider the direction of motion of a *positive* charge and indicate this motion by placing outward-pointing arrows on the lines. A negatively charged sphere has a field of the same shape, but its lines are directed inward.

Consider now the field about two adjacent charges, one + and the other −. Here again we ask, how would a small charge move? Between the charges it would move directly from + toward −; off to one side, repelled by one and attracted by the other, it would move in a curved path (Fig. 10.7).

Fields of magnetic or gravitational force may be similarly described. The lines used to describe a field—gravitational, magnetic, or electric—are called *lines of force*. Although purely fictitious, lines of force enable us to describe concisely how an object will tend to move in any part

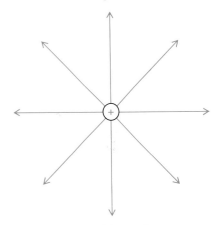

10.6 The lines of force about an isolated positive charge are radial lines whose direction is outward from the charge. *Lines of force do not actually exist, but are merely aids to our thinking.*

lines of force

of a field. Where many lines are close together the field is strong; where they spread apart the field is weak. Note that lines of force do not cross or branch and can end only on one of the exciting objects or at infinity.

GLOSSARY

A body possessing *electric charge* is capable of producing a spark and of attracting small, light objects. *Negative charge* is that produced on a rubber rod rubbed with fur; *positive charge* is that produced on a glass rod rubbed with silk. The *coulomb* is the unit of charge.

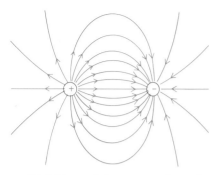

10.7 Lines of force around nearby electric charges. *Each line of force represents the path that would be taken by a positive charge.*

A *conductor* is a substance through which charges can move freely.

An *insulator* is a substance through which charges can move only with difficulty.

A body is *grounded* when its charge is permitted to flow to the earth.

Coulomb's law states that the force between two charges is directly proportional to the product of their charges and inversely proportional to the distance between them; the force is repulsive when the charges have the same sign, attractive when they have different signs.

Electrons, *protons*, and *neutrons* are the elementary particles of which all matter is composed. Electrons have charges of -1.6×10^{-19} coul, protons have charges of $+1.6 \times 10^{-19}$ coul, and neutrons are uncharged.

A *field of force* is a region of altered space (for example, around a mass, an electric charge, a magnet) that exerts a force on appropriate bodies placed in that region.

Lines of force are imaginary lines used to describe a field of force; their direction is that of the force which the field would exert on a positive test particle, and their density is proportional to the strength of the field (that is, they are closest together where the field is strongest, and farthest apart where the field is weakest).

EXERCISES

1 Describe simple experiments that demonstrate that there are two kinds of electricity and no more.

2 Describe what happens when a positively charged electroscope is touched with
   a A rubber rod that has been stroked with fur.
   b A glass rod that has been rubbed with silk.

3 In the eighteenth century both heat and electricity were thought to be weightless fluids. To what extent is this still a useful notion? What

ELECTRICITY

experimental evidence in each case indicates that this hypothesis is inadequate?

4 List the similarities and differences between electric and gravitational fields.

5 How do we know that the force holding the earth in its orbit about the sun is not an electric force, since both gravitational and electric forces vary inversely with the square of the distance between centers of force?

6 Explain why lines of force can never cross one another.

7 A rod has a charge of $+q$ at one end and a charge of $-q$ at the other. How will the rod move when it is placed in a uniform electric field whose direction is parallel to the rod? Whose direction is perpendicular to the rod?

8 A charge of $+3 \times 10^{-9}$ coul is located 0.5 m from a charge of -5×10^{-9} coul. What is the magnitude and direction of the force between them?

9 Two small spheres are given identical positive charges. When they are 1 cm (0.01 m) apart the repulsive force between them is 0.002 newton. What would the force be if
a The distance is increased to 3 cm?
b One charge is doubled?
c Both charges are tripled?
d One charge is doubled and the distance is increased to 2 cm?

10 (*a*) A metal sphere with a charge of $+1 \times 10^{-5}$ coul is 10 cm from another metal sphere with a charge of -2×10^{-5} coul. Find the magnitude of the attractive force between the spheres. (*b*) The two spheres are bought in contact and again separated by 10 cm. Find the magnitude of the new force between the spheres.

11 How many electrons of charge -1.6×10^{-19} coul must be added to a pith ball to give it a charge of -1 coul? What would be the total mass of these electrons? Why is it unlikely that this would ever be done?

12 Two charges, one of $+3 \times 10^{-5}$ coul and the other of $+6 \times 10^{-5}$ coul, are 15 cm apart. A test charge of $+5 \times 10^{-5}$ coul is placed between the two so that it is 5 cm from the $+3 \times 10^{-5}$ coul charge and 10 cm from the $+6 \times 10^{-5}$ coul charge. What are the magnitude and direction of the force on the test charge?

13 Find the electric field intensity 40 cm from a charge of $+7 \times 10^{-6}$ coul.

14 A particle of mass 10^{-6} kg is suspended in equilibrium by an electric field of 9.8 newtons/coul intensity. Find the magnitude of the charge on the particle.

15 (*a*) Find the force acting on a particle of charge 10^{-5} coul when it is in an electric field of intensity 50 newtons/coul.
(*b*) If the particle is released from rest in this field, what will its kinetic energy be after it has traveled 1 m?

16 Two electrons exert forces on each other equal in magnitude to the weight of an electron. How far apart are the electrons?

chapter 11
Electricity and Matter

THE ELECTRON
cathode rays
the electron
discovery of the electron

ELECTRICAL PROPERTIES OF MATTER
ions and electrons
induction

ELECTRIC CURRENT
circuits
the ampere
potential difference
the electron volt
ohm's law
electric power

On the crude level of our personal experience, matter and electric charge both are continuous, apparently capable of being subdivided into smaller and smaller bits without limit. The contrary is true of matter: There is another level, beyond the direct reach of our senses, where matter would be revealed as aggregates of atoms, distinct building blocks which cannot be further fragmented by ordinary means. Electric charge, too, has a microscopic structure, whose basic unit, both positive and negative, is 1.6×10^{-19} coul, a quantity abbreviated as e. This charge always appears associated with specific particles, notably the electron ($q = -e$) and the proton ($q = +e$), and all charges are multiples of this fundamental quantity. In this chapter we shall be introduced to the evidence for e and to some of the electrical phenomena in matter that are consequences of the quantization of charge. Finally, we shall glance at some of the basic considerations involved in the flow of electric current.

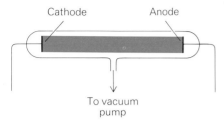

11.1 A cathode-ray discharge tube. *A colored discharge occurs as the air pressure in the tube is reduced. The negative electrode is called the cathode, and the positive electrode is called the anode.*

THE ELECTRON

cathode rays

In the repertory of physics demonstrations, few experiments are more spectacular than an exhibition of cathode rays. On the lecture table is a long glass tube, with a small metal disk sealed into each end (Fig. 11.1). A side tube is connected with a vacuum pump. The room is darkened, and the experimenter throws a switch, giving one metal disk a strong positive charge, the other a strong negative charge. Nothing is visible yet except a faint glow at each end of the tube, since too much air separates the charges for a spark to jump between them. Now the vacuum pump is turned on, reducing slowly the air pressure in the tube. Suddenly a bright discharge leaps between the metal disks, like a long spark but wider and more diffuse, at first branching, irregular, dancing from side to side. Widening as the air pressure falls, the discharge becomes a wavering, purplish column filling the tube completely. Now the pressure has dropped to less than 1 mm of mercury (less than a thousandth of normal atmospheric pressure): the purple column breaks near one end, the dark gap grows wider, and soon the tube is once more nearly invisible. Still the pump pulls traces of air from the tube, reducing its pressure to 0.001 mm of mercury or less. Now at last, on the walls of the tube opposite the negative disk, appears a greenish glow, faint at first, but spreading and brightening until much of the glass is softly luminous. The interior of the tube remains dark, but the green radiance of the glass shows that the tube is filled with the invisible and once-mysterious *cathode rays*.

Electric discharges in tubes of this sort attracted the serious attention of physicists in the 1870s, but not until 1897 was the green glow of the glass at very low pressures explained. Discovery of the true nature of cathode rays we owe to many workers, in this country and in Europe, but one name stands out above the rest—that of the great English physicist, J. J. Thomson. With Thomson's work on cathode rays begins that extraordinary scientific development we call *modern physics*.

an anode is a positive electrode, a cathode is a negative electrode

The two metal disks of the evacuated tube are called its *electrodes*—a general term applied to any pair of adjacent conductors on which unlike charges are maintained. The negative electrode is named the *cathode*, the positive one the *anode*. Cathode rays are so named because they stream from the negative metal plate. We can demonstrate that they originate in the cathode by using a tube equipped with an opaque object that can be set before the negative electrode; the object is silhouetted sharply against the green glow at the far end of the tube (Fig. 11.2), showing that the invisible rays in this direction have been blocked off.

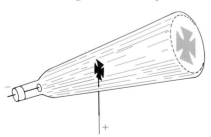

11.2 A cathode-ray discharge tube with an obstruction in the path of the rays. *This experiment demonstrates that the rays originate in the cathode, or negative terminal, of the tube.*

the electron

Opinion was divided in the early days of cathode-ray experimentation, one group maintaining that the radiation was a form of invisible light, the other group that the radiation consisted of material particles. Experiments designed to settle the issue brought out many properties of the mysterious rays. Under their influence not only glass but many other substances glow softly, or *fluoresce.* Experiments showing that obstacles in the path of the rays produce sharp shadows proved that the radiation travels approximately in straight lines. The rays can be focused on a point by using a cathode with a concave surface, somewhat as light is focused by a concave mirror. A piece of platinum placed at the point where the rays converge quickly becomes red-hot, proving that the cathode radiation carries considerable energy. All these properties could be explained satisfactorily either by supposing that the rays consisted of particles or by supposing that they were akin to ordinary light.

But other experiments favored the particle hypothesis. Most convincing were experiments showing that the rays carry an electric charge. One way to show this is to make a tube with a second pair of metal plates sealed into its side, so that the cathode rays must pass between them (Fig. 11.3). When these plates are given opposite charges, the greenish fluorescence caused by the rays moves from its position at the end of the tube to a spot on the side, as if the rays had been deflected. The direction of the deflection indicates that the rays are repelled by the negative charge and attracted by the positive charge—in other words, that the rays themselves are negatively charged. No radiation like light has ever been found to carry a charge, so this property is good evidence that cathode rays consist of moving particles. These tiny particles, each possessing a minute charge of negative electricity, are *electrons.*

The chief properties of the electron are as follows:

1. The electron is exceedingly small. Its mass is 9.1×10^{-31} kg, which is roughly $\frac{1}{1,800}$ of the mass of a hydrogen atom.
2. The charge of the electron is 1.60×10^{-19} coul. Small as it is, this charge is enormous for so minute an object as an electron. A gram of electrons placed 1 cm away from another gram of electrons would be repelled with a force of over 10^{26} tons!
3. Because they are so light and so highly charged, it is easy to accelerate electrons to very high speeds. In almost all devices employing electron streams (for instance, television picture tubes) the electron velocities are very close to the velocity of light—186,000 mi/sec.

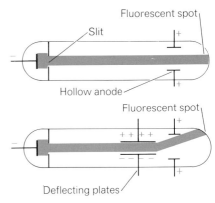

11.3 Cathode rays consist of particles. *A beam of cathode rays is repelled by a negative charge and attracted by a positive one, evidence that these rays consist of streams of negatively charged particles.*

cathode rays are streams of electrons

electron properties

discovery of the electron

The early years of the twentieth century saw an extensive program of careful experiments to measure the properties of the electron. In England, J. J. Thomson was able to determine accurately the ratio q/m between the charge and mass of an individual electron by observing the deflection of a beam of electrons in an ingenious combination of electric and magnetic fields. His figure was

$$\frac{q}{m} = 1.76 \times 10^{11} \text{ coul/kg}$$

This information firmly established the particle nature of cathode rays, since electrons have mass, but it did not, by itself, demonstrate that all electrons are alike.

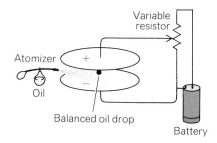

11.4 The Millikan oil-drop experiment. *The variable resistor makes it possible to vary the electric field between the plates.*

The final step in the "discovery" of the electron was taken by an American, Robert A. Millikan, in his famous oil-drop experiment. As shown in Fig. 11.4, Millikan sprayed a mist of fine oil drops between a pair of electrodes. Friction between the drops and the atomizer nozzle caused many of the drops to become electrically charged. The electrodes were connected to a battery, which charged them oppositely to produce a uniform electric field E. Then, by adjusting the magnitude and direction of this field, the upward electric force

$$F = qE$$

that it exerts on an oil drop carrying the charge q could be made to precisely counterbalance the weight of the drop,

$$W = Mg$$

When this happened, that particular drop was stationary, while others would drift up or down, depending upon their mass and charge. For the stationary drop,

$$W = F$$
$$Mg = qE$$

and the charge of the drop is

$$q = \frac{Mg}{E}$$

Since the mass M of the drop could be determined by independent means and both g and E were known, q could be found.

Millikan obtained several different values for q, but in every case they were multiples of the single charge

$$e = 1.60 \times 10^{-19} \text{ coul}$$

and no smaller charge has ever been found. Therefore e is considered the basic unit of electric charge.

Since Thomson had shown that the charge-to-mass ratio q/m of electrons is always 1.76×10^{11} coul/kg, the constancy of e meant that all electrons are indeed identical. The mass of the electron is, since $q = e$,

$$m = \frac{e}{q/m} = \frac{1.60 \times 10^{-19} \text{ coul}}{1.76 \times 10^{11} \text{ coul/kg}}$$
$$= 9.1 \times 10^{-31} \text{ kg}$$

ELECTRICAL PROPERTIES OF MATTER

all matter contains electrons

All ordinary matter contains electrons, but most of the time it is electrically neutral. This suggests that the negative charge of the electrons is balanced by positive charges somewhere within the atoms and molecules.

ELECTRICITY AND MATTER

It is possible to isolate the particles that carry the positive charges, to set them free as electrons are set free in a cathode-ray tube, but the experiments are much more difficult because the positively charged particles are very tightly bound together. The properties of the positively charged particles (which are called protons) will occupy us later on (Chap. 16), and for the moment we shall concentrate on the behavior of electrons in matter.

ions and electrons

A solid conductor, in terms of electrons, is a substance whose electrons are relatively free to move; an insulator is a substance whose electrons are firmly bound in their atoms. If a copper wire is placed between a negatively charged sphere and an electroscope, some of the movable electrons in the wire, repelled by the negative charge, move into the electroscope and give it a negative charge. Some of the excess electrons on the sphere, so to speak, crowd into one end of the wire, pushing before them the movable electrons of the wire so that some at the other end spill over onto the electroscope. If silk thread is used in place of wire,

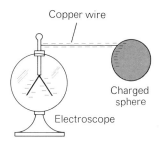

11.5 Electric current is a flow of electrons. *Substances whose electrons are relatively free to move are conductors, while those whose electrons are tightly bound in place are insulators. Copper, like other metals, is a good conductor.*

electrons can move freely in a solid conductor, whereas they are bound firmly to the atoms of an insulator

A dense beam of electrons is used in a vacuum chamber for the precision machining of metals. The concentrated energy of the electron beam causes the metal to vaporize where the beam strikes it.

all solid conductors are metals

its electrons, although repelled by the charge, are so firmly fixed in their atoms that almost none of them move to the electroscope. A positive charge on the sphere would draw electrons out of the wire or out of any piece of metal connected to it. Thus, electrons move easily along a metal away from a negative charge or toward a positive charge—which is what we mean by saying that all metals are conductors.

Conduction of electricity through gases and liquids—in a cathode-ray tube, a fluorescent light, or the acid of a storage battery—is somewhat more complicated. Electricity moves through most gases and liquids not as simple flow of electrons but, for the most part, as motions of charged atoms and molecules called *ions*. An atom or molecule gains a positive charge (becomes a positive ion) if it loses one or more electrons, and it gains a negative charge (becomes a negative ion) by attaching to itself electrons in excess of its normal number.

an ion is a charged atom or molecule

The process of forming ions, or *ionization*, may take place in a number of ways. A gas like ordinary air becomes ionized when X-rays, ultraviolet light, or radiation from a radioactive material passes through it, when an electric spark is produced, or even when a flame burns in it. Air molecules are sufficiently disturbed by these processes so that electrons are ripped off some of them; the electrons thus set free may attach themselves to adjacent molecules, so both positive and negative ions are formed (Fig. 11.6). Ordinarily the ions last no more than a few seconds, because movement of the gas molecules bring oppositely charged ions together and the electrons redistribute themselves to give neutral particles. Liquids, in contrast with gases, may contain positive and negative ions that persist indefinitely; the ions may be formed when the liquid itself is formed or when other substances are dissolved in it. We shall have much more to say about ions in liquids in Chap. 24.

electric current in gases and liquids involves the motion of ions

Electric charge can move through a gas or liquid (at ordinary pressures and temperatures and at moderate fields) only if ions are present. Then the flow of electricity is a movement of ions in two directions, positive ions moving toward the cathode and negative ions (plus electrons in the case of gases) toward the anode, as in Fig. 11.7. When no ions are present,

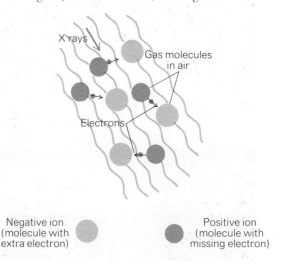

11.6 Ionization of air. *X-rays can disturb air molecules so that electrons are torn from some of them and become attached to others, which results in the formation of both positive and negative ions.*

gases and liquids are excellent insulators, as is shown by the fact that oppositely charged spheres may be placed close to each other in air without any discharge occurring. If the charges are made large enough, however, they are capable themselves of ionizing some of the gas molecules nearby; then a current is possible, and we say that a spark jumps between them. If the air pressure is reduced, as it is in a cathode-ray tube, ions can travel farther and the discharge can jump over a longer gap. The light produced in a spark or in the glowing column of a cathode-ray tube comes from atoms whose internal structures are disturbed when fast-moving ions collide with other molecules. Cathode rays themselves consist of electrons knocked out of the cathode when it is struck by positive ions.

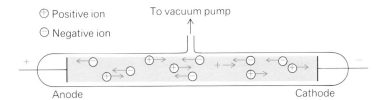

11.7 Conduction in a gas. *The flow of electricity in a gas consists of the motion of positive charges to the cathode and of negative charges to the anode.*

The simple experiments with which we began our study of electricity—stroking a rubber rod with fur or a glass rod with silk—probably involve movement of both ions and electrons. Contact between rubber and fur apparently permits electrons to move from the fur to the rubber, making the rod negative and leaving the fur positive. But both substances are poor conductors, so we would not expect electrons to move in great numbers. Probably the separation of charge is accomplished in part by a movement of ions, chiefly ions in films of water and air on their surfaces. Thus the complete explanation of these elementary experiments becomes very complicated. For most purposes, however, an explanation in terms of electrons is adequate. In discussions to follow, when we are considering movements of electricity in solids we shall speak only of moving electrons; for nonconductors this represents an oversimplification but not a serious one unless we are concerned with intimate details of the movement.

To summarize, all matter contains both electrons and positive charges. Flow of electricity in solids is largely a movement of electrons alone, ease of movement in a given solid determining how good a conductor it is. A negative charge on a solid object is a concentration of electrons in excess of the normal number its atoms contain; a positive charge is a deficiency of electrons. In most liquids and gases conduction of electricity takes place by the movement in two directions of much larger particles called ions, which are atoms or molecules containing either an excess or a deficiency of electrons. These generalizations have exceptions, for free electrons can move through some fluids and ions can move along the surface of some solids. But the rules apply to so many processes that we shall find them frequently useful, and they serve as a good illustration of how effectively the discovery of electrons has brought order into the study of electrical phenomena.

mechanisms of current flow

induction

We are now in a position to consider a question that we have managed to avoid thus far. One sign that a body possesses an electric charge is that it causes other objects to move toward it. A charged rubber rod readily attracts a pith ball; but the ball was not itself charged originally, and so it is not clear where the attractive force comes from that pulls it to the rod.

The explanation is not difficult if we think in terms of electrons. When the rod, given its usual negative charge by stroking with a piece of fur, is brought near the pith ball, electrons in the ball, repelled by the negative charge, move as far away as they can—which is to the far side of the ball (Fig. 11.8). The side of the ball near the rod is left with a positive charge, and the ball is accordingly attracted to the rubber. If the rod is removed without actually touching the ball, the disturbed electrons resume their normal positions and the ball is unchanged. But, if contact is made, some of the rod's electrons flow onto the ball, giving the ball as a whole a negative charge and causing the violent repulsion we have observed before.

The + and − charges on opposite sides of the pith ball, produced without actual contact with another charged object, are *induced charges*. The induced charges are temporary, disappearing as soon as the negative rod is removed. But a permanent charge may be fixed on the ball by induction, provided that the ball is grounded while the rubber rod is near (Fig. 11.9). Grounding the ball (say, by touching it with a finger) makes it and the earth temporarily part of a single huge conductor; in this conductor the negative rod drives electrons as far away as possible, that is, from the ball down through the body into the earth. If, now, the ground connection is broken while the rod is still held near by, the electrons have no means of returning to the ball, and the ball therefore has a positive charge that will remain when the rod is taken away.

In the same fashion a negative charge may be given to another pith ball: hold near it a positively charged glass rod; ground the ball, letting electrons attracted by the rod flow up from the ground into the ball; break the ground connection and remove the rod, thus leaving the excess electrons stranded on the ball. Note that the induced charge on the ball is always opposite in sign to the original charge.

In general, any object with a negative charge induces positive charges on all objects near it, simply because it repels away from itself the movable negative charges in these objects. Insulated objects near the negative charge will have a concentration of positive electricity on the side toward the charge, a concentration of negative electricity on the opposite side. Grounded objects will have lost some of their electrons to the earth. The amount of the induced charge on any object depends on the size and shape of the object and on its nearness to the negative charge. In a similar way a positive charge induces negative charges on objects in its immediate vicinity.

ELECTRIC CURRENT

Until now we have been concerned primarily with stationary electric charges. These charges sometimes moved from one place to another—from

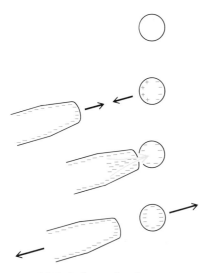

11.8 A charged rod attracts an uncharged object by first causing a separation of charge within the object. *If contact is made, charge flows into the object, and the two bodies then repel.*

the proximity of a charge induces a separation of charge in nearby objects

11.9 Charging an object by induction. *No contact is made between the charged rod and the object. When the latter is grounded momentarily, it assumes a charge of the opposite sign to that of the rod.*

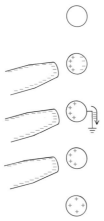

When a high potential difference (here 60,000 volts) is applied to an inadequately insulated electric cable, the surrounding air becomes ionized and a corona discharge occurs.

208   ELECTRICITY AND MAGNETISM

a rubber rod to a pith ball, along a wire to an electroscope—but our principal interest was in the causes and results of the motion of charges, not in the motion itself. Further, in the cases we considered, the motion took place in a very short period of time, nearly instantaneously as far as we were concerned. Static electricity is interesting enough in itself, but it is not the kind of electricity that lights our lamps, runs our motors, and permits us to talk to people many miles away. The electricity of everyday use is in the form of *electric current*—electric charges in continuous motion. There is no fundamental difference, of course, between a moving charge and a charge at rest, but, as we shall see, the fact that the former is in motion gives it properties not shared by its stationary twin.

an electric current consists of moving charges

circuits

We shall start with the simple type of current produced when a wire is connected between the terminals of a battery. A battery is a device for maintaining a positive charge on one terminal and a negative charge on the other. This is accomplished by a chemical reaction in the battery, with chemical energy being continuously transformed into electric energy whenever the battery is in use. Connecting a wire between the terminals provides a path for the excess electrons at the negative electrode to move toward the positive electrode. The flow of electrons tends to cancel out the two charges, but chemical processes in the battery build up the charges as fast as they are depleted. Thus the current in the wire consists of a movement of electrons from one end to the other. Note that we do not say, "The electrons carry the current," or "The motion of electrons produces a current"; the moving electrons *are* the current.

Suppose the wire is cut and its free ends are attached to a switch; then by closing and opening the switch we can start or stop the current at will. The current flows whenever a continuous conducting path is provided between the electrodes and stops whenever the path is broken. When the conducting path is complete (that is, when the switch is closed), we say the circuit is *closed*; when the path is interrupted, we say the circuit is *open* (Fig. 11.10).

Given a battery, a switch, and wires of different sizes and materials, let us examine electric currents in some detail. First of all, how do we know when there is an electric current present? A wire looks the same whether the circuit is closed or open; how can we tell whether electrons are moving along it?

One way in which a current may betray its presence is by the spark produced if a small air gap is left somewhere in the circuit—for instance, if the wire is detached from one terminal and its end held near but not touching the terminal. A large gap would break the circuit, but a spark will jump readily across a small air space.

Another easily detectable effect of the passage of electrons is the heating produced. The amount of heating for a given flow of electrons depends greatly on the wire used: small wires become hotter than large ones; wires of certain metals, like tungsten and iron, become much hotter than silver

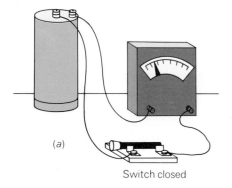

(a) Switch closed

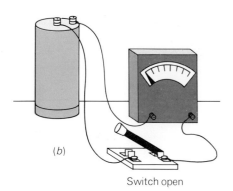

(b) Switch open

11.10 *A closed circuit is one in which there is a complete conducting path; an open circuit is one in which there is an interruption in the conducting path.*

heat is produced when a current flows through a material

or copper wires of similar size. The heating is explained by the resistance the wires offer to the movement of electrons along them—much as the heat due to friction is explained by resistance to mechanical movement. We take advantage of the heating effect of currents in a great many common electrical devices. Electric-light bulbs are an example, with a slender tungsten filament made hot enough to emit light.

the ampere

Every conductor exhibits a certain amount of resistance to the flow of electric current. Before we can examine the way in which the current in an electric circuit is affected by the resistance present, we must consider current itself from a quantitative point of view.

In a number of ways the flow of electricity along a wire is analogous to the flow of water in a pipe. The analogy is often helpful because we can visualize water much more readily than electrons. When we wish to describe the rate at which water is moving through a pipe, we might express the flow in terms of, say, gallons per second. If 1.5 gal of water pass through a given pipe each second, the flow is 1.5 gal/sec. The quantitative description of electric current is very similar. Quantity of electric charge is measured in coulombs, as quantity of water is measured in gallons, and a natural way of referring to the flow of charge in a wire is in terms of the number of coulombs per second going past a given point in the wire, just as a natural way of referring to the flow of water is in terms of gallons per second. This unit of electric current is called the *ampere* (amp), after the French physicist André Marie Ampère. That is,

1 amp = 1 coul/sec

the ampere is the unit of electric current

potential difference

Suppose that a gallon of water is poised at the brink of a waterfall. We say that it possesses here a certain potential energy, since it is capable of moving under gravitational attraction. When it drops to the base of the fall, its potential energy decreases. The work obtainable from the gallon of water during its fall is equal to this decrease in potential energy.

Now consider a coulomb of negative charge on the negative terminal of a battery. It is acted upon by both the repulsion of adjacent negative charges and the attraction of charges on the other terminal. We say, therefore, that it possesses a certain potential energy by reason of its position on the negative electrode. When it has moved along a wire to the positive terminal, its potential energy is smaller, since here it can no longer move spontaneously. The amount of work the coulomb can perform in flowing from the negative to the positive terminal is measured by its decrease in potential energy.

The decrease in its potential energy brought about by the motion of 1 coul from the negative to the positive terminal is a quantity called the *potential difference* between the two terminals. It is a quantity analogous to difference of elevation in the case of water (Fig. 11.11). The potential difference between two points is thus equal to the corresponding energy difference per unit charge. We measure difference of elevation in feet;

potential difference is electrical potential energy per unit charge

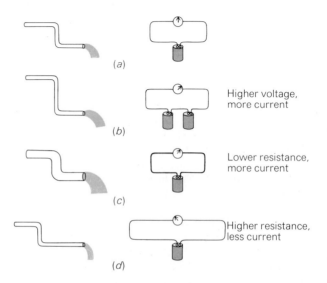

11.11 The flow of electric current in a wire is analogous in many ways to the flow of water in a pipe. Thus at (b) having the water fall through a greater height than at (a) yields a greater flow of water, which corresponds to using two batteries to obtain a higher potential difference and thus a greater flow of current. At (c) a larger pipe yields a greater flow of water, which corresponds to using a larger wire that offers less resistance to the flow of current. At (d) a longer pipe yields a smaller flow of water, which corresponds to using a longer wire that offers more resistance to the flow of current.

we measure difference of potential in *volts* (named for the Italian physicist Alessandro Volta). When 1 coul of charge travels through 1 volt of potential difference, the work that it does is equal to 1 joule by definition:

the volt is the unit of potential difference

1 volt = 1 joule/coul

(As we recall from Chap. 4, the joule is the unit of work and energy in the metric system.) Because the volt is defined in this way, it is easy to make transitions from mechanical to electrical quantities.

The maximum difference of potential between the terminals of an automobile storage battery is about 12 volts, of a dry cell about 1.5 volts. Every coulomb of electricity at the negative terminal of the storage battery, therefore, is capable of doing eight times as much work as a coulomb at the negative electrode of a dry cell—just as a gallon of water at the brink of a waterfall 120 ft high is capable of doing eight times as much work as a gallon at the brink of a 15-ft fall. If a storage battery and a dry cell are connected in exactly similar circuits, the battery will push eight times as many electrons around its circuit in a given time as the dry cell, giving a current (in amperes) eight times as great. Very crudely, we may speak of the potential difference between two points as the amount of "push" effective in moving charges between the points.

the electron volt

the electron volt is a unit of energy

A widely used unit of energy in atomic physics is the *electron volt*, which is the amount of energy gained by an electron that is accelerated by a

A Cockroft-Walton generator at Brookhaven National Laboratory accelerates protons to energies of 750,000 ev. These protons are then further accelerated to 50 Mev in a linear accelerator before being injected into the 33-Bev alternating gradient synchrotron.

potential difference of 1 volt. The abbreviation for electron volt is *ev*. Since the charge on the electron is 1.6×10^{-19} coul,

$$1 \text{ ev} = 1.6 \times 10^{-19} \text{ coul} \times 1 \text{ volt}$$
$$= 1.6 \times 10^{-19} \text{ joule}$$

The electron volt is a very small unit, and frequently its multiples the *Mev* and the *Gev* (sometimes *Bev*) are used, where

$$1 \text{ Mev} = 1 \text{ million ev} = 10^6 \text{ ev} = 1.6 \times 10^{-13} \text{ joule}$$
$$1 \text{ Gev} = 1 \text{ billion ev} = 10^9 \text{ ev} = 1.6 \times 10^{-10} \text{ joule}$$

mev and gev

The M and G respectively stand for the prefixes mega ($=10^6$) and giga ($=10^9$); other examples of their use are the megabuck and the gigawatt.

Some examples of energies normally quoted in electron volts are the following:

The energy required to separate a hydrogen atom into its constituent electron and proton is 13.6 ev.

A hydrogen molecule consists of two hydrogen atoms bound together by forces whose nature we shall investigate in Chap. 20. The energy required to separate a hydrogen molecule into two hydrogen atoms is 4.5 ev.

The energy liberated when the nucleus (central core) of a uranium atom is split into two fragments, a process called nuclear fission, averages about 200 Mev.

A *proton synchrotron* is a device for accelerating protons to extremely high energies for research into the properties of elementary particles. The most powerful synchrotron currently operating in the United States is at Brookhaven National Laboratory near New York and produces 33-Gev protons. A 70-Gev synchrotron is in operation in Serpukhov in Russia, and work is about to start on a 200-Gev synchrotron in Illinois.

ohm's law

If a circuit is set up so that different voltages can be applied to the ends of the same piece of wire and if the temperature of the wire is maintained constant, it is found experimentally that the current flowing in the wire is proportional to the potential difference. The greater the voltage of the battery across the wire, the more current flows. This generalization is called *Ohm's law* after its discoverer, the German physicist Georg Simon Ohm (1787–1854).

the resistance of a conductor is a measure of its opposition to the flow of current

Evidently there is some property of the wire that opposes the flow of current. The more current we want, the more potential difference we must apply. This property is called *resistance*, and, as we have said, it is somewhat similar to friction. A large current requires a large potential difference; moving a heavy box across a room requires more force than moving a light box.

According to Ohm's law, the current in a circuit, usually denoted by the letter i, is proportional to the potential difference, denoted by V, across the circuit. Clearly, the more the resistance, the less the current, so that, if we write R for resistance, Ohm's law may be expressed in the form

ohm's law $i = \dfrac{V}{R}$

When i is in amperes and V in volts, the unit of R is the *ohm*. That is,

the ohm is the unit of electrical resistance $1 \text{ amp} = \dfrac{1 \text{ volt}}{1 \text{ ohm}}$

The great usefulness of Ohm's law is that, once we know the resistance R of a given electric circuit, we can determine immediately how much potential difference V is needed to cause a given amount of current i

to flow. Or, given the potential difference and resistance, we can compute the current.

Let us see how Ohm's law can be applied in a practical situation. A typical automobile has a 12-volt battery whose capacity is 60 ampere-hours (amp-hr)—which means that it can provide a current of 60 amp for 1 hr, or 6 amp for 10 hr, or 1 amp for 60 hr, etc., before it becomes "dead." How long can such an automobile have its headlights, of total resistance 4 ohms, left on without the engine running (which would recharge the battery) before the battery runs down? To solve the problem, we first calculate the current (Fig. 11.12), which is

$$i = \frac{V}{R} = \frac{12 \text{ volts}}{4 \text{ ohms}} = 3 \text{ amp}$$

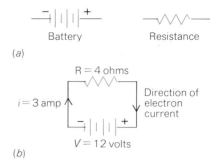

11.12. (a) Conventional symbols for a battery and a resistance. (b) *A current of 3 amp flows in a circuit whose resistance is 4 ohms when a potential difference of 12 volts is applied. The current consists of a flow of electrons, so its direction is from the − terminal of the battery to the + terminal.*

Since the battery's capacity is 60 amp-hr, the lights can be left on for 20 hr before the battery runs down.

Ohm's law can also be used to find the resistance of an electric appliance when its voltage and current ratings are known. An electric toaster draws a current of 4 amp when it is plugged into a 120-volt supply line. To find its resistance, we rewrite Ohm's law in the form $R = V/i$ and substitute the values given:

$$R = \frac{V}{i} = \frac{120 \text{ volts}}{4 \text{ amp}} = 30 \text{ ohms}$$

The resistance of the toaster is 30 ohms.

Despite its name, Ohm's law is not a basic physical principle such as the law of conservation of energy. Ohm's law is actually a relationship that is obeyed only by metallic conductors, but not by gaseous or liquid conductors and not by such electronic devices as transistors and vacuum tubes.

electric power

Electric energy is so very useful because it is readily converted into other kinds of energy. Electric energy in the form of electric current is converted into radiant energy in a light bulb, into chemical energy when a storage battery is charged, into mechanical energy in an electric motor, into heat energy in an electric oven. In each case the current performs work on whichever device it passes through, and the device then transforms this work into another kind of energy. A very important quantity in any discussion of electric current, therefore, is the rate at which a current is doing work—in other words, the *power* of the current.

electric power is the rate at which an electric current performs work

Earlier in this chapter we learned that, when 1 coul of charge is pushed through a circuit by a potential difference of 1 volt, the amount of work done is equal to 1 joule. In general, then,

Electrical work = charge × potential difference

$$W = qV$$

where W is expressed in joules, q in coulombs, and V in volts. Power P is defined as the rate at which work is being done; so

$$\text{Power (watts)} = \frac{\text{work (joules)}}{\text{time (sec)}}$$

$$P = \frac{W}{t}$$

If we substitute for W its electrical equivalent qV, we find that

$$P = \frac{W}{t} = \frac{qV}{t}$$

But, by definition,

$$\text{Current (amp)} = \frac{\text{charge (coul)}}{\text{time (sec)}}$$

or, in symbols,

$$i = \frac{q}{t}$$

This means that we are able to express electrical power as

electric power

$$P = \frac{q}{t} V = iV$$

Power (watts) = current (amp) × potential difference (volts)

Now we can see why nearly all electric appliances are rated in terms of watts: this designation expresses the rate at which the appliance uses electric energy. A 60-watt light bulb requires twice the power of a 30-watt light bulb, and one-tenth the power of a 600-watt electric fan. Because

$$P = iV$$

we can readily determine the current requirements of an appliance that is rated in watts when it is connected across the power main whose voltage is specified. A 60-watt bulb connected across a 120-volt power line needs a current of

$$i = \frac{P}{V} = \frac{60 \text{ watts}}{120 \text{ volts}} = \frac{1}{2} \text{ amp}$$

fuses and circuit breakers

Electric fuses and circuit breakers are designed to protect a power line by opening the circuit whenever an unsafe amount of current passes through them. The fuses used in homes are often rated at 15 amp. Since the power-line voltage is 120 volts, the greatest power that the line can provide without blowing the fuse is

$P = iV = 15 \text{ amp} \times 120 \text{ volts} = 1{,}800 \text{ watts}$

Users of electricity pay for the quantity of energy they consume. The usual commercial unit of electric energy is the *kilowatt-hour* (kw-hr), which is the energy supplied per hour when the power level is 1 kilowatt (1 kilowatt = 1 kw = 1,000 watts). Thus, if electricity is sold at 0.4 cent per kilowatt-hour, the cost of operating a 1,500-watt (1.5 kw) electric heater for 7 hr would be

the kilowatt-hour is a unit of energy

Cost = price per unit of energy × energy used
  = (0.4 cent/kw-hr) × (1.5 kw × 7 hr)
  = 4.2 cents

GLOSSARY

An *electron* is a tiny, negatively charged particle found in matter. *Cathode rays* are streams of electrons produced in an evacuated tube when a strong electric field is set up in it.

An *ion* is an atom or molecule possessing an electric charge. The process of forming ions is called *ionization*.

Electric charges in motion constitute an *electric current*. The unit of electric current is the *ampere*, equal to a flow of 1 coul/sec.

The *potential difference* between two points in an electric circuit is the amount of potential energy lost by 1 coul of charge flowing from one point to the other. The unit of potential difference is the *volt*, equal to a potential energy of 1 joule/coul of charge.

The *electron volt* is a unit of energy equal to 1.6×10^{-19} joule, which is the amount of energy acquired by an electron accelerated by a potential difference of 1 volt. A *Mev* is 10^6 ev and a *Gev* is 10^9 ev.

Ohm's law states that the current in a circuit is equal to the potential difference between the ends of the circuit divided by the *resistance* of the circuit, symbolically, $i = V/R$. Resistance is expressed in units called *ohms*.

Electric power is the rate at which an electric current is doing work. The unit of power is the *watt*, equal to 1 joule/sec.

EXERCISES

1 How could you charge an electroscope negatively with only the help of a positively charged object?

2 Name five good conductors of electricity and five good insulators. How well do these substances conduct heat? What general relationship between the ability to conduct heat and the ability to conduct electricity could you infer from this information?

3 Explain in terms of electrons why the production of electricity by friction always yields equal amounts of positive and negative electricity.

4 How is the movement of electricity through air different from its movement through a copper wire?

5 What properties of cathode rays demonstrate that they are streams of particles rather than a wave phenomenon like light?

6 Contact between road and tires often produces an appreciable electric charge in the body of an automobile or truck. Enough charge may accumulate to produce a spark. Trucks carrying gasoline or other flammable materials sometimes have metal chains attached to their frames which touch the road. How does this chain prevent dangerous sparks?

7 A person can be electrocuted while taking a bath if he touches a poorly insulated light switch. Why is the electric shock he receives under these conditions so much more dangerous than usual?

8 An electric appliance is sometimes said to "use up" electricity. What does it actually use in its operation?

9 Explain why bending a wire does not affect its electrical resistance, even though a bent pipe offers more resistance to the flow of water than a straight one.

10 A fuse prevents more than a certain amount of current from flowing in a particular circuit. What might happen if too much current were to flow? What determines how much is too much?

11 If a 75-watt light bulb is connected to a 120-volt power line, how much current flows through it? What is the resistance of the bulb? How much power does the bulb consume?

12 How many coulombs of electric current pass through an electric appliance in 20 min if the current through the appliance is 0.4 amp? If the potential difference across the appliance is 120 volts, how much power does it consume? How much energy in joules does it draw from the circuit in 20 min?

13 If your home has a 120-volt power line, how much power in watts can you draw from the line before a 30-amp fuse will burn out? How many 100-watt light bulbs can you put in the circuit before the fuse will burn out?

ELECTRICITY AND MATTER

14 A hot-water heater employs a 2,000-watt resistance element. If all the heat from the resistance element is absorbed by the water in the heater, how much water per hour can be warmed from 10 to 70°C?

15 If electricity costs 3 cents/kw-hr, how much does it cost to warm a kilogram of water from 10 to 70°C?

16 How many electrons per second flow past a point in a wire carrying a current of 2 amp?

17 A power rating of 1 horsepower (hp) is equivalent to 746 watts. How much current does a $\frac{1}{4}$-hp electric motor require when it is operated at 120 volts?

18 Find the speed of an electron whose energy is 26 ev.

19 Find the energy (in electron volts) of an electron whose speed is 10^6 m/sec.

20 About 10^{10} electrons in each cm participate in carrying a 1-amp current in a typical wire. Find the average speed of these electrons.

21 Two resistors R_1 and R_2 are connected in series, which means that current flows consecutively from one to the other. Show that their combined resistance is $R_1 + R_2$ by making use of the fact that the same current flows through both of them and then adding together the potential differences iR_1 and iR_2 across each one.

22 Two resistors R_1 and R_2 are connected in parallel, which means that they are connected together at both terminals so that the current in any circuit they are part of is divided between them. Show that their combined resistance is $R_1 R_2 / (R_1 + R_2)$ by making use of the fact that the same potential difference exists across each resistor and then adding together the currents that flow through each one.

chapter 12

Magnetism

PERMANENT MAGNETS
- magnetic poles
- permanent magnets
- magnetic lines of force

THE MAGNETIC FIELD
- oersted's experiment
- the nature of magnetism
- configurations of currents

MAGNETIC FORCES
- magnetic forces on currents
- galvanometers and motors

ELECTROMAGNETIC INDUCTION
- the dynamo effect
- the transformer

Although it is not immediately obvious, there is an intimate relationship between electricity and magnetism. By the middle of the nineteenth century the chief features of this relationship were understood, and they not only furnished the basis for the industrial utilization of electricity, but also led directly to the discovery of the electromagnetic nature of light. Before we begin to explore the connection between electricity and magnetism, which is one of the most significant in all of physics, we shall inquire into magnetism and magnetic phenomena in order to have a background from which to work.

PERMANENT MAGNETS

magnetic poles

Ordinary magnets are familiar to everybody. The simplest kind consists of a bar of iron that has been magnetized in one of a variety of ways,

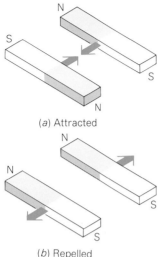

(a) Attracted

(b) Repelled

12.1 Like magnetic poles repel each other; unlike magnetic poles attract.

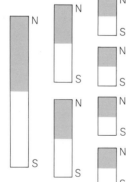

12.2 Cutting a magnet in half produces two other magnets. *There is no such thing as a single free magnetic pole.*

say by having been stroked by another magnet. A magnetized bar of iron is recognized, of course, by its ability to attract and hold other pieces of iron. Another property, illustrated by the familiar compass needle, is its tendency, when freely suspended, to turn so that one end points north and the other south. We call the north-pointing end the *north pole* of the magnet, and the south-pointing end the *south pole*. Here, near the ends, the greater part of the magnetization is concentrated, as can easily be shown by testing the attraction of various parts of the bar for small iron nails.

If two magnets are brought near to each other, the two poles are found to behave quite differently. Laid end to end so that the two north poles are near together, the magnets repel each other; if a north pole is brought near a south pole the two attract each other (Fig. 12.1). We may formulate a simple rule analogous to that for electric charges: **Like magnetic poles repel each other; unlike poles attract.**

It would be convenient for experimental purposes if a single north pole could be isolated unencumbered by the south pole at the other end of the magnet. Seemingly, isolation of the north pole should not be difficult; we need only saw the magnet in half. Unfortunately this method will not work. The resulting half magnets have each a north pole and a south pole, two new poles appearing where the middle of the former magnet was, as shown in Fig. 12.2. We may cut the resulting magnets in two again, with the same results, and continue as long as we have tools fine enough for the cutting; still each magnet that we prepare, however small, will have a north pole and a south pole. **There is no such thing as a single free magnetic pole.**

permanent magnets

Since a magnet can be cut into smaller and smaller fragments indefinitely, each fragment acting as a small magnet, we may reasonably assume that magnetism is a property of the smallest particles of a substance. Thus each particle of iron (or any other material that can be magnetized) behaves as if it had a north pole and a south pole. In ordinary iron the particles are haphazardly arranged, and adjacent north and south poles neutralize each other's effect. When the iron is magnetized, we imagine that many or all of the particles are aligned with their north poles in the same direction, so that the strengths of all the tiny magnets are added together (Fig. 12.3).

The ultimate magnetic particles in iron have been identified as the iron atoms themselves. The magnetic properties of an atom are determined by the behavior within it of the electrons it contains, since *all* magnetic fields have their origins in moving charges. The units that line up when a bar of iron is magnetized are usually microscopic "domains" containing many atoms in more or less uniform alignment. The metal iron is distinguished from most other substances by this tendency of its atoms to act cooperatively, which permits iron objects to be strongly magnetized. A "permanent" magnet may be demagnetized by heating it strongly or by hammering it, both processes that agitate the atoms and restore them to their normal state of random orientation.

Certain alloys (mixtures) of other metals are also strongly magnetic; the metals cobalt and nickel, certain compounds of iron, and liquid oxygen are less magnetic. With delicate instruments, in fact, all substances can be shown to have magnetic properties. Some are attracted to a magnet, as iron is; most are very slightly repelled.

Let us try an experiment with magnets analogous to the electrical experiment with conductors and insulators. Place a magnet a foot or so from an unmagnetized piece of iron and connect the two by various materials. No substance will be found to carry the magnetism from the original magnet to the unmagnetized iron except a third piece of iron (or a magnetic alloy). If iron is used for the connection, the unmagnetized piece becomes temporarily able to attract small nails or filings, but all or nearly all of its magnetism vanishes when the connection is severed. Meanwhile the strength of the original magnet remains practically unimpaired. That is, when a magnet touches a piece of iron, its magnetism in effect extends along the iron but is not conducted permanently through it in the sense in which an electric charge is carried through a conductor.

iron is not the only substance with magnetic properties

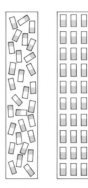

12.3 *In an unmagnetized iron bar the tiny magnetic domains have random field directions, while in a magnetized bar their field directions are aligned.*

magnetic lines of force

Since magnets can exert forces on one another when the magnets themselves are not in actual contact, we can speak of the *magnetic field* of a single magnet even though there is no other magnet nearby for it to

magnetic field

The configuration of the magnetic field around a bar magnet and a nearby key is revealed by the pattern of iron filings shaken on a card. It is convenient to think of the pattern in terms of "lines of force," but such lines do not actually exist since the field is a continuous property of the region of space it occupies.

attract or repel. As we know, it is often convenient to picture fields of force in terms of imaginary *lines of force* whose direction at any point is the same as the direction in which a test body would move if placed at that point. In the case of a gravitational field the test body is a particle; in the case of an electric field it is a small charge. Lines of force are also helpful in describing magnetic fields, particularly because they are easily exhibited with help of iron filings. An iron filing, just like a compass needle, tends to line up in a magnetic field along the direction of the field. If we scatter iron filings on a card placed over a magnet and shake the card slightly, the filings form into a pattern indicative of the configuration of the magnetic field, with the filings gathering most thickly where the field is most intense. Figure 12.4 shows the patterns obtained

iron filings and compass needles tend to line up in the direction of a magnetic field

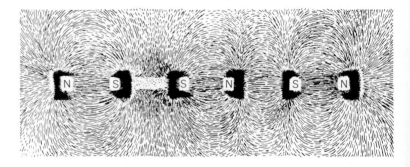

12.4 Patterns formed by iron filings sprinkled near three bar magnets. *The filings align themselves in the direction of the lines of force passing through them.*

near bar magnets in various combinations, and Fig. 12.5 shows the patterns for the same magnets obtained by calculation; they are essentially the same.

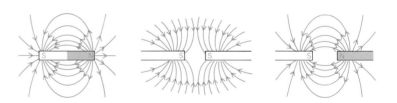

12.5 Lines of force around magnets. *These lines of force were obtained by calculation and may be compared with the patterns of iron filings near bar magnets.*

THE MAGNETIC FIELD

oersted's experiment

Perhaps unfamiliar to us as sources of magnetic field are electric currents, yet every current is surrounded by such a field. To repeat a famous experiment first performed in 1820 by the Danish physicist Hans Christian Oersted, let us connect a horizontal wire to a battery and hold beneath the wire a small compass needle (Fig. 12.6); the needle swings into a position at right angles to the wire. When the compass is placed just above the wire, the needle swings completely around until it is again perpendicular to the wire but pointing in the opposite direction.

MAGNETISM

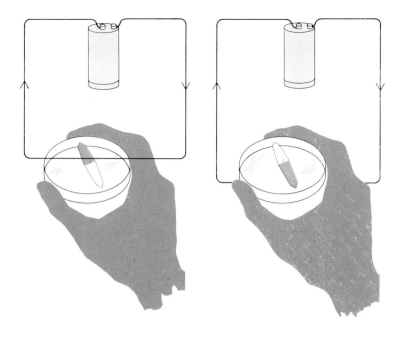

12.6 Oersted's experiment, showing that a magnetic field surrounds every electric current. *The field direction above the wire is opposite to that below the wire.*

We can use iron filings to determine the magnetic field pattern around a wire carrying a current. When we do this, we find that the lines of force near the wire consist of concentric circles (Fig. 12.7). The direction of the lines of force (that is, the direction in which the north pole of the compass points) depends on the direction of flow of electrons through the wire; when one is reversed, the other reverses also.

In general, the direction of the magnetic field around a wire can be found by encircling the wire with the fingers of the *left hand*, so that the extended thumb points along the wire *in the direction in which the electrons move*; the fingers then point in the direction of the field. (Engineers traditionally think in terms of a hypothetical positive current that flows from the + terminal of a battery to its − terminal, the opposite of electron current. In terms of a positive current, the above rule becomes a right-hand rule instead. Either way, the current and the field are perpendicular to each other.)

left-hand rule for direction of magnetic field

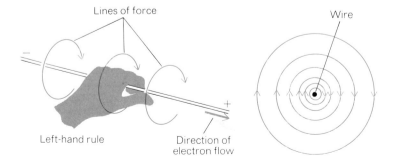

12.7 Magnetic lines of force around a wire carrying an electric current. *The direction of the lines may be found by placing the thumb of the left hand in the direction of electron flow; the curled fingers then point in the direction of the lines of force. In the right-hand diagram the electron current flows up from the paper.*

A magnetic field and the current that produces it are related in a rather involved way, and we shall not go into the details here. However, a few specific items are worth knowing. In Chap. 10 we learned that the electric field intensity **E** at a point is defined in terms of the force it exerts on a unit electric charge at that point. The corresponding definition of the *magnetic induction* (symbol **B**, a vector quantity) at a point is in terms of the force the magnetic field exerts on a unit length of wire carrying a unit current perpendicular to the field. The unit of magnetic induction is the *tesla*. (The tesla was formerly known as the *weber per square meter*.) In order to give an idea of the magnitude of this unit, we note that the magnetic induction of the earth's magnetic field averages about 3×10^{-5} tesla at sea level; the magnetic induction near a typical permanent magnet is about 1 tesla; and the strongest electromagnets can produce fields whose magnetic induction is as much as 100 teslas.

Oersted's discovery was the first positive proof that a connection exists between electricity and magnetism; it was also the first demonstration of the principle on which the electric motor is based. Magnetism and electricity are related, but magnetic poles and electric charges *at rest* have no effect on one another. This is an important fact. A magnet is completely uninfluenced by a stationary electric charge near it, and vice versa.

the nature of magnetism

When a current passes through a wire bent into a circle, the resulting magnetic field, shown in Fig. 12.8, is exactly the same as that produced by a bar magnet. One end of the loop acts as a north pole, the other as a south pole; if suitably suspended, the loop swings to a north-south position. A current loop attracts pieces of iron just as a bar magnet does. Indeed, the magnetic properties of iron and other substances can be traced to minute currents within their atoms. Thus it is correct to say that **all magnetic fields originate in electric currents.**

An electric charge at rest is surrounded by an electric field, and when it is in motion it is surrounded by a magnetic field as well. Suppose we travel alongside a moving charge, in the same direction and at the same speed. All we detect now is an electric field—the magnetic field has disappeared. Let us try another experiment in this vein and move past a stationary charge with our instruments. We find both an electric and

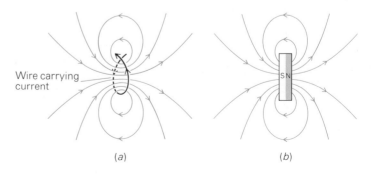

12.8 *The magnetic field of a loop of electric current is the same as that of a bar magnet.*

MAGNETISM

a magnetic field! Clearly the *relative motion* between charge and observer is crucial to the production of a magnetic field: no relative motion, no magnetic field.

The proper interpretation of what we perceive as separate electric and magnetic fields is that they are both manifestations of a single electromagnetic field that surrounds every electric charge. The electric field is always there, but the magnetic field only appears when relative motion is present. In the case of a wire that carries an electric current, there is only a magnetic field because the wire itself is electrically neutral. The electric field of the electrons is cancelled out by the opposite electric field of the positive ions in the wire, but the latter are at rest and therefore have no magnetic field to cancel the magnetic field of the moving electrons. If we simply move a wire that has no current flowing in it, the electric and magnetic fields of the electrons are cancelled by the electric and magnetic fields of the positive ions.

electric and magnetic fields are different aspects of a single electromagnetic field

In Chap. 5 we learned of some other curious effects that appear when measurements are made of the properties of things in motion relative to an observer: lengths are shorter, time intervals are longer, and masses are greater. These effects are predicted by the theory of relativity, and they actually occur, but they are only appreciable when the speed of the relative motion is near the speed of light (3×10^8 m/sec, which is 186,000 mi/sec). In magnetism, however, we have a relativistic phenomenon which occurs at much lower speeds; after all, the effective speeds of the electrons that carry current in a wire are only about 1 in./min. There are two reasons why magnetic effects are so conspicuous despite the low relative speed. First, electric forces are extremely strong (as we saw in Chap. 10, the electrical attraction between the electron and proton in a hydrogen atom is 10^{39} times stronger than the gravitational attraction between them), so even a small alteration in their character, which is one way to regard magnetic forces, may have appreciable consequences. Second, although the individual charges that produce a magnetic field usually move slowly, there may be so many of them that the result is substantial. For example, even a modest current in a wire may involve the motion of 10^{20} electrons in every centimeter of the wire.

magnetism is a relativistic effect

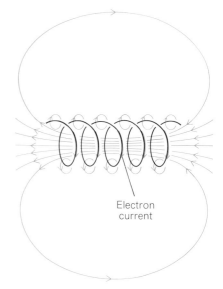

configurations of currents

When several wires all carrying currents in the same direction are placed side by side, as in Fig. 12.9, the magnetic fields of the individual wires add together to give a resultant magnetic field. The combined field that results is just like that of a single wire with a current in it equal to the sum of all of the separate currents. The effect is often employed to increase the magnetic field of a current loop. Instead of one loop many loops of wire are wound into a coil, and the resulting magnetic field is as many times stronger than the field of one turn as there are turns in the coil; a coil with 50 turns produces a field 50 times greater than a coil with just one turn.

12.9 *The magnetic field of a coil is like that of a single loop but is stronger.*

The magnetic strength of a coil of wire is enormously increased if a rod of soft iron is placed inside it (Fig. 12.10). This combination of coil

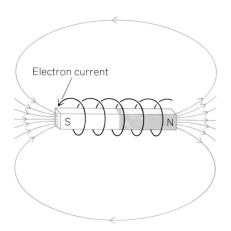

12.10 An electromagnet consists of a coil with an iron core, which considerably enhances the magnetic field produced.

and iron core is called an *electromagnet*. An electromagnet exerts magnetic force only when current flows through its turns, and so its action can be turned on and off. Also, by using many turns and a sufficient amount of current, it can be made far more powerful than a permanent magnet of similar size. These two properties make electromagnets among the most widely employed devices in the technical world. They range in size from the tiny coils in telephone receivers to the huge coils that load and unload scrap iron.

MAGNETIC FORCES

magnetic forces on currents

Suppose that a horizontal wire connected to a battery is suspended as in Fig. 12.11, so that it is free to move from side to side; and suppose

Adjustable electromagnets like this one are used in many laboratories. Here the magnetic behavior of a chemical compound is studied at low temperatures; the metal container is filled with liquid helium at −269°C to keep the sample cold.

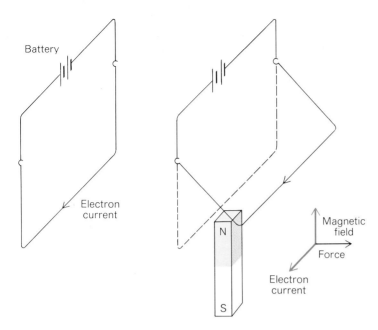

12.11 The motor effect. *The wire moves to one side in a direction perpendicular to both the magnetic field and the current.*

that the north pole of a strong bar magnet is placed directly beneath it. This arrangement is the reverse of Oersted's experiment: Oersted placed a movable magnet near a wire fixed in position, whereas here we have a movable wire near a fixed magnet. We might predict, from Oersted's results and Newton's third law of motion, that in this case the wire will move. It fulfills the prediction, jumping out quickly to one side as soon as the circuit is closed. The direction of its motion is perpendicular to the lines of force of the bar magnet's field. Whether it jumps to one side or the other depends on the direction of flow of electrons in the wire and on which pole of the magnet is used.

The push that a magnetic field exerts on a wire carrying a current can be called the *motor effect*, since the running of an electric motor depends on this force. We should keep in mind the nature of the force: it is not attraction or repulsion, but a *sidewise push*. The wire does not move toward the magnet or away from it, but perpendicular to its field.

The motor effect can be demonstrated with a cathode-ray discharge tube. In this tube (Fig. 12.12) an electric current is reduced to bare essentials: a flow of electrons, unencumbered by a wire, moving across a vacuum away from the cathode. By placing a fluorescent screen at the end of the tube, we can make the invisible electrons leave a trace of their movements as a bright line. If now we bring near the side of a tube so equipped one end of a bar magnet or place the tube between the poles of a horseshoe magnet, the line bends either upward or downward. That is, the swiftly moving electrons are pushed in a direction at right angles to the magnetic field and so are forced out of their accustomed straight path. Since the amount of deflection depends, among other things, on the speed, mass, and charge of the electrons, experiments of this sort are what gave physicists the basic data about these particles.

a magnetic field exerts a force on a current

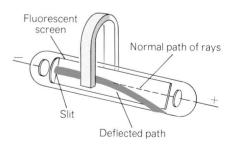

12.12 Deflection of cathode rays in a magnetic field. *The deflection is perpendicular to the direction of both the magnetic field and the electron movement.*

galvanometers and motors

Figure 12.13 shows a small coil suspended between the poles of a horseshoe magnet. The coil hangs limp until a current is passed through it; then it snaps into a position such that its north pole is as near as possible to the south pole of the horseshoe, its south pole near the north pole of the horseshoe. The spring suspending the coil resists somewhat the coil's turning, and the coil comes to rest at an intermediate position. How far the coil turns toward a direct alignment with the magnet depends on two things: the resistance of the spring to twisting, and the strength of the coil's north and south poles. The magnetic strength of the coil depends in turn on the amount of current passing through it. Hence, for a given supporting thread, the angle through which the coil turns is a measure of the strength of the electric current applied to it. Devices built so that this angle can be measured accurately are our most convenient instruments for detecting and measuring electric currents. They are called *galvanometers* (after the Italian scientist Luigi Galvani). The ordinary direct-current *voltmeters* and *ammeters* used by electricians are galvanometers of this type adapted for measuring potential difference and current, respectively.

A galvanometer has nearly all the essential parts of a simple direct-current electric motor. In a motor the coil must be supported on an axle instead of a sensitive spring, and some device must be used for changing the direction of current through the coil every time it aligns itself with the magnet. Changing the current reverses the poles of the coil, so that every time it swings into and a little past the position of alignment, it receives a new impulse to turn into the opposite position. Thus its motion becomes continuous. The device used for automatically changing the current direction is called a *commutator;* it may be seen on the shaft of a motor as a copper sleeve divided into two or more segments (Fig. 12.14).

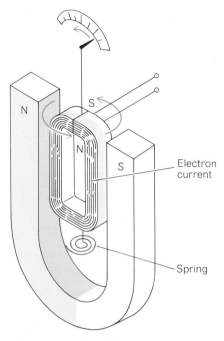

12.13 A simple galvanometer. *When a current passes through the coil, it rotates so that its north pole is close to the south pole of the magnet and its south pole close to the north pole of the magnet. The stronger the current, the farther the coil can turn against the torque of the spring.*

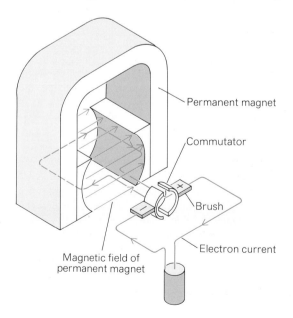

12.14 A simple direct-current electric motor. *The purpose of the commutator is to reverse the current in the loop periodically so that it always rotates in the same direction.*

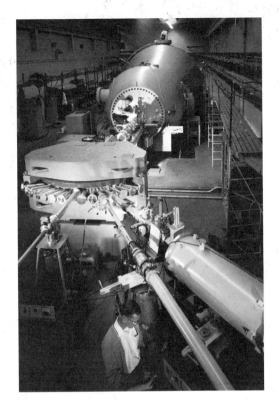

Van de Graaff particle accelerator uses a static electric field to produce a beam of high-speed ions. The magnet in the center of the picture, acting together with an electrostatic analyzer, channels particles of selected masses and speeds in different directions.

Practical electric motors are not built on quite so simple a pattern. There are many designs, adapted to different uses and different kinds of electric currents. Ordinarily electromagnets are employed rather than the simple permanent magnet of a galvanometer. In some motors the coil is fixed in position, and the magnet or magnets rotate within it. Motors built for alternating rather than direct current do not need commutators. But regardless of design, electric motors without exception utilize in their operation the force between a magnet and an electric current in its vicinity.

ELECTROMAGNETIC INDUCTION

Energy in any of its usual forms can be converted directly into electric energy. The battery is a device for changing chemical energy into electric energy. Instruments called *thermocouples*, used in the measurement of temperature, convert heat directly into electric energy. Even radiant energy can produce small electric currents, as we shall see later.

But the electric energy that is supplied so copiously to our homes and factories comes neither from light nor from heat nor from chemical reactions, but from mechanical energy. The great dynamos in power plants that supply electricity to cities are driven by water power or steam turbines; the boilers that supply the steam obtain heat from coal, oil, or

This giant set of coils provides the magnetic field in a modern generator. A large electric motor would appear nearly the same. In a generator, a current is induced in the coils of the turning armature, while in a motor, current in the coils of the armature causes it to turn.

natural gas, or from nuclear reactors (Chap. 16). Ships and isolated farms may have small generators operated by gasoline or diesel motors. In all cases the energy that is turned into electricity is the mechanical energy of moving machinery.

the dynamo effect

Commercial generators are based upon a very interesting principle. The story goes back to some famous observations by the English physicist Michael Faraday. Intrigued by the researches of Ampère and Oersted on the magnetic fields around electric currents, Faraday reasoned that, if a current can produce a magnetic field, then somehow a magnet should be capable of generating an electric current. Now a wire placed in a magnetic field and connected to a galvanometer shows no sign of a current. What Faraday discovered is that **if there is relative motion between a wire and a magnet, a current is produced.** As long as the wire continues to cut across lines of force in the magnetic field, the current persists; when the motion stops, the current stops. Because it is produced by motion in the presence of a magnetic field, without any direct contact with electric charges, this sort of current is called an *induced current*. The entire phenomenon is known as *electromagnetic induction*.

Let us repeat Faraday's experiment in very simple form. Suppose that the copper wire of Fig. 12.15 is moved back and forth across the lines of force of the bar magnet. The galvanometer will indicate a current flowing first in one direction, then in the other. Note that the wire is held approximately at right angles to the lines of force; thus the motion of electrons along the wire is at right angles to the magnetic field. The direction along the wire in which the induced current will flow depends on the direction of its motion and the direction of the lines of force; reverse the direction of motion, or use the opposite magnetic pole, and the current is reversed. The strength of the current depends on the rapidity of movement and on the strength of the field.

Electromagnetic induction, often called the *dynamo effect*, is related to the motor effect. A motor runs because electrons flowing along a wire are pushed sidewise in a magnetic field. In Faraday's experiment we again cause electrons to move through a magnetic field, but this time by moving the wire as a whole. The electrons are pushed sidewise as before and, in response to the push, move along the wire as an electric current.

To intensify the induced current produced by moving a conductor near a magnet, commercial dynamos employ a large coil rather than a single wire and several electromagnets instead of a bar magnet. Turned rapidly between the electromagnets by a steam or water turbine, the wires of the coil cut lines of force first in one direction, then in the other. Operation of the dynamo is illustrated in simplified form in Fig. 12.16, where a coil is shown turning between two magnets. Evidently, during half a revolution, each side of the coil cuts the field in one direction, then, during the other half revolution, cuts the field in the opposite direction. Hence the induced current flows alternately one way and the other. We call such a current an *alternating current*.

in electromagnetic induction, a current is induced in a wire moved across a magnetic field

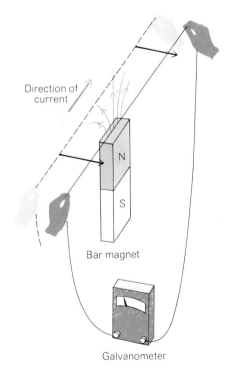

12.15 The dynamo effect. *The direction of the induced current is perpendicular both to the magnetic lines of force and to the direction in which the wire is moving. No current is induced when the wire is at rest.*

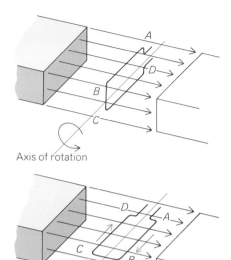

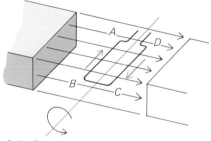

12.16 An alternating-current dynamo. *As the loop rotates, current is induced in it first in one direction (ABCD), and then in the other (DCBA). No current flows at those times when the loop is moving parallel to the magnetic field.*

Currents produced by batteries, thermocouples, and similar devices are one-way, or *direct*, currents, flowing steadily in one direction unless we arbitrarily change the connections. In alternating currents electrons move first one way, then the other—not from one end of the circuit to the other but back and forth over short distances. In the 60-cycle current that we ordinarily use in our homes, electrons change their direction of motion 120 times each second.

By using commutators similar to the commutators used on direct-current motors, dynamos can be constructed to produce direct current, but alternating current is usually produced because it is more economical for transmission over long distances.

the transformer

To generate an induced current requires that magnetic lines of force be made to move across a conductor. Two ways of accomplishing this motion have been mentioned: the wire may be moved past a magnet, or the wire may be held stationary while the magnet moves. We come now to a third, less obvious method, which involves no visible motion at all.

Let us connect coil A in Fig. 12.17 to a switch and a battery, and connect coil B to a galvanometer. When the switch is closed, a current flows through A, building up a magnetic field around it. The current and field do not start instantaneously; during a tiny fraction of a second the current increases from zero to its normal value, and its magnetic influence expands from a weak field close to the wire to a strong field perceptible at some distance. We may imagine the lines of force expanding and moving outward across the wires of coil B. This motion of the lines across coil B produces in it a momentary current, recorded by a sharp deflection of the galvanometer needle. Once the current in A reaches its normal, steady value, the field becomes stationary and the induced current in B stops.

Next, let us open the switch. Again in a small fraction of a second the current in A drops to zero, and its magnetic field collapses back around the coil. Once more lines of force cut across B, and the galvanometer responds with another deflection, this time in the opposite direction since the motion of the lines of force has changed. Thus starting and stopping the current in A has the same effect as moving a magnet in and out of B. An induced current is generated whenever the switch is opened or closed.

Suppose that A is connected not to a battery but to a 60-cycle alternating current. Now we need no switch; automatically, 120 times each second, the current comes to a complete stop and starts off again in the other direction. Its magnetic field expands and contracts at the same rate, and the lines of force cutting B first in one direction, then the other, induce an alternating current similar to that in A. An ordinary galvanometer will not respond to these rapid alternations, but an instrument built to measure alternating currents will show the induced current readily.

Thus an alternating current in one coil will produce an alternating current in a second coil, even though a considerable distance separates

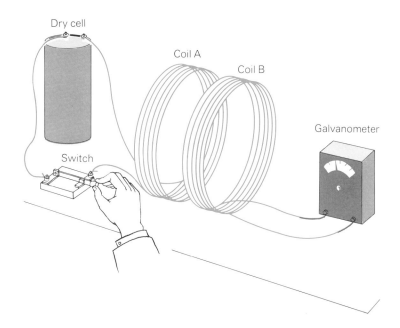

12.17 A simple transformer.
Momentary currents are detected by the galvanometer when the current in coil A is started or stopped.

them. However, to generate an induced current most efficiently, the two coils should be close together and wound on a core of soft iron. Such a combination of two coils and an iron core constitutes a *transformer*. The coil into which electricity is fed from an outside source is the primary coil, that in which an induced current is generated is the secondary coil.

Transformers are useful because the voltage of the induced current can be made any desired multiple or fraction of the primary voltage by suitable winding of the coils. If the number of turns of wire in the secondary coil is the same as the number of turns in the primary, the induced voltage will be the same as the primary voltage. If the secondary has twice as many turns, its voltage is twice that of the primary; if it has one-third as many turns its voltage is one-third that of the primary, and so on. With the help of a suitably designed transformer, then, we may secure any desired voltage, high or low, from a given alternating current.

In a similar way the current in the secondary depends inversely on the ratio of turns in the two coils. It is greater in the secondary than in the primary if the number of turns is less, and vice versa; that is, a transformer that reduces voltage produces a greater current, and one that increases voltage produces a smaller current. As an illustration, suppose that a transformer has 100 turns on its primary coil, 1,000 on its secondary. If the voltage applied to the primary is 110 volts, the secondary voltage is 1,100 volts; and if the secondary circuit draws a current of 0.3 amp, then the primary current must be 3 amp (neglecting heat losses in the coils). The primary current automatically adjusts itself to supply the power needed in the secondary circuit.

The relation between current and voltage in the two coils of a transformer is a necessary consequence of the law of conservation of energy. If we neglect heat losses, the electric power generated in the secondary

the transformer

the ratio of voltages in the coils of a transformer is equal to the ratio of turns in the coils *when the secondary coil of a transformer has a higher voltage than the primary, its current is lower than that in the primary, and vice versa*

coil should equal that supplied to the primary. Now electric power (in watts) is equal to voltage (in volts) × current (in amperes); hence in a perfectly efficient transformer, volts × amperes for the primary should equal volts × amperes for the secondary. So an increased voltage in the secondary must mean a decreased current, and vice versa.

For a multitude of purposes, in homes and factories and laboratories, it is useful to change the voltage of alternating currents. But perhaps the most valuable service transformers render is in making possible long-distance transmission of power. Currents in long-distance transmission must be as small as possible, since large currents mean energy losses in heating the transmission wires. Hence at the powerhouse electricity from the generator is led into a "step-up" transformer, which increases the potential difference and decreases the current, each by several hundred times. On high-voltage, or "high-tension," lines (sometimes carrying currents at potential differences exceeding 500,000 volts) this current is carried to local substations, where other transformers "step down" its voltage to make it safe for transmission along city streets.

The desirability of changing the voltage of the current at least twice between powerhouse and consumer explains our customary use of alternating currents rather than direct: as yet there is no device for changing the voltage of direct currents which can compare with the transformer in simplicity and efficiency, although such devices are under development.

GLOSSARY

A *magnet* is a body with the property of attracting iron objects. When freely suspended, the *north pole* of a magnet points north while its *south pole* points south. Like poles repel each other; unlike poles attract.

The *left-hand rule* states that when a current-carrying wire is grasped so that the thumb of the left hand points in the direction in which the electrons move, the fingers of that hand point in the direction of the magnetic field around the wire.

An *electromagnet* is a coil of wire with an iron core to enhance its magnetic field.

The *motor effect* is the sidewise force a magnetic field exerts on a current-carrying wire; the force is perpendicular to both the direction of the current and the direction of the field.

A *galvanometer* is a device for measuring electric currents with the help of the motor effect.

Electromagnetic induction refers to the production of a current in a wire moving through a magnetic field. It also is known as the *dynamo effect*.

A *direct current* is one that always flows in one direction; an *alternating current* periodically reverses its direction of flow.

MAGNETISM

A *transformer* is a device that transfers electric energy in the form of alternating current from one coil to another by means of electromagnetic induction.

EXERCISES

1. What kind of observations would you have to make in order to prepare a map showing the lines of force of the earth's magnetic field?

2. In what fundamental respect are the motor effect and the dynamo effect similar?

3. Would you expect to find direct or alternating current in
   a. The filament of a light bulb in your home?
   b. The filament of a light bulb in an automobile?
   c. The secondary coil of a transformer?

4. Given a coil of wire and a small light bulb, how can you tell whether the current in another coil is direct or alternating without touching the second coil or its connecting wires?

5. A long coil is suspended by a thread at its midpoint between the poles of a strong magnet. What (if anything) happens when a direct current is sent through the coil? When an alternating current is sent through the coil?

6. Why is a piece of iron attracted to either pole of a magnet?

7. All atoms contain moving electrons. What connection would you suspect between this fact and the fact that all atoms exhibit magnetic properties?

8. What are the similarities and differences between galvanometers and simple direct-current motors?

9. A current-carrying wire is in a magnetic field. What angle should the wire make with the direction of the field for the force on it to be zero? What should the angle be for the force to be a maximum?

10. An electric motor requires more current when it is started than when it is running continuously. Why?

11. When a wire loop is rotated in a magnetic field, the direction of the current induced in the loop reverses itself twice per rotation. Why?

12. The shaft of a dynamo is much easier to turn when the dynamo is not connected to an outside circuit than when such a connection is made. Why?

13 An electron current flows east along a power line. Find the directions of the magnetic field above and below the power line. (Neglect the earth's magnetic field.)

14 A strong current is passed through a loop of flexible wire lying on a table. What shape does the loop assume? Why? (Neglect the earth's magnetic field.)

15 A transformer has a 120-turn primary winding and an 1,800-turn secondary winding. A current of 10 amp flows in the primary winding when a potential difference of 550 volts is placed across it. Find the current in the secondary winding and the potential difference across it.

16 A transformer has a primary coil with 1,000 turns and a secondary coil with 200 turns. If the primary voltage is 660, what is the secondary voltage? What primary current is required if 1,000 watts is to be drawn from the secondary?

chapter 13

Light

ELECTROMAGNETIC WAVES

faraday and maxwell
origin of electromagnetic waves
types of electromagnetic waves
the speed of light
the cerenkov effect

LIGHT

light "rays"
reflection
refraction

WAVE NATURE OF LIGHT

color
interference
diffraction

We are so accustomed to light—which is, after all, what our eyes respond to—that we seldom stop to wonder about its true nature. But something enters our eyes when we experience the sensation of vision, and it is by no means obvious just what that something is. The search for an explanation for light and its effects in terms of known physical principles was pursued for several centuries, but not until the essentials of electricity and magnetism were known was a satisfactory explanation found. It seems odd on first acquaintance that light should be related to electrical and magnetic phenomena, and the discovery of the details of this relationship is one of the most noteworthy in all physical science.

ELECTROMAGNETIC WAVES

In 1864 the British physicist James Clerk Maxwell made the remarkable suggestion that an electric charge moving with a changing velocity (in other words, an *accelerated* charge) generates combined electrical and

electromagnetic waves consist of coupled electric and magnetic fields that vary periodically

magnetic disturbances capable of traveling indefinitely through empty space. These disturbances are called *electromagnetic waves*. Such waves are difficult to visualize, because they represent periodic changes in fields which are themselves difficult to imagine.

faraday and maxwell

Michael Faraday, the son of a blacksmith, was apprenticed to a bookbinder in his youth, and schooled himself in chemistry and physics from the books he was learning to bind. At the age of twenty-one he obtained the humble position of bottle washer for Sir Humphrey Davy at that great chemist's laboratory in the Royal Institution of London. Within 20 years the blacksmith's son succeeded Davy as head of the institution. During those 20 years Faraday's experiments, particularly in chemistry, had won him wide recognition in scientific circles. To the later years of his life belong the more famous investigations into electricity and magnetism that we discussed briefly in the last chapter. Never adept at mathematics, Faraday is remembered as one of physical science's greatest experimental geniuses. Like many of us, he felt the necessity of working with real, tangible things, such as the coils and magnets of his laboratory. To make real the electric and magnetic forces that he could not see or feel, Faraday invented *lines of force*—lines which do not exist, which give at best a crude picture of fields of force, but which students ever since have found useful in their early attempts to visualize the abstractions of electricity.

James Clerk Maxwell was born into an old and distinguished Scottish family in 1831, when Faraday was forty years old. Given the best education that England could provide, Maxwell became a precocious scientist. At fifteen he published his first paper; at twenty-five he was made professor of physics and astronomy at Cambridge. Maxwell was gifted with extraordinary mathematical ability; for him fields of force could be better expressed by equations than in terms of Faraday's lines. With his abstract equations, based largely on Faraday's experiments, Maxwell at length not only expressed the interconnections between electricity and magnetism but discovered the nature of light as well.

Many a moral could be drawn from the lives of these two men, so different in talent and in training, who stand out above the host of nineteenth-century scientists who were seeking an explanation for the mysterious phenomena of electricity. Their work is an especially good illustration of the progress of science from experiment to broad generalization. Sometimes experimental ability and theoretical insight are combined in the same person, but more often we find these faculties in different men. As Faraday's work paved the way for Maxwell, so the observations of Tycho Brahe lay behind Kepler's laws of planetary behavior and Galileo's experiments with falling bodies lay behind Newton's laws of motion.

origin of electromagnetic waves

Maxwell was led to the discovery of electromagnetic waves by an argument based on the symmetry of nature. We know that a changing magnetic field can give rise to a current in a suitable conductor, from which we conclude that such a field has an electric field associated with it. Maxwell proposed the converse, that a changing electric field always has a magnetic field associated with it whose strength is proportional to the rate of change of the electric field. The mathematical formulation of this hypothesis is a persuasive one, but it must nevertheless meet the test of experiment. The weak electric fields produced by electromagnetic induction are easy to detect because metals offer so little resistance to the flow of charges. There is no such thing as a magnetic current, and it is accordingly very difficult to measure the feeble magnetic fields Maxwell predicted.

However, if Maxwell was right, then electromagnetic waves must exist in which changing electric and magnetic fields are coupled together by both electromagnetic induction and the converse mechanism that he proposed. The argument is straightforward. Since the velocity of an accelerated charge is changing, the magnetic field its motion produces is also changing. A changing magnetic field means that an electric field comes into being, according to electromagnetic induction. In turn, the electric field is changing (since the electric field due to electromagnetic induction represents an alteration in the existing field), and by Maxwell's hypothesis a changing electric field gives rise to a magnetic field. The latter field represents a change in the preexisting magnetic field, so another electric field is induced, and so on. The linked fields created in this way do not remain stationary in space but spread out, much as ripples spread out when a stone is dropped in a body of water. <u>The energy contained in an electromagnetic wave is constantly being interchanged between the fluctuating electric field and the fluctuating magnetic field.</u>

Figure 13.1 shows how a pair of opposite charges that vibrate back and forth give rise to the coupled, constantly changing electric and magnetic field that electromagnetic waves consist of. (This is not the only way in which such waves can be generated, since *every* accelerated charge

a magnetic field is associated with a changing electric field, just as an electric field is associated with a changing magnetic field

→ Electric line of force
• Magnetic line of force into paper   × Magnetic line of force out of paper

13.1 How vibrating electric charges give rise to the coupled electric and magnetic fields that constitute electromagnetic waves. *The waves spread out from their source with the speed of light.*

produces them, but it provides an especially easy means for visualizing how this occurs.) In *a* the charges are moving apart. Some of the electric lines of force between them are shown as color lines; the magnetic lines of force are in the form of circles perpendicular to the page. In *b* the charges have stopped, so there is no magnetic field being produced at this moment, but the existing magnetic field continues to spread out along with the electric field. Both fields travel with the speed of light. In *c* the charges are moving toward each other, so the new magnetic field is opposite in direction to the old one although the electric field is in the same direction as before. In *d* the charges have passed each other and are now moving apart. The magnetic field is still in the same direction as in *c*, but now the electric field has reversed itself.

The result of the above sequence is that the outermost electric and magnetic lines of force form closed loops which are no longer connected to the oscillating charges. The loops move freely through space and constitute an electromagnetic wave. As the charges continue to oscillate back and forth, additional closed loops of electric and magnetic lines of force are produced that similarly expand outward.

Figure 13.2 shows a way to visualize the relationship between the electric and magnetic fields in an electromagnetic wave that does not involve lines of force. In this drawing the fields are represented by a series of vectors that indicate the magnitude and direction of the fields in the path of the wave. The fields are perpendicular to each other and to the direction of the wave, and they remain in step as they periodically reverse their directions.

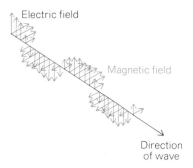

13.2 The electric and magnetic fields in an electromagnetic wave vary simultaneously. *The fields are perpendicular to each other and to the direction of the wave.*

types of electromagnetic waves

During Maxwell's lifetime electromagnetic waves remained a theoretical prediction. Finally, in 1887, the German physicist Heinrich Hertz showed experimentally that electromagnetic waves exist and behave exactly as Maxwell expected them to.

Hertz was not aware of the commercial possibilities of his experimental production of electromagnetic waves, and it remained for others to develop what we now call radio. Radio transmission is accomplished by means of electromagnetic waves produced by electrons oscillating hundreds of thousands to millions of times per second in the antenna of the sending station. When these waves pass the antenna of a receiving station, they cause the electrons there to vibrate in unison with those of the transmitter. By means of electronic devices of various kinds the receiver can be *tuned* so that it responds to a specific frequency of electromagnetic waves only, and thus, since each transmitter operates on a different frequency, a receiver can pick up the signals sent out by whatever station it wishes. The original current set up by the radio waves is very feeble, but it can be amplified in the receiver so that its variations are strong enough to produce sounds in a loudspeaker.

Radio waves are particularly useful because they have the property of being reflected from layers of ionized gas high in the earth's atmosphere. If it were not for this property, the direct reception of radio programs would be limited to short distances, for electromagnetic waves travel in straight lines and would be shielded from more distant receivers by the curvature of the earth. However, since the waves can bounce repeatedly between the upper atmosphere and the earth's surface, transmission is possible for long distances, even to the opposite side of the earth.

In recent times methods have been found of generating electric currents with more rapid oscillations and, hence, of producing electromagnetic waves of higher frequency (shorter wavelength). Frequencies of ordinary radio waves extend up to more than 10^6 Hz and those of waves used in shortwave broadcasting up to 3×10^7 Hz; waves with frequencies in the range from 3×10^7 to 3×10^{10} Hz have found widespread use in television and radar. These shorter, high-frequency waves are not able to bounce back from ionized layers in the atmosphere, so direct reception of television is limited by the horizon. On the other hand, such waves are especially useful because they can be channeled easily into beams in particular directions and because the beams are reflected by solid objects like ships and airplanes; this is the principle behind radar detection.

in radar, a narrow beam of high-frequency electromagnetic waves bounces off an obstacle and is picked up on its return to the instrument that sent it out

Electromagnetic waves have been produced by causing charges to oscillate more than a million million times a second, and the limit has probably not yet been reached. But Maxwell's theory did not stop with predicting waves of these various frequencies; it went on to propose that <u>light itself consists of electromagnetic waves.</u> It was known before Maxwell's time that light showed characteristics of wave motion, but his work gave the first intimation that light, electricity, and magnetism were closely related. How close the relationship is and how intimately all three are connected with the innermost structure of matter physicists have learned only during the present century.

light waves are electromagnetic in character

Not only light but many other kinds of radiation have been shown to consist of electromagnetic waves. Infrared radiation has wavelengths longer than those of visible light, and ultraviolet radiation has shorter wavelengths. Still shorter are the electromagnetic waves of X-rays, and shorter yet those of gamma radiation from atomic nuclei. The electromagnetic nature of these radiations is shown only indirectly, but the demonstration that they are similar in nature to radio waves is beyond all doubt. Electromagnetic waves thus include an enormous range of frequencies and wavelengths—from radio waves a few miles in length, with frequencies less than 100,000 Hz, to waves of gamma radiation less than 10^{-10} cm long and having frequencies greater than 10^{20} Hz (Fig. 13.3). Waves in various parts of this range have particular characteristics, but all have the same basic properties, notably a speed in vacuum of 186,000 mi/sec (3×10^8 m/sec).

13.3 The electromagnetic spectrum. *All electromagnetic waves have the same fundamental character and the same speed in vacuum, but many aspects of their behavior depend upon their frequency.*

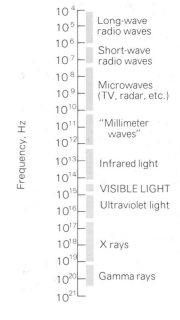

It is possible to measure in a direct way how long the waves of light are. Red light turns out to have a wavelength of 0.00007 cm, and violet light about half as much. If these numbers and the speed of light are substituted in the equation $v = \lambda f$, the frequencies of vibration can be calculated. They turn out to be prodigious, ranging from 4×10^{14} to 8×10^{14} Hz. If these vibrations are produced by oscillating electric charges, like the longer electromagnetic waves of radio and television, then somewhere in a luminous material charges must be moving back and forth at these unimaginable rates. Maxwell had no way of knowing what these moving charges might be, but today we can identify them as the electrons inside atoms.

the speed of light

If light consists of electromagnetic waves that transport energy from one place to another, it should be possible to detect its motion and to measure its speed. However, this turns out to be a matter of great difficulty in practice. We switch on a lamp, and instantly the room is filled with light; there is no impression of anything spreading out from the lamp to the walls. We say that light comes to the earth from the sun, but nothing seems to move; we are simply bathed in a sea of solar luminosity. If we try to measure how fast light travels by arranging to have a distant light turned on at a specified time and then noting its arrival with a stopwatch, we find no detectable lag between the turning on of the light and the time it reaches our eyes (an experiment first tried by Galileo). Evidently light moves so very rapidly that measurement of its speed requires more than simple experiments.

The first measurement of the speed of light happened almost by accident. The Danish astronomer Olaf Rømer, a contemporary of Isaac Newton, was making a careful study of the times when Jupiter eclipsed one of its satellites, in the hope that a table of these times would be useful to navigators in determining longitude at sea. Rømer predicted the exact times of these eclipses, based upon the constant 42.5-hr period of satellite revolution, and then compared the figures with observations. He found that the predicted times were about 22 min in error after six months had elapsed, but the discrepancy disappeared after a year. Apparently the satellite decreased in speed when the earth was moving away from Jupiter so as to be "late" for its eclipse when the earth was farthest distant, and increased in speed in the next half of the earth's orbital circuit to be on time again when the earth was near Jupiter.

Rømer correctly supposed that the time differences were due to the difference in the paths which light from Jupiter had to follow to reach the earth (Fig. 13.4); as the earth circles the sun away from Jupiter, light takes longer and longer to reach us, while this time decreases as the earth approaches Jupiter on its return. Rømer's estimate of the speed of light was not very accurate since neither the exact time delay nor the diameter of the earth's orbit was known correctly in those days, but modern figures for these quantities give the same value for the speed of light as more direct measurements. Today it is known that the maximum apparent retardation of an eclipse of a Jovian satellite is just about 1,000 sec, and the earth's orbit has a diameter of twice the 93-million-mile sun-earth distance, or 186 million miles. Hence the speed of light, symbol c, is given by

$$c = \frac{\text{distance}}{\text{time}} = \frac{186{,}000{,}000 \text{ miles}}{1{,}000 \text{ sec}} = 186{,}000 \text{ mi/sec}$$

The reflection of reeds from the surface of a still pond.

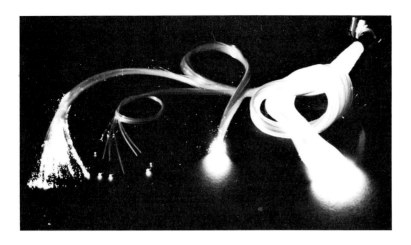

Light can be "piped" from one place to another by means of successive reflections from the wall of a glass rod.

The metric equivalent of this speed is 3×10^8 m/sec.

Accurate laboratory measurements of the speed of light became possible only in the mid-nineteenth century. The first successful attempt, by the French physicist Fizeau in 1849, was made by using a rapidly moving toothed wheel to interrupt a narrow beam of light (Fig. 13.5). The light was reflected from a mirror placed about 5 miles away, and the reflected beam was observed through the same toothed wheel. As long as the wheel turned slowly, the light passing through any gap between two teeth could reach the mirror, be reflected, and return through the same gap, so that the mirror would be visible to an observer behind the wheel. However, if the wheel was going faster, the light originally passed through a gap would return to the wheel in time to strike the following tooth; thus it would be blocked off, and the mirror could no longer be seen through the wheel. When the wheel's speed was increased still further, the light reappeared, this time passing through the gap following the one it went through on the outward journey. The speed of the wheel gave the time needed for a tooth to move into the position of the adjacent gap. In this tiny fraction of a second the light had traversed a distance of about 10 miles, and these two figures provided the necessary data for computing the speed of light.

a laboratory method for measuring the speed of light

Fizeau's method was improved the following year by his colleague Foucault, with the introduction of a rotating mirror in place of the toothed

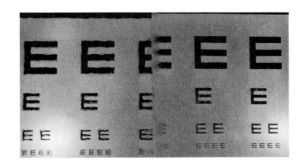

Irregularities in a lacquer film are revealed by the distorted images that result from refraction of light through the film.

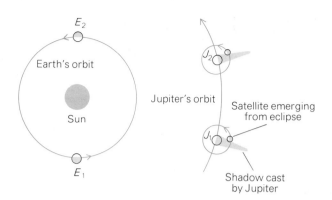

13.4 Rømer's measurement of the speed of light. *When the earth is at E_1 its distance from Jupiter is decreasing and the eclipses of the satellite thereafter occur earlier than predicted. Six months later, with the earth at E_2 and Jupiter at J_2, the distance is increasing and the eclipses thereafter occur later than predicted.*

13.5 Fizeau's method of determining the speed of light. *A beam of light is directed into a partly silvered glass plate, which reflects part of the beam through the edge of a toothed wheel to a distant mirror. The mirror reflects the light back along the same path, and part of it is transmitted to the observer through the lightly silvered glass. When the wheel is turned rapidly, light moving through a gap in the wheel is reflected from the mirror in time to meet the adjacent tooth on its return journey, so that no light reaches the observer.*

wheel. This procedure was greatly refined by the American physicist Michelson in a long series of experiments which gave a very accurate figure for the speed of light. In recent years other methods have been devised for checking Michelson's results. Today the accepted figure for the speed of light in vacuum is

$$c = 2.998 \times 10^8 \text{ m/sec}$$

For most purposes it is sufficiently accurate to let $c = 3 \times 10^8$ m/sec.

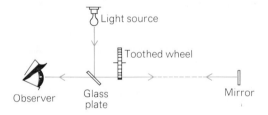

the cerenkov effect

The relativity of mass places a limit on the speed that a body can have relative to an observer. As the body's speed v approaches the speed of light c, the ratio v^2/c^2 approaches 1 and the quantity $\sqrt{1 - v^2/c^2}$ approaches 0. Hence the mass m of the body as measured by the observer, which is given by

$$m = \frac{m_0}{\sqrt{1 - v^2/c^2}}$$

approaches infinity as v approaches c. The relative speed v therefore must be less than c. An infinite force would be needed to accelerate a body to a speed at which its mass is infinite, and there are neither infinite forces nor infinite masses in the universe.

The speed of light c in special relativity is the *speed of light in free space*, 3×10^8 m/sec. In all material media, such as water, glass, or air, visible light travels more slowly than this, and atomic particles are capable of moving with higher speeds in such media than the speed of light *in them*—though never faster than the speed of light in free space. When an electrically charged particle moves through a substance at a speed exceeding that of light in the substance, a cone of light waves is emitted in a process roughly similar to that in which a ship produces a bow wave as it moves through the water at a speed greater than that of water waves. These light waves are known as *Cerenkov radiation*.

cerenkov radiation

LIGHT

light "rays"

Early in life we learn that light travels in straight lines. As direct evidence we could point to the sharp shadows cast by objects illuminated by small light sources and to the straight beams produced when light from small holes penetrates the recesses of a dusty basement. Actually, our entire physical orientation to the world about us, our sense of the location of things in space, depends on *assuming* that light follows straight-line paths. In everyday life we use this property to define straightness, rather than Euclid's definition about "the shortest distance between two points"; if we want to know whether a given line is straight or not, we sight along it instead of trying to determine whether it is the shortest of all possible lines between its end points.

Just as familiar, however, is the fact that light does not always follow straight lines. We see most objects by *reflected* light, light that has been turned abruptly on striking a surface. The distorted appearance of objects seen through a glass of water, through glass objects of irregular shape, or through the heated air rising above a flame, all testify to the ability of light to be bent from a straight path. In these cases we say that the light is *refracted*, and we note that this occurs when light moves from one transparent material to a different one.

Although we recognize, with the intellectual part of our minds, that light can be reflected and refracted, our eyes are so accustomed to finding light traveling along straight paths that we are often deceived about the true position of objects. When we look in a mirror, for example, we actually see light that has traveled in a broken path to the mirror and back again, but our eyes tell us that the light has followed a straight path from our image behind the mirror. When we look at someone standing in shallow water, his legs appear foreshortened; actually light coming from water into air is bent, but our eyes have no way to take this into account and give us the illusion of foreshortening.

A good deal can be learned about the behavior of light and our reactions to it simply from a study of the paths that light follows in various situations. Since light appears to travel in straight paths, we can represent its motion by geometrically straight lines called *rays*. In nature there are no such things as rays, for light sources and light beams always have finite size. But rays are a convenient abstraction, and we can visualize what we mean by thinking of a narrow pencil of light in a darkened room.

light "rays" provide a convenient way of thinking for many purposes, although light actually consists of waves

Note that, in making this abstraction, we exclude many familiar properties of light—its energy content and its color, for example. Nor does

Radioactive $_{27}Co^{60}$ sources immersed in a pool of water. Gamma rays from the decay of $_{27}Co^{60}$ nuclei impart energy to nearby electrons, which in turn produce Cerenkov radiation as they move through the water at speeds faster than the speed of light in water.

this abstraction imply any hypothesis about the nature of light; we could suppose that it consisted of particles or waves or even streams of continuous fluid, and rays would still be convenient for describing its behavior. As a matter of fact, in using rays we do not even need to know which way light moves; the first man to investigate these phenomena carefully, the Greek geometer Euclid, thought that something traveled from the eye to the object seen rather than vice versa, but this faulty hypothesis did not invalidate his conclusions. Rays of light are a beautiful example of how far we can go in science by using a simple idealized model having no actual counterpart in nature and by ignoring those aspects of a phenomenon with which we are not at the moment concerned.

reflection

When we look at ourselves in a mirror, light from all parts of the body (this is reflected light, of course, but we may treat it as if it originated in the body) is reflected from the mirror back to our eyes (Fig. 13.6). Light from the foot, for example, follows the path $CC'E$; our eyes, looking along the ray $C'E$ and unconsciously projecting it in a straight line, see the foot at the proper distance but apparently behind the mirror at C''. A ray from the top of the head is similarly reflected at A', and our eyes see the point A as if it were behind the mirror at A''. Rays from other points of the body are similarly reflected, and in this manner a complete *image* is formed behind the mirror.

The fact that the image is a perfect replica of the body (provided the glass of the mirror has no imperfections) means that each ray must be reflected from the mirror at the same angle as the angle of approach; in other words, the angle made by CC' with the mirror must equal the angle made by $C'E$. This property of rays is expressed by the statement that **the angle of incidence equals the angle of reflection.** In order to make this rule applicable to curved as well as plane surfaces, the angles are usually measured from a perpendicular to the surface (Fig. 13.7) rather than from the surface itself.

Why do we not see images of ourselves in walls and furniture as well as in mirrors? This is largely a question of the relative roughness of surfaces. Rays of light are reflected from walls just as they are from mirrors, but the reflected rays are scattered in all directions by the many

in reflection, what is seen is an image of the actual object

angle of incidence equals angle of reflection

The indexes of refraction of most substances vary with the wavelength of the light sent through them. Here a beam of sunlight is separated into its component wavelengths (each of which causes the sensation of a different color) by a glass prism.

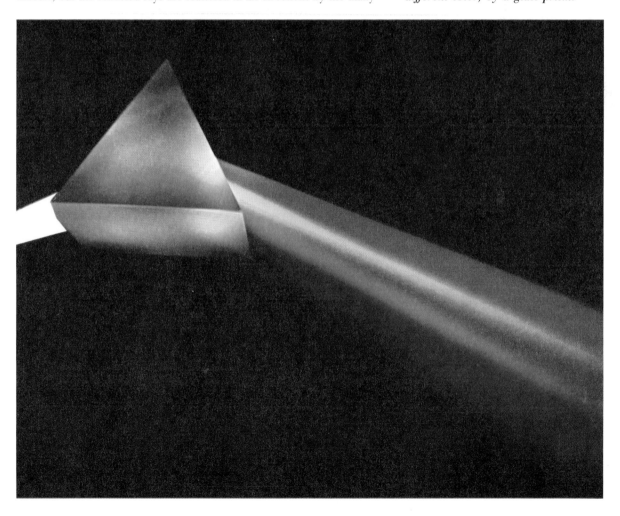

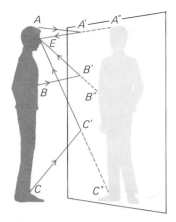

13.6 Formation of an image in a plane mirror. *The image appears to be behind the mirror because we instinctively respond to light as though it travels in straight lines.*

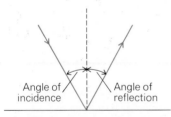

13.7 When light is reflected, the angle of incidence equals the angle of reflection.

13.8 Light is refracted when it travels obliquely from one medium to another. *Here the effect of refraction is to make the water appear shallower than it actually is.*

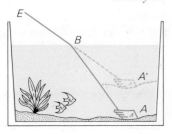

surface irregularities. We see the wall by the light scattered from its surface.

refraction

Next, consider what happens when light reaches our eyes from an object under water. If we look obliquely through a water surface at a stone lying on the bottom, the stone appears to be lying at a higher point than it actually is (Fig. 13.8): a body of water is always deeper than it appears to be. Another aspect of the same phenomenon is the apparent shortening of a friend's body when he is standing in water and we look at him from the side. The explanation depends on the change in direction of light as it moves from water into air: a ray of light from the stone in Fig. 13.8 follows the broken path ABE to our eyes, but our eyes deceive us by telling the brain that the segment BE is part of the straight-line path originating at A'.

Note that rays of light from water are bent *toward* the water surface as they emerge into air. A ray starting in air and going obliquely into water would follow the same path in reverse—as can be shown experimentally by letting a slender pencil of light fall on the surface of slightly turbid water in a darkened room. We may generalize the behavior of light during refraction by saying that rays going obliquely from one medium into another are bent away from a perpendicular to the surface if the second medium is less dense than the first, toward the perpendicular if the second medium is more dense (Fig. 13.9). Light moving from air into glass, for example, is bent toward the perpendicular; light moving from glass into air or from glass into water is bent away from the perpendicular. The explanation for the bending lies in different speeds of light in media of different densities. Light that enters another medium along the perpendicular is not deflected.

The phenomenon of refraction occurs whenever light passes from one medium into another in which its speed is different. Figure 13.10 shows two rays of light, I and II, that are part of a beam of light going from air into a glass plate. Ray I reaches the glass first, at point A; at this time ray II is at A'. After some time interval t ray II enters the glass at B', ray I having meanwhile proceeded to B. The distance $A'B'$ is equal to $v_a t$, where v_a is the speed of light in air, and the distance AB is equal to $v_g t$, where v_g is the speed of light in glass. Since light travels more slowly in glass than in air, AB is less than $A'B'$, and the two rays again proceed parallel to each other but in a direction different from their original one.

WAVE NATURE OF LIGHT

Convenient though the ray model undoubtedly is, light is too complex and many-sided a phenomenon to be described entirely in terms of so drastic a simplification. The basic wave nature of light must be invoked in order to understand such aspects of its behavior as color, interference, and diffraction.

color

When white light passes through simple lenses, color fringes appear that tend to blur the outlines of images. Similar color fringes appear in field glasses and telescopes, unless the lenses are specially constructed to eliminate this effect. Something in the process of refraction through glass seems to produce color, and we can make the colors especially strong by using a prism instead of a lens. In this case white light appears to spread itself out into a rainbow band of color, or *spectrum*, with the colors showing the familiar sequence from red through orange and yellow to green, blue and violet. If we use the ray representation of light, we draw a single ray for the incident white light and a succession of rays for the refracted colors (Fig. 13.11).

This experiment is interpreted today by saying that white light consists of a mixture of electromagnetic waves of different wavelengths, each of which produces the visual sensation of a particular color. The prism separates the wavelengths by refracting each one through a different angle, with red light showing the least refraction and violet light the most. When the experiment was first studied intensively nearly 300 years ago by Isaac Newton, none of these conclusions was obvious. Colors were thought to be *qualities* that light could assume under various conditions; one color could supposedly be changed to another, and white light was considered no different from the rest. Only by a long series of careful experiments was Newton able to establish the permanence of the different colors and their relation to white light—concepts that now seem to us so easy and so reasonable because we have grown up with them in our intellectual background.

Detailed analysis of the wavelengths present in any kind of light is accomplished by means of a *spectroscope*, an instrument that has played a central role in the development of modern physics and astronomy. The simplest form of spectroscope consists of a prism placed between two cylindrical tubes (Fig. 13.12). One of the tubes has a narrow slit at one end, through which the light to be examined is admitted. The light is spread out into its separate colors by the prism and then passes down the second tube to the eyepiece. The tubes also contain a system of lenses that focuses an image of the slit on the eyepiece, so that each separate color appears as an image of the slit. If all wavelengths in the visible part of the electromagnetic spectrum are present, the slit images are next to one another and the result is a continuous rainbow. If only a few wavelengths are present, the spectrum consists of a few bright lines.

Any solid or liquid material, or any gas if sufficiently compressed, when it is heated until it glows brightly, gives out light of all colors. At lower temperatures the most intense radiation is at the red end of the spectrum ("red heat"); as the temperature rises, the greatest intensity shifts to the middle of the spectrum, and the light appears white ("white heat"); at still higher temperatures the light becomes bluish. Because of this relation between color and temperature, astronomers can estimate the temperatures of stars from the intensity of various parts of their spectra.

Luminous gases at low or moderate pressures ordinarily produce light of only a few wavelengths, giving spectra that consist of a few bright lines. These are *discontinuous spectra*, contrasted with the *continuous*,

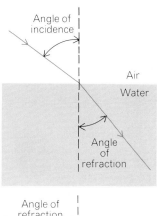

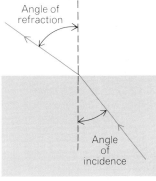

13.9 Properties of refraction. *Light rays are bent toward the perpendicular when they enter a dense medium and away from the perpendicular when they leave a dense medium. The paths taken by light rays are always reversible.*

13.10 The phenomenon of refraction occurs whenever light passes from one medium into another in which its speed is different. *Here two rays of light, I and II, pass from air, in which their speed is v_a, to glass, in which their speed is v_g. Because v_g is less than v_a, $A'B'$ is longer than AB, and the beam of which I and II are part changes in direction when it enters the glass.*

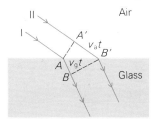

rainbowlike spectra of heated solids. Thus, the sodium-vapor lights used to illuminate foggy roads give a spectrum restricted largely to two lines close together in the orange-yellow region; the neon lights that adorn our city streets have a spectrum consisting of a few lines near the red end.

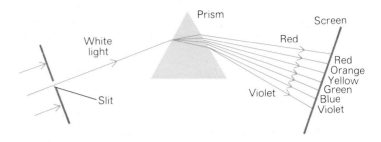

13.11 Formation of a spectrum by the refraction of white light in a glass prism. *The different colors blend into one another smoothly.*

A colored pattern occurs when an uneven thin film of oil floats on water because light of each wavelength undergoes constructive interference at different oil thicknesses.

Colors of objects that we see by reflected light depend on the kind of light falling on them and on their composition. If an object can reflect red light but absorbs other colors, it will appear red in sunlight; if it reflects chiefly green light, it will appear green in sunlight, and so on. Most objects reflect or absorb more than one color, and the color we see is a combination of those that are reflected. Obviously, an object can reflect a certain color only if that color is present in the light falling on it; thus, in sodium

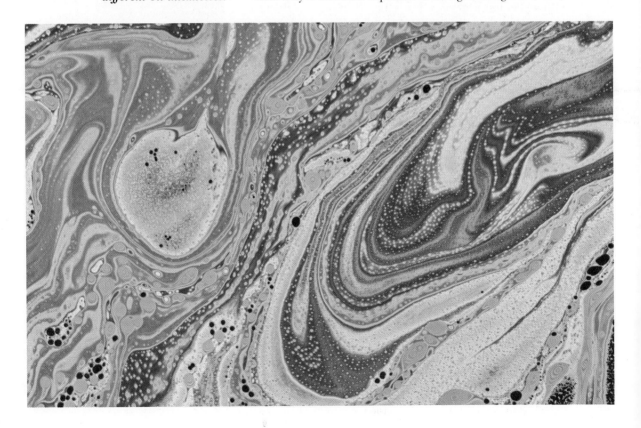

light, a red or green object will appear black. The ghastly hues that even the most carefully made-up complexions assume under sodium or mercury light are due simply to the absence in these lights of the colors that skin and cosmetics normally reflect.

Some colors have a different origin. The blue of the sky, for example, is due to scattering of the sun's light by molecules and dust particles in the atmosphere. Blue light is scattered more effectively than red; hence the sky, which we see only by scattered light, has an excess of blue, and the sun itself is a little more yellowish or reddish than it would appear if the earth had no atmosphere. At sunrise or sunset, when the sun's light must traverse great thicknesses of the atmosphere, the scattering of its blue light is more pronounced, and the sun is often a brilliant red. Above the atmosphere, the sky appears black, and the moon, stars, and planets are visible in the daytime.

A laser is a device that produces an intense beam of monochromatic, coherent light from the cooperative radiation of excited atoms. The light waves in a coherent beam are all in step with one another, which greatly increases their effectiveness. A laser beam is virtually nondivergent and hence can be focused into a minute spot. Here the energy in a laser beam is used for machining high-carbon steel by creating temperatures exceeding 6000°C over a very small area. The greenish plume is steel that has been vaporized by the beam.

Diffraction pattern formed by monochromatic light passing through a small aperture in a screen. Diffraction limits the ability of optical systems to produce sharp images.

a thin film appears black where light waves reflected from its upper and lower surfaces are out of step

13.12 *A simple spectroscope, showing the arrangement of slit, prism, and lenses. Modern research spectroscopes employ "diffraction gratings," whose operation involves both diffraction and interference, to separate a beam of light into its component wavelengths.*

13.13 *Interference bands produced by reflections of monochromatic light from a soap film. The film is formed across a loop of wire.*

interference

As we learned in Chap. 9 *interference* refers to the adding together of two or more waves of the same kind that pass the same point in space at the same time. The interference of light is readily exhibited by shining light obliquely on a soap film (Fig. 13.13) or a film of air between two glass plates (Fig. 13.14). The results are clearest if we use light of only one color, say the yellow light of a sodium-vapor lamp; a thin film in this kind of light shows a succession of yellow and black bands, in straight lines or circles or irregular figures, depending upon the character of the film.

The wave picture gives a direct explanation of these bands, as illustrated by Fig. 13.15. This drawing shows a greatly enlarged cross section of a part of the film. Light falling on the film is reflected twice, once from the upper surface and once from the lower surface. Light rays from the two reflections travel upward along nearly the same path, and their waves interfere. At some places on the film wave crests in the reflected rays are out of step, so that one cancels the effect of the other, while at other places wave crests are in step and reinforce each other. Where cancellations occur, the film appears black because no light propagates; where reinforcement occurs, it is bright yellow. What effect interference will have at any point on the film depends on the thickness of the film at that point, since the thickness determines how far one reflected ray lags behind the other.

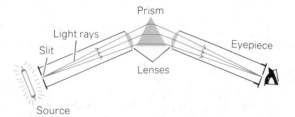

When white light is used instead of monochromatic light, the reflected rays of one color will be in step at a given point while rays of other colors will not. Hence, rather than white and black bands, the film shows a succession of brilliant colors. These are the rainbow effects we see so often in bubbles and in oil films on water.

diffraction

Diffraction, too, occurs in light waves. As we recall from Chap. 9, diffraction refers to the "bending" of waves around the edges of an obstacle in their path. Because of diffraction, a shadow is never completely dark, although the wavelengths of light waves are so short that the effects of diffraction are largely limited to the border of the shadow region.

Diffraction limits the useful magnification of optical instruments such as those of microscopes and telescopes. The larger the diameter of a lens

LIGHT

13.14 Interference of light after reflection from air-glass interfaces. *In each drawing the two glass plates are touching at one edge and very slightly separated at the opposite edge, leaving a wedge-shaped film of air between them. Light reflected from one surface of the film interferes with light reflected from the other surface. In the left-hand drawing the surfaces are plane, while in the right-hand drawing they are uneven.*

(or curved mirror which acts like a lens), the less significant is diffraction in its performance. For this reason it is impossible to construct a small telescope capable of high magnification, since the result would be a blurred image instead of a sharp one. The largest telescope in the world, the Hale telescope at Mt. Palomar in California, employs a 200-in.-diameter concave mirror, and diffraction limits the ability of this telescope to view detail on the moon's surface, for example, to objects about 150 ft across.

diffraction limits the useful magnification of optical instruments; the larger the lenses or mirrors used, the sharper the image

13.15 Enlarged cross section of soap film in Fig. 13.13. *AO and BQ are two rays of monochromatic light falling on the film. AO is reflected in the two rays OA′ and PA″ and BQ is reflected in the rays QB′ and RB″. In each case the reflected rays interfere. OA′ and PA″ cancel each other because their waves are out of step, while QB′ and RB″ reinforce each other.*

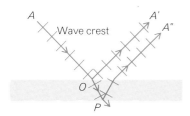
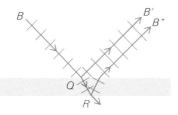

GLOSSARY

Electromagnetic waves are coupled periodic electrical and magnetic disturbances that spread out from accelerated electric charges. Among the various kinds of electromagnetic waves, distinguished only by their frequencies, are gamma rays, X-rays, ultraviolet radiation, visible light, infrared radiation, millimeter waves, microwaves, and radio waves. Electromagnetic waves all travel in vacuum with the *speed of light*.

A *spectrum* is the band of different colors produced when light from a particular source passes through a glass prism or other device that disperses light of different wavelengths.

A *spectroscope* is a device for analyzing a beam of light into its component colors.

EXERCISES

1 We say that light is a transverse wave. What is it that varies at right angles to the direction in which a light wave travels?

The intensity and high degree of collimation of a laser beam are demonstrated here. A $\frac{1}{16}$-in. diameter beam of 6.328×10^{-7} m wavelength light from a laser at the Spectra-Physics plant in California is photographed from the Lick Observatory, 25 miles away, where the beam has diverged to a diameter of only 6 ft. The camera is not in line with the beam to avoid overexposing the film.

2 The light-year is an astronomical unit of length equal to the distance light travels in a year. What is the length of a light-year in miles?

3 Devise an experiment to show that light is a form of energy.

4 Why is light sometimes spoken of in terms of rays and sometimes in terms of waves?

5 Some stars appear red, some yellow, and some blue. Which has the highest temperature? Which the lowest?

6 Why do large bodies of water appear bluish? How does the sky appear to an astronaut above the earth's atmosphere during the day? During the night?

7 What color would red cloth appear if it were illuminated by light from
  a The sun?
  b A neon sign?
  c A sodium-vapor lamp?

8 When a beam of white light passes perpendicularly through a flat pane of glass, it is not dispersed into a spectrum. Why not?

9 The period of daylight is increased by a small amount because of the refraction of sunlight by the earth's atmosphere. Show with the help of a diagram how this effect comes about.

10 A diamond shows flashes of color when held in white light. Why? If the diamond were exposed to red light, what would happen?

11 Why do you think stars twinkle when seen from the ground but not when seen by astronauts in space vehicles?

12 Light waves carry both energy and momentum. Why doesn't the momentum of the sun diminish with time as its energy content does?

LIGHT

13. Diffraction was discussed in Chap. 9. Radio waves are able to diffract readily around buildings, as anybody with a portable radio receiver can verify. However, light waves, which are also electromagnetic waves, undergo no discernible diffraction around buildings. Why not?

14. If the earth were moving toward a star instead of the star toward the earth, would lines in the star's spectrum appear to be shifted? If so, toward which end of the spectrum?

15. An opera performance is being broadcast by radio. Who will hear a certain sound first, a member of the audience 100 ft from the stage or a listener with his ear glued to a radio receiver in a city 3,000 miles away?

16. What is the wavelength of radio waves whose frequency is 500 kHz? 1,500 kHz? 10 mHz? (1 kHz = 10^3 Hz, 1 mHz = 10^6 Hz.)

17. It is desired to repeat Fizeau's determination of the speed of light using a mirror 7 miles from the light source. If the toothed wheel has 100 teeth, how fast must it turn for light leaving through one gap to return through the next gap?

18. A radar pulse is aimed at the moon from a station on the earth, 238,000 miles away. How long does it take for the radar echo to return?

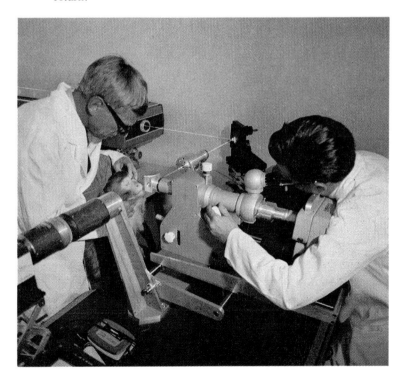

The light beam from a laser is so narrow and intense that it provides an excellent way to repair a detached retina by "welding" it in place. In this technique a laser beam is directed at a small region of the retina, which becomes hot and adheres to the tissue behind it. Here researchers at Stanford University are aiming the blue-green light from an argon laser at a monkey's eye.

19 What is the height of the smallest mirror in which you could see yourself at full length? Use a diagram to explain your answer. Does it matter how far away you are?

20 When a fish looks up through the water surface at an object in the air, will the object appear to be its normal size and distance above the water? Use a diagram to explain your answer, and assume that the fish's eye, like the human eye, is accustomed to interpreting light rays as straight lines.

part four
THE ATOM

Matter is composed of atoms: of what are atoms themselves composed? The puzzle of atomic structure eventually required for its solution contributions from mechanics, electromagnetism, relativity, and quantum theory besides ingenious experiments, and it is interesting to see how many separate physical principles are involved in the modern picture of the atom.

The quantum theory contains some of the least expected relationships in physics: all waves have a certain particle character as well, and all moving particles exhibit wave behavior.

Quantum phenomena seem unfamiliar only because they occur on a scale of size far too small to be visible, but for things on this scale—atoms and molecules, for instance—such phenomena are commonplace.

Our study of the atom itself begins with its division into nucleus and electron cloud, and then goes on to examine the nucleus itself in detail. The reader will encounter a host of curious elementary (that is, indivisible) particles, learn about the composition of nuclei, discover isotopes—atoms of the same element that are nevertheless slightly different—and find out where "atomic energy" resides and how it can be liberated. Finally we pass to the atom as a whole, which means considering the arrangement of its electrons and why they are able to exist in a stable arrangement despite the contrary predictions of electromagnetic theory.

chapter 14

Waves and Particles

QUANTUM THEORY OF LIGHT

the photoelectric effect
the quantum theory of light
what is light?
x-rays
the compton effect
photons and gravity

MATTER WAVES

de broglie wavelength
the electron microscope
probability density

THE UNCERTAINTY PRINCIPLE

the uncertainty principle
a hypothetical experiment
waves versus particles

To most of us there is nothing mysterious or ambiguous about the concepts of *particle* and *wave*. A stone dropped into a lake and the ripples that spread out from its point of impact apparently have in common only the ability to carry energy and momentum from one place to another. Classical physics, which describes scientifically the "physical reality" of our sense impressions, treats particles and waves as separate aspects of that reality. The mechanics of particles and the optics of waves are, by tradition, independent subjects, each with its own chain of experiments and hypotheses.

But the physical reality we experience has its roots in phenomena that occur in the microscopic world of atoms and molecules, electrons and nuclei, and in this world there are neither particles nor waves in our sense of these terms. We regard electrons as particles because they possess charge and mass and behave according to the laws of particle mechanics in such familiar devices as television picture tubes. However, there is as much evidence in favor of interpreting a moving electron as a wave manifestation as there is in favor of interpreting it as a particle manifestation. We regard electromagnetic waves as waves because under suitable circumstances they exhibit such characteristic wave behavior as interference. However, under other circumstances electromagnetic waves behave as though they consist of streams of particles. Together with the theory of special relativity, the wave-particle duality is central to an understanding of modern physics, and we shall explore its meaning in this chapter.

QUANTUM THEORY OF LIGHT

the photoelectric effect

light causes metal surfaces to emit electrons

Late in the nineteenth century a series of experiments were performed which revealed that electrons are emitted from a metal surface when light of sufficiently high frequency falls on it (Fig. 14.1). For most metals ultraviolet light is necessary for electron emission to occur, but very active metals, such as potassium and cesium, respond to visible light as well. The emission of electrons by a substance being irradiated with light is known as the *photoelectric effect*.

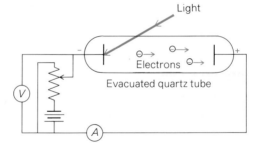

14.1 Experimental observation of the photoelectric effect. *Electrons are emitted from the surface of the target metal when a light beam is directed on it.*

Evidently electromagnetic waves can concentrate their energy on single electrons, setting them free from the positive charge of their parent atoms. In itself this observation does not seem surprising; as a mechanical analogy, we can think of pebbles dislodged from a beach by water waves. Light waves also carry energy, and some of the energy absorbed by the metal may somehow concentrate on individual electrons and give them enough kinetic energy to escape. When we look more closely at the data, however, we find that the photoelectric effect can hardly be interpreted so simply.

the photoelectric effect cannot be explained by the electromagnetic theory of light

One of the features of the photoelectric effect that particularly puzzled its discoverers is that the maximum energy of the emitted electrons (called *photoelectrons*) is independent of the intensity of the light. A bright light beam yields more photoelectrons than a faint one of the same frequency, but the maximum electron energy is the same. If water waves followed the same rule, which they do not, pebbles would move out from shore with the same average speed no matter how high the waves might be;

the number of pebbles would change with the height of the waves, but not the violence of their motion. Also, within the limits of experimental accuracy, there is no time lag between the arrival of light at a metal surface and the emission of photoelectrons. These observations cannot be understood from the electromagnetic theory of light, which predicts that the electron energy should vary with the intensity of the light and that, under the conditions of the actual experiments, an average of nearly a year would be required for an individual electron to accumulate the energy needed for escape.

Equally odd from the point of view of the wave theory is the fact that the maximum photoelectron energy depends upon the *frequency* of the light employed. At frequencies below a certain critical frequency characteristic of each particular metal, no electrons whatever are emitted. Above this threshold frequency the photoelectrons have a range of energies from zero to a certain maximum value, and this maximum energy increases linearly with increasing frequency. High frequencies result in high maximum photoelectron energies, low frequencies in low maximum photoelectron energies. Thus a faint blue light produces electrons with more energy than those produced by a bright red light, although the latter yields a greater number of them.

the quantum theory of light

The electromagnetic theory of light, one of the immortal achievements of the human intellect, accounts so well for such a variety of phenomena that it must contain some measure of truth. Yet this well-founded theory is completely at odds with the photoelectric effect.

In 1905 Albert Einstein found that the paradox presented by the photoelectric effect could be understood by taking seriously a notion proposed five years earlier by the German theoretical physicist Max Planck. Planck was seeking to explain the characteristics of the radiation emitted by bodies hot enough to be luminous, a problem notorious at the time for its resistance to solution. He was able to derive a formula for the spectrum of this radiation (that is, the relative brightness of the various colors present) as a function of the temperature of the body that was in agreement with experiment, provided he assumed that the radiation is emitted *discontinuously* as little bursts of energy. These bursts of energy are called *quanta*, or alternatively, *photons*. Planck found that the quanta associated with a particular frequency f of light all have the same energy, and that this energy E is directly proportional to f; that is,

radiant energy propagates in bursts called quanta or photons

$$E = hf$$

photon energy is proportional to frequency

where h, today known as *Planck's constant*, has the value

$$h = 6.63 \times 10^{-34} \text{ joule-sec}$$

planck's constant

Planck was not altogether happy about this idea, which, though it led to agreement with certain experimental data, nevertheless was contrary to the orthodox electromagnetic theory of light. He took the position that,

All light-sensitive detectors (including the eye) are based upon the absorption by target electrons of energy from photons of light. The electrical resistance of the photoconductive cells shown here decreases with increasing illumination.

although light is emitted discontinuously in little bursts, it nevertheless is propagated in waves exactly as everybody thought.

Einstein, however, thought that, if light is emitted in little packets, it should travel through space and ultimately be absorbed in the same little packets. His idea fitted the experiments on the photoelectric effect perfectly. He supposed that some specific minimum amount of energy, which we might call w, is required in order to pull an electron away from a metal surface. If the frequency of the light is too low, so that E, the energy of the photons, is less than w, no photoelectrons are emitted. When E is greater than w, a photon of light striking an electron can impart to the electron enough energy for it to emerge from the metal with a certain amount of kinetic energy. Einstein's formula for the process is very simple:

photoelectric equation $hf = \text{KE} + w$

where hf = energy of a photon of light whose frequency is f
KE = kinetic energy of emitted electron
w = energy needed to extract the electron from the metal

The quantity w for a particular metal is known as its *work function*, and is a measure of the force that holds the electrons in the metal.

Not all photoelectrons have the same amount of kinetic energy, but come out of the metal with energies up to the KE of the above formula. This is because w is the work that must be done to take an electron through the metal surface from just beneath it, and more work is required when the electron originates deeper in the metal.

what is light?

The view that light propagates as a series of little packets of energy is directly opposed to the wave theory of light. And the latter, which provides the sole means of explaining a host of optical effects, notably diffraction and interference, is one of the most securely established of physical theories. Planck's suggestion that a hot object emits light in separate quanta led to no more than raised eyebrows among physicists in 1900 since it did not apparently conflict with the propagation of light as a wave. Einstein's suggestion in 1905 that light travels through space in the form of distinct photons, on the other hand, astonished many of his colleagues.

According to the wave theory, light waves spread out from a source in the way ripples spread out on the surface of a lake when a stone falls into it. The energy carried by the light in this picture is distributed continuously throughout the wave pattern. According to the quantum theory, however, light spreads out from a source as a series of localized concentrations of energy, each sufficiently small to be capable of absorption by a single electron. Curiously, the quantum theory of light, which treats it as a strictly particle phenomenon, incorporates the light frequency f, a strictly wave concept (Fig. 14.2).

The quantum theory of light is strikingly successful in explaining the photoelectric effect. It predicts correctly that the maximum photoelectron energy should depend upon the frequency of the incident light and not upon its intensity, contrary to what the wave theory suggests, and it is able to explain why even the feeblest light can lead to the immediate emission of photoelectrons, again contrary to the wave theory. The wave theory can give no reason why there should be a threshold frequency such that when light of lower frequency is employed, no photoelectrons are observed, no matter how strong the light beam, something that follows naturally from the quantum theory.

Which theory are we to believe? A great many physical hypotheses have had to be modified or discarded when they were found to disagree with experiment, but seldom before have we arrived at two totally different theories to account for a single physical phenomenon. The situation here is fundamentally different from what it is, say, in the case of relativistic versus newtonian mechanics, where the latter turns out to be an approximation to the former. There is no way of deriving the quantum theory of light from the wave theory of light or vice versa.

In a specific event light exhibits *either* a wave or a particle nature, never both simultaneously. The same light beam that will show interference in a suitable optical device can cause the emission of photoelectrons from a suitable surface, but these processes occur independently. The wave theory of light and the quantum theory of light complement each other. Electromagnetic waves provide the sole possible explanation for certain experiments involving light and optical phenomena, while photons provide the sole possible explanation for all the other experiments in this field. We have no alternative to regarding light as something that manifests itself as a stream of discrete photons in certain experiments and as a wave train in others. The "true nature" of light is no longer a meaningful concept in terms of everyday experience, and we must accept both wave and quantum theories, apparent contradictions and all, as the closest we can get to a complete description of light.

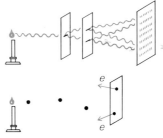

14.2. (a) The wave theory of light is able to account for interference patterns of bright and dark bands in terms of light waves that are respectively in step and out of step with one another. (b) The quantum theory of light is able to account for the photoelectric effect by assuming that light consists of tiny bundles of energy.

the wave theory and the quantum theory of light complement each other

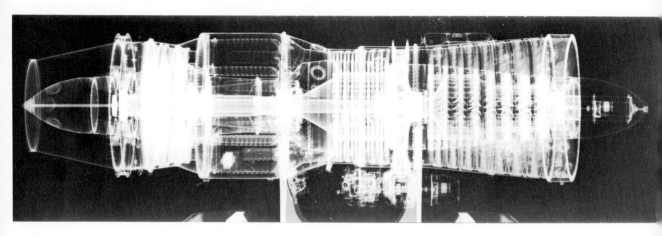

The structure of a jet aircraft engine, laid bare in a single x-ray photo. (Courtesy Eastman Kodak Company.)

x-rays

The photoelectric effect provides convincing evidence that photons of light can transfer energy to electrons. Is the inverse process also possible? That is, can part or all of the kinetic energy of a moving electron be converted into a photon? As it happens, the inverse photoelectric effect not only does occur, but also had been discovered (though not at all understood) before the theoretical work of Planck and Einstein.

In 1895, in his laboratory at Würzburg, Wilhelm Roentgen accidentally observed that a screen coated with a fluorescent salt glowed every time he switched on a nearby cathode-ray tube. Roentgen knew that cathode rays themselves could not escape through the glass walls of his tube, yet it was clear some sort of invisible radiation was falling on the screen. The radiation was very penetrating; thick pieces of wood, glass, and even metal could be interposed between tube and screen, and still the screen glowed. At length Roentgen found that his mysterious rays pass readily through human flesh and leave a photographic record of bones beneath the flesh. With these observations Roentgen announced his discovery to the world, christening the radiations X-rays after the algebraic symbol for an unknown quantity.

X-rays are produced where cathode rays (which are fast electrons) stop. Rapidly moving electrons must be brought to rest suddenly; the more abruptly they are stopped, the more powerful are the resulting X-rays. An X-ray tube is simply a cathode-ray tube designed to produce and to bring to a sudden stop a dense beam of fast electrons. Usually the stopping takes place in the metal of the anode (Fig. 14.3). Crudely, we might compare the production of X-rays to the stopping of a bullet by a tree; just as the bullet buries itself in the wood and gives up its kinetic energy as heat, so an electron buries itself in the anode and gives up some of its energy as X-rays.

x-rays are high-frequency electromagnetic waves

What are X-rays? After many attempts had been made to determine their nature, Max von Laue and his students in 1912 were able to show definitely by means of an interference-type experiment that they are

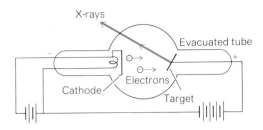

14.3 An X-ray tube. *High-frequency electromagnetic waves called X-rays are emitted by a metal target when it is bombarded by fast electrons.*

electromagnetic waves of extremely high frequency. X-rays contain frequencies higher than those in ultraviolet light, but somewhat lower than those of the gamma rays produced by radioactive atomic nuclei.

Even the early workers with X-rays noted that increasing the voltage applied to the tube—which means faster electrons striking the anode—led to the production of X-rays of greater penetrating power. Experiments showed that the more penetrating X-rays were of higher frequency than the less penetrating ones, so it was clear that there was a direct relationship between the speed of the electrons and the frequency of the resulting X-rays. Furthermore, the more intense the beam of electrons, the more intense the beam of X-rays, but there was no correlation between the electron intensity and the X-ray energy; many slow electrons would produce many low-frequency X-rays, and a few fast electrons would produce a few high-frequency X-rays.

The quantum theory of light is in complete accord with these observations. Most of the electrons striking the target lose their kinetic energy gradually in numerous collisions, their energy merely going into heat. (It is for this reason that the targets in X-ray tubes are normally of high-melting-point metals, and an efficient means of cooling the target is often employed.) A few electrons, though, lose most or all of their energy in single collisions with target atoms, and this is the energy that is evolved as X-rays. X-ray production, then, represents an inverse photoelectric effect. Instead of photon energy being transformed into electron kinetic energy, electron kinetic energy is being transformed into photon energy. The energy of an X-ray photon of frequency f is hf, and therefore the minimum kinetic energy of the electron that, in stopping, produced the X-ray photon must be equal to hf; that is,

x-rays are produced by an inverse photoelectric effect

$$KE_{min} = hf = \text{photon energy}$$

which agrees with the experimental data.

the compton effect

A further triumph for the quantum theory of light came in 1923. The American physicist Arthur H. Compton noticed that when a beam of X-rays passed through a gas, some of the X-rays were scattered off to the side. The scattered X-rays were found to be of lower frequency than the original beam. The quantum theory provides a straightforward explanation for this unexpected observation.

According to the quantum theory of light, photons behave like particles except for the absence of any rest mass. If this is true, then it should be possible for us to treat collisions between photons and, say, electrons in the same manner as billiard-ball collisions are treated in elementary mechanics. Figure 14.4 shows how such a collision might be represented, with an X-ray photon striking an electron and being scattered away from its original direction of motion, while the electron receives an impulse and begins to move. When this happens the photon must give up some of its energy to the electron, just as in any other collision. If the electron energy after the collision is KE, then, from the law of conservation of energy,

photons behave like particles in collisions

Loss in photon energy = gain in electron energy

$$hf - hf' = \text{KE}$$

where f is the initial frequency of the X-ray, and f' is its frequency after having been scattered. Compton was able to verify his idea through careful measurements, and subsequently received a Nobel prize for this work.

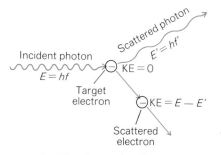

14.4 The Compton effect. *When an X-ray photon is scattered by an electron, its new frequency is lower than before because part of its original energy is imparted to the electron in the collision.*

photons and gravity

Although a photon has no rest mass, since light cannot be "at rest" in any frame of reference but always travels in free space with the speed c, it nevertheless behaves as though it possesses inertial mass. According to the quantum theory of light, the energy of a photon is hf. According to the theory of relativity, the total energy of a photon is mc^2, where m is its inertial mass. Equating these two values, we find

$$mc^2 = hf$$
$$m = \frac{hf}{c^2}$$

photons behave gravitationally as though they possess mass

Does a photon possess gravitational mass as well? Since light is deflected by a gravitational field, as we learned in Chap. 5, it would seem natural to assume that photons have the same gravitational behavior as other particles. This assumption turns out to be valid and has been verified by measurements both made on starlight and carried out entirely in the laboratory.

Let us consider first what happens when a rapidly moving particle leaves the surface of a star. As it travels outward, the particle gives up some of its energy in overcoming the gravitational attraction of the star—the particle must do work to leave the star. Hence the final kinetic energy of the particle when it is out in space is *less* than its initial kinetic energy at the star's surface.

Now we transfer the same reasoning to the case of a photon of frequency f emitted by the star. The photon's initial energy is hf at the star's surface. When it is a large distance away, for instance at the earth, the photon has given up some of its original energy to the gravitational field of the star and now has the lower energy hf'. Since hf' must be less than hf, the photon has a *lower* frequency than it had when it was emitted (Fig. 14.5). (The gravitational attraction of the earth for the photon will, of

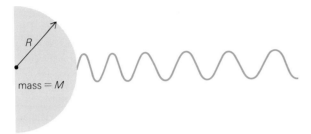

14.5 The frequency of a photon emitted from the surface of a star decreases as it moves away from the star, hence its wavelength increases.

course, increase its energy from hf', but by a negligible amount since the earth's mass is only about a millionth that of a typical star.) A photon in the visible region of the spectrum is thus shifted toward the red (low-frequency) end, and this phenomenon is accordingly known as the *gravitational red shift*. It must be distinguished from the Doppler red shift observed in the spectra of distant galaxies due to their apparent recession from the earth, a recession attributed to the general expansion of the universe. The change in frequency of a photon due to the gravitational red shift is calculated to be

$$\frac{\Delta f}{f} = \frac{f - f'}{f} = \frac{GM}{c^2 R}$$

where M is the star's mass and R is its radius.

As we shall learn in Chap. 18, the atoms of every element, when suitably excited, emit photons of certain specific frequencies only. The existence of the gravitational red shift can therefore be checked by comparing the frequencies found in stellar spectra with those in spectra obtained in the laboratory. Because the gravitational constant G is a very small quantity (6.67×10^{-11} newton-m^2/kg^2) and the square of the speed of light is a very large quantity (9×10^{16} m^2/sec^2), the ratio G/c^2 is extremely small and consequently the gravitational red shift can be observed only in radiation from very dense stars. In the case of the sun, $\Delta f/f$ amounts to only 1 part in 10^6, and is undetectable.

However, there is a class of stars in the final stages of their evolution called *white dwarfs* that are composed of atoms whose electron structures have "collapsed," and such stars have quite enormous densities—typically about 5 tons/in.^{3} A white dwarf might have a radius about 1 percent that of the sun while its mass might be about 60 percent that of the sun, and in this case $\Delta f/f$ would be about 1 part in 10^4, which is measurable under favorable circumstances. Actual observations of white dwarf spectra reveal red shifts that agree reasonably well with the predicted values, which confirms that photons possess gravitational mass. Recently it has become possible to measure the gravitational change in frequency of extremely high frequency electromagnetic waves that have "fallen" through a certain height in the laboratory, and the results again agree with the predictions.

MATTER WAVES

The distinction between particles and waves, a distinction that seems so obvious in our everyday world, becomes hazy in the world of the very small when we learn that light has both wave and particle aspects. The distinction grows still more ambiguous when we find that objects with rest mass—things like electrons, protons, and alpha particles, whose particle nature we have "proved"—sometimes act just like waves. In retrospect it may seem odd that two decades passed between the 1905 discovery of the particle properties of waves and the speculation that the converse might also be true. However, it is one thing to suggest a revolutionary hypothesis to explain otherwise mysterious data and quite another to advance an equally revolutionary hypothesis in the absence of a strong experimental mandate. The latter is what Louis de Broglie did in 1924 when he proposed that matter possesses wave as well as particle characteristics. So different was the intellectual climate at the time from that prevailing at the turn of the century that de Broglie's notion soon received respectful attention, while the earlier quantum theory of light of Planck and Einstein created hardly any stir despite its striking empirical support.

de broglie wavelength

moving bodies exhibit wave properties

Although it is hard enough to believe that light sometimes behaves as if it has wave properties and sometimes as if it has particle properties, it is harder still to accept the idea that a particle can sometimes behave as if it were a wave. Yet de Broglie's hypothesis could not be ignored, for in the next few years a number of experiments were performed with beams of fast electrons that demonstrated that moving particles do have dual characters. The wavelengths involved are too short to be significant in the large-scale world to which our senses respond, but in the atomic world that lies beneath what we see, the wave nature of particles and the particle nature of waves are of the utmost importance.

According to the quantum theory of light, the momentum of a photon whose wavelength is λ is h/λ, where h is Planck's constant. Drawing upon an intuitive feeling for symmetry in nature, de Broglie asserted that this relationship is a completely general one that applies to material particles as well as to photons. The momentum of a particle of mass m and velocity v is mv, and consequently its *de Broglie wavelength* is

de broglie wavelength
$$\lambda = \frac{h}{mv}$$

There is nothing imaginary about these *matter waves*. They are perfectly real, just as light waves or sound waves are. However, matter waves are not necessarily detectable in every situation. A 3,000-lb car moving at 30 mi/hr has a de Broglie wavelength of only about 10^{-38} ft, which is so small relative to its size that no wave behavior can be observed. On the other hand, an electron whose speed is 10^7 m/sec has a de Broglie wavelength of 10^{-8} m, which is comparable with atomic dimensions, and

WAVES AND PARTICLES

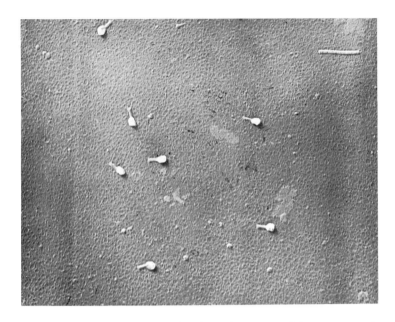

An electron micrograph of bacteriophages, magnified 80,000 ×. (Percy Brooks)

the wave behavior of electrons has been demonstrated in many experiments.

Not only electrons but also all other particles can be shown to behave like waves under suitable circumstances. All matter may be said to consist of waves just as correctly as it is said to consist of particles. In fact for many aspects of the behavior of small particles their wave nature provides a better explanation than their particle nature, as we shall see in subsequent chapters. As with electromagnetic waves, the wave and particle aspects of moving bodies can never be simultaneously observed, so that we cannot determine which is the "correct" description. All we can say is that sometimes a moving body exhibits wave properties and at other times it exhibits particle properties.

the wave and particle pictures of moving bodies complement each other

the electron microscope

Shortly after their discovery, matter waves were used for a practical purpose similar to a use we often make of light waves: to illuminate objects in a microscope. The difficulty with light waves for microscopic work is that they are too long—the magnifying power of a microscope is limited by the fact that objects cannot be distinguished unless they are somewhat larger than the waves of light reflected from them. Shorter waves, like the waves of ultraviolet radiation or X-rays, would improve the performance of a microscope, but to find materials for lenses that will both transmit and refract these radiations has proved difficult. Electron waves are an excellent solution to this problem, for they have wavelengths similar to those of X-rays and at the same time carry a charge, which permits them to be controlled by electric and magnetic fields. To construct an electron microscope, then, we must let a beam of electrons pass through an object,

because the wavelengths of the fast electrons in an electron microscope are shorter than those of the light waves in an ordinary microscope, the electron microscope can produce sharp images at higher magnifications

then direct it through a system of fields that act like lenses in focusing the beam, and, finally, let it fall on a fluorescent screen or photographic plate so that the image may become visible (Fig. 14.6). The remarkable photographs obtained with electron microscopes are evidence of the physical reality of electron waves.

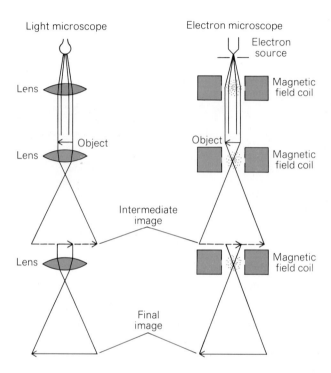

14.6 *Comparison of the principles of the conventional optical microscope and the electron microscope.*

probability density

wave function

The variable quantity that characterizes de Broglie waves is called the *wave function*, usually denoted by the symbol ψ (the Greek letter *psi*). The value of the wave function associated with a moving particle at a particular point in space at a particular time is related to the likelihood of finding the body there at that time.

The wave function ψ itself has no direct physical significance in terms of an experiment. There is a simple reason for this. The probability that something be somewhere at a given time can have any value from 0 (the particle is definitely not there) to 1 (the particle is definitely there). A probability of 0.2, for instance, signifies a 20 percent chance of finding the body. But the amplitude of any wave may be negative as well as positive, and a negative probability is meaningless—a quite different situation from that of a negative electric field in the case of a light wave or from that of a negative amplitude (meaning a trough) in the case of a water wave. Hence ψ itself cannot be an observable quantity.

This objection does not apply to ψ^2, the square of the wave function, which is known as the *probability density*. The probability of experimentally finding the body described by a wave function ψ at a particular place at a particular time is proportional to the value of ψ^2 there at that time. A large value of ψ^2 means the strong possibility of the body's presence, whereas a small value of ψ^2 means the slight possibility of its presence.

The de Broglie waves associated with a moving body are in the form of a group, or packet, of waves, as in Fig. 14.7. This wave packet travels with the same speed v as the body does.

probability density

the wave packet associated with a moving body moves with the same speed as the body

THE UNCERTAINTY PRINCIPLE

The fact that a moving particle must be regarded as a wave packet rather than as a localized entity suggests that there is a fundamental limit to the accuracy with which we can measure its particle properties, such as position and momentum. This limit is expressed in the *uncertainty principle*, one of the most significant of physical laws. We shall encounter the uncertainty principle many times in the chapters to come, since it provides the sole means of interpreting many otherwise puzzling phenomena in the world of atoms, molecules, and nuclei.

14.7 *The packet of de Broglie waves that corresponds to a certain particle moving with the speed v in the x direction.*

the uncertainty principle

The particle whose wave packet is shown in Fig. 14.7 may be located anywhere within the packet. Of course, the probability density ψ^2 is a maximum in the middle of the packet, but we may still find the particle somewhere else, in fact anywhere that ψ^2 is not actually 0. If the wave packet of a certain moving particle is very narrow, as in Fig. 14.8a, the position of the particle is readily established, since ψ^2 is 0 outside an extremely small region. However, in this case it is impossible to find the wavelength accurately, and, since $\lambda = h/mv$, this means that we cannot determine the particle's momentum. At the other extreme a wide wave packet, as in Fig. 14.8b, permits a satisfactory wavelength estimate, but where is the particle located?

A straightforward calculation based upon the mathematical properties of wave packets makes it possible to relate the inherent uncertainty Δx in a measurement of particle position with the inherent uncertainty Δmv in a simultaneous measurement of its momentum. The result is usually written in the form

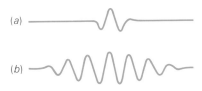

14.8 *(a) A narrow wave packet. The position of the particle can be accurately determined, but the wavelength (and hence the particle's momentum) cannot be established. (b) A wide wave packet. The wavelength can be accurately determined, but not the position of the particle.*

$$\Delta x\, \Delta mv \geq \frac{h}{2\pi}$$

the uncertainty principle

where h is Planck's constant. The sign $\geq$ (which means "equal to or greater than") is a consequence of the fact that wave packets may have different shapes: under the best of circumstances, $\Delta x\, \Delta mv = h/2\pi$, while under other circumstances $\Delta x\, \Delta mv$ is greater than $h/2\pi$. The Δx and

it is impossible to determine both the exact position and exact momentum of any body

Δmv are irreducible minima that follow from the wave natures of moving bodies. Any experimental errors that arise in the actual conduct of a measurement only augment the product $\Delta x \, \Delta mv$.

The above formula is one form of the uncertainty principle first obtained by Werner Heisenberg in 1927. It states that the product of the uncertainty Δx in the position of a particle at some instant and the uncertainty Δmv in its momentum at the same instant is at best equal to $h/2\pi$. If we arrange matters so that Δx is small, corresponding to the narrow wave packet of Fig. 14.8a, then Δmv will be large. If we reduce Δmv in some way, corresponding to the wide wave group of Fig. 14.8b, then Δx will be large. These uncertainties are not in our apparatus but in nature.

a hypothetical experiment

Let us approach the uncertainty principle from another point of view, basing the argument now on the particle nature of waves instead of the wave nature of particles. Suppose we could construct a microscope powerful enough to see an electron. We cannot use ordinary light to illuminate the electron on the stage of this microscope, because ordinary light has a wavelength about a hundred million times the diameter of the electron, and a microscope will show details only down to a size near the wavelength of the light employed. But perhaps we can arrange to use some sort of high-frequency gamma rays, with wavelengths about the same as the electron's diameter. These rays would not be visible to our eyes, but we might design an appropriate device that would respond to them.

So we arrange to get an electron on the stage of our microscope, and we illuminate it by turning on the gamma-ray light. We see a momentary flash, and then the field is blank; our electron has disappeared. Then we recall that gamma rays of short wavelength are photons of very high energy. At least one of these photons must be reflected from the electron in order for us to see it, but when the electron is struck by the photon it recoils out of the microscope field (Fig. 14.9). If the electron is to stay in our field of view, we shall have to be content with gamma rays

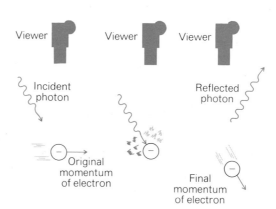

14.9 The uncertainty principle. *An electron cannot be observed without changing its momentum by an unknown amount; thus position and momentum cannot both be precisely determined.*

of longer wavelength and lower energy. Making this change to longer waves, we again look down the microscope tube. Now the electron moves more slowly, so that we can actually see it; but we see it only as a hazy blur, because with longer waves our microscope loses its power to show details clearly. No matter how we alter the microscope, we cannot improve matters. It is impossible to obtain a clear look at the electron without disturbing it.

In the first of the above experiments we tried to see the electron as a distinct image in a definite spot, and the electron at once moved out of our field of view. In the second experiment we kept it in the field long enough to see it, but we did not locate it as a clearly defined image. If we calculate the relationship between the uncertainties Δx and Δmv in these experiments, we again find

$$\Delta x\, \Delta mv \geq \frac{h}{2\pi}$$

We can get rough values for both position and velocity, but if we try to get the position more accurately, the measurement of velocity becomes very inaccurate, and if we try to determine the velocity precisely, the position becomes uncertain. The indeterminateness is not due to faulty experimental technique. Whether we think in terms of the impossibility of observing electrons without disturbing them or in terms of the wave nature of moving particles, we get exactly the same result, which is expressed in the uncertainty principle.

the uncertainty principle represents a fundamental limit to experimental accuracy to which instrumental errors and inaccuracies must be added

waves versus particles

Well then, what *is* an electron? It has mass, and so we have spoken of it as a particle in the sense of being a very tiny object that obeys the same laws of motion as the larger objects around us. When it is moving it seems to be a type of wave motion and exhibits certain properties that lie exclusively in the domain of waves. To us the two are mutually exclusive: a wave is a wave; a particle is a particle. How can we determine the true nature of the electron? Or, perhaps, is the question itself meaningless, because we are trying to use models based on ordinary experience to describe entities so small that we have only indirect evidence of them? To put it another way, just because we do not find things that sometimes seem like waves and at other times like particles in everyday life, we might not be justified in assuming that such dual personalities are not truly characteristic of electrons, protons, neutrons, and the like.

The uncertainty principle makes it impossible to decide experimentally whether a moving body is a wave or a particle in the sense that no experiment can be devised to give us a definite answer.

The clearest picture we can get of an electron is to regard it as a particle and to think of the associated waves as "waves of probability." According to the uncertainty principle, if we know anything at all about the velocity of an electron we cannot know its position exactly, but we can express its position as a probability of its being in any particular spot. Not only the present position but future positions and velocities of the electron may be expressed as probabilities. We cannot say just where an electron will be 2 sec hence or how fast it will be moving, but we can say that it will be more likely to be in one place than another and that its speed will be more likely to have one value than another.

Thus an essential unpredictability is inherent in nature. We describe the behavior of large objects like stones or rifle bullets in terms of definite laws. Knowing the position and speed of a falling stone, we can predict accurately where it will be at future times by using the law of falling bodies. We are confident in our predictions, because all falling stones of our experience have behaved similarly. Stones and other objects of everyday life obey the principle of *causality*, that similar events always follow similar causes. But electrons (and other submicroscopic particles) are different; we cannot make accurate predictions about an electron's future motion, and we cannot be sure that two electrons with roughly similar positions and velocities will behave at all alike. The principle of causality, expressing our faith in the uniformity of natural processes, does not hold for subatomic particles.

causality is valid only for macroscopic bodies

All bodies, of any size whatever, are affected by the uncertainty principle, which means that their motions likewise can be expressed only as probabilities. Since the objects of everyday life are made up of small particles, we might conjecture that even their behavior should be a matter of probability rather than of fixed laws. There is a chance, for example, that this book will someday defy the law of gravity and rise straight up in the air. But for objects this large—in fact, even for objects the size of molecules—the relevant probabilities are so large as to be practically certainties. The likelihood that this book will continue to obey the law of gravity is so large that we might wait a trillion years without seeing any sign of deviation. Only for electrons and other particles of similar size is there appreciable chance for differences from expected behavior.

GLOSSARY

The *photoelectric effect* is the emission of electrons from a metal surface when light is shone upon it.

Quanta, or *photons*, are tiny packets of energy whose motion constitutes the propagation of light; under certain circumstances light appears as an electromagnetic wave phenomenon, and under others as a quantum phenomenon. The energy of each quantum of light is hf, where h is *Planck's constant* and f is the frequency of the light.

X-rays are high-frequency electromagnetic waves produced whenever fast electrons are brought to rest quickly.

The *Compton effect* occurs in the scattering of X-rays by electrons; the scattered X-rays have lower frequencies than they had originally, corresponding to the loss of energy by their photons upon colliding with electrons.

WAVES AND PARTICLES

A *matter wave* is associated with rapidly moving bodies, whose behavior in certain respects resembles wave behavior. The wavelength of such matter waves is h/mv, where h is Planck's constant, m is the mass of the body, and v is its velocity. The matter waves associated with a moving body are in the form of a group, or packet, of waves which travel with the same speed as the body.

The probability of finding a moving particle at a certain place at a certain time is proportional to the *probability density* ψ^2, where ψ, the *wave function* of the particle, is the variable that characterizes matter waves.

The *uncertainty principle* states that it is impossible simultaneously to determine the position and momentum of a body with complete accuracy. Hence, in dealing with such problems as those involving the physics of electrons, all we can consider are probabilities rather than specific positions and states of motion.

EXERCISES

1. List as many differences as you can between a photon and an electron.

2. Compare the evidence for the wave nature of light with the evidence for its particle nature.

3. Give a reason why the wave aspect of light was established long before its particle aspect.

4. Energy is transported in light by means of separate photons, yet even the faintest light that we can perceive does not appear as a series of flashes. Explain.

5. Explain the relationship between the photoelectric effect and the Compton effect.

6. The uncertainty principle applies to *all* bodies, yet its consequences are only significant for such extremely small particles as electrons, protons, and neutrons. Explain.

7. The atoms in a solid have a certain minimum "zero-point" energy even at $0°K$, while the molecules in an ideal gas would be at rest at $0°K$. Use the uncertainty principle to explain these statements.

8. Find the energy of the photons in light whose wavelength is 2.5×10^{-7} m.

9. Calculate the amount of energy in a photon of ultraviolet light whose frequency is 2×10^{16} Hz. Do the same for a photon of radio waves whose frequency is 2×10^5 Hz.

10 Yellow light has a frequency of about 5×10^{14} Hz. How many photons are emitted per second by a lamp that radiates yellow light at a power of 100 watts?

11 The eye can detect as little as 10^{-18} joule of energy in the form of light. How many photons of frequency 5×10^{14} Hz does this amount of energy represent?

12 How many photons per second are emitted by a 10,000-watt radio transmitter whose frequency is 880 kHz?

13 The radiant energy reaching the earth from the sun is about 1,400 watts/m². If this energy is all green light of wavelength 5.5×10^{-7} m, how many photons strike each square meter per second?

14 An energy of 4×10^{-19} joule is required to remove an electron from the surface of a particular metal. What is the frequency of the light that will just dislodge photoelectrons from the surface? What is the maximum energy of photoelectrons emitted through the action of light of wavelength 2×10^{-7} m?

15 A photon whose energy is 2×10^5 ev strikes a free electron which acquires an energy of 1.5×10^5 ev in the collision. Find the frequency of the scattered photon.

16 Electrons are accelerated through potential differences of approximately 10,000 volts in television picture tubes. Find the maximum frequency of the X-rays that are produced when these electrons strike the screen of the tube.

17 Calculate the de Broglie wavelengths of
  a. an electron of mass 9.1×10^{-31} kg and velocity 1.5×10^8 m/sec*
  b. an automobile of mass 1,500 kg and velocity 30 m/sec

18 A 10^4-kg truck has a de Broglie wavelength of 2.7×10^{-39} m. Find its speed. Will its wave properties be observable?

19 An electron microscope uses 40-kev (4×10^4 ev) electrons. Find its ultimate resolving power on the assumption that this is equal to the wavelength of the electrons.

20 The position and momentum of a 1,000-ev electron are simultaneously determined. If the position is located to within 10^{-10} m, what is the percentage of the uncertainty in its momentum?

---

*The relativistic increase in mass of the electron must be taken into account. See Chap. 5 and the footnote on p. 105.

chapter 15

Atoms and Molecules

ELEMENTS AND COMPOUNDS

chemical change

elements, compounds, and mixtures

combustion

oxygen

conservation of mass

ATOMS AND MOLECULES

the atomic theory

symbols and formulas

avogadro's hypothesis

ATOMIC STRUCTURE

rutherford's experiment

atomic number

failure of classical physics

Far in the past people began to suspect that matter, despite its appearance of being continuous, actually possesses a definite structure on a microscopic level beyond the direct reach of our senses. This suspicion did not take on a more concrete form until a little over a century and a half ago. Since then the existence of atoms and molecules, the ultimate particles of matter in its common forms, has been amply demonstrated, and their own ultimate particles have been identified and studied as well. In this chapter and in several others to come our chief concern will be the structure of the atom, since it is this structure that is responsible for nearly all the properties of matter that have shaped the world around us.

ELEMENTS AND COMPOUNDS

Chemistry as a science is relatively new, since its early history was occupied almost exclusively with a search for a way of converting ordinary metals into gold. This fruitless quest, called *alchemy* by the Arabs, was not abandoned until the seventeenth century, when men like John Mayow and Robert Boyle in England, Jean Rey in France, and Georg Stahl in Germany began a realistic inquiry into the properties of matter and their changes during chemical reactions.

chemical change

Properties of materials undergo profound changes in many familiar processes. We have already discussed the melting of solids and the vaporization of liquids, which are changes brought about by alterations in molecular motions and separations. Other processes, such as the rusting of iron, the burning of wood, and the explosion of dynamite, involve more drastic changes in properties and alterations in the molecules themselves rather than merely in their motions. Processes of this sort that consist of molecular changes are called *chemical reactions*.

<small>a chemical reaction involves fundamental changes in the properties of the interacting substances</small>

Let us examine a specific chemical reaction in a qualitative way. Suppose that we mix some powdered zinc metal with a somewhat greater amount of powdered sulfur on a sheet of asbestos and ignite the mixture with the flame from a bunsen burner. The result is an explosion, and light and heat are given off. When the fireworks have died down and any excess sulfur has burned away, we are left with a brittle, white substance that resembles neither the original zinc nor the original sulfur. What has happened?

Further experiment would quickly show that neither zinc nor sulfur alone gives any such reaction on heating, that explosion of the mixture takes place as well in a vacuum as in air, and that a metal or porcelain surface may be substituted for the asbestos without influencing the reaction. In other words, the process requires the presence of both zinc and sulfur, but no other materials. We conclude that the two substances have *combined* to form the new material. To convince the most skeptical, we could find ways to change the white, brittle product back into zinc and sulfur, but to accomplish this change would require considerable time and resourcefulness.

<small>a homogeneous substance is one whose constituent particles are all identical</small>

This is a simple example of the kind of alteration in the properties of matter that we call a chemical reaction. Let us analyze it by examining the properties of the three substances concerned. Sulfur is a yellow solid, rather soft, with a low density, melting at about $114°C$, not dissolved by water or acid but easily dissolved by a liquid called carbon disulfide. Every grain of the yellow powder exhibits these properties, and if we should crush the grains their fragments would still show the same properties. Because every particle of sulfur is like every other particle, it is a *homogeneous* substance. Zinc, too, is homogeneous, with such characteristic properties as a light gray color, fairly high density, a melting point of $419.4°C$, solubility in dilute acids, and insolubility in carbon disulfide. The brittle, white product of reaction is a third homogeneous substance, called zinc sulfide, with properties different from those of either zinc or sulfur: a much higher melting point, an intermediate density, and insolubility in both very dilute acids and carbon disulfide.

Corrosion is an example of chemical change.

Now suppose we prepare two identical zinc-sulfur mixtures, heating one until it reacts but leaving the second unheated. Both the resulting zinc sulfide and the unheated mixture contain zinc and sulfur, and both differ from pure zinc and sulfur in such properties as color and density. But the mixture is a *heterogeneous* material: its properties change from one particle to the next. With a microscope and a needle we can separate particles of zinc from particles of sulfur. Carbon disulfide will dissolve part of the mixture, and dilute acid will dissolve the rest. Properties of individual particles in the mixture have not changed at all.

In the zinc sulfide, on the other hand, every particle has the same properties as every other particle, and these properties are quite different from those of zinc and sulfur particles. Two substances can be mixed,

a heterogeneous substance is one whose constituent particles are of two or more different kinds

often very intimately, simply by stirring them together, but their properties are radically altered only if they undergo a chemical reaction.

Most familiar chemical changes do not involve so spectacular a liberation of energy as the zinc-sulfur reaction. Rust, for example, is the product of a slow combination of iron with certain gases of the atmosphere. Silverware slowly tarnishes because the metal combines with small amounts of sulfur contained in certain foods and in compounds in the air. Cooking involves many complex but hardly spectacular reactions. In the growth and decay of living things there occur even more complex types of slow chemical change. Among commonplace chemical processes only the reactions involved in burning result in rapid liberation of energy.

elements, compounds, and mixtures

elements are simple substances that cannot be decomposed or transformed into one another by chemical means

From the long, disheartening search of the alchemists ultimately emerged the idea that certain materials, like iron, mercury, gold, sulfur, were simple substances that could be neither decomposed nor transformed into one another. Such simple substances were called *elements*. The belief slowly grew that the earth contains only a limited number of these elements and that all other materials are combinations of them in various proportions. The formation of a new substance from others by chemical change, then, is possible only if its elements are present in the other substances. Never expressed as a formal law, this statement is nevertheless the fundamental axiom that distinguishes chemistry from alchemy.

These modern elements are concrete, tangible substances. Our belief that they are the building materials of all matter is founded on the fact that every other variety of matter that we can bring into the laboratory can be broken down into two or more of them. Of the 103 known elements, 11 are gases, 2 are liquids, and the remaining 90 are solids at room temperature and atmospheric pressure. Hydrogen, oxygen, chlorine, and neon are familiar gaseous elements; bromine and mercury are the two liquids; iron, zinc, tin, aluminum, copper, lead, silver, gold, carbon, and sulfur are among the solid elements.

two or more elements may combine to form a compound with characteristic properties of its own

Elements are put together to form the other materials of the earth in a variety of ways. Some materials contain two or more elements united in a chemical *compound*, as zinc and sulfur are united in the compound zinc sulfide. Other materials consist of *mixtures* of elements, or mixtures of compounds. The distinction between compounds and mixtures is of fundamental importance, since chemical reactions always involve the formation or the breaking down of one or more compounds.

Let us consider first the kinds of material that can be formed from two elements only. The elements may form a *heterogeneous mixture*, like the mixture of zinc and sulfur we have been considering; in such a mixture each element retains its own distinguishing properties, and small fragments of each can be separated mechanically from the mixture. The elements may be mixed more intimately to form a *homogeneous mixture*, or *solution;* thus the gases hydrogen and oxygen when placed in the same container will mix so thoroughly that no ordinary means will show that any one part of the mixture is different from any other part. Finally, the elements may form a compound; if an electric spark is allowed to jump through a mixture of hydrogen and oxygen, the two react violently, and liquid droplets of the compound water are formed.

The distinction between compounds and heterogeneous mixtures is easy, since the separate elements are still recognizable in the mixtures. It is not always a simple matter, however, to tell whether two elements have formed a solution or have undergone a chemical reaction to produce a compound. We may say in general that their properties are more profoundly altered if they have united chemically, and we may put the matter to experimental test in one of the three following ways:

1 Measure the freezing point or boiling point of the material. For a compound the freezing point or boiling point is a constant temperature, but for a solution the temperature changes during both boiling and freezing. Thus water, a compound of hydrogen and oxygen, boils at precisely $100°C$ (if the pressure is 1 atm); liquid air, a homogeneous mixture of nitrogen and oxygen, starts to boil at about $-192°C$ and continues to boil up to $-182°C$ as vaporization proceeds.

a compound boils and freezes at definite, constant temperatures

2 See whether the material can be separated into its elements by boiling or freezing. Ordinarily the composition of a compound is not altered by a change of state, but a solution is wholly or partially separated into its constituents. Water shows no tendency to decompose into its elements at $100°C$ or even far above this temperature; when liquid air boils, however, the vapor that comes off first is largely nitrogen, and the vapor that comes off in the last stages is largely oxygen.

a compound is not altered by a change of state

3 Add more of one of the constituent elements to the material, and see whether the material remains homogeneous. Elements in a compound are combined in a definite invariable proportion (the *law of definite proportions*), whereas the composition of a solution or mixture is variable. In water every gram of oxygen is combined with precisely 0.126 g of hydrogen; if more oxygen or more hydrogen is added, it does not mix with water but forms a heterogeneous mixture of gas and liquid. With liquid or gaseous air, on the other hand, more nitrogen or oxygen will mix readily, and the material remains homogeneous (Fig. 15.1).

according to the law of definite proportions, the elements in a compound are in specific ratios by weight

65.4 g Zinc + 32.1 g Sulfur = 97.5 g Zinc sulfide

75.4 g Zinc + 32.1 g Sulfur = 97.5 g Zinc sulfide + 10.0 g Zinc

15.1 An example of the law of definite proportions: Zinc and sulfur always combine in the same ratio of weights when they form zinc sulfide even though the actual weights may vary. *Any excess of either constituent is left over after the reaction.*

By a proper choice of reactant substances, fireworks of all colors can be produced. Usually the oxygen needed for the combustion of fireworks is incorporated in one of the ingredients.

ATOMS AND MOLECULES

Experiments of this sort will determine whether any unknown material is a compound or a solution. The experiments, of course, reflect the fact that in a compound the elements have combined to form a new substance, with characteristic properties of its own, whereas in a mixture each element retains its identity.

To sum up, all matter in bulk may be classified first into homogeneous and heterogeneous matter, and then further as follows:

Heterogeneous matter—mixtures of elements, compounds, or both
Homogeneous matter—pure substances (elements and compounds)
—homogeneous mixtures (solutions)

classification of matter

combustion

The first chemical change to be studied intensively by quantitative methods was the process of burning, or combustion. The transformation from wood to smoke and ashes, with its accompanying heat and dancing flames, is by all odds the most spectacular chemical change with which men of earlier times had immediate contact, and it has piqued the curiosity of thoughtful individuals from remotest antiquity to the present. Primitive man based his explanation on ever-present demons and spirits. In the religions of many advanced civilizations the fire god has a respected place. The more scientifically minded Greeks gave the first rational explanation in nonsupernatural terms, recorded in the writings of Aristotle: every inflammable material was suppose to contain the elements "earth" and "fire," the fire escaping while the material burned and the earth (ashes) remaining behind. In various guises this explanation of Aristotle's persisted through the centuries of alchemy down even to the time of the French Revolution.

The particular form this explanation took in the eighteenth century was the *phlogiston hypothesis* developed by two Germans, Johann Joachim Becher and Georg Stahl. The essential idea was the same as Aristotle's, but Becher and Stahl showed how it could be extended to many other reactions besides burning; and for the substance that supposedly escaped during burning, they abandoned Aristotle's simple word *fire* and coined the more esoteric term *phlogiston*. The story of the overthrow of the phlogiston hypothesis and the establishment of modern conceptions of chemical change is an impressive chapter in the history of ideas.

Today we never hear the word phlogiston, but in its day there was no more respected concept in all of chemistry. Early scientists explained combustion in the following way. All substances that can be burned contain phlogiston, and the phlogiston escapes as the burning takes place. We observe that air is necessary for combustion, but this is explained by assuming that phlogiston can leave a substance only when air is present to absorb it. Many metals when heated in air change slowly to soft powders; zinc and tin give white powders, mercury a reddish powder, iron a black scaly material. These changes, like the changes in ordinary burning, were ascribed to the escape of phlogiston; a metal was assumed to be a compound of powder plus phlogiston, and heating the metal simply

phlogiston is a substance invented by early chemists to account for the chemical and physical changes that occur in combustion; it does not exist

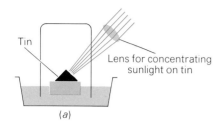

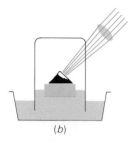

15.2 Lavoisier's experiment showing that tin, upon heating, combines with a gas from the air. (a) Before heating; (b) after heating. The tin is partly changed to a white powder, and the water level rises until only four-fifths as much air is left as there was at the start. Further heating causes no additional change.

caused the compound to decompose. Now, many of the powders can be changed back to metal by heating with charcoal, an observation interpreted to mean that charcoal must be a form of nearly pure phlogiston that simply reunited with the powder to form the compound (the metal). When hydrogen was discovered in 1766, its ability to burn without leaving any ash suggested that it was another form of pure phlogiston; one could predict, then, that heating an ore or a powder with hydrogen would form a metal, and this prediction was confirmed by experiment.

So far so good, but soon the phlogiston hypothesis ran into serious trouble. When wood burns its ashes weigh less than the original wood, and the decrease in weight can reasonably be ascribed to the escape of phlogiston. But, when a metal is heated until it turns into a powder, the powder weighs *more* than the original metal! The believers in phlogiston were forced to assume that phlogiston sometimes could have negative weight, so that if it left a substance the remaining material could weigh more than before. To us this notion of negative weight is nonsense, but in the eighteenth century it was taken quite seriously.

The French scientist Antoine Lavoisier conducted a series of experiments in the latter part of the eighteenth century that effectively demolished the phlogiston hypothesis. Son of a wealthy lawyer, Lavoisier was given a good education and had more than ample means for carrying on his scientific work. For many years of his busy life he served as a public official and showed himself keenly aware of the acute social problems that France was facing. But neither an immense scientific reputation nor long public service could save him during the Revolution; denounced by Marat, he was sent to the guillotine in 1794.

Lavoisier knew that tin could be converted to a white powder when it was heated, and the powder definitely weighed more than the original metal. To study the change in greater detail, Lavoisier placed a piece of tin on a wooden block floating in water (Fig. 15.2), covered the block with a glass jar, and heated the tin by focusing the sun's rays upon it with a magnifying glass—a common method of heating before gas burners and electric heaters were invented. The tin was partly changed to white powder, and the water level rose in the jar until only four-fifths as much air was left as there had been at the start. Further heating caused no detectable change. In another experiment Lavoisier heated tin in a sealed flask until as much as possible was converted to powder. The flask was weighed accurately before and after heating, and the two weights proved to be identical. Then the flask was opened, and air rushed in. With the additional air, the weight of the flask was greater than it had been at the start—and by the same amount as the increase in weight of the tin.

oxygen

To Lavoisier these experiments suggested clearly that the tin had absorbed a gas from the air. We need only imagine that one-fifth of the air consists of a gas that can combine with tin; then the powder is a compound of this gas with the metal, and the increase in weight is the weight of the gas. Water rises in the jar to take the place of the gas that has been removed; when the sealed flask is broken, air rushes in to replace the gas that the tin has adsorbed. This explanation is simple and direct, involving no phlogiston but only substances that have definite weights.

At about the time these experiments were completed, Lavoisier learned that an English experimenter, Joseph Priestley, had prepared a new gas with strange properties. Priestley was the poverty-stricken minister of a small church, with only limited time and equipment; yet his experimental talents led him to a number of significant discoveries. The gas that he had found caused lighted candles to flare up brightly and glowing charcoal to burst into flames, and a mouse kept in a closed jar of the gas lived longer than one kept in a closed jar of air. Lavoisier gave the gas its modern name, *oxygen*, and found it useful not only for explaining the changes in metals on heating but for explaining the processes of combustion as well. The burning of candles, wood, coal, according to Lavoisier, involves a combination of their materials with oxygen. They appear to lose weight because the products of the reaction are gaseous; actually, as shown by experiment, the gaseous products weigh more than the original material. Thus burning and the reactions of metals in air both received a complete explanation in terms of real substances, and the necessity for assuming the mysterious phlogiston disappeared. — oxygen

Oxygen under ordinary conditions is a colorless, odorless, tasteless gas. Air owes its ability to support combustion to the free oxygen that it contains; it cannot support combustion as well as pure oxygen because in air the element is so diluted with inactive gases. Air is composed of about one-fifth oxygen and four-fifths nitrogen, with small amounts of other gases.

When oxygen combines chemically with another substance, the process is called *oxidation*, and the other substance is said to be *oxidized*. Rapid oxidation accompanied by the liberation of noticeable heat and light is the process of *combustion*, or burning. In the experiments of Lavoisier, tin and mercury oxidized slowly when heated. A lighted candle oxidizes rapidly in air, still more rapidly in pure oxygen. Slow oxidation is involved in many familiar processes, such as the rusting of iron, the decay of wood, the hardening of paint. The energy to maintain life comes from the slow oxidation of food in our bodies, by oxygen breathed in through the lungs and transported by the blood stream. — oxidation is the process by which oxygen combines with another substance; combustion is rapid oxidation accompanied by the release of considerable energy

A substance formed by the union of another element with oxygen is called an *oxide*. The white powder that Lavoisier obtained by heating tin is tin oxide. Rust is largely iron oxide. In general, oxides of metals are solids. Oxides of other elements may be solid, liquid, or gaseous. The oxide of sulfur is an evil-smelling gas called sulfur dioxide; carbon forms two gaseous oxides, called carbon monoxide and carbon dioxide; the oxide of silicon is found in nature as the solid called quartz, the chief constituent of ordinary sand; the oxide of hydrogen is water. Oxides of nearly all the elements can be prepared, most of them simply by heating the elements with oxygen. A few oxides (mercuric oxide, lead oxide, barium peroxide) are easily decomposed by heating, giving one convenient laboratory method for preparing oxygen; other oxides, such as lime (calcium oxide), are not decomposed even at the temperature of an electric arc, $3000°C$. — an oxide is a compound of oxygen and another element

A fire can be extinguished either by cooling it below the ignition temperature of whatever is burning or by smothering it with an agent that keeps air, and hence oxygen, from the burning material. Here a gasoline fire is being put out by the latter method with a spray of Freon FE-1301, which is widely used in aircraft fire extinguishers.

conservation of mass

Lavoisier's discovery of the true nature of combustion was made possible by his use of the balance and by his insistence on the importance of weights in studying chemical reactions. This emphasis on weights marked a profound change in viewpoint and is one of Lavoisier's great contributions to chemistry. From his day to ours the balance has remained the chemist's most valuable tool.

The balance not only enabled Lavoisier to overthrow the phlogiston hypothesis; it also led him to a generalization as fundamental to modern chemistry as the law of energy conservation is to physics. Accurate weighing of many substances before and after undergoing chemical reaction convinced him that the total mass of the products of a chemical reaction is always the same as the total mass of the original materials, no matter how startling the chemical change may be. This is the *law of conservation of mass*. It may be stated more tersely: Matter can be neither created nor destroyed. When wood burns, mass seems to disappear because some of the products of reaction are gases; if the mass of the original wood is added to the mass of the oxygen that combined with it and if the mass of the resulting ash is added to the mass of the gaseous products, the two sums will turn out exactly equal. Iron increases in weight on rusting because it combines with gases from the air, and the increase in weight is exactly equal to the weight of gas consumed. In the thousands of reactions that have been tested with accurate chemical balances, no deviation from the law has ever been found.

law of conservation of mass

As we learned in Chap. 5, relativity has altered our ideas about the conservation of mass and energy. We can no longer assert that matter and energy are indestructible, for many processes involve the transformation of matter into energy, or vice versa. To take account of these processes we must for accuracy combine the two conservation laws in the single statement: The total amount of energy, including its mass equivalent, in the universe is constant. Transformations of matter into energy will be discussed in future chapters; in chemical reactions, the original separate conservation laws hold within the limits of experimental error.

ATOMS AND MOLECULES

At the start of the nineteenth century nobody had any defensible ideas about the structure of matter, of how elements go to make up compounds. The theory that answered these questions came from a most unlikely individual, an awkward, colorless, poorly educated teacher named John Dalton. Dalton was a plodding, literal thinker who had had no formal instruction in physics or chemistry, but these liabilities turned out to be assets, because they meant that he constantly sought simple explanations for complex phenomena without being hampered by the misconceptions of other people.

the atomic theory

The atomic theory of matter emerged from Dalton's crude attempts at picturing the ultimate particles of which gases were composed. He began with the notion of Democritus and Lucretius that everything was composed of *atoms* and made this idea quantitative. Some of Dalton's original ideas have had to be modified, and rather than go into the full story of the evolution of the atomic theory we shall simply summarize the modern picture of atoms and molecules.

molecules are the ultimate particles of a compound

In brief, the ultimate particles of any compound are called *molecules*. Molecules may be further broken down, but when this happens they are no longer representative of the essential characteristics of the compound.

atoms are the ultimate particles of an element

The molecules of a compound consist of the *atoms* of its constituent elements. Although the ultimate particles of any element are atoms, many elemental gases consist of molecules instead of individual atoms. Thus gaseous oxygen contains oxygen molecules, each of which is a pair of oxygen atoms bound together by forces whose nature we shall explore in Chap. 20. Other elemental gases, for instance helium and neon, consist of single atoms. Elemental solids and liquids are assemblies of individual atoms.

The molecules of a compound have definite, invariable compositions and structures, as Fig. 15.3 shows. Each water molecule contains two hydrogen atoms and one oxygen atom with the hydrogen atoms 105° apart, for example, while each ammonia molecule contains three hydrogen atoms 107.5° apart.

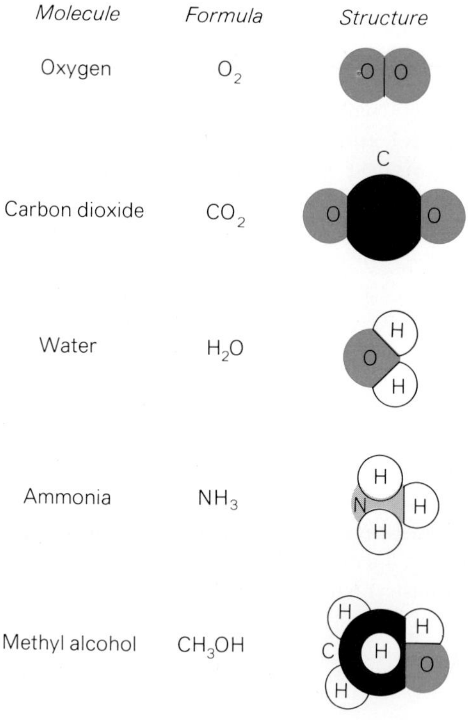

Molecule	Formula	Structure
Oxygen	O_2	
Carbon dioxide	CO_2	
Water	H_2O	
Ammonia	NH_3	
Methyl alcohol	CH_3OH	

15.3 Structures of several common molecules.

symbols and formulas

By convention an atom of an element is represented by an abbreviation of the element's name. For many elements the first letter is used; an atom of oxygen is O, an atom of hydrogen H, an atom of carbon C. When the names of two elements begin with the same letter, two letters are used in the abbreviation for one or both: Cl stands for an atom of chlorine, He for helium, Zn for zinc. For some elements abbreviations of Latin names are used: a copper atom is Cu (cuprum), an iron atom Fe (ferrum), a mercury atom Hg (hydrargyrum). These abbreviations are called *symbols* for the elements. Table 15.1 is a list of the elements in alphabetical order together with their chemical symbols.

chemical symbols

Two or more atoms joined to form a molecule are represented by writing their symbols side by side; a carbon monoxide molecule is CO, a zinc sulfide molecule ZnS, a mercuric oxide molecule HgO. When a molecule contains two or more atoms of the same kind, a small subscript indicates the number present; the familiar expression H_2O means that a molecule of water contains two H atoms and one O atom; a molecule of oxygen, containing two O atoms, is written O_2; a molecule of carbon tetrachloride (CCl_4) contains one C atom and four Cl atoms; a molecule of nitrogen pentoxide (N_2O_5) contains two N (nitrogen) atoms and five O atoms. Each subscript applies only to the symbol immediately before it. These expressions for molecules are called *formulas*.

chemical formulas

Symbols and formulas give the chemist a convenient shorthand for expressing the joining together and separating of atoms, which are the fundamental processes of chemical change.

avogadro's hypothesis

In the year 1808 the French scientist Joseph Gay-Lussac concluded, on the basis of very careful experiments, that, when gases undergo chemical reactions, the *volumes* of the reacting and product gases are related by simple whole numbers. Thus, when water is decomposed into hydrogen and oxygen, the volume of the hydrogen is exactly twice the volume of the oxygen; when nitrogen and oxygen unite to form the colorless gas nitric oxide, one volume each of nitrogen and oxygen produce two volumes of nitric oxide. Here were some numerical relations that the atomic theory ought to explain but could not. According to Dalton, each molecule of nitric oxide should contain an atom each of oxygen and nitrogen. Since the molecules in a gas are far apart, their size should make no difference in the volume occupied by gas, and so Dalton reasoned that one volume of nitrogen plus one volume of oxygen should yield one volume of nitric oxide. Dalton therefore dismissed Gay-Lussac's measurements as incorrect.

The young Italian physicist Amadeo Avogadro was not so hasty. Three years after Gay-Lussac's work, he put forward two important ideas. First, he asserted that **equal volumes of all gases, under the same conditions of pressure and temperature, contain the same number of molecules.** He suggested further that the molecules of some gaseous elements might not consist simply of individual atoms of the

avogadro's hypothesis

TABLE 15.1 Chemical symbols of the elements.

Element	Symbol	Element	Symbol
Actinium	Ac	Lithium	Li
Aluminum	Al	Lutetium	Lu
Americium	Am	Magnesium	Mg
Antimony	Sb	Manganese	Mn
Argon	Ar	Mendelevium	Md
Arsenic	As	Mercury	Hg
Astatine	At	Molybdenum	Mo
Barium	Ba	Neodymium	Nd
Berkelium	Bk	Neon	Ne
Beryllium	Be	Neptunium	Np
Bismuth	Bi	Nickel	Ni
Boron	B	Niobium	Nb
Bromine	Br	Nitrogen	N
Cadmium	Cd	Nobelium	No
Calcium	Ca	Osmium	Os
Californium	Cf	Oxygen	O
Carbon	C	Palladium	Pd
Cerium	Ce	Phosphorus	P
Cesium	Cs	Platinum	Pt
Chlorine	Cl	Plutonium	Pu
Chromium	Cr	Polonium	Po
Cobalt	Co	Potassium	K
Copper	Cu	Praseodymium	Pr
Curium	Cm	Promethium	Pm
Dysprosium	Dy	Protactinium	Pa
Einsteinium	Es	Radium	Ra
Erbium	Er	Radon	Rn
Europium	Eu	Rhenium	Re
Fermium	Fm	Rhodium	Rh
Fluorine	F	Rubidium	Rb
Francium	Fr	Ruthenium	Ru
Gadolinium	Gd	Samarium	Sm
Gallium	Ga	Scandium	Sc
Germanium	Ge	Selenium	Se
Gold	Au	Silicon	Si
Hafnium	Hf	Silver	Ag
Helium	He	Sodium	Na
Holmium	Ho	Strontium	Sr
Hydrogen	H	Sulfur	S
Indium	In	Tantalum	Ta
Iodine	I	Technetium	Tc
Iridium	Ir	Tellurium	Te
Iron	Fe	Terbium	Tb
Krypton	Kr	Thallium	Tl
Lanthanum	La	Thorium	Th
Lawrencium	Lr	Thulium	Tm
Lead	Pb	Tin	Sn

ATOMS AND MOLECULES

TABLE 15.1 Chemical symbols of the elements (cont.).

Element	Symbol	Element	Symbol
Titanium	Ti	Ytterbium	Yb
Tungsten	W	Yttrium	Y
Uranium	U	Zinc	Zn
Vanadium	V	Zirconium	Zr
Xenon	Xe		

element, but that they might be composed of two identical atoms joined together.

Let us see how Avogadro's ideas fit Gay-Lussac's measurements into the framework of the atomic theory. Consider, for example, the formation of nitric oxide, in which two volumes of product are obtained from one volume each of nitrogen and oxygen (Fig. 15.4). Since each volume represented contains the same number of molecules (Avogadro's hypothesis), a given number of oxygen molecules must react with the *same* number of nitrogen molecules to give *twice* as many molecules of nitric oxide. Thus 1,000 oxygen molecules would give 2,000 nitric oxide molecules; 10 oxygen molecules would give 20 of nitric oxide; 1 oxygen molecule would give 2 of nitric oxide. Now each nitric oxide molecule contains some oxygen; hence the oxygen molecule must have split, half of it going to each molecule of product. Avogadro interpreted this deduction to mean that an oxygen molecule consists of at least two atoms.

the molecules of some gaseous elements each consist of two atoms

Nitrogen   Oxygen   Nitric oxide

15.4 Formation of nitric oxide from nitrogen and oxygen. *Because one volume each of nitrogen and oxygen combine to yield two volumes of nitric oxide, nitrogen and oxygen must consist of molecules each containing two atoms.*

Similar reasoning applied to other reactions shows that oxygen molecules apparently often split into two parts, but never into more than two. We may safely infer that each molecule has no more than a pair of atoms. Also, for the other common gaseous elements—nitrogen, hydrogen, chlorine—Gay-Lussac's volume relationships suggest molecules made up of two atoms apiece.

How does the formation of water from its constituent elements fit into this picture? Gay-Lussac had shown that, at temperatures above 100°C, two volumes of hydrogen reacting with one volume of oxygen produce two volumes of water vapor (Fig. 15.5). In terms of molecules, then, a given number of hydrogen molecules react with half as many oxygen molecules to give the same number of water molecules. Two hydrogen molecules plus one oxygen molecule give two water molecules. Now, each

15.5 How the composition of water molecules may be determined. *Two volumes of hydrogen reacting with one volume of oxygen yield two volumes of water vapor. Hydrogen and oxygen molecules each contain two atoms, and therefore water molecules must each contain two hydrogen atoms and one oxygen atom.*

Hydrogen   Oxygen   Water vapor

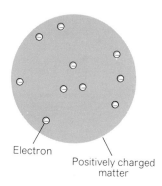

15.6 The Thomson model of the atom. *Experiment shows it to be incorrect.*

the thomson model of the atom does not agree with experiment

15.7 The Rutherford scattering experiment. *The distribution of scattering angles reveals the structure of the atoms in the foil.*

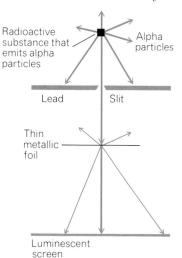

of the original molecules contains two atoms apiece; hence four hydrogen atoms and two oxygen atoms go into the building of the two water molecules. Accordingly, each water molecule must contain two hydrogen atoms and one oxygen atom, a 2:1 ratio.

Avogadro's idea that the volume of a gas at any given temperature and pressure is determined only by the number of molecules present, in his day merely a hypothesis, has been so well confirmed that today it is regarded as an established fact.

ATOMIC STRUCTURE

Although most of the scientists of the nineteenth century accepted the idea that the chemical elements consist of atoms, they knew virtually nothing about the atoms themselves. The discovery of the electron and the realization that all atoms contain electrons provided the first important insight into atomic structure. Electrons contain negative electric charges, but atoms themselves are electrically neutral: every atom must therefore contain enough positively charged matter to balance the negative charge of its electrons. Furthermore, electrons are thousands of times lighter than whole atoms, which suggests that the positively charged constituent of atoms is what provides them with nearly all their mass. When J. J. Thomson proposed in 1898 that atoms are uniform spheres of positively charged matter in which electrons are embedded, his hypothesis then seemed perfectly reasonable. Thomson's plum-pudding model of the atom—so-called from its resemblance to the raisin-studded delicacy—is sketched in Fig. 15.6. Despite the importance of the problem, thirteen years passed before a definite experimental test of the plum-pudding model was made. This experiment, as we shall see, compelled the abandonment of this apparently plausible model, leaving in its place a concept of atomic structure incomprehensible in the light of classical physics.

rutherford's experiment

The most direct way to find out what is inside a plum pudding is to plunge a finger into it, a technique not very different from that used by Geiger and Marsden to find out what is inside an atom. In their classic experiment, performed in 1911 at the suggestion of Ernest Rutherford, they employed the fast alpha particles spontaneously emitted by certain radioactive elements as probes (see Chap. 17). Geiger and Marsden placed a sample of an alpha-particle-emitting substance behind a lead screen that had a small hole in it, as in Fig. 15.7, so that a narrow beam of alpha particles was produced. This beam was directed at a thin gold foil. A movable zinc sulfide screen, which gives off a visible flash of light when struck by an alpha particle, was placed on the other side of the foil.

It was anticipated that most of the alpha particles would go right through the foil, while the remainder would suffer at most slight deflections. This behavior follows from the Thomson atomic model, in which the charges within an atom are assumed to be uniformly distributed throughout its volume. If the Thomson model is correct, only weak electric forces are exerted on alpha particles passing through a thin metal foil, and their

ATOMS AND MOLECULES

initial momenta should be enough to carry them through with only minor deflections from their original paths.

What Geiger and Marsden actually found was that, although most of the alpha particles indeed emerged without deviation, some were scattered through very large angles. A few were even scattered in the backward direction. Since alpha particles are relatively heavy (over 7,000 times more massive than electrons) and those used in this experiment traveled at high speed, it was clear that strong forces had to be exerted upon them to cause such marked deflections. To explain the results, Rutherford was forced to picture an atom as being composed of a tiny *nucleus*, in which its positive charge and nearly all its mass are concentrated, with its electrons some distance away (Fig. 15.8). With an atom largely empty space, it is easy to see why most alpha particles go right through a thin foil. However, when an alpha particle approaches a nucleus, it encounters an intense electric field, and is likely to be scattered through a considerable angle. The atomic electrons, being so light, do not appreciably affect the motion of incident alpha particles.

in the rutherford model of the atom, the positive charge and most of the mass of an atom are concentrated in a tiny nucleus at its center

Numerical estimates of electric field intensities within the Thomson and Rutherford models emphasize the difference between them. If we assume with Thomson that the positive charge within a gold atom is evenly spread throughout its volume and we neglect the electrons completely, the electric field intensity at the atom's surface (where it is a maximum) is about 10^{13} volts/m. On the other hand, if we assume with Rutherford that the positive charge within a gold atom is concentrated in a small nucleus at its center, the electric field intensity at the surface of the nucleus exceeds 10^{21} volts/m—a factor of 10^8 greater. Such a strong field can deflect or even reverse the direction of an energetic alpha particle that comes near the nucleus, while the feebler field of the Thomson atom cannot.

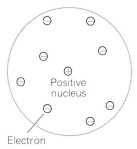

Suppose, as a rough analogy, that a small star approaches the solar system from outer space. With sufficient speed, the chances are excellent that it will pass through undeflected; even direct collision with a planet would not affect its motion appreciably. Only if it happened to approach closely the great central mass of the sun would its direction be radically altered. Similarly, said Rutherford, an alpha particle plows straight through the electric field of an atom, undeterred even by striking an electron now and then; only close approach to the heavy central core turns it aside.

15.8 The Rutherford model of the atom. *This conception of atomic structure is supported by experiment; almost all of the atom consists of empty space.*

Ordinary matter, then, is mostly empty space. The solid wood of a table, the steel that supports a skyscraper, the hard rock underfoot, all are but myriads of moving electric charges, isolated from one another by greater distances, comparatively, than the distance of the earth from its sister planets. If all the actual matter, all the electrons and nuclei, in our bodies could somehow be packed closely together, we should shrivel to specks just visible with a magnifying glass.

the electrons in an atom are relatively far from the nucleus, so most of an atom is empty space

atomic number

The concept of an atom as a positive nucleus surrounded by moving electrons accounts for many familiar observations. Since an ordinary atom is electrically neutral, the total positive charge on the nucleus must equal the negative charge of all the electrons in the electron cloud. If one of the electrons in the cloud is temporarily lost, the atom as a whole will show a positive charge; if an extra electron is temporarily added to the cloud, the atoms as a whole will have a negative charge. Ions produced by high-speed particles are just such charged atoms or charged molecules that have lost or gained electrons as the particles moved through them. Ordinary positive charges on pith balls or electrodes imply a deficiency of electrons in the atoms present, and negative charges imply an excess. The easy movement of electric currents through metals suggests that some electrons in metal atoms are loosely held and can jump from one atom to the next.

Rutherford's work supplied also more precise information about the makeup of different atoms. The deflection that an alpha particle undergoes as it approaches an atomic nucleus depends on the amount of positive charge in the nucleus, so measurements of the deflection by atoms of different elements provide a means of estimating the amounts of nuclear charge. Rutherford found that all atoms of any one element have the same nuclear charge and that this charge is different for different elements. The nucleus of hydrogen has a positive charge equal in magnitude to the negative charge of a single electron, that is, $+e$; the nucleus of helium has a charge equal to $+2e$; and so on up to the most complex element known to occur naturally, plutonium, whose nucleus has a charge equal to $+94e$. **The number of unit positive charges on the atomic nuclei of an element is called the atomic number of the element.** Thus the atomic number of hydrogen is 1, of helium 2, of lithium 3, and of plutonium 94. Atomic numbers of the elements are listed in Table 16.1.

every atom of a given element has the same atomic number, which is the number of unit positive charges its nucleus contains

Since in normal atoms the positive nuclear charge must be equaled by the total negative charge of the electrons in the cloud, atomic number can also be defined as the number of electrons in the uncharged atoms of an element. For example, the atomic number of phosphorus is 15; this means that the nucleus of the phosphorus atom has a positive charge of $15e$ and that around this nucleus is an electron cloud of 15 electrons.

The atomic number of an element is its most fundamental property. Atoms with the same atomic number may have somewhat different masses and still have almost identical other properties. But no change in atomic number is possible without a radical change in properties. The amount of positive charge on the nucleus of an atom serves to determine the fundamental nature of the atom and to distinguish it from all others.

failure of classical physics

Rutherford's finding that an atom consists of a tiny, positively charged nucleus surrounded at a great distance by negatively charged electrons led immediately to this question: What keeps the electrons out there? As we know, oppositely charged particles attract, and yet the electrons do not fall into the nucleus. By analogy with the planets of the solar system, where the centripetal force needed for rotation is furnished by the gravitational force of the sun, we might suppose that the electrons within an atom are in constant motion, circling around the nucleus at just the proper speed to avoid being sucked into the nucleus.

This is not a bad notion, but there is a serious difficulty associated with it. According to Maxwell's theory, an electron so moving should emit electromagnetic radiation continuously. In giving out radiation it should continuously lose energy, its orbit should become steadily smaller, and in a very short time (perhaps 10^{-10} sec) it should collide with the nucleus (Fig. 15.9). But atomic electrons do not behave in this way, regardless of Maxwell's theory. Under ordinary conditions atoms do not emit any radiation, and, needless to say, atoms do not spontaneously collapse.

Whenever they have been directly tested outside the atomic domain, the predictions of electromagnetic theory have always agreed with experiment—yet atoms do not collapse. This contradiction supports the conclusions of Chap. 14: The laws of physics that are valid in the macroscopic world do not hold true in the microscopic world of the atom. In Chap. 18 we shall see how the radical changes in the fundamental concepts of physics introduced in Chap. 14 enable us to understand the structure of the atom.

classical physics predicts that atoms should collapse, but quantum physics is able to account for their stability

GLOSSARY

A *homogeneous substance* can be subdivided indefinitely into identical portions, whereas when a *heterogeneous substance* is subdivided, the portions ultimately have different properties.

Elements are substances that can be neither decomposed nor transformed into one another. There is a limited number of elements, and all other substances are combinations of them in various proportions.

A *compound* is a homogeneous combination of elements in definite proportions. The properties of a compound are generally very different from those of its constituent elements.

A *solution* is a homogeneous combination of elements or compounds without any definite proportions. The constituents of a solution retain most of their original properties.

Oxidation is the chemical combination of a substance with oxygen; *combustion* is rapid oxidation accompanied by the evolution of noticeable heat and light.

The *law of conservation of mass* in chemical reactions states that the total mass of the products of a chemical reaction is always the same as the total mass of the original materials.

The *law of definite proportions* states that the elements that make up a chemical compound are always combined in the same definite proportions by weight.

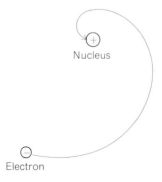

15.9 *An atomic electron should, according to classical physics, spiral rapidly into the nucleus as it radiates energy due to its acceleration.*

A *chemical symbol* is a one- or two-letter abbreviation of the name of an element. A *chemical formula* expresses the composition of the molecules of a compound as a list of the symbols for the elements in the compound with subscripts indicating the number of atoms of each element present in each molecule.

According to the *Rutherford model of the atom,* an atom consists of a tiny, positively charged *nucleus* with the negatively charged electrons some distance away. The atom is largely empty space.

The *atomic number* of an element is the number of unit positive charges on each of the nuclei of its atoms.

EXERCISES

1 List as many properties as you can of ice and of water. Why do we call the change from water to ice a physical change?

2 List the properties of iron and of rust. Why do we call the change from iron to rust a chemical change?

3 How can you show that water is a compound rather than a homogeneous mixture of hydrogen and oxygen?

4 How would you determine which of a pair of otherwise unknown liquid samples is a compound and which a mixture?

5 Air was long considered an element. How could you show that this idea is false?

6 When a large jar is placed over a lighted candle, the candle burns for a few minutes and then goes out. Can you explain why this occurs?

7 On what kind of experimental evidence is the law of conservation of mass based? Can you suggest a reason why this law was not formulated before the time of Lavoisier?

8 Can iron be prepared from rust? Sulfur from zinc sulfide? Lead from iron? Hydrogen from oxygen?

9 Compare the history of the phlogiston hypothesis with that of the ptolemaic hypothesis. In what ways was each a successful hypothesis? What were the chief flaws in each? What special assumptions had to be introduced to rescue each hypothesis, and how were these avoided in the hypotheses that replaced them? Is it really proper to group the phlogiston and ptolemaic hypotheses together?

ATOMS AND MOLECULES

10 The formula for liquid water is H_2O, for solid zinc sulfide ZnS, and for gaseous nitrogen dioxide NO_2. Precisely what information do these formulas convey? What information do they *not* convey?

11 Which of the following substances is homogeneous and which is heterogenous? Blood, carbon dioxide gas, solid carbon dioxide, rock, steak, iron, rust, concrete, air, oxygen, salt, milk.

12 Which of the following homogeneous liquids are elements, which are compounds, and which are solutions? Alcohol, mercury, liquid hydrogen, pure water, sea water, rum.

13 Compare the Thomson and Rutherford models of the atom. In what ways do they agree? In what ways do they disagree?

14 Alpha particle tracks through gases and thin metal foils are nearly always straight lines. To what conclusion regarding atomic structure does this observation lead?

15 How many molecules of water vapor could be made from two molecules of hydrogen?

16 Nitrogen and oxygen are present in air in the virtually constant ratio of 3.2 g of nitrogen to every 1 g of oxygen. These elements can also combine chemically to form the compound nitric oxide, which contains 7 g of nitrogen to every 8 g of oxygen. Both air and nitric oxide are colorless gases. Suggest methods for showing experimentally that one is a compound and one a mixture.

17 One volume of hydrogen reacts with one volume of chlorine to give two volumes of the gas hydrogen chloride. Show from these figures that each hydrogen molecule must contain at least two atoms.

18 The analysis of ammonia, a compound of nitrogen and hydrogen, shows that 14 g of nitrogen is combined with every 3 g of hydrogen. A nitrogen atom is approximately fourteen times heavier than a hydrogen atom. What is the ratio of nitrogen to hydrogen atoms in the ammonia molecule? How many grams of nitrogen would 51 g of ammonia contain? How many grams of ammonia could be prepared from 100 g of hydrogen?

chapter 16

The Nucleus

NUCLEAR STRUCTURE

>   cloud and bubble chambers
>   the proton
>   the neutron
>   isotopes
>   nuclear nomenclature

NUCLEAR ENERGY

>   atomic masses
>   binding energy
>   nuclear fission
>   nuclear reactors
>   nuclear fusion

The behavior of atomic electrons is responsible for the chief properties (except mass) of atoms, molecules, solids, and liquids. But the atomic nucleus itself is far from insignificant in the grand scheme of things. For instance, the elements exist because of the ability of nuclei to possess multiple electric charges, and explaining this ability is the central problem of nuclear physics. Furthermore, the energy that powers the continuing evolution of the universe apparently can all be traced to nuclear reactions and transformations. And, of course, there are plenty of applications of nuclear energy in the industrial world of today.

NUCLEAR STRUCTURE

There are a number of ways we can trace the paths followed by alpha particles (and other ions) after they have been emitted from a radioactive substance. The earliest way is based on the fact that water droplets tend to condense on the ions left by the passage of an alpha particle in a very moist atmosphere. By shining a light from the side on these droplets, we can make them gleam brightly against a dark background. The cloud chamber, as such a device is called, played an important part in the discovery of the proton and the neutron, the two particles found in atomic nuclei.

cloud and bubble chambers

A cloud chamber is a device for rendering visible the path of a moving charged particle. Invented by C. T. R. Wilson in 1907, the cloud chamber makes use of the fact that a supersaturated vapor condenses into liquid droplets around any ions that may be present within it. In its simplest version, a cloud chamber consists of a glass-fronted cylinder containing a mixture of air and water vapor (Fig. 16.1). When the piston is moved back rapidly, the vapor expands and cools to a supersaturated state. If a charged particle passes through the chamber at precisely this time, the ions it leaves behind serve as the nuclei of water droplets that condense from the vapor. This trail can be observed and photographed by illuminating the chamber from the side. The identity and initial energy of a particle that stops in the chamber can be determined from the length of the track (the longer the track, the greater the energy), from the density of the track (the greater the density, the heavier the particle and the slower it is), and from the character of the track (the straighter the track, the heavier the particle). Cloud chambers are especially useful in studying the interactions between an incoming particle and the nuclei of gas atoms it may strike.

moving charged particles leave trails of water droplets in a cloud chamber

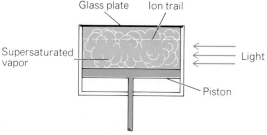

16.1 A cloud chamber. *The trail of ions left by an alpha particle or other rapidly-moving atomic particles is made visible because water droplets condense on the ions when a supersaturated vapor is present. The piston is used to expand the vapor suddenly, which cools it into a supersaturated state.*

Ordinary cloud chambers are not ideal because they are sensitive only for a fraction of a second after the expansion and because the gas density is so low that high-energy particles usually neither stop within the chamber nor experience collisions there. Another type of cloud chamber eliminates the first objection by using a container kept cold at its bottom. If a heavy gas fills the chamber and a lighter one is allowed to diffuse downward, the latter becomes supersaturated as it nears the cooler bottom and thus establishes a layer of continuous sensitivity. The density objection is overcome in the bubble chamber, a device that is essentially a cloud chamber operating in reverse. A bubble chamber contains a superheated liquid (a liquid heated beyond its normal boiling point), and bubbles of vapor form around any ions created within it. Because a liquid rather than a gas is involved, the likelihood of finding interesting events along the path of an incident particle is increased; modern high-energy accelerators make extensive use of bubble chambers in investigating nuclear and

moving charged particles leave trails of tiny bubbles in a bubble chamber

the proton

In 1919 Rutherford found in a cloud-chamber photograph of alpha particles traveling through nitrogen an alpha track that ended in a peculiar way. From the end of the wide track, where the alpha particle came to a stop, a thinner track started off in a different direction. Another particle, less effective in ionizing the gas than alpha particles, evidently had come out of the collision between the alpha particle and a nitrogen atom. From the characteristics of the track, Rutherford was able to determine that the new particle had a mass about one-fourth that of an alpha particle and a positive charge equal in magnitude to the negative charge of the electron. This particle is the *proton*.

In later photographs of alpha-particle tracks through substances other than nitrogen Rutherford found occasional collisions showing the telltale thin fog tracks of protons. Since protons could be knocked out of many kinds of atoms, he guessed that these particles made up a part of all atomic nuclei. This guess has been amply confirmed by later work. In ordinary atoms, protons are the fundamental positive charges, as electrons are the fundamental negative charges. **The atomic number of an element is the number of protons in the nuclei of its atoms.**

protons provide the positive charge in an atom

An ordinary hydrogen atom consists of a proton with a single electron revolving about it. The two particles have charges of the same magnitude (though with opposite sign), but the mass of the proton is 1.6725×10^{-27} kg, which is 1,836 times that of the electron. Practically all the mass of the hydrogen atom, then, is concentrated in the proton.

the hydrogen atom consists of a proton circled by an electron

the neutron

In the late 1920s, several observers reported a different kind of collision involving alpha particles and atoms of very light elements, particularly atoms of the metal beryllium. In these collisions no fog tracks of emitted particles appeared from the actual sites of the collisions, but the emission of some kind of radiation was suggested by tracks that appeared to start spontaneously in other parts of the cloud chamber. Evidently the radiation itself was uncharged, since it produced no ionization, but was capable of interacting with atoms at a distance to produce ionizing particles. This effect might be produced by high-energy gamma rays, and for some time this explanation was favored.

Finally, in 1932, James Chadwick, working in Rutherford's laboratory, found that the radiation consisted not of gamma rays, but of tiny *uncharged* particles that the alpha particles had knocked out of beryllium nuclei. These uncharged particles, termed *neutrons*, have approximately the same mass as protons but lack positive charge. The neutron mass is 1.6748×10^{-27} kg, which is 1,839 times the electron mass. Because of their lack of charge and their small size, they can travel through the

the neutron has no charge and has slightly more mass than the proton

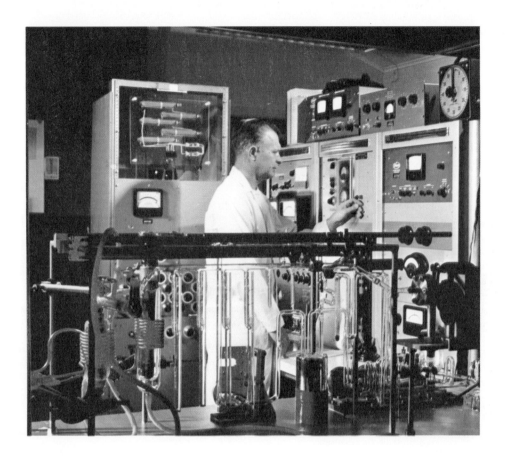

A mass spectrometer being used to study the composition of meteorites.

electron clouds of atoms without producing enough disturbance to cause appreciable ionization, but when they happen to strike a nucleus, charged particles may be emitted which cause the fog tracks observed at a distance from the original collision.

Free neutrons outside of nuclei are unstable and undergo radioactive decay into a proton and an electron each. The half-life of this decay is about 11 min according to the latest measurements; that is, if we start with 100 neutrons, in 11 min 50 of them will have turned into protons and electrons, in another 11 min 25 of the remaining 50 neutrons will have decayed, and so on. (Radioactive decay is discussed in Chap. 17.) It is not correct to think of a neutron as a composite of a proton and an electron, despite its decay into these particles: a neutron is an independent entity with unique properties. If we were to try to create a neutron by bringing together a proton and an electron, we would merely get a hydrogen atom as the result, not a neutron.

neutrons decay into protons and electrons outside of nuclei

isotopes

neutrons and protons are called nucleons

The neutron and the proton (together often called *nucleons*) have masses about the same as the hydrogen atom and nearly 2,000 times the mass of the electron. Neutrons and protons make up almost the entire mass of an atom; the electron is so much lighter that even in the heaviest atoms the combined mass of all the electrons is only a small fraction of the mass of a single nucleon.

Table 16.1 is a list of the atomic masses of the various elements expressed in *atomic mass units* (amu), where, by definition,

$$1 \text{ amu} = 1.660 \times 10^{-27} \text{ kg}$$

the atomic mass unit

The energy equivalent of the mass unit is 931 Mev. The figures in the table are based upon measurements made on large numbers of atoms of each element, not on individual atoms.

If atomic nuclei consist of protons and neutrons, it should follow that the mass of any nucleus is approximately an even multiple of the mass of the hydrogen nucleus, since protons (which are hydrogen nuclei) and neutrons have about the same masses. Because electrons contribute very little to an atom's mass, this means, further, that the atomic mass of any element ought to be an integral multiple of the atomic mass of hydrogen. In other words, since hydrogen has an atomic mass in amu of approximately 1, all other atomic masses should be whole numbers or nearly so. Now it is a striking fact that atomic masses in amu for many elements are indeed close to whole numbers; thus carbon has a mass of 12.01, fluorine 19.00, nitrogen 14.01. On the other hand, some atomic masses in amu are conspicuously different from whole numbers; for example, chlorine 35.45, copper 63.54, magnesium 24.31. How can we possibly construct atoms with these masses out of protons and neutrons?

The answer to this question is that the atoms of any one element are not necessarily all exactly alike. The experiment that verified this radical suggestion was carried out in 1919 and consisted of sending positively charged ions (atoms missing an electron) of known speed through a magnetic field in a device called a *mass spectrometer* (Fig. 16.2). In their passage through the field heavy ions were deflected somewhat less than light ions, and the amount of the deflection provided an accurate means of determining just how heavy different atoms were. When ions of chlorine were passed through this instrument, the ions did not all form a single beam corresponding to a mass of 35.45, but instead were divided into two beams, the stronger beam corresponding to a mass of 35 and the

the mass spectrometer is a device for measuring the masses of individual atoms

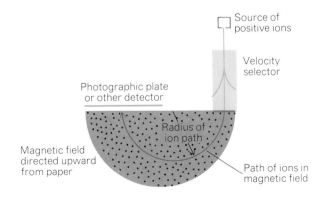

16.2 An idealized mass spectrometer. *Only ions of a single speed can pass through the velocity selector. In the magnetic field, the ions are bent into circular paths whose radii are proportional to the ion masses; heavy ions are deflected least, light ones most.*

TABLE 16.1 Atomic masses of the elements.

Atomic number	Element	Symbol	Atomic mass
1	Hydrogen	H	1.008
2	Helium	He	4.003
3	Lithium	Li	6.939
4	Beryllium	Be	9.012
5	Boron	B	10.81
6	Carbon	C	12.01
7	Nitrogen	N	14.01
8	Oxygen	O	16.00
9	Fluorine	F	19.00
10	Neon	Ne	20.18
11	Sodium	Na	22.99
12	Magnesium	Mg	24.31
13	Aluminum	Al	26.98
14	Silicon	Si	28.09
15	Phosphorus	P	30.97
16	Sulfur	S	32.06
17	Chlorine	Cl	35.45
18	Argon	Ar	39.95
19	Potassium	K	39.10
20	Calcium	Ca	40.08
21	Scandium	Sc	44.96
22	Titanium	Ti	47.90
23	Vanadium	V	50.94
24	Chromium	Cr	52.00
25	Manganese	Mn	54.94
26	Iron	Fe	55.85
27	Cobalt	Co	58.93
28	Nickel	Ni	58.71
29	Copper	Cu	63.54
30	Zinc	Zn	65.37
31	Gallium	Ga	69.72
32	Germanium	Ge	72.59
33	Arsenic	As	74.92
34	Selenium	Se	78.96
35	Bromine	Br	79.91
36	Krypton	Kr	83.80
37	Rubidium	Rb	85.47
38	Strontium	Sr	87.62
39	Yttrium	Y	88.91
40	Zirconium	Zr	91.22
41	Niobium	Nb	92.91
42	Molybdenum	Mo	95.94
43	Technetium	Tc	(99)
44	Ruthenium	Ru	101.0
45	Rhodium	Rh	102.9
46	Palladium	Pd	106.4
47	Silver	Ag	107.9
48	Cadmium	Cd	112.4
49	Indium	In	114.8
50	Tin	Sn	118.7
51	Antimony	Sb	121.8
52	Tellurium	Te	127.6

TABLE 16.1 Atomic masses of the elements (cont.).

Atomic number	Element	Symbol	Atomic mass
53	Iodine	I	126.9
54	Xenon	Xe	131.3
55	Cesium	Cs	132.9
56	Barium	Ba	137.4
57	Lanthanum	La	138.9
58	Cerium	Ce	140.1
59	Praseodymium	Pr	140.9
60	Neodymium	Nd	144.2
61	Promethium	Pm	(147)
62	Samarium	Sm	150.4
63	Europium	Eu	152.0
64	Gadolinium	Gd	157.3
65	Terbium	Tb	158.9
66	Dysprosium	Dy	162.5
67	Holmium	Ho	164.9
68	Erbium	Er	167.3
69	Thulium	Tm	168.9
70	Ytterbium	Yb	173.0
71	Lutetium	Lu	175.0
72	Hafnium	Hf	178.5
73	Tantalum	Ta	181.0
74	Tungsten	W	183.9
75	Rhenium	Re	186.2
76	Osmium	Os	190.2
77	Iridium	Ir	192.2
78	Platinum	Pt	195.1
79	Gold	Au	197.0
80	Mercury	Hg	200.6
81	Thallium	Tl	204.4
82	Lead	Pb	207.2
83	Bismuth	Bi	209.0
84	Polonium	Po	(209)
85	Astatine	At	(210)
86	Radon	Rn	(222)
87	Francium	Fr	(223)
88	Radium	Ra	(226)
89	Actinium	Ac	(227)
90	Thorium	Th	232.0
91	Protactinium	Pa	(231)
92	Uranium	U	238.0
93	Neptunium	Np	(237)
94	Plutonium	Pu	(244)
95	Americium	Am	(243)
96	Curium	Am	(247)
97	Berkelium	Bk	(247)
98	Californium	Cf	(251)
99	Einsteinium	Es	(254)
100	Fermium	Fm	(257)
101	Mendelevium	Md	(256)
102	Nobelium	No	(254)
103	Lawrencium	Lr	(257)

masses in parentheses are those of the most stable isotopes of the elements

weaker to a mass of 37. This means that natural chlorine must contain two kinds of atoms, normally mixed in constant proportions of about 76 percent of mass 35 and 24 percent of mass 37, giving the apparent atomic mass of 35.45.

Not only chlorine but many other elements are found to have atoms with slightly different weights when examined in a mass spectrometer. These varieties of an element with different atomic weights are called *isotopes*. Extensive research has shown that fractional atomic masses always mean mixtures of isotopes; once the atoms are ionized and separated, each isotope proves to have an atomic mass very nearly a whole number. In other words, all atoms have masses that are nearly integral multiples of the mass of the hydrogen atom. This conclusion, of course, is consistent with the idea that all atomic nuclei are built wholly of neutrons and protons.

the isotopes of an element have the same atomic number but different atomic masses

Even the element hydrogen has isotopes. Ordinary hydrogen nuclei, of course, consist of a single proton. In addition, there is an isotope of hydrogen called *deuterium*, of atomic mass 2, whose nuclei are each composed of a proton plus a neutron, and an isotope called *tritium*, of atomic mass 3, whose nuclei are each composed of a proton plus two neutrons (Fig. 16.3). Both of these heavy isotopes are found in nature, but only in relatively small amounts. So-called *heavy water* is water in which deuterium atoms instead of hydrogen atoms are present in combination with oxygen. Tritium is unstable, and undergoes radioactive decay with the emission of an electron; the process has a half-life of about $12\tfrac{1}{2}$ years.

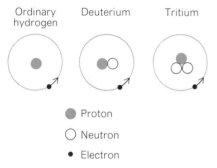

16.3 The isotopes of hydrogen.

Experiments involving knocking protons and neutrons out of atoms and the existence of isotopes make it necessary for us to reexamine our definition of a chemical element. In the previous chapter we defined an element as a substance not decomposable into other substances, and the atomic theory in its simplest form assumes that all the atoms of an element are identical. But, if alpha particles can change an atom of one element into an atom of another by ejecting one or more nucleons, the above definition is not valid, and the discovery of isotopes renders the latter assumption untenable. The proper statement is that **an element is a substance all of whose atoms have the same atomic number.**

the atoms of an element all have the same atomic number, that is, the same number of protons in their nuclei

Atoms of two isotopes have the same number of protons in their nuclei but different numbers of neutrons. The light isotope of chlorine, for example, has nuclei containing 17 protons and 18 neutrons, and the heavy isotope has nuclei containing 17 protons and 20 neutrons. The electron clouds, whose structure is determined by the total positive charge on the nucleus, are practically identical. In other words, the different isotopes of an element have the same atomic number but different atomic masses. These relations are illustrated for several elements in Table 16.2.

most of the properties of an element are determined by the electron clouds of its atoms

The ordinary physical and chemical properties of an element are determined almost wholly by the electron clouds of its atoms and, therefore, by its atomic number. Since different isotopes have the same atomic numbers and almost identical electron structures, their properties are very nearly the same. The two isotopes of chlorine, for instance, have the same yellow color, the same suffocating odor, the same efficiency as poisons and bleaching agents, and the same readiness to combine with metals and hydrogen. Their densities, boiling points, freezing points, and rates

THE NUCLEUS

TABLE 16.2 The isotopes of hydrogen, chlorine, and zinc. The atomic mass of an element is the average of the atomic masses of its isotopes, taking into account their relative abundances. The mass number of a nucleus is the total number of protons and neutrons it contains. Atomic masses are in amu.

Element	Properties of element		Properties of isotope				
	Atomic number	Atomic mass	Protons in nucleus	Neutrons in nucleus	Mass number	Atomic mass	Relative abundance, percent
Hydrogen	1	1.008	1	0	1	1.008	99.985
			1	1	2	2.014	0.015
			1	2	3	3.016	very small
Chlorine	17	35.45	17	18	35	34.97	75.53
			17	20	37	36.97	24.47
Zinc	30	65.37	30	34	64	63.93	48.89
			30	36	66	65.93	27.81
			30	37	67	66.92	4.11
			30	38	68	67.92	18.56
			30	40	70	69.93	0.62

of diffusion depend somewhat on the masses of the atoms and thus are very slightly different. The extreme similarity in properties explains why isotopes are not appreciably separated in natural processes and why chemists failed to detect them before the mass spectrometer was developed.

nuclear nomenclature

The following terms and symbols are widely used to describe a nucleus:

Z = atomic number = number of protons
N = neutron number = number of neutrons
$A = Z + N$ = mass number = total number of neutrons and protons

The term nucleon refers to both protons and neutrons, so that the mass number A is the number of nucleons in a particular nucleus. Nuclear species (sometimes called *nuclides*) are identified according to the scheme

$$_Z X^A$$

where X is the chemical symbol of the species. Thus the arsenic isotope of mass number 75 is denoted by

$$_{33}As^{75}$$

since the atomic number of arsenic is 33. Similarly, a nucleus of ordinary hydrogen, which is a proton, is denoted by

$$_1H^1$$

Here the atomic and mass numbers are the same because no neutrons are present. Sometimes the mass number is placed to the left of the symbol, so that the above nuclides would be denoted $^{75}_{33}$As and $^{1}_{1}$H.

NUCLEAR ENERGY

atomic masses

We say that the atomic masses of isotopes are "very nearly" integral multiples of the mass of a hydrogen atom. Why must we make this statement indefinite? Why should not atomic masses be *exactly* whole numbers relative to the atomic mass of hydrogen? If we examine the matter more closely, and add up the masses of the individual neutrons and protons in a complex nucleus, we find that the sum of the particle masses turns out to be almost equal to the mass of the nucleus, but invariably a little more. If our ideas about the structure of the nucleus are correct, why should such discrepancies crop up?

the mass of an atom is always slightly less than the sum of the masses of its constituent particles

We might very well suspect that the answer is somehow connected with the mass-energy relationship of the special theory of relativity. Let us consider a specific example. We know that a helium atom consists of two neutrons, each with an atomic mass of 1.0087 amu, and two hydrogen atoms (each a proton plus an electron), each set with an atomic mass of 1.0078 amu. The total mass of these four particles is 4.0330 amu. The mass of a helium atom, however, is only 4.0026 amu, smaller by 0.0304 amu. In other words, the mass of a helium atom is *less* by almost 1 percent than the combined masses of its constituent particles.

In view of the mass-energy relationship

$$E = mc^2$$

this mass discrepancy is not so outrageous as it seems. All that we need assume is that, in the process of combining to form a helium atom, the two neutrons and two hydrogen atoms evolve a certain amount of energy. This energy comes from their mass, so the total mass left in the helium atom is a little less than the original mass of the constituents.

nuclei are stable because they lack enough mass to break up into separate nucleons

The fact that the mass of the helium nucleus is less than the masses of the individual particles that went into it is, in fact, capable of explaining in a general way another question that had troubled physicists: Why does the cluster of protons in a nucleus, all repelling one another since all are positively charged, stay together? The nucleus does not contain enough mass to form free neutrons and protons, and so it cannot fall apart into these particles unless enough *outside* energy is imparted to it to make up for the missing mass. Therefore the helium nucleus is ordinarily a very stable affair, but, if we can give it energy through a collision of some kind, we might then be able to break it up.

Another example will illustrate the ideas in the preceding paragraph. When a piece of lithium is bombarded by protons, the reaction pictured in Fig. 16.4 takes place. The masses of the various particles are as follows:

THE NUCLEUS

H nucleus + Li nucleus ⟶ 2 He nuclei
  1.0073   7.0144   2 × 4.0015

The total mass on the left side is 8.0217 amu, that on the right side is 8.0030 amu; 0.0187 mass unit has disappeared. This missing mass is equivalent to an energy of 17 Mev, to use the energy unit of nuclear physics, which we might assume to be carried off as kinetic energy by the two alpha particles. (1 amu = 931 Mev.) When this reaction takes place in the laboratory, careful measurements show that the energies of the alpha particles indeed add up to 17 Mev. Sometimes, as in this reaction, the energy formed when mass disappears takes the form of kinetic energy of moving particles; sometimes it appears as gamma radiation, or as both kinetic energy and gamma radiation. Whatever form the energy takes, its total amount is always found to be related to the loss in mass by Einstein's equation $E = mc^2$.

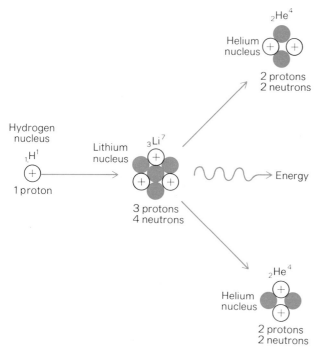

16.4 An example of transmutation. A lithium nucleus struck by a proton yields two helium nuclei plus evolved energy. The total mass of the products is less than the total mass of the reactants; the "missing" mass appears in the form of energy.

binding energy

The energy equivalent of the mass discrepancy in a nucleus is called its *binding energy* and is a measure of the stability of the nucleus. The greater its binding energy, the more the energy that must be supplied from the outside to break up a nucleus.

Nuclear binding energies are strikingly large. To appreciate their magnitude, it is helpful to convert the figures from million electron volts per nucleon to more familiar units, say kilocalories per kilogram. Since 1 Mev = 1.60×10^{-13} joule and 4,185 joules = 1 kcal (the mechanical equivalent of heat), we find that 1 Mev = 3.83×10^{-17} kcal. One

the greater the binding energy of a nucleus, the more stable it is

atomic mass unit is equal to 1.66×10^{-27} kg, and each nucleon in a nucleus has a mass of very nearly 1 amu. Hence

$$1 \frac{\text{Mev}}{\text{nucleon}} = \frac{3.83 \times 10^{-17} \text{ kcal}}{1.66 \times 10^{-27} \text{ kg}} = 2.31 \times 10^{10} \frac{\text{kcal}}{\text{kg}}$$

A binding energy of 8 Mev per nucleon, a typical value, is therefore equivalent to 1.85×10^{11} kcal/kg. By contrast, the heat of vaporization of water is a mere 540 kcal/kg, and even the heat of combustion of gasoline, 1.13×10^4 kcal/kg, is 10 million times smaller.

The binding energy *per nucleon*, arrived at by dividing the total binding energy of a nucleus by the number of nucleons it contains, is a most interesting quantity. The binding energy per nucleon is plotted versus mass number A in Fig. 16.5; we recall that A is the total number of neutrons and protons in a particular nucleus, so that Fig. 16.5 shows

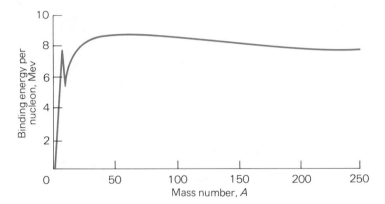

16.5 The binding energy per nucleon is a maximum for nuclei of mass numbers near $A = 56$. *Such nuclei are the most stable.*

large and small nuclei are less stable than those of intermediate size

how the binding energy per nucleon varies with the size of the nucleus. The curve rises steeply at first until it reaches a maximum in the vicinity of $A = 56$, corresponding to iron nuclei, and then drops slowly. Evidently nuclei of intermediate size are the most stable, since the greatest amount of energy must be supplied to liberate their constituent neutrons and protons. This fact suggests that energy will be evolved if heavy nuclei can somehow be split into lighter ones, or if light nuclei can somehow be joined to form heavier ones. The former process is known as nuclear *fission* and the latter as nuclear *fusion*, and both indeed occur under the proper circumstances and do evolve energy as predicted.

If energy can be produced by appropriate nuclear reactions, why was there a lapse of a quarter of a century between the pioneer experiments and the construction of the first nuclear reactor? The explanation lies in the low efficiency of the bombardment experiments. In all these experiments tiny "bullets" are sent in the general direction of the "target" atoms; a very few bullets score direct hits and set free momentary bursts of energy, but the great majority pass through the target without encountering atomic nuclei. The reactions we have been considering are collisions between *single* atoms and *single* particles, not between large amounts of matter. Although each individual reaction releases energy,

High Flux Beam Research Reactor (HFBR) has a power of 40 million watts and a neutron flux of 1.6×10^{15} neutrons per square centimeter per second.

so many particles pass through the target without interacting that the energy that must be supplied to make the particles move is far greater than the total energy produced. It is as if someone had scattered small sticks of dynamite over a hillside and we were trying to explode them by firing a machine gun at the hill from a long distance away. Once in a while one of our bullets would hit a stick and we would see a small explosion, but most of the bullets would plow harmlessly into the ground. The dynamite has an abundance of stored energy, but with this technique of releasing it we must supply more energy with our gun than we can obtain from the dynamite.

nuclear fission

In 1939 it was discovered that nuclei of the uranium isotope $_{92}U^{235}$ sometimes split into two smaller nuclei when bombarded with neutrons. The most striking aspect of this fission reaction is the magnitude of the energy liberated, which is easy to compute with the help of Fig. 16.5. Nuclei of U^{235} have binding energies of about 7.6 Mev per nucleon, while the fission fragments (as the new, smaller nuclei are called) have binding energies of about 8.5 Mev per nucleon. Hence 0.9 Mev per nucleon is released during fission, which for the 235 nucleons involved means over 200 Mev! Ordinary chemical reactions, such as those that participate in the combustion of coal and oil, liberate only a few electron volts per individual reaction, and even nuclear reactions (other than fission) liberate no more than several million electron volts.

in nuclear fission, an incident neutron causes a large nucleus to split into two smaller ones

Most of the energy that is released in nuclear fission goes into the kinetic energy of the fission fragments. These fragments are usually radioactive, and further energy is given off in their beta and gamma decays. In addition, two or three neutrons are emitted during fission, and they carry with them a certain amount of energy as well. The latter property of nuclear fission has a significant implication: under suitable conditions, the neutrons that emerge from one uranium nucleus might cause the splitting of other uranium nuclei, with these in turn producing further neutrons that might split still other uranium nuclei, and so on, with a succession of fission reactions spreading through an entire mass of uranium. A *chain reaction* of this kind was first demonstrated in Chicago in 1942 under the direction of Enrico Fermi; Fig. 16.6 is a schematic diagram of the events that occur in a chain reaction in U^{235}.

chain reaction

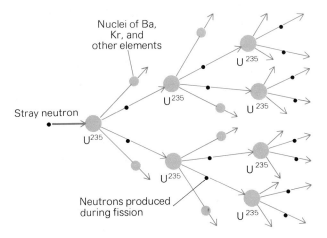

16.6 Sketch of a chain reaction as it might occur in pure U^{235}. The reaction is self-sustaining if at least one neutron from each fission event on the average induces another fission event. If more than one neutron per fission on the average induces another fission, the reaction is explosive.

The condition for a chain reaction to occur in an assembly of fissionable material (U^{235} is only one of several isotopes capable of undergoing fission) is simple: At least one neutron produced during each fission must, on the average, initiate another fission instead of either escaping from the assembly or being absorbed in a nonfissionable nucleus. If too few neutrons initiate fissions, the reaction will slow down and stop. If precisely one neutron per fission causes another fission, energy will be released at a constant rate; this is the case in a *nuclear reactor*, which is a device for producing controlled power from nuclear fission. If the frequency of fissions increases, with more than one neutron from each fission leading to others, the energy release will be so rapid that an explosion will occur; this is the case in an atomic bomb.

in a nuclear reactor, fission occurs at a controlled rate

nuclear reactors

A major obstacle to constructing a nuclear reactor lies in the fact that only a particular kind of uranium readily undergoes the fission reaction. Ordinary uranium consists of three isotopes; atoms with a mass number of 238 are by far the most abundant, about 1 atom in 140 has a mass number of 235, and about 1 in 16,000 a mass number of 234. Only the nuclei of U^{235} are appreciably fissionable by neutrons. Despite the small amount of this isotope, a chain reaction might still take place in natural uranium if a neutron could be depended on to bounce away from collisions with other atoms until it met one of the fissionable variety; unfortunately, most neutrons do not bounce, but are absorbed into U^{238} nuclei, which means that in natural uranium the chain reaction is quickly interrupted. Evidently one method of helping the chain reaction to take place would be to separate the two isotopes.

Separating out the U^{235} from natural uranium is a very difficult and expensive process. However, there is another way to solve the problem of causing a chain reaction to take place in uranium. It happens that the neutrons produced in U^{235} fission are high-speed neutrons, which are readily absorbed by U^{238} atoms but which are not so effective as slow neutrons in splitting atoms of U^{235}. Fast neutrons can be slowed down if they are passed through material consisting of light atoms that will not absorb them, like carbon atoms; their energy is gradually lost as they collide with one light nucleus after another. Now if small pieces of uranium could be scattered through a large block of graphite (a dense form of pure carbon), some of the fast neutrons originating by fission in one piece should escape from the piece, be slowed as they pass through the graphite, and thus be ready to produce fission in the U^{235} atoms of another piece (Fig. 16.7). This arrangement permits the chain reaction to take place in spite of the presence of U^{238}. Of course, many neutrons are still absorbed by U^{238} atoms, and these atoms quickly change to plutonium. Such a uranium-graphite reactor, first set up at the University of Chicago in 1942, showed by its successful operation that a controlled chain reaction was possible.

U^{235} is the only isotope of uranium that undergoes fission readily

slow neutrons are more effective in producing fission in U^{235} than fast ones

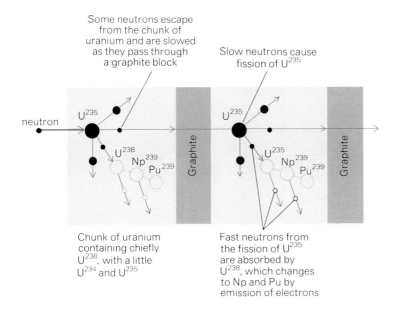

16.7 Principle of a uranium-graphite nuclear reactor. *Energy is evolved at each step of the process.*

A nuclear reactor generates prodigious amounts of heat, and this heat serves quite as well as a coal or an oil fire to make water boil and so provide power for a steam engine. Many different designs have been tried, but nearly all have three features in common: (1) a source of energy, generally uranium that has been somewhat enriched in the U^{235} isotope; (2) a *moderator*, or material to slow down fast neutrons, which is usually heavy water; and (3) a material to control the flow of neutrons by absorbing some of them, generally a rod of cadmium or boron steel that can be moved in and out of the reactor. The radiations produced by nuclear processes in the reactor are extremely dangerous, so elaborate shielding with lead or a similar material is necessary. Furthermore, automatic safety devices are essential, so that the control rods will immediately be inserted far enough to stop the reaction when anything goes wrong.

The necessity for heavy shields to prevent radiation damage has so far limited applications of nuclear power to situations where weight has little importance. Stationary power plants have been built to produce electricity commercially, and in an increasing number of places such plants can provide power cheaply enough to compete with electricity from coal, oil, and falling water. Many submarines are operating with reactor-powered steam turbines, and similar engines have been installed in surface ships. But in automobiles and airplanes, where limitation of weight is essential, the use of nuclear reactors as power sources will remain impractical until the problem of adequate but lightweight shielding has been solved.

Symmetric Tokamak is a device used in research into thermonuclear energy. Located at Princeton University, it is based upon a Russian design.

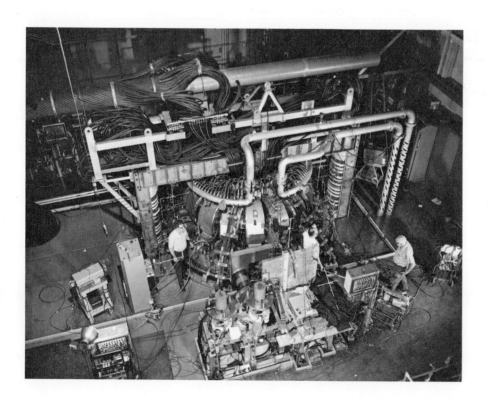

The powerful and deadly radiation from reactors, although a serious handicap in power production, has proved very useful in other ways. The radiation consists chiefly of neutrons and gamma rays, but may contain secondary particles produced when these strike other atoms. When other substances are introduced into a reactor and subjected to this radiation, a variety of nuclear reactions is possible. In particular, many radioactive isotopes of common elements can be prepared, and these have found wide use in chemical, medical, and biological research. Some of the artificially radioactive substances have proved more effective than radium or X-rays in the treatment of disease. One spectacular result of experiments with the radiation from nuclear reactors has been the artificial preparation of new elements—not only plutonium and neptunium, which were found in the course of work on the atomic bomb, but several others both lighter and heavier than these as well.

The chain reaction in natural uranium moderated with graphite liberates great quantities of energy, chiefly in the form of heat. The reaction is not rapid enough for an effective bomb, although it is an entirely practical source of energy. In the reactors operated during World War II, however, energy production was incidental; the importance of the reactors lay in the gradual enrichment of their uranium fuel with plutonium. By 1945, enough of both U^{235} and plutonium was available for the construction of bombs, which were made from both materials.

But difficulties were by no means ended with successful production of the fissionable elements in pure form. These new explosives could not be simply cut up into sticks and exploded by percussion caps, like dynamite. If a large enough mass of one of the elements were to be heaped together, it would explode by itself; there would be no way to keep it from exploding, since always in the surrounding air there would be enough stray neutrons to start the chain reaction. On the other hand, a small piece could not be exploded at all, since so many neutrons would escape from it that the chain reaction could not persist. An atomic bomb, then, must consist of small pieces of fissionable material sufficiently separated so that they will not explode, together with a device for bringing them together at the proper moment. The mechanical problem is severe, for if the separate pieces are not brought together very rapidly, the bomb will start to explode prematurely and will fly apart before it can do appreciable damage. It was only when this final problem was solved that the bomb was completed.

the atomic bomb

nuclear fusion

Enormous as the energy production in uranium fission is, there are other nuclear processes which can give out energy in even larger amounts. One of these is the reaction of four hydrogen nuclei to produce helium, a process that actually takes place in the sun and other stars and is one of the important sources of stellar energy (see Chap. 42 for the details of this reaction). As we saw at the beginning of this chapter, the mass of a helium atom is less than the sum of the masses of its constituent particles before their combination, and a corresponding amount of energy is therefore released each time a helium nucleus is synthesized. The energy produced during the formation of every kilogram of helium is equivalent to the heat produced by the combustion of 3,000 tons of coal.

More feasible for duplication on the earth than the reaction of four hydrogen nuclei to form a helium nucleus is a reaction between two nuclei of heavy hydrogen (deuterium), or even better, a reaction between deuterium and the still heavier artificial isotope of hydrogen, tritium:

$$_1H^2 + {_1H^3} \longrightarrow {_2He^4} + {_0n^1} + \text{energy}$$
$$\text{Deuterium Tritium} \qquad \text{Helium Neutron}$$

in nuclear fusion, two small nuclei unite to form a larger one

A reaction of this sort, in which two particles unite to form a heavier particle, is often referred to as a *fusion* process, in contrast to the *fission* of uranium or plutonium. Maintaining a fusion reaction requires temperatures of several million degrees, temperatures attained on the earth during the explosion of a uranium bomb. Addition of deuterium and tritium to a uranium bomb, therefore, makes possible a still more efficient instrument of destruction, the hydrogen bomb, in which the uranium fission process serves to produce the necessary conditions for the more violent deuterium-tritium fusion. Such a bomb was exploded for the first time in 1952. The destructive capacity of a hydrogen bomb is limited only by the mechanical difficulties involved in getting its ingredients together rapidly enough in order to prevent a premature explosion.

Today laboratories throughout the world are working on means for controlling the production of energy in fusion processes. "Magnetic bottles" have been devised which employ strong magnetic fields to contain the reacting nuclei at very high temperatures, and a variety of ingenious schemes have been proposed for heating the nuclei to temperatures such that the fusion of light nuclei to form heavier ones will take place. *Thermonuclear power* is the name that has been given to the energy evolved during nuclear fusion, and it seems likely that this will be the ultimate source of power on the earth when fossil fuels such as coal and oil have been exhausted; nuclear reactors employing uranium and plutonium will almost certainly, in the long run, prove more expensive to operate than thermonuclear power plants.

thermonuclear power is derived from nuclear fusion

GLOSSARY

The *proton* is a positively charged elementary particle constituting the nucleus of the hydrogen atom.

The *neutron* is an electrically neutral elementary particle whose mass is approximately that of the proton. Atomic nuclei consist of neutrons and protons.

The *mass number* of a nucleus is the total number of neutrons and protons it contains.

The *atomic number* of an element is the number of protons in the nuclei of its atoms. Thus an element is a substance all of whose atoms have the same atomic number.

THE NUCLEUS

The *isotopes* of an element are varieties of the element whose atoms have the same atomic number but different mass numbers; that is, their nuclei contain the same number of protons but different numbers of neutrons.

When a heavy nucleus splits into two or more lighter nuclei it is said to undergo *nuclear fission*. Considerable energy is evolved each time fission occurs.

A *chain reaction* is a succession of nuclear fissions in which excess neutrons produced by each fission induce further fissions in other atoms.

A *nuclear reactor* is a device in which fissions occur at a controlled rate.

When two light nuclei unite to form a heavier one, they are said to undergo *nuclear fusion*. Considerable energy is evolved in such processes; this energy is called *thermonuclear energy*.

EXERCISES

1 The following statements were thought to be correct in the nineteenth century. Which of them are now known to be incorrect? For those that are incorrect, indicate the experimental information that proved the statement wrong, and modify the statement appropriately so that it is in accordance with modern views.
   a. Atoms are indivisible and indestructible.
   b. Energy can be neither created nor destroyed.
   c. The acceleration of a body is directly proportional to the force applied to it and inversely proportional to its mass.
   d. An element is a substance that cannot be decomposed into other substances.
   e. All atoms of any one element are identical.

2 We have seen instances in which introducing new concepts to rescue an old theory has proved futile. How does the successful notion of isotopes differ from such unsuccessful notions as phlogiston?

3 For each of the following elements, the mass number of whose chief isotope is given at right, state (a) the number of protons in its nucleus, (b) the number of neutrons in its nucleus, and (c) the number of electrons surrounding its nucleus.

Element	Atomic number	Mass number
Oxygen	8	16
Iron	28	56
Iodine	53	127
Bismuth	83	209

4 Find the number of neutrons and protons in each of the following nuclei: $_3L^6$, $_6C^{13}$, $_{15}P^{31}$, $_{40}Zr^{94}$, $_{56}Ba^{137}$.

5 What are the differences and similarities between fusion and fission processes?

6 Why are magnetic fields rather than solid containers used to confine atomic nuclei at very high temperatures? Would this scheme work for electrically neutral atoms and molecules?

7 What advantage is there in using neutrons as bombarding particles for investigating nuclear reactions?

8 Can you give a reason why it is so difficult to produce new isotopes in the laboratory?

9 (a) One kilogram of U^{235} loses about 1 g in mass when it undergoes fission. How much energy is released when this mass disappears? (b) If 1 ton of TNT releases 4×10^9 joules when detonated, how many tons of TNT are equivalent to a bomb containing 1 kg of U^{235}?

10 A nucleus of $_3Li^6$ is struck by a nucleus of $_1H^2$, and a nuclear reaction occurs in which two identical nuclei are produced. State the atomic number, mass number, and chemical name of these nuclei.

11 A nucleus of $_4Be^9$ is struck by an alpha particle, and a neutron is emitted in the nuclear reaction that occurs. State the atomic number, mass number, and chemical name of the resulting nucleus.

12 In some stars three alpha particles join together in a single reaction to form a $_6C^{12}$ nucleus. The alpha-particle mass is 4.0015 amu, and that of $_6C^{12}$ is 11.9967 amu; find the energy liberated in this reaction in Mev (1 amu = 931 Mev).

13 The hydrogen isotope $_1H^3$ has a nuclear mass of 3.0155 amu, and the helium isotope $_2He^3$ has a nuclear mass of 3.0149 amu. Find the mass discrepancy for each nucleus in atomic mass units, and calculate the energy equivalent of this discrepancy in Mev. (The proton mass is 1.0073 amu and the neutron mass is 1.0087 amu.)

14 Would you expect the gravitational attractive force between two protons in a nucleus to counterbalance their electrical repulsion? Calculate the ratio between the electric and gravitational forces acting between two protons. Does this ratio depend upon how far apart the protons are?

chapter 17

Elementary Particles

RADIOACTIVITY

    radium
    alpha, beta, and gamma rays
    radioactive decay
    half-life

NUCLEAR STABILITY

    energy levels
    instability and decay
    meson theory of nuclear forces

ELEMENTARY PARTICLES

    antiparticles
    the neutrino
    the four fundamental interactions

Ordinary matter is composed of protons, neutrons, and electrons. Because these particles cannot be broken down further, they are called *elementary particles*. Another elementary particle we have encountered is the photon. Although the photon has no rest mass and it always moves with the speed of light, it has all the other properties usually associated with the term particle—it is localized in a small region of space, it has energy and momentum, and it interacts with other particles in more or less the same way as a billiard ball interacts with other billiard balls.

    A large number of other elementary particles have been discovered in addition to the proton, neutron, electron, and photon. Most of these particles are unstable and become transformed into other elementary particles in a fraction of a second after they come into being. Some of them, such as the *pion* and the *neutrino*, have fairly clear roles to play in the structure and behavior of ordinary matter. The place of many elementary particles in the scheme of things is still obscure, however, although great progress has been made in the past few years in understanding the interrelationships among them.

It is appropriate to preface our examination of elementary particles with a look at radioactivity, since there are many links between these two aspects of physics.

RADIOACTIVITY

In the year 1896 Henri Becquerel made a curious discovery in his laboratory in Paris. A small amount of a yellow salt containing uranium had been placed for several days on a photographic plate in his darkroom; when he developed the plate, he found the silhouette of the pile of salt as a developed image. This was extraordinary—a substance that could take a photograph of itself without emitting any visible light whatever. More of this uranium salt placed on other photographic plates produced similar behavior, no matter how carefully the plates were covered and protected from outside influences. The uranium salt was unquestionably giving out some sort of unknown radiation, a property Becquerel named *radioactivity*.

radium

Experiments following his initial discovery convinced Becquerel that the part of the salt responsible for darkening his photographic plates was the element uranium. Any compound of this metal produced a similar darkening, and the amount of darkening was roughly dependent on the amount of uranium present. Having discovered these few facts about the new phenomenon, Becquerel turned the problem over to a young woman working in the laboratories at the Sorbonne. Her name was Marie Curie.

The story of Madame Curie and her husband has been told often. Early in their work on radioactivity came the surprising discovery that a piece of pitchblende, a black mineral from Joachimsthal in what is now Czechoslovakia, produced a darkening of photographic plates out of all proportion to its uranium content. This suggested that the mineral contained traces of some other element far more powerful than uranium. The Curies set themselves the difficult task of isolating this unknown element. Two years of labor followed in which they managed to extract from a ton of black ore a fraction of a gram of each of *two* new elements. The first to be discovered was named polonium, after Madame Curie's native Poland; the other more highly radioactive element was the famous metal radium.

Radium somewhat resembles calcium and barium. Like them, it is a soft, silvery metal, tarnishing rapidly in air and dissolving in water with the evolution of hydrogen. Unlike calcium and barium, radium shows an astonishing ability to produce the radiations Becquerel associated with uranium. Thousands of times more active than uranium, radium and its salts not only blacken photographic plates, but, when mixed with certain compounds such as zinc sulfide, cause them to glow softly in a darkened room.

alpha, beta, and gamma rays

The radiation from radium and other radioactive elements causes fluorescence, darkens a photographic plate, and ionizes gases, and at least part of it is highly penetrating. The harshest physical or chemical treatment cannot stop the radiation or even slow it down; whether the elements are free or combined in salts, cooled in liquid air or heated in an electric arc, their ability to radiate remains unchanged. Spontaneously and continuously, year after year, these substances emit their radiation. Evidently the emission is somehow associated with the atoms themselves—and with a part of each atom that is not affected by ordinary physical and chemical changes.

In electric and magnetic fields the radiation shows a complex behavior. It splits into three parts; one part is deflected to one side, a second part to the other side, and a third part continues straight through (Fig. 17.1). Before their nature was known, the three parts were labeled provisionally with the first three letters of the Greek alphabet: *alpha rays* were those that were deflected as if they carried a positive charge; *beta rays*, those that seemed to have a negative charge; and *gamma rays*, those that passed undeflected through a field. Investigation has shown that gamma rays are electromagnetic waves shorter than X-rays and that beta rays are electrons.

Alpha particles ultimately proved to be atoms of helium. More correctly, we should say nuclei of helium—atoms that have been stripped of their two electrons apiece, so that they carry a double positive charge. Seven thousand times as heavy as electrons, moving much more slowly than beta or gamma rays (only about 10,000 mi/sec), the particles of alpha rays cannot penetrate nearly so far through gases or metals; but their relatively enormous mass makes them highly destructive to atoms that get in their way. Collisions between alpha particles and other atoms, as we saw in Chap. 15, gave physicists their earliest clues in deciphering atomic structures.

One type of alpha-particle collision can be observed in a rather simple manner. Take a watch or clock with a luminous dial into a darkened room, give your eyes time to become thoroughly accustomed to the darkness (at least 15 min), and then look at the glowing paint through a low-power magnifying glass. You will find that the glow is made up of myriads of tiny flashes, like the sparks from a skyrocket, each flash lasting only a moment. In the paint is a small bit of some radioactive substance together with a material that emits light under its influence; each flash is the disturbance created when a single alpha particle is stopped by molecules of the luminous substance. You do not see the helium nuclei themselves, of course, but you see the effect that each one produces.

alpha, beta, and gamma rays are emitted by radioactive substances

gamma rays are short wavelength electromagnetic waves

beta rays are electrons

alpha rays are helium nuclei

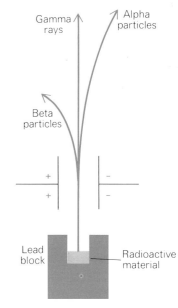

17.1 Three types of radiation from radioactive materials. *In the presence of an electric field, alpha particles behave as though positively charged, beta particles as though negatively charged, and gamma rays as though uncharged.*

radioactive decay

The fraction of a gram of paint on your watch dial, with its incredibly small amount of actual radioactive material, has been giving off helium ions in immense numbers since the watch was made. Now it is fairly easy to show that no helium is present in the materials of the paint when it is mixed, and equally easy to show that the helium is produced just as rapidly when all the materials of the paint are removed except the infinitesimal speck of some radioactive element that it contains. *The helium is evidently being produced spontaneously from the atoms of the radioactive element.* This is further evidence for abandoning the

The effects of nuclear radiation on living things are studied in a 50-acre forest on Long Island. A gamma-radioactive source is located at the end of the path in the picture, and nearby trees have died as a result of their exposure to these rays.

idea that elements are substances that cannot be broken down into simpler substances.

We may think of the expulsion of an alpha particle from a radium atom as a sort of minute explosion: a fragment of the atom bursts out of its interior, and part of the energy of the explosion appears as gamma radiation. The structure remaining after the explosion is no longer a radium atom; it has lost four atomic mass units and is, accordingly, an atom of an element with altogether different chemical properties. This is the element *radon*, a gas which, like radium, is intensely radioactive. Each of its atoms emits another alpha particle, becoming in the process an atom of polonium which likewise is radioactive. Thus the decay of a radium atom sets afoot a long series of radioactive changes.

Radium itself is the product of a series of changes starting with uranium. Table 17.1 and Fig. 17.2 show this complete series of radioactive elements, one of four such series. Each element in the series is produced

radioactive series

ELEMENTARY PARTICLES

TABLE 17.1 Radioactive transformations in the uranium series.

Element	Mass number	Half-life	Particle emitted during transformation
Uranium	238	4.4×10^9 years	Alpha
Thorium	234	24.5 days	Beta
Protactinium	234	1.14 min	Beta
Uranium	234	3×10^5 years	Alpha
Thorium	230	8×10^4 years	Alpha
Radium	226	1,590 years	Alpha
Radon	222	3.82 days	Alpha
Polonium	218	3.05 min	Alpha
Lead	214	26.8 min	Beta
Bismuth	214	19.7 min	Beta
Polonium	214	10^{-6} sec	Alpha
Lead	210	22 years	Beta
Bismuth	210	4.9 days	Beta
Polonium	210	140 days	Alpha
Lead	206	Stable	

by the disintegration of the one above it and produces in its turn the one below it. Note that some of the changes take place by the emission of an alpha particle, some by emission of a beta particle. Gamma radiation does not of itself produce a change from one element to another, but consists of ultrahigh-energy X-rays given out in the course of alpha- and beta-particle emission.

Three other disintegration series are known that involve other atoms of high atomic number. They are the *thorium* and *actinium* series, both of which occur in nature and end in lead, like the uranium series, and the *neptunium* series, which begins with an atom that must be artificially produced (although it probably existed in nature early in the history of the universe) and ends in bismuth. Elements whose atomic numbers are

most naturally radioactive elements have atomic numbers higher than that of lead

A scintillation counter detects gamma rays and high-speed charged particles by the flashes of light they cause in a luminescent crystal. A sensitive photoelectric cell responds to these flashes, and its electric signals are then fed into a suitable recorder. Here the counter is inside a lead shield to keep out stray radiation.

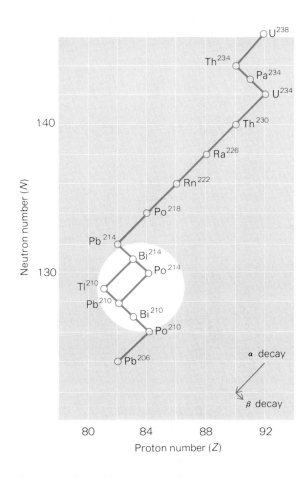

17.2 The uranium decay series. The decay of $_{83}Bi^{214}$ may proceed either by alpha emission and then beta emission or in the reverse order. There are three other decay series known.

less than that of lead, with only a few exceptions, are not naturally radioactive. Radioactive isotopes of the light elements, however, can be prepared by subjecting ordinary atoms to various kinds of radiation from accelerators or nuclear reactors.

half-life

half-life

The *half-life* of a radioactive isotope is the length of time necessary for half of any quantity of the isotope to decay. If we start with a certain number of radium atoms, whose half-life is 1,600 years, 50 percent of them will have decayed to radon by 1,600 years from now. In another 1,600 years half the remainder will have decayed, so only 25 percent of the original sample of radium will be left. In a further 1,600 years only $12\frac{1}{2}$ percent of the original sample will be left, and so on (Fig. 17.3). Every radioactive isotope has a characteristic and unchanging half-life; some have half-lives of a millionth of a second, others have half-lives that range up to billions of years.

The fact that radioactive decay follows the above pattern is strong evidence that this phenomenon is statistical in nature: every nucleus in a sample of radioactive material has a certain likelihood of decaying, but there is no way of knowing in advance *which* nuclei will actually decay in a particular time span. The situation is like that of a country's population, whose death rate can be accurately calculated even though it is impossible to predict exactly when a given individual will die.

If the sample of radioactive material is large enough, the actual fraction of it that decays in a particular time span will be the same as the probability for any individual nucleus to decay. The statement that radium has a half-life of 1,600 years, then, signifies that every radium nucleus has a 50 percent chance of decaying in any 1,600-year period. This does *not* mean a 100 percent chance of decaying in 3,200 years; a nucleus does not have a memory, and its likelihood of decaying stays the same until it actually does decay.

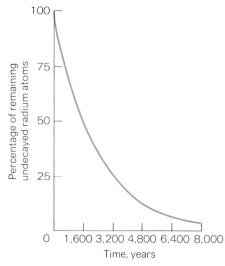

17.3 The decay of radium. *The number of remaining undecayed radium atoms decreases by one-half in each 1,600-year period. This time span is accordingly known as the "half-life" of radium.*

NUCLEAR STABILITY

Not all combinations of neutrons and protons form stable nuclei. In general, light nuclei contain approximately equal numbers of neutrons and protons, while in heavier nuclei the proportion of neutrons becomes progressively greater. This is evident from Fig. 17.4, which is a graph of N (number of neutrons) versus Z (number of protons) for stable nuclei.

energy levels

The tendency for N to equal Z follows from the existence of nuclear *energy levels*. A neutron or proton in a nucleus can have only certain specific energies. At most two neutrons and two protons can have identical energies in a nucleus; in the language of physics, each energy level can "contain" two neutrons and two protons. Energy levels are filled in sequence to achieve configurations of minimum energy and hence maximum stability. A nucleus with, say, three neutrons and one proton above filled lower levels will have more energy than a nucleus with two neutrons and two protons in the same situation, since in the former case one of the neutrons must go into a higher level while in the latter case all four nucleons fit into the lowest available level. Figure 17.5 shows how this picture accounts for the absence of a stable $_5B^{12}$ isotope while permitting $_6C^{12}$ to exist. In Chap. 19 we will see how atomic energy levels are filled by electrons in a similar way, and how this is related to the chemical properties of the elements.

the tendency for a nucleus to have equal numbers of neutrons and protons is due to the existence of nuclear energy levels

The preceding argument is only part of the story. Protons are positively charged and repel one another. This repulsion is so great in nuclei with more than 10 or so protons that an excess of neutrons, which produce only attractive forces between themselves and nucleons of either kind, is required for stability. Thus the experimental points of Fig. 17.4 depart more and more from the $N = Z$ line as the atomic number Z increases. Even in light nuclei N may exceed Z, but is never smaller; $_5B^{11}$ is stable, for instance, but not $_6C^{11}$.

the mutual repulsion between protons is the reason for the excess of neutrons in heavy nuclei

A common technique in contemporary chemical research involves the use of radioactive atoms as "tracers" in experiments. The details of a reaction can be studied by keeping track of the "tagged" substances with radiation detectors.

instability and decay

Nuclear forces are limited in range, and as a result nucleons interact strongly only with their nearest neighbors in a nucleus. Because the electrical repulsion of the protons is appreciable throughout the entire nucleus, there is a limit to the ability of neutrons to hold together a large nucleus. This limit is represented by the bismuth isotope $_{83}\text{Bi}^{209}$, which is the heaviest stable nuclide. All nuclei with more than 83 protons and more than 126 neutrons spontaneously transform themselves into lighter ones by the emission of one or more alpha particles, which as we know

ELEMENTARY PARTICLES

0.75-Mev protons from the Cockroft-Walton generator pictured on page 211 are boosted to energies of 50 Mev in this linear accelerator at Brookhaven National Laboratory. Electrodes within the accelerators are fed by alternating current in such a way that successive groups of protons experience electric fields that accelerate them.

are $_2\text{He}^4$ nuclei. Since an alpha particle consists of two protons and two neutrons, an alpha decay reduces the Z and the N of the original nucleus by two each.

If the resulting daughter nucleus has either too small or too large a neutron/proton ratio to be stable, it may beta decay to a more appropriate configuration. In negative beta decay, a neutron is transformed into a proton and an electron within the nucleus:

$$n \longrightarrow p + e^-$$

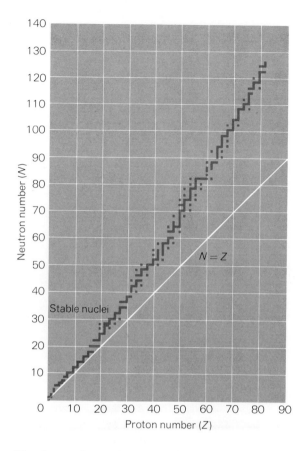

17.4 Stable nuclei. *The lighter nuclei tend to have equal numbers of neutrons and protons, while the proportion of neutrons is greater in heavy nuclei.*

The electron leaves the nucleus and is observed as a "beta particle." In positive beta decay, a proton becomes a neutron and a positron (positive electron) is emitted:

$$p \longrightarrow n + e^+$$

beta decay alters the neutron/proton ratio in a nucleus

Thus negative beta decay decreases the proportion of neutrons and positive beta decay increases it. Figure 17.6 shows how alpha and beta decays enable stability to be achieved. Positrons are discussed in more detail later in this chapter.

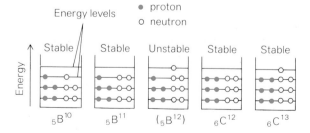

17.5 Each nuclear energy level in the light nuclei can contain two protons and two neutrons. *Energy levels are filled in sequence so that the nuclei have minimum energies.*

ELEMENTARY PARTICLES

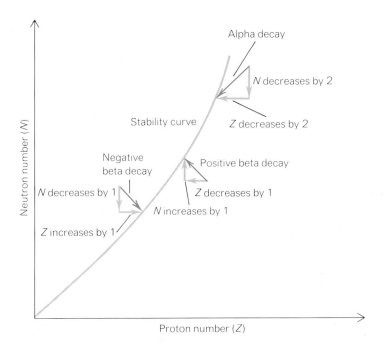

17.6 Alpha and beta decays tend to bring an unstable nucleus to a stable state by reducing its size (alpha decay) or by altering its neutron/proton ratio (beta decay).

Why are alpha particles emitted instead of, say, individual protons or $_2\text{He}^3$ nuclei? The answer follows from the extremely high binding energy of the alpha particle, which is responsible for the peak at the left-hand side of the binding energy per nucleon curve of Fig. 16.5. To escape from a nucleus, a particle must have kinetic energy, and the alpha-particle mass is sufficiently smaller than that of its constituent nucleons for such energy to be available. For example, alpha decay in $_{92}\text{U}^{232}$ is accompanied by the release of 5.4 Mev, while 6.1 Mev would somehow have to be supplied from the outside if a proton is to be emitted and 9.6 Mev if a $_2\text{He}^3$ nucleus is to be emitted.

Where do gamma rays fit into this picture? Often the nucleons in a daughter nucleus after alpha or beta decay are not all in their lowest energy levels. When such nucleons "fall" to these normal levels, the energy they lose is given off as photons of electromagnetic energy. These photons constitute gamma rays.

origin of gamma decay

meson theory of nuclear forces

At this point we know of the existence of electrons and positrons, neutrons and protons, and it seems possible to explain the behavior of atomic nuclei without going any further. However, if we were to go into the question of what holds nuclei together in a really deep way, we should want to know something about the nature of the forces involved. Because nuclei weigh less than their component particles, they are stable, but what is it in the interactions between nucleons that leads to this loss of mass? After considering this matter in detail, the Japanese physicist Hideki

The 50-Mev proton beam from the linear accelerator at Brookhaven enters the alternating-gradient synchrotron there through the 4-in. pipe at lower right. A changing magnetic field in the synchrotron imparts a final energy of 33 Gev to the protons by electromagnetic induction.

nuclear forces arise through meson exchange

Yukawa in 1935 proposed that nuclear forces arise from the constant exchange of particles (called π *mesons*) back and forth between nearby nucleons. There were precedents for this kind of analysis: We shall see in Chap. 20 how molecules are held together by the circulation of electrons among their component atoms, and the electrical forces between charges can be formally interpreted as the result of the exchange of electromagnetic photons between them.

According to the meson theory of nuclear forces, all nucleons consist of identical cores surrounded by a "cloud" of one or more mesons. Mesons may be neutral or carry either charge, and the sole difference between neutrons and protons is supposed to lie in the composition of their respective meson clouds. The forces that act between one neutron and another and between one proton and another are the result of the exchange of neutral mesons (designated π^0) between them. The force between a neutron and a proton is the result of the exchange of charged mesons (π^+ and π^-) between them. Thus a neutron emits a π^- meson and is converted into a proton

$$n \longrightarrow p + \pi^-$$

while the absorption of the π^- by the proton that the neutron was interacting with converts it into a neutron:

$$p + \pi^- \longrightarrow n$$

In the reverse process, a proton emits a π^+ meson whose absorption by a neutron converts it into a proton:

$$p \longrightarrow n + \pi^+$$
$$n + \pi^+ \longrightarrow p$$

Although there is no simple mathematical way of demonstrating how the exchange of particles between two bodies can lead to attractive and repulsive forces, a rough analogy may make the process intuitively meaningful. Let us imagine two boys exchanging basketballs (Fig. 17.7). If they throw the balls at each other, they each move backward, and when they catch the balls thrown at them, their backward momentum increases further. Thus this method of exchanging the basketballs yields the same effect as a repulsive force between the boys. However, if the boys snatch the basketballs from each other's hands, the result will be equivalent to an attractive force acting between them.

It is possible to prove, by more advanced mathematical techniques than we are using in this book, that the exchange of mesons between nucleons can indeed lead to mutually attractive forces, but a fundamental problem presents itself. If nucleons constantly emit and absorb mesons, why do we never find neutrons or protons with other than their usual masses?

The answer is based upon the uncertainty principle. The laws of physics refer exclusively to experimentally measurable quantities, and the uncertainty principle limits the accuracy with which we can make certain combinations of measurements. The emission of a meson by a nucleon which does not change in mass—a clear violation of the law of conservation of energy—can occur provided that the nucleon absorbs a meson emitted by the neighboring nucleon that it is interacting with so soon afterward that we cannot *even in principle* determine whether or not any mass change actually was involved.

A decade after Yukawa's work the π meson (or *pion*) was indeed discovered, and today there is no doubt that it is closely related to nuclear structure. Pions may carry $+$ or $-$ charges equal to the charge of the electron or may be uncharged. The charged pions have masses 273 times that of the electron; the uncharged ones are a little lighter.

Curiously enough, the pion is not a stable particle outside of a nucleus, but tends to "decay" in a fraction of a second. The charged pion decays into another kind of meson called the muon, whose mass is 207 times that of the electron. The uncharged pions decay into two gamma rays. Muons are also unstable and decay after brief existences, positive muons into positrons, negative muons into electrons. The mass differences between pions and muons and between muons and electrons, of course, appear as kinetic energy. All the mass of the uncharged pion is converted into electromagnetic energy in the two gamma rays (Fig. 17.8).

Repulsive force due to particle exchange

Attractive force due to particle exchange

17.7 Attractive and repulsive forces both can arise from particle exchange. *This is the concept underlying the meson theory of nuclear forces.*

meson exchange can occur without violating conservation of energy provided the process occurs rapidly enough

17.8 Decay schemes of the positive, negative, and neutral pi mesons (pions).

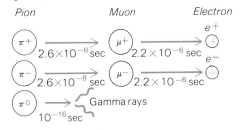

TABLE 17.2 Thirty-seven elementary particles. The photon, π^0 meson, and η^0 meson are their own antiparticles. Only the chief mode or modes of decay are listed.

Name	Particle	Anti-particle	Mass, in electron masses	Stability and decay	Average lifetime, sec
Photon	γ	(γ)	0	Stable	
Neutrino	ν_e	$\bar{\nu}_e$	0	Stable	
	ν_μ	$\bar{\nu}_\mu$	0	Stable	
Electron	e^-	e^+	1	Stable	
Muon	μ^-	μ^+	207	Unstable; decays into electron plus two neutrinos	2.2×10^{-6}
Pion	π^+	π^-	273	Unstable; decays into muon plus neutrino	2.6×10^{-8}
	π^0	(π^0)	264	Unstable; decays into two gamma rays	8.9×10^{-17}
Kaon	K^+	K^-	966	Unstable; decays into muon plus neutrino or into two or three pions	1.2×10^{-3}
	K_1^0	$\overline{K_1^0}$	974	Unstable; decays into two pions	8.7×10^{-11}
	K_2^0	$\overline{K_2^0}$	974	Unstable; decays into three pions or into pion and neutrino plus muon or electron	5.3×10^{-8}
Eta meson	η^0	(η^0)	1,073	Unstable; decays into three pions or two gamma rays	10^{-18}
Proton	p^+	p^-	1,836	Stable	
Neutron	n^0	$\bar{n}^0$	1,839	Unstable in free space; decays into electron, proton, and neutrino	1×10^3
Lambda hyperon	Λ^0	$\overline{\Lambda^0}$	2,182	Unstable; decays into pion plus neutron or proton	2.5×10^{-10}
Sigma hyperon	Σ^+	$\overline{\Sigma^-}$	2,328	Unstable; decays into pion plus neutron or proton	8×10^{-11}
	Σ^-	$\overline{\Sigma^+}$	2,341	Unstable; decays into neutron plus pion	1.6×10^{-10}
	Σ^0	$\overline{\Sigma^0}$	2,332	Unstable; decays into lambda hyperon plus gamma ray	10^{-14}
Xi hyperon	Ξ^-	$\overline{\Xi^+}$	2,583	Unstable; decays into lambda hyperon plus pion	1.7×10^{-10}

ELEMENTARY PARTICLES

TABLE 17.2 Thirty-seven elementary particles. The photon π^0 meson, and η^0 meson are their own antiparticles. Only the chief mode or modes of decay are listed (cont.).

Name	Particle	Anti-particle	Mass, in electron masses	Stability and decay	Average lifetime, sec
	Ξ^0	$\overline{\Xi^0}$	2,571	Unstable; decays into lambda hyperon plus pion	2.9×10^{-10}
Omega hyperon	Ω^-	$\overline{\Omega^+}$	3,290	Unstable; decays into lambda hyperon plus kaon or into xi hyperon plus pion	10^{-10}

Is the meson theory of nuclear forces correct? We are entitled to expect of any hypothesis that it make quantitative predictions in agreement with experiment, yet the meson theory cannot account for nuclear properties in the detailed manner that, as we shall see in the next two chapters, quantum theory can account for atomic properties. However, particles *have* been discovered whose mass and behavior are exactly in accord with meson theory; hence there can be little doubt that Yukawa was on the right track, and it is possible, perhaps likely, that an elaboration of the meson theory will prove successful.

ELEMENTARY PARTICLES

In addition to pions and muons, a large number of other unstable fundamental particles have been discovered in the past twenty years in experiments involving the bombardment of atomic nuclei with very high energy particles. Some of these new particles have only moderately short lifetimes (a millionth of a second is a long time in this branch of physics), while others exist only for the briefest of instants. Table 17.2 lists the 37 longest-lived of the known elementary particles, which is only a fraction of the total. The very term "elementary particle" has lost much of its meaning with the discovery of all these entities, most of which now seem to be closely related to one another though not in a simple way. The observed regularities in the properties of these particles give hope that a comprehensive theory to account for them may be forthcoming.

antiparticles

Nearly all the particles listed in Table 17.2 have *antiparticles:* The antiparticle of a given particle has the same mass and behaves similarly in most respects, but its electrical charge is opposite in sign. Thus the positron e^+ is the antiparticle of the electron e^-, and the negatively

the charge of a particle is opposite to that of its antiparticle

The 80-in. hydrogen bubble chamber at Brookhaven. In the chamber, liquid hydrogen at $-412°F$ is superheated by a sudden reduction in pressure timed to coincide with the arrival of particles from the alternating-gradient synchrotron. The bubbles that form along the ion trails of charged particles are photographed at the lower left for subsequent analysis.

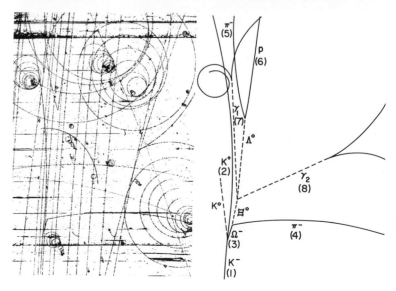

Bubble-chamber photograph of the track of the first omega-minus hyperon to be observed. This particle was predicted to exist by a new theory of elementary particles. Track 1 is an incoming K^- meson, whose collision with one of the protons in the liquid hydrogen results in the production of a K^0 meson, a K^+ meson (track 2), and a negative omega hyperon (track 3). The omega-minus hyperon decays in about 10^{-10} sec into several other particles (tracks 4–8).

charged *antiproton* p^- is the antiparticle of the proton p^+. Certain uncharged elementary particles, such as the neutron, have antiparticles because they have properties other than charge which are different in the particle and its antiparticle.

If a particle and its antiparticle happen to come together, they simultaneously vanish in a process called *annihilation*. The missing mass reappears in the form of gamma rays when electrons and positrons are annihilated and in the form of mesons when nucleons and antinucleons are annihilated (Fig. 17.9). In annihilation, matter is converted into energy and electric charge disappears—but the "lost" matter becomes precisely mc^2 of energy, and the total charge of the universe is unaltered, since the amounts of $+$ and $-$ charge that vanish are equal.

Just as matter can be converted into energy and electric charge disappear, so energy can be converted into matter and electric charge be created. The latter occurs in the process of *pair production*, which is the reverse of annihilation and can take place when a high-energy gamma ray passes near a nucleus (Fig. 17.10). The energy equivalent of the electron mass is about 0.5 Mev, so that the production of an electron-positron pair requires a gamma ray with an energy of about 1 Mev. Gamma rays of several million electron volts energy are easily obtained, and the positron was discovered as long ago as 1932.

The mass energy of a proton is nearly 1 Gev, and gamma rays of the 2 Gev or more energy needed for a proton-antiproton pair are less readily produced. In the past decade huge machines have been developed that can generate gamma rays of the required energies, and antiprotons (and antineutrons) have been definitely identified in the debris that results when they impinge on nuclei. An antiproton can exist for only a brief period, since upon contact with a proton in ordinary matter both vanish into mesons.

There seems no reason why atoms could not be composed of antiparticles—antiprotons, antineutrons, and positrons. Such *antimatter* would behave exactly like ordinary matter in every way, except that upon contact with ordinary matter, equal amounts of both would be annihilated in a violent explosion. It is entirely possible that elsewhere in the universe are stars, planets, and living things that consist of antimatter.

matter is annihilated when a particle and its antiparticle come together

matter is created from energy in pair production

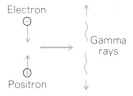

17.9 *The mutual annihilation of an electron and a positron results in a pair of gamma rays whose total energy is equal to mc^2, where m is the total mass of the electron and positron.*

antimatter

the neutrino

Another particle of great interest is the *neutrino*. This particle, which has no mass, is nevertheless able to possess energy and momentum, just as the photon does, although neutrinos and photons are quite different in other respects.

The electron energies observed in the beta decay of a particular nuclear species are found to vary *continuously* from 0 to a maximum value KE_{max} characteristic of the species. Figure 17.11 shows the energy spectrum of the electrons emitted in the beta decay of $_{83}Bi^{210}$; here $KE_{max} = 1.17$ Mev. In every case the maximum energy

$$E_{max} = m_0 c^2 + KE_{max}$$

17.10 *Pair production. The presence of a nucleus is required in order that both momentum and energy be conserved.*

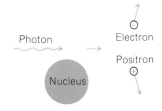

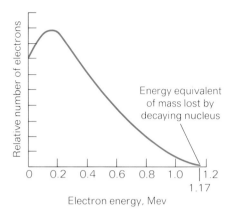

17.11 Energy distribution of electrons from the beta decay of $_{83}Bi^{210}$. The "missing" energy in each decay is carried off by a massless, changeless neutrino. If there were no neutrino, all the electrons would have 1.17 Mev of energy.

carried off by the decay electron is equal to the energy equivalent of the mass difference between the parent and daughter nuclei. Only seldom, however, is an emitted electron found with an energy of KE_{max}.

In 1930 Wolfgang Pauli proposed that, if an uncharged particle of zero mass is emitted in beta decay together with the electron, the energy discrepancy discussed above would be removed. It was supposed that this particle, later christened the *neutrino*, carries off an energy equal to the difference between KE_{max} and the actual electron kinetic energy (the recoil nucleus carries away negligible kinetic energy) and, in so doing, has a momentum exactly balancing those of the electron and the recoiling daughter nucleus. Subsequently, it was found that there are *two* kinds of neutrino involved in beta decay, the neutrino itself (symbol ν) and the antineutrino (symbol $\bar{\nu}$). In ordinary beta decay it is an antineutrino that is emitted:

$$n \longrightarrow p + e^- + \bar{\nu}$$

Positron emission corresponds to the conversion of a nuclear proton into a neutron, a positron, and a neutrino:

$$p \longrightarrow n + e^+ + \nu$$

While a neutron outside a nucleus can undergo negative beta decay into a proton because its mass is greater than that of the proton, the lighter proton cannot be transformed into a neutron except within a nucleus.

The neutrino hypothesis has turned out to be completely successful. The reason neutrinos were not experimentally detected until recently is that their interaction with matter is extremely feeble. Lacking charge and mass, and not electromagnetic in nature as is the photon, the neutrino can pass unimpeded through vast amounts of matter. A neutrino must pass through 130 *light-years* of solid iron on the average before interacting! (A light-year is the distance light travels in a year.) An immense flux of neutrinos is produced in the sun and other stars in the course of the nuclear reactions that occur within them, and this flux moves freely through the universe. The energy these neutrinos carry is—apparently—lost forever in the sense of being unavailable for conversion into other forms.

neutrinos interact only feebly with matter, and hence their energy is lost from the rest of the universe

In 1962 it was discovered that the neutrinos emitted when charged pions decay into muons are different from those emitted during beta decay. The existence of these two classes of neutrinos is indicated in Table 17.2 by the symbols ν_μ for the neutrino that accompanies the production of a muon and ν_e for the neutrino that accompanies beta decay. The corresponding antineutrinos are denoted by the same symbols with bars over them.

the four fundamental interactions

There are apparently only four ways in which elementary particles can interact with one another. These four types of interaction give rise to all the physical processes in the universe. The fundamental interactions are the following, listed in order of increasing strength:

1. The **gravitational interaction,** which is responsible for the attractive force one mass exerts upon another.
2. The **weak nuclear interaction,** which is responsible for particle decays in which neutrinos are involved, notably beta decays.
3. The **electromagnetic interaction,** which is responsible for the electric forces that charges at rest exert upon each other and for the magnetic forces that charges in motion exert upon each other.
4. The **strong nuclear interaction,** which is responsible for holding nucleons together to form atomic nuclei.

the fundamental interactions that give rise to all physical processes

The relative strengths of the strong, electromagnetic, weak, and gravitational interactions are in the ratios $1:10^{-2}:10^{-13}:10^{-38}$. Of course, the distances through which the corresponding forces act are very different. While the strong force between nearby nucleons is many powers of 10 greater than the gravitational force between them, when they are a meter apart the proportion is the other way. The structure of nuclei is determined by the properties of the strong interaction, while the structure of atoms is determined by those of the electromagnetic interaction. Matter in bulk is electrically neutral, and the strong and weak forces are severely limited in their range. Hence the gravitational interaction, utterly insignificant on a small scale, becomes the dominant one on a large scale. The role of the weak force in the structure of matter is apparently to see to it that nuclei with inappropriate neutron/proton ratios undergo corrective beta decays.

As mentioned earlier in this chapter, it is possible in a certain sense to think of the strong nuclear interaction as being "carried" from one nucleon to another by mesons and of the electromagnetic interaction as being "carried" from one charge to another by photons. Since mesons and photons actually exist with the properties expected of them on this basis, the question arises as to whether the gravitational and weak interactions, too, might not have characteristic particles associated with them. These hypothetical particles, called the *graviton* and the *intermediate boson* respectively, are being actively searched for, but as yet without success.

the graviton and intermediate boson

Another kind of particle which, according to an apparently plausible theory, ought to exist is the *quark*. There are three varieties of quark, plus their antiparticles, and all elementary particles are supposed to consist of combinations of quarks and antiquarks. The really revolutionary thing about quarks is that two of them should have charges of $-\frac{1}{3}e$ and the third should have a charge of $+\frac{2}{3}e$. According to this theory, nucleons and heavier particles are each composed of three quarks, and the various mesons are composed of quark-antiquark pairs. Despite much effort, no experimental evidence in support of the existence of quarks has been found thus far, but the ideas that underlie their prediction are so persuasive that the hunt continues. Elementary-particle physics is evidently one of the most active and exciting branches of science.

the quark

GLOSSARY

Radioactivity is the property possessed by certain atomic nuclei of spontaneously emitting charged particles or very high-frequency electromagnetic waves, or both.

Alpha particles are the nuclei of helium atoms; each one has a charge of $+2e$.

Beta particles are electrons.

Gamma rays are very high-frequency electromagnetic waves.

A *radioactive series* is a group of certain elements arranged in an order such that each element is produced by the radioactive decay of the one before it and decays in turn into the one below it. The four principal radioactive series are those of uranium, thorium, actinium, and neptunium.

The *half-life* of a radioactive element is the period of time required for one-half of an original sample of the element to decay.

An *elementary particle* is one of the various indivisible particles found in nature.

Pions are elementary particles involved in holding atomic nuclei together. Their masses are between those of the electron and proton, and they may have either charge or be neutral. Charged pions decay into the lighter *muons*, which in turn decay into electrons. Uncharged pions decay into a pair of gamma rays.

The *neutrino* is a massless, uncharged particle that is emitted during beta decay and in the decay of various elementary particles.

Almost every elementary particle has an *antiparticle* counterpart of the same mass whose electric charge and certain other properties have the opposite sign. Thus the antiparticle of the electron is the *positron*, whose charge is $+e$. When a particle and an antiparticle of the same kind come together, they *annihilate* each other, with the vanished mass reappearing in the form of energy.

Pair production is the materialization of an electron and a positron (or a proton and an antiproton) when a sufficiently energetic gamma ray passes near an atomic nucleus.

The four *fundamental interactions* are the gravitational, weak nuclear, electromagnetic, and strong nuclear interactions. They give rise to all the physical processes in the universe.

EXERCISES

1. Radium decomposes spontaneously into radon and helium. Why is radium considered an element rather than a compound of radon and helium?

ELEMENTARY PARTICLES

A spark chamber. The space between the aluminum plates is filled with a helium-neon gas mixture, and the plates are maintained at high relative potentials. When charged particles pass through the chamber, sparks jump between the plates along the paths of the particles.

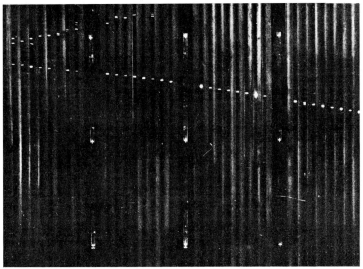

Photograph of a neutrino-induced event in a 10-ton spark chamber shielded from all particles except neutrinos by 7,600 tons of steel and 9,100 tons of concrete. The long track is that of a mu meson created by an incident neutrino. After the passage of 10^{14} synchrotron-produced neutrinos, 50 photographs were found to record mu mesons and none to record electrons. This experiment showed that the neutrinos produced in the decay of pi mesons are different from those produced in beta decay.

2 What happens to the atomic number and atomic mass of a radioactive isotope when it emits
   a. A negative beta particle?
   b. A positive beta particle?
   c. A gamma ray?

3 Of the following particles, which are emitted by nuclei in radioactive decay? Neutrons, electrons, positrons, mesons, neutrinos, protons, antiprotons, alpha particles, photons.

4 Both $_8O^{14}$ and $_8O^{19}$ undergo beta decay in order to become stable nuclei. Which of them would you expect to emit a positron and which an electron? Why?

5 Does a gamma-ray photon require more or less energy to materialize into a neutron-antineutron pair than to materialize into a proton-antiproton pair?

6 What is the physical principle that makes it possible for the law of conservation of energy to be violated under certain specific circumstances? Give an example of such circumstances.

7 When the nucleus of a radium atom, which has an atomic number of 88 and an atomic weight of 226, undergoes alpha decay, what happens to its atomic number and mass number?

8 The carbon isotope $_6C^{11}$ decays radioactively by the emission of a positive electron. Find the atomic number, mass number, and chemical name of the resulting isotope.

9 The thorium isotope $_{90}Th^{233}$ undergoes two successive negative beta decays. Find the atomic number, mass number, and chemical name of the resulting isotope.

10 The hydrogen isotope $_1H^3$ (tritium) decays radioactively by the emission of a negative electron. Find the atomic number, mass number, and chemical name of the resulting isotope. The half-life of tritium is 12.5 years. What percentage of an original quantity of tritium will still be present as tritium 50 years later?

11 If 1 kg of radium is sealed into a container, how much of it will remain as radium after 1,600 years? After 4,800 years? If the container is opened after a period of time, what gases would you expect to find inside it?

12 One-sixteenth of a sample of $_{11}Na^{24}$ remains undecayed after 60 hrs. Find the half-life of this radioisotope.

chapter 18

Theory of the Atom

ATOMIC SPECTRA

emission and absorption spectra

spectral series

BOHR MODEL OF THE ATOM

energy levels

waves and orbits

atomic excitation

QUANTUM THEORY OF THE ATOM

quantum mechanics

quantum numbers

electron probability density

electron spin

the exclusion principle

In Chap. 15 we learned that an atom consists of a tiny positively charged nucleus surrounded at some distance by electrons. This picture raised the important and vexing problem of how the electrons are able to stay away from the nucleus without circling it in orbits and thereby radiating energy continuously. To make sense of atomic structure the two key concepts discussed in Chap. 14 are needed: The quantum theory of light, according to which radiant energy is transported by individual photons of energy hf; and the wave theory of moving particles, according to which a particle with the momentum mv behaves in certain respects like a wave of wavelength h/mv. We shall now see how these concepts when linked to Rutherford's atomic model give rise to a theory of the atom in complete agreement with experiment. Our starting point will be the hydrogen atom, the simplest of all with but one electron outside a singly charged nucleus that consists of just a proton.

ATOMIC SPECTRA

The study of the radiation emitted by atoms was largely responsible for the growth of modern ideas about atomic structure. Most of this radiation is in the form of visible light, and the late nineteenth century saw many investigations of *atomic spectra*, as the characteristic series of wavelengths emitted by the atoms of particular elements are called. Since light is an electromagnetic phenomenon that can originate in the motion of electrons, the discovery that atoms contain electrons led to the speculation that atomic spectra can somehow be traced to these electrons.

emission and absorption spectra

continuous spectra contain all wavelengths

line spectra contain only specific wavelengths

We have already mentioned spectra briefly and distinguished between the rainbow bands produced when light from a glowing metal passes through a spectroscope (*continuous spectra*) and the sharp, bright lines from the light of a heated gas (*discontinuous*, or *line*, *spectra*). The continuous colored band from red to violet means that all visible wavelengths are present in the light; the discontinuous, bright-line spectrum indicates that only a few wavelengths are represented.

The observed features of the radiation from heated solids can be explained on the basis of the quantum theory of light independently of the details of the radiation process itself or of the nature of the solid. From this fact it follows that, when a solid is heated to incandescence, we are witnessing the collective behavior of a great many interacting atoms rather than the characteristic behavior of the individual atoms of a particular element.

emission and absorption spectra

At the other extreme, the atoms or molecules in a rarefied gas are so far apart on the average that their only mutual interactions occur during occasional collisions. Under these circumstances we should expect any emitted radiation to be characteristic of the individual atoms or molecules present, an expectation that is realized experimentally. When an atomic gas or vapor at somewhat less than atmospheric pressure is suitably "excited," usually by the passage of an electric current through it, the emitted radiation has a spectrum which contains certain discrete wavelengths only. An idealized laboratory arrangement for observing such atomic spectra is sketched in Fig. 18.1. Figure 18.2 shows the atomic spectra of several elements; they are called *emission line spectra*.

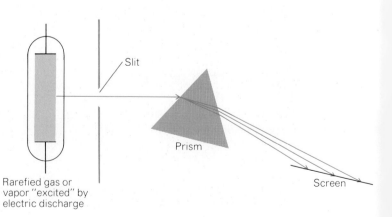

18.1 An idealized spectrometer. *Each wavelength produces a different image of the slit on the screen owing to dispersion in the prism.*

THEORY OF THE ATOM

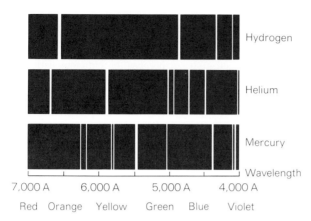

18.2 Portions of the emission spectra of hydrogen, helium, and mercury ($1\ A = 10^{-10}\ m$).

Spectra of a different sort, *absorption spectra*, are produced when light from an incandescent source passes through a cool gas before entering the spectroscope. The light source alone would give a continuous spectrum, but the gas absorbs certain wavelengths out of the light that passes through it. Hence the continuous spectrum appears to be crossed by dark lines, each line representing one of the wavelengths absorbed by the gas. If the bright-line spectrum of an incandescent gas is compared with the absorption spectrum of the same gas, the dark lines in the latter are found

Some typical spectra in the visible region. (a) Molecular hydrogen. (b) Atomic hydrogen. (c) Sodium vapor. (d) Helium. (e) Neon. (f) Lithium.

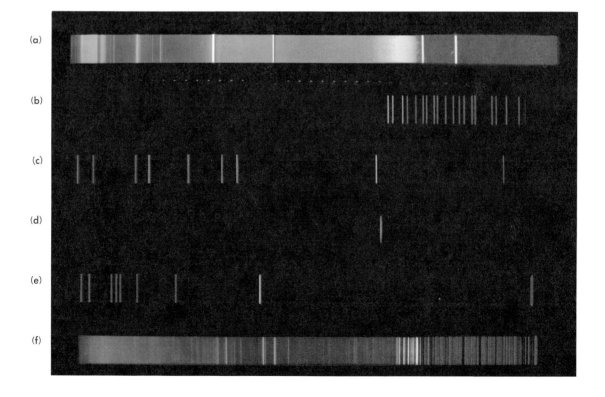

to correspond in wavelength to a number of the bright lines in the emission spectrum. Thus a cool gas absorbs wavelengths of light that it is capable of emitting when heated to incandescence. The dark Fraunhofer lines in the solar spectrum occur because the luminous part of the sun, which radiates in a manner approximately like that of any object heated to 6000°K (see Fig. 42.1), is surrounded by an envelope of cooler gas which absorbs light of certain wavelengths only.

spectral series

the line spectrum of every element contains certain characteristic wavelengths

The line spectrum of each element (either its bright-line spectrum or its absorption spectrum) contains lines of certain wavelengths that are characteristic of that element. These lines can be recognized in complex spectra so that the element can be identified either in a light source or in an absorbing gas. This fact makes the spectroscope a valuable tool in chemical and metallurgical analysis; even minute traces of most elements are readily identified by the lines in their spectra.

The number, intensity, and position of the lines in the spectrum of an element vary somewhat with temperature, with pressure, with the presence of electric and magnetic fields, and with the method by which the element is made incandescent. Thus an expert can tell by examination of spectra not only what elements are present in a light source, but much about their physical condition as well. An astronomer, for example, can deduce from the spectrum of a star the composition of its atmosphere, whether it is approaching or receding from the earth, and what substances in its atmosphere are ionized.

In the latter part of the nineteenth century it was discovered that the wavelengths present in atomic spectra fall into definite sets called *spectral series*. The wavelengths in each series can be specified by a simple empirical formula, the formulas for the various series that comprise the complete spectrum of an element being remarkably similar. The first such spectral series was found by J. J. Balmer in 1885 in the course of a study of the visible part of the hydrogen spectrum. Figure 18.3 shows the *Balmer series*. The line with the longest wavelength is designated H_α, the next is designated H_β, and so on. As the wavelength decreases, the lines are found closer together and weaker in intensity until the *series limit* is reached, beyond which there are no further separate lines but only a faint continuous spectrum. Balmer's formula for the wavelengths of this series is quite simple:

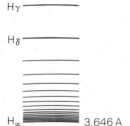

18.3 The Balmer series of hydrogen. *The spectral lines become closer and closer together near the short-wavelength limit $(1 A = 10^{-10} m)$.*

$$\frac{1}{\lambda} = R\left(\frac{1}{2^2} - \frac{1}{n^2}\right) \qquad n = 3, 4, 5, \ldots$$

the wavelengths in each spectral series are related by a simple formula

The quantity R, known as the *Rydberg constant*, has the value 1.097×10^7 m^{-1}. The H_α line corresponds to $n = 3$, the H_β line to $n = 4$, and so on. The series limit corresponds to $n = \infty$, so that it occurs at a wavelength of $4/R$, in agreement with experiment.

The Balmer series contains only those wavelengths that are in the visible portion of the hydrogen spectrum. The spectral lines of hydrogen in the

THEORY OF THE ATOM

ultraviolet and infrared regions fall into several other series, as Fig. 18.4 shows. The Brackett series evidently overlaps the Paschen and Pfund series.

The existence of such remarkable regularities in the hydrogen spectrum, together with similar regularities in the spectra of more complex elements, poses a definitive test for any theory of atomic structure. Here once more we find the familiar pattern of scientific reasoning, from observations to simple mathematical relations between the observations, and then from these relations to an inclusive theory. As Tycho Brahe's observations of the planets were correlated by Kepler's simple formulas and as these paved the way for Newton's idea of universal gravitation, so in the past hundred years observations of spectra have led to simple relations among the wavelengths of spectral lines and these in turn have led to modern concepts of atomic structure.

BOHR MODEL OF THE ATOM

The first theory of the hydrogen atom to succeed in accounting for the more conspicuous aspects of its behavior was presented by Niels Bohr, a Dane, in 1913. Bohr applied quantum ideas to atomic structure to obtain a model which, despite its serious inadequacies and later replacement by a theory of greater accuracy and usefulness, nevertheless persists as the mental picture many scientists have of the atom.

energy levels

Bohr applied the basic concept of the quantum theory of light—that energy comes in small packets rather than in a continuous stream—to the problem of atomic structure. He began by postulating that there were certain specific orbits in which an electron inside an atom could circle the nucleus without losing energy. Because these orbits are each a different distance from the nucleus, electrons in them have different amounts of energy. In other words, Bohr suggested that electrons within an atom can have only certain particular energy values. An electron in the innermost orbit has the least energy; an electron in an outer orbit has more. Thus each orbit may be called, alternatively, an *energy level*. The orbits (or energy levels) are identified by their *quantum number*, n, which is $n = 1$ for the innermost orbit, $n = 2$ for the next, and so on.

The emission and absorption by atoms of only particular frequencies of light (which we observe as spectral lines) fits Bohr's atomic model perfectly. An atomic electron in some stable orbit can absorb only those photons of light whose energy hf will permit it to "jump" into another stable orbit farther out. Emission of radiation takes place when an electron jumps from one orbit to another of smaller energy, with the difference in energy between the two orbits being equal to hf, where f is the observed frequency of the emitted light.

Figure 18.5 shows schematically the arrangement of possible orbits for the single electron in a hydrogen atom. Here the orbit pursued by the

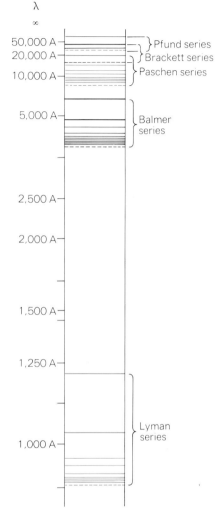

18.4 The spectral series of hydrogen. *The wavelengths in each series can be related by simple formulas ($1\ A = 10^{-10}\ m$).*

solitary electron under ordinary conditions is represented by the circle nearest the nucleus. The other circles are possible orbits in which the electron would possess greater energy than in its normal orbit, since it would then be farther from the nucleus (much as a stone at the top of a building has more potential energy than on the ground, since it is farther from the earth's center).

Suppose that the electron at first is in the normal orbit. If the atom is supplied with energy—by strong heating, by an electrical discharge, or by radiation—the electron may be induced to jump to a larger orbit. This jump means that the atom has absorbed some energy. It retains the added energy as long as it remains in the *excited state*, that is, as long as its electron stays in the larger orbit. But the excited state is unstable, and in a small fraction of a second the electron jumps spontaneously back to its original orbit (or to another orbit smaller than the original one). In this second jump radiation is emitted, the energy of the radiation representing the difference in energy between the excited and normal states of the atom. The amount of energy given out evidently depends on which of the outer orbits the electron followed in the excited state.

To gain a rough picture of the emission of radiation during an electron jump, we may imagine the electron vibrating for a moment as it subsides into the smaller orbit, much as a rubber ball will bounce repeatedly when it loses potential energy by dropping from a table to the floor. The vibrating electron sends out electromagnetic waves, just as the vibrating electrons in a radio antenna send out waves. The electron in the atom vibrates much more rapidly than those in the antenna, so its waves have far higher frequency and shorter wavelength than radio waves.

The energy and, therefore, the frequency of the radiation emitted from a hydrogen atom are determined, according to Bohr's hypothesis, by the particular jump that its electron makes. If the electron jumps from orbit $n = 4$ to orbit $n = 1$ (Fig. 18.5), the energy (and frequency) of the radiation will be greater than if it jumps from 3 to 1 or 2 to 1. Starting from orbit 4, it may return to 1 not by a single leap but by stopping at 2 and 3 on the way; corresponding to these jumps will be radiations with energies (and frequencies) determined by the energy differences between 4 and 3, 3 and 2, 2 and 1. Each of these several jumps gives radiation of a single frequency and will, therefore, be represented in the hydrogen spectrum by a single bright line. Further, the frequencies of the different lines will be simply related one to the other, since they correspond to different possible jumps in the same set of orbits. And the relations between the lines that Bohr *predicted* by this mechanism precisely matched the *observed* relations between lines in the spectrum of hydrogen (Fig. 18.6).

an electron in an orbit larger than its normal one is said to be in an excited state

a photon of electromagnetic energy is emitted when an electron in an excited state returns to its normal state

18.5 Electron orbits in the Bohr model of the hydrogen atom (not to scale). *The radius of each orbit is proportional to n^2, the square of its quantum number. The inner orbit is the electron's normal path and the outer orbits represent states of higher energy. If the electron absorbs enough energy for it to jump to an outer orbit, it may return to its normal one by a suitable jump or combination of jumps; each jump is accompanied by the emission of radiation.*

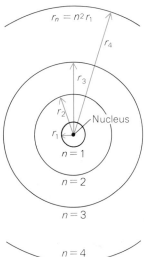

waves and orbits

It is now necessary to inquire into why an atomic electron can exist in certain orbits only. The clue we need comes from an analysis of the wave properties of an electron revolving around a hydrogen nucleus: it turns out that the de Broglie wavelength of the electron is exactly equal to

THEORY OF THE ATOM

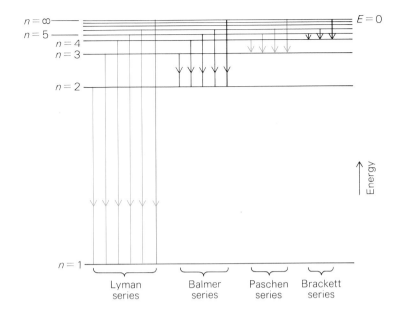

18.6 Spectral lines are the result of transitions between energy levels.

the circumference of its normal (that is, innermost) orbit. Thus the normal orbit of the electron in a hydrogen atom corresponds to one complete electron wave joined on itself (Fig. 18.7).

This fact provides us with the information we need to construct a theory of the atom. If we consider the vibrations of a wire loop (Fig. 18.8), we find that their wavelengths always fit an integral number of times into the loop's circumference, so that each wave joins smoothly with the next. If the wire were rigid enough, these vibrations would continue indefinitely. Why are these the only vibrations possible in a wire loop? If a fractional number of wavelengths are placed around the loop, as in Fig. 18.9, destructive interference will occur as the waves travel around the loop, and vibrations will die out rapidly. By regarding the behavior of electron waves in an atom as being analogous to the vibrations of a wire loop, then, we might suppose that **an electron can circle a nucleus without radiating energy provided that its orbit contains a whole number of de Broglie wavelengths.**

The above hypothesis is the decisive one in our understanding of the atom. It combines both the particle and wave characters of the electron into a single statement, since the electron wavelength is computed from the orbital speed required to balance the electrostatic attraction of the nucleus. Although we can never experimentally verify these contradictory characters simultaneously, they are inseparable aspects of the natural world.

It is a simple matter to express the condition that an electron orbit contain an integral number of de Broglie wavelengths. The circumference of a circular orbit of radius r is $2\pi r$, and so we may write the condition for orbit stability as

$$n\lambda = 2\pi r_n \qquad n = 1, 2, 3, \ldots$$

18.7 The condition for a stable electron orbit. *The orbit of the electron in a hydrogen atom corresponds to a complete electron de Broglie wave joined on itself.*

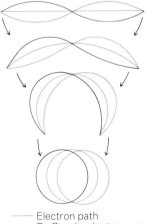

— Electron path
— De Broglie electron wave

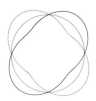

Circumference = 2 wavelengths

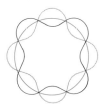

Circumference = 4 wavelengths

Circumference = 8 wavelengths

18.8 Vibration of a wire loop. *In each case an integral number of wavelengths fit into the circumference of the loop.*

where r_n is the radius of the nth orbit, that is, the one that contains n whole wavelengths. The integer n is called the *quantum number* of the orbit. A detailed calculation (which takes into account the speed the electron must have to circle the nucleus without being drawn into it by electrostatic attraction) shows that the stable electron orbits are restricted to those having the radii

$$r_n = n^2 r_1 \qquad n = 1, 2, 3, \ldots$$

The orbit closest to the nucleus in the hydrogen atom has the radius

$$r_1 = 5.3 \times 10^{-11} \text{ m}$$

The spacing between adjacent orbits increases rapidly with increasing quantum number n.

The various permitted orbits involve different electron energies. The general formula for them is

$$E_n = -\frac{E_1}{n^2} \qquad n = 1, 2, 3, \ldots$$

The energies E_n given by this formula are plotted in Fig. 18.6. The energies are all negative, which means that the electron is bound to the nucleus. The lowest energy level E_1 (equal to -2.2×10^{-18} joule, which is -13.6 ev) is that of the *ground state* of the atom, and the higher levels E_2, E_3, E_4, and so on, are those of the excited states. With increasing n, the corresponding energy E_n approaches closer and closer to 0; in the limit of $n = \infty$, $E_\infty = 0$, and the electron is no longer bound to the nucleus to form an atom. (A positive energy for a nucleus-electron combination means that the electron is not bound to the nucleus and has no quantum conditions to fulfill; such a combination does not constitute an atom, of course.)

atomic excitation

18.9 *A fractional number of wavelengths cannot persist in a wire loop because destructive interference will occur.*

There are two principal mechanisms that can excite an atom to an energy level above its ground state, thereby enabling it to radiate. One mechanism is a collision with another particle during which part of their joint kinetic energy is absorbed by the atom. An atom excited in this way will return to its ground state in an average of 10^{-8} sec by emitting one or more photons (Fig. 18.10). To produce an electric discharge in a rarefied gas, an electric field is established which accelerates electrons and atomic ions until their kinetic energies are sufficient to excite atoms they happen to collide with. Neon signs and mercury-vapor lamps are familiar examples of how a strong electric field applied between electrodes in a gas-filled tube leads to the emission of the characteristic spectral radiation of that gas, which happens to be reddish light in the case of neon and bluish light in the case of mercury vapor.

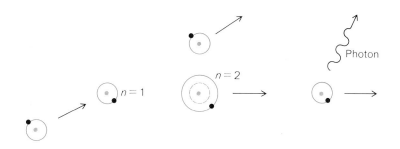

18.10 Excitation by collision. *Some of the available energy is absorbed by one of the atoms, which goes into an excited energy state. The atom subsequently emits a photon in returning to its ground (normal) state.*

A different excitation mechanism is involved when an atom absorbs a photon of light whose energy is just the right amount to raise the atom to a higher energy level. For example, a photon of wavelength 1.026×10^{-7} m is emitted when a hydrogen atom in the $n = 3$ state drops to the $n = 1$ state; the absorption of a photon of wavelength 1.026×10^{-7} m by a hydrogen atom initially in the $n = 1$ state will therefore bring it up to the $n = 3$ state. This process accounts for the origin of absorption spectra. When white light, which contains all wavelengths, is passed through hydrogen gas, photons of those wavelengths that correspond to transitions between energy levels are absorbed. The resulting excited hydrogen atoms reradiate their excitation energy almost at once, but these secondary photons come off in random directions with only a few in the same direction as the original beam of white light (Fig. 18.11). The dark lines in an absorption spectrum are therefore never

an atom can be excited by absorbing a photon

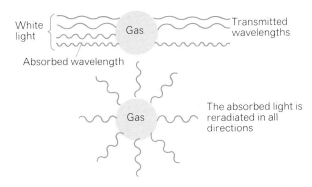

18.11 Origin of an absorption spectrum. *The dark lines in such a spectrum appear that way because the absorbed wavelengths are reradiated in random directions, so only a small part of the original intensity of these wavelengths continues in the direction of the primary beam.*

completely black, but only seem so by contrast with the bright background. We expect the lines in the absorption spectrum of any element to have counterparts in its emission spectrum, then, which agrees with observation (Fig. 18.12).

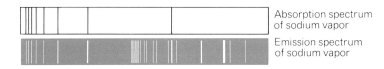

18.12 The dark lines in the absorption spectrum of an element correspond to bright lines in its emission spectrum.

QUANTUM THEORY OF THE ATOM

The above theory of the hydrogen atom, which is essentially that developed by Bohr in 1913 (although he did not have the de Broglie hypothesis of electron waves to guide his thinking), is able to account for certain experimental data in a convincing manner, but it has a number of severe limitations. While the Bohr theory correctly predicts the spectral series of hydrogen, it is incapable of being extended to treat the spectra of complex atoms having two or more electrons each; it can give no explanation of why certain spectral lines are more intense than others (that is, why certain transitions between energy levels have greater probabilities of occurrence); and it cannot account for the observation that many spectral lines actually consist of several separate lines whose wavelengths differ slightly. And, perhaps most important, it does not permit us to obtain what a really successful theory of the atom should make possible: an understanding of how individual atoms interact with one another to endow macroscopic aggregates of matter with the physical and chemical properties we observe.

The above objections to the Bohr theory are not put forward in an unfriendly way, for it was one of those seminal achievements that transform scientific thought, but rather to emphasize that an approach to atomic phenomena of greater generality is required. Such an approach was developed in 1925–1926 by Erwin Schrödinger, Werner Heisenberg, and others, under the apt name of *quantum mechanics*. By the early 1930s the application of quantum mechanics to problems involving nuclei, atoms, molecules, and matter in the solid state made it possible to understand a vast body of otherwise puzzling data and—a vital attribute of any theory—led to predictions of remarkable accuracy.

quantum mechanics

The fundamental difference between newtonian mechanics and quantum mechanics lies in what they describe. Newtonian mechanics is concerned with the motion of a particle under the influence of applied forces, and it takes for granted that such quantities as the particle's position, mass, velocity, and acceleration can be measured. This assumption is, of course, completely valid in our everyday experience, and newtonian mechanics provides the "correct" explanation for the behavior of moving bodies in the sense that the values it predicts for observable magnitudes agree with the measured values of those magnitudes.

the uncertainty principle is an integral part of quantum mechanics, which is concerned with probabilities

Quantum mechanics, too, consists of relationships between observable magnitudes, but the uncertainty principle radically alters the definition of "observable magnitude" in the atomic realm. According to the uncertainty principle, the position and momentum of a particle cannot be accurately measured at the same time, whereas in newtonian mechanics both are assumed to have definite, ascertainable values at every instant. The quantities whose relationships quantum mechanics explores are *probabilities*. Instead of asserting, for example, that the radius of the electron's orbit in a ground-state hydrogen atom is always exactly 5.3×10^{-11} m, quantum mechanics states that this is the *most probable* radius; if we conduct a suitable experiment, many trials will yield a different value, either larger or smaller, but the value most likely to be found will be 5.3×10^{-11} m.

Thus in quantum mechanics there is no attempt to force nature into the arbitrary molds of our personal experience. The theory is completely abstract instead of being based on a more or less mechanical model, and its only contact with reality is through observable quantities. That is, we can measure the mass of the electron and its electric charge, we can measure the frequencies of spectral lines emitted by excited atoms, and so on, and the theory must be able to relate them all. But we *cannot* measure the diameter of an electron's orbit or watch it jump from one orbit to another, and these notions, therefore, are not incorporated in the theory.

quantum mechanics deals only with quantities that can be measured

The quantum-mechanical theory of the atom starts with Schrödinger's equation, a formula which requires a background in advanced mathematics to understand. The procedure is to substitute into Schrödinger's equation certain facts about the atom under consideration, such as the number of protons in its nucleus, the mass and charge of the electron, and so on, and then to solve the resulting equation for the wave function ψ (see Chap. 14) of whichever electrons we are concerned with. From this wave function can be calculated the probability density of the electron, which is the likelihood that it be found in any particular place whose coordinates we specify. Because of the mathematical properties of Schrödinger's equation, a classification of atomic electrons in terms of energy levels follows naturally. This classification is the analog of our previous picture of stable orbits, but it has a much more secure theoretical background and can be extended to cover many more situations accurately than can the simpler Bohr theory.

Quantum mechanics represents a complete abandonment of the traditional approach to physics in which models capable of being visualized are the starting points of theories. But although quantum mechanics does not provide any glimpses into the inner world of the atom, it does tell us almost everything we need to know about the measurable properties of atoms. And close inspection reveals a striking fact: **Newtonian mechanics is no more than an approximate version of quantum mechanics.** The certainties proclaimed by Newton are illusory, and their agreement with experiment is a consequence of the fact that macroscopic bodies consist of so many individual atoms that departures from average behavior are unnoticeable. Instead of two sets of physical principles, one for the macroscopic universe and one for the microscopic universe, there is only a single set, and quantum mechanics represents our best effort to date at formulating it.

newtonian mechanics is an approximation of quantum mechanics valid only for large assemblies of atoms

quantum numbers

The quantum-mechanical theory of the atom reveals that a total of three quantum numbers, n, l, and m_l, is needed to specify the physical state of the electron in a hydrogen atom. It is interesting to consider the

the energy of an atomic electron is quantized (quantum number *n*)

the angular momentum of an atomic electron is quantized in both magnitude and direction (quantum numbers *l* and m_l)

18.13 Right-hand rule for angular momentum L. *The magnetic quantum number m_l specifies the direction of* **L** *by determining the projection of* **L** *in the field direction. The possible values of m_l for a given value of orbital quantum number l range from +l to 0 to −l, so that the number of possible orientations of the angular-momentum vector* **L** *in a magnetic field is $2l + 1$. We must regard an atom characterized by a certain value of m_l as standing ready to assume a certain orientation of its angular momentum* **L** *relative to an external magnetic field, in the event it finds itself in such a field.*

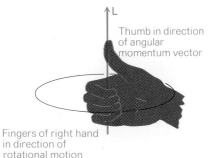

interpretation of these quantum numbers in terms of the Bohr model of the atom. This model, as we saw earlier in this chapter, corresponds to planetary motion in the solar system, except that the inverse-square force holding the electron to the nucleus is electrical rather than gravitational. Two quantities are *conserved*—that is, maintain a constant value at all times—in planetary motion, as Newton was able to show from Kepler's three empirical laws: these are the total energy and the angular momentum of each planet. In the quantum-mechanical theory of the atom the energy of each electron is also a constant, with its value determined by the quantum number n of its "orbit." The theory of planetary motion can also be worked out from Schrödinger's equation, and it yields an identical energy restriction. However, the total quantum number n for any of the planets turns out to be so immense that the separation of permitted energy levels is far too small to be observable. For this reason classical physics provides an adequate description of planetary motion, but it fails within the atom. The total quantum number n refers to the quantization of electron energy in the hydrogen atom.

Quantum mechanics shows that the angular momentum of an atomic electron is also both quantized and conserved. (Angular momentum, as we learned in Chap. 4, is the rotational analog of linear momentum.) According to quantum mechanics, the possible values of angular momentum are determined by l, the *orbital quantum number*. An electron whose total quantum number is n can have an orbital quantum number of zero or any integer up to $n − 1$.

In macroscopic planetary motion the quantum number describing angular momentum is so large that the separation into individual angular momentum states cannot be experimentally observed, but such a separation is detectable in atomic electron structures.

The orbital quantum number l determines the *magnitude* of the electron's angular momentum. However, angular momentum, like linear momentum, is a vector quantity, and so to describe it completely requires that its *direction* be specified as well as its magnitude. What possible significance can a direction in space have for an atom? The answer becomes clear when we reflect that an electron revolving about a nucleus is a minute current loop and has a magnetic field like that of a tiny bar magnet. In an external magnetic field a bar magnet has an amount of energy that depends both upon how strong it is and upon its orientation with respect to the field. It is therefore not surprising to learn that the *direction* of the angular-momentum vector (that is, the direction of the axis about which the electron may be thought to revolve) is also quantized with respect to an external magnetic field. The quantum number that governs this direction is m_l (Fig. 18.13).

In a magnetic field, then, the energy of a particular atomic state depends upon the value m_l as well as upon that of n. A state of total quantum number n breaks up into several substates when the atom is in a magnetic field, and their energies are slightly more or slightly less than the energy of the state in the absence of the field. This phenomenon is one of the effects that leads to a "splitting" of individual spectral lines into separate lines when atoms radiate in a magnetic field, with the spacing of the lines dependent upon the magnitude of the field. The splitting of

THEORY OF THE ATOM

spectral lines by a magnetic field is called the *Zeeman effect* after the Dutch physicist Zeeman, who first observed it in 1890 (Fig. 18.14). The Zeeman effect is a vivid confirmation of space quantization, and has proved invaluable to astronomers in studying the properties of stars and galaxies.

The various atomic states are conventionally labeled according to their quantum numbers n (which governs the total energy) and l (which governs the magnitude of the angular momentum) according to the scheme shown in Table 18.1. There are no states such as $1p$ or $2d$ because l must always be smaller than n.

the zeeman effect is the splitting of spectral lines when the emitting atoms are in a magnetic field

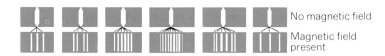

No magnetic field
Magnetic field present

18.14 Some examples of the Zeeman effect.

electron probability density

In Bohr's model of the hydrogen atom the electron is visualized as revolving around the nucleus in a circular path. The quantum theory of the atom modifies the straightforward Bohr picture in two ways. First, no definite value for the electron's position at any time can be given, only the relative probabilities for finding the electron at various locations. This imprecision is, as mentioned earlier, a consequence of the wave nature of the electron. Second, we cannot even think of the electron as moving around the nucleus in any conventional sense because the probability density ψ^2 is independent of time and may vary considerably from place to place.

Figure 18.15 shows how the electron probability density varies in space for various states of the hydrogen atom. The larger the quantum number n, the farther away the electron is, on the average, from the nucleus. For a given value of n, the less angular momentum the electron has, the farther away it is from the nucleus, again on the average. Or we can just as well think of the electron as though it is smeared out in space, so that instead of probability density the shaded patterns of Fig. 18.15 represent negative charge density. Evidently all three quantum numbers n, l, and m_l affect ψ^2. The probability densities for s states are spherically

TABLE 18.1 The symbolic designation of atomic states.

	s $l=0$	p $l=1$	d $l=2$	f $l=3$	g $l=4$	h $l=5$
$n=1$	$1s$					
$n=2$	$2s$	$2p$				
$n=3$	$3s$	$3p$	$3d$			
$n=4$	$4s$	$4p$	$4d$	$4f$		
$n=5$	$5s$	$5p$	$5d$	$5f$	$5g$	
$n=6$	$6s$	$6p$	$6d$	$6f$	$6g$	$6h$

18.15 Cross-sectional views of the electron probability density ψ^2 for several states of the hydrogen atom. symmetric, while those of the others are not. The pronounced lobe patterns characteristic of many of the states turn out to be significant in chemistry since these patterns help determine the manner in which adjacent atoms in a molecule interact with each other, as we shall see in Chap. 20.

electron spin

Despite the accuracy with which quantum mechanics accounts for certain of the properties of the hydrogen atom, and despite the elegance and essential simplicity of this theory, it cannot approach a complete description of this atom nor of other atoms without the further hypothesis of electron spin and the exclusion principle associated with it. In the next several chapters we shall examine the role of electron spin in atomic phenomena and why the exclusion principle is the key to understanding the structures of complex atomic systems.

In an effort to account for a number of small but definite peculiarities in otherwise regular atomic spectra, S. A. Goudsmit and G. E. Uhlenbeck proposed in 1925 that the electron possesses an intrinsic angular momentum called *spin* independent of any orbital angular momentum it might have, and in addition behaves like a tiny bar magnet. What Goudsmit and Uhlenbeck had in mind was a classical picture of an electron as a charged sphere spinning on its axis. The rotation involves angular momentum, and because the electron is negatively charged, it is in effect a loop of electric current and hence has a magnetic field around it. The notion of electron spin proved to be successful in explaining a wide variety of atomic effects. Of course, the idea that electrons are spinning charged spheres is hardly in accord with quantum mechanics, but in 1928 P. A. M. Dirac was able to show, on the basis of a relativistic quantum-theoretical treatment, that particles having the charge and mass of the electron must have just the behavior attributed to them by Goudsmit and Uhlenbeck.

electrons behave as though they are spinning charged spheres

The spin angular momentum of an electron is restricted to a single value, but the spin magnetic quantum number m_s can have the two values $m_s = +\frac{1}{2}$ and $m_s = -\frac{1}{2}$, corresponding to two possible orientations of the axis of spin ("up" and "down") in a magnetic field.

direction of electron spin is quantized (quantum number m_s)

the exclusion principle

In the normal configuration of a hydrogen atom, the electron is in its lowest quantum state. What are the normal configurations of more complex atoms? Are all 92 electrons of a uranium atom in the same quantum state, to be envisioned perhaps as circling the nucleus crowded together in a single Bohr orbit? Many lines of evidence make this hypothesis unlikely. An example is the great difference in chemical behavior exhibited by certain elements whose atomic structures differ by just one electron: for instance, the elements having atomic numbers 9, 10, and 11 are respectively the halogen gas fluorine, the inert gas neon, and the alkali metal sodium. Since the electronic structure of an atom controls its interactions with other atoms, it would be hard to understand why the chemical properties of the elements should change so abruptly with a small change in atomic number if all the electrons in an atom were to exist together in the same quantum state.

In 1925 Wolfgang Pauli discovered the fundamental principle that governs the electronic configurations of atoms having more than one

each electron in an atom has a different set of quantum numbers

electron. His *exclusion principle* states that **no two electrons in an atom can exist in the same quantum state.** Each electron in an atom must have a different set of quantum numbers n, l, m_l, m_s, which determine its physical state, that is, its energy, angular momentum, and spin.

Pauli was led to the exclusion principle by a study of atomic spectra. It is possible to determine the various states of an atom from its spectrum, and the quantum numbers of these states can be inferred. In the spectra of every element but hydrogen a number of lines are *missing* that correspond to transitions to and from states having certain combinations of quantum numbers. Thus no transitions are observed in helium to or from the ground-state configuration in which the spins of both electrons are in the same direction to give a total spin of 1, although transitions *are* observed to and from the other ground-state configuration in which the spins are in opposite directions to give a total spin of 0. In the absent state the quantum numbers of *both* electrons would be $n = 1$, $l = 0$, $m_l = 0$, $m_s = \frac{1}{2}$, while in the state known to exist, one of the electrons has $m_s = \frac{1}{2}$ and the other $m_s = -\frac{1}{2}$. Pauli showed that every nonexistent atomic state involves two or more electrons with identical quantum numbers, and the exclusion principle is a statement of this empirical finding.

GLOSSARY

An *emission spectrum* is one produced by a light source alone; it may be a *continuous spectrum*, with all colors present, or a *bright-line spectrum*, in which only a few specific wavelengths characteristic of the source appear.

An *absorption spectrum* is one produced when light from an incandescent source passes through a cool gas; it is also called a *dark-line spectrum* because it appears as a continuous band of colors crossed by dark lines corresponding to characteristic wavelengths absorbed by the gas.

The lines in the spectrum of an element fall into several *spectral series*, in which the frequencies of the lines are related by simple formulas.

In the *Bohr model of the atom* electrons are supposed to move around nuclei in circular orbits of definite size; when an electron jumps from one orbit to another, a photon of light is either emitted or absorbed whose energy corresponds to the difference in the electron's energy in the two orbits.

Quantum mechanics is a highly abstract theory of atomic phenomena which treats only experimentally measurable quantities and does not invoke mechanical models that contradict the uncertainty principle.

Four *quantum numbers* are required to specify completely the physical state of an atomic electron. These are the *total quantum number n*, which

THEORY OF THE ATOM

determines the energy of the electron; the *orbital quantum number l*, which determines the magnitude of the electron's angular momentum; the *magnetic quantum number* m_l, which determines the direction of the electron's angular momentum; and the *spin magnetic quantum number* m_s, which determines the orientation of the electron's spin.

The Pauli *exclusion principle* states that no more than one electron in an atom can have the same set of quantum numbers.

EXERCISES

1. What kind of spectrum would you expect to observe if you used a spectroscope to analyze: (*a*) light from the sun; (*b*) light from the tungsten filament of a light bulb; (*c*) light from a sodium-vapor highway lamp; (*d*) light from an electric-light bulb that has passed through cool sodium vapor?

2. In terms of the Bohr theory, explain why the hydrogen spectrum contains many lines, even though the hydrogen atom has only a single electron.

3. Why is the Bohr theory incompatible with the uncertainty principle?

4. On the basis of the Bohr model of the atom, explain why the dark (absorption) lines in the spectrum of hydrogen have the same wavelengths as the bright (emission) lines of the same element.

5. The Bohr theory permits us to visualize the structure of the atom, whereas quantum mechanics is very complex and concerned with such ideas as wave functions and probabilities. What reasons would lead to the replacement of the Bohr theory by quantum mechanics?

6. A beam of electromagnetic radiation that contains all wavelengths passes through a tube containing hydrogen gas. The hydrogen atoms are all in their ground states initially. What spectral series will be present in the absorption spectrum that results?

7. The four quantum numbers needed to describe an atomic electron are n, l, m_l, and m_s. What quantity is governed by each of them?

8. The earth has a mass of 6×10^{24} kg, and it circles the sun at 3×10^4 m/sec in an orbit 1.5×10^{11} m in radius. How many earth de Broglie wavelengths fit into this orbit?

9. Calculate the speed of the electron in the innermost ($n = 1$) Bohr orbit of a hydrogen atom. Is this speed great enough for the relativistic mass change to be appreciable? [Hint: Equate the centripetal force on the electron with the electrical attraction of the proton it circles around.]

10 How much energy is required to remove the electron from a hydrogen atom when it is in the $n = 3$ state? The $n = 5$ state?

11 Of the following transitions in a hydrogen atom (*a*) which emits the photon of highest frequency, (*b*) which emits the photon of lowest frequency, and (*c*) which absorbs the photon of highest frequency? $n = 1$ to $n = 2$, $n = 2$ to $n = 1$, $n = 2$ to $n = 6$, $n = 6$ to $n = 2$.

12 A proton and an electron, both initially at rest and far away from each other, combine all at once to form a hydrogen atom in its ground state. Find the wavelength of the photon that would be created in this hypothetical event.

13 A sample of hydrogen gas is to be heated until the average molecular energy is equal to the binding energy of the hydrogen atom. What temperature would this be?

part five
ATOMS IN COMBINATION

Armed with an introduction to the principles that underlie atomic structure, we are in a position to appreciate why and how atoms interact with one another.

After first examining the origin of the periodic law in the electron shells of atoms, we go on to consider how bonds between atoms arise. A continuation of this theme carries us into the realm of solids and the various bonding processes that hold them together.

The next few chapters, then, constitute a bridge between physics and chemistry: until now we have been studying what is usually regarded as physics, and afterward we shall study what is usually regarded as chemistry.

chapter 19

The Periodic Law

THE PERIODIC LAW

    metals and nonmetals
    active and inactive elements
    families of elements
    the periodic table
    mendeleev
    groups and periods

ATOMIC STRUCTURE

    shells and subshells
    explaining the periodic table

There is a very intimate relationship between the electronic structure of an atom and its chemical behavior. This is hardly surprising, for it is the outermost electrons of atoms that interact when atoms join together to form molecules. About the only familiar physical or chemical property of matter traceable to nuclei rather than to their surrounding electrons, in fact, is atomic mass. Thus the first step in any serious inquiry into the fundamentals of chemistry must begin with the architecture of the electron cloud around each atom, an architecture whose understanding is one of the triumphs of modern physics.

THE PERIODIC LAW

Before we discuss how the quantum theory of the atom, the notion of electron spin, and the exclusion principle all fit together to account for the chemical properties of complex atoms, it is appropriate to turn to empirical chemistry for background information on the periodic law. This law, now a century old, is one of the great generalizations that helped chemistry to complete its transition to a true science, and the explanation of the periodic law is one of the most significant aspects of modern physical science.

metals and nonmetals

We shall start our approach to the periodic law by noting that one of the more obvious distinctions between different elements is that which divides metals from nonmetals. So generally familiar is the idea of a metal that we have used this distinction already without trying to make it precise. Iron, mercury, gold, aluminum, sodium, and tin are examples of metallic elements; carbon, sulfur, hydrogen, chlorine, and helium are nonmetals.

metals have a characteristic luster and conduct heat and electricity readily

The outstanding physical properties that differentiate metals from other substances are (1) their characteristic sheen, or *metallic luster,* and (2) the ease with which they conduct heat and electricity. Instinctively we associate also the qualities of hardness and toughness with metals, but a moment's glance at the softness of gold, lead, and sodium shows that these are not general characteristics. Nonmetals in the solid state are usually brittle materials without metallic luster (graphite and one form of silicon are exceptions) and are very poor conductors of heat and electricity (graphite is an exception, but its conductivity is small compared with that of most metals). In some other physical properties nonmetals have an extreme range: in melting point from helium ($-272°C$) to carbon ($3550°C$), and in hardness from diamond to soft white phosphorus.

In chemical behavior metals show considerable differences among themselves. Sodium, for example, is extremely active, whereas gold and platinum are highly resistant to chemical change. In general: (1) Metals combine with nonmetals much more readily than with one another. All metals combine directly with fluorine and chlorine, and most combine directly with oxygen. Many metals mix readily to form alloys, but definite compounds between metals are few and unstable. (2) All the more active metals react with dilute acids to liberate hydrogen, and very active metals liberate hydrogen from water. (3) Oxides of the more active metals react with water to form bases. Thus sodium oxide placed in water yields sodium hydroxide, which is a base. (Acids and bases are discussed in Chap. 25.)

properties of nonmetals

Nonmetals show an even greater variety of chemical properties than do the metals. Some (argon, helium, neon) form hardly any compounds, whereas others (chlorine, fluorine) are highly active. In general: (1) Nonmetals (except for those like argon) combine readily with active metals, somewhat less readily with one another. Thus, chlorine and fluorine react violently with active metals but will not combine directly with oxygen. Sulfur and phosphorus, on the other hand, burn brightly in oxygen. (2) The nonmetals do not react with dilute acids. Several are attacked by bases, but the reactions do not follow a single pattern. (3) Oxides of nonmetals, if soluble, combine with water to form acids.

hydrogen has both metallic and nonmetallic properties

The nonmetal hydrogen is unique: although distinctly nonmetallic in its physical properties and much of its chemical behavior, some of its reactions suggest similarities with the metals. For example, hydrogen combines more readily and more violently with nonmetals than with metals, and its compounds with atom groups like SO_4 and CO_3 are similar in formula to metallic compounds.

We shall learn later a more sophisticated general definition of metals and nonmetals. Even this better definition, however, does not make the distinction sharp, for a few elements show properties in some measure characteristic of both groups. Metals far outnumber the nonmetals; only 20 of the 103 elements known today are considered definitely nonmetallic.

active and inactive elements

Sodium we call an *active* metal, gold a very *inactive* one. Precisely what do these terms mean?

We know that sodium is tarnished by a few seconds' exposure to air, whereas a gold ring keeps its luster after years of exposure to air and perspiration. We think of the spectacular combustion of sodium in chlorine, accompanied by much heat and light energy; gold combines sluggishly with chlorine, setting free little energy. We recall that sodium liberates hydrogen rapidly from dilute acids, even from water; gold is unaffected by ordinary acids, dissolving only in a mixture of concentrated hydrochloric (HCl) and nitric (HNO_3) acids. In reactions like these we say that sodium exhibits its greater activity.

active elements liberate more heat when they react than inactive elements do

There are a number of other criteria with whose help we can establish the relative activity of the various elements. We might measure the amount of heat liberated in similar chemical reactions undergone by two elements. Suppose, for instance, that we combine a given amount of chlorine with gold metal and the same amount with sodium metal. We find that the process of forming sodium chloride gives off over 15 times as much heat as the process of forming gold chloride, and so we conclude that sodium is much more active than gold.

Or we might start out with similar compounds and ask how easily they can be separated into their component elements or element groups. In the case of gold chloride and sodium chloride, the results are that gold chloride decomposes when it is heated to about 300°C, but sodium chloride must be heated to well over 1000°C for its decomposition to take place. Accordingly, we say that gold chloride is a relatively *unstable* compound, but sodium chloride is a *stable* compound. In general, the more active an element is, the more difficult it is to decompose its compounds, and so active elements tend to form stable compounds whereas inactive ones tend to form unstable compounds.

active elements usually form compounds more stable than those formed by inactive elements

There are still other means of establishing the relative activity of the elements, but those mentioned above are probably the most useful. Both metals and nonmetals can be arranged in order of their activities.

In the partial listing at right, the most active elements are at the top of each series, the least active at the bottom.

families of elements

The resemblances among some elements are so striking that these elements seem to be members of the same natural family. As examples of families of elements whose grouping together is almost inevitable to the chemist, we shall discuss a family of active nonmetals called the *halogens*, a group of active metals called the *alkali metals*, and a group of gases that undergo almost no chemical reactions, the *inert gases*.

Metals		*Nonmetals*
Potassium		Fluorine
Sodium		Chlorine
Calcium	More active ↑	Bromine
Magnesium		Oxygen
Aluminum		Iodine
Zinc		Sulfur
Iron	Less active ↓	
Lead		
Copper		
Mercury		
Silver		
Gold		

the halogens

The *halogens* consist of the highly active elements fluorine (F), chlorine (Cl), bromine (Br), and iodine (I) listed in order of increasing atomic weight. They are responsible for some of the vilest odors (*bromos* is Greek for "stink") and most brilliant colors (*chloros* is Greek for "green") to be found in the laboratory. The name *halogen* means "salt former," a token of the fact that these elements produce white, crystalline solids when they combine with many metals. Fluorine is a pale yellow gas and chlorine a greenish-yellow gas at room temperature, bromine is a reddish-brown liquid, and iodine is a steel-gray solid. Bromine evaporates readily, a property known as *volatility*; iodine sublimates.

What are the similarities among the halogens? For one thing, all their molecules contain two atoms at ordinary temperatures: F_2, Cl_2, Br_2, I_2. Also, the compounds they form with metals have similar formulas. Here are three examples:

NaF	ZnF_2	AlF_3
NaCl	$ZnCl_2$	$AlCl_3$
NaBr	$ZnBr_2$	$AlBr_3$
NaI	ZnI_2	AlI_3

Note that, in all compounds with a specific metal, the same number of halogen atoms combine with each metal atom, though this number may vary for different metals.

All the halogens react with hydrogen to form, as the case may be, HF, HCl, HBr, or HI. These compounds can be dissolved in water to form acids, of which hydrochloric acid is a familiar example. The halogens are readily soluble in carbon tetrachloride, giving solutions colored in the same way as their vapors, but they are only slightly soluble in water.

Although the halogens are all active elements, their activity declines markedly with increasing atomic number. Fluorine is the most active of all nonmetals—so active that it is difficult to prepare, difficult to keep, dangerous to work with. Chlorine is somewhat less active; bromine and iodine, still less. All metals combine directly with fluorine and chlorine, but the less active ones are not affected by bromine and iodine. Table 19.1 lists the amounts of energy liberated when one atom of potassium reacts with the various halogen atoms to form KF, KCl, KBr, and KI, respectively; these energies of formation decrease with increasing atomic number, indicating less and less stability. Also evident in this table is

TABLE 19.1   Properties of the halogens.

Name	Symbol	Atomic number	Energy of formation per molecule, ev	Melting point, °C	Boiling point, °C
Fluorine	F	9	KF, 5.8	−220	−188
Chlorine	Cl	17	KCl, 4.6	−101	−35
Bromine	Br	35	KBr, 4.1	−7	59
Iodine	I	53	KI, 3.5	114	184

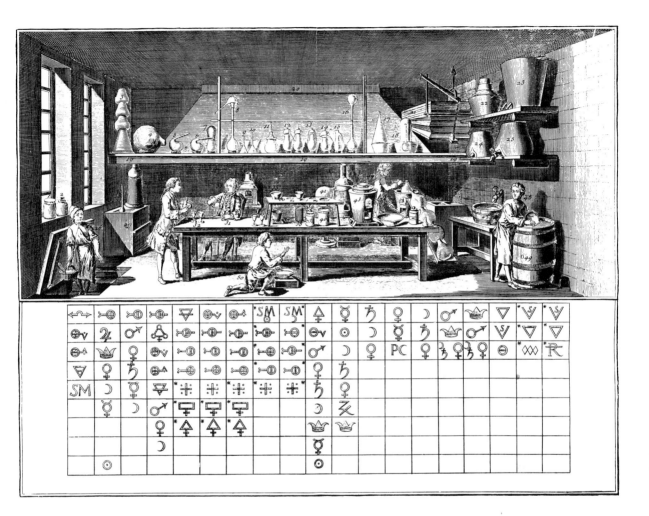

a steady increase in both melting and boiling points with atomic number. Thus the halogens constitute a group of elements with many similar properties, some of which change progressively with increasing atomic number.

The *alkali metals* are all soft, light, and extremely active metals. In order of increasing atomic number they are lithium (Li), sodium (Na), potassium (K), rubidium (Rb), and cesium (Cs). Like sodium, the other alkali metals tarnish quickly in air, liberate hydrogen from water and dilute acids, combine energetically with active nonmetals to form very stable compounds, and form oxides that combine with water to make bases. Formulas of their compounds are strikingly similar:

Bromides	LiBr	NaBr	KBr	RbBr	CsBr
Sulfides	Li_2S	Na_2S	K_2S	Rb_2S	Cs_2S
Hydroxides	LiOH	NaOH	KOH	RbOH	CsOH

This eighteenth century table of chemical symbols, published in Diderot's Encyclopedia, represents an early attempt at showing relationships among the elements in schematic form. A chemical laboratory of the time is pictured in the engraving.

the alkali metals

Among the other properties the alkali metals have in common are rather low melting points for metals: cesium melts in a warm room, and even lithium, which has the highest melting point of the group, liquefies at only 179°C. Table 19.2 lists a few of the properties of the alkali metals; evidently their densities increase steadily with atomic number (potassium is an exception), while their melting and boiling points decrease.

In general, the chemical activity of the alkali metals increases as atomic number increases. The three heavier metals liberate so much energy in their reactions with cold water that the hydrogen produced ignites spontaneously, while from lithium and sodium, hydrogen is evolved without burning. Cesium forms the most stable compounds with chlorine and bromine; lithium, the least stable compounds. Thus, like the halogens, the alkali metals show striking similarities in their chemical and physical properties, several of which change progressively in magnitude with increasing atomic number.

TABLE 19.2 Properties of the alkali metals.

Name	Symbol	Atomic number	Density, 10^3 kg/m^3	Melting point, °C	Boiling point, °C
Lithium	Li	3	0.53	179	1317
Sodium	Na	11	0.97	98	892
Potassium	K	19	0.87	64	774
Rubidium	Rb	37	1.53	39	688
Cesium	Cs	55	1.87	29	690

the inert gases

The *inert gases*, in marked contrast to the active halogens and alkali metals, are so inactive that they form only a handful of compounds with other elements. In fact, their atoms are so inactive that they do not even join together into molecules, as the atoms of other gaseous elements do. The family of inert gases includes, once more in order of increasing atomic number, helium (He), neon (Ne), argon (Ar), krypton (Kr), xenon (Xe), and radon (Rn). All the inert gases are found in small amounts in the atmosphere, with argon making up nearly 1 percent of the air and the others much less. Their scarcity and inactivity prevented their discovery until the very end of the nineteenth century. The physical properties of the inert gases, outlined in Table 19.3, show the same general similarity and regular gradations with increasing atomic number that we have found in the other groups.

the periodic table

the periodic law

When the elements are listed in order of increasing atomic number, elements with similar chemical and physical properties recur at regular intervals. This empirical observation, known as the *periodic law*, was first formulated by Dmitri Mendeleev about a century ago. A tabular arrangement of the elements exhibiting this recurrence of properties is called a *periodic table*. Table 19.4 is perhaps the simplest form of periodic table; though more elaborate ones have been devised, this table is adequate for our purposes.

TABLE 19.3   Properties of the inert gases.

Name	Symbol	Atomic number	Density of liquid, 10^3 kg/m^3	Melting point, °C	Boiling point, °C
Helium	He	2	0.15	−272	−269
Neon	Ne	10	1.21	−249	−246
Argon	Ar	18	1.40	−189	−186
Krypton	Kr	36	2.16	−156	−152
Xenon	Xe	54	3.52	−112	−107
Radon	Rn	86	4.4	−71	−62

To see how the periodic table organizes our knowledge of the elements, we shall trace the ideas that underlie its construction. In order of increasing atomic number, the first element is hydrogen. Next is the inert gas helium; then the alkali metal lithium; then a rare metal called beryllium, less active than lithium; then boron, a relatively inactive nonmetal, which forms the chloride BCl_3; then carbon, a nonmetal that forms both CCl_4 and CH_4; then nitrogen, another nonmetal; then oxygen, a more active nonmetal; and fluorine, most active of all nonmetals. From lithium to fluorine is a complete transition from a highly active metal to a highly active nonmetal. After fluorine comes neon, another inert gas like helium; then sodium, an alkali metal like lithium. To suggest these resemblances, we break off the rows of elements at helium and neon and start new rows with lithium and sodium. In the seven elements beyond neon, we find again a transition from active metals to active nonmetals.

how the periodic table is constructed

After calcium, in the fourth row, more difficulties appear. Scandium, the next element, is similar to aluminum in some properties, different in others. Titanium (Ti) is even less like carbon and silicon. Then follow ten metals (including iron, copper, zinc), quite similar among themselves but differing conspicuously from the nonmetals at the end of the first three rows. Only after the ten metals do three relatives of these nonmetals appear, arsenic (As), selenium (Se), and bromine. Thus, between the first inert gas (He) and the second (Ne), is a sequence of eight elements; between neon and argon is another sequence of eight; but between argon and krypton the sequence includes eighteen. Beyond krypton is a second sequence of eighteen, including again a dozen metals somehow related. From xenon to the last inert gas, radon, is a yet more complex sequence of 32 elements.

mendeleev

The periodic table of the elements is another of those great intellectual achievements that cannot be separated from the name of the man responsible for it. The planetary laws bring to mind Kepler; the laws of motion, Newton; the atomic theory, Dalton. The Russian Dmitri Ivanovich Mendeleev is similarly associated with the periodic table.

Mendeleev, who was born in Siberia, was for many years professor of chemistry at the University of St. Petersburg. He devoted himself to

TABLE 19.4 The periodic table of the elements.

Group	I	II										III	IV	V	VI	VII	VIII
Period 1	1.008 H 1																4.00 He 2
2	6.94 Li 3	9.01 Be 4										10.82 B 5	12.01 C 6	14.01 N 7	16.00 O 8	19.00 F 9	20.18 Ne 10
3	22.99 Na 11	24.31 Mg 12										26.98 Al 13	28.09 Si 14	30.98 P 15	32.06 S 16	35.46 Cl 17	39.95 Ar 18
4	39.10 K 19	40.08 Ca 20	44.96 Sc 21	47.90 Ti 22	50.94 V 23	52.00 Cr 24	54.94 Mn 25	55.85 Fe 26	58.93 Co 27	58.71 Ni 28	63.54 Cu 29	65.37 Zn 30	69.72 Ga 31	72.59 Ge 32	74.92 As 33	78.96 Se 34	79.91 Br 35	83.8 Kr 36
5	85.47 Rb 37	87.62 Sr 38	88.91 Y 39	91.22 Zr 40	92.91 Nb 41	95.94 Mo 42	(99) Tc 43	101.1 Ru 44	102.91 Rh 45	106.4 Pd 46	107.87 Ag 47	112.40 Cd 48	114.82 In 49	118.69 Sn 50	121.76 Sb 51	127.61 Te 52	126.90 I 53	131.30 Xe 54
6	132.91 Cs 55	137.34 Ba 56	* 57–71	178.49 Hf 72	180.95 Ta 73	183.85 W 74	186.2 Re 75	190.2 Os 76	192.2 Ir 77	195.09 Pt 78	196.97 Au 79	200.59 Hg 80	204.37 Tl 81	207.19 Pb 82	208.98 Bi 83	(209) Po 84	(210) At 85	(222) Rn 86
7	(223) Fr 87	226.05 Ra 88	† 89–103															

The number above the symbol of each element is its atomic mass, and the number below the symbol is its atomic number. The elements whose atomic masses are given in parentheses do not occur in nature, but have been prepared artificially in nuclear reactions. The atomic mass in such a case is the mass number of the most long-lived radioactive isotope of the element.

* *Rare earths*	138.91 La 57	140.12 Ce 58	104.91 Pr 59	144.24 Nd 60	(147) Pm 61	150.35 Sm 62	151.96 Eu 63	157.25 Gd 64	158.92 Tb 65	162.50 Dy 66	164.93 Ho 67	167.26 Er 68	168.93 Tm 69	173.04 Yb 70	174.97 Lu 71
† *Actinides*	(227) Ac 89	232.04 Th 90	(231) Pa 91	238.03 U 92	(244) Np 93	(244) Pu 94	(243) Pu 95	(247) Cm 96	(247) Bk 97	(251) Cf 98	(254) Es 99	(257) Fm 100	(256) Md 101	(254) No 102	(257) Lr 103

government service as well as to scientific work, although his outspoken liberal ideas were frequently embarrassing to the tsarist regime. Mendeleev was a gifted teacher, an able experimenter, but above all a dreamer, a scientific visionary. If some of his speculations seem fantastic, for one vision at least chemistry owes him a great debt, the vision that gave him the key to the classification of the elements. In this vision he was not alone, for a few of his contemporaries reached the same conclusion independently; but Mendeleev was the first to apply such ideas to all the known elements and to predict from them the existence of elements then unknown.

The relationships brought out by the periodic table are somewhat vague in places, but on the whole the table brings together similar elements

THE PERIODIC LAW

with considerable accuracy. Mendeleev's achievement is all the more remarkable if we recall that in 1869, when the periodic law was formulated, the notion of atomic number had not been discovered and only 63 elements in all were known. Mendeleev actually used atomic mass, not atomic number, in setting up the periodic table, and he and later chemists found it necessary to deviate from the strict sequence of atomic masses for certain elements. Potassium, for example, has a smaller atomic mass than argon, yet ending the third period with potassium would put this active metal under the inert gases helium and neon, whereas argon

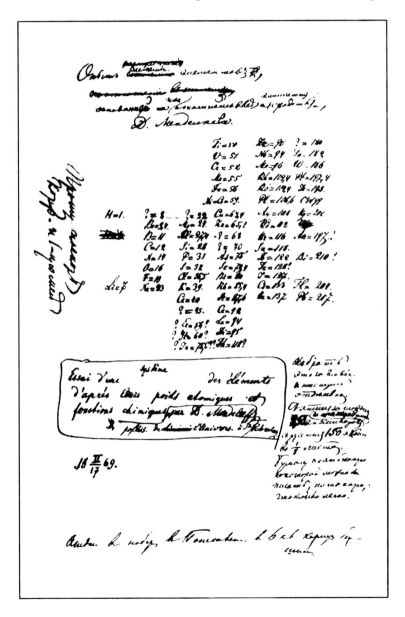

Mendeleev's periodic table in manuscript form, as presented for publication in 1869.

mendeleev predicted the existence and properties of several elements which were later discovered

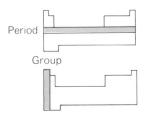

19.1 *The elements in a group (or column) of the periodic table have similar properties, while those in a period have different properties.*

the elements in each horizontal row of the periodic table have different properties and constitute a period

the elements in each vertical column of the periodic table have similar properties and constitute a group

would go beneath sodium and lithium. To avoid this discrepancy, the order is reversed to bring argon before potassium, which, of course, is consistent with the order of their atomic numbers.

Because so few elements were known in his time, Mendeleev had to leave many gaps in his table of elements in order to make similar elements fall one under the other. Sure of the correctness of his classification, he suggested that these gaps represented undiscovered elements. Further, from the position of each gap, from the properties of the elements around it, and from the variation of these properties across the periods and down the columns, he predicted the properties of the unknown elements. His predictions included not only general chemical activity, but precise numerical values for densities, melting points, and so on. As the unknown elements were discovered one by one and as their properties were found to agree closely with Mendeleev's predictions, the correctness and usefulness of the periodic classification became firmly established. Perhaps its greatest triumph came at the end of the century, when the inert gases were discovered: here were six new elements whose existence Mendeleev was not aware of, but they fitted perfectly, as one more family of similar elements into the periodic table.

Today all gaps in the table have been filled. The usefulness of the periodic law in predicting new elements is past, but its usefulness in coordinating chemical knowledge is greater today than in Mendeleev's time. A chemist need not learn in detail the properties of all the elements, or even of a large proportion of them; if he knows thoroughly the chemical behavior of a few elements and if he knows how properties vary in the periods and groups of the periodic table, then for any other element he need only glance at its position in the table to learn its chief physical and chemical characteristics.

groups and periods

Most periodic tables show similar elements in vertical rows, called *columns* or *groups* of elements. The horizontal rows, containing elements with widely different properties, are called *periods* (Fig. 19.1). Across each period is a more or less steady transition from an active metal through less active metals and weakly active nonmetals to highly active nonmetals and finally to an inert gas. Within each column there is also a steady change in properties, but much less rapid and less conspicuous than within the periods. Thus, increasing atomic number brings increasing activity in the alkali metal family and decreasing activity among the nonmetals of the halogen family. These changes are typical; chemical properties within each group change from top to bottom in the direction of increased metallic activity or decreased nonmetallic activity—which amount to the same thing (Fig. 19.2).

A complete periodic table should display not only these major changes in properties but also the more subtle relationships among elements in the long periods. Many ingenious arrangements of the complete table have been suggested in order to show such relationships, but for present purposes the comparatively simple form of Table 19.4 is sufficiently informative. The eight principal groups (indicated by Roman numerals

THE PERIODIC LAW

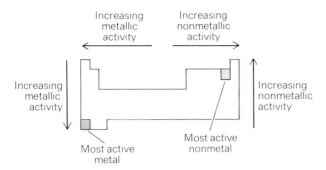

19.2 *How chemical activity varies in the periodic table.*

at the top) form vertical columns, as usual. The inert gases (Group VIII) are placed at the right since this arrangement has the advantage of grouping them with the other nonmetals at the right side of the table (Fig. 19.3). Each of the eight-element periods (periods 2 and 3) is broken after the second element, in order to keep the elements in columns with the most closely related elements of the long periods below.

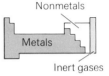

19.3 *The majority of the elements are metals.*

The *transition elements* in each long period are metals showing considerable chemical resemblance to one another but no pronounced resemblance to elements in the major groups (Fig. 19.4). Period 6 contains 32 elements altogether, but 15 of these are taken out and placed in the box below the table; these *rare-earth metals* resemble one another very closely, so much so that their separation is a matter of extreme difficulty, and hence they are all lumped together in the spot (marked by an asterisk) below yttrium (Y). A similar group of closely related elements (*actinide metals*) appears in the same position in period 7, and these elements are shown with the rare earths in the box below the table. The position of the nonmetal hydrogen in Group I with the very active alkali metals can be justified on the grounds that its chemical behavior is more like that of these metals than that of any other group.

ATOMIC STRUCTURE

Now we return to atomic structure to seek the basis of the periodic law. There are two basic principles that determine (together with the quantum theory of the atom) just how the electrons in a complex atom are arranged. The first is the exclusion principle, which prevents all the various electrons from occupying the lowest quantum state. Each electron in the complex atom must have a different set of quantum numbers n, l, m_l, and m_s. The second is the general rule that a system of particles of any kind is stable when its total energy is a minimum, which means here that the quantum states of lowest energy in an atom are occupied to the extent permitted by the exclusion principle.

the electrons in an atom must be in the lowest energy levels permitted by the exclusion principle

shells and subshells

Let us examine how electron energy varies with quantum state. All the electrons in an atom that have the same total quantum number n are,

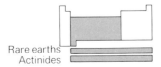

19.4 *The transition elements are metals.*

19.5 *The binding energy of an atomic electron increases with increasing quantum number n and, to a smaller extent, with increasing quantum number l. The symbol for an l = 0 state is s, for an l = 1 state is p, for an l = 2 state is d, and for an l = 3 state is f.*

the electrons in a shell have the same total quantum number n

on the average, about the same distance from the nucleus. These electrons therefore interact with virtually the same electric field and have similar energies. It is convenient to speak of such electrons as occupying the same atomic *shell*.

The energy of an electron in a particular shell also depends to a certain extent upon its orbital quantum number l, though this dependence is not so great as that upon n. In a complex atom the degree to which the full nuclear charge is shielded from a given electron by intervening shells of other electrons varies with its probability-density distribution. As we can see from Fig. 18.15, an electron of small l is more likely to be found near the nucleus (where it is poorly shielded by the other electrons) than one of higher l, which results in a lower total energy (that is, higher binding energy) for it. The electrons in each shell accordingly increase in energy with increasing l. This effect is illustrated in Fig. 19.5, which shows how the binding energies of various atomic electrons vary with n and l.

the electrons in a subshell have the same orbital quantum number l

Electrons that share a certain value of l in a shell are said to occupy the same *subshell*. All of the electrons in a subshell have almost identical energies, since the dependence of electron energy upon m_l and m_s is comparatively minor.

explaining the periodic table

The concept of electron shells and subshells fits perfectly into the pattern of the periodic table, which turns out to mirror the atomic structures of the elements. Let us see how this pattern arises.

the exclusion principle limits the occupancies of shells and subshells

The exclusion principle places definite limits on the number of electrons that can occupy a given shell or subshell. A subshell is characterized by a certain total quantum number n and an orbital quantum number l, where

THE PERIODIC LAW

$$l = 0, 1, 2, \ldots, n - 1$$

Thus the $n = 3$ shell has three subshells, corresponding to $l = 0$, $l = 1$, and $l = 2$. These subshells are respectively denoted $3s$, $3p$, and $3d$.

There are $2l + 1$ different values of the magnetic quantum number m_l for any l, since

$$m_l = 0, \pm 1, \pm 2, \ldots, \pm l$$

and two possible values of the spin magnetic quantum number m_s ($+\frac{1}{2}$ and $-\frac{1}{2}$) for any m_l. Hence each subshell can contain a maximum of $2(2l + 1)$ electrons. The maximum number of electrons in a shell of quantum number n is $2n^2$.

An atomic shell or subshell containing its full quota of electrons is said to be *closed*. A closed subshell with $l = 0$ holds 2 electrons; a closed subshell with $l = 1$, 6 electrons; a closed subshell with $l = 2$, 10 electrons; and so on.

a closed shell or subshell contains its full quota of electrons

Table 19.5 which is illustrated in Fig. 19.6 shows the occupancy of the various shells in a number of elements. The figure is arranged in the same manner as the periodic table to emphasize their relationship. To interpret Table 19.5 we note that the electrons in a closed shell are all very tightly bound, since the positive nuclear charge that attracts them is large relative to the negative charge of the inner electrons (Fig. 19.7). An atom containing only closed shells or subshells has its electric charge

19.6 Electron configurations. *A highly schematic representation of part of Table 19.5.*

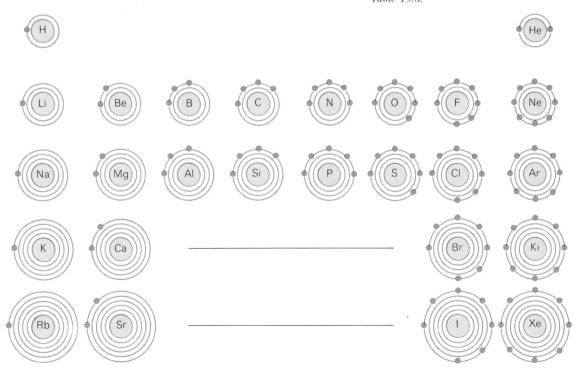

TABLE 19.5 Electron configurations of the elements.

		n = 1	n = 2		n = 3			n = 4				n = 5				n = 6			n = 7
		1s	2s	2p	3s	3p	3d	4s	4p	4d	4f	5s	5p	5d	5f	6s	6p	6d	7s
1	H	1																	
2	He	2	Inert gas																
3	Li	2	1	Alkali metal															
4	Be	2	2																
5	B	2	2	1															
6	C	2	2	2															
7	N	2	2	3															
8	O	2	2	4															
9	F	2	2	5	Halogen														
10	Ne	2	2	6	Inert gas														
11	Na	2	2	6	1	Alkali metal													
12	Mg	2	2	6	2														
13	Al	2	2	6	2	1													
14	Si	2	2	6	2	2													
15	P	2	2	6	2	3													
16	S	2	2	6	2	4													
17	Cl	2	2	6	2	5	Halogen												
18	A	2	2	6	2	6	Inert gas												
19	K	2	2	6	2	6		1	Alkali metal										
20	Ca	2	2	6	2	6		2											
21	Sc	2	2	6	2	6	1	2											
22	Ti	2	2	6	2	6	2	2											
23	V	2	2	6	2	6	3	2											
24	Cr	2	2	6	2	6	5	1											
25	Mn	2	2	6	2	6	5	2											
26	Fe	2	2	6	2	6	6	2											
27	Co	2	2	6	2	6	7	2											
28	Ni	2	2	6	2	6	8	2											
29	Cu	2	2	6	2	6	10	1											
30	Zn	2	2	6	2	6	10	2											
31	Ga	2	2	6	2	6	10	2	1										
32	Ge	2	2	6	2	6	10	2	2										
33	As	2	2	6	2	6	10	2	3										
34	Se	2	2	6	2	6	10	2	4										
35	Br	2	2	6	2	6	10	2	5	Halogen									
36	Kr	2	2	6	2	6	10	2	6	Inert gas									
37	Rb	2	2	6	2	6	10	2	6			1	Alkali metal						
38	Sr	2	2	6	2	6	10	2	6			2							
39	Y	2	2	6	2	6	10	2	6	1		2							
40	Zr	2	2	6	2	6	10	2	6	2		2							
41	Nb	2	2	6	2	6	10	2	6	4		1							
42	Mo	2	2	6	2	6	10	2	6	5		1							
43	Tc	2	2	6	2	6	10	2	6	5		2							
44	Ru	2	2	6	2	6	10	2	6	7		1							
45	Rh	2	2	6	2	6	10	2	6	8		1							
46	Pd	2	2	6	2	6	10	2	6	10									
47	Ag	2	2	6	2	6	10	2	6	10		1							
48	Cd	2	2	6	2	6	10	2	6	10		2							
49	In	2	2	6	2	6	10	2	6	10		2	1						
50	Sn	2	2	6	2	6	10	2	6	10		2	2						
51	Sb	2	2	6	2	6	10	2	6	10		2	3						
52	Te	2	2	6	2	6	10	2	6	10		2	4						

THE PERIODIC LAW

TABLE 19.5 Electron configurations of the elements (cont.).

	$n=1$	$n=2$		$n=3$			$n=4$				$n=5$				$n=6$			$n=7$	
	$1s$	$2s$	$2p$	$3s$	$3p$	$3d$	$4s$	$4p$	$4d$	$4f$	$5s$	$5p$	$5d$	$5f$	$6s$	$6p$	$6d$	$7s$	
53 I	2	2	6	2	6	10	2	6	10		2	5							Halogen
54 Xe	2	2	6	2	6	10	2	6	10		2	6							Inert gas
55 Cs	2	2	6	2	6	10	2	6	10		2	6			1				Alkali metal
56 Ba	2	2	6	2	6	10	2	6	10		2	6			2				
57 La	2	2	6	2	6	10	2	6	10		2	6	1		2				
58 Ce	2	2	6	2	6	10	2	6	10	2	2	6			2				
59 Pr	2	2	6	2	6	10	2	6	10	3	2	6			2				
60 Nd	2	2	6	2	6	10	2	6	10	4	2	6			2				
61 Pm	2	2	6	2	6	10	2	6	10	5	2	6			2				
62 Sm	2	2	6	2	6	10	2	6	10	6	2	6			2				
63 Eu	2	2	6	2	6	10	2	6	10	7	2	6			2				Rare
64 Gd	2	2	6	2	6	10	2	6	10	7	2	6	1		2				earths
65 Tb	2	2	6	2	6	10	2	6	10	9	2	6	1		2				
66 Dy	2	2	6	2	6	10	2	6	10	10	2	6			2				
67 Ho	2	2	6	2	6	10	2	6	10	11	2	6			2				
68 Er	2	2	6	2	6	10	2	6	10	12	2	6			2				
69 Tm	2	2	6	2	6	10	2	6	10	13	2	6			2				
70 Yb	2	2	6	2	6	10	2	6	10	14	2	6			2				
71 Lu	2	2	6	2	6	10	2	6	10	14	2	6	1		2				
72 Hf	2	2	6	2	6	10	2	6	10	14	2	6	2		2				
73 Ta	2	2	6	2	6	10	2	6	10	14	2	6	3		2				
74 W	2	2	6	2	6	10	2	6	10	14	2	6	4		2				
75 Re	2	2	6	2	6	10	2	6	10	14	2	6	5		2				
76 Os	2	2	6	2	6	10	2	6	10	14	2	6	6		2				
77 Ir	2	2	6	2	6	10	2	6	10	14	2	6	7		2				
78 Pt	2	2	6	2	6	10	2	6	10	14	2	6	9		1				
79 Au	2	2	6	2	6	10	2	6	10	14	2	6	10		1				
80 Hg	2	2	6	2	6	10	2	6	10	14	2	6	10		2				
81 Tl	2	2	6	2	6	10	2	6	10	14	2	6	10		2	1			
82 Pb	2	2	6	2	6	10	2	6	10	14	2	6	10		2	2			
83 Bi	2	2	6	2	6	10	2	6	10	14	2	6	10		2	3			
84 Po	2	2	6	2	6	10	2	6	10	14	2	6	10		2	4			
85 At	2	2	6	2	6	10	2	6	10	14	2	6	10		2	5			Halogen
86 Rn	2	2	6	2	6	10	2	6	10	14	2	6	10		2	6			Inert gas
87 Fr	2	2	6	2	6	10	2	6	10	14	2	6	10		2	6		1	Alkali metal
88 Ra	2	2	6	2	6	10	2	6	10	14	2	6	10		2	6		2	
89 Ac	2	2	6	2	6	10	2	6	10	14	2	6	10		2	6	1	2	
90 Th	2	2	6	2	6	10	2	6	10	14	2	6	10		2	6	2	2	
91 Pa	2	2	6	2	6	10	2	6	10	14	2	6	10	2	2	6	1	2	
92 U	2	2	6	2	6	10	2	6	10	14	2	6	10	3	2	6	1	2	
93 Np	2	2	6	2	6	10	2	6	10	14	2	6	10	4	2	6	1	2	
94 Pu	2	2	6	2	6	10	2	6	10	14	2	6	10	5	2	6	1	2	
95 Am	2	2	6	2	6	10	2	6	10	14	2	6	10	6	2	6	1	2	
96 Cm	2	2	6	2	6	10	2	6	10	14	2	6	10	7	2	6	1	2	Actinides
97 Bk	2	2	6	2	6	10	2	6	10	14	2	6	10	8	2	6	1	2	
98 Cf	2	2	6	2	6	10	2	6	10	14	2	6	10	10	2	6		2	
99 Es	2	2	6	2	6	10	2	6	10	14	2	6	10	11	2	6		2	
100 Fm	2	2	6	2	6	10	2	6	10	14	2	6	10	12	2	6		2	
101 Md	2	2	6	2	6	10	2	6	10	14	2	6	10	13	2	6		2	
102 No	2	2	6	2	6	10	2	6	10	14	2	6	10	14	2	6		2	
103 Lr	2	2	6	2	6	10	2	6	10	14	2	6	10	14	2	6	1	2	

inert gas atoms have closed shells or subshells

uniformly distributed, so it does not attract other electrons and its electrons cannot be readily detached. Such atoms we expect to be passive chemically, like the inert gases—and the inert gases all turn out to have closed-shell or closed-subshell electron configurations!

Those atoms with but a single electron in their outermost shells tend to lose this electron, which is relatively far from the nucleus and is shielded by the inner electrons from all but an effective nuclear charge of $+e$. Hydrogen and the alkali metals are in this category, and accordingly show very similar behavior. In the sodium atom, for instance, the total nuclear charge of $+11e$ acts on the two innermost electrons, which are held very tightly. These electrons shield part of the nuclear charge from the eight electrons in the second shell, which are attracted by a net charge of $+9e$. All ten electrons in the first and second shells act to shield the outermost electron, which therefore "sees" a net nuclear charge of only $+e$ and is held much less securely to the atom than any of the other electrons (Fig. 19.7).

a metal atom has one or several electrons outside closed shells or subshells and combines chemically by losing these electrons to nonmetal atoms

Atoms whose outer shells lack a single electron from being closed tend to acquire such an electron through the strong attraction of the imperfectly shielded nuclear charge, which accounts for the chemical behavior of the halogens. In the chlorine atom, for instance, there are the same ten electrons in the inner two shells that there are in the sodium atom, but the nuclear charge of chlorine is $+17e$. Hence the net charge "seen" by each of the seven outer electrons is $+7e$, and the attractive force is accordingly seven times greater than in the case of the lone outer electron in sodium. By analogous reasoning the similarities of the members of the various groups of the periodic table may be accounted for in a wholly straightforward manner.

a nonmetal atom lacks one or several electrons of having closed shells or subshells and combines chemically by picking up electrons from metal atoms or by sharing electrons with other nonmetal atoms

To summarize, subshells are filled up for atoms having 2, 10, 18, 36, 54, and 86 electrons, which explains why these elements are the inert gases with no chemical activity. The great activity of the alkali metals, which have a single electron outside a closed subshell, is accounted for by their strong tendency to get rid of this electron and thereby achieve a stable electronic structure. Similarly, the halogens, which are just one electron short of a closed subshell, have a strong tendency to pick up an additional electron. Corresponding arguments hold for the other groups of the periodic table.

origin of chemical activity

As Table 19.5 shows, in several places the electron shells and subshells are not always filled in consecutive order. Thus the transition elements in any period have the same outer electron shells and add electrons successively to inner shells, which accounts for their similar properties. Another consequence of the delayed filling of inner shells is the strongly magnetic behavior of iron, nickel, and cobalt (atomic numbers 26, 27, and 28). Several of the electrons in the $3d$ subshells of these atoms have parallel spins without having their magnetic properties cancelled out by electrons with oppositely directed spins. (In a filled subshell, half the electrons have spins that point one way and half have spins that point the opposite way.) An iron atom, for instance, has five of its six $3d$ electrons with their spins parallel, which gives it a strongly magnetic character since an electron is much like a tiny bar magnet.

GLOSSARY

Metals possess a characteristic sheen (metallic luster) and are good conductors of heat and electricity. They combine with nonmetals more readily than with one another. The more active metals liberate hydrogen from dilute acids and their oxides react with water to form bases.

Nonmetals have an extreme range of physical properties; in the solid state they are usually lusterless and brittle and are poor conductors of heat and electricity. Some nonmetals form no compounds whatever; the others combine more readily with active metals than with one another. Soluble nonmetal oxides react with water to form acids.

When comparing two elements, the more *active* one is more rapid in comparable chemical reactions and liberates more energy than the less active one. The compounds of the more active elements are also more stable.

The *halogens* are a family of highly active nonmetals with similar chemical properties. The halogens are fluorine, chlorine, bromine, and iodine in order of atomic mass.

The *alkali metals* are a family of soft, light, extremely active metals with similar chemical properties. The alkali metals are lithium, sodium, potassium, rubidium, and cesium in order of atomic mass.

The *inert gases* are a family of almost totally inactive elements consisting of helium, neon, argon, krypton, xenon, and radon in order of atomic mass.

The *periodic law* states that if the elements are listed in the order of their atomic number, elements with similar properties recur at definite intervals. A tabular arrangement of the elements showing this recurrence of properties is called a *periodic table*.

An electron *shell* in an atom consists of all the electrons having the same principal quantum number n. When a particular shell contains all the electrons possible it is called a *closed shell*.

An electron *subshell* in an atom consists of all the electrons having both the same principal quantum number n and the same orbital quantum number l. When a particular subshell contains all the electrons possible it is called a *closed subshell*.

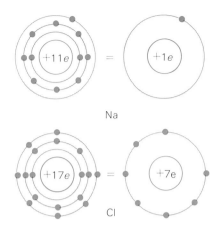

19.7 Electron shielding in sodium and chlorine. *Each outer electron in a Cl atom is acted upon by an effective nuclear charge seven times greater than that acting upon the outer electron in a Na atom, even though the outer electrons in both cases are in the same shell.*

EXERCISES

1 Sodium never occurs in nature as the free element, and platinum seldom occurs in combination. How are these observations related to the chemical activities of the two metals?

2 From what physical and chemical characteristics of iron do we conclude that it is a metal? From what physical and chemical characteristics of sulfur do we conclude that it is a nonmetal?

3 The element astatine (At), which appears at the bottom of the halogen column in the periodic table, has been prepared artificially in minute amounts but has not been found in nature. Using the periodic law and your knowledge of the halogens, predict the properties of this element, as follows:
   a. At room temperature, is it solid, liquid, or gaseous?
   b. How many atoms does a molecule of its vapor contain?
   c. Is it very soluble, moderately soluble, or slightly soluble in water?
   d. What is the formula for its compound with hydrogen?
   e. What are the formulas for its compounds with potassium and calcium?
   f. Is its compound with potassium more or less stable than potassium iodide?

4 The following metals are listed in order of decreasing chemical activity: potassium, sodium, calcium, magnesium. How does this order agree with their positions in the periodic table? Where would you place cesium in the above list?

5 What is the number of electrons in the outermost shells of the elements in Group II of the periodic table?

6 How many elements would there be if atoms having filled electron shells up to and including $n = 6$ were able to exist?

7 Chemists refer to each quantum state of given n, l, and m_l as an *orbital*. How many electrons can occupy each orbital?

8 The energy needed to remove an electron from an atom is called its *ionization energy*. The ionization energies of the elements of atomic numbers 20 through 29 are very nearly the same, although wide variations are found in the ionization energies of other sequences of elements. Explain this observation with the help of Table 19.5.

chapter 20

The Chemical Bond

MOLECULAR FORMATION
   types of bond
   molecules and crystals

THE IONIC BOND
   electron transfer
   chemical activity
   the ionic bond

THE COVALENT BOND
   the hydrogen molecule
   covalent substances
   molecular orbitals

What is the nature of the forces that bond atoms together to form molecules? This question, of fundamental importance to the chemist, is hardly less important to the physicist, whose theory of the atom cannot be correct unless it provides a satisfactory answer. The ability of the quantum theory of the atom to explain chemical bonding is further testimony to the power of this approach.

MOLECULAR FORMATION

A molecule is a stable combination of two or more atoms. By "stable" is meant that a molecule must be given energy from an outside source in order to break up into its constituent atoms. In other words, a molecule exists because the energy of the joint system is less than that of the system of separate atoms. If the interactions among a certain group of atoms reduce their total energy, a molecule can be formed; if the interactions increase their total energy, the atoms repel one another.

a molecule has less total energy than the separate atoms of which it is composed

types of bond

Let us consider what happens when two atoms are brought closer and closer together. Three extreme situations may occur:

a covalent bond consists of one or more pairs of shared electrons

1. A *covalent bond* is formed. One or more pairs of electrons are shared by the two atoms. As these electrons circulate between the atoms, they spend more time between the atoms than elsewhere, which produces an attractive force. An example is H_2, the hydrogen molecule, whose two electrons belong jointly to the two protons (Fig. 20.1a).

an ionic bond is formed by the attraction of ions

2. An *ionic bond* is formed. One or more electrons from one atom may transfer to the other, and the resulting positive and negative ions attract each other. An example is NaCl, where the bond exists between Na^+ and Cl^- ions and not between Na and Cl atoms (Fig. 20.1b).

3. No bond is formed. When the electron structures of two atoms overlap, they constitute a single system, and according to the exclusion principle no two electrons in such a system can exist in the same quantum state. If some of the interacting electrons happen thereby to be forced into higher energy states than they occupied in the separate atoms, the system may have much more energy than before and be unstable. To visualize this effect, we may regard the electrons as fleeing as far away from one another as possible to avoid forming a single system, which leads to a repulsive force between the nuclei. (Even when the exclusion principle can be obeyed with no increase in energy, there will be an electrostatic repulsive force between the two nuclei; this is a much less significant factor than the exclusion principle in influencing bond formation, however.)

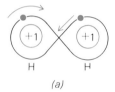

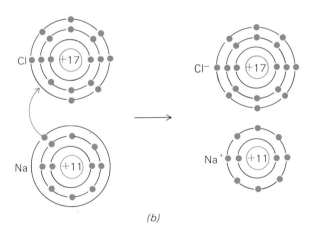

20.1 (a) *Covalent bonding.* The shared electrons spend more time on the average between their parent nuclei and therefore lead to an attractive force (b) *Ionic bonding.* Sodium and chlorine combine chemically by the transfer of electrons from sodium atoms to chlorine atoms; the resulting ions attract electrostatically.

bonds may be intermediate between covalent and ionic in character

In H_2 the bond is purely covalent and in NaCl it is purely ionic, but in many other molecules an intermediate type of bond occurs in which the atoms share electrons to an unequal extent. An example is the HCl molecule, where the Cl atom attracts the shared electrons more strongly than the H atom. A strong argument can be made for thinking of the ionic bond as no more than an extreme case of the covalent bond, but it is customary to analyze them separately.

THE CHEMICAL BOND

molecules and crystals

Ionic bonds usually do not result in the formation of molecules. Strictly speaking, a molecule is an electrically neutral aggregate of atoms that is held together strongly enough to be experimentally observable as a particle. Thus the individual units that constitute gaseous hydrogen each consist of two hydrogen atoms, and we are entitled to regard them as molecules.

definition of molecule

On the other hand, the crystals of rock salt (NaCl) are aggregates of sodium and chlorine ions which, although invariably arranged in a certain definite structure, do not pair off into discrete molecules consisting of one Na^+ ion and one Cl^- ion; rock salt crystals may in fact be of almost any size. While there are always equal numbers of Na^+ and Cl^- ions in rock salt, so that the formula NaCl correctly represents its composition, these ions form molecules rather than crystals only in the gaseous state. Despite the absence of individual NaCl molecules in solid NaCl, there is a very specific interaction between adjacent Na^+ and Cl^- ions in the latter, and it is the properties of this interaction that make NaCl as characteristic an example of chemical bonding as H_2.

THE IONIC BOND

electron transfer

The simplest example of a chemical reaction involving electron transfer is the combination of a metal and a nonmetal. For a specific case, let us consider the burning of sodium in chlorine to give sodium chloride. From Fig. 19.7 it is evident that the lone outer electron in an Na atom is weakly attached, whereas there is a vacancy in the outer shell of a Cl atom where electrons are strongly held. An Na atom readily gives up its outer electron, and a Cl atom readily adds another electron to the seven already present. Thus Na and Cl are perfect mates; one has an electron to lose, the other an electron to gain. In the process of combination, an electron may be thought of as being transferred from Na to Cl (Fig. 20.1b).

The great stability of the resulting closed electron shells within both reacting atoms is indicated by the large amount of energy in the form of heat and light given out when this reaction takes place. The combination NaCl is extremely unreactive since each of its constituent atoms has a stable electronic structure. To break it apart—which means to return the electron to Na, destroying both stable eight-electron structures—requires the expenditure of the same considerable amount of energy that the combination set free.

The Na atom in the compound, shorn of its electron, is no longer a normal atom, since it consists of a nucleus with 11 positive charges and only 10 electrons. The structure as a whole therefore has a positive charge; it is called a *sodium ion*, with the symbol Na^+. The Cl atom has one electron in excess of its normal number and so is charged negatively; this structure is called a *chlorine ion*, Cl^-. The solid salt NaCl has a crystalline structure made up of alternate sodium and chloride ions (Fig. 20.2) which attract one another electrostatically.

the compound NaCl consists of Na^+ and Cl^- ions

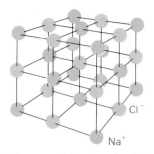

20.2 *The crystal lattice of solid NaCl consists of sodium and chlorine ions in a regular array.*

The fundamental characteristic of all metal atoms is their tendency to lose their outer electrons, like sodium in the above example. Nonmetal atoms, on the other hand, tend to gain electrons so as to fill in gaps in their outer shells. In most reactions of this sort a metal atom loses all its outer electrons, and a nonmetal atom fills all the gaps in its structure. Thus, when sodium combines with sulfur, each S atom has two spaces to fill (Fig. 20.3), but each Na atom has only one electron to give; hence two Na atoms are required for each S atom, and the resulting compound is Na_2S. When calcium combines with oxygen, each Ca atom contributes two electrons to each O atom, and the formula of the compound is CaO.

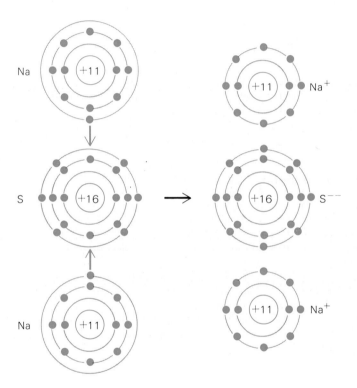

20.3 *The formation of sodium sulfide, Na_2S.*

ionic compounds are formed by electron transfer and usually are crystalline in nature

certain groups of atoms act as units in chemical reactions

Compounds formed by electron transfer are called *ionic compounds*. In addition to simple compounds like NaCl, $MgBr_2$, and K_2S, ionic compounds include substances with more complex formulas like Na_2SO_4, KNO_3, and $CaCO_3$, in which electrons from the metal atoms have been transferred to nonmetal atom *groups* (SO_4, NO_3, CO_3) instead of to single nonmetal atoms. Ionic compounds in general contain a metal and one or more nonmetals, and their crystal structures are made up of alternate positive and negative ions. Most of them are crystalline solids with high melting points, as might be expected, since melting involves a separation of the ions. Another important characteristic of ionic compounds that we shall discuss presently is their ability in the molten state or in solution to conduct electricity.

THE CHEMICAL BOND

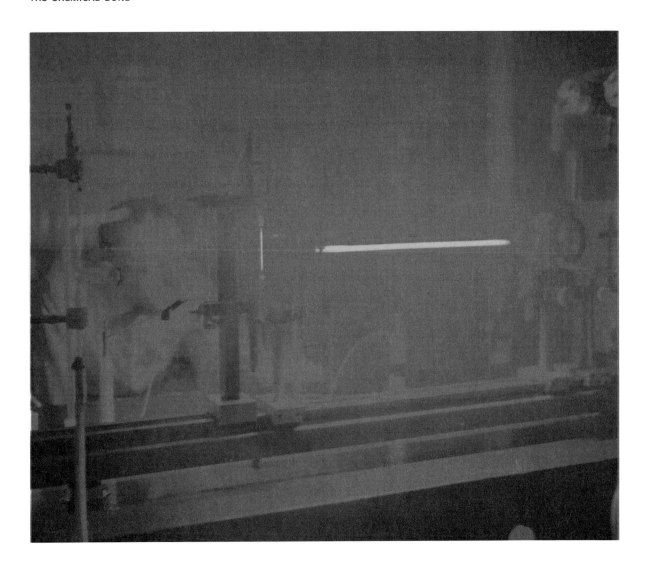

chemical activity

Since metals are characterized by the tendency of their atoms to give up outer electrons to other atoms, we may describe an active metal as a metal whose atoms give up their electrons comparatively readily. Thus magnesium is a more active metal than platinum because the magnesium atom loses its outer electrons more readily than the platinum atom does. Similarly we may describe an active nonmetal as a nonmetal whose atoms readily take electrons from metal atoms to complete their outer shells. Thus chlorine reacts more violently with metals than iodine does because its atoms have a greater attraction for the outer electrons of metal atoms.

When we discussed the periodic table in the previous chapter, we saw that elements within each horizontal row and each vertical column show

An intense flash of light is used here to promote gas-phase chemical reactions by disrupting existing molecular bonds.

why some elements are more active than others

ATOMS IN COMBINATION

a progressive change in metallic or nonmetallic activity. In a horizontal row properties change from those of an active metal (alkali group) at one side to those of an active nonmetal (halogen group) at the other. The change takes place through a series of progressively less active metals and then a series of progressively more active nonmetals. In the vertical columns there is a much less striking change, tending as a rule toward more active metals (and less active nonmetals) at the bottom of the table. Can we find an explanation for these progressive changes in properties in terms of atomic structure? And, a related problem, can we correlate chemical activity with the strengths of chemical bonds, as exhibited by the exceptional stability of compounds composed of active elements? The notions of ionization energy and electron affinity make it possible to answer these questions quantitatively.

the energy needed to remove an electron from an atom is its ionization energy

The *ionization energy* of an atom is the energy needed to remove one of its electrons and is therefore a measure of how tightly bound its outermost electrons are. The smaller its ionization energy, the more readily a metal atom can contribute an electron to an ionic bond and the more chemically active that metal is. Ionization energy is sometimes called *ionization potential*.

Table 20.1 shows the ionization energies of the elements. An atom of any of the alkali metals of Group I has a single electron outside a closed subshell. The electrons in the inner shells partially shield the outer electron from the nuclear charge $+Ze$, so that the effective charge holding the outer electron to the atom is just $+e$ rather than $+Ze$ (Fig. 20.4). Relatively little work must be done to detach an electron from such an atom, and the alkali metals form positive ions readily. The larger the atom, the farther the outer electron is from the nucleus and the weaker is the electrostatic force on it, which is why the ionization energy generally decreases as we go down any group. Potassium, for example, has 18

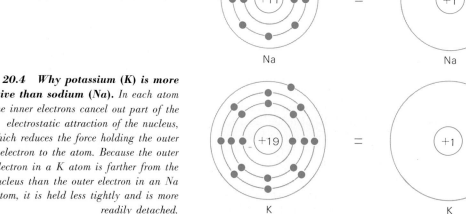

20.4 Why potassium (K) is more active than sodium (Na). *In each atom the inner electrons cancel out part of the electrostatic attraction of the nucleus, which reduces the force holding the outer electron to the atom. Because the outer electron in a K atom is farther from the nucleus than the outer electron in an Na atom, it is held less tightly and is more readily detached.*

TABLE 20.1 Ionization energies of the elements in electron volts.

1 H 13.6																	2 He 24.6
3 Li 5.4	4 Be 9.3											5 B 8.3	6 C 11.3	7 N 14.5	8 O 13.6	9 F 17.4	10 Ne 21.6
11 Na 5.1	12 Mg 7.6											13 Al 6.0	14 Si 8.1	15 P 11.0	16 S 10.4	17 Cl 13.0	18 Ar 15.8
19 K 4.3	20 Ca 6.1	21 Sc 6.6	22 Ti 6.8	23 V 6.7	24 Cr 6.8	25 Mn 7.4	26 Fe 7.9	27 Co 7.9	28 Ni 7.6	29 Cu 7.7	30 Zn 9.4	31 Ga 6.0	32 Ge 7.9	33 As 9.8	34 Se 9.8	35 Ar 11.8	36 Kr 14.0
37 Rb 4.2	38 Sr 5.7	39 Y 6.5	40 Zr 7.0	41 Nb 6.8	42 Mo 7.1	43 Tc 7.3	44 Ru 7.4	45 Rh 7.5	46 Pd 8.3	47 Ag 7.6	48 Cd 9.0	49 In 5.8	50 Sn 7.3	51 Sb 8.6	52 Te 9.0	53 I 10.5	54 Xe 12.1
55 Cs 3.9	56 Ba 5.2	*	72 Hf 5.5	73 Ta 7.9	74 W 8.0	75 Re 7.9	76 Os 8.7	77 Ir 9.2	78 Pt 9.0	79 Au 9.2	80 Hg 10.4	81 Tl 6.1	82 Pb 7.4	83 Bi 7.3	84 Po 8.4	85 At —	86 Rn 10.7
87 Fr —	88 Ra 5.3	†															

*	57 La 5.6	58 Ce 6.9	59 Pr 5.8	60 Nd 6.3	61 Pm —	62 Sm 5.6	63 Eu 5.7	64 Gd 6.2	65 Tb 6.7	66 Dy 6.8	67 Ho —	68 Er 6.1	69 Tm 5.8	70 Yb 6.2	71 Lu 5.0
†	89 Ac —	90 Th 7.0	91 Pa —	92 U 6.1	93 Np —	94 Pu 5.1	95 Am 6.0	96 Cm —	97 Bk —	98 Cf —	99 Es —	100 Fm —	101 Md —	102 No —	103 Lr —

electrons inside the orbit of the outermost electron, but sodium has only 10. This means that the potassium atom is larger and that its outer electron is farther from the positive charge on the nucleus. Hence potassium atoms should lose their outer electrons more easily than sodium atoms. This checks with the experimental fact that potassium is a more active metal than sodium.

The increase in ionization energy from left to right across any period is accounted for by the increase in nuclear charge while the number of inner shielding electrons stays constant. There are two electrons in the common inner shell of Period 2 elements, and the effective nuclear charge acting on the outer electrons of these atoms is therefore reduced by two electron charges. The outer electron in a lithium atom is held to the atom by an effective charge of $+e$, while each outer electron in an atom of beryllium, boron, carbon, and so on, is held to its parent atom by effective charges of $+2e$, $+3e$, $+4e$, and so on. Hence a lithium atom loses its outer electron more easily than a beryllium atom, a boron atom loses its outer electron more easily than a carbon atom, and so on.

At the other extreme from alkali metal atoms, which tend to lose their outermost electrons, are halogen atoms, which tend to complete their outer subshells by picking up an additional electron each. The *electron affinity* of an atom is defined as the energy released when an electron is added to it. The greater the electron affinity, the more tightly bound is the added electron, and the more chemically active that nonmetal is in its role as an electron acceptor. Table 20.2 shows the electron affinities of the halogens. In general, electron affinities decrease going down any group of the periodic table and increase going from right to left across any period. The experimental determination of electron affinities is quite difficult, and those for only a few elements are accurately known.

electron affinity

TABLE 20.2 Electron affinities of the halogens in electron volts.

Fluorine	3.45
Chlorine	3.61
Bromine	3.36
Iodine	3.06

the ionic bond

conditions for ionic bond

An ionic bond between two atoms can exist when one of them has a low ionization energy, and hence a tendency to become a positive ion, while the other has a high electron affinity, and hence a tendency to become a negative ion. Sodium, with an ionization energy of 5.1 ev, is an example of the former, and chlorine, with an electron affinity of 3.6 ev, is an example of the latter. When an Na^+ ion and a Cl^- ion are in the same vicinity and are free to move, the attractive electrostatic force between them brings them together. The condition that a stable unit of NaCl result is simply that the total energy of the system of the two ions be less than the total energy of a system of two atoms of the same elements; otherwise the surplus electron on the Cl^- ion would transfer to the Na^+ ion, and the neutral Na and Cl atoms would no longer be bound together. Let us see how this criterion is met by NaCl.

We begin with an Na atom and a Cl atom infinitely far apart. We expend 5.1 ev of work to remove the outer electron from the Na atom, leaving it an Na^+ ion; that is,

$$Na + 5.1 \text{ ev} \longrightarrow Na^+ + e^-$$

When this electron is brought to the Cl atom, the latter absorbs it to complete its outer electron subshell and thereby becomes a Cl^- ion. Since the electron affinity of chlorine is 3.6 ev, signifying that this amount of work must be done to liberate the odd electron from a Cl^- ion, the formation of Cl^- *gives off* 3.6 ev of energy. Hence

$$Cl + e^- \longrightarrow Cl^- + 3.6 \text{ ev}$$

The net result of these two events is the sum of the preceding two equations,

$$Na + Cl + 1.5 \text{ ev} \longrightarrow Na^+ + Cl^-$$

Our net expenditure of energy to form Na^+ and Cl^- ions from Na and Cl atoms has been only 1.5 ev.

Now we permit the electrostatic attraction between the Na^+ and Cl^- ions to bring them together to 2.4×10^{-10} m of each other. The energy given off when this occurs is equal to the potential energy of the system of a $+e$ charge that distance from a $-e$ charge. This energy is 6 ev. Schematically,

$$Na^+ + Cl^- \underset{\underset{\infty}{\longleftarrow}}{\longrightarrow} Na^+ + Cl^- + 6 \text{ ev} \atop {\underset{2.4 \times 10^{-10} \text{ m}}{\longleftarrow\longrightarrow}}$$

If we shift an electron from an Na atom to an infinitely distant Cl atom and allow the resulting ions to come together, then the entire process in the absence of other effects gives off an energy of 6 ev − 1.5 ev = 4.5 ev:

$$Na + Cl \underset{\underset{\infty}{\longleftarrow}}{\longrightarrow} Na^+ + Cl^- + 4.5 \text{ ev} \atop {\underset{2.4 \times 10^{-10} \text{ m}}{\longleftarrow\longrightarrow}}$$

Evidently an Na^+Cl^- combination with this interatomic spacing formed by the electrostatic attraction of Na^+ and Cl^- ions is stable, since 4.5 ev of work must be done on such a molecule in order to separate it into Na and Cl atoms.

We may well ask what stops the two ions from continuing to approach each other until their electron structures mesh together. The reasons were indicated at the beginning of this chapter. First of all, it is easy to see that, if such a meshing were to take place, the positively charged nuclei of the ions would no longer be completely shielded by their surrounding electrons and would simply repel each other electrostatically.

There is another phenomenon that is even more effective in keeping the ions apart. According to the Pauli exclusion principle, no two electrons in the same atomic system can exist in the same quantum state. If the electron structures of Na^+ and Cl^- overlap, they constitute a single atomic system rather than separate, independent systems. If such a system is to obey the exclusion principle, some electrons will have to go to higher quantum states than they would otherwise occupy. These states have more energy than those corresponding to the normal electron configuration of the ions, and so the total energy of an Na^+Cl^- unit increases when the Na^+ and Cl^- ions approach each other too closely.

the exclusion principle plus nuclear repulsion prevents ions from getting too close together

An increasing potential energy when two bodies are brought together means that the force between them is repulsive, just as a decreasing potential energy under the same circumstances means an attractive force. (The gravitational potential energy of an object decreases the closer it is to the earth, for instance.) Figure 20.5 contains a curve showing how the potential energy V of the system of an Na^+ ion and a Cl^- ion varies with separation distance r. At $r = \infty$, an infinite separation, $V = +1.5$ ev. When the ions are closer together, the attractive electrostatic force between them takes effect and the potential energy drops. Finally, when they are about 3×10^{-10} m apart, their electron structures begin to interact, a repulsive force comes into play, and the potential energy levels off. The minimum in the potential energy curve occurs at $r = 2.4 \times 10^{-10}$ m, where $V = -4.2$ ev. (The difference between this figure and the

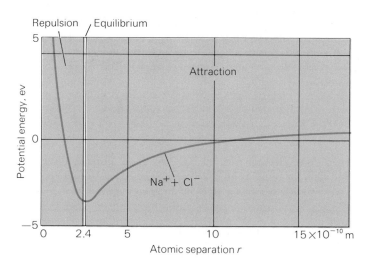

20.5 *Variation of the potential energy of the system Na⁺ and Cl⁻ with their distance apart.* Attractive and repulsive forces balance at the lowest point in the curve; at greater separations the attractive force predominates, while at smaller separations the repulsive force predominates.

-4.5 ev calculated on the basis of electrostatic forces only is due to the effect of the repulsive forces.) At this separation distance the mutually attractive and repulsive forces on the ions exactly balance, and the system is in equilibrium. To *dissociate* an NaCl molecule into Na and Cl atoms thus requires an energy of 4.2 ev.

THE COVALENT BOND

The theory of the ionic bond between atoms of opposite valence is unable to explain such molecules as H_2 or Cl_2, whose constituent atoms are identical. Pure ionic binding involves the transfer of one or more electrons from one atom to another, and the resulting electrostatic forces are responsible for the stability of the molecule. In *covalent binding*, on the other hand, electrons may be thought of as being *shared* by the atoms composing a molecule. In the course of circulating among these atoms, the electrons spend more of their time between atoms than on the outside of the molecule, which leads to a net attractive electrostatic force that holds the molecule together (Fig. 20.1a). In some molecules more than one pair of electrons is shared; for instance, N_2 has three shared pairs of electrons.

the shared electrons in a covalent bond spend more time between the atoms, which serves to hold the atoms together

the hydrogen molecule

The simplest possible molecular system is H_2^+, the hydrogen molecular ion, in which a single electron bonds two protons. Let us look in a general way into how it is possible for two protons to share an electron and why such sharing should lead to a lower total energy and hence to a stable system.

One of the consequences of the wave nature of a moving particle (Chap. 14) is that it is impossible to permanently trap such a particle in a box

quantum mechanics shows that it is possible for a trapped particle to escape from its "box"

THE CHEMICAL BOND

of any kind: the particle can "leak" out of the box even though it does not have enough energy to break through the wall because the particle's wave function extends beyond it. Only if the wall is infinitely strong—which is never the case—is the wave function wholly inside the box. The electric field around a proton is in effect a box for an electron, and two nearby protons correspond to a pair of boxes with a wall between them (Fig. 20.6). There is no mechanism in classical physics by which the electron in a hydrogen atom can transfer spontaneously to a neighboring proton. In quantum physics, however, the mechanism is straightforward. There is a certain probability that an electron trapped in one box will tunnel through the wall and get into the other box, and once there it has the same probability for tunneling back. This situation can be described by saying that the electron is shared by the protons.

To be sure, the likelihood that an electron will pass through the region of high potential energy—the "wall"—between two protons depends strongly upon how far apart the protons are. If the proton-proton distance is 10^{-10} m, the electron may be regarded as going from one proton to the other about every 10^{-15} sec, which means that we can legitimately consider the electron as being shared by both. If the proton-proton distance is 10×10^{-10} m, however, the electron shifts across an average of only about once per second, which is practically an infinite time on an atomic scale. Since the effective radius of the $1s$ wave function in hydrogen is 0.53×10^{-10} m, we conclude that electron sharing can take place only between atoms whose wave functions overlap appreciably.

Granting that two protons can share an electron, there is a simple argument that shows why the energy of such a system could be less than that of a separate hydrogen atom and proton. According to the uncertainty principle, the smaller the region to which we restrict a particle, the greater must be its momentum and hence kinetic energy. An electron shared by two protons is less confined than one belonging to a single proton, which means that it has less kinetic energy. The total energy of the electron in H_2^+ is therefore less than that of the electron in $H + H^+$, and since the magnitude of the proton-proton repulsion in H_2^+ is not too great, H_2^+ is stable.

The H_2 molecule contains two electrons instead of the single electron of H_2^+. According to the exclusion principle, both electrons can be in the same quantum state, and hence be described by the same wave function, provided that their spins are antiparallel (Fig. 20.7). With two electrons to contribute to the bond, H_2 ought to be more stable than H_2^+—at first glance, twice as stable, with a bond energy of 5.3 ev compared with 2.65 ev for H_2^+. However, the H_2 wave functions are not quite the same as those of H_2^+ because of the electrostatic repulsion between the two electrons in H_2, a factor absent in the case of H_2^+. The latter repulsion weakens the bond in H_2, so that the actual bond energy is 4.72 ev instead of 5.3 ev.

covalent substances

Substances whose atoms are joined by shared electron pairs are called *covalent substances*. In general, they are nonmetallic elements or com-

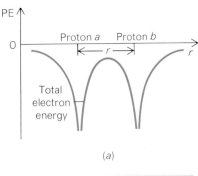

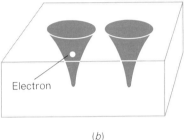

20.6 (a) *The solid line shows the potential energy of an electron in the electric fields of two nearby protons.* The total energy of a ground-state electron in a hydrogen atom is indicated. (b) *Two nearby protons correspond in quantum mechanics to a pair of boxes separated by a wall.*

20.7 *Contours of electron probability for two adjacent hydrogen atoms whose electron spins are parallel and antiparallel.* The closer the spacing of the lines, the greater the likelihood of finding the electron. A repulsive force is present in the former situation, an attractive one in the latter.

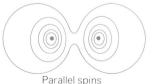

Parallel spins

Antiparallel spins

pounds of one nonmetal with another, although some compounds of metals belong to this class. Since a pair of electrons shared between two atoms remain in between the respective nuclei, neither atom acquires an electric charge; hence the crystals of covalent compounds are made up of atoms or molecules rather than ions. In general, covalent substances are poor conductors of electricity.

The distinction between covalent and ionic compounds is not sharp. In both kinds one or more electrons are held between two atoms; the distinction depends on whether the electrons are held chiefly by one atom (ionic) or equally by the two atoms (covalent). Some covalent compounds have an intermediate character: the electrons are somewhat closer to one atom than to the other. Two examples are HCl and H_2O:

$$H : Cl \qquad H : O$$
$$ \cdot\cdot$$
$$ H$$

polar covalent compounds

These substances are called *polar* covalent compounds, because one part of the molecule is relatively negative and another part positive. All gradations can be found between symmetric covalent molecules at one extreme, through polar covalent molecules, to ionic compounds at the other extreme. For example,

Covalent	Cl	:	Cl
Polar covalent	H	:	Cl
Ionic	Na		:Cl

electronegativity

The relative tendency of an atom to attract shared electrons when it is part of a molecule is called its *electronegativity*. In the HCl molecule, for instance, Cl is more electronegative than H. A scale of electronegativities has been devised, principally through the work of Linus Pauling, on the basis of observed bond energies. A partially ionic character to a bond means a higher bond energy—greater stability—than if the bond were purely covalent with equal electron sharing, since the attractive force due to the negative charge density between the atoms is supplemented by a direct mutual attraction between the more positive atom and the more negative one. Thus the difference between the actual bond energy in a molecule and the bond energy calculated by assuming equal electron sharing is a measure of the difference between the electronegativities of the atoms involved.

Table 20.3 is a list of atomic electronegativities. The values in this table permit us to predict how the electrons in a molecule are shared. Thus the electronegativity of H is 2.2 and that of Cl is 3.1, in agreement with the fact that the shared electrons favor the Cl atom. On the other hand, the electronegativity of Li is 1.0, so in the LiH molecule the shared electrons favor the H atom. The electronegativities of Na and Cl are quite far apart, being respectively 0.9 and 3.1, in accord with the ionic character of the bond in NaCl.

THE CHEMICAL BOND

TABLE 20.3 Electronegativities of the elements.

1 H 2.2																	2 He —
3 Li 1.0	4 Be 1.6											5 B 2.0	6 C 2.6	7 N 3.0	8 O 3.4	9 F 4.0	10 Ne —
11 Na 0.9	12 Mg 1.3											13 Al 1.6	14 Si 1.9	15 P 2.2	16 S 2.6	17 Cl 3.1	18 Ar —
19 K 0.8	20 Ca 1.0	21 Sc 1.4	22 Ti 1.5	23 V 1.6	24 Cr 1.7	25 Mn 1.6	26 Fe 1.8	27 Co 1.9	28 Ni 1.9	29 Cu 1.9	30 Zn 1.7	31 Ga 1.8	32 Ge 2.0	33 As 2.2	34 Se 2.6	35 Br 3.0	36 Kr —
37 Rb 0.8	38 Sr 1.0	39 Y 1.2	40 Zr 1.3	41 Nb 1.6	42 Mo 2.2	43 Tc 1.9	44 Ru 2.2	45 Rh 2.3	46 Pd 2.2	47 Ag 1.9	48 Cd 1.7	49 In 1.8	50 Sn 2.0	51 Sb 2.0	52 Te 2.1	53 I 2.7	54 Xe —
55 Cs 0.8	56 Ba 0.9	57–71 — 1.1–1.3	72 Hf 1.3	73 Ta 1.5	74 W 2.4	75 Re 1.9	76 Os 2.2	77 Ir 2.2	78 Pt 2.3	79 Au 2.5	80 Hg 2.0	81 Tl 2.0	82 Pb 2.3	83 Bi 2.0	84 Po 2.0	85 At 2.2	86 Rn —
87 Fr 0.7	88 Ra 0.9	89–94 — 1.1–1.4															

molecular orbitals

In chemistry it is customary to refer to the wave function characterized by specific quantum numbers n, l, and m_l as an *orbital*; each orbital can describe (or "contain") two electrons of opposite spins. In discussing covalent bonding, it is helpful to be able to visualize the distributions in space of the various atomic orbitals, which resemble those of hydrogen. The pictures in Fig. 18.15 are limited to two dimensions and hence are not suitable for this purpose. It is more appropriate here to draw boundary surfaces that outline the regions within which the total probability of finding the electron described by the orbital has some definite value, say 90 or 95 percent. Figure 20.8 contains boundary surface diagrams for s and p orbitals.

an orbital can contain two electrons of opposite spins

When two atoms come together, their orbitals overlap and the result may be an increased electron probability density between them that signifies a *bonding molecular orbital* or it may be a decreased electron probability density that signifies an *antibonding molecular orbital*. Figure 20.9 shows how s and p atomic orbitals can form bonding molecular orbitals. There are two kinds of bonding molecular orbital that can result from s and p atomic orbitals. In one of these, the σ orbital (σ is the Greek letter *sigma*), the electron probability density is concentrated on the bond axis, a line joining the two parent nuclei. The bond axis is conventionally called the z axis, so s and p_z atomic orbitals can form σ molecular orbitals. In a π orbital, the electron probability density is

bonding and antibonding molecular orbitals

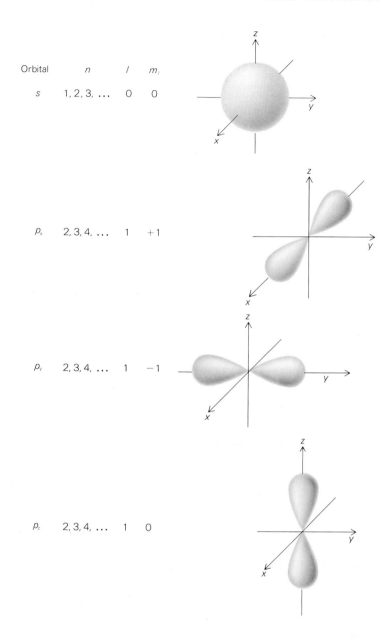

20.8 The configurations of s and p atomic orbitals. Each orbital can "contain" two electrons. There is a high probability of finding an electron described by one of these orbitals in the shaded region.

concentrated on two sides of the bond axis, so p_x and p_y atomic orbitals both form π molecular orbitals. In Chap. 28 we shall learn of the important role played by π orbitals in the chemistry of carbon compounds. As in the case of an atomic orbital, a molecular orbital can contain two electrons.

The configurations of the three p atomic orbitals together with their ability to join with s orbitals to form bonding molecular orbitals makes it possible to understand the geometries of many molecules. The water molecule H_2O is an example. Offhand we might expect a linear molecule,

THE CHEMICAL BOND

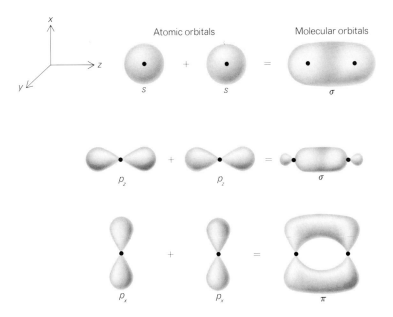

20.9 The formation of σ and π bonding molecular orbitals. *Two p_y atomic orbitals can combine to form a π molecular orbital in the same way as shown for two p_x atomic orbitals but with a different orientation.*

H—O—H (each dash represents a shared electron pair), since oxygen is more electronegative than hydrogen and each H atom in H_2O accordingly exhibits a small positive charge. The resulting repulsion between the H atoms should keep them as far apart as possible, namely on opposite sides of the O atom. In reality, however, the water molecule has a structure closer to O—H, with an angle of 104.5° between the two O—H bonds.
|
H

The bent shape of the water molecule is not hard to account for. The $2p_y$ and $2p_z$ orbitals in O are only singly occupied, so that each can join with the 1s orbital of an H atom to form a σ bonding orbital (Fig. 20.10). The y and z axes are 90° apart, and the larger 104.5° angle that is actually found may plausibly be attributed to the mutual repulsion of the H atoms.

A similar argument is able to explain the pyramidal shape of the ammonia molecule NH_3. The $2p_x$, $2p_y$, and $2p_z$ atomic orbitals in N are singly occupied, which means that each of them can form a σ bonding orbital with the 1s orbital of an H atom. The bonding molecular orbitals in NH_3 should therefore be centered along the x, y, and z axes (Fig. 20.11) with N—H bonds 90° apart. As in H_2O, the actual bond angles in NH_3 are somewhat greater than 90°, in this case 107.5°, owing to repulsions among the H atoms.

To sum up, the likelihood of finding a particular electron near its parent nucleus depends both upon the distance from the nucleus and upon direction. The electron probabilities that can be computed from the quantum theory of the atom are not symmetric, but instead show "lobes" of high probability in certain directions for electrons in certain quantum states. Covalent binding occurs when the appropriate probability lobes

the water molecule

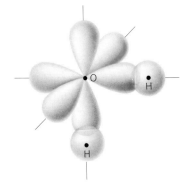

20.10 The H_2O molecule. *The atomic orbitals shown as overlapping form σ bonding molecular orbitals. The angle between the O—H bonds is actually 104.5°.*

of adjacent atoms overlap. When an atom combines with two or more other atoms, the latter are therefore generally "attached" at specific places on the atoms corresponding to the lobes of greatest magnitude, instead of being located at random or in some universal configuration. Figure 20.12 shows some examples of directed bonds in complex molecules.

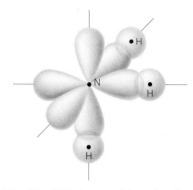

20.11 The NH_3 (ammonia) molecule. *The atomic orbitals shown as overlapping form σ bonding molecular orbitals. The angles between the N—H bonds are actually 107.5°.*

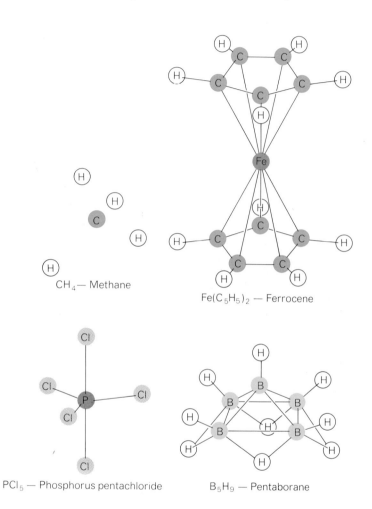

20.12 Directed bonds in complex molecules. *The geometrical arrangements of the atoms in a number of molecules are shown here.*

CH_4 — Methane

$Fe(C_5H_5)_2$ — Ferrocene

PCl_5 — Phosphorus pentachloride

B_5H_9 — Pentaborane

GLOSSARY

A *molecule* is an electrically neutral combination of two or more atoms held together strongly enough to be experimentally observable as a particle.

A *covalent bond* occurs when one or more pairs of electrons are shared between two atoms.

THE CHEMICAL BOND

An *ionic bond* occurs when one or more electrons are transferred from one atom to another, so that the resulting ions attract each other electrostatically.

The *ionization energy* of an atom is the amount of work required to remove an electron from it. The *electron affinity* of an atom is the amount of energy that it releases when it picks up an extra electron beyond its normal complement.

The *electronegativity* of an atom refers to its tendency to attract shared electrons when it participates in a chemical bond.

An *orbital* is the wave function that describes an electron in an atom or molecule; an orbital can contain two electrons, one of either spin. A *bonding molecular orbital* is one in which there is an increased electron probability density between two atoms, leading to a covalent bond.

EXERCISES

1. Illustrate with electronic diagrams (*a*) the reaction between a lithium atom and a fluorine atom, and (*b*) the reaction between a magnesium atom and a sulfur atom. Would you expect lithium fluoride and magnesium sulfide to be ionic or covalent compounds?

2. What part of the atom is chiefly involved in each of the following processes?
   a. the burning of charcoal
   b. radioactive disintegration
   c. the production of X-rays
   d. the ionization of air
   e. the emission of spectral lines
   f. the explosion of a hydrogen bomb
   g. the rusting of iron

3. Electrons are much more readily liberated from metals than from nonmetals when irradiated with visible or ultraviolet light. Can you explain why this is true? From metals of what group would you expect electrons to be liberated most easily?

4. The rare element selenium has the following arrangement of electrons: 2 in the first shell, 8 in the second, 18 in the third, and 6 in the fourth. Would you expect selenium to be a metal or a nonmetal? To what group in the periodic table would it belong?

5. What is the difference in atomic structure between the two isotopes of chlorine? How would you account for the great chemical similarity of the two isotopes?

6 Would you expect magnesium or calcium to be the more active metal? Explain your answer in terms of atomic structure.

7 Lithium is directly below hydrogen in the periodic table, yet lithium atoms do not join together to form Li_2 molecules the way hydrogen atoms do. Why not?

8 Which of the following compounds do you expect to be ionic and which covalent? IBr, NO_2, SiF_4, Na_2S, CCl_4, $RbCl$, Ca_3N_2.

9 Distinguish between orbits and orbitals.

10 Why do sodium atoms exhibit more pronounced chemical activity than sodium ions?

11 Why do chlorine atoms exhibit more pronounced chemical activity than chlorine ions?

12 How much energy is required to form an ion pair from each of the following pairs of atoms? $K + Cl$; $K + I$; $Li + Br$.

13 Use the electronegativity scale to predict which atom in each of the following compounds the electrons that participate in the bonding are closest to on the average: KCl, CuS, CO, NO, HgO. Arrange these compounds in order of increasing ionic character.

chapter 21

The Solid State

CRYSTAL STRUCTURE
   solids and liquids
   x-ray diffraction
   crystal defects

BONDING IN SOLIDS
   ionic crystals
   covalent crystals
   the metallic bond
   molecular crystals
   the hydrogen bond

BAND THEORY OF SOLIDS
   conductors and insulators
   semiconductors
   optical properties of solids

A solid consists of atoms very close together in a fixed arrangement. Most solids are crystalline in nature or nearly so, though a few, such as glass, have no definite ordered structures. The ionic and covalent bonds that hold molecules together have important counterparts in the solid state. In addition there are the *van der Waals* and *metallic* bonds that provide the cohesive forces in, respectively, molecular crystals (such as ice) and metals. All these bonds are electrical in origin, so that the chief distinctions among them lie in the distribution of electrons around the atoms or molecules whose regular arrangements constitute crystals.

CRYSTAL STRUCTURE

The majority of solids are crystalline, with the ions, atoms, or molecules of which they are composed falling into regular, repeated three-dimensional patterns. The presence of *long-range order* is thus the defin-

long-range order is characteristic of a crystal

ing property of a crystal. Other solids lack long-range order in the arrangements of their constituent particles, and may properly be regarded as supercooled liquids whose stiffness is due to an exceptionally high viscosity. Glass, pitch, and many plastics are examples of such amorphous ("without form") solids.

amorphous solids exhibit short-range order only

Amorphous solids do exhibit *short-range order* in their structures, however. The distinction between the two kinds of order is nicely exhibited in boron trioxide (B_2O_3), which can occur in both crystalline and amorphous forms. In each case every boron atom is surrounded by three oxygen atoms, which represents a short-range order. In a B_2O_3 crystal the oxygen atoms are present in hexagonal arrays, as in Fig. 21.1, which is a long-range ordering, while amorphous B_2O_3, a vitreous or "glassy" substance, lacks this additional regularity.

solids and liquids

The resemblance between an amorphous solid and a liquid is worth pursuing as a means of better understanding both states of matter. Although liquids are like gases in certain respects—both are fluids and the molecules of both are randomly arranged—nevertheless there is more in common between liquids and solids. For instance, a gas sample has no fixed volume, unlike liquid and solid samples, but expands to fill its container. The density of a given liquid is usually almost the same as that of the corresponding solid, which suggests that their molecules are about the same distance apart in both states. (We will refer to the structural units of liquids and solids as molecules for the moment, though they may be either atoms or molecules depending upon the substance.) Furthermore, X-ray studies indicate that liquids have definite short-range structures at any instant, quite similar to those of amorphous solids except that the groupings of liquid molecules are continually shifting.

liquids resemble amorphous solids in structure

Now let us see what happens when a liquid is cooled to its freezing point. The transformation from liquid to solid begins with the appearance of minute crystals called nuclei, which proceed to grow in size until the transformation is complete. The nuclei may appear spontaneously, or foreign particles such as dust grains or impurity crystals may act as nuclei. If the liquid is pure and undisturbed, it may be possible to cool it below the freezing point without crystallization. Ordinarily such a supercooled

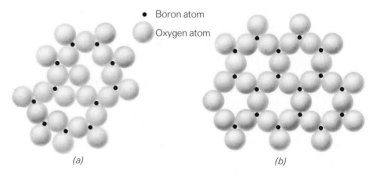

21.1 (a) Amorphous B_2O_3 exhibits only short-range order. (b) Crystalline B_2O_3 exhibits long-range order as well.

liquid is unstable and will crystallize at the slightest disturbance, but in some cases the molecules are unable to move about at the lowered temperature to arrange themselves into a crystalline pattern. The result is a solid material whose internal configuration is that of a liquid; no change of state has occurred, but instead one of the transient molecular arrangements of the liquid has been preserved by the reduction in molecular mobility.

A vitreous solid is thus basically unstable, since its internal energy would be reduced if it became a crystalline solid at the same temperature. Crystallization from the vitreous state is so sluggish that it ordinarily does not occur, but is not unknown: glass may devitrify when heated until it has not quite begun to soften, and extremely old glass specimens are sometimes found to have crystallized.

Since amorphous solids are essentially liquids, they have no sharp melting points. We can interpret this behavior on a microscopic basis by noting that, since an amorphous solid lacks long-range order, the bonds between its molecules vary in strength. When the solid is heated, the weakest bonds rupture at lower temperatures than the others, so that it softens gradually. In a crystalline solid the transition between long-range and short-range order (or no order at all) involves the breaking of bonds whose strengths are more or less identical, and melting occurs at a precisely defined temperature.

It is worth noting that most actual solids are not huge single crystals but instead are aggregates of a great many tiny crystals, and the behavior of a polycrystalline bulk solid may be quite different from that of one of its component units.

x-ray diffraction

A valuable method for determining crystal structure is based upon the use of X-rays. As we know, X-rays consist of electromagnetic waves of short wavelength and considerable penetrating power. When a beam of X-rays is directed at a crystal, it is observed that, while most of them simply pass right through, some are scattered at certain specific angles that vary with the wavelength of the X-rays and with the type of crystal (Fig. 21.2). This phenomenon (customarily referred to as *X-ray diffraction*) is an interference effect and, originally employed to verify the wave nature of X-rays, is today widely used to study the arrangements of atoms in crystals.

Figure 21.3 shows a narrow beam of X-rays of the same wavelength that fall obliquely on one face of a crystal. The atoms in the crystal, because of their regular arrangement, can be thought of as lying in a series of parallel planes, and a portion of the X-ray beam is, in effect, reflected by each of the planes. This situation corresponds very closely

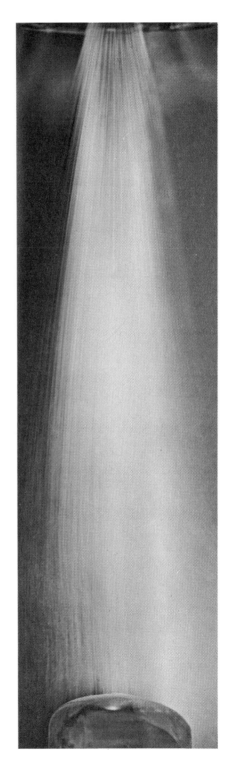

A synthetic ruby rod being "grown" in a Verneuil furnace. Powdered corundum (Al_2O_3) with a small percentage of chromic oxide (Cr_2O_3) is fed from above through an oxyhydrogen flame and solidifies into a single crystal.

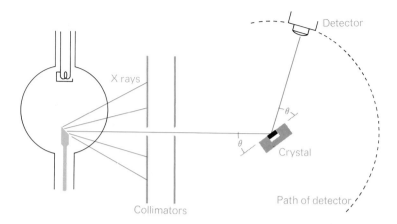

21.2 X-ray scattering experiment. *At certain angles constructive interference occurs in the X-rays scattered by the various layers of atoms in the crystal.*

to that of Fig. 13.15, where the light waves reflected by two surfaces that are close together reinforce each other in certain directions and cancel each other in intermediate directions through the process of interference.

The scattering of X-rays by crystals was first analyzed in a similar way by W. L. Bragg in 1912. He considered a beam of X-rays striking two

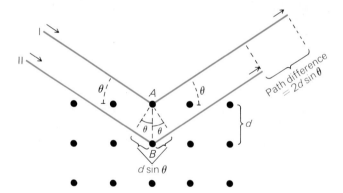

21.3 X-ray scattering from a cubic crystal. *When the path difference is an integral number of wavelengths, constructive interference occurs and a "reflected" ray emerges.*

successive atomic planes a distance d apart at an angle θ, as in Fig. 21.3. The beam goes past atom A in the first plane and atom B in the next, and each of them scatters part of the beam in random directions. Constructive interference takes place only between those scattered rays that are parallel and whose paths differ by exactly λ, 2λ, 3λ, and so on. That is, the path difference must be $n\lambda$, where n is an integer. The only rays scattered by A and B for which this is true are those labeled I and II in the figure. If the path difference is not a whole number of wavelengths, the scattered rays will be out of step and will cancel each other out.

The first condition upon I and II is that their common scattering angle be equal to the angle of incidence θ of the original beam. (This condition is the same as that for ordinary optical reflection: angle of incidence = angle of reflection.) The second condition is that

THE SOLID STATE

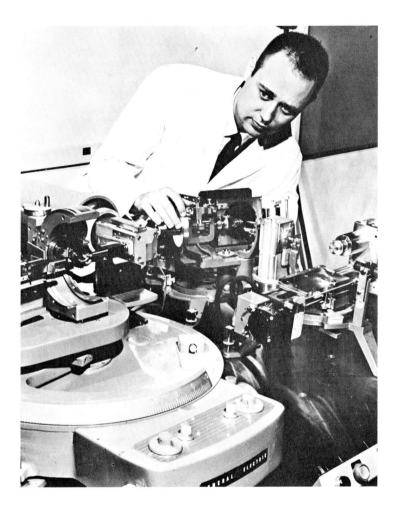

A modern X-ray diffraction apparatus being used here to study the molecular structures of cell membranes.

Path difference = whole number of wavelengths

$$2d \sin \theta = n\lambda \qquad n = 1, 2, 3, \ldots$$

If we know the wavelength λ of the X-rays in an experimental beam, then we can ascertain the angles θ at which the scattered rays emerge from a crystal under study and solve the above equation for the atomic spacing d.

Variations of Bragg's method have been devised which have been used to determine the exact arrangements of atoms in crystals with more complex structures than the simple one of Fig. 21.3.

the interference of x-rays scattered by atoms in a crystal permits determining the crystal's structure

crystal defects

An ideal crystal is one in which each atom occupies a definite location in a regular array. Actual crystals are far from meeting this specification.

The interference pattern produced by the reflection of X rays from particles in a crystal of sodium chloride. The bright spots are the places where X rays reflected from various layers in the crystal come in step. The cubic pattern of the sodium chloride lattice is suggested by the fourfold symmetry of the X-ray photograph.

Defects in the structure of a crystal—missing atoms, atoms out of place, irregularities in the spacing of the rows of atoms, and so forth—often have a considerable bearing on the physical properties of solids. In particular, the mechanical strength of a solid is largely determined by the nature and concentration of defects in its structure.

A *dislocation* is a type of crystal defect in which a line of atoms is not in its proper position. Dislocations are of two basic kinds. Figure 21.4 shows an *edge dislocation,* which can be visualized in terms of the removal of part of one layer of atoms and the subsequent accommodation of the array to the defect. The dislocation itself is indicated by the symbol ⊥, and in its immediate neighborhood the crystal structure is severely distorted. The bonds between atoms in this crystal are represented by lines.

dislocations in a solid

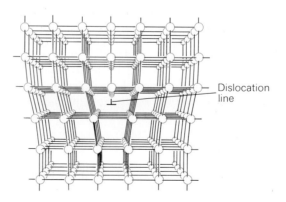

21.4 *An edge dislocation.*

The other kind of dislocation is the *screw dislocation*. We can visualize the formation of a screw dislocation by imagining that a cut is made a certain distance into a perfect crystal and one side of the cut then displaced relative to the other (Fig. 21.5). The atomic layers spiral around the dislocation, which accounts for its name. Actual dislocations in crystals are usually combinations of the edge and screw varieties.

Dislocations are important because they provide an explanation for the plastic behavior of solids. When an applied stress exceeds its elastic limit, a material no longer returns to its original shape but is permanently deformed (Chap. 7). Most metals can undergo substantial plastic deformation before fracture, a property described as ductility; in other solids the plastic range is usually smaller. The elastic response of a solid is readily interpreted in terms of the bonding forces within it, which act like Hooke's law restoring forces for small displacements from the equilibrium configuration. But this direct approach fails to account for plastic behavior, since calculations of the force required to slide one layer of atoms in a crystal past another yield figures about a thousand times higher than those actually observed.

dislocations make possible plastic flow

The presence of dislocations makes it possible to understand why solids are only 0.1 percent or so as strong as they ought to be on the basis of perfect crystal structures. Figure 21.6 shows how a crystal that contains an edge dislocation can be permanently deformed by a relatively modest pair of forces. The line of atoms below and to the right of the dislocation

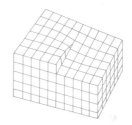

21.5 A screw dislocation.

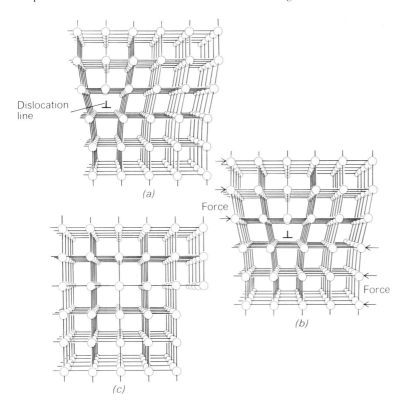

21.6 Slip results from the motion of a dislocation through a crystal under stress. (a) Initial configuration of crystal. (b) The dislocation moves to the right as the atoms in the layer under it successively shift their bonds with those of the upper layer one line at a time. (c) The crystal has taken on a permanent deformation.

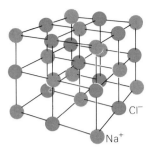

21.7 *The face-centered-cubic crystal structure of sodium chloride.*

shift their bonds to the line of atoms directly above it when the forces are applied, which causes the dislocation to move one atom spacing to the right. The process is repeated until the dislocation arrives at the edge of the crystal, and the deformation is then permanent. The entire process is called *slip*, and the plane along which the dislocation moves is called the *slip plane*. In slip, the atomic bonds holding one layer to the next are broken only one line at a time, whereas in a perfect crystal all the bonds between two layers would have to be broken more or less simultaneously for plastic flow to occur, a much more formidable proposition.

Because dislocations multiply during plastic flow, the continued deformation of a solid increases its dislocation content. Eventually the dislocations become so numerous and so tangled together that they impede one another's motion, which *decreases* the plasticity of the material. This phenomenon is known as *work hardening*. If a work-hardened crystal is heated, its disordered atomic array tends to return to regularity; the rise in temperature, which increases the vibrational energy of the atoms, initiates the release of the energy stored in the dislocations. The process of heating a work-hardened crystal to restore its ductility is called *annealing*. Steel bars and sheets formed by cold-rolling are much harder than those formed by hot-rolling.

foreign atoms reduce slip in steel, which is why steel is stronger than pure iron

Besides work hardening, another means of impeding slip in a metal is the deliberate introduction of foreign atoms to interfere with the motion of dislocations. Thus the addition of small amounts of such other elements as carbon, chromium, manganese, and tungsten to iron converts it into the vastly stronger steel by reducing slip.

BONDING IN SOLIDS

ionic crystals

Ionic bonds in crystals are very similar to ionic bonds in molecules. Such bonds come into being when atoms which have low ionization energies, and hence lose electrons easily, interact with other atoms which have high electron affinities. The former atoms give up electrons to the latter, and they thereupon become positive and negative ions respectively. In an ionic crystal these ions come together in an equilibrium configuration in which the attractive forces between positive and negative ions predominate over the repulsive forces between similar ions.

the repulsion between atoms that are very close together is a manifestation of the exclusion principle

As in the case of molecules, crystals of all types are prevented from collapsing under the influence of the cohesive forces present by the action of the exclusion principle, which requires the occupancy of higher energy states when electron shells of different atoms overlap and mesh together.

Figure 21.7 shows the arrangement of Na^+ and Cl^- ions in a sodium chloride crystal. The ions of either kind may be regarded as being located at the corners and at the centers of the faces of an assembly of cubes, with the Na^+ and Cl^- assemblies interleaved. Each ion thus has six nearest neighbors of the other kind. Such a crystal structure is called *face-centered cubic*. In NaCl crystals the distance between like ions is 5.63×10^{-10} m; the distance between adjacent ions is half this.

A different structure is found in cesium chloride crystals, where each ion is located at the center of a cube at whose corners are ions of the other kind (Fig. 21.8). This structure is called *body-centered cubic*, and each ion has eight nearest neighbors of the other kind. In CsCl crystals the distance between like ions is 4.11×10^{-10} m.

Most ionic solids are hard, owing to the strength of the bonds between their constituent ions, and have high melting points. They are usually brittle as well, since the slipping of atoms past one another that accounts for the ductility of metals is prevented by the ordering of positive and negative ions imposed by the nature of the bonds.

properties of ionic solids

covalent crystals

The cohesive forces in covalent crystals arise from the presence of electrons between adjacent atoms. Each atom participating in a covalent bond contributes an electron to the bond, and these electrons are shared by both atoms rather than being the virtually exclusive property of one of them as in an ionic bond. Diamond is an example of a crystal whose atoms are linked by covalent bonds. Figure 21.9 shows the structure of a diamond crystal. Each carbon atom has four nearest neighbors and shares an electron pair with each of them. The length of each bond is 1.54×10^{-10} m.

the bonds in a covalent crystal arise from shared electrons

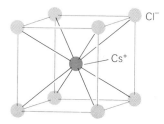

Purely covalent crystals are relatively few in number. In addition to diamond, some examples are silicon, germanium, and silicon carbide; in SiC each atom is surrounded by four atoms of the other kind in the same tetrahedral structure as that of diamond. All covalent crystals are hard (diamond is the hardest substance known, and SiC is the industrial abrasive Carborundum), have high melting points, and are insoluble in all ordinary liquids, behavior which reflects the strength of the covalent bonds.

21.8 *The body-centered-cubic crystal structure of cesium chloride.*

There are several ways to ascertain whether the bonds in a given nonmetallic, nonmolecular crystal are predominantly ionic or covalent. In general, a compound of an element from Group I or II of the periodic table with one from Group VI or VII exhibits ionic bonding in the solid state. Another guide is the coordination number of the crystal, which is the number of nearest neighbors about each constituent particle. A high coordination number suggests an ionic crystal, since it is hard to see how an atom can form purely covalent bonds with six other atoms (as in a face-centered-cubic structure like that of NaCl) or with eight other atoms (as in a body-centered-cubic structure like that of CsCl). A coordination number of four, however, as in the diamond structure, is compatible with an exclusively covalent character. To be sure, as with molecules, it is not always possible to classify a particular crystal as being wholly ionic or covalent: AgCl, whose structure is the same as that of NaCl, and CuCl, whose structure resembles that of diamond, both have bonds of intermediate character, as do a great many other solids.

21.9 *The crystal lattice of diamond. The carbon atoms are held together by covalent bonds.*

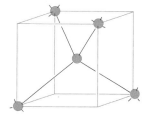

the metallic bond

The basic concept that underlies the modern theory of metals is that the valence (outermost) electrons of the atoms comprising a metal may be

Forging increases the strength of a metal object by increasing the number of dislocations present in its crystal structure.

common to the entire atomic aggregate, so that a kind of "gas" of free electrons pervades it. The interaction between this gas and the positive metal ions constitutes a strong cohesive force. Further, the presence of such free electrons accounts very nicely for the high electrical conductivities and other unique properties of metals. To be sure, no electrons in any solid at ordinary temperatures, even a metal, are able to move about its interior with complete freedom. All of them are influenced to some extent by other particles present, and when the theory of metals is refined to include these perturbations, a comprehensive picture emerges that is in excellent agreement with experiment.

a "gas" of free electrons is present in a metal

A convenient way of understanding the metallic bond is to view it as an unsaturated covalent bond. Let us compare the binding process in hydrogen and in lithium, both members of Group I of the periodic table. An H_2 molecule contains two electrons with opposite spins, the maximum number of $n = 1$ electrons that can be present. The H_2 molecule is therefore saturated, since the exclusion principle requires that any additional electrons be in states of higher energy and the stable attachment of further H atoms cannot occur unless their electrons are in $n = 1$ states.

a saturated covalent bond involves a pair of electrons shared by two adjacent atoms

Superficially lithium might seem obliged to behave in a similar way, having two $n = 1$ electrons and one $n = 2$ electron. However, there are six *unfilled* 2p ($n = 2, l = 1$) quantum states in every Li atom whose energies are only very slightly greater than those of the two 2s ($n = 2$, $l = 0$) states. When an Li atom comes near an Li_2 "molecule," it readily becomes attached with a covalent bond without violating the exclusion principle, and the resulting Li_3 "molecule" is stable since all its valence electrons remain in $n = 2$ states. There is no limit to the number of Li atoms that can join together in this way, since lithium forms body-centered-cubic crystals (Fig. 21.8) in which each atom has eight nearest neighbors. With only one electron per atom available to enter into bonds, each bond involves one-fourth of an electron on the average, instead of two electrons as in ordinary covalent bonds. Hence the bonds are far from being saturated, which is true of the bonds in other metals as well.

the bonds between metal atoms involve less than two electrons on the average

One consequence of the unsaturated nature of the metallic bond is the weakness of metals as compared with perfect ionic and covalent crystals having saturated bonds. Another is the ease with which metals can be deformed. Having neither definite localized bonds between adjacent atoms, as in a covalent crystal, nor a configuration of alternating positive and negative ions, as in an ionic crystal, the atoms of a metal can be rearranged in position without rupturing the crystal. For the same reason the properties of a mixture of different metal atoms do not depend critically on the proportion of each kind of atom provided that their sizes are similar. Thus the characteristics of an alloy often vary smoothly with changes in its composition, in contrast to the specific atomic proportions found in ionic solids and in covalent solids such as SiC.

The most striking consequence of the unsaturated bonds in a metal is the ability of the outermost atomic electrons to wander freely from atom to atom. To understand this phenomenon intuitively, we can think of each such electron as constantly moving from bond to bond. In solid Li, each electron participates in eight bonds, so it only spends a short

the electron gas model of metals accounts for their ability to conduct electric current

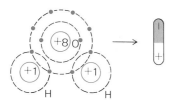

21.10 *Water consists of polar molecules which behave as if negatively charged at one end and positively charged at the other. See also Fig. 20.10.*

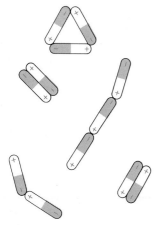

21.11 *Polar molecules attract each other.*

van der waals forces bond molecules together to form liquids and solids

21.12 *Polar molecules attract normally nonpolar molecules.*

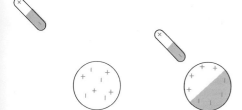

time between any pair of Li$^+$ ions. The electron cannot remember (so to speak) which of the two ions it really belongs to, and it is just as likely to move on to a bond that does not involve its parent ion at all. The outermost atomic electrons in a metal therefore behave in a manner quite similar to that of molecules in a gas. Ohm's law (Chap. 11) can be derived on the basis of this "electron gas" picture of metals.

As in the case of any other solid, metal atoms stick together because their collective energy is lower when they are bound together than when they exist as separate atoms. To understand why this reduction in energy occurs in a metallic crystal, we note that, because the ions are near one another, each floating electron is on the average closer to one nucleus or another than it would be if it belonged to an isolated atom. Consequently, the potential energy of the electron is less in the crystal than in the atom, and it is this decrease in potential energy that is responsible for the metallic bond.

molecular crystals

There are many substances whose molecules are so stable that, when brought together, they have no tendency to lose their individuality by joining together in a collective lattice with multiple linkages like those found in ionic and covalent crystals. Most organic compounds are examples of such noninteracting substances. However, even they can exist as liquids and solids through the action of the attractive *van der Waals* intermolecular forces.

We begin by noting that many molecules, which we called polar molecules in the previous chapter, behave as though they are negatively charged at one end and positively charged at the other. An example is the H$_2$O molecule, in which the concentration of electrons around the oxygen atom makes that end of the molecule more negative than the end where the hydrogen atoms are (Fig. 21.10). Such molecules tend to align themselves so that ends of opposite sign are adjacent, as in Fig. 21.11, and in this orientation the molecules strongly attract one another.

A polar molecule is also able to attract nonpolar molecules whose charges are normally uniformly distributed. The process is illustrated in Fig. 21.12: The electric field of the polar molecule induces a separation of charge in the other molecule, and the two now attract each other electrostatically.

More remarkably, two nonpolar molecules can attract each other by the above mechanism. Even though the electron distribution in a nonpolar molecule is symmetric *on the average*, the electrons themselves are in constant motion and at *any given instant* one part or another of the molecule has an excess of them. Instead of the fixed charge asymmetry of a polar molecule, a nonpolar molecule has a constantly shifting asymmetry. When two nonpolar molecules are close enough, their fluctuating charge distributions tend to shift together, adjacent ends always having opposite sign (Fig. 21.13) and so always causing an attractive force. This kind of force is named after the Dutch physicist van der Waals, who suggested its existence nearly a century ago to explain observed departures from the ideal-gas law; the explanation of the actual mechanism of the force, of course, is more recent, since it is based upon the modern theory of the atom.

Van der Waals forces are present, not only between all molecules, but also between all atoms, including those of the inert gases which do not otherwise interact. Van der Waals bonds are much weaker than ionic and covalent bonds, and as a result molecular crystals, whose lattices are composed of whole molecules rather than ions or atoms, generally have low melting and boiling points and little mechanical strength. Ordinary ice and Dry Ice (solid CO_2) are examples of molecular solids.

the hydrogen bond

Certain compounds, notably water, have much higher melting and boiling points than would be expected if they followed the general trend of these figures for other compounds of the same types. For example, in order of decreasing molecular mass, the nonmetallic hydrogen compounds (called *hydrides*) H_2Te, H_2Se, and H_2S have the respective boiling points -2, -42, and $-60°C$, while H_2O, with a still smaller molecular mass, boils at $100°C$. Similar though less marked anomalies are exhibited by ammonia (NH_3) and hydrogen fluoride (HF). The intermolecular forces in H_2O, NH_3, and HF are due to *hydrogen bonds*, which are stronger than ordinary van der Waals bonds but weaker than ionic or covalent bonds.

Hydrogen bonds occur in the hydrides of elements so electronegative that we may think of the molecules as having virtually bare protons present on the outside. More precisely, the electron distribution in such a molecule is so distorted by the affinity of the "parent" atom for electrons that each hydrogen atom in essence has donated most of its negative charge to the parent atom, leaving behind a poorly shielded proton. In the case of HF, the result is a molecule with a strong, localized positive charge at one end which can link up with the less concentrated negative charge at the opposite end of another HF molecule. The key factor here is the small effective size of the poorly shielded proton, since electrostatic forces vary as $1/r^2$ where r is measured from the centers of the charged bodies involved. So strong are the hydrogen bonds in hydrogen fluoride that even in the gas phase such supermolecules as H_2F_2, H_3F_3, H_4F_4, H_5F_5, and H_6F_6—or, better, HF·HF, HF·HF·HF, and so on—occur frequently.

Water molecules are exceptionally prone to form hydrogen bonds because the four pairs of electrons around the O atom occupy orbitals that project outward as though toward the vertexes of a tetrahedron (Fig. 21.14). (A tetrahedron is a pyramid with four corners, or *vertexes*.) Hydrogen atoms are at two of these vertexes, which accordingly exhibit localized positive charges, while the other two vertexes exhibit somewhat more diffuse negative charges. Each H_2O molecule can therefore form hydrogen bonds with *four* other H_2O molecules; in two of these bonds the central molecule provides the bridging protons, and in the other two the attached molecules provide them.

In the liquid state, the hydrogen bonds between adjacent H_2O molecules are continually being broken and reformed owing to thermal agita-

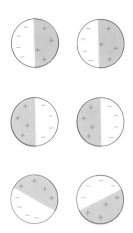

21.13 *The fluctuating charge distributions of nearby molecules tend to shift together, leading to an attractive force between them, even though neither is a polar molecule on the average.*

a hydrogen bond is an especially strong type of van der waals bond

21.14 *In an H_2O molecule, the four pairs of outer (valence) electrons around the oxygen atom, six contributed by the O atom and one each by the H atoms, occupy orbitals that form the pattern shown. Each H_2O molecule can form hydrogen bonds with four other H_2O molecules.*

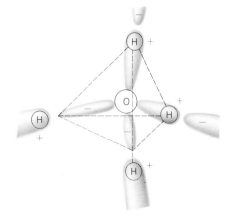

TABLE 21.1. Crystal types. The cohesive energy is equal to the work needed to remove an atom (or molecule) from the crystal.

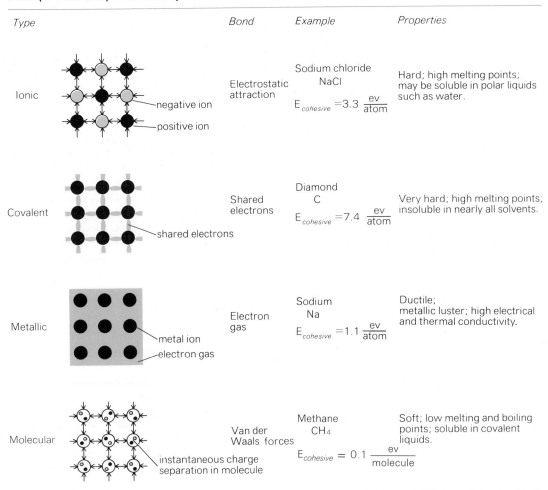

Type	Bond	Example	Properties
Ionic	Electrostatic attraction	Sodium chloride NaCl $E_{cohesive} = 3.3 \frac{ev}{atom}$	Hard; high melting points; may be soluble in polar liquids such as water.
Covalent	Shared electrons	Diamond C $E_{cohesive} = 7.4 \frac{ev}{atom}$	Very hard; high melting points; insoluble in nearly all solvents.
Metallic	Electron gas	Sodium Na $E_{cohesive} = 1.1 \frac{ev}{atom}$	Ductile; metallic luster; high electrical and thermal conductivity.
Molecular	Van der Waals forces	Methane CH_4 $E_{cohesive} = 0.1 \frac{ev}{molecule}$	Soft; low melting and boiling points; soluble in covalent liquids.

tion, but even so at any instant the molecules are combined in definite clusters. In the solid state, these clusters are large and stable and constitute ice crystals.

The characteristic hexagonal pattern (Fig. 21.15) of an ice crystal arises from the tetrahedral arrangement of the four hydrogen bonds each H_2O molecule can participate in. With only four nearest neighbors around each molecule, ice crystals have extremely open structures, which is the reason for the exceptionally low density of ice. Because the molecular clusters are smaller and less stable in the liquid state, water molecules on the average are packed more closely together than are ice molecules, and water has the higher density: hence ice floats. (The density of ice is only about 92 percent of that of water.) The density of water increases from

ice has a more open structure than water, hence its density is less than that of water and it floats

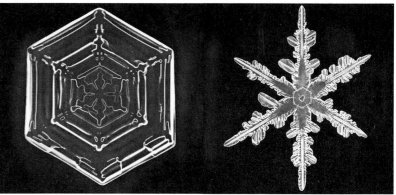

0°C to a maximum at 4°C as large clusters of H_2O molecules are broken up into smaller ones that occupy less space in the aggregate; only above 4°C does the normal thermal expansion of a liquid manifest itself in a decreasing density with increasing temperature (Fig. 21.16).

The water molecules in a snowflake are held together by van der Waals bonds.

BAND THEORY OF SOLIDS

The atoms in almost every crystalline solid, whether a metal or not, are so close together that their valence electrons constitute a single system of electrons common to the entire crystal. The exclusion principle is obeyed by such an electron system because the energy states of the outer electron shells of the atoms are all altered somewhat by their mutual interactions. In place of each precisely defined characteristic energy level of an individual atom, the entire crystal possesses an *allowed energy band* composed of myriad separate levels very close together. Since there are as many of these separate levels as there are atoms in the crystal, the band cannot be distinguished from a continuous spread of permitted energies.

conductors and insulators

The allowed energy bands in a solid correspond to the energy levels in an atom, and an electron in a solid can possess only those energies that fall within these energy bands. The various energy bands in a solid may overlap, as in Fig. 21.17, in which case its electrons have a continuous distribution of permitted energies. In other solids the bands may *not* overlap (Fig. 21.18), and the intervals between them represent energies which their electrons cannot possess. Such intervals are called *forbidden bands*. The electrical behavior of a crystalline solid is determined both by its energy-band structure and by how these bands are normally filled by electrons.

Figure 21.19 is a simplified diagram of the energy levels of a sodium

21.15 Top view of an ice crystal, showing open hexagonal arrangement of H_2O molecules. Each molecule has four nearest neighbors to which it is attached by hydrogen bonds.

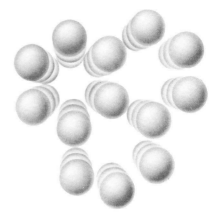

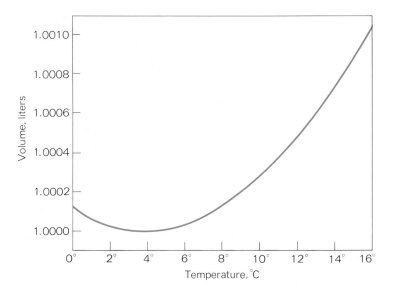

21.16 *The variation with temperature of the volume of 1 kg of water.* The volume is a minimum, corresponding to maximum density, at 4°C.

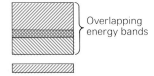

21.17 *The energy bands in a solid may overlap.*

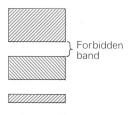

21.18 *A forbidden band separates nonoverlapping energy bands.*

atom and the energy bands of solid sodium. A sodium atom has a single electron in its outer shell. This means that the highest occupied energy band in a sodium crystal is only half-filled, since each level in the band, like each level in the atom, is able to contain *two* electrons. When an electric field is set up across a piece of solid sodium, electrons easily acquire additional energy while remaining in their original energy band. The additional energy is in the form of kinetic energy, and the moving electrons constitute an electric current. Sodium is therefore a good conductor of electricity, as are other crystalline solids with allowed energy bands that are only partially filled. These solids constitute the metals.

Figure 21.20 is a simplified diagram of the energy bands of diamond. The two lower energy bands are completely filled with electrons, and there is a gap of 6 ev between the top of the higher of these bands and the empty band above it. This means that at least 6 ev of additional energy must be provided an electron in a diamond crystal if it is to have any kinetic energy, since it cannot have an energy lying in the forbidden band. An energy increment of this magnitude cannot readily be given to an electron in a crystal by an electric field. Diamond is therefore a very poor conductor of electricity, and is accordingly classed as an insulator.

semiconductors

Silicon has a crystal structure resembling that of diamond, and as in diamond, a gap separates the top of a filled allowed energy band from a vacant higher allowed band. The forbidden band in silicon is only 1.1 ev wide, however. At low temperatures silicon is little better than diamond as a conductor, but at room temperature a small proportion of its electrons have sufficient kinetic energy of thermal origin to jump the forbidden band and enter the allowed band above it. These electrons are sufficient to permit a limited amount of current to flow when an electric field is applied. Thus silicon has an electrical resistivity intermediate between those of conductors and those of insulators, and is termed a *semiconductor*.

The resistivity of semiconductors can be significantly affected by small amounts of impurity. Let us incorporate a few arsenic atoms in a silicon crystal. Arsenic atoms have five electrons in their outermost shells, while silicon atoms have four. When an arsenic atom replaces a silicon atom in a silicon crystal, four of its electrons are incorporated in covalent bonds with its nearest neighbors. The fifth electron requires little energy to be detached and move about in the crystal. Such a substance is called an *n-type* semiconductor because electric current in it is carried by negative charges (Fig. 21.21).

If we alternatively incorporate gallium atoms in a silicon crystal, a different effect occurs. Gallium atoms have only three electrons in their outer shells, and their presence leaves vacancies called "holes" in the electron structure of the crystal. An electron needs relatively little energy to enter a hole, but as it does so it leaves a new hole in its former location. When an electric field is applied across a silicon crystal containing a trace of gallium, electrons move toward the anode by successively filling holes. The flow of current here is conveniently described with reference to the holes, whose behavior is like that of positive charges since they move toward the negative electrode (Fig. 21.22). A substance of this kind is called a *p-type* semiconductor. The unusual properties of *n-* and *p-*type semiconductors have made possible the development of such compact and efficient electronic devices as the transistor.

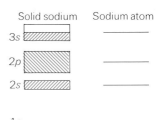

21.19 Energy levels in the sodium atom and the corresponding bands in solid sodium (not to scale). *The upper band is only half filled, and so electrons need little additional energy in order to move through the metal. Sodium is therefore a good electrical conductor.*

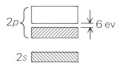

21.20 Energy bands in diamond (not to scale). *A large gap separates the upper filled band from the empty band above it. Electrons must acquire 6 ev of energy in order to move through the crystal, which is therefore a poor electrical conductor.*

21.21 An n-type semiconductor. *Current is carried by excess electrons that do not fit into the electron structure of the crystal.*

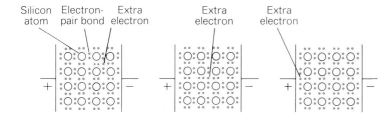

optical properties of solids

There is a close connection between the optical properties of solids and their energy-band structures. Photons of visible light have energies between about 1 and 3 ev. Such amounts of energy are readily absorbed by a "free" electron in a metal, since its allowed energy band is only partly filled, and metals are accordingly opaque. The electrons in an insulator, on the other hand, need more than 3 ev of energy to jump across the forbidden band to the next allowed band. Insulators therefore cannot absorb photons of visible light, and so they are transparent. Of course, most samples of insulating materials do not appear transparent, but this is because of the scattering of light by irregularities in their

21.22 *A p-type semiconductor. Current is carried by the motion of sites of missing electrons. Such "holes" in the electron structure of the crystal migrate toward the negative electrode as electrons successively fill them.*

conductors are opaque to visible light, insulators are transparent

structures. Thus ground glass is practically opaque, but becomes transparent when it is polished. Insulators are opaque to ultraviolet light, whose higher frequencies mean high enough photon energy to permit electrons to cross the forbidden band. Because the forbidden bands in semiconductors are about the same in width as the energies of photons of visible light, they are usually opaque to visible light but are transparent to infrared light, whose lower frequencies mean lower photon energies that cannot be absorbed.

GLOSSARY

A *crystal* is a solid whose constituent ions, atoms, or molecules are arranged in a regular, repeated pattern. An *amorphous solid* exhibits only short-range order in its structure.

An *ionic crystal* consists of individual ions in an equilibrium array in which the attractive forces between ions of opposite charge balance the repulsive forces present.

A *covalent crystal* consists of individual atoms which share electrons with their neighbors.

The *metallic bond* has its origin in a "gas" of freely moving electrons that pervades every metal.

Van der Waals forces have their origin in the electrostatic attraction between asymmetrical charge distributions in atoms and molecules. In a *hydrogen bond*, a nearly bare proton on the outside of a molecule exerts an exceptionally strong van der Waals force on a neighboring molecule.

The proximity of the atoms in a crystal alters slightly the energy states of their outer electron shells. As a result the crystal as a whole has *energy bands* of myriad separate levels close together. Gaps between the bands are called *forbidden bands*, and electrons in the crystal cannot have energies that fall in these gaps.

Semiconductors are intermediate between metals and insulators in their ability to carry electric current. Electrons carry current in an *n-type* semiconductor, whereas vacancies in its electron structure called *holes* carry current in a *p-type* semiconductor.

Cross section of a transistor, a device that capitalizes upon the properties of n- and p-type semiconductors to control electric currents. Here a p region is sandwiched between two n regions (the latter are silvery in appearance in the picture).

EXERCISES

1. State the four principal types of binding in solids and give an example of each. What is the fundamental physical origin of all of them?

2. Electrons and neutrons are used to study crystal structure in addition to X-rays. What property of these particles do you think makes them useful for this purpose?

3. What properties do liquids have in common with gases? With solids?

4. Distinguish between the plastic and the elastic behavior of solids. How is each accounted for in terms of the structure of solids?

5. Van der Waals forces are strong enough to hold inert gas atoms together to form liquids at low temperatures, yet they do not lead to inert gas molecules at higher temperatures. Why?

6. Metal crystals are said to be weaker than ionic or covalent crystals, yet in our experience metals are stronger than most other substances. Can you think of any reasons why ionic and covalent crystals might not be able to exhibit their theoretical strength as well as metals can?

7. What property of the hydrogen atom is responsible for the strength of hydrogen bonds in molecular solids? What common compound has an especially strong tendency to form hydrogen bonds?

8. Would you expect metals to be transparent to infrared light? Would you expect insulators to be transparent to infrared light?

9. A 1-kg copper bar is rolled out into a thin sheet. The energy stored in the dislocations that result in its crystal structure total 10^7 joules/m^3. The copper sheet is then "annealed" by being heated to $200°C$, at which temperature it recrystallizes and the stored energy is released. What is its final temperature before it is allowed to cool? (The density of copper is 8.9×10^3 kg/m^3 and its specific heat is 0.093 kcal/kg-°C; 1 kcal = 4,185 joules.) Is it necessary to know the mass of the copper?

part six
BASIC CHEMISTRY

The four chapters to come are concerned with some of the properties of matter and some of the ideas that are basic to chemistry.

Chemistry, like all true sciences, is quantitative in character, so it is not out of place to have an entire chapter devoted to such topics as valence, chemical equations, and stoichiometry.

Next the chemical behavior is sketched of eight elements, either important in their own rights or typical of specific classes of elements. Then liquid solutions in general and ionic solutions in particular are discussed. The ions of a substance are usually very different from the atoms or molecules of that substance, which makes it necessary for us to adopt a special point of view with respect to chemical reactions that occur in an ionic solution.

Finally we look into the three familiar classes of compounds that form ions when dissolved in water: acids, bases, and salts.

chapter 22
Chemical Calculations

VALENCE
saturation
metal valences
nonmetal valences
how valence is applied

CHEMICAL EQUATIONS
balancing an equation

STOICHIOMETRY
avogadro's number
the mole
molarity
gas volumes
the universal gas constant

The work of Lavoisier (1743–1794) marks the transition from alchemy to chemistry because of his reliance on the balance as an essential tool in investigating chemical reactions. In this chapter we shall examine some of the quantitative aspects of chemistry: the useful idea of valence, the equations that express what happens in a reaction, and finally how to translate formulas into definite masses, and vice versa.

VALENCE

The interatomic forces that lead to chemical bonds exhibit *saturation*: An atom joins with only a limited number of other atoms. This is true in crystals as well as in molecules. The chemist describes saturation in terms of *valence*, a very useful concept although somewhat limited in scope.

saturation

saturation

We have seen that two H atoms can combine to form an H_2 molecule, and indeed hydrogen molecules in nature always consist of two H atoms. Let us now review how the exclusion principle prevents molecules such as He_2 and H_3 from existing while permitting such other molecules as H_2O to be stable.

Every He atom in its ground state has an electron of each spin in the $n = 1$ shell. If it is to join with another He atom by exchanging electrons, each atom will have two electrons with the same spin for part of the time as the electrons circulate between them; that is, one atom will have both electron spins up and the other will have both spins down. The exclusion principle, of course, prohibits two electrons with otherwise the same quantum numbers in an atom from having the same spins, which is manifested by a repulsion between He atoms. Hence the He_2 molecule cannot exist.

the exclusion principle is responsible for saturation

A similar argument holds in the case of H_3. An H_2 molecule contains two electrons whose spins are antiparallel. Should another H atom approach whose electron spin is, say, up, the resulting molecule would have two spins up and one down, and this is impossible if all three electrons are to be in the lowest energy state. Hence the existing H_2 molecule repels the additional H atom. The exclusion-principle argument does not apply if one of the three electrons in H_3 is in an excited state. All such states are of higher energy than the $n = 1$ state, however, and the resulting configuration therefore has more energy than $H_2 + H$ and so will decay rapidly to $H_2 + H$.

The water molecule H_2O achieves stability because the O atom lacks two electrons of having a completed outer electron shell. This lack is remedied when the O atom forms covalent bonds with two H atoms so that the latter's electrons are shared by the O atom without violating the exclusion principle. The H_2O structure has less energy than the separate atoms because of the electron affinity of O, and so is favored.

The phenomenon of saturation leads in a natural way to a kind of electronic bookkeeping system in which valence is the medium of exchange. By definition, **the valence of an element in a compound is the number of electrons each of its atoms has gained, lost, or shared; elements in the free state have a valence of 0.**

definition of valence

An atom that has only a few electrons in its outermost subshell tends to combine with one or more other atoms whose outermost subshells lack a total of the same number of electrons. An atom that donates n electrons to a bond is said to have a valence of $+n$, while one that receives m electrons from a bond is said to have a valence of $-m$. Hence metals have positive valences, nonmetals negative valences: The valence of a metal is the number of electrons in the outer shell of each of its atoms, while the valence of a nonmetal is the number of electrons needed to complete the outer shell of each of its atoms.

metals have positive valences

nonmetals have negative valences

positive and negative valences balance in simple compounds

For stability, **the total number of positive and negative valences in a simple compound must be equal.** Thus H has a valence of $+1$ and O a valence of -2, and H_2O has a net valence of $1 + 1 - 2 = 0$. In the case of H_2, we may think of one H atom as having a valence of $+1$ and the other of -1. Atoms such as carbon, which form bonds by sharing four electrons, may be regarded as having valences of either $+4$ or -4, depending upon the atoms they combine with. For example, in CH_4 (methane) carbon has a valence of -4, while in CO_2 (carbon dioxide) its valence is $+4$.

To be sure, in both CH_4 and CO_2 the bonds are largely covalent and electrons circulate between the atoms of each molecule instead of being transferred as in purely ionic bonds. For this reason positive and negative valences are often not distinguished in covalent compounds, and the valence of an element in such a compound is just the number of electron pairs which each of its atoms shares with other atoms. Thus carbon has a valence of 4 in either CH_4 or $CHCl_3$, H has a valence of 1, and Cl has a valence of 1. In SO_2, sulfur has a valence of 4 and oxygen of 2, since each sulfur atom shares four electron pairs and each oxygen atom shares two.

in a covalent compound, valence refers to the number of shared electron pairs

Since the number of electrons in the outer shell of an atom determines its normal valence, these outer electrons are often called *valence electrons*. Thus H has one valence electron, C has four valence electrons, Cl has seven, Ca has two, and so on.

outer-shell electrons are called valence electrons

metal valences

A simple way to look at valence is to regard the valence of a metal as equal to the number of chlorine atoms per metal atom in the formula for the chloride of that metal. The chlorides of the alkali metals are an example:

Lithium chloride	LiCl
Sodium chloride	NaCl
Potassium chloride	KCl
Rubidium chloride	RbCl
Cesium chloride	CsCl

The atoms of the alkali metals all combine with a single chlorine atom to form their respective chlorides, and so these metals all have the valence $+1$.

Barium, calcium, and magnesium, among other metals, form chlorides whose formulas are

Barium chloride	$BaCl_2$
Calcium chloride	$CaCl_2$
Magnesium chloride	$MgCl_2$

The atoms of these metals require two chlorine atoms each to form their respective chlorides, and so they all have the valence $+2$.

In the same way we find that aluminum and chromium have valences of $+3$, hafnium has the valence $+4$, and tantalum the valence $+5$. Some elements may have two or more valences, depending upon the specific circumstances surrounding the formation of their compounds; thus

some elements exhibit more than one valence

iron may have the valence $+2$, in which case its compounds are called *ferrous* [for example, ferrous chloride ($FeCl_2$)] or the valence $+3$, in which case its compounds are called *ferric* [for example, ferric chloride ($FeCl_3$)].

nonmetal valences

For nonmetals, valence may be thought of as equal to the number of hydrogen atoms per nonmetal atom in the formula for the hydrogen compound of that nonmetal.

The halogens, which form

Hydrogen fluoride	HF
Hydrogen chloride	HCl
Hydrogen bromide	HBr
Hydrogen iodide	HI

when combined with hydrogen, all have the valence -1. Oxygen, which forms the compound

Water H_2O

with hydrogen, has the valence -2. Nitrogen, which forms the compound

Ammonia NH_3

with hydrogen, has the valence -3.

The inert gases, which form virtually no compounds, stand in a class by themselves with the normal valence of zero.

how valence is applied

the formula of a simple compound can be determined from the valences of its constituent atoms

The notion of valence is chiefly useful in considering simple compounds that consist of a metal (or hydrogen) and a nonmetal or nonmetal group. Because the sum of the valences of the constituents of a simple compound must be zero, these numbers supply information needed to write the formula of the compound with correct subscripts when the name of the compound is given. Thus sodium has a valence of $+1$ and oxygen a valence of -2, so sodium oxide must have the formula Na_2O in order that positive and negative valences balance out. Similarly aluminum has a valence of $+3$ and bromine a valence of -1, and the formula of aluminum bromide is $AlBr_3$.

atom groups

Certain groups of atoms are observed to act as units, appearing as parts of many compounds and remaining intact through chemical reactions. An example is the group SO_4, which consists of a sulfur atom joined to four oxygen atoms. This *sulfate group* is a constituent of a number of similar compounds:

CHEMICAL CALCULATIONS

Sodium sulfate	Na_2SO_4
Potassium sulfate	K_2SO_4
Copper sulfate	$CuSO_4$
Zinc sulfate	$ZnSO_4$
Magnesium sulfate	$MgSO_4$

To demonstrate that the sulfate group enters into chemical reactions as an entity, we might simply mix solutions of magnesium sulfate and barium chloride. The result is a solution of magnesium chloride that contains solid barium sulfate, an insoluble compound:

$$MgSO_4 + BaCl_2 \longrightarrow BaSO_4 + MgCl_2$$

When two or more groups of a single kind are present in each molecule of a compound, the formula is written with a pair of parentheses around the group:

Calcium nitrate	$Ca(NO_3)_2$
Aluminum sulfate	$Al_2(SO_4)_3$

To establish the valences of atom groups, it is only necessary to compare compounds that contain them with corresponding compounds that contain atoms of a single kind in their place. The nitrate group NO_3 appears in the compound HNO_3, and since Br has the valence -1 in the corresponding compound HBr, we conclude that the NO_3 group also has the valence -1. The compound aluminum nitrate therefore must have the formula $Al(NO_3)_3$. Table 22.1 is a list of the valences of some common elements and atom groups.

TABLE 22.1 The valences of some common elements and atom groups.

Name	Symbol	Valence	Name	Symbol	Valence
Lithium	Li	+1	Fluorine	F	−1
Sodium	Na	+1	Chlorine	Cl	−1
Cesium	Cs	+1	Bromine	Br	−1
Potassium	K	+1	Iodine	I	−1
Silver	Ag	+1	Nitrate group	NO_3	−1
Ammonium group	NH_4	+1	Permanganate group	MnO_4	−1
Beryllium	Be	+2	Chlorate group	ClO_3	−1
Magnesium	Mg	+2	Hydroxide group	OH	−1
Calcium	Ca	+2	Cyanide group	CN	−1
Strontium	Sr	+2	Sulfate group	SO_4	−2
Barium	Ba	+2	Carbonate group	CO_3	−2
Zinc	Zn	+2	Chromate group	CrO_4	−2
Cadmium	Cd	+2	Silicate group	SiO_3	−2
Aluminum	Al	+3	Phosphate group	PO_4	−3

valence is a concept of limited usefulness

It is important to be aware that, for all the convenience they offer, valences have certain severe limitations. For one thing, they are of no help in considering compounds of nonmetals with one another, and cannot be applied consistently to many complex compounds and even to some simple ones that contain both nonmetals and metals. Also, some elements have two or more different valences, for instance copper with $+1$ and $+2$ and iron with $+2$ and $+3$; the formulas for "copper oxide" and "iron oxide" cannot be determined merely by knowing what their constituent elements are with no further information.

In Chap. 27 we shall see how the more general concept of oxidation number is of help in many situations where valences fail to account for the observed reactions.

CHEMICAL EQUATIONS

As a shorthand method of expressing the results of a chemical change, the formulas of the substances involved can be combined into a *chemical equation*. An equation includes the formulas of all the substances entering the reaction on the left-hand side with the formulas of all the products on the right-hand side. The formulas may be written in any order and are connected by $+$ signs; between the two sides of the equation is placed an arrow. Thus, when carbon burns, the two substances that react are carbon (C) and oxygen (O_2), and the only product is carbon dioxide (CO_2):

a chemical equation describes the results of a chemical change

$$C + O_2 \longrightarrow CO_2$$

This equation means, in words: "carbon reacts with oxygen to form carbon dioxide."

balancing an equation

In an equation the number of atoms of any one kind must be the same on both sides of the equation. This provision ensures that the law of conservation of mass is obeyed in every chemical equation, corresponding to the fact that this law is obeyed in the actual reaction being described. For example, the decomposition of water that occurs when an electric current is passed through a water sample (a process called electrolysis) might be written in words as

$$\text{Water} \longrightarrow \text{hydrogen} + \text{oxygen}$$

Using the formulas for these substances, we might write

unbalanced equation
$$H_2O \longrightarrow H_2 + O_2$$

Here two atoms of oxygen are shown on the right-hand side, only one atom on the left; in chemical terms, the equation is *unbalanced*. We cannot help matters by simply writing O instead of O_2 on the right, for

CHEMICAL CALCULATIONS

we know that gaseous oxygen has the formula O_2. Nor is it legitimate to write a subscript "2" after the O in H_2O, for H_2O_2 is the formula for hydrogen peroxide, not water. The remedy is to show two molecules of H_2O on the left, giving two molecules of hydrogen and one of oxygen:

$$2H_2O \longrightarrow 2H_2 + O_2 \qquad \text{balanced equation}$$

Now the equation is *balanced*, for there are two O atoms and four H atoms on each side. Note that a number placed in front of a formula multiplies everything in the formula, whereas a subscript applies only to the atom immediately in front of it.

Balancing an equation consists in making the number of atoms of each kind the same on both sides by writing the proper numbers in front of various formulas. For simple equations, balancing involves no more than careful inspection. Let us consider three example (Fig. 22.1):

in a balanced equation, the number of atoms of each kind is the same on both sides

1 Mercury heated in oxygen is changed to red mercuric oxide:

Mercury + oxygen $\longrightarrow$ mercuric oxide

Unbalanced equation:

$Hg + O_2 \longrightarrow HgO$

Balanced equation:

$2Hg + O_2 \longrightarrow 2HgO$

2 Zinc added to a solution of hydrochloric acid liberates hydrogen and forms a solution of zinc chloride:

Zinc + hydrochloric acid $\longrightarrow$ hydrogen + zinc chloride

Unbalanced equation:

$Zn + HCl \longrightarrow H_2 + ZnCl_2$

Balanced equation:

$Zn + 2HCl \longrightarrow H_2 + ZnCl_2$

3 When hydrogen sulfide is burned in oxygen, water and sulfur dioxide are formed:

Hydrogen sulfide + oxygen $\longrightarrow$ water + sulfur dioxide

Unbalanced equation:

$H_2S + O \longrightarrow H_2O + SO_2$

Balanced equation:

$$2H_2S + 3O_2 \longrightarrow 2H_2O + 2SO_2$$

Two facts about equations need emphasis: (1) An equation shows simply what chemical change has taken place; it tells nothing about the conditions (temperature, pressure, etc.) that are necessary to bring about the change. (2) An equation is not a means of predicting what chemical change *will take place*, but is a concise summary of a change that *has taken place*. To write an equation, the formulas of all products as well as those of all the substances that react must be known.

(1)

$$2Hg + O_2 \longrightarrow 2HgO$$

(2)

$$Zn + 2HCl \longrightarrow H_2 + ZnCl_2$$

(3)

$$2H_2S + 3O_2 \longrightarrow 2H_2O + 2SO_2$$

22.1 Schematic diagrams of three chemical reactions described in the text.

An equation in itself is not a means of predicting chemical changes, but if a chemist is familiar enough with the chemical behavior of an element, he can often express by means of an equation his inference about the reaction of the element in a new set of circumstances.

STOICHIOMETRY

Now that we know something of the atomic basis of chemistry, it is appropriate to take a look at its large-scale aspects. Our first concern

CHEMICAL CALCULATIONS

will be mass relationships: How much of element A is present in compound B? How much of compound C must be added to how much of compound D in order that their reaction yield a given amount of substance E? Questions of this kind fall into the province of *stoichiometry*, a branch of chemistry of major importance in both industry and the laboratory.

stoichiometry refers to mass and volume relationships in chemical reactions

avogadro's number

No matter how small may be the samples of matter involved in a chemical process, they contain enormous numbers of atoms. However, if the numbers of atoms in two samples of different elements are the same, the ratio of the masses of the samples is equal to the ratio of the atomic masses of the corresponding elements. A sample of helium that contains as many helium atoms as there are oxygen atoms in a sample of oxygen always has a mass

$$\frac{4.003}{16.00} = 0.2502$$

of that of the oxygen, while a sample of lead whose atoms are equal in number to the oxygen atoms has a mass

$$\frac{207.2}{16.00} = 12.95$$

times greater, since the atomic masses of helium, oxygen, and lead are, respectively, 4.003, 16.00, and 207.2 amu. (We recall that 1 amu = 1 atomic mass unit = 1.66×10^{-27} kg.)

Since it is impossible to count the atoms in a sample directly, while determining its mass is easy, the above argument is most useful stated in reverse: When the masses of two samples of different elements are in proportion to their atomic masses, they contain the same number of atoms. Because of this, it is convenient to define a quantity called the *gram atom*. **A gram atom of any element is that amount of it whose mass is equal to its atomic mass expressed in grams.** A gram atom of oxygen is 16.00 g; a gram atom of helium is 4.003 g; and a gram atom of lead is 207.2 g.

the gram atom

A gram atom of any element contains the same number of atoms as a gram atom of any other element. This number is a constant of nature known as *Avogadro's number* (N_0), whose value is

$$N_0 = 6.02 \times 10^{23} \text{ atoms/g atom}$$

avogadro's number is the number of atoms in a gram atom of any element

A knowledge of Avogadro's number permits us to calculate the number of atoms in a sample of any element. For example, we might wish to know how many atoms there are in a pound of copper. There are two steps in this calculation:

(1) We begin by noting that a weight of 1 lb is equivalent to a mass of 454 g and the atomic mass of copper is 63.54 amu. Hence a gram atom of copper has a mass of 63.54 g, and

$$\text{Gram atoms of Cu} = \frac{\text{mass of Cu}}{\text{atomic mass of Cu}}$$

$$\frac{454 \text{ g}}{63.54 \text{ g/g atom}} = 7.15 \text{ g atoms}$$

There are 7.15 g atoms of copper per pound.

(2) We now multiply the number of gram atoms by the number of atoms per gram atom, which is N_0, to find the number of atoms:

$$\text{Atoms of Cu} = \text{gram atoms of Cu} \times \frac{\text{atoms}}{\text{gram atom}}$$

$$= 7.15 \text{ g atoms} \times 6.02 \times 10^{23} \frac{\text{atoms}}{\text{g atom}}$$

$$= 4.3 \times 10^{24} \text{ atoms}$$

There are 4.3×10^{24} atoms in a pound of copper. This calculation is shown schematically in Fig. 22.2.

the mole

formula mass

The *formula mass* of a compound is the sum of the atomic masses of its constituent elements, each multiplied by the number of times it appears in the formula of that compound. Thus the formula mass of sodium chloride (NaCl) is the atomic mass of Na, which is 22.99 amu, plus that of Cl, which is 35.45 amu, for a total of 58.44 amu.

The formula mass of the more complex compound sodium sulfate (Na_2SO_4) may be calculated as follows:

$$
\begin{aligned}
2\text{Na} &= 2 \times 22.99 = 45.98 \\
1\text{S} &= 1 \times 32.06 = 32.06 \\
4\text{O} &= 4 \times 16.00 = 64.00 \\
&\text{Total} = \overline{142.04} \text{ amu}
\end{aligned}
$$

The formula mass of sodium sulfate is 142.04 amu.

The reason for speaking of formula mass rather than of molecular mass is that individual molecules do not exist as such in most solids. The notion of formula mass can be employed with *all* compounds, while that of molecular mass has meaning only for compounds that occur as aggregates of individual molecules. Water is an example of the latter type of compound, since it consists of H_2O molecules in its solid, liquid, and gaseous states. The mass of an H_2O molecule is

$$
\begin{aligned}
2\text{H} &= 2 \times 1.008 = 2.016 \\
1\text{O} &= 1 \times 16.00 = 16.00 \\
&\text{Total} = \overline{18.02} \text{ amu}
\end{aligned}
$$

CHEMICAL CALCULATIONS

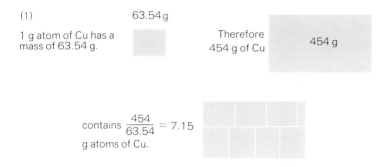

22.2 *How the number of atoms in 1 lb (454 g) of copper is calculated.*
(1) First the number of g atoms is found.
(2) Then the total number of atoms present is determined.

and this is the same as the formula mass of H_2O, of course. However, it is not correct to refer to the molecular mass of NaCl (although this is often done), because NaCl crystals consist of continuous lattices of Na^+ and Cl^- ions and not of NaCl molecules.

The useful concept of the gram atom can be extended to compounds by defining the *gram mole*, usually called the *mole*. **A mole of any compound is that amount of it whose mass is equal to its formula mass expressed in grams.** A mole of sodium chloride has a mass of 58.44 g, a mole of sodium sulfate has a mass of 142.04 g, and a mole of water has a mass of 18.02 g.

the mole

A mole of any substance contains the same number of formula units as a mole of any other substance, and this is the same as the number of atoms in a gram atom. Hence Avogadro's number N_0 may also be written

$$N_0 = 6.02 \times 10^{23} \text{ formula units/mole}$$

avogadro's number is also the number of formula units in a mole

In the case of sodium chloride, a formula unit is one atom of Na and one of Cl; in the case of sodium sulfate, a formula unit is two atoms of Na, one of S, and four of O; in the case of water, a formula unit is two atoms of H and one of O. Only for water is the formula unit the same as a molecule, since the other compounds do not normally occur as individual molecules.

As an example of a calculation facilitated by the use of the mole we shall find the percentage by mass of the sodium that is incorporated in the compound sodium sulfate (Na_2SO_4). According to its formula, each mole of Na_2SO_4 contains 2 g atoms of Na, namely 45.98 g. Hence a mole of Na_2SO_4, which is 142.04 g since the formula mass of Na_2SO_4 is 142.04, contains

$$\text{Percent sodium} = \frac{\text{mass of Na per mole}}{\text{total mass per mole}} \times 100$$

$$= \frac{45.98 \text{ g}}{142.04 \text{ g}} \times 100$$

$$= 32.4$$

Once we know this percentage, of course, we can immediately establish the mass of sodium in any given quantity of sodium sulfate. Thus 2,000 kg of sodium sulfate contains

$$0.324 \times 2{,}000 \text{ kg} = 648 \text{ kg of sodium}$$

molarity

It is often more convenient to describe the concentration of a solution in terms of the number of moles of solute per liter it contains instead of in terms of the number of grams of solute per liter. (As discussed in Chap. 24, when a solid or a gas is dissolved in a liquid, the liquid is called the *solvent*, and the solid or gas is called the *solute*. When a liquid is dissolved in another liquid, the one present in the larger quantity is considered the solvent. The *liter*, equal to 1,000 cm^3, or 10^{-3} m^3, is a metric unit of volume; it is a little larger than a quart.)

solvent and solute

molarity of a solution refers to the concentration of solute present

The *molarity* of a solution is the number of moles of solute per liter of solution and is designated by M. Thus a solution that contains 2 moles of NaCl per liter would be called $2M$ NaCl, and one that contains 0.3 moles of HCl (hydrochloric acid) per liter would be called $0.3M$ HCl.

It is not necessary that there actually be exactly 1 liter of a solution described in this way: the molarity of a solution refers only to the concentration of the solute, not to the total amount present. To prepare 0.4 liter of $2M$ NaCl, we dissolve 0.8 mole of NaCl in enough water to make 0.4 liter of solution. Since the formula mass of NaCl is $22.99 + 35.45 = 58.44$, 1 mole of NaCl has a mass of 58.44 g, and 0.8 mole has a mass of 46.75 g.

An advantage of designating a solution by its molarity is that we can then obtain a given number of moles of the solute simply by measuring out the equivalent volume of solution, which is usually easier and quicker than first figuring out the formula mass of the solute and then weighing out the required quantity. If we should require 0.1 mole of NaCl, for instance, all we need do is pour out 50 ml (1 ml = 1 milliliter = 10^{-3} liter) of $2M$ NaCl solution, since

$$\frac{0.1 \text{ mole}}{2 \text{ moles/liter}} = 0.05 \text{ liter} = 50 \text{ ml}$$

gas volumes

According to the kinetic theory of gases (Chap. 8), equal volumes of all gases, under the same conditions of pressure and temperature, contain the same number of molecules. Careful measurements show that, at 0°C and atmospheric pressure, 1 mole of any gas occupies a volume of exactly 22.4 liters. This fact makes it possible to deal quantitatively with gas volumes in chemical reactions.

one mole of any gas occupies 22.4 liters at 0°C and atmospheric pressure

As an illustration, let us compute how much potassium chlorate ($KClO_3$) must be decomposed to yield exactly 1 liter of oxygen at 0°C and atmospheric pressure. When $KClO_3$, a white solid, is heated, it turns into KCl, another white solid, and evolves oxygen as a gas. The balanced equation of this reaction is

$$2KClO_3 \longrightarrow 2KCl + 3O_2$$

In terms of moles, this equation states that

2 moles of $KClO_3 \longrightarrow$ 2 moles of KCl + 3 moles of O_2

Dividing through by 3,

$\frac{2}{3}$ mole of $KClO_3 \longrightarrow \frac{2}{3}$ mole of KCl + 1 mole of O_2

To obtain 1 mole of O_2, we must decompose $\frac{2}{3}$ mole of $KClO_3$. But we want 1 liter of O_2, whereas 1 mole of O_2 occupies 22.4 liters. Hence we must decompose $\frac{1}{22.4}$ of that amount of $KClO_3$ which yields 1 mole of O_2,

$$\frac{1}{22.4} \times \frac{2}{3} \text{ mole of } KClO_3 = 0.0298 \text{ mole of } KClO_3$$

What mass of $KClO_3$ does 0.0298 mole represent? First we determine the formula mass of $KClO_3$:

$$1 K = 1 \times 39.10 = 39.10$$
$$1 Cl = 1 \times 35.45 = 35.45$$
$$3 O = 3 \times 16.00 = 48.00$$
$$\text{Total} = 122.55$$

The formula mass of $KClO_3$ is 122.55, and a mole of this compound has a mass of 122.55 g. Our conclusion, then, is that

$$0.0298 \times 122.55 \text{ g} = 3.65 \text{ g}$$

of potassium chlorate will yield 1 liter of oxygen at 0°C and atmospheric pressure when it is decomposed.

the universal gas constant

equal volumes of gases at the same temperature and pressure contain equal numbers of molecules

Molecules are so small that the space they occupy in a gas sample is minute compared with the volume of the sample. For this reason, equal volumes of all gases, under the same conditions of temperature and pressure, contain equal numbers of molecules. At 0°C and atmospheric pressure, 1 mole of any gas occupies a volume of 22.4 liters, which is 22.4×10^3 cm^3. This fact makes it possible to obtain a numerical value for the constant factor in the ideal-gas law (Chap. 7),

$$\frac{pV}{T} = \text{const}$$

Let us evaluate pV/T for N moles of an ideal gas at 0°C and atmospheric pressure. We have

p = atmospheric pressure = 10.13 newtons/cm^2

$V = 22.4 \times 10^3 \dfrac{\text{cm}^3}{\text{mole}} \times N$ moles

$\quad = 2.24 \times 10^4 \, N$ cm^3

$T = 0°\text{C} = 273°\text{K}$

and so

$$\frac{pV}{T} = NR$$

universal gas constant

where R, the *universal gas constant*, has the value

$$R = \frac{(10.13 \text{ newtons/cm}^2) \times 2.24 \times 10^4 \text{ cm}^3}{273°\text{K}}$$

$\quad = 8.31 \times 10^3$ joules/mole-°K

$\quad = 1.99 \times 10^{-3}$ kcal/mole-°K

ideal gas law

The ideal-gas law is often written in the form

$$pV = NRT$$

where N is the number of moles of gas present.

GLOSSARY

The *valence* of an element or atomic group refers to its combining ability. An atom that donates n electrons to a chemical bond is said to have a valence of $+n$, and an atom that receives m electrons from a chemical bond is said to have a valence of $-m$. Metals have positive valences,

CHEMICAL CALCULATIONS

nonmetals negative valences. In covalent compounds the valence of an element or atomic group is the number of electron pairs that are shared.

A *chemical equation* is a shorthand method of expressing the results of a chemical change; it includes the formulas of all the substances entering the reaction on the left-hand side of an arrow and the formulas of all the products on the right-hand side. A *balanced equation* is one in which the number of atoms of each kind is the same on both sides of the arrow.

A *gram atom* of an element is that amount of it whose mass is equal to its atomic mass expressed in grams. The *formula mass* of a compound is the sum of the atomic masses of its constituent elements, each multiplied by the number of times it appears in the formula of that compound. A *gram mole* (or just *mole*) of any compound is that amount of it whose mass is equal to its formula mass expressed in grams.

Avogadro's number is the number of atoms in a gram atom of any element; it is also the number of formula units in a gram mole of any compound.

The *molarity* of a solution is the number of moles of a dissolved substance it contains per liter.

EXERCISES

1 From the following formulas, determine the valences of the underlined elements: $\underline{Cd}O$, $\underline{As}H_3$, $\underline{U}O_2$, $\underline{Cr}Cl_3$, $\underline{Pb}(OH)_2$, $\underline{Ni}(NO_3)_2$, $Na_2\underline{Si}O_3$.

2 What are the valences of the underlined atom groups in the formulas that follow? $Na_2\underline{SiO_3}$, $K_2\underline{CrO_4}$, $\underline{NH_4}Cl$, $Cu(\underline{NH_3})_4SO_4$.

3 Quicklime (CaO) is prepared by heating limestone ($CaCO_3$). The gas CO_2 is evolved during the reaction. Write the equation for the process.

4 The rusting of iron is a complex reaction in detail, but it is essentially the oxidation of iron to form ferric oxide (Fe_2O_3). Write the equation for this reaction.

5 When an electric current is passed through a solution of hydrogen chloride, hydrogen is liberated at one electrode and chlorine at the other. Write an equation for the reaction. Would you expect the volumes of hydrogen and chlorine liberated to be equal? Why or why not?

6 When an acetylene flame is used in welding, the intense heat is produced by the burning of acetylene gas (C_2H_2) to form carbon dioxide and water. Write the equation.

7 Which of the following equations are balanced?
 a. $Zn + H_2SO_4 \longrightarrow H_2 + ZnSO_4$

b. $Al + 3O_2 \longrightarrow Al_2O_3$
c. $H_2CO_3 \longrightarrow H_2O + CO_2$
d. $3CO + Fe_2O_3 \longrightarrow 3CO_2 + 2Fe$
e. $N_2 + H_2 \longrightarrow 2NH_3$
f. $6Na + Fe_2O_3 \longrightarrow 2Fe + 3Na_2O$
g. $MnO_2 + 4HCl \longrightarrow MnCl_2 + 2H_2O + Cl_2$

8 Write balanced equations for the following reactions:
   a. Aluminum reacts with ferric oxide (Fe_2O_3) to form iron and aluminum oxide (Al_2O_3).
   b. Aluminum reacts with hydrochloric acid solution (HCl) to liberate hydrogen and form a solution of aluminum chloride ($AlCl_3$).
   c. Cesium reacts with bromine gas to form cesium bromide (CsBr).
   d. Potassium chlorate ($KClO_3$) is decomposed into potassium chloride (KCl) and oxygen.

9 Write balanced equations for the following reactions:
   a. Hydrogen and oxygen combine to form water.
   b. Carbon burns in air to form carbon monoxide (CO).
   c. Sulfur trioxide (SO_3) combines with water to form sulfuric acid (H_2SO_4).
   d. Potassium and sulfur combine to form potassium sulfide (K_2S).
   e. Barium reacts with water to liberate hydrogen.

10 What is the percentage by weight of the oxygen in sugar, $C_{12}H_{22}O_{11}$?

11 How many atoms are there in a ton of lead?

12 Find the mass of 2 moles of Fe_2O_3.

13 Does Ag_5SbS_4 or Ag_3AsS_3 contain more silver per ton?

14 How much aluminum is required to react with 10 g of oxygen to form Al_2O_3? How much Al_2O_3 will be formed?

15 Find the number of molecules in 0.0042 g of gaseous nitrogen. What is the volume of the sample at 0°C and atmospheric pressure?

16 Find the mass of a 0.2-mole sample of gaseous oxygen. How many molecules are there in the sample? How many atoms? What volume does the sample have at 0°C and atmospheric pressure?

chapter 23

Eight Elements

EIGHT ELEMENTS

    hydrogen

    oxygen

    nitrogen

    carbon

    silicon

    sulfur

    chlorine

    sodium

THE LANGUAGE OF CHEMISTRY

    naming compounds

    acids and bases

A number of elements are of especial interest to the chemist, either because their behavior is unique, as in the case of hydrogen or carbon, or because they are typical of an entire family of elements, as in the case of chlorine. We shall outline the properties of eight elements here—hydrogen, oxygen, nitrogen, carbon, silicon, sulfur, chlorine, and sodium—in preparation for further work in the fundamentals of chemistry. (The chemical behavior of another element, iron, is described in Chap. 27.) Then we go on to consider the specialized vocabulary of chemistry, which permits relevant information to be presented clearly and concisely. We shall learn no new concepts or principles in this chapter, but instead expand our store of basic chemical knowledge.

EIGHT ELEMENTS

hydrogen

Hydrogen is the lightest of all substances, having a density of only 0.09 kg/m^3 at 0°C and atmospheric pressure. It is a colorless, tasteless, and odorless gas above its boiling point of −253°C (20°K); its freezing point is $-259°C$ ($14°K$). At ordinary temperatures hydrogen molecules contain two atoms each, and its formula is accordingly H_2.

Hydrogen is a fairly abundant element, making up about 1 percent by weight of the earth's crust. Most of it is combined with oxygen in water. In compounds with carbon and oxygen, hydrogen is present in all animal and vegetable tissue. Free hydrogen, uncombined with other elements, is very scarce on the earth; it sometimes occurs as a minor constituent of volcanic gases and of natural gas.

In the laboratory, hydrogen is commonly prepared by the reaction between certain metals and water or acids and by electrolysis, which is the decomposition of water when an electric current is passed through it. A convenient apparatus for the first method is shown in Fig. 23.1. Acid is poured down the thistle tube onto pieces of metal in the flask, and the gas bubbles off steadily as long as both acid and metal are present. Zinc is often used as the metal, and sulfuric acid as the acid; the reaction between these two may be written

$$\underset{\text{Zinc}}{Zn} + \underset{\text{Sulfuric acid (soln)}}{H_2SO_4} \longrightarrow \underset{\text{Hydrogen}}{H_2} + \underset{\text{Zinc sulfate (soln)}}{ZnSO_4}$$

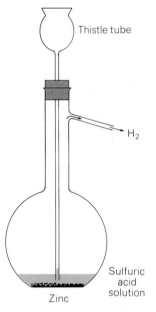

laboratory preparation of hydrogen

23.1 Preparation of hydrogen by the reaction between a metal and an acid. *Here sulfuric acid dissolved in water reacts with zinc to evolve gaseous hydrogen.*

After the reaction, the zinc sulfate is not visible, but may be obtained as a white, crystalline solid by evaporating the remaining liquid.

The second method for preparing hydrogen, electrolysis, involves the passage of an electric current through a fluid, with resulting decomposition of the fluid (Fig. 23.2). Here the fluid is water, made a conductor by addition of a little acid or alkali, and the current passes between small platinum plates (marked "Pt") connected to a battery or generator. Hydrogen bubbles collect at one plate, oxygen bubbles collect at the other, and each gas rises to the top of its tube. As the gases accumulate, the volume of hydrogen remains always twice as great as the volume of oxygen. This reaction may be summarized

$$\underset{\text{Water}}{2H_2O} \longrightarrow \underset{\text{Hydrogen}}{2H_2} + \underset{\text{Oxygen}}{O_2}$$

Such a statement says nothing about how the reaction was carried out, but merely describes the chemical change that has occurred and indicates the relative quantities of the materials undergoing change.

Hydrogen burns readily in air or oxygen with a hot, colorless flame. The gas produced in the flame is water vapor, as may be shown by condensing some of it. This reaction is the reverse of the decomposition of water by electrolysis:

$$\underset{\text{Hydrogen}}{2H_2} + \underset{\text{Oxygen}}{O_2} \longrightarrow \underset{\text{Water}}{2H_2O}$$

A mixture of hydrogen and oxygen will not react at ordinary temperatures, but once the mixture is ignited by the heat of a flame or an electric spark, the reaction generates sufficient heat to keep itself going. If the gases are mixed in the proper proportions, an explosion results, because of the sudden expansion of the mixture as it is heated by the reaction. Miniature explosions produced by bringing a flame near hydrogen-air mixtures in a test tube give a convenient method of testing for the element.

Another compound of hydrogen and oxygen is *hydrogen peroxide* (H_2O_2), an unstable liquid which is commonly used in dilute solution as a disinfectant and bleaching agent. Its formula is written H_2O_2 rather than HO because its molecular mass, determined by comparing the density of its vapor with that of oxygen (and in other ways), is 34 and not 17. In H_2O_2 there is a covalent bond between the two O atoms as well as between each O atom and an H atom. Thus each oxygen atom participates in two covalent bonds, as the single oxygen atom in H_2O does. However, the bonds in H_2O_2 are not as strong as those in either H_2O or O_2 (gaseous oxygen), and so it readily decomposes into them:

hydrogen peroxide

$$\underset{\text{Hydrogen peroxide}}{2H_2O_2} \longrightarrow \underset{\text{Water}}{2H_2O} + \underset{\text{Oxygen}}{O_2}$$

Because light accelerates the decomposition of H_2O_2, it is usually stored in dark bottles.

Hydrogen is an active element at moderately high temperatures, combining directly with a number of other nonmetallic elements besides oxygen and less readily with several of the metals. Thus with nitrogen it forms *ammonia* (NH_3); with chlorine, *hydrogen chloride* (HCl); with calcium, *calcium hydride* (CaH_2). With many metallic oxides hydrogen reacts to form water and the free metal; thus

$$\underset{\text{Copper oxide}}{CuO} + \underset{\text{Hydrogen}}{H_2} \longrightarrow \underset{\text{Copper}}{Cu} + \underset{\text{Water}}{H_2O}$$

In a reaction of this sort, where oxygen is removed from combination with a metal, the oxide is said to be *reduced*. Reduction is the opposite process to oxidation.

The low density of hydrogen (7 percent of the density of air) makes it useful for filling high-altitude research balloons, although its flammability is a constant source of danger. The intense heat produced when hydrogen burns in oxygen is made use of in cutting and welding metals with the oxyhydrogen blowtorch. Hydrogen is a principal constituent of artificial gas fuels. The hardening of oils to form solid fats and the production of synthetic ammonia and methyl alcohol are among the other commercial uses of the element.

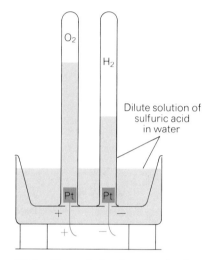

23.2 Electrolysis of water. *An electric current decomposes water into gaseous hydrogen and oxygen. The volume of the hydrogen evolved is twice that of the oxygen, since both are diatomic gases and water contains twice as many hydrogen atoms as oxygen atoms. A trace of sulfuric acid is used to enable the water to conduct electricity.*

oxygen

Oxygen, under ordinary conditions, is a colorless, odorless, tasteless gas. If cooled sufficiently, it condenses to a clear blue liquid with the property of being strongly attracted to a magnet. The boiling point of liquid oxygen is $-183°C$, and its freezing point is $-218°C$. Oxygen is chemically active even in the liquid state, its increased density offsetting the very low temperature. Iron, for example, will burn so vigorously in liquid oxygen that the metal is melted by the heat of the reaction, in spite of the intensely cold liquid surrounding it.

Oxygen is by far the most abundant of the elements that make up the earth's crust, its total amount (by weight) being nearly equal to that of all the rest put together. Most of the oxygen is in compounds, compounds which are the chief constituents of rocks, soil, and living things. Water is a compound of hydrogen and oxygen. The free element is one of the chief constituents of the atmosphere.

Air owes its ability to support combustion to the free oxygen which it contains; it cannot support combustion as well as pure oxygen because the element is so diluted with inactive gases. (Air is about one-fifth oxygen, four-fifths nitrogen, with small amounts of other gases.) That air is a mixture (or solution) of oxygen and nitrogen rather than a compound may be readily shown by liquefying it and allowing the liquid to boil. Boiling commences near the boiling point of nitrogen ($-196°C$), and the vapor that comes off first consists mostly of this element; as boiling proceeds the temperature rises, and toward the end practically pure oxygen is left in the liquid. The ease with which oxygen can be separated from nitrogen by letting liquid air boil makes this a convenient method for preparing oxygen for commercial use.

When oxygen combines chemically with another substance, the process is called *oxidation*, and the other substance is said to be *oxidized*. Rapid oxidation accompanied by the liberation of noticeable heat and light is the process of *combustion*, or burning. In the experiments of Lavoisier, tin and mercury oxidized slowly when heated. A lighted candle oxidizes rapidly in air, still more rapidly in pure oxygen. Slow oxidation is involved in many familiar processes, such as the rusting of iron, the decay of wood, the hardening of paint. The energy to maintain life comes from the slow

oxygen is the most abundant element in the earth's crust

oxidation is the combination of oxygen with another substance

Anhydrous ammonia is sometimes applied directly to the soil to supply it with nitrogen. Soil shown here is chernozem in the American Middle West (see page 605).

Rough diamonds from South Africa.

oxidation of food in our bodies by oxygen breathed in through the lungs and transported by the bloodstream.

A substance formed by the union of another element with oxygen is called an *oxide*. The white powder that Lavoisier obtained by heating tin is tin oxide. Rust is largely iron oxide. In general, oxides of metals are solids. Oxides of other elements may be solid, liquid, or gaseous. The oxide of sulfur is an evil-smelling gas (the odor of burning sulfur) called sulfur dioxide; carbon forms two gaseous oxides, called carbon monoxide and carbon dioxide; the oxide of silicon is found in nature as the solid called quartz, the chief constituent of ordinary sand; the oxide of hydrogen is water. Oxides of nearly all the elements can be prepared, most of them simply by heating the elements with oxygen. A few oxides (mercuric oxide, lead oxide, barium peroxide) are easily decomposed by heating, whereas others, such as lime (calcium oxide), are not decomposed even at the 3000°C temperature of an electric arc.

oxides

nitrogen

Nitrogen, the most abundant constituent of the atmosphere, is (like oxygen) a colorless, odorless, tasteless gas under ordinary conditions. Nitrogen boils at $-196°C$ and freezes at $-210°C$. The formula of gaseous nitrogen is N_2, and the atoms in each molecule are tightly held together by three covalent bonds. The bond energy in N_2 is accordingly quite high, namely 9.8 ev, as compared with 5.1 ev in O_2 in which there are two covalent bonds per molecule. Thus almost twice as much work must be done to break apart an N_2 molecule as is required to break apart

nitrogen molecules are very stable because three covalent bonds hold each of them together

an O_2 molecule. Nitrogen molecules are very stable, as we might expect, and N_2 is a comparatively unreactive gas.

The *amino acids* of which proteins are composed are nitrogen compounds, and since the solid flesh of our bodies is largely protein, nitrogen is one of the most important elements to man. The ultimate source of all our protein material is plants, although much of it comes to us secondhand in such animal proteins as those in meat, eggs, and milk. Plants in turn manufacture their proteins from simpler nitrogen compounds which enter their roots from the soil in which they grow.

nitrogen compounds are essential for plant and animal life

Green plants are unable to utilize the stable molecules of free nitrogen in the air around them; all their nitrogen, and therefore all the nitrogen that goes into animal bodies as well, comes from nitrogen compounds in the soil. All the nitrogen molecules we breathe can do us no good either, for the atoms in these molecules of the free element are united by bonds which our body processes are unable to break. Like a shipwrecked mariner surrounded by water but dying of thirst, mankind is surrounded by an ocean of nitrogen but would perish except for the combined nitrogen which plants can absorb through their roots.

The formation of plant and animal proteins continually removes nitrogen compounds from the soil. Just as continually fixed nitrogen is returned to the soil by the decay of animal excrement and of dead plants and animals, the nitrogen of proteins being converted by decay into ammonia and ammonium salts which are then oxidized to nitrates by soil bacteria. But the replenishment is never complete: either in decay or during combustion, some of the nitrogen atoms of the proteins manage to join together into molecules and escape into the air as nitrogen gas. Some nitrogen is also lost permanently from the soil by solution of nitrates and ammonium salts in streams and rainwash, and by bacteria which decompose nitrates into free nitrogen.

atmospheric nitrogen is fixed in nature by bacteria and during thunderstorms

Nature makes good these losses in two ways: another kind of soil bacteria, the "nitrogen-fixing" bacteria, have the unique ability to break down the stable nitrogen molecule and to manufacture compounds from the atoms; and electric discharges during thunderstorms cause some combination of atmospheric nitrogen and oxygen into nitrogen oxides, which are carried to the soil in solution in rain water. So in nature nitrogen goes through a continuous, rather complicated cycle (Fig. 23.3), which keeps the amount of fixed nitrogen in the soil approximately constant.

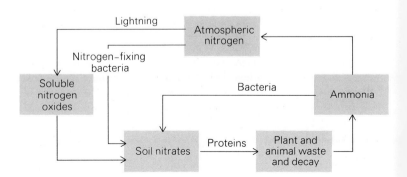

23.3 The nitrogen cycle.

Civilized man drastically disturbs this natural cycle. Much of the protein material that goes into his body is never returned to the soil, and his habit of using plant material for fires greatly accelerates the conversion of combined nitrogen into unavailable free nitrogen. One of the problems of modern nations is to keep enough nitrogen supplied to agricultural lands to make up for this steady depletion. The problem is partially solved by the use of manure and by planting crops on which nitrogen-fixing bacteria can grow, but these expedients cannot supply all the needed combined nitrogen.

One place to look for additional supplies of combined nitrogen should seemingly be in rocks: compounds of almost all other elements are found as mineral deposits, and there is no immediately obvious reason why nitrogen should be an exception. The difficulty is that nearly all inorganic nitrogen compounds are soluble in water. They cannot accumulate under ordinary conditions, because rain water, streams, and underground seepage carry them away too rapidly. Only in the world's most arid regions are nitrogen compounds found in appreciable quantities; in only one place, the desert of northern Chile, where rain falls less than once a decade, have deposits formed that are large enough for commercial use. The chief compound here is sodium nitrate ($NaNO_3$, Chile saltpeter), a material useful directly as a fertilizer and easily convertible into nitric acid for the manufacture of explosives. Exploitation of the Chilean deposits began in the early nineteenth century, and for more than 100 years much of the world's fixed nitrogen came from this little patch of desert.

By the beginning of the twentieth century the need for another source of fixed nitrogen was becoming acute, and scientists in several countries set to work to find one. Some nitrogen compounds could be obtained from coal when it was heated to form coke, but not nearly enough to satisfy industrial requirements. The only possible source seemed to be the atmosphere itself; the great difficulty was to make nitrogen atoms leave their partners and combine with atoms of another element. Several processes were tried, the most successful one being the union of nitrogen with hydrogen in the Haber process (Chap. 26) to form ammonia,

$$N_2 + 3H_2 \longrightarrow 2NH_3$$
Nitrogen Hydrogen   Ammonia

the synthesis of ammonia

The ammonia gas produced may be converted to ammonium salts for fertilizers by combination with sulfuric or nitric acid, or it may be oxidized by atmospheric oxygen to nitric acid for use in making explosives. Much of the productivity of American agriculture is the result of the intensive use of artificial fertilizers.

carbon

Diamond and graphite are two naturally occurring forms of the pure element carbon. Diamond is the hardest known natural substance, clear and colorless when strictly pure, not breaking easily in any direction, and a very poor conductor of electricity. Graphite is soft, opaque, steel

diamond and graphite are both pure carbon but have different crystal structures

Synthetic diamonds were first made a decade ago. The ones shown are intended for industrial applications, and are magnified several times in the photograph.

23.4 The arrangement of carbon atoms in a diamond crystal. *The unit of structure is a tetrahedron consisting of four atoms around a central atom; one of these units is shown linked by colored lines.*

23.5 The arrangement of carbon atoms in the crystal lattice of graphite. *The unit of structure is a hexagonal ring of six atoms; one of these units is shown in color. The layers are held together by weak van der Waals forces.*

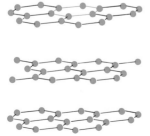

gray to black, composed of tiny flakes that split apart easily, and a fairly good electrical conductor. Ordinary carbon in the form of coke, soot, and charcoal is impure graphite in the form of minute crystals. Both diamond and graphite are extremely resistant to heat; they vaporize at atmospheric pressure at temperatures near 3500°C. Carbon liquefies only at extremely high pressures and temperatures.

That two materials as different in properties as diamond and graphite can be forms of a single element seems at first incredible. That this is so may be proved by burning each in oxygen at temperatures above 700°C; both give carbon dioxide gas as the only product. The differences between the two arise from a difference in crystal structure. The carbon atoms of diamond are arranged in a compact framework in which each atom is surrounded by four others at the corners of a tetrahedron (Fig. 23.4), whereas the carbon atoms of graphite lie in parallel planes, each plane made up of hexagonal rings (Fig. 23.5).

The bonding within each graphite layer is partly covalent and partly metallic in nature. A graphite sample has a substantial electrical conductivity parallel to its layers but a much smaller one across the layers, which agrees with the foregoing picture. The van der Waals forces that bond the graphite layers together are quite weak, in contrast to the strong bonding forces within the layers. The layers consequently can slide past one another readily and are easily flaked apart, which is why graphite is so useful as a lubricant and in pencils. The internuclear separation in a graphite layer is 1.42×10^{-10} m and the spacing between layers is 3.40×10^{-10} m, while in diamond the internuclear spacing is 1.54×10^{-10} m throughout the crystal. The more open structure of graphite is reflected in its density of 2.25 g/cm^3, as compared with 3.51 g/cm^3 for diamond. Many other solids besides carbon, both elements and compounds, can exist in two or more different forms, the properties of each depending on the particular arrangement of particles in its crystal lattice.

Inactive at ordinary temperatures, carbon at high temperatures reacts readily with many substances. It burns in air or oxygen to form the gas *carbon dioxide* (CO_2) if abundant oxygen is available, or the gas *carbon monoxide* (CO) if the oxygen supply is limited. Carbon combines slowly with a few other nonmetallic elements: with sulfur to form the volatile, flammable liquid carbon disulfide (CS_2), with hydrogen to form the gas methane (CH_4) and related compounds. Some metals react with carbon to form solids called *carbides*, such as calcium carbide (CaC_2) and iron carbide (Fe_3C). Like hydrogen, hot carbon reduces the oxides of many metals; for instance,

$$\text{ZnO} + \text{C} \longrightarrow \text{Zn} + \text{CO}$$
Zinc oxide   Carbon   Zinc   Carbon monoxide

The two oxides of carbon, carbon monoxide and carbon dioxide, are colorless, odorless, tasteless gases. One or the other, depending on the amount of oxygen available, is formed, not only from the burning of carbon, but from the burning of any carbon compound. The slow oxidation of carbon compounds in the bodies of animals produces carbon dioxide,

The zone refining of a silicon rod. The rod is passed through an induction coil which melts a segment of it, and as the rod moves, impurities stay in the molten zone and are thereby carried to the end of the rod. Less than 1 part of impurity in 10^{10} remains.

which is exhaled from the lungs. Carbon monoxide is flammable and forms an important constituent of artificial gas fuels. It is also a deadly poison, especially dangerous because it has no odor. Carbon dioxide is heavy, nonpoisonous, and neither flammable nor able to support combustion. Dissolved in water under pressure, it forms ordinary soda water. It is the gas responsible for the "rising" of bread and cakes during baking. Solid carbon dioxide, Dry Ice, is sometimes used in refrigeration.

Carbon is a relatively scarce element in the earth's crust, making up only about 0.03 percent by weight of the crust's materials. But its importance to humanity is out of all proportion to its abundance. In the form of coal, carbon is an important industrial fuel. Coal and coke are indispensable in winning many of the common metals from their ores. The compound carbon dioxide is an important minor constituent of air, since plant growth depends on its presence. Combined with calcium and oxygen in calcium carbonate ($CaCO_3$), carbon is a constituent of the common and useful rock limestone. Compounds of carbon with hydrogen (hydrocarbons) make up natural gas, gasoline, and lubricating oils. More complex carbon compounds are the chief constituents of our bodies, of the food we eat, of the clothes we wear, of the wood from which our houses are built. Artificially produced carbon compounds include an endless variety of dyes, perfumes, explosives, drugs, and plastics. In the number, variety, and importance of its compounds, carbon outranks all other elements. Chapter 28 is devoted exclusively to the chemistry of carbon compounds.

silicon

Next to oxygen, the most abundant element in the earth's crust is silicon. Silicon melts at 1410°C and boils at 2355°C. Silicon never occurs free in nature, and the pure element is a rarity even in chemical laboratories, but its compounds make up some 87 percent of the rocks and soil which compose the earth's solid outer portion. In the chemistry of naturally occurring inorganic materials, silicon plays the same sort of central role that carbon plays in the chemistry of living things.

most of the rocks and soil of the earth's crust contain silicon

Silicon is just below carbon in the fourth group of the periodic table. Its atoms, therefore, should be somewhat larger than carbon atoms and should have similar outer shells of four electrons. Silicon should form compounds in which it has a valence of 4, and in these compounds its behavior should be less actively nonmetallic than that of carbon. These general predictions are fulfilled, but in detail the chemistry of silicon does not resemble that of carbon as closely as we might expect.

Free silicon finds commercial use as a semiconductor and in the manufacture of hard, resistant steels. It is made by reducing the dioxide (SiO_2) at high temperatures with active metals or carbon. Two forms may be prepared: an "amorphous" brown powder probably made up of minute crystals, and the more common gray, crystalline solid. The latter has a structure like that of diamond, with each Si atom linked to four others. It is hard enough to scratch glass and has a high melting point and boiling point. If an excess of carbon is used in the reduction of SiO_2, the liberated silicon combines with carbon to form the important abrasive *carborundum* (SiC, silicon carbide); this compound has a diamondlike structure, with Si and C atoms alternating, and has a hardness only slightly less than that of diamond itself.

silicon carbide is an extremely hard substance

Like carbon, silicon is inactive at room temperatures but combines directly with active metals and nonmetals at moderately high temperatures. Again like carbon, silicon enters chemical combination by sharing the four outer electrons of its atoms and nearly always shows a valence of 4. It forms two oxides (SiO and SiO_2) analogous to CO and CO_2; a tetrachloride ($SiCl_4$), analogous to CCl_4; and compounds with metals and oxygen called *silicates*, some of which (for instance Na_2SiO_3) resemble in their formulas the carbonates (Na_2CO_3). That silicon atoms can share

silicates

electrons with each other is shown by their ability to fit into the diamond structure; the bond is not so strong as the bond between carbon atoms, however, and silicon atoms show only in slight degree the ability of carbon atoms to form chains.

Silicon dioxide, commonly called *silica*, resembles CO_2 in its formula but in few other properties. CO_2 is a gas, moderately soluble in water; SiO_2 is a hard, insoluble solid with a high melting point. Silica is extremely common in nature, occurring when pure as clear crystals of *quartz* or rock crystal, and with minor impurities as amethyst, agate, onyx, flint, jasper, opal, and so on. It is the chief component of most sands and sandstones. Clear quartz is transparent not only to visible light but to much of the ultraviolet as well, a property that makes it valuable for optical instruments. Heated above $1700°C$ silica becomes a viscous liquid, which on rapid cooling solidifies to a hard, amorphous solid resembling ordinary glass; since this "fused quartz" has much of the ultraviolet transparency of crystalline quartz and is far cheaper, it has found wide use in instruments for the production and study of ultraviolet radiation. Fused silica is also highly prized for laboratory use for its ability to withstand sudden temperature changes without cracking.

silica is silicon dioxide

Nearly all the earth's silicon is either combined with oxygen in silica or combined with oxygen and one or more metals in the various silicates. In number and variety the silicates will hardly stand comparison with the carbon compounds, but their complexities are sufficient to make the chemistry of silicon an intricate and difficult subject. The structures of silicon compounds are discussed in some detail in Chap. 30.

sulfur

Sulfur, a yellow, odorless, and tasteless solid, is the brimstone of ancient times. It melts at $113°C$ and boils at $445°C$, and has a density about twice that of water. The common occurrence of sulfur near active volcanoes, together with the blue flame and sharp odor produced when it burns, probably explains its long literary association with the subterranean abode of deceased sinners. The free element is also found among the deposits of long extinct volcanoes and in marine deposits associated with gypsum and rock salt. Gypsum is a familiar naturally occurring sulfur compound ($CaSO_4 \cdot 2H_2O$); the sulfides pyrite (FeS_2), galena (PbS), and cinnabar (HgS) are others.

Liquid sulfur has the peculiar property of becoming highly viscous as its temperature is raised toward $200°C$ and then losing viscosity as the boiling point is approached. When sulfur near its boiling point is quickly cooled by pouring into water, it solidifies into a brown, pliable, elastic material called *amorphous* sulfur which is quite different from the familiar yellow, crystalline form.

amorphous sulfur

Chemically, sulfur is a moderately active element. It burns readily in air or oxygen to form *sulfur dioxide* (SO_2), a gas whose strong odor is that often described as "the odor of sulfur." Sulfur unites with many metals on heating to form compounds called *sulfides*, for example, copper sulfide (CuS) and silver sulfide (Ag_2S). With active metals this reaction

Liquid sulfur emerging from a pipe during the refining process.

may liberate considerable energy, as in the zinc-sulfur reaction described in Chap. 15. With hydrogen, sulfur combines to form hydrogen sulfide (H_2S), a gas whose odor is like that of rotten eggs. Other common sulfur compounds contain both metals and oxygen: "Epsom salts," magnesium sulfate ($MgSO_4 \cdot 7H_2O$); "blue vitriol," copper sulfate ($CuSO_4 \cdot 5H_2O$); gypsum, calcium sulfate ($CaSO_4 \cdot 2H_2O$). The H_2O in these formulas indicates water, loosely held in the solid crystals, which may be removed on mild heating.

Industrially the most important sulfur compound is sulfuric acid (H_2SO_4), a heavy, colorless, viscous liquid that is highly corrosive and dissolves readily in water with evolution of much heat. A solution of the acid is a common laboratory reagent, and among its countless industrial uses the most important are in petroleum refining and in the manufacture of fertilizers and explosives.

chlorine

Chlorine is a poisonous, greenish-yellow gas with a disagreeable odor. Chlorine melts at $-102°C$ and boils at $-35°C$, and its density at $0°C$

and atmospheric pressure is 3.16 kg/m³. At ordinary temperatures chlorine molecules contain two atoms each, and so its chemical formula is Cl_2.

Chlorine is far too active to exist free in nature. Its most abundant compound is sodium chloride (NaCl), familiar to us as ordinary salt, which occurs in solution in the ocean and in salt lakes and in solid form as deposits of rock salt. In the laboratory, chlorine may be prepared from sodium chloride by heating it with sulfuric acid and manganese dioxide, while industrially the gas is prepared by the electrolysis of a concentrated sodium chloride solution.

Chemically, chlorine is one of the most active of all elements, combining directly with all metallic elements and many nonmetallic elements. Oxygen is one of the few with which it will not react, but unstable oxides can be prepared indirectly. The more active metals burn brilliantly in chlorine just as they do in oxygen, forming *chlorides* instead of oxides. Thus copper foil burns to form *copper chloride* ($CuCl_2$) and sodium burns with an intense yellow flame to form white clouds of tiny *sodium chloride* crystals. This spectacular reaction between the active gas chlorine and the active metal sodium, both highly poisonous, to produce the harmless substance salt is one of the most impressive examples of the profound changes in properties that chemical reactions can bring about.

A jet of hydrogen burns in chlorine as readily as in oxygen, forming a colorless gas with a sharp, unpleasant odor called *hydrogen chloride* (HCl). A mixture of hydrogen and chlorine will react at room temperature, provided that the mixture is exposed to light; if the mixture is prepared

chlorine is an extremely active element

Chlorine being produced by the mercury cell process at a plant of PPG Industries in Louisiana.

in darkness and then brightly illuminated, the reaction is explosive. For either this *photochemical* ("caused by light") reaction or the burning of hydrogen, the chemical change may be summarized

photochemical reactions are induced by the action of light

$$\underset{\text{Hydrogen}}{H_2} + \underset{\text{Chlorine}}{Cl_2} \longrightarrow \underset{\text{Hydrogen chloride}}{2HCl}$$

A solution of hydrogen chloride in water is a common strong acid called *hydrochloric acid* (the "muriatic acid" of commerce).

Chlorine reacts with many colored compounds of carbon, changing them to colorless compounds. This property accounts for the extensive use of the element in bleaching. It is used also as a disinfectant and in the preparation of dyes, explosives, and poison gases.

sodium

Sodium is a silver-gray solid that rapidly tarnishes when exposed to air. It melts at 98°C and boils at 892°C. It is one of the more abundant elements, making up nearly 2.5 percent of the earth's crust, but its extreme chemical activity prevents it from occurring free in nature. Its compounds are widely distributed in rocks, soil, and, in solution, in bodies of water.

Sodium is so soft that it can be cut like cheese; so light that it will float on water; so easily corroded by air and water that it must be kept under oil, lest a violent reaction occur. By ordinary standards these properties suggest that sodium should be called anything but a metal, yet to the chemist sodium is strongly metallic in every way. It has a silvery luster on freshly cut surfaces, and it is an excellent conductor of electricity—two characteristically metallic properties. Its chemical behavior shows in an exaggerated form certain properties common to all the more active metals. Among these are:

1 Its ability to burn brightly, with the evolution of much heat, in both oxygen [forming *sodium peroxide* (Na_2O_2)] and chlorine. Some nonmetallic elements will burn in oxygen or chlorine, a few in both, but the energy liberated is in general not so great as for the metals.
2 Its ability to liberate hydrogen from acids. Many common metals, for instance zinc, iron, and aluminum, liberate hydrogen slowly from acids, but with sodium the reaction is violent and the liberation of gas exceedingly rapid.
3 Its ability to liberate hydrogen from water. This reaction produces, in addition to hydrogen, a solution of a compound called *sodium hydroxide* (NaOH):

sodium combines directly with water to liberate hydrogen

$$\underset{\text{Sodium}}{2Na} + \underset{\text{Water}}{2H_2O} \longrightarrow \underset{\text{Hydrogen}}{H_2} + \underset{\text{Sodium hydroxide (soln)}}{2NaOH}$$

Several of the commoner metals, even iron, can be made to react slowly with hot water, but a little chunk of sodium need only be dropped on cold water to start a reaction that sets it skimming over the surface and generates enough heat to melt the metal; an explosion may even occur.

Sodium combines readily with most of the nonmetallic elements, but not in general with other metals. Like hydrogen and carbon, it reduces many metallic oxides to the pure metal. Of the simple compounds of sodium, the more familiar are ordinary salt (NaCl), washing soda or sodium carbonate (Na_2CO_3), baking soda or sodium bicarbonate ($NaHCO_3$), caustic soda or sodium hydroxide (NaOH).

THE LANGUAGE OF CHEMISTRY

Every element has its own special name, but when it is part of a compound this name may undergo a change or may even not be mentioned specifically in the name of the compound. The way in which compounds are named is quite ingenious, and as a prelude to further work in chemistry, it is useful for us to learn something about this special language. We shall find that the order in which the chemical symbols representing the various elements in a compound are placed in the formula for that compound follows a special pattern. The first time we face all the various rules for naming and symbolizing compounds, they may seem complicated, and we must not lose sight of the fact that these rules were devised to bring some order into the complex world of chemistry.

naming compounds

The first rule we consider is straightforward: In a compound containing a metal and one or more elements that are not metals, the name of the metal comes first and its symbol appears first in the formula. Thus common salt is

if a compound contains a metal, its name comes first

Sodium chloride   NaCl

For compounds containing only nonmetallic elements there is no simple rule. If carbon or hydrogen is present, it usually stands first:

Carbon dioxide CO_2
Hydrogen sulfide H_2S
Carbon tetrachloride CCl_4

An exception that has become standard usage is

Ammonia NH_3

If oxygen or chlorine is present, it is usually written last:

Carbon monoxide   CO
Phosphorus trichloride PCl_3
Sulfur dioxide SO_2

In compounds containing more than two elements, oxygen is very often one of the constituents. In such cases oxygen usually does not appear explicitly in the name of the compound and it is last in the formula:

Calcium carbonate	$CaCO_3$
Sodium sulfate	Na_2SO_4

hydroxides contain the OH atom group

An exception occurs in the case of certain compounds containing both hydrogen and oxygen, called hydroxides, in whose formulas hydrogen comes last, for instance,

Sodium hydroxide	$NaOH$
Calcium hydroxide	$Ca(OH)_2$

-ide

Names of compounds made up of only two elements always end in *-ide*, which is used as a suffix to the name of the second element in the name. Thus we have

Hydrogen chloride	HCl
Zinc chloride	$ZnCl_2$
Aluminum chloride	$AlCl_3$
Hydrogen sulfide	H_2S
Iron carbide	Fe_3C

A number of other compounds also have names that end in *-ide*, for instance, the hydroxides, which contain hydrogen and oxygen in the combination OH.

When two or more compounds contain the same pair of elements, they are distinguished by one of two methods:

mono-, di-, tri-, . . .

1 A prefix (*mono-*, *di-*, *tri-*, etc.) may be added to the name of the second element in each, indicating the number of its atoms per molecule:

Carbon monoxide	CO
Carbon dioxide	CO_2
Phosphorus trichloride	PCl_3
Phosphorus pentachloride	PCl_5

-ic and -ous

2 The suffixes *-ic* and *-ous* may be added to the name of the first element, the *-ic* referring to a compound containing more atoms of the second element relative to the first (for iron, the suffixes are added to the Latin name, ferrum, and for copper, to cuprum):

Ferric chloride	$FeCl_3$
Ferrous chloride	$FeCl_2$
Mercuric oxide	HgO
Mercurous oxide	Hg_2O
Cupric sulfide	CuS
Cuprous sulfide	Cu_2S

In general, the second method is used for compounds of a metal with a nonmetal and the first for compounds of two nonmetals, but this rule has several exceptions.

The term *peroxide,* often applied to dioxides, refers properly only to a group of dioxides with the special property of having oxygen-oxygen bonds. Such bonds are relatively weak and peroxides are accordingly unstable at high temperatures. The only common peroxides are

peroxides

Hydrogen peroxide	H_2O_2
Sodium peroxide	Na_2O_2
Barium peroxide	BaO_2

Most compounds with three elements, as we said, contain oxygen, another nonmetallic element, and a metal. Normally, the suffix *-ate* is added to the name of the nonmetal to indicate the presence of oxygen:

-ate

Sodium sulfate	Na_2SO_4
Calcium sulfate	$CaSO_4$
Potassium nitrate	KNO_3
Magnesium carbonate	$MgCO_3$

acids and bases

We shall not attempt rigorous definitions of these important classes of compounds at this point. In general, acids are characterized experimentally by the facts that (1) their solutions have a sour taste, and (2) their solutions will change the color of certain dyes; for example, they will turn blue litmus paper red. Their formulas are characterized by the presence of hydrogen combined with one or more nonmetallic elements. If only one element besides hydrogen is present, the acid is named by adding the prefix *hydro-* and the suffix *-ic* to the name of the second element:

all acids contain hydrogen

Hydrochloric acid (hydrogen chloride)	HCl
Hydrosulfuric acid (hydrogen sulfide)	H_2S

If the acid contains oxygen in addition to another nonmetal, the prefix *hydro-* is omitted from the name:

Sulfuric acid (hydrogen sulfate)	H_2SO_4
Nitric acid (hydrogen nitrate)	HNO_3
Carbonic acid (hydrogen carbonate)	H_2CO_3

The hydroxides, for instance,

Sodium hydroxide   NaOH

bases are hydroxides

violate the rules of naming that have been set up for other compounds and are exceptional also in their chemical properties. They contain the OH atom group. Soluble hydroxides of metals are characterized by (1) their bitter taste, and (2) their ability to reverse the changes in the color of dyes brought about by acids; for instance, a soluble hydroxide will turn red litmus paper blue. In many respects opposite in behavior to acids, the hydroxides of metals are called collectively *bases* even though the term does not appear in their names.

EXERCISES

1 In what ways is the chemical behavior of sodium similar to that of hydrogen? In what ways is the chemical behavior of oxygen similar to that of chlorine?

2 Name the following acids and bases: H_2S, HNO_3, $Ca(OH)_2$, $Al(OH)_3$, NaOH, HF, H_2SO_4, H_3BO_3.

3 Name the following compounds: $MgCO_3$, $HgSO_4$, SiO_2, $AgNO_3$, AgCl, Na_3N, K_2CO_3, NiS, $Al_2(SO_4)_3$, $Zn(NO_3)_2$, UF_6.

4 Match each of the chemical compounds listed below with the appropriate formula from the list at the right.

    a. ammonia HNO_3
    b. ferrous chloride H_2O_2
    c. ferric hydroxide   HCl
    d. nitric acid $FeCl_2$
    e. potassium hydroxide   KOH
    f. hydrochloric acid $Fe(OH)_3$
    g. sodium oxide NH_3
    h. hydrogen peroxide Na_2O

5 Name (*a*) the two oxides of phosphorus, P_2O_3 and P_2O_5; (*b*) the two chlorides of mercury, Hg_2Cl_2 and $HgCl_2$; (*c*) the two hydroxides of iron, $Fe(OH)_2$ and $Fe(OH)_3$.

6 Write formulas for aluminum oxide, magnesium iodide, lithium carbonate, calcium sulfide, sodium nitride, rubidium hydroxide, potassium sulfate, and barium nitrate.

7 Which of the following elements does not react with oxygen to form an oxide? Hydrogen, carbon, sulfur, chlorine, silicon.

8 What element is most abundant in the earth's crust? In the earth's atmosphere?

9 Below silicon in the fourth group of the periodic table is the rare element germanium. Would you expect germanium to be more metallic or less metallic than silicon? Would you expect it to form compounds with hydrogen analogous to the hydrocarbons? Write the formula you would expect for (a) the most common oxide of germanium, (b) the most common chloride, (c) sodium germanate.

chapter 24

Ions and Solutions

ELECTROLYSIS
electrolysis
faraday's laws of electrolysis

SOLUBILITY
solvent and solute
polar liquids

IONS IN SOLUTION
electrolytes
properties of ions
arrhenius
ionic equations

As we have seen, the forces that bind atoms together to form molecules, solids, and liquids are electric in origin. In this chapter we shall further explore the role of electricity in chemical processes, with special emphasis on the behavior of ions in solution.

ELECTROLYSIS

electrolysis

Electricity is conducted through gases and nonmetallic liquids by a mechanism quite different from that characteristic of metals. The current almost always consists not of moving electrons but of moving *ions*. Ions, as we have mentioned before, are electrically charged atoms or groups of atoms—structures resembling ordinary atoms and molecules but possessing either too few or too many electrons to neutralize the positive nuclear charges present. Electrical conduction through a gas or liquid usually involves the movement of both positive and negative ions.

electric currents in gases and nonmetallic liquids consist of moving ions

Ionic compounds like ordinary salt are good conductors when liquefied. These compounds, as we have seen, at ordinary temperatures are solids with crystal structures made up of positive and negative ions. The solids are nonconductors, since their ions are held firmly in position, but in the liquid state the ions are free to move about and hence are free to conduct electricity from one electrode to another.

Molten salt is made up of the two ions Na^+ (sodium ion, a sodium atom with its outer electron missing) and Cl^- (chloride ion, a chlorine atom with one excess electron). When two electrodes connected to the terminals of a battery are immersed in the liquid, Na^+ ions are attracted to the cathode (negative electrode) and Cl^- ions to the anode (positive electrode), as in Fig. 24.1. At the anode each Cl^- is *neutralized*; it gives up its extra electron to the electrode and becomes a normal chlorine atom. At the cathode each Na^+ is neutralized by the addition of an electron from the electrode and becomes an atom of ordinary metallic sodium. Thus electrons move from the electrode to the liquid at the cathode, from the liquid to the electrode at the anode. In effect, a current passes through the liquid, with the current in the liquid consisting of a movement of ions.

neutralization

The current sent through the molten salt breaks up the compound NaCl into its constituent elements:

$$2NaCl \longrightarrow 2Na + Cl_2$$

electrolysis is the process by which the passage of an electric current through a liquid liberates a free element

The sodium, a liquid at the temperature of molten salt, collects around the cathode, and chlorine gas bubbles up from the anode. As we know, a process of this sort in which free elements are liberated from liquids by an electric current is called *electrolysis*. In the preparation of many elements and in electroplating, electrolysis finds important commercial applications; for example, the procedure just outlined is commonly used to prepare metallic sodium.

pure water is a poor conductor of electric current

Electrolysis is usually carried out in water solutions rather than in molten salts. Pure water contains an exceedingly small number of ions, and hence is practically a nonconductor. But ionic compounds dissolve in water as separate ions: thus a solution of ordinary salt contains Na^+ and Cl^-; a solution of copper bromide ($CuBr_2$) contains Cu^{++} (a Cu atom with its two valence electrons missing) and Br^- (a Br atom with one electron added to its normal outer shell of seven). The ability of these dissolved ions to move through a solution makes the solution a good conductor.

Details of electrolysis are much the same in solutions and in molten salts. In the electrolysis of a $CuBr_2$ solution, for example, Cu^{++} ions move to the cathode and Br^- ions to the anode. The cathode gives two electrons to each Cu^{++}, changing the ion to a Cu atom; the anode takes one electron from each Br^-, changing the ion to neutral Br. Thus copper is plated out at the cathode, and free bromine is liberated and goes into solution at the anode. Every time one Cu^{++} ion and two Br^- ions are neutralized, two electrons are in effect transferred from cathode to anode, so that a continuous current is maintained across the solution.

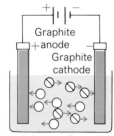

24.1 The electrolysis of molten sodium chloride. *The current in the liquid consists of moving ions.*

faraday's laws of electrolysis

Today, with our knowledge of electricity and atomic structure, it is easy to understand electrolysis. A century ago, however, a great deal of mystery surrounded this phenomenon, and Michael Faraday's discovery of the two laws of electrolysis in 1834 is further testimony to his remarkable ability as an experimenter. After many measurements, Faraday found that the mass of any particular substance liberated at an electrode is directly proportional to

1. The total charge Q that has passed through the liquid
2. The atomic mass of the substance divided by its valence

faraday's laws of electrolysis

These are *Faraday's laws of electrolysis*.

A charge of 96,500 coul was found to liberate 1 g atom of elements of valence number ± 1, $\frac{1}{2}$ g atom of elements of valence number ± 2, and so on. We can combine this latter observation with Faraday's laws in the formula

$$m = \frac{Q}{96{,}500 \text{ coul}} \times \frac{\text{atomic mass}}{\text{valence}}$$

mass liberated in electrolysis

where m is the liberated mass. The name *faraday* has been given to 96,500 coul of charge in honor of Faraday's pioneer work.

the faraday is a certain quantity of electric charge

The above equation is a direct consequence of the atomic nature of matter and charge. A gram atom of any element contains Avogadro's number N_0 of atoms. If the element has unit valence, one electron must be added to or removed from it (depending upon whether it is a metal or nonmetal, respectively) in order to convert its ions to neutral atoms. Thus the amount of charge Q_0 that must be transferred to liberate a gram atom of a univalent element is equal to N_0, the number of individual atoms, multiplied by e, the charge of the electron, and hence the charge that is absorbed or given off per atom. Inserting the values of N_0 and e to four significant figures,

$$\begin{aligned} Q_0 &= N_0 e \\ &= (6.024 \times 10^{23} \text{ atoms/g atom}) \times 1.602 \times 10^{-19} \text{ coul/atom} \\ &= 96{,}500 \text{ coul/g atom} \\ &= 1 \text{ faraday} \end{aligned}$$

Actually, since Q_0 and e can both be measured more or less directly, electrolysis provides a valuable method for determining Avogadro's number.

When the element involved has a valence higher than 1, more than one electron per atom must be transferred, and proportionately less than a gram atom of the element will be liberated by the passage of a faraday during electrolysis. This factor is taken into account by dividing the atomic mass by the valence.

As an example, let us calculate how much sodium metal is produced by sending a current of 10 amp through molten NaCl for 1 hr. Since current i is the rate of flow of electric charge,

$$Q \text{ (coul)} = i \text{ (amp)} \times t \text{ (sec)}$$

Electrolytic cells used in the production of caustic soda (NaOH) and chlorine from salt brine (a solution of NaCl in water). Hydrogen rather than sodium is liberated, since H^+ ions have greater affinity for electrons than Na^+ ions. The leftover Na^+ and OH^- ions remain in solution, and solid NaOH is obtained by evaporation.

Here

$$t = 1 \text{ hr} \times 60\,\frac{\text{min}}{\text{hr}} \times 60\,\frac{\text{sec}}{\text{min}} = 3{,}600 \text{ sec}$$

and

$$Q = 10 \text{ amp} \times 3{,}600 \text{ sec} = 36{,}000 \text{ coul}$$

The atomic mass of sodium is 22.99, and its valence is $+1$; hence

IONS AND SOLUTIONS

$$m = \frac{36,000 \text{ coul}}{96,500 \text{ coul}} \times \frac{22.99 \text{ g}}{1}$$

$$= 8.58 \text{ g}$$

The mass of sodium liberated is 8.58 g.

What volume of chlorine is evolved in the above process? At 0°C and atmospheric pressure, 1 mole of any gas occupies 22.4 liters. The valence of chlorine is -1, so 1 faraday will produce 1 g atom of chlorine. However, chlorine molecules contain *two* atoms of chlorine each, so 2 faradays is needed to produce 1 mole of Cl_2. Here the charge involved is 36,000 coul, which is

$$\frac{36,000 \text{ coul}}{96,500 \text{ coul/faraday}} = 0.373 \text{ faraday}$$

We therefore can write

$$\text{Volume evolved} = \frac{22.4 \text{ liters/mole}}{2 \text{ faradays/mole}} \times 0.373 \text{ faraday}$$

$$= 4.18 \text{ liters}$$

SOLUBILITY

Chemical combination, we have learned, takes place either by the transfer of electrons from one atom to another or by the sharing of pairs of electrons between atoms. The distinction is not sharp but serves as the basis for a convenient separation of two kinds of compounds: ionic compounds formed by electron transfer and covalent compounds formed by electron sharing. An ionic compound nearly always contains a metal and one or more nonmetals; its crystal lattice is made up of ions; it has a high melting point, since melting involves separating its ions; and in the liquid state or in solution it conducts an electric current, since its

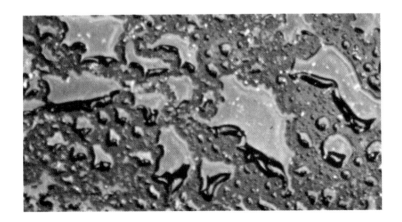

Oil and water do not mix because water molecules are highly polar and accordingly tend to cluster together, thereby squeezing out the nonpolar oil molecules.

ions are free to move about. A covalent compound most commonly contains two nonmetals; the crystal lattice of its solid state consists of molecules or of atoms; and it is a poor conductor either in the liquid state or in solution.

With these ideas in the background, let us inquire into the properties of solutions and the processes by which they are formed.

solvent and solute

A solution is a very intimate mixture of two or more different substances. Solutions can be formed of any of the three states of matter; air is a solution of several gases, sea water is a solution of various solids and gases in a liquid, and many alloys are "solid solutions" of two or more metals. Here our chief concern will be solutions in liquids.

In a solution containing two substances, the substance present in larger amount is called the *solvent*, the other the *solute*. When solids or gases dissolve in liquids, the liquid is always considered the solvent. Thus, when sugar is stirred into water, the sugar is the solute and the water is the solvent. Water is by far the commonest and most active of all solvents.

Solutions, like compounds, are homogeneous, but unlike compounds they do not have fixed compositions. To a solution of 10 g of salt in 100 g of water, for example, a little more salt or a little more water may be added; the composition of the solution is altered, but it remains homogeneous. Some pairs of liquids form solutions in all proportions; any amount of alcohol may be mixed with any amount of water to form a homogeneous liquid. More commonly, however, a given liquid will dissolve only a limited amount of another substance. Common salt can be stirred into water at 20°C until the solution contains 36 g of salt for every 100 g of water; further additions of salt will not dissolve, no matter how prolonged the stirring. This figure, 36 g per 100 g of water, is called the *solubility* of salt in water at 20°C. **The solubility of a substance is the maximum amount that can be dissolved in a given quantity of solvent at a given temperature.** It is most often expressed in grams of solute per 100 g of water but may be given as grams per liter, ounces per quart, and so on. A solution that contains the maximum amount of solute is called a *saturated* solution.

Solubilities of gases decrease as the temperature rises; solubilities of most solids increase (Fig. 24.2). If a solution of a solid is saturated at a high temperature and allowed to cool to a temperature at which its solubility is smaller, some of the solid usually crystallizes out. Thus the solubility of KNO_3 is 136 g per 100 g at 70°C and 31 g per 100 g at 20°C; cooling 236 g of saturated solution through this 50° range would force 105 g of solid KNO_3 to crystallize out.

Sometimes, if the cooling is allowed to take place slowly and without disturbance, a solute may remain in solution even though its solubility is exceeded to form a *supersaturated* solution. Supersaturated solutions are in general unstable, with the solute crystallizing out suddenly when the solution is jarred or otherwise disturbed.

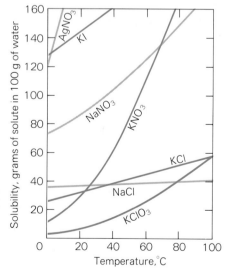

24.2 Variation of solubility with temperature for various substances dissolved in water.

polar liquids

Solubilities of different materials vary widely. Water, for example, readily dissolves such diverse substances as salt, sugar, alcohol, and ammonia, but it will not dissolve materials like camphor, fat, sulfur, or diamond. Gasoline, on the other hand, dissolves fat, but will not affect salt or sugar. Can we account for these relationships in terms of molecular and atomic structure?

The explanation in part depends on the electrical structures of different kinds of molecules. Water, for example, has polar molecules which behave as if negatively charged at one end and positively charged at the other. This occurs because, as we saw in a previous chapter, oxygen is more electronegative than hydrogen, so the electrons shared between O and H are considerably closer to the oxygen. We call water a *polar liquid*; a liquid like gasoline, whose molecules have positive and negative charges symmetrically arranged, is *nonpolar*.

Water and other strongly polar liquids consist in large part of molecular aggregates rather than simple molecules. The molecules join together in groups, positive charges against negative charges (Fig. 24.3). Water molecules can pair up similarly with polar molecules of other substances, such as alcohol and sugar (Fig. 24.4), so water dissolves these substances readily.

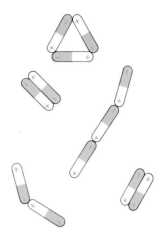

24.3 Aggregates of water molecules.
Water molecules cluster together because of electric forces that arise from their polar character.

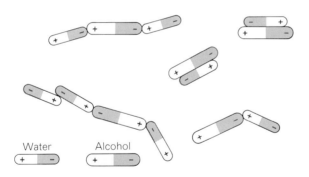

24.4 Alcohol dissolved in water.
Polar compounds readily dissolve in water because their molecules can pair up with water molecules.

Molecules of fats and oils are less strongly polar or are nonpolar and are not so strongly attracted by water molecules. If oil is shaken with water, the strong attraction of the polar molecules of water for one another "squeezes out" the oil molecules from between them, so the liquids separate quickly into layers. Oil or fat molecules mix readily, however, with the similar nonpolar molecules of gasoline (Fig. 24.5).

The solubility of a covalent substance that has distinct molecules, then, depends primarily on the electrical structure of its molecules. It dissolves only in liquids whose molecules have similar electrical structures.

Ionic compounds are highly polar in the sense that negative charges are concentrated on some atoms, positive charges on others, although in the solid state they do not consist of distinct molecules. They dissolve only in highly polar liquids. The process of solution, somewhat different

"like dissolves like"

ionic compounds dissolve only in highly polar liquids

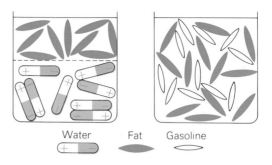

24.5 Gasoline dissolves fat; water does not. Nonpolar compounds dissolve only in nonpolar liquids.

from that for polar covalent compounds, can be visualized with the aid of Fig. 24.6, which represents the solution of NaCl in water. At the surface of the salt crystal water molecules cluster around the ions, positive ends toward negative ions and negative ends toward positive ions. The attraction of so many water molecules is sufficient to overcome the electric forces within the crystal lattice, and each ion moves off into the solution with its retinue of solvent molecules. As each layer is removed, the next is attacked until the salt is completely dissolved or until the solution becomes saturated. Since this process depends on the overcoming of forces within the crystal by electric forces exerted by the solvent molecules, it is understandable that ionic compounds will not dissolve in any but the most polar solvents.

IONS IN SOLUTION

water is a highly polar liquid

Since water is the most polar of all ordinary liquids, it has a unique capacity for dissolving ionic compounds according to the mechanism just described. This ability to dissolve materials *as ions* is a major reason why water is such an immensely important substance on our planet.

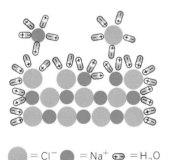

electrolytes

24.6 Solution of sodium chloride crystal in water. Water molecules exert electric forces on the Na^+ and Cl^- ions which are strong enough to remove them from the crystal lattice.

The ions that a compound releases on dissolving are, as we know, simply its constituent atoms or atom groups, with electric charges determined by their excess or deficiency in electrons. In the example of the last section, every sodium ion is a sodium atom minus its outer electron, and every chloride ion is a chlorine atom with eight electrons instead of the normal seven in its outer shell—precisely the structures, of course, that are present in the original crystal lattice. A potassium nitrate crystal contains positive potassium ions, which are potassium atoms shorn of their outer electrons, and negative nitrate ions, which are nitrate groups with one electron in excess of the total number of protons; when the crystal dissolves, these same two ions are free in the solution. It is commonly said that an ionic compound "ionizes" when it dissolves, but ionization in this sense means only that ions already present in the crystal are set free to move about independently in the solution. A more appropriate

dissociation term is *dissociation*.

IONS AND SOLUTIONS

Formulas for ions are the symbols of the atoms or atom groups with enough + or − signs attached to indicate their net charges. The two ions of NaCl are Na$^+$ (Na minus an electron) and Cl$^-$ (Cl plus an electron), and those of KNO$_3$ are K$^+$ and NO$_3^-$. CaCl$_2$ dissociates to form Ca^{++} and 2Cl$^-$ (two chlorine ions for every one calcium ion), Na$_2$SO$_4$ to form 2Na$^+$ and SO$_4^=$. Since the charge on an ion is determined by how many electrons it has gained or lost in electron transfer, the charge is the same as the valence of the atom or atom group.

Substances that separate into free ions on solution in water are called *electrolytes*. Electrolytes include all ionic compounds that are soluble in water and some covalent compounds containing hydrogen (for example, HCl) that form ions by reaction with water. Other soluble covalent compounds, like sugar and alcohol, that do not ionize in solution are *nonelectrolytes*.

> electrolytes separate into free ions on solution in water

A property of electrolytes by which they may be recognized quickly —the property, in fact, that gives them their name—is the ability of their solutions to conduct an electric current. Conduction is possible because the ions are free to move; positive ions migrate through the solution toward the negative electrode, negative ions toward the positive electrode. The migration of ions, together with the reactions that occur at the electrodes, made up the process of electrolysis, as we saw.

> electrolytes in solution conduct electric current by the motion of ions

properties of ions

One of the early objections to the ionic theory of solutions was that sodium chloride was assumed to break down into separate particles of sodium and chlorine and yet the solution remains colorless. Why, if chlorine is present as free ions, should we not find the characteristic greenish-yellow color of chlorine in the solution? We might reply that chloride ion has altogether different properties from chlorine gas—a different color, a different taste, different chemical reactions. This answer is straightforward enough, but its full significance is not easy to grasp.

We must regard a solution of sodium chloride not as a solution of NaCl but as a solution of two substances with the formulas Na$^+$ and Cl$^-$. Further, each of these substances has its own set of properties, properties that are quite different from those of the active metal Na and the poisonous gas Cl$_2$. We must think of each ion in an electrolytic solution as a new and separate material with characteristic properties of its own.

> dissociation is a type of chemical change

By the *properties of an ion*, we mean, of course, the properties of solutions in which the ion occurs. A solution of a single kind of ion, all by itself, cannot be prepared; positive ions and negative ions must always be present together, so that the total number of charges of each sign will be the same. But each ion gives its own characteristic properties to all solutions containing it, and these properties can be recognized whenever they are not masked by other ions. For example, a property of the copper ion Cu^{++} is its blue color, and all solutions of this ion are blue (unless some other ion is present that has a stronger color). A characteristic of the hydrogen ion H$^+$ is its sour taste, and all solutions containing this ion (acids) are sour. The silver ion Ag$^+$ shows the property

> ions in solution have their own characteristic properties

of forming a white precipitate, AgCl, when mixed with solutions of the chloride ion Cl^-; any solution of an electrolyte containing silver when mixed with a solution of any chloride will give this precipitate.

To emphasize the difference between the properties of an ion and the properties of the corresponding neutral substance, we may set down in tabular form a comparison of the properties of the two substances Cl^- and Cl_2:

Cl_2	Cl^-
Greenish-yellow color	Colorless
Strong, irritating taste and odor	Mild, pleasant taste
Combines with all metals	Does not react with metals
Combines readily with hydrogen	Does not react with hydrogen
Does not react with Ag^+	Forms AgCl with Ag^+
Very soluble in CCl_4	Insoluble in CCl_4

In general, Cl_2 is much more active—as might be expected, since its atoms have only seven valence electrons, whereas chloride ions have eight.

For each ion we may write down a list of properties, which means a list of the properties common to all its solutions. In general, the properties of a solution of an electrolyte are the sum of the properties of the ions that it contains. The properties of sodium chloride are the properties of Na^+ and the properties of Cl^-; the properties of copper sulfate are the properties of Cu^{++} and $SO_4^=$. This is one of the reasons why the ionic picture is so valuable in the study of solutions: Instead of learning the individual properties of several hundred different electrolytes, we need only learn the properties of a few ions to be able to predict the properties of an electrolytic solution that contains them.

arrhenius

The hypothesis that many substances exist in solution as ions was proposed in 1887 by a young Swedish chemist, Svante Arrhenius. So radical was the idea that older chemists derided it, and Arrhenius all but ruined his career at its beginning by defending his hypothesis vigorously. Later, as the weight of accumulating evidence supported him more and more strongly, Arrhenius won worldwide acclaim.

Today the idea of ions in solution follows naturally from the electrical structure of matter. We know that some compounds are formed by the loss of electrons from one kind of atom to another, so that some of the atoms gain a positive charge and the others a negative charge, and we find no difficulty in imagining that a liquid like water can separate these electrically charged atoms. But in 1887 this modern picture of the atom was not even dreamed of. Without this modern knowledge Arrhenius's contemporaries had reason to consider his idea of neutral substances breaking up into electrically charged fragments farfetched.

IONS AND SOLUTIONS

The only direct experimental evidence that suggested the existence of charged particles in solution was that provided by electrolysis. When a current passes through a solution, one kind of material patently migrates to the anode and another to the cathode, just as if the two kinds had opposite electric charges. Faraday had supposed that passage of the current *caused* the substance in solution to break down into ions, and this explanation of electrolysis was generally accepted.

Arrhenius instead proposed that in solution ions are set free, not by the passage of a current, but *whenever an electrolyte dissolves.* Positive and negative charges are formed in equal numbers, so that both the solution and the original solute appear to be electrically neutral. Of the detailed evidence that Arrhenius presented in support of his theory, we shall mention only two major points:

ions are set free whenever an electrolyte dissolves

1 Reactions between electrolytes which take place instantaneously in solution often are very slow or do not occur at all if the electrolytes are dry. An example is the reaction between silver nitrate and sodium chloride. If solutions of the two electrolytes are mixed, a white *precipitate* (ppt)—a solid that forms as a result of a chemical reaction in solution—of the insoluble substance silver chloride appears immediately, at first as a general cloudiness through the solution, later as tiny crystals which settle to the bottom. If the liquid is separated from these crystals by filtration and then evaporated to dryness, another solid, sodium nitrate, will be left. The equation for the reaction may be written

a precipitate is a solid that forms as a result of a chemical reaction in solution

$$AgNO_3 + NaCl \longrightarrow AgCl(ppt) + NaNO_3 \text{ (in soln)}$$

The silver and sodium seem to have "exchanged partners." But if dry salt is mixed with dry silver nitrate, no reaction of this sort occurs. Apparently, the exchange of partners is aided by solution, which suggests that the "partners" are already free in the solutions, ready to react as soon as they are mixed.

2 Solutions of electrolytes have abnormally low freezing points. Any substance dissolved in water lowers its freezing point somewhat: We take advantage of this fact, for instance, by dissolving various materials in the water of automobile radiators during cold weather to keep the water from freezing. For nonelectrolytes, careful study shows that the amount by which the freezing point is lowered depends only on the concentration of molecules of dissolved material present, not on the nature of the material. Equal numbers of sugar, alcohol, and glycerin molecules dissolved in the same quantity of water lower the freezing point by almost exactly the same amount. But the same number of "molecules" of salt (assuming the molecule to be NaCl) lowers the freezing point nearly twice as much; so also do the same number of molecules of $MgSO_4$, KBr, and $AgNO_3$, to give a few more examples. Arrhenius concluded that, in solutions of these substances, twice as many particles were actually present as would be indicated by the simple formulas because the substances were broken down into ions.

dissolved substances lower the freezing point of water by an amount that depends upon the concentration of ions or molecules

Similarly a substance like $CaCl_2$ lowers the freezing point nearly three times as much as sugar does, because each "molecule" is broken down into three particles, one calcium ion (Ca^{++}) and two chloride ions (Cl^- + Cl^-). Boiling points and vapor pressures of electrolytic solutions show similar abnormalities when compared with solutions of nonelectrolytes, and these abnormalities also find ready explanation in terms of ions.

Arrhenius's hypothesis, based on the careful study and evaluation of these and other experimental observations, has been confirmed by modern research into the electrical structure of the atom. Not all Arrhenius's ideas are still accepted; he believed, for instance, that electrolytes consist of molecules rather than ions in the solid state, which has been disproved by X-ray studies of crystals. But the basic notion that electrolytes dissolve in water to form free ions capable of moving about in solution remains the foundation of all modern work on solutions in water. Thus the 1923 Debye-Hückel theory of electrolytes, a largely successful quantitative treatment that takes into account the electrostatic attractive forces between ions in solution, is an elaboration of Arrhenius's hypothesis and does not represent a change in his essential argument.

ionic equations

Reactions involving ions can be expressed in equation form by following the basic principle that a chemical equation is a summary of an actual chemical change. We may use the formula of a solid compound, as we have done previously, to indicate the relative number of atoms present. Thus the formula of solid sodium chloride is NaCl, even though the crystal actually consists of sodium ions and chloride ions. In solution, we write the formula of an electrolyte in terms of its ions, signifying that the ions are now independent substances rather than parts of a crystal.

The change from solid salt to salt in solution, for example, may be represented as

dissolving NaCl in water liberates Na^+ and Cl^- ions

$$NaCl \longrightarrow Na^+ + Cl^-$$

The equation means that sodium ions and chloride ions have been set free to move about in the solution. If the solution is cooled or evaporated so that solid sodium chloride crystallizes out, the re-formation of the crystal structure may be written

$$Na^+ + Cl^- \longrightarrow NaCl$$

ions in solution act independently of one another

Once an electrolyte is dissolved, each ion is independent and its reactions need not involve the other ions present. For example, let us consider the reaction between silver nitrate and sodium chloride mentioned earlier. Both $AgNO_3$ and NaCl, when dissolved, are completely dissociated into ions. Of the products, $NaNO_3$ is soluble and completely ionized, but AgCl is a solid precipitate not appreciably ionized. Hence we might write the equation

$$Ag^+ + NO_3^- + Na^+ + Cl^- \longrightarrow AgCl + Na^+ + NO_3^-$$

On each side of this equation appear the formulas Na^+ and NO_3^-; these substances have not changed during the reaction but have merely remained in solution (Fig. 24.7). Actually, the only chemical change that has occurred is the disappearance of Ag^+ and Cl^- and the formation of solid AgCl:

$$Ag^+ + Cl^- \longrightarrow AgCl$$

This brief equation is a complete summary of the reaction. The other ions are present but take no part. If KCl had been used instead of NaCl, or $AgC_2H_3O_2$ (silver acetate) instead of $AgNO_3$, the other ions would have been different but the actual chemical change would still have been the union of Ag^+ and Cl^- to form AgCl. The first equation is not wrong, but the second expresses more clearly that the reaction with chloride ion is a property of Ag^+ and not of some particular silver compound.

A similar example is furnished by the precipitation of the white solid calcium carbonate ($CaCO_3$) when a solution of calcium ion is added to

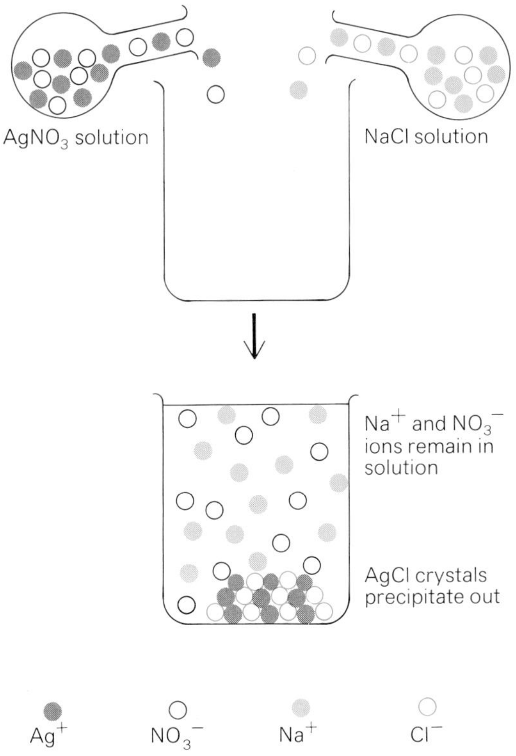

AgCl is an insoluble compound and precipitates out of a solution that contains Ag^+ and Cl^- ions

an ionic equation summarizes the actual reaction

24.7 *When silver nitrate ($AgNO_3$) and sodium chloride (NaCl) are dissolved in water, a precipitate of the insoluble compound silver chloride is produced.* The sodium and nitrate ions remain in solution.

CaCO₃ is an insoluble compound and precipitates out of a solution that contains Ca^{++} and $CO_3^=$ ions

a solution of carbonate ion. If $CaCl_2$ is used to supply Ca^{++} and Na_2CO_3 to supply $CO_3^=$, the equation may be written

$$Ca^{++} + 2Cl^- + 2Na^+ + CO_3^= \longrightarrow CaCO_3 + 2Cl^- + 2Na^+$$

Sodium ions and chloride ions take no part in the reaction, so we may simplify the equation to

$$Ca^{++} + CO_3^= \longrightarrow CaCO_3$$

no reaction occurs between two electrolytes in the same solution unless an undissociated substance is formed

When two electrolytes are mixed, no reaction occurs unless some undissociated substance like AgCl and $CaCO_3$ can form. If solutions of $Cu(NO_3)_2$ and NaCl are mixed, for example, a possible reaction might be an exchange of partners to form $NaNO_3$ and $CuCl_2$. But both of these are soluble salts, completely ionized in solution, so no reaction can occur. In symbols,

$$Cu^{++} + 2NO_3^- + Na^+ + Cl^- \longrightarrow Cu^{++} + Cl^- + Na^+ + 2NO_3^-$$

The same substances are represented on each side of the "equation," and no chemical change has taken place.

For a further example of an ionic process we shall consider the reaction between an active metal and an acid, such as

$$Zn + H_2SO_4 \longrightarrow H_2 + ZnSO_4$$

for the reaction between zinc and sulfuric acid. Let us try to express this equation in terms of ions. Zinc is a solid metal, and hence is not ionized; H_2 represents the covalent molecule of a gas, and is also not ionized. Sulfuric acid and zinc sulfate, however, are ionized in solution, and so we have

$$Zn + 2H^+ + SO_4^= \longrightarrow H_2 + Zn^{++} + SO_4^=$$

This is one way to write the ionic equation, but inspection shows that it can be simplified. $SO_4^=$ appears on both sides of the equation, and so may be omitted. The actual reaction is therefore

$$Zn + 2H^+ \longrightarrow Zn^{++} + H_2$$

This shorter statement implies that the only chemical changes are the formation of zinc ions from zinc metal and the formation of hydrogen gas from hydrogen ions.

GLOSSARY

Electrolysis is the liberation of free elements from a liquid by the passage through it of an electric current.

IONS AND SOLUTIONS

In a solution containing two substances, the one present in the larger amount is the *solvent* and the other the *solute*. When solids or gases are dissolved in a liquid, the liquid is always considered the solvent.

A *saturated solution* is one that contains the maximum amount of solute at a given temperature.

A *supersaturated solution* is one in which more solute is dissolved than normally possible at that temperature; it is unstable and the excess solute readily crystallizes out.

A *polar molecule* is one that behaves as if it were negatively charged at one end and positively charged at the other. A *polar liquid* is a liquid whose molecules are polar, while a *nonpolar liquid* has molecules whose charge is symmetrically arranged.

An *electrolyte* is a substance that separates (or *dissociates*) into free ions on solution in water.

A *precipitate* is a solid that forms as a result of a chemical reaction in solution.

EXERCISES

1 Into what kind of energy is electric energy converted during electrolysis? In what common device is this energy transformation reversed?

2 Why is the solubility of one gas in another unlimited?

3 How can you tell whether a sugar solution is saturated or not?

4 Give examples of polar and nonpolar liquids, and state several substances soluble in each.

5 How could you distinguish experimentally between an electrolyte and a nonelectrolyte?

6 Contrast the properties and electron structures of Na and Na^+. Which would you expect to be more active chemically?

7 Name one property by which you could distinguish
   a. Cl^- from NO_3^-
   b. Ag^+ from Na^+
   c. Ca^{++} from Na^+
   d. Cu^{++} from Ca^{++}

8 Why do so many substances dissolve in water? Why do oils and fats not dissolve in water?

9 With the help of Fig. 24.2, predict what will happen when a concentrated solution of sodium nitrate at 50°C is added to a saturated solution of potassium chloride at the same temperature.

10 At 10°C, which is more concentrated, a saturated solution of potassium nitrate or a saturated solution of potassium chloride? At 60°C?

11 A current of 50 amp flows through molten NaCl for 10 min. What mass of metallic sodium will be deposited at the negative electrode?

12 A current of 1 amp is passed through a solution that contains 0.5 mole of $CuSO_4$. How long will it take for all the copper to be removed from the solution?

13 One faraday of electricity is passed through a water bath. How many grams of water are decomposed?

14 A total of 400,000 coul is passed through a sample of molten lithium chloride. What volume of chlorine gas at 0°C and atmospheric pressure is liberated?

chapter 25

Acids, Bases, and Salts

ACIDS

    hydronium
    strong and weak acids

BASES

    strong and weak bases
    the pH scale

NEUTRALIZATION

    neutralization
    indicators
    salts
    normality of solutions
    strong base and weak acid
    weak base and strong acid
    ammonia and ammonium ion
    acidic and basic oxides

We continue our study of the behavior of ions in solution by considering now the three important classes of electrolytes: acids, bases, and salts.

ACIDS

Acids have been described earlier as substances containing hydrogen whose water solutions taste sour and change the color of a dye called litmus from blue to red. Strong acids like hydrochloric acid (HCl) and sulfuric acid (H_2SO_4) are poisonous, cause painful burns if allowed to remain on the skin, and are injurious to clothing. Weak acids like carbonic acid (H_2CO_3) and citric acid, far from being harmful, add a pleasant sour taste to foods and drinks.

hydronium

What is it that gives an acid its characteristic properties? Following the line of reasoning in the previous chapter we note (1) that solutions of acids conduct electricity and hence must contain ions, and (2) that acids contain hydrogen united with one or more nonmetals. It is reasonable, then, to express the dissociation of acids into ions with equations such as

$$HCl \longrightarrow H^+ + Cl^-$$

and

$$H_2SO_4 \longrightarrow 2H^+ + SO_4^=$$

and to conclude that the characteristic properties of acids are the properties of the free hydrogen ion H^+.

There are difficulties with this simple picture. For one thing, pure acids when liquefied do not conduct current, as NaCl does, so we cannot properly speak of a pure acid as made up of ions. It is as if acids formed free ions not by simple separation of the positively and negatively charged atoms already present but only by reaction with water. (In more general terms, acids are covalent rather than ionic substances, and they can form ions only by reaction with a polar liquid like water or alcohol.)

acids are covalent substances that dissociate only when they react with a polar liquid

A second difficulty is the nature of the ion H^+, which is nothing more than a single proton—the nucleus of a hydrogen atom shorn of its lone electron. All other ions are particles of the same general size as atoms, particles whose structures consist of nuclei and electron clouds. This one ion would be entirely different, a naked positive charge with a volume only about 10^{-15} as great. It seems very unlikely that such a particle could have an independent existence in a liquid; undoubtedly it would attach itself immediately to some other atom or molecule.

To eliminate these difficulties, we could write more correctly

$$HCl + H_2O \longrightarrow H_3O^+ + Cl^-$$

the hydronium ion

Here the acid is shown taking part in a specific reaction with water instead of simply splitting up into ions, and the proton is considered to be attached to a water molecule (H_3O^+) rather than free in solution. The ion H_3O^+ is called the *hydronium ion;* it is a combination of H^+ with H_2O, or a *hydrated* hydrogen ion (Fig. 25.1). The characteristic properties of acids are described more correctly as properties of the hydronium ion than as properties of the simple hydrogen ion.

Nevertheless, it is customary in chemistry to write the characteristic ion of acids with the formula H^+ rather than H_3O^+. This is chiefly for convenience. H_3O^+ is more correct than H^+, but it is still not entirely correct; if we strove for strict accuracy we should have to write sometimes $H_5O_2^+$, $H_7O_3^+$, and $H_9O_4^+$ as well as H_3O^+, and we should have to use hydrates for other ions also—$Na(H_2O)^+$ instead of Na^+, $Cl(H_2O)^-$ instead of Cl^-, and so on. Our equations would quickly become too cumbersome for easy use. In most chemical reactions the water of hydration does not play an important role, and the reactions can be adequately represented by equations from which it is omitted.

For our purposes, then, **an acid may be defined as a substance containing hydrogen whose solution in water increases the number of hydrogen ions present.** We can think of free hydrogen ions as being present in all acid solutions and give these solutions their common properties. When we say that acid solutions taste sour, turn litmus pink, and liberate hydrogen gas by reaction with metals, we mean that the hydrogen ion does these things.

definition of acid

strong and weak acids

Acids differ greatly in their degree of dissociation. Some acids, called *strong acids*, dissociate completely. For example, when HCl dissolves in water it breaks down completely into H^+ and Cl^-, with no molecules of undissociated HCl. Other acids, called *weak acids*, dissociate only slightly. The greater dissociation of strong acids means that, in solutions of similar total concentration, a strong acid has a much larger proportion of hydrogen ions than a weak acid: It has a much sourer taste, it is a better conductor of electricity, and if the two acids are poured on zinc, the evolution of hydrogen gas is much faster from the reaction with strong acid.

strong acids dissociate completely in solution, weak acids only slightly

An interesting example of a weak acid is carbonic acid (H_2CO_3), which is formed by the solution of CO_2 gas in water:

$$CO_2 + H_2O \longrightarrow H_2CO_3$$

The carbonic acid then dissociates into H^+ and HCO_3^-:

$$H_2CO_3 \longrightarrow H^+ + HCO_3^-$$

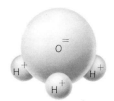

25.1 *A model of the hydronium ion, H_3O^+.*

Very little H^+ is found in a solution of CO_2 in water, which means that we can think of carbonic acid as being weak. Actually, it seems unlikely on the basis of experimental evidence that much H_2CO_3 ever exists as such, so that perhaps a better way to express the "dissociation of carbonic acid" is simply to write

$$CO_2 + H_2O \longrightarrow H^+ + HCO_3^-$$

Thus the "weakness" of carbonic acid resides in the unlikelihood of its formation, rather than in any tendency it may have to remain as undissociated H_2CO_3. However, it is convenient to speak of carbonic acid as being formed when CO_2 is dissolved in water and to regard it as a weak acid, which is an acceptable summary of the situation provided that the correct picture of what is happening is kept in mind.

hydrochloric, sulfuric, and nitric acids are strong acids

The three most common strong acids are HCl (hydrochloric), H_2SO_4 (sulfuric), and HNO_3 (nitric). Some familiar weak acids besides carbonic acid are acetic acid ($HC_2H_3O_2$), the acid in vinegar; boric acid (H_3BO_3),

acetic acid, boric acid, and citric acid are weak acids

found in every medicine cabinet; and citric acid, the acid of citrus fruits.

The stronger an acid, the weaker is the attachment of hydrogen in its molecules. In a strong acid like HCl the attachment is so weak that the H^+ and Cl^- are split apart completely and go their separate ways in solution. In a weak acid like "H_2CO_3," on the other hand, the attachment is strong enough so that most of the molecules remain undissociated. For instance, in a saturated solution of CO_2, only 1 CO_2 molecule in every 1,000 combines with H_2O to yield H^+. Acetic acid, which is not quite so weak as carbonic and which therefore holds its hydrogen ions less securely, has about 10 molecules out of every 1,000 broken up into ions in a saturated solution. In hydrochloric acid, where the attachment of hydrogen in the molecule is very weak, every one of the 1,000 molecules would be split up.

the covalent bonds in the molecules of strong acids are easily broken to liberate H^+

Other substances besides CO_2 that do not contain hydrogen in their formulas are capable of giving acid solutions because they react with water to liberate H^+ from H_2O molecules. Solutions of many iron and copper salts (compounds like $CuSO_4$, $FeCl_2$, $FeCl_3$) give slight acid reactions, as do solutions of nonmetal oxides like SO_2 and NO_2.

BASES

We have already described bases as substances whose solutions in water have a bitter taste and an ability to turn red litmus to blue. Their formulas, for example, NaOH and $Ba(OH)_2$, show that bases consist of a metal united with one or more hydroxide groups (OH). On dissolving in water, bases dissociate into ions according to reactions such as

bases contain the hydroxide (OH) atom group

$$NaOH \longrightarrow Na^+ + OH^-$$

and

$$Ba(OH)_2 \longrightarrow Ba^{++} + 2OH^-$$

Just as H^+ is the characteristic ion of acid solutions, so OH^- is the characteristic ion in water solutions of bases. The bitter taste and the ability to change the color of dyes are properties of the OH^- ion. We may therefore say that **a base is a substance that contains an OH group and that increases the number of OH^- ions present upon dissolving in water.**

definition of base

strong and weak bases

Like acids, bases may be grouped according to their degree of dissociation as strong bases and weak bases. Thus NaOH is a strong base because it breaks up completely into Na^+ and OH^- on dissolving.

strong bases dissociate completely in solution, weak bases only slightly

Bases differ from acids, however, in that soluble weak bases with simple formulas are rare. The most common weak base is ammonium hydroxide (NH_4OH), in which the NH_4 group plays the role of a metal. Ammonium

ACIDS, BASES, AND SALTS

hydroxide, like carbonic acid, is considered a compound only for the sake of convenience, since a "solution of ammonium hydroxide" is more correctly just a solution of ammonia gas (NH_3) in water. The "dissociation of ammonium hydroxide" is really a reaction between the dissolved NH_3 and H_2O:

$$NH_3 + H_2O \longrightarrow NH_4^+ + OH^-$$

The former belief that this process first goes through the intermediate step

$$NH_3 + H_2O \longrightarrow NH_4OH$$

was a consequence of the highly polar nature of both ammonia and water molecules, which leads to the great solubility of ammonia. The "weakness" of ammonium hydroxide means that the reaction of the first equation is relatively infrequent; we can take the liberty of referring to NH_4OH as a compound as a kind of verbal shorthand, but we must remember that the direct combination of NH_3 and H_2O to yield NH_4^+ and OH^- is what is really involved.

The three strong bases most widely used are KOH (ordinary lye or caustic potash), NaOH (soda lye or caustic soda), and $Ba(OH)_2$. These substances are quite as poisonous and quite as destructive to flesh and clothing as the strong acids and, like the strong acids, are extensively used in chemical industry. Another hydroxide that gives a fairly basic solution is $Ca(OH)_2$, which makes ordinary limewater. In so far as it dissolves, $Ca(OH)_2$ dissociates completely, but the amount of OH^- obtainable is limited because the compound is not very soluble. All other common hydroxides, like $Al(OH)_3$, $Fe(OH)_3$, $Cu(OH)_2$, are practically insoluble in water. These substances contribute only negligible amounts of OH^- to solution, both because they are weak bases and because they dissolve to such a slight extent. *[potassium hydroxide, sodium hydroxide, and barium hydroxide are strong bases]*

Many substances that do not contain OH in their formulas gives basic solutions because they are capable of reacting with water to produce OH^- from H_2O molecules. Familiar examples are ammonia water (NH_3 dissolved in water), washing soda (sodium carbonate, Na_2CO_3), and borax (sodium tetraborate, $Na_2B_4O_7$). These will be recognized as common household items, useful as cleaning agents because of their ability to dissolve grease. Ordinary soap also gives a slightly basic solution, but its cleaning ability is due more to its emulsifying action than to its basic nature. *[solutions of ammonia, washing soda, and borax are basic although these substances are not bases]*

A name often used for any substance that dissolves to give a basic solution is *alkali*, an old Arabic term that referred originally to the bitter extract obtained by leaching the ashes of a desert plant. The alkali that crystallizes in desert basins is chiefly Na_2CO_3. Because NaOH and KOH are strong alkalies, sodium and potassium are often called *alkali metals*. An *alkaline solution* is any solution with appreciable quantities of OH^-; the terms *alkaline* and *basic* are practically synonymous. *[alkali]*

More general definitions of acids and bases cover systems from which water may be absent. Thus a "Brønsted-Lowry acid" is a substance that acts as the proton donor in a chemical reaction and a "Brønsted-Lowry base" is a substance that acts as the proton acceptor. A still broader definition was suggested by G. N. Lewis: a "Lewis acid" is a substance that acts as an electron-pair acceptor in sharing electron pairs provided by a "Lewis base."

the pH scale

Even pure water dissociates to a small extent. The reaction can be written

water dissociates into H^+ and OH^- ions to a very small extent

$$H_2O \longrightarrow H^+ + OH^-$$

The hydroxide ion OH^- attracts protons much more strongly than does the neutral water molecule H_2O, and the reverse reaction

H^+ and OH^- ions in solution recombine readily to form water molecules

$$H^+ + OH^- \longrightarrow H_2O$$

is accordingly preferred.

The dissociation of water means that there will be some H^+ (actually H_3O^+, of course) and OH^- ions present in pure water at all times, and the strong tendency for these ions to recombine means that their concentration will be small. As dominant as recombination is over dissociation in water, it can never be complete since new ions come into being all the time and on the average a certain period elapses until they neutralize each other. In pure water the concentration of H^+ and OH^- ions is 10^{-7} mole of each per liter. Since there are 1,000 g of water per liter, the concentration of H_2O is 55 moles/liter, and only 0.0000002 percent is dissociated: two molecules out of every billion.

In an acid solution the concentration of H^+ is greater than 10^{-7} mole/liter and that of OH^- is lower, while in a basic solution the concentration of OH^- is greater than this figure and that of H^+ is lower. Thus we can characterize a water solution with more than 10^{-7} mole/liter of H^+ as acid and one with less than 10^{-7} mole/liter of H^+ as basic; a concentration of exactly 10^{-7} mole of H^+ per liter means a neutral solution, since in this event the OH^- concentration is the same.

an acid solution contains more than 10^{-7} mole/liter of H^+ ions, a basic solution contains less

The pH scale is an ingenious method for expressing the exact degree of acidity or basicity of a solution in terms of its H^+ ion concentration. By definition, the pH of a solution is the negative logarithm (to the base 10) of its H^+ concentration. **If the H^+ concentration is 10^{-x}, the pH is x.** As an equation,

definition of pH

$$pH = -\log [H^+]$$

where $[H^+]$ represents the H^+ concentration in moles per liter.

In a neutral solution, $[H^+] = 10^{-7}$ mole/liter, and its pH is

the pH of a neutral solution is exactly 7

$$pH = -\log 10^{-7} = 7$$

In a strongly acid solution, there might be 0.1 mole/liter of H^+, and

ACIDS, BASES, AND SALTS

$$pH = -\log 0.1 = -\log 10^{-1} = 1$$

A basic solution, on the other hand, has an excess of OH⁻ that suppresses the H⁺ concentration by combining with the H⁺ ions, and a strongly basic solution might have an H⁺ concentration of only 10^{-12} mole/liter. Here

$$pH = -\log 10^{-12} = 12$$

Evidently a pH of 7 signifies a neutral solution, a smaller pH than 7 signifies an acid solution, and a higher pH than 7 signifies a basic solution. Figure 25.2 shows the pH scale.

the pH of an acid solution is less than 7

the pH of a basic solution is more than 7

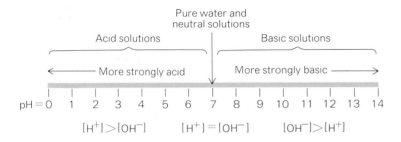

25.2 The pH scale. *The hydrogen ion concentration is symbolized by [H⁺] and that of the hydroxide ion by [OH⁻].*

NEUTRALIZATION

When sodium hydroxide solution is added slowly to hydrochloric acid, there is no visible sign that a reaction is taking place. Both the original solutions are colorless, and the resulting solution is colorless also. That a reaction does occur can be shown in several ways:

1. The mixture becomes warm, showing that chemical energy is changing to heat energy.
2. The taste of the acid becomes less and less sour as the base is added.
3. If small samples of the acid are taken out while the base is being added, the samples show progressively less and less active evolution of hydrogen gas when poured on zinc.

The base evidently destroys, or *neutralizes*, the characteristic acid properties, and the reaction is accordingly called *neutralization*. In the same way the characteristic properties of a base can be neutralized by adding a strong acid.

acids and bases neutralize each other

neutralization

What is the chemical change in the neutralization of HCl by NaOH? We could write the preliminary equation

$$HCl + NaOH \longrightarrow H_2O + NaCl$$

An inexpensive way to obtain approximate pH values involves matching a sensitized paper tape moistened with the solution under test against a color code on the spool. Here a pH of 2 turns the paper purple, pH 4 orange, pH 6 tan, pH 8 green, and pH 10 blue. A pH 7 solution, which is neutral, leaves the paper unchanged.

neutralization involves the formation of H_2O molecules from H^+ and OH^- ions

To make the equation more precise, we consider the ions involved. HCl, a strong acid, dissociates completely in water to give H^+ and Cl^-; NaOH, a strong base, dissociates into Na^+ and OH^-; the product NaCl, likewise a soluble electrolyte, remains dissociated in solution. Of the four substances shown, only water is a nonelectrolyte, so it alone should appear undissociated in the equation; hence

$$H^+ + Cl^- + Na^+ + OH^- \longrightarrow H_2O + Na^+ + Cl^-$$

Since Na^+ and Cl^- appear on both sides, they may be omitted, leaving

$$H^+ + OH^- \longrightarrow H_2O$$

This is the actual chemical change, stripped of all nonessentials. The neutralization of a strong acid by a strong base in water solution is essentially a reaction between hydrogen ions and hydroxide ions, forming water.

indicators

To carry out a neutralization reaction in the laboratory, some method is necessary to determine when just enough acid has been added to neutralize all the base present, or vice versa. Otherwise the resulting solution will be either acidic or basic, depending on which is in excess. One convenient method for determining the "neutral point" is to add a few drops of litmus solution to the base before the reaction is carried out, giving the basic solution a blue color. As the acid is added, the blue color persists until all the base is used up, then changes to pink. Similarly, if a base is to be added to an acid, the acid may first be made pink with litmus solution; then its color will change to blue as soon as enough base has been added to neutralize the acid.

A substance like litmus, whose color enables a chemist to tell whether a solution is acidic or basic and whose sharp change in color shows when neutralization has occurred, is called an *indicator*. Many different indicators are used in chemical laboratories, some being more useful than litmus because changes in their color during neutralization are more abrupt. Two common ones are phenolphthalein, which is pink in basic solution and colorless in acid, and methyl orange, which is yellow in basic solution and salmon pink in dilute acid. While the color change in litmus occurs at a pH of 7, corresponding to a neutral solution, the color changes in phenolphthalein and methyl orange occur at pH values of 9 and 4, respectively, corresponding to slightly basic and slightly acid solutions. In most situations the fact that the latter indicators do not change color at a pH of exactly 7 is of little importance.

All indicators are complex compounds that contain carbon, hydrogen, and oxygen; several are commercial dyes. One familiar "indicator" is the

the color of an indicator shows whether a solution is acid or basic

An electronic pH meter permits the degree of acidity or basicity of a solution to be determined rapidly and accurately.

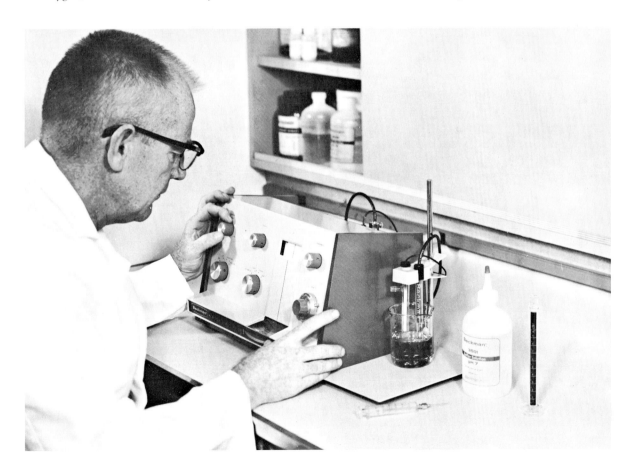

red coloring matter of cherry juice: When a red cherry stain is washed with soap (which gives a weakly alkaline solution), its color changes abruptly to blue, showing that the acid of the fruit juice has been neutralized.

salts

When an NaOH solution is neutralized with HCl, the resulting solution should contain nothing but the ions Na^+ and Cl^-. If the solution is evaporated to dryness, the ions combine to form the white solid NaCl. This substance, ordinary *salt*, gives its name to an important class of compounds, most of which are crystalline solids at ordinary temperatures and most of which consist of a metal combined with one or more nonmetals. Typical salts are KBr, $MgSO_4$, $Al(NO_3)_3$, and $ZnCO_3$. Crystal structures of salts consist of alternate positive and negative ions, which means that practically all soluble salts dissociate into free ions completely in a water solution. No salt is completely insoluble in water, but some, like AgCl and $CaCO_3$, are so very slightly soluble that we often refer to them as insoluble; such salts, of course, can give only a very few ions in solution.

The solubilities of most common salts can be summarized as follows:

solubilities of common salts

All nitrates (salts containing the NO_3 group) are soluble.
All acetates (salts containing the $C_2H_3O_2$ group) are soluble.
All chlorides are soluble, except AgCl and a very few others.
All sulfates are soluble, except $BaSO_4$ and a few others.
All carbonates are insoluble, except Na_2CO_3, K_2CO_3, and $(NH_4)_2CO_3$.
All sulfides are insoluble, except Na_2S, K_2S, $(NH_4)_2S$, CaS, and BaS.
All salts of Na, K, and NH_4 are soluble.

Any salt can be formed by mixing the appropriate acid and base and evaporating the solution to dryness. Thus KNO_3 is formed when solutions of KOH and HNO_3 are mixed and evaporated; $CuSO_4$ is formed when H_2SO_4 is poured on the insoluble hydroxide $Cu(OH)_2$ and the resulting solution is evaporated. We could say, in general, that neutralization reactions give water and a solution of a salt. It is important to remember, however, that the salt itself is not produced directly by the neutralization. Neutralization is essentially a reaction between hydrogen ion and hydroxide ion; as a result of this process ions may be left in solution that on evaporation will unite to form the crystals of a salt.

normality of solutions

a gram-equivalent of an acid provides 1 mole of H^+; a gram-equivalent of a base provides 1 mole of OH^-

A quantity called the *gram-equivalent* is widely used in chemical calculations. A gram-equivalent of an acid is the amount of that acid needed to furnish 1 mole of H^+; a gram-equivalent of a base is the amount of that base needed to furnish 1 mole of OH^-. Thus a gram-equivalent of an acid or base provides Avogadro's number N_0 of H^+ or OH^- ions respectively. One mole (36.46 g) of HCl can provide 1 mole of H^+, hence 1 gram-equivalent of HCl is 36.46 g. One mole of H_2SO_4 (98.08 g) can provide 2 moles of H^+, since

$$H_2SO_4 \longrightarrow 2H^+ + SO_4^=$$

Hence 1 gram-equivalent of H_2SO_4 is half of 98.08 g or 49.04 g.

The *normality* of an acid solution is the number of gram-equivalents of acid per liter of solution, and the normality of a basic solution is the number of gram-equivalents of base per liter of solution. The normality of a solution is designated by N and is a measure of the concentration of the ions (H^+ or OH^-) in the solution. Thus a $1N$ HCl solution contains 1 mole of HCl per liter, and a $1N$ H_2SO_4 solution contains 0.5 mole of H_2SO_4 per liter. Although their concentrations are different in terms of moles, a given volume of either solution can neutralize an equal volume of a $1N$ basic solution.

normality of a solution

strong base and weak acid

Neutralization is a more complicated process when the acid or the base is weak. As an example, suppose that we add NaOH slowly to acetic acid ($HC_2H_3O_2$), which is a weak acid. (The formula for acetic acid is written $HC_2H_3O_2$ because only one of the four H's has any tendency to be set free as H^+; the other three remain always a part of the acetate ion $C_2H_3O_2^-$.) As the base is added, the vinegarlike odor of $HC_2H_3O_2$ decreases and finally disappears. If we use amounts of solution containing equal numbers of molecules of base and acid, we find that the resulting mixture is basic (that is, it contains an excess of OH^-), whereas equal amounts of NaOH and a strong acid such as HCl would give a strictly neutral solution.

Why is there any difference? We can see the reason by writing the equation in terms of ions. To do this, we recall that NaOH, as a strong base, must be completely dissociated in solution; that $NaC_2H_3O_2$, as a soluble salt, must also be completely dissociated; that water is practically undissociated; and that acetic acid, as a weak acid, is present chiefly as undissociated molecules. So we write

$$Na^+ + OH^- + HC_2H_3O_2 \longrightarrow Na^+ + C_2H_3O_2^- + H_2O$$

which reduces to

$$OH^- + HC_2H_3O_2 \longrightarrow C_2H_3O_2^- + H_2O$$

Evidently H^+ is taken away from the acetate ion by OH^-, the H^+ and OH^- uniting to form water and the acetate ion being left free in solution. We can picture the OH^- and the $C_2H_3O_2^-$ as competing for the H^+; the $C_2H_3O_2^-$ ions, although they are attached to the H^+ ions originally, lose most of them because the attraction of OH^- ions is stronger. Now the important point is that the $C_2H_3O_2^-$ ions lose most of the H^+ ions, *but not all of them*. When we say that acetic acid is weak, we mean that there is a strong attraction between H^+ and $C_2H_3O_2^-$; it is not as strong as the attraction between H^+ and OH^- in water, but it is nevertheless strong enough to hold *some* of the H^+ out of reach of the OH^-. Hence some of the OH^- is not neutralized but remains free in the solution, and the solution is therefore basic.

<small>the neutralization of a weak acid by a strong base</small>

Generalizing from this reaction, we could say that the neutralization of a weak acid by a strong base is essentially a reaction of OH^- with undissociated acid molecules to form water and set free the negative ion of the acid. If equal numbers of acid molecules and OH^- ions are used, some of the OH^- will be left over because the acid retains some of its H^+ ions.

Suppose, now, we try to reverse the above reaction by mixing $C_2H_3O_2^-$ with H_2O. The most direct way to do this is to dissolve solid sodium acetate in water. The salt dissociates as follows:

$$NaC_2H_3O_2 \longrightarrow Na^+ + C_2H_3O_2^-$$

We now have a mixture of $C_2H_3O_2^-$ and water. Some Na^+ is present also, of course, but this ion is relatively inert and will not interfere. Next we add litmus or another indicator to the solution, and we find that the solution is slightly basic. Evidently some OH^- has been set free:

$$C_2H_3O_2^- + H_2O \longrightarrow HC_2H_3O_2 + OH^-$$

This reaction has not gone very far, since the concentration of OH^- is small, but apparently acetate ion has enough attraction for hydrogen ion so that it can take a little away from water and leave the OH^- free.

The above kind of reaction is called *hydrolysis*, and we say in general that any salt consisting of the ion of a strong base and the ion of a weak acid will hydrolyze on dissolving in water to give a basic solution. This is one of the reactions mentioned earlier by which substances whose formulas do not contain OH can react with water to give basic solutions. The alkalinity of solutions of washing soda (Na_2CO_3), borax ($Na_2B_4O_7$), water glass (Na_2SiO_3), and soap are all explained by similar hydrolysis reactions.

weak base and strong acid

We might guess that reactions will take place involving weak bases and their ions that are analogous to those described in the preceding section. When ammonium hydroxide is neutralized with hydrochloric acid, for example, and equal numbers of molecules of the acid and base are used, the resulting solution is slightly acid. This can be explained by writing the ionic equation

$$NH_4OH + H^+ + Cl^- \longrightarrow NH_4^+ + Cl^- + H_2O$$

in which HCl is written in terms of its ions because it is a strong acid, NH_4Cl is written in terms of ions because it is a soluble salt, and NH_4OH is written in molecular form because it is a weak base. The chloride ion may be omitted because it appears on both sides of the equation. Hence the true reaction is

$$NH_4OH + H^+ \longrightarrow NH_4^+ + H_2O$$

The essential process is the taking away of OH^- from NH_4OH by reaction with H^+, with the NH_4^+ ion then being left free. Thus the NH_4^+ and H^+ are in competition for OH^-, the H^+ getting the lion's share because its attraction for OH^- is greater. Most of the OH^- goes with H^+, but not all; some remains with NH_4^+ since ammonium hydroxide is a weak base. Because some of the H^+ is not neutralized, the resulting mixture is acid.

the neutralization of a weak base by a strong acid

The reverse reaction takes place when NH_4Cl is dissolved in water to give

$$NH_4Cl \longrightarrow NH_4^+ + Cl^-$$

Such a solution gives an acid reaction with litmus, indicating that the reaction

$$NH_4^+ + H_2O \longrightarrow NH_4OH + H^+$$

occurs. The ion of any other weak base will react similarly with water to set free a small amount of H^+. Such reactions account in part for the fact mentioned earlier, that many substances whose formulas contain no hydrogen can give acid solutions by reacting with water.

Reactions like those that occur when NH_4Cl and $FeCl_3$ are dissolved in water are further examples of hydrolysis. **Thus hydrolysis refers either to the formation of free OH^- by reaction of water with the negative ion of a weak acid or to the formation of free H^+ by reaction of water with the positive ion of a weak base.** (In the chemistry of carbon compounds it has a wider significance, but this definition is sufficient for our present purposes.)

hydrolysis

ammonia and ammonium ion

The weak base ammonium hydroxide deserves special attention. It is a common substance in the laboratory, in chemical industry, and around the house ("ammonia water"). It differs from most other common bases in that its positive ion is a combination of two nonmetal atoms rather than a metal.

Ammonium hydroxide may be thought of as being formed when a strong base is added to an ammonium salt or when ammonia (NH_3) is dissolved in water:

$$NH_3 + H_2O \longrightarrow NH_4OH$$

ammonium hydroxide is a weak base

Ammonia is a gas with a strong odor that is extremely soluble in water. Just how much of the NH_3 dissolved in water actually reacts to form NH_4OH is uncertain, and in all likelihood the latter only exist as such momentarily. As we said earlier, for most purposes it is immaterial whether we regard a solution of ammonia as containing NH_4OH or as a mixture of NH_3 and H_2O. In any event, the solution is a weak base in the sense that the concentration of OH^- ions is only a tiny fraction of the total amount of dissolved ammonia.

NH_4OH is neutralized by a solution of HCl, and a similar reaction takes place between gaseous NH_3 and gaseous HCl:

$$NH_3 + HCl \longrightarrow NH_4Cl$$

This reaction is conspicuous when bottles of HCl and NH_4OH solutions are opened side by side: The two gases, escaping from the solutions, mix and form a dense white cloud consisting of tiny particles of NH_4Cl.

ammonium salts

Ammonium chloride (commercially called "sal ammoniac"), although made up of three nonmetals, is a white crystalline substance that behaves like a salt. Its crystals are made up of the ions NH_4^+ and Cl^-, the NH_4^+ (ammonium ion) playing the role of a metal ion like K^+. Similar ionic compounds, called in general *ammonium salts*, are formed by the reaction of ammonium or ammonium hydroxide with the other acids: ammonium sulfate, $(NH_4)_2SO_4$, ammonium nitrate, NH_4NO_3, ammonium carbonate, $(NH_4)_2CO_3$, and so on.

Ammonium sulfate and ammonium nitrate, like ammonium chloride, give slightly acid solutions because the ammonium ion hydrolyzes. Ammonium carbonate, on the other hand, gives a basic solution because both of its ions hydrolyze and the hydrolysis of carbonate ion is more effective than that of ammonium ion:

$$(NH_4)_2CO_3 \longrightarrow 2NH_4^+ + CO_3^=$$
$$NH_4^+ + H_2O \longrightarrow NH_4OH + H^+$$
$$CO_3^= + 2H_2O \longrightarrow H_2CO_3 + 2OH^-$$

Ammonia is readily driven out of solution by heating, which reduces the solubility of NH_3. If we wish, we can regard the process as the decomposition of NH_4OH:

$$NH_4OH \longrightarrow NH_3 + H_2O$$

This reaction is the basis of a delicate test for the presence of ammonium salts in any solution: one need only add a strong base to give NH_4OH, and then heat the solution to produce the characteristic odor of NH_3. Similarly, the presence of a solid ammonium salt in a mixture can be tested for by heating with a solid hydroxide:

$$(NH_4)_2SO_4 + Ca(OH)_2 \longrightarrow CaSO_4 + 2H_2O + 2NH_3$$

acidic and basic oxides

In an earlier chapter we learned that one important distinction between metals and nonmetals is the fact that oxides of the former often react with water to form bases, whereas oxides of the latter often react with water to form acids. For example,

$$CaO + H_2O \longrightarrow Ca(OH)_2$$
$$SO_3 + H_2O \longrightarrow H_2SO_4$$
$$CO_2 + H_2O \longrightarrow H_2CO_3$$

Not only do these oxides dissolve in water to give acidic and basic solutions, but also in some reactions they may themselves play the roles of acids or bases. For example, Na_2O can "neutralize" H_2SO_4 quite as effectively as NaOH can, in the sense that it destroys its acidic properties:

$$Na_2O + 2H^+ + SO_4^= \longrightarrow 2Na^+ + H_2O + SO_4^=$$

or

$$Na_2O + 2H^+ \longrightarrow 2Na^+ + H_2O$$

Some oxides (CuO, Fe_2O_3, SiO_2) will not dissolve in water. Their relationship with acids and bases is shown, however, by the fact that they can be prepared by reactions of the type

$$Cu(OH)_2 \longrightarrow CuO + H_2O$$
$$2Fe(OH)_3 \longrightarrow Fe_2O_3 + 3H_2O$$
$$H_4SiO_4 \longrightarrow SiO_2 + 2H_2O$$

These equations show what happens when copper hydroxide, iron hydroxide, and silicic acid, respectively, are heated.

Because any oxide may be regarded as derived from an acid or base (although sometimes the acid or base is purely hypothetical), oxides of metals in general are referred to as *basic oxides* and oxides of nonmetals as *acidic oxides*. Thus SiO_2 (the chief constituent of ordinary sand) is called an acidic oxide, although it is insoluble in water and neither tastes sour nor turns litmus red, while Fe_2O_3 (the chief constituent of hermatite ore) is called a basic oxide, although it neither tastes bitter nor turns litmus blue.

basic and acidic oxides

Certain oxides have the ability to neutralize both acids and bases. An example is zinc oxide (ZnO), which can undergo both of the following reactions:

$$ZnO + 2H^+ \longrightarrow Zn^{++} + H_2O$$
$$ZnO + 2OH^- + H_2O \longrightarrow Zn(OH)_4^=$$

An oxide of this kind is called *amphoteric*.

amphoterism

GLOSSARY

The *hydronium ion* consists of a water molecule with a hydrogen ion attached; its symbol is H_3O^+. (Sometimes more than one water molecule may be attached to a single hydrogen ion; the notion of the hydronium ion is for convenience only.)

An *acid* is any substance whose molecules contain hydrogen and whose water solution contains hydrogen ions.

A *base* is any substance whose molecules contain OH groups and whose water solution contain OH^- ions.

A *salt* is one of a class of compounds most of which are crystalline solids at ordinary temperatures and most of which consist of a metal combined with one or more nonmetals. Any salt can be formed by mixing the appropriate acid and base and evaporating the solution to dryness.

The *neutralization* of a strong acid by a strong base in water solution is a reaction between hydrogen and hydroxide ions to form water.

An *indicator* is a substance that changes color when neutralization occurs.

The pH of a solution is a method for expressing the exact degree of its acidity or basicity in terms of its hydrogen-ion concentration. A pH of 7 signifies a neutral solution, a smaller pH than 7 signifies an acid solution, and a higher pH than 7 signifies a basic solution.

Hydrolysis refers to both the formation of a basic solution when the negative ion of a weak acid reacts with water and the formation of an acidic solution when the positive ion of a weak base reacts with water.

EXERCISES

1 How can you tell whether an unknown solution is acidic, basic, or neutral?

2 Which of the following are weak acids, and which are weak bases? H_2SO_4, $HC_2H_3O_2$, NH_4OH, H_2CO_3, HCl, NaOH, H_3BO_3.

3 Would you expect HBr to be a weak or strong acid? Why?

4 Write the ionic equation for the neutralization of KOH by HNO_3. What actual chemical changes does this equation show?

5 How could you tell whether an unknown mixture of salts contains (*a*) a salt of ammonium ion, (*b*) a salt of carbonate ion?

6 What reaction would take place if $FeCl_3$ solution were added to KOH solution?

ACIDS, BASES, AND SALTS

7 Boric acid (H_3BO_3) is a weaker acid than carbonic acid. What would happen if solutions of Na_3BO_3 (sodium borate) and HCl were mixed? Would you expect a solution of Na_3BO_3 to be acidic, basic, or neutral?

8 If the following salts are dissolved in water, which would give acidic solutions, which alkaline solutions, which neutral solutions? Na_2CO_3, KCl, $KC_2H_3O_2$, $BaCl_2$, $(NH_4)_2SO_4$, $NaNO_3$.

9 How could you prepare the weak acid H_2S from the salt Na_2S (sodium sulfide)?

10 From the fact that H_2S is a weak acid, would you predict that a solution of Na_2S would be acidic, basic, or neutral? Explain.

11 Which hydrolyzes more, $CO_3^=$ or $C_2H_3O_2^-$? What is the general relation between the extent of hydrolysis of a negative ion and the weakness of the corresponding acid?

12 One common type of baking powder ("alum" baking powder) contains a salt of aluminum, such as $Al_2(SO_4)_3$. Explain how such a salt can furnish acid to liberate CO_2 (from $NaHCO_3$) when the baking powder is mixed with water. [Use the fact that $Al(OH)_3$ is insoluble and a weak base.]

13 What reaction takes place when a solution of $Ca(C_2H_3O_2)_2$ is added to a solution of H_2SO_4?

part seven

CHEMISTRY IN ACTION

We now come to what most of us think of as the fundamental subject matter of chemistry: chemical reactions and their products.

First we shall consider chemical energy—the energy that powers and heats our bodies, for instance, is chemical energy obtained from the oxidation of food. An important example of a reaction for which external energy is necessary is photosynthesis, in which solar radiant energy is absorbed by plants during their production of carbohydrates from water and carbon dioxide. While in the original sense oxidation and reduction refer to reactions involving respectively the gain and loss of oxygen by a substance, the chemist finds these terms valuable in describing a broad category of reactions.

Organic chemistry, which treats of carbon compounds, is the last topic in our study of chemistry. The profusion and variety of carbon compounds result from the ability of carbon atoms to form strong bonds with other carbon atoms, so that chains and rings of carbon atoms occur in organic molecules. Industrial organic chemistry with its synthetic fibers, elastomers, and plastics and the organic chemistry of life are important aspects of this branch of chemistry.

chapter 26
Chemical Reactions

CHEMICAL ENERGY
- exothermic and endothermic reactions
- nature of chemical energy
- activation energies
- fuels
- explosives

REACTION RATES
- temperature
- concentration and surface area
- catalysis

CHEMICAL EQUILIBRIUM
- reversible reactions
- chemical equilibrium
- le châtelier's principle
- synthetic ammonia

Chemical reactions have significant aspects quite apart from the changes that occur when the reactants combine to yield the products. For instance, some reactions evolve energy whereas others require energy to be supplied externally if they are to take place. Even those reactions that liberate energy may not occur unless an initial amount of energy is furnished to start the process. Chemical changes may be almost instantaneous or may take years to be completed, depending upon many factors. And not all reactions can ever actually be "completed": often an intermediate equilibrium situation exists with the products undergoing reverse reactions to form the starting substances just as fast as the primary reaction proceeds. These and still other considerations are involved in actual chemical reactions, and they form the subject of this chapter.

CHEMICAL ENERGY

Ever since our ancestors learned the value of fire, mankind has been putting chemical energy to practical use. Today we transform it not only into heat and light but into mechanical energy and electric energy as well. Locked up in the atoms of matter, chemical energy long remained a mystery. Modern theories of the atom and of the chemical bond, however, have given us a great deal of insight into the origin of this energy.

exothermic and endothermic reactions

exothermic reactions liberate energy

Chemical changes that *liberate* energy are called *exothermic reactions*. Familiar examples are the burning of coal and the explosion of a mixture of hydrogen and oxygen. The energy liberated is often stated in the equation for the process:

$$C + O_2 \longrightarrow CO_2 + 94.4 \text{ kcal}$$
$$2H_2 + O_2 \longrightarrow 2H_2O + 117 \text{ kcal}$$

These figures represent the heat produced when an amount of each substance is used equal to its formula mass expressed in grams multiplied by its coefficient in the equation; in other words, by the number of moles given by the coefficient. When 1 mole (12 g) of carbon is burned, 94.4 kcal is produced; when 2 moles (4.032 g) of hydrogen is burned, 117 kcal is produced. These particular amounts are chosen so that the energies liberated for similar numbers of molecules may be compared for different reactions.

endothermic reactions absorb energy

Chemical changes that take place only when heat or some other kind of energy is *supplied* are called *endothermic reactions*. Thus water can be decomposed into hydrogen and oxygen only by heating to very high temperatures or by supplying electric energy during electrolysis:

$$2H_2O + 117 \text{ kcal} \longrightarrow 2H_2 + O_2$$

The formation of nitric oxide (NO) from its elements is an endothermic reaction, which takes place only at high temperatures:

$$N_2 + O_2 + 43.2 \text{ kcal} \longrightarrow 2NO$$

From the law of conservation of energy we might predict that, if a given reaction is exothermic, the reverse reaction will be endothermic and, further, that the amount of heat liberated by one reaction must be equal to the amount absorbed by the other. This prediction is borne out in the case of water, as we can see above, and might be checked by any number of other reactions. For example, sodium burning in chlorine liberates 197 kcal for every 2 moles (46 g) of sodium:

$$2Na + Cl_2 \longrightarrow 2NaCl + 197 \text{ kcal}$$

and NaCl is decomposed in an endothermic process requiring the absorption of this same amount of heat:

$$2NaCl + 197 \text{ kcal} \longrightarrow 2Na + Cl_2$$

Energy changes accompanying ionic reactions are represented in equations in the same manner as energy changes for other reactions. The dissociation of most salts is an endothermic process; for example, when KNO_3 is dissolved in water, the container becomes cold, since dissociation of the salt absorbs heat from its surroundings:

dissociation is usually an endothermic process

$$KNO_3 + 8.5 \text{ kcal} \longrightarrow K^+ + NO_3^-$$

Neutralization, on the other hand, is an exothermic ionic process. If concentrated solutions of NaOH and HCl are mixed, for instance, the mixture quickly becomes too hot to touch:

neutralization is an exothermic process

$$H^+ + OH^- \longrightarrow H_2O + 13.7 \text{ kcal}$$

The neutralization of any strong acid by any strong base liberates almost precisely this same amount of heat for each mole (1.008 g) of H^+—as might be expected, since the actual chemical change in all cases is simply the joining together of hydrogen ions and hydroxide ions.

nature of chemical energy

The heat given out or absorbed in a chemical change is an approximate measure of the chemical energy possessed by the substances which react, and hence also a measure of their stability. If much energy is required to decompose a substance, that is, if its decomposition is strongly endothermic, the substance is (with rare exceptions) relatively stable; if its decomposition is weakly endothermic or exothermic, the substance is in general unstable. From the reactions above, we can see at a glance that CO_2, H_2O, and NaCl are stable compounds, since the formation of each is strongly exothermic and its decomposition endothermic. NO, on the other hand, is unstable, since its decomposition liberates heat. We can say further that the combinations H_2 and O_2, Na and Cl_2, H^+ and OH^- are relatively unstable, since they react with evolution of much energy, while N_2 and O_2 form a stable mixture.

stable and unstable substances

The general interpretation of chemical-energy changes in terms of molecular structure follows readily from our earlier discussion of the chemical bond. When sodium reacts with chlorine, for example, an electron from each sodium atom is transferred to the outer shell of a chlorine atom, a position in which it has a smaller amount of potential energy with respect to the atomic nuclei. When carbon reacts with oxygen, the atoms are joined by electron pairs, with the formation of covalent bonds also involving a decrease in the potential energy of the electrons. Thus, **chemical energy is electron potential energy;** when electrons

chemical energy is electron potential energy

CHEMISTRY IN ACTION

An oxyacetylene welding torch. Note the two supply hoses, one for acetylene (C_2H_2) and the other for compressed oxygen. Part of the heat generated comes from the decomposition of the acetylene molecules, and part from the oxidation of the resulting carbon and hydrogen atoms.

move to new positions in an exothermic reaction, some of their potential energy is liberated. The surplus energy of the shifted electrons causes a violent disturbance in the new-formed molecules: It may give the molecules themselves motions that correspond to heat energy or it may excite outer electrons into higher energy levels from which they return with the emission of radiant energy. In endothermic reactions, some other form of energy must be supplied to increase the potential energies of electrons in order that the reactions occur.

heats of formation are greatest for molecules whose atoms are farthest apart in electronegativity

The energy change in a chemical reaction is evidently closely connected with the precise nature of the bonds between the atoms in the substances involved. In general, bond strengths are greatest between atoms far apart on the electronegativity scale (Chap. 20), least between atoms close together. Thus we might expect that compounds between metals and nonmetals, which in general are at opposite ends of the scale, have considerable heats of formation, while compounds between nonmetals of similar electronegativity have small heats of formation. The relationship between the heat of formation of a compound and the electronegativity difference between its atoms is not always a simple one, to be sure: for example, the presence of multiple bonds between atoms instead of single ones, as in O_2 with two covalent bonds and N_2 with three covalent bonds, means additional stability and so a high heat of formation.

Both the energy of chemical reactions and the energy of nuclear processes are in a sense "atomic" energy, in that both involve the elementary particles which make up atoms. But the parts of the atoms concerned are different. The chemical energy evolved in combustion and other exothermic reactions depends on rearrangements in the outer part of the electron clouds of atoms; the energy of nuclear reactors and weapons, popularly called "atomic energy," is produced by rearrangements of the protons and neutrons of atomic nuclei. So much more energy is contained in the nuclei of atoms than in their electron clouds that atomic energy in the future is likely to supplant chemical energy for many purposes.

activation energies

Coal burns in air to give great quantities of heat; how, then, can coal be kept indefinitely at ordinary temperatures in contact with air? The decomposition of nitric oxide liberates considerable energy; why does it not, therefore, break up spontaneously? How can the compound exist at all? A mixture of hydrogen and oxygen will produce a violent explosion; why should heat or an electric spark be necessary to start the explosion? Why, in general, do not all exothermic reactions take place instantaneously of their own accord?

Experience indicates that many exothermic processes occur only if some energy is provided to start them. A mixture of hydrogen and oxygen may be likened to the car of Fig. 26.1, whose potential energy may be converted into kinetic energy if it moves down into the deep valley. However, it can do this only if it is first given sufficient energy to climb to the top of the intervening hill. Similarly, the chemical energy stored in the

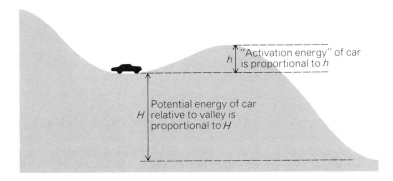

26.1 Activation energy. *The potential energy of the car will be converted into kinetic energy if it moves down into the valley. However, it requires an initial kinetic energy in order to climb the hill between it and the valley, analogous to the activation energy required in many exothermic reactions.*

electrons of hydrogen and oxygen can be liberated as heat only if the molecules have sufficient energy, or are sufficiently *activated*, to make the reaction start. The energy necessary for activation, corresponding to the energy required to move the car up the first hill, is called the *activation energy* of the reaction.

The electronic picture of chemical combination gives a plausible reason why activation should be necessary. The combination of oxygen and hydrogen involves the formation of electron-pair bonds between O and H atoms, a process that gives out energy, but, before these bonds can be formed, the covalent bonds between hydrogen atoms in H_2 molecules and oxygen atoms in O_2 molecules must be broken. To break these bonds requires energy. The energy is in part supplied by the thermal energy of the molecules and, after the reaction starts, by the energy it gives out, but initially some additional energy must be supplied to loosen the bonds.

an activated molecule has enough energy to participate in a reaction

A molecule with sufficient energy above the average to enable it to react is called an *activated molecule*. In some gas reactions an activated molecule may be actually broken down into atoms. Other activated molecules may simply have high kinetic energy, or they may have one or two electrons in excited states. In reactions that take place spontaneously at room temperatures (for example, the reaction between hydrogen and fluorine), enough of the molecules have sufficient heat energy to break the necessary bonds without further activation. In many ionic reactions no bonds need be broken; ions are, so to speak, already activated and react almost instantaneously. But a great number of exothermic reactions require the preliminary activation of some molecules before they can take place at appreciable rates.

Once an exothermic reaction is started, it usually supplies its own activation energy. In other words, the energy given out when some of the molecules react supplies neighboring molecules with sufficient energy to activate them, so the reaction spreads quickly. Thus, a mixture of hydrogen and oxygen need only be touched with a flame for the reaction to spread so rapidly that an explosion results. When a bed of coal is set on fire it continues to burn, since the heat liberated in one part is sufficient to ignite the coal in adjacent parts.

fuels

The first requirement of a good fuel is naturally that its combination with oxygen should be a strongly exothermic reaction. Other requirements are that it should be cheap, that it should be easy to store, that the products of combustion should be easily disposed of. Many substances satisfy the first requirement, but only a few fulfill the other three as well. Sodium, for example, would be an excellent fuel as far as its heat-producing qualities go, but it is expensive, it must be stored under oil, and getting rid of its oxide would be a problem.

The substances that best fulfill the requirements for fuels are carbon and some of its compounds. These occur in nature as wood, coal, petroleum, and natural gas, materials which may themselves be used as fuels or which may be converted into artificial fuels for special purposes. Fuels containing carbon are abundant and cheap, and their inertness at ordinary temperatures makes them easy to store. The principal products of combustion are two gases, CO_2 and water vapor, which escape into the air, and the ashes from wood and coal are inactive materials that can be easily removed.

Liquid Fuels. Most liquid fuels are obtained from the black, oily liquid called *petroleum*, a complex mixture of compounds of carbon and hydrogen. Distillation separates petroleum into simpler mixtures: Some of the more volatile constituents form the mixture called gasoline, those with slightly higher boiling points kerosene, those with still higher boiling points diesel oil, lubricating oil, petroleum jelly, and paraffin. By far the most important liquid fuel is gasoline, which is widely used to produce mechanical energy in internal combustion engines.

petroleum

The ability of gasoline to produce mechanical energy depends on the large amount of heat (11 kcal/g) given out when it burns. The gaseous products of burning expand very rapidly because they are intensely heated by the reaction, and this expansion forces down the pistons of the engine which in turn cause the crankshaft to turn (see p. 158). The equation for the burning of heptane, one of the constituents of gasoline, is

$$C_7H_{16} + 11O_2 \longrightarrow 7CO_2 + 8H_2O$$

Fifteen molecules result from the reaction of 12, which means that the volume of the products is not much greater than that of the original substances at ordinary temperatures, and so their expansion is due almost entirely to the considerable rise in temperature.

Gas Fuels. Gas fuels, like liquid fuels, leave no solid ash on burning but are converted entirely to CO_2 and H_2O. *Natural gas* is often found with petroleum, and like petroleum consists of carbon-hydrogen compounds; its chief constituent is methane (CH_4). Artificial gas fuels are produced from coal or coke. The one most widely used is *water gas*, a mixture of hydrogen and carbon monoxide produced by passing steam over hot coke:

methane is the chief constituent of natural gas

$$H_2O + C \longrightarrow CO + H_2$$

Hydrogen itself would be an excellent gas fuel, since its heat of combustion is very high (34.5 kcal/g). It is too expensive for ordinary purposes but is widely used in the oxyhydrogen blowtorch and as a fuel in advanced rocket engines.

Solid Fuels. *Coal*, long the most important source of industrial energy, now takes second place to petroleum. The heat of combustion of good bituminous (soft) coal is about 7.8 kcal/g, that of good anthracite coal about 7.4 kcal/g. These values are close to the heat of combustion of

coal

pure carbon (7.9 kcal/g), but the similarity is partly accidental: most of the heat produced by burning coal comes from the combustion of carbon compounds, and coal always contains a fairly large percentage of unburnable ash. *Coke* is a fuel derived from coal by heating it in the absence of air; volatile constituents are driven off, leaving free carbon and ash in the coke. Coke is used in place of coal where a hotter and less smoky fire is desirable. *Wood* is chiefly cellulose, one of the carbohydrates that plants produce by photosynthesis (see p. 503). Its heat of combustion is less than that of coal, ranging from 2.5 to 4.5 kcal/g. The exact structure of cellulose is unknown, but its burning may be represented roughly by the equation

$$C_6H_{10}O_5 + 6O_2 \longrightarrow 6CO_2 + 5H_2O + 671 \text{ kcal}$$

This reaction is the reverse of photosynthesis: Radiant energy from sunlight, stored in wood as chemical energy, is released by burning as heat energy. The energy of burning coal comes ultimately from the same source, for coal consists of plant material buried beneath layers of sediment and altered by long ages of slow chemical change.

explosives

Explosives are unstable substances which react to liberate large quantities of gas suddenly, the expansion of the gas producing the desired mechanical energy. We refer to mixtures of hydrogen and oxygen, or of gasoline vapor and oxygen, as "explosive mixtures," since the gases produced by their reactions are heated enough to expand suddenly; but the term *explosive* usually means a solid or liquid material which reacts to form gas when appropriately disturbed.

most explosives contain nitrogen

Most explosives contain compounds of nitrogen, and the chemistry of explosives is in large part the chemistry of this element. Nitrogen itself is an extremely inactive gas, uniting with other elements only when strongly heated. The chief reason for this inactivity is the strength of the triple electron-pair bonds between nitrogen atoms in N_2 molecules. If these bonds are broken momentarily by passing an electric discharge through nitrogen, the resulting N atoms are much more active than the ordinary molecules. Because the N_2 molecule is so stable, nitrogen atoms unite to form it very readily, even when they are already joined with other atoms in compounds. In other words, many compounds containing nitrogen are unstable and decompose readily to form nitrogen gas. This property makes nitrogen compounds valuable in explosives.

gunpowder

The earliest widely used explosive was *gunpowder,* made by mixing potassium nitrate (KNO_3) with charcoal and sulfur. Gunpowder explodes because the charcoal combines with oxygen from KNO_3 to form CO and CO_2, and nitrogen gas is set free; most of the sulfur unites with potassium to form solid particles of smoke. The reaction is highly exothermic and takes place very suddenly, its expanding gaseous products exerting enormous pressures in every direction.

Modern explosives give gaseous products entirely, and contain carbon, oxygen, and nitrogen united in the same molecule so that the explosion

CHEMICAL REACTIONS

The largest dynamite explosion ever produced blasts an underwater obstruction at Ripple Rock in the coastwise route from Seattle to Alaska. 1,375 tons of explosive were detonated.

can be even more rapid. They are made by the action of nitric and sulfuric acids on various carbon compounds, such as glycerin, cellulose, and toluene; from these three, respectively, come the explosives *nitroglycerin* [$C_3H_5(NO_3)_3$], *nitrocellulose* [$C_6H_7O_2(NO_3)_3$], and *trinitrotoluene* or *TNT* [$C_7H_5(NO_2)_3$]. These are alike in that they contain the same four elements, and that their products are nitrogen, the oxides of carbon, and water vapor. For example, when nitroglycerin explodes,

decomposition of nitroglycerin

$$4C_3H_5(NO_3)_3 \longrightarrow 12CO_2 + 10H_2O + 6N_2 + O_2$$

Nitroglycerin is dangerous to handle, for it explodes at the slightest shock. Nitrocellulose and TNT require a fairly strong shock, often provided by a small amount of another explosive in "percussion caps." *Dynamite* is a mixture of nitrocellulose and nitroglycerin, with various amounts of sawdust, flour, and salts added to adjust the speed of explosion and to make the material safe to work with.

REACTION RATES

Some chemical changes are practically instantaneous. In neutralization, for example, acid and base react as soon as they are stirred together; silver chloride is precipitated immediately when solutions of silver ion and chloride ion are mixed; the reaction involved in a dynamite explosion is exceedingly rapid. Other chemical changes, like the formation of ammonia and the rusting of iron, take place slowly. For many of these slow reactions, we can set up experiments to measure exactly how fast they are going, that is, what fractions of the original substances have disappeared at various times after the reactions start.

the speed of a reaction depends upon temperature, concentration, surface area, and the presence of a catalyst

Reaction rates depend first of all on the nature of the reacting substances. Obviously some materials undergo chemical change more rapidly than others: Meat decays faster than wood; iron corrodes faster than copper. For any particular reaction the rate is influenced by four principal factors: (1) temperature, (2) concentrations of the reacting substances, (3) amount of surface exposed (in reactions involving solids), and (4) catalysts.

temperature

a 10°C rise in temperature approximately doubles the speed of a chemical reaction

Reaction rates are always increased by a rise in temperature. We use this simple fact, of course, when we put food in the refrigerator to retard its decay and when we use hot water rather than cold for washing. As a rule, reaction rates are approximately doubled for every 10°C rise in temperature.

The kinetic theory of matter suggests one obvious reason for the increase of rate with temperature: Most reactions depend on collisions between particles, and the number of collisions increases with rising temperature because molecular speeds are increased. But a 10°C rise at ordinary temperatures is very far from enough to double the number of collisions in a particular sample. To find an adequate explanation, we must go back to the idea of activation energy.

If molecules must be activated before they can react, reaction rates should depend not on how many ordinary collisions occur each second but on the number of collisions between *activated* molecules. Now activated molecules in a fluid may be produced by ordinary molecular motion as a result of exceptionally energetic collisions; these remain activated for only a small fraction of a second, losing their excess energy by further collisions unless they react in the meantime. So, in any fluid, provided the temperature is not too low, a fraction of the particles should be activated at any one instant. The fraction may be very small at ordinary temperatures, but it increases rapidly as the temperature rises and molecular motion speeds up. Reaction rates increase with temperature chiefly because the number of activated molecules grows larger.

reaction rates increase with temperature because of an increase in the number of activated molecules

A mixture of hydrogen and oxygen, for example, contains at room temperature very few molecules with sufficient energy to react, and the reaction is so exceedingly slow that the gases may remain mixed for years without appreciable change. Even at 400°C the rate is negligibly small, but at 600° enough of the molecules are activated to make the reaction fast, and at 700° so many are activated that the mixture explodes. This kind of behavior is typical of many reactions, especially those involving molecules whose bonds must be broken: At low temperatures the chemical change is so slow that for all practical purposes it does not occur; in a range of intermediate temperatures the reaction is moderately rapid; and at high temperatures it becomes instantaneous. Reactions between ions, on the other hand, are practically instantaneous even at room temperatures, for the ionic state itself is a form of activation.

ionic reactions are practically instantaneous

concentration and surface area

The general effect of concentration on reaction speed is well shown by rates of burning in air and in pure oxygen; the pure gas has almost five times as many oxygen molecules per cubic centimeter as air has, and rates of burning are very much greater. The concentration effect appears even more spectacularly in the burning of an iron wire in liquid air; despite the low temperature, oxygen molecules are so abundant and so close together in the liquid that the metal burns brightly.

As a general rule, the rate of a simple chemical reaction is directly proportional to the concentration of each reacting substance. This is an experimental result, for which the kinetic theory gives a reasonable explanation: The number of collisions between activated molecules, which determines the reaction speed, should depend on the total number of collisions and this, in turn, on how many molecules each cubic centimeter contains.

reaction rates are proportional to the concentrations of the reacting substances

When a reaction takes place between two solids or between a fluid and a solid, its rate depends markedly on the amount of solid surface exposed. A finely powdered solid presents vastly more surface than a few large chunks, and reactions of powders are accordingly much faster. Granulated sugar dissolves more rapidly in water than lump sugar; finely

the greater the surface area, the faster the reaction

A culture of the unicellular alga Chlorella is used here in the study of photosynthesis.

divided zinc is attacked by acid quickly, large pieces only slowly; ordinary iron rusts slowly, but, if the metal is very finely powdered, its oxidation is fast enough to produce a flame. A kinetic explanation is obvious enough: The greater the surface, the more quickly molecules can get together to react. For a similar reason, efficient stirring speeds up reactions between fluids.

catalysis

Harder to explain is the action of catalysts. These are substances with the remarkable property of changing the rate of a chemical reaction without being themselves altered or used up.

A catalyst may either speed up or retard a reaction. For a simple example, hydrogen peroxide solutions are unstable at ordinary temperatures, slowly decomposing into water and oxygen. If a little of the black powder called manganese dioxide is added to hydrogen peroxide, the decomposition becomes much more rapid, and oxygen bubbles from the solution in large quantities. At the end of the reaction the manganese dioxide catalyst can be recovered unchanged. Ordinary solutions of hydrogen peroxide contain a little of a carbon compound called acetanilid, a catalyst of the opposite sort, that slows down the decomposition.

catalysts are substances that change the rate of a chemical reaction

An extremely important example of *catalysis* (the name given to the action of a catalyst) occurs in nature. This is the formation of carbohydrates in the leaves of green plants, which involves a direct conversion of radiant energy from sunlight into chemical energy. Carbohydrates are complex compounds of carbon, including sugar, starch, and cellulose—the first two vitally important foods, the last a major constituent of wood, paper, and many kinds of cloth.

Plants are able to manufacture carbohydrates out of water, which enters through their roots, and CO_2, which is taken from the air, according to the reaction

photosynthesis is the manufacture by plants of carbohydrates from water and carbon dioxide

$$6CO_2 + 5H_2O + 671 \text{ kcal} \longrightarrow C_6H_{10}O_5 + 6O_2$$

The reaction is highly endothermic, with the necessary energy coming from sunlight. The energy is absorbed not by the CO_2 and H_2O directly but instead by a substance called *chlorophyll*, which is part of the green coloring matter of leaves; the catalyst chlorophyll is not changed by the reaction but serves to pass on the sun's energy to the reacting molecules. This reaction is called *photosynthesis*, since light is necessary for its occurrence. Mankind depends on photosynthesis not only for the chemical energy in carbohydrates but also for the constant replenishment of oxygen in the atmosphere (Fig. 26.2). Approximately 70 billion tons of carbon dioxide are removed from the atmosphere each year by plants and returned by them as oxygen, and a corresponding amount of oxygen is simultaneously being converted by plants and animals into carbon dioxide.

chlorophyll is the catalyst for photosynthesis

Catalysts may affect reaction rates in several ways. In some cases, the catalyst is known to form an unstable intermediate compound with one of the reacting substances, which decomposes again as the reaction pro-

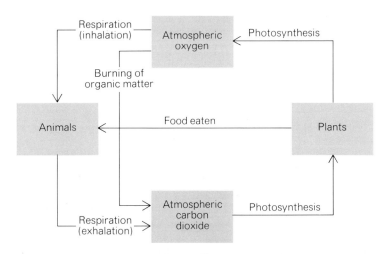

26.2 The oxygen-carbon dioxide cycle in the atmosphere.

ceeds. Other catalysts, notably certain metals, can affect reaction rates by producing activated molecules at their surfaces. For the action of many others no adequate explanation has been suggested. In general, a given reaction is influenced only by a few catalysts, and these may or may not affect other reactions. Catalysts are highly important in many industrial processes, but in searching for new ones a chemist must usually rely more on experience and trial-and-error methods than on any comprehensive knowledge of how catalysts work.

CHEMICAL EQUILIBRIUM

reversible reactions

most chemical reactions are reversible

Most chemical reactions are *reversible*. That is, the products of a chemical change, under suitable conditions, can usually be made to react to give the original substances. We have discussed many examples in other connections. Hydrogen and oxygen combine to form water when ignited, and water can be decomposed into its elements at extreme temperatures or by electrolysis. Mercury and oxygen combine when heated moderately, and mercuric oxide decomposes when heated more strongly. Carbon dioxide dissolves in water to form carbonic acid, and carbonic acid decomposes into carbon dioxide and water.

There is no reason, of course, why the forward and backward processes of a chemical change cannot take place simultaneously, provided their rates are approximately equal. If CO_2 is kept over water in a stoppered bottle, for instance, some of the gas reacts with water. But a number of the resulting ions join together to give CO_2 and H_2O as soon as an appreciable concentration is built up. The recombination rate increases with concentration until, finally, as many ions recombine each second as are formed by the dissolving of CO_2. At this point the rates of the forward and backward reactions are the same, and no further change occurs in the amounts of the various substances. We may represent this situation by the single equation

$$H_2O + CO_2 \rightleftharpoons H^+ + HCO_3^-$$

in which the double arrow indicates that the reactions in both directions occur together.

chemical equilibrium

A situation of this kind is called *chemical equilibrium*. It is a state of balance determined by two opposing processes. The two processes do not reach equilibrium and stop, but continue indefinitely to maintain a balance because one constantly undoes what the other accomplishes. As a crude analogy, we might imagine a man walking down an escalator while the escalator is moving upward; if he walks as fast in one direction as the escalator carries him in the other, the two motions will be in equilibrium and he will remain at the same place indefinitely.

a chemical equilibrium is a balance between forward and reverse reactions

A great many chemical changes reach a state of equilibrium instead of going to completion in one direction or the other. Equilibrium may be established when a reaction is very nearly complete, or when it is only just starting, or when both products and reacting substances are present in comparable amounts. At what point equilibrium occurs depends entirely on the rates of the opposing reactions; one reaction takes place until a sufficient concentration of products is built up for the reverse reaction to go at the same rate. The ionizations of different acids offer good examples. HCl dissociates completely into H^+ and Cl^-; here there is no reverse reaction and no equilibrium, except in very concentrated solutions. On the other hand, $HC_2H_3O_2$ ionizes to only a slight extent, and when a small concentration of ions is built up, the recombination goes at the same rate as the dissociation. The extent to which an acid is ionized depends on how fast its molecules break down into ions compared with how fast the ions recombine.

many chemical reactions do not proceed to completion but reach an intermediate equilibrium state instead

Frequently a chemist encounters this problem: he wishes to prepare a compound but finds that the reaction that produces it reaches equilibrium before much of the compound has been formed. Once equilibrium is established, waiting for more of the product to form is futile, for its amount thereafter does not change. How can the equilibrium conditions be altered, so that the yield of the product will be larger?

Since equilibrium depends on a balance between two rates, a solution to this problem depends on finding a way to change the speed of one reaction or the other. Speeding up or retarding one of the reactions in an equilibrium is not quite so simple as changing the rate of a single reaction, but we might expect that the same general factors that affect reaction rates would influence equilibrium also. The chemist has three chief methods at his disposal for shifting an equilibrium in one direction or another:

1 Changing the concentration of one or more substances
2 Changing the temperature
3 Changing the pressure (especially in gas reactions)

methods for adjusting a chemical equilibrium

Since, generally, the two opposite reactions involved in a chemical equilibrium are affected differently by these three factors, their proper adjustment often gives a yield of the desired reaction product larger than would otherwise be possible. We may note that neither using a catalyst nor changing the amount of exposed surface influences equilibrium, since these factors always affect forward and backward actions alike.

le châtelier's principle

The above methods for adjusting an equilibrium are illustrative of *Le Châtelier's principle,* which was proposed by the French chemist H. L. Le Châtelier in 1884. In its most general form, this principle states that **when the conditions of a system in equilibrium are altered, the equilibrium will shift in such a way as to attempt to restore the initial conditions.**

a system in equilibrium tends to maintain its initial conditions

As a simple example, we might consider the decomposition of mercuric oxide by heating in a closed container:

$$2HgO \rightleftharpoons 2Hg + O_2$$

The reaction is reversible in a closed container because mercury and oxygen are able to combine directly to form mercuric oxide. To increase the yield of Hg all we need do is pump out the oxygen produced, which alters the equilibrium to favor the decomposition reaction as the system attempts to maintain the original oxygen concentration. Since the oxygen is being removed as soon as it is formed, the attempt must fail, and all the HgO will be decomposed.

an equilibrium can be shifted by changing the concentration of one of the reactants or products

As a rule, changing the concentration of one of the products in an equilibrium affects the speed of the backward reaction without affecting that of the forward one, whereas a change in the concentration of one of the original substances affects the speed of the forward reaction only. Thus we may change the rate of either reaction separately and hence shift the equilibrium in one direction or the other.

If one reaction in an equilibrium is exothermic, the opposite reaction must be endothermic. An increase in temperature, of course, accelerates both reactions, but by Le Châtelier's principle the endothermic reaction—the one which absorbs heat—is favored more than the other. As an example, we can consider the formation of NO from N_2 and O_2, a moderately endothermic reaction. At fairly high temperatures NO decomposes exothermically, so that an equilibrium is established:

$$N_2 + O_2 + 44 \text{ kcal} \rightleftharpoons 2NO$$

The forward reaction absorbs heat, the backward reaction evolves heat. Hence a rise in temperature favors the formation of NO, and for good yields of this gas the reaction must be carried out at very high temperatures.

an equilibrium can be shifted by changing the temperature

Thus a knowledge of the energy change in an equilibrium reaction gives us another method of causing the equilibrium to shift in one direction

CHEMICAL REACTIONS

The first step in the manufacture of artificial fertilizers is the fixation of atmospheric nitrogen in ammonia. Liquid fertilizer hardens into pellets as it is dropped from the top of the square tower in this fertilizer plant in Tennessee.

or the other. If the desired product is formed by an endothermic process, the reaction should be carried out at high temperatures; if it is formed by an exothermic reaction, low temperatures are more favorable.

an equilibrium involving a gas can be shifted by changing the pressure

If a gas reaction involves a change in volume, its equilibrium position may be shifted by changing the total pressure. An increase in pressure causes the reaction to take place in a direction which gives the smaller number of molecules, whereas a decrease in pressure favors the opposite reaction. High pressure, so to speak, squeezes the reaction mixture into as small a volume as possible. For example, at fairly high temperatures equilibrium is established among the three gases oxygen (O_2), sulfur dioxide (SO_2), and sulfur trioxide (SO_3):

$$2SO_2 + O_2 \rightleftharpoons 2SO_3$$

If the pressure is constant, the SO_3 occupies a volume only two-thirds as great as the volume of the gases from which it formed. Hence an increase in pressure favors the forward reaction: To obtain large yields of SO_3, the reaction should be carried out at as high a pressure as possible.

synthetic ammonia

As we have learned, there are several means which a chemist can use to increase the yields of substances involved in equilibrium systems. Let us see how the use of these means led the German chemist Fritz Haber to a successful commercial method for the preparation of ammonia from nitrogen and hydrogen.

The reaction is reversible, forming an equilibrium represented by the equation

synthesis of ammonia

$$N_2 + 3H_2 \rightleftharpoons 2NH_3$$

The forward reaction is exothermic, so that the yield of ammonia is greatest at low temperatures. But the reaction has a high activation energy and is exceedingly slow at ordinary temperatures; only above 700°C does it become rapid. At 700°, however, the yield of ammonia is far too small for practical use, as the second column of Table 26.1 shows. So Haber's problem was to find a method either to speed up the reaction at moderate temperatures or to increase the yield at high temperatures.

A possible way to increase the rate of reaction was to discover a catalyst. Extensive search brought to light several fairly good ones: the rare metals osmium and uranium, and a mixture of the more common metals iron and molybdenum. But even the best catalyst could not make the reaction proceed at a reasonable speed below 500°, and at this temperature the yield of ammonia is still negligible.

One other possibility remained: forcing the equilibrium to shift in the direction of the forward reaction by a change in pressure. The equation shows that ammonia occupies only half the volume of the gases from which it is made (when all are measured at the same pressure), so that its yields should be improved by increasing the pressure. How the yield actually changes with pressure is indicated by the experimental results in Table

CHEMICAL REACTIONS

TABLE 26.1 The yield of ammonia from the reaction of hydrogen and nitrogen at various combinations of temperature and pressure. The speed of the reaction increases with temperature, although the yield decreases.

Temperature, °C	Yield of ammonia, percent		
	At 1 atm	At 100 atm	At 200 atm
500	0.13	10.2	17.6
600	0.05	4.5	8.2
700	0.02	2.0	4.0
800	0.01	1.1	2.2

26.1. For temperatures near 500° the yields at high pressures proved large enough for commercial use. A mixture of nitrogen and hydrogen could be allowed to reach equilibrium, the 10 to 20 percent of ammonia which had formed could be removed by freezing or by solution in water, then more nitrogen and hydrogen could be added to the remaining gases and the process repeated.

This brief account tells nothing of the technical difficulties that Haber faced in putting the ammonia process into actual operation; it merely outlines his solution for the purely chemical problems of speeding up a sluggish reaction and increasing the yield from an equilibrium. Commercial production of "synthetic" ammonia by Haber's process began in Germany in 1913.

GLOSSARY

An *exothermic reaction* is one that liberates energy.

An *endothermic reaction* is one that must be supplied with energy in order to occur.

The *activation energy* of a reaction is the energy that must be supplied initially for the reaction to start.

A *catalyst* is a substance that can alter the rate of a chemical reaction without itself being changed by the reaction.

Chlorophyll is the catalyst that makes possible the reaction of water and carbon dioxide in green plants in order to produce carbohydrates. The reaction is called *photosynthesis*.

A *chemical equilibrium* occurs when a chemical reaction and its reverse reaction both take place at the same rate.

Le Châtelier's principle states that when the conditions of a system in equilibrium are altered, the equilibrium will shift in such a way as to attempt to restore the initial conditions.

EXERCISES

1. Which of the following are exothermic reactions and which endothermic?
   a. The explosion of dynamite
   b. The burning of methane
   c. The decomposition of water into its elements
   d. The dissociation of water into ions
   e. The burning of iron in chlorine
   f. The combination of zinc and sulfur to form zinc sulfide

2. From the observation that the slaking of lime [addition of water to CaO to form $Ca(OH)_2$] gives out heat, would you conclude that the following reaction is endothermic or exothermic?

   $$Ca(OH)_2 \longrightarrow CaO + H_2O$$

3. In what fundamental way is the explosion of an atomic bomb different from the explosion of dynamite?

4. Give two examples of reactions that are
   a. practically instantaneous at room temperatures
   b. fairly slow at room temperatures

5. Suggest three ways to increase the rate at which zinc dissolves in sulfuric acid.

6. Under ordinary circumstances coal burns slowly, but the fine coal dust in mines sometimes burns so rapidly as to cause an explosion. Explain the difference in rates.

7. Explain why a reaction with high activation energy is slow at room temperature.

8. Ammonia gas dissolves in water and reacts according to the equation

   $$NH_3 + H_2O \rightleftharpoons NH_4^+ + OH^-$$

   How would the amount of ammonium ion in solution be affected by
   a. increasing the pressure of NH_3?
   b. pumping off the gas above the solution?
   c. raising the temperature?
   d. adding a solution of HCl?

9. Hydrogen sulfide gas dissolves in water and ionizes very slightly: $H_2S \rightleftharpoons 2H^+ + S^=$. How would the acidity of the solution (concentration of H^+) be affected by
   a. increasing the pressure of H_2S?
   b. raising the temperature?
   c. adding a solution of KOH?

d. adding a solution of silver nitrate? [Silver sulfide (Ag_2S) is insoluble.]

10 Limestone ($CaCO_3$) dissolves in carbonic acid to form calcium bicarbonate. The latter dissociates readily, so that an equilibrium is set up

$$CaCO_3 + H_2CO_3 \rightleftharpoons Ca^{++} + 2HCO_3^-$$

How would this equilibrium be affected by
a. raising the temperature?
b. allowing the solution to evaporate?
c. increasing the pressure of CO_2, thereby increasing the concentration of H_2CO_3?

Under what natural conditions, then, is limestone most soluble? Under what conditions will it be precipitated from solution?

11 The reaction $2SO_2 + O_2 \longrightarrow 2SO_3$ is exothermic. How will a rise in temperature affect the yield of SO_3 in an equilibrium mixture of the three gases? Will an increase in pressure raise or lower this yield? In what possible way can the speed of the reaction be increased at moderate temperatures?

12 The three gases H_2, O_2, and H_2O are in equilibrium at temperatures near 2000°C. Write the equation for the equilibrium. Would the yield of H_2O be increased or decreased by raising the temperature? By raising the pressure?

chapter 27
Oxidation and Reduction

OXIDATION-REDUCTION REACTIONS
  displacement reactions
  order of activity
  electrochemical cells

OXIDATION NUMBER
  valence
  oxidation number
  changes in oxidation number
  iron compounds
  rust
  atmospheric oxygen

METALLURGY
  ores
  copper and tin
  iron
  aluminum

Earlier in this book we employed the term *oxidation* to mean the chemical combination of a substance with oxygen. A related term also in common use is *reduction*, which refers to the removal of oxygen from a compound. Whenever oxygen reacts with another substance (except fluorine), the oxygen atoms acquire electrons donated by the atoms of that substance; when the resulting compound is reduced, its atoms regain the electrons initially lost to the oxygen atoms. Hence **oxidation involves the loss of electrons by the atoms of an element and reduction involves the gain of electrons.**

It is convenient to generalize oxidation and reduction to refer to *any* chemical process that involves the transfer of electrons from one substance to another, whether oxygen itself is involved or not. The oxidation of one element is always accompanied by the reduction of another. For example, the burning of zinc in chlorine means the reduction of chlorine as the

oxidation involves the loss of electrons, reduction the gain of electrons

zinc is oxidized: the electrons lost by the zinc atoms are gained by the chlorine atoms. Since in chemical reactions atoms cannot lose, gain, or share electrons by themselves, oxidation and reduction must take place together. Reactions involving electron transfer (again, whether or not oxygen itself is involved) are called *oxidation-reduction reactions*, and they constitute a large and important category.

oxidation-reduction reactions involve electron transfer

OXIDATION-REDUCTION REACTIONS

displacement reactions

If a piece of copper wire is covered with a solution of silver nitrate and allowed to stand for a few hours, the wire becomes coated with gray crystals and the solution turns pale blue. The crystals are metallic silver, and the telltale blue color shows the presence of copper ions in the solution. Evidently some copper has become ionized, and at the same time silver has been deposited. This reaction is summarized by the equation

$$Cu + 2Ag^+ + 2NO_3^- \longrightarrow Cu^{++} + 2Ag + 2NO_3^-$$

Or, since the nitrate ion is not affected, we may write

copper displaces silver from a solution

$$Cu + 2Ag^+ \longrightarrow Cu^{++} + 2Ag$$

This is an oxidation-reduction reaction in the simplest form. Each copper atom has lost two electrons, to become a copper ion; each silver ion has gained one electron, to become a neutral silver atom. Copper is oxidized; silver is reduced. We may describe this reaction by saying that copper has *displaced* silver from solution.

A similar reaction takes place when a steel knife blade is held in a solution of copper sulfate. After a few moments the blade is coated with a reddish film of copper, and chemical tests would show the presence of iron ions in solution. The equation for this reaction is therefore

$$Fe + Cu^{++} \longrightarrow Fe^{++} + Cu$$

Iron atoms reduce copper ions to free copper and are themselves oxidized to positive ions. Or we may say that iron has displaced copper from solution.

order of activity

Iron atoms give up electrons to copper ions; copper atoms give up electrons to silver ions. Thus we might arrange the three metals in the order Fe, Cu, Ag, showing their relative abilities to give up electrons. By studying other displacement reactions, we should find that other metals may be added to this series, each metal being capable of giving electrons to the ions of metals that follow it:

OXIDATION AND REDUCTION

Na	Sodium	
Ca	Calcium	
Mg	Magnesium	
Al	Aluminum	order of activity of
Zn	Zinc	metals
Fe	Iron	
Pb	Lead	
(H)	(Hydrogen)	
Cu	Copper	
Hg	Mercury	
Ag	Silver	
Au	Gold	

This sequence is in order of decreasing ability to lose electrons or of decreasing ability to reduce the ions of other metals. Magnesium placed in a solution of copper chloride gives its electrons to the copper ions; lead placed in a silver nitrate solution reduces the silver ions. This is precisely the order we described earlier as the *order of activity* of the metals. Oxidation-reduction reactions, in fact, furnish a precise measure for the activities of different metals.

Hydrogen may be placed in the above sequence, for the solution of a metal in acids is a typical displacement reaction:

$$Zn + 2H^+ \longrightarrow Zn^{++} + H_2$$

Zinc gives up electrons to hydrogen ions, going into solution as zinc ions and setting hydrogen free. To find where hydrogen belongs in the series, we need only see which metals will dissolve in acids and which will not. We find that all the metals from Na to Pb will reduce H^+, but those from Cu to Au are unaffected by ordinary acids. Hence H belongs between Pb and Cu.

Displacement reactions among nonmetals are shown especially well by the halogens. If chlorine is added to the solution of potassium bromide, the solution turns brownish because bromine is set free:

$$Cl_2 + 2Br^- \longrightarrow Br_2 + 2Cl^-$$

Chlorine oxidizes bromide ion; in other words, each chlorine atom takes an electron from Br^-, setting free a bromine atom and itself becoming ionized. Similarly bromine displaces iodine, and fluorine displaces any one of the other halogens. By means of these reactions, nonmetals also may be arranged in an activity series:

F	Fluorine	
Cl	Chlorine	order of activity of
Br	Bromine	nonmetals
O	Oxygen	
I	Iodine	
S	Sulfur	

This sequence is in order of decreasing ability to gain electrons, or of decreasing ability to oxidize the ions of other nonmetals. Note that the activity of nonmetals is measured by their oxidizing ability, that of metals by their reducing ability—which is just another way of stating the fundamental facts that metals combine chemically by losing electrons to other elements and that nonmetals combine chemically by gaining electrons.

electrochemical cells

oxidation-reduction reactions underlie the operation of electrochemical cells

An oxidation-reduction reaction can be used to produce an electric current if we arrange matters so that the electrons transferred during the reaction do not move directly between the reactants but instead pass through an external wire. The dry cells of a flashlight, the storage battery in a car, and the fuel cell in a spacecraft are all based upon oxidation-reduction reactions.

A simple example of an *electrochemical cell* is the Daniell cell, which was once widely used in telegraphy. If we put a zinc bar into a copper sulfate solution, metallic copper is deposited on the bar while zinc ions go into solution because copper has the greater ability to hold electrons. The overall reaction is therefore

$$Zn + Cu^{++} \longrightarrow Zn^{++} + Cu$$

Actually, there are two separate so-called *half-reactions* that take place here:

$$Zn \longrightarrow Zn^{++} + 2e^-$$
$$Cu^{++} + 2e^- \longrightarrow Cu$$

In the first half-reaction, Zn is oxidized, and in the second, Cu^{++} is reduced. This pair of half-reactions can be used as the basis of an electrochemical cell if they occur at different locations, so that the electrons flow from the Zn atoms to the Cu^{++} ions through an external circuit rather than directly.

the displacement of copper by zinc in a solution is the basis of the daniell cell

In a Daniell cell, a zinc electrode is immersed in a zinc sulfate solution that is separated by a porous barrier from a copper electrode immersed in a copper sulfate solution (Fig. 27.1). The porous barrier allows ions to pass through it when the cell is in operation, but it prevents the mixing of the two solutions. When the electrodes are connected together, electrons flow from the zinc electrode through the wire to the copper electrode, where they combine with copper ions to form metallic copper via the half-reaction

$$Cu^{++} + 2e^- \longrightarrow Cu$$

The electrons are produced in the conversion into ions of zinc atoms from the zinc electrode via the other half-reaction

$$Zn \longrightarrow Zn^{++} + 2e^-$$

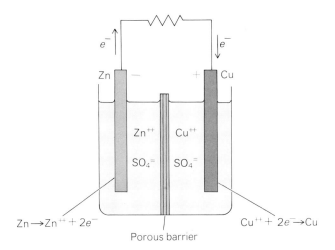

27.1 The Daniell electrochemical cell. The half-reactions that occur at each electrode are shown.

The solution of zinc and the plating out of copper are thus accompanied by a flow of current outside the cell. The current will continue to flow as long as there are zinc atoms left on the electrode and Cu^{++} ions left in the solution. The potential difference across a Daniell cell is about 1.1 volts.

In the storage battery of a car, plates of lead and of lead dioxide, PbO_2, are immersed in a solution of sulfuric acid (Fig. 27.2). No porous barrier is needed. The sulfuric acid is dissociated into H^+ and $SO_4^=$ ions. When the battery is in operation, electrons leave the lead electrode and the Pb^{++} ions thus formed combine with $SO_4^=$ ions from the solution to give lead sulfate, $PbSO_4$, which is insoluble. Thus the half-reaction at the lead electrode is

the lead-acid storage battery

$$Pb + SO_4^= \longrightarrow PbSO_4 + 2e^-$$

At the other electrode, lead dioxide, H^+ ions, $SO_4^=$ ions, and incoming electrons from the wire combine to give lead sulfate and water:

$$PbO_2 + 4H^+ + SO_4^= + 2e^- \longrightarrow PbSO_4 + 2H_2O$$

Thus lead sulfate is deposited on *both* electrodes when current is drawn from the battery.

The concentrations of H^+ and $SO_4^=$ ions in the solution evidently decrease as the battery is discharged, and the Pb and PbO_2 in the electrodes are progressively converted to $PbSO_4$. After a while there is not enough of the reactants to sustain the reaction, and the battery is "dead." To recharge the battery, a current is passed through it in the opposite direction. This current causes the two half-reactions to proceed in reverse, which restores the plates and the acid bath to their original compositions.

The state of charge of a storage battery can be ascertained by measuring the specific gravity of its acid bath. (The *specific gravity* of something is its density relative to that of water.) A fully charged battery has a specific gravity of about 1.26, which means that its density is 1.26 times

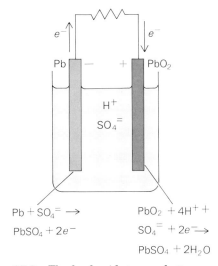

27.2 The lead-acid storage battery. The half-reactions that occur at each electrode are shown.

Technicians completing assembly of the prototype of a hydrogen-oxygen fuel cell. The actual cell is used in the Apollo spacecraft to provide electricity and drinking water.

that of water. When the battery is fully discharged, its specific gravity drops to about 1.11. The potential difference across a fully charged storage battery is 2.1 volts; a "12-volt" battery contains six cells connected in series.

fuel cells

In a *fuel cell*, the reacting substances are fed in continuously, so that the cell can provide current indefinitely without having to be replaced or recharged. Fuel cells are already used in spacecraft since they are very light in proportion to the electric power they can supply. In the future it seems likely that fuel cells will be perfected to the point where they are economical sources of power for individual homes and for electric cars. The combination of 1 lb of hydrogen and 8 lb of oxygen in a hydrogen-oxygen fuel cell produces about 10^8 joules of electric energy, which is enough to power a 100-watt light bulb for almost two weeks. The overall reaction in such a cell is

$$2H_2 + O_2 \longrightarrow 2H_2O + \text{flow of 4 electrons}$$

OXIDATION NUMBER

valence

As we know, the valence of an element in a simple compound is the number of electrons that each atom has gained, lost, or shared. In ionic compounds, the valence of the metal is positive, that of the nonmetal (or atomic group acting as the nonmetal) negative. Here the valence of an atom represents its combining ability in the sense of the number of single bonds it can form, with the sign of the valence indicating whether it normally acts as an electron donor (positive valence) or as an electron acceptor (negative valence) in these bonds. By establishing the valences of common elements and atomic groups, as was done in Chap. 22, it is possible to predict the formulas for many compounds and to understand their reactions.

This straightforward picture fails for compounds in which covalent rather than ionic bonding is dominant and is hardly capable of being applied in the cases of such elements as iron, tin, and copper which have different "valences" under different conditions; indeed, even oxygen can behave erratically in forming such compounds as F_2O and H_2O_2. Rather than extend the definition of valence to include the possibility that it may be different in both significance and value for the same element in different compounds, many chemists prefer to avoid the word completely and replace it with a variety of other, more precisely defined terms such as ionic valence, oxidation number, coordination number, and so on. However, just as it is convenient to refer to "hydrogen ions" in a solution even though they do not exist as such, so the "valence" of an element remains in the vocabulary of chemistry despite its ambiguous aspects. What we shall do here is use valence in its most elementary interpretation as the number of, say, chlorine atoms that can combine with a particular metal atom or the number of hydrogen atoms that can combine with a particular nonmetal atom, and introduce another quantity, *oxidation number*, to assist in electronic bookkeeping in certain of the circumstances in which valence as so defined is of no help.

oxidation numbers are useful in keeping track of the electrons in oxidation-reduction reactions

oxidation number

The oxidation number of an atom in a given situation is established on the basis of a series of specific if sometimes rather artificial rules. These rules prescribe the manner in which the electrons involved are assigned, and the oxidation number of each of the atoms is the electric charge (in units of *e*, the magnitude of the charge on the electron) it has when this assignment is made. The more general rules are as follows:

rules for assigning oxidation numbers

1. The oxidation number of an atom in the elemental state is 0. Thus the oxidation numbers of each hydrogen atom in H_2, each copper atom in metallic Cu, and each sulfur atom in S_8 are all 0.
2. The oxidation number of a monatomic ion is equal to its actual charge.

Thus the oxidation number of H^+ is $+1$, that of Al^{3+} is $+3$, and that of $O^=$ is -2.

3 In a covalent compound, each shared electron pair is arbitrarily assigned to the more electronegative of the two atoms involved, that is, the one with the higher electron affinity. If the two atoms are of the same element, the pair is divided between them. An example of the former is ClF, where each atom contributes an electron to the covalent bond between them and F is given the oxidation number -1 because, as the more electronegative partner, it is assumed to hold both electrons; Cl then lacks an electron, and so has the oxidation number $+1$ in the compound. In other words, what is in reality a case of electron sharing is regarded as one of electron transfer for convenience in analysis.

4 The sum of the oxidation numbers in a compound must be zero, and this sum in an atomic group must equal the charge on the group. In H_2SO_4 the oxidation number of each H is $+1$, of each O is -2, so the oxidation number of the S must be $+6$.

In addition to these general rules there are a number of specific assignments that are worth keeping in mind. Some of them are that, in a compound, the elements in Group I of the periodic table (the alkali metals lithium, sodium, potassium, and so on) always have the oxidation number $+1$; the elements in Group II (beryllium, magnesium, calcium, and so on) always have the oxidation number $+2$; fluorine always has the oxidation number -1, while the other halogens have the oxidation number -1 except in compounds with other, more electronegative halogens or with oxygen; the element oxygen usually has the oxidation number -2, with exceptions in the case of F_2O and in the various peroxides; and when hydrogen is combined with a nonmetal, its oxidation number is $+1$, whereas when combined with a metal it is -1.

changes in oxidation number

In the process of oxidation, which involves electron loss, the oxidation number of an element increases in the sense of becoming more positive; if the number was originally negative, it becomes less so. In the reverse process of reduction, which involves electron gain, the oxidation number decreases; if the number was originally negative, it becomes more so (Fig. 27.3). Let us examine a few examples.

When zinc is oxidized by burning in oxygen to form ZnO,

$$2Zn + O_2 \longrightarrow 2ZnO$$

the oxidation number of an element increases when it is oxidized and decreases when it is reduced

27.3 *Oxidation means an increase in oxidation number, reduction means a decrease in oxidation number.* An atom in the elemental state has an oxidation number of 0.

OXIDATION AND REDUCTION

and its oxidation number increases from 0 to +2. The same change in oxidation number occurs when zinc burns in chlorine,

$$Zn + Cl_2 \longrightarrow ZnCl_2$$

and this too is an oxidation process for the zinc. The elemental oxygen and chlorine are both reduced in the process, with the oxidation number of the former going from 0 to -2 and that of the latter going from 0 to -1.

In the decomposition of mercuric oxide by heating,

$$2HgO \longrightarrow 2Hg + O_2$$

the oxygen is itself oxidized because its oxidation number goes from -2 to 0. The mercury is reduced since its oxidation number changes from $+2$ to 0.

In the reduction of iron by heating one of its oxides in a stream of hydrogen,

$$Fe_2O_3 + 3H_2 \longrightarrow 2Fe + 3H_2O$$

the hydrogen is simultaneously oxidized, but the oxidation number of the oxygen is unchanged and it does not participate directly in the exchange of electrons that constitutes the oxidation-reduction reaction.

The gas propane, C_3H_8, is often used as a fuel in cooking stoves and for blowtorches. When it burns in air, the reaction is

$$C_3H_8 + 5O_2 \longrightarrow 3CO_2 + 4H_2O$$

Is the hydrogen or the carbon content of propane being oxidized, or both? To answer the question we need only investigate the oxidation number changes that occur. In C_3H_8, carbon has the oxidation number $-8/3$ and hydrogen the oxidation number $+1$, while in CO_2 carbon has the oxidation number $+4$ and in H_2O hydrogen has the oxidation number $+1$. Hence it is the carbon which is oxidized.

iron compounds

Many elements show two or more different oxidation numbers in their compounds. Mercury, for example, forms compounds in which it has an oxidation number of $+1$, such as Hg_2O, Hg_2Cl_2, $Hg_2(NO_3)_2$, Hg_2SO_4 (mercur*ous* oxide, mercur*ous* chloride, etc.), and other compounds in which it has an oxidation number of $+2$, such as HgO, $HgCl_2$, $Hg(NO_3)_2$, $HgSO_4$ (mercur*ic* oxide, etc.). Tin in some compounds ($SnCl_2$, SnO) has an oxidation number of $+2$; in others, an oxidation number of $+4$ ($SnCl_4$, SnO_2). Carbon forms the two oxides CO and CO_2; sulfur, the two oxides SO_2 and SO_3.

an element may have two or more oxidation numbers

Changes from one oxidation number to another are, of course, oxidation-reduction reactions. For example, carbon is oxidized when carbon monoxide is burned:

$$2CO + O_2 \longrightarrow 2CO_2$$

and sulfur is reduced when sulfur trioxide is strongly heated:

$$2SO_3 \longrightarrow 2SO_2 + O_2$$

ferrous and ferric compounds

Especially noteworthy are changes in oxidation state among the compounds of iron. This element has two principal oxidation numbers, $+2$ and $+3$; thus it forms the two oxides FeO (ferr*ous* oxide) and Fe_2O_3 (ferr*ic* oxide), the two chlorides $FeCl_2$ and $FeCl_3$, and the two sulfates $FeSO_4$ and $Fe_2(SO_4)_3$. Many compounds of each type are soluble in water and ionize to give, respectively, the ions Fe^{++} and Fe^{3+}. Most ferrous compounds in solution are pale green in color, most ferric compounds pale yellow. (Solutions of ferric compounds often show stronger tints of yellowish to reddish brown because ferric ion has a tendency to form complex ions by attaching itself to water molecules and to other ions.) As a general rule, iron compounds are strongly colored, both in solution and in the solid state: ferric compounds have shades ranging from yellow to brown and red; ferrous compounds are gray, green, or black.

ferrous and ferric compounds have characteristic colors

Changes from one oxidation state of iron to the other are easily brought about. If chlorine, for example, is bubbled into a solution of ferrous sulfate, the greenish color of the solution changes quickly to yellow, showing that Fe^{++} has been converted to Fe^{3+}:

$$Cl_2 + 2Fe^{++} \longrightarrow 2Cl^- + 2Fe^{3+}$$

Each chlorine atom oxidizes a ferrous ion by taking one electron away from it. One way of effecting the opposite change is to add an iodide, say KI, to a solution of a ferric compound. Now the yellow solution changes to deep brown as iodine is liberated:

$$2I^- + 2Fe^{3+} \longrightarrow I_2 + 2Fe^{++}$$

Each iodide ion gives an electron to a ferric ion, thereby reducing it to ferrous ion. Many other oxidizing and reducing substances besides Cl_2 and I^- can be used in these reactions.

within the earth, ferrous compounds are more common; at the surface, ferric compounds are more common

In the presence of air, most ferrous compounds are slowly oxidized to ferric compounds by atmospheric oxygen. Deep in the earth, however, at high temperatures and out of reach of the atmosphere, ferrous compounds are more stable than ferric compounds. Most of the iron in rocks formed at these depths, therefore, goes into ferrous compounds, chiefly ferrous silicates. When such rocks appear at the surface, either brought up by volcanic activity or exposed by long erosion, their iron compounds are no longer stable. Slowly, oxygen attacks the iron, aided by the slight solvent action of carbonic acid in streams and rain water. The complex silicates are broken down, and the iron is oxidized as in ordinary rusting to various hydrates of Fe_2O_3. The rusty stains on the surfaces and in cracks of so many common rocks are due to this iron oxide, formed by slow oxidation of ferrous compounds present in the rocks.

Many of the bright and somber hues in nature we owe to the colorful compounds of this one element. Not only the rusty surface stains but

OXIDATION AND REDUCTION

A refinery worker ladles molten lead from one vessel to another in the refining process.

yellow-brown and reddish-brown colors in rocks themselves are nearly always due to ferric compounds. Many black rocks and green rocks get their colors from ferrous compounds. Sand, clay, and soil likewise owe their brown tints to ferric compounds, their gray and black shades often to ferrous compounds. The colors of ferric compounds are so strong that only a little iron is needed to color a rock or a soil conspicuously.

Reduction of ferric compounds at the earth's surface is accomplished chiefly by carbon compounds produced by plants and animals. Soils containing much organic matter are usually black, since their iron is in the form of ferrous compounds whose relatively weak colors do not obscure the dark carbonaceous materials. Often we find a thin layer of black soil resting on brown soil, the dark color showing the depth to which decaying organic matter keeps the iron of the soil reduced.

Thus in nature we find continuous transformations of ferric compounds to ferrous compounds and back again, an unending series of oxidation-reduction reactions.

rust

Rust is a hydrated ferric oxide which can be formed from iron or steel only in the presence of both water and oxygen. Although the formula

both water and oxygen are needed to cause iron to rust

of rust is often written $2Fe_2O_3 \cdot 3H_2O$, the relative proportions of ferric oxide and water are to some extent variable. The formation of rust can be summarized as

$$4Fe + 3O_2 + 3H_2O \longrightarrow 2Fe_2O_3 \cdot 3H_2O$$

The actual reaction apparently takes place in a series of steps, however, rather than all at once. The first step is thought to be the oxidation of metallic iron to ferrous ions (Fe^{++}), with the lost electrons being picked up by hydrogen ions:

$$Fe + 2H^+ \longrightarrow Fe^{++} + 2H$$

That this reaction actually occurs is substantiated by the accelerated rusting of iron when in an acid environment. The resulting hydrogen atoms combine with oxygen molecules to form water, a reaction for which iron is known to be a catalyst:

$$4H + O_2 \longrightarrow 2H_2O$$

In the final step the ferrous ions are further oxidized to yield both rust and hydrogen ions:

$$4Fe^{++} + O_2 + 7H_2O \longrightarrow 2Fe_2O_3 \cdot 3H_2O + 8H^+$$

The hydrogen ions are therefore not consumed in the formation of rust but are available to continue the process.

Interestingly enough, rusting seems to be promoted by a moderate restriction in the supply of oxygen. If the preceding sequence of reactions takes place fast enough, the rust acts to some extent as a protective coating and impedes further corrosion. (Because rust is itself a catalyst for the production of rust, it does not provide as much protection as do oxide films on other metals, for instance zinc and aluminum.) However, if the sequence is slow, the ferrous ions diffuse away from where they were formed and turn into rust nearby, leaving behind a pit in the metal that continues to be eaten away.

atmospheric oxygen

Oxygen is a powerful oxidizing agent, but most of its reactions under ordinary conditions are so slow that it seems relatively inactive. Moderately high temperatures speed its reactions greatly, and the pure gas is more active than the dilute mixture with nitrogen which makes up ordinary air. The presence of water containing hydrogen ions also increases the rate at which oxygen reacts with other substances; even the minute concentration of hydrogen ions formed by the solution of carbon dioxide in water is enough to speed up oxidations enormously. How great the effect of carbonic acid can be is readily shown by exposing to the air two polished iron surfaces, one damp and the other dry: while the dry surface retains its polish for weeks, the damp one begins to rust in a few hours.

Oxidation of Metals. All active metals react slowly with atmospheric oxygen, just as iron does. Some form simple oxides, others react further with water or carbon dioxide to form hydroxides or carbonates. Sodium and potassium "rust" so quickly on exposure to air that they must be kept under oil. Some active metals, like zinc and aluminum, show no visible corrosion because their oxides, formed quickly on fresh surfaces, cling to the metal as an impervious film that prevents further oxidation.

Spontaneous Combustion. Carbon and most of its compounds oxidize so very slowly under ordinary conditions that we commonly think of them as stable substances. Some carbon compounds, however, do react with oxygen at perceptible rates. Usually the heat given out by these reactions is dissipated into the surrounding air, but if free circulation of air is prevented, the heat may gradually accumulate. Thus when coal is piled too deeply, heat formed by slow oxidation of coal within the pile cannot escape and may ultimately raise the temperature sufficiently to ignite the pile. Piles of oily rags are another source of danger, for many oils oxidize slowly in air. Burning started in this manner, by accumulation of heat from slow oxidation, is often called "spontaneous combustion."

Decay. The decay of organic matter is principally a slow oxidation by atmospheric oxygen. The processes of decay may be extremely complex, but the chief ultimate products are the two which nearly always result from the complete oxidation of carbon compounds—CO_2 and H_2O. Nitrogen in the original compounds is changed principally to ammonia. Decay is greatly accelerated by moisture and also by bacteria which can use the energy of the slow oxidation for their life processes.

Oxidation of Food. The complex carbon compounds that form the principal part of our food are broken down by the processes of digestion into simpler compounds that can be transported by the bloodstream and stored in various parts of the body. To obtain energy from food, our bodies must oxidize these stored compounds. The necessary oxygen is taken in through the lungs and is carried to the tissues by a substance in the blood called *hemoglobin*; the oxidation process is aided by complex organic catalysts. The products of oxidation, besides energy, are CO_2 and H_2O, the former being carried by the blood to the lungs and exhaled from there.

To sum up, oxygen in the air is a good oxidizing agent but under ordinary conditions reacts slowly. Its rates of reaction in different circumstances may be speeded up by heat, by the presence of dilute carbonic acid, by bacteria, and by catalysts. Active metals, organic substances, and other oxidizable materials are always subject to slow attack by oxygen when exposed to the air.

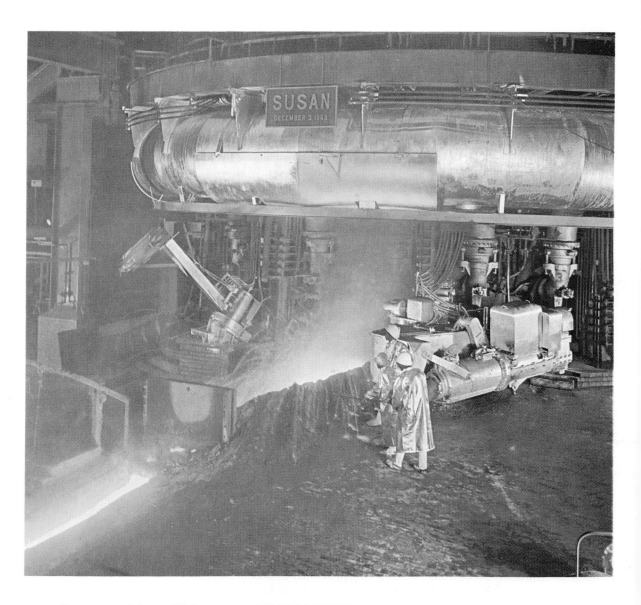

Susan, one of the world's largest blast furnaces, reduces iron ore to metallic iron.

METALLURGY

Oxidation and reduction reactions are highly important in the various processes by which metals are obtained from their ores. These processes are among those studied in the science of *metallurgy*.

ores

ore and gangue

An *ore* of a metal is any naturally occurring material from which the metal can be extracted profitably. In an ore, a metal may occur as the free element or in a compound, mixed usually with large amounts of valueless material called *gangue*. For example, a common gold ore consists of particles of free metal in a gangue of sand and gravel. The first problem of metallurgy is to concentrate the ore, that is, to remove as much of the gangue as possible. This is usually a physical problem rather than a chemical one. A common method of concentration is a large-scale application of the gold miner's pan: the crushed ore is shaken, usually under water, and the heavier metallic particles settle to the bottom. More complicated methods are necessary when the metallic part of the ore has nearly the same density as the gangue.

The metallurgist must then separate the metal from the remainder of the gangue and from any elements that are combined with it. This is simple for the few metals which occur as free elements—platinum, gold, silver, more rarely copper. These may be separated from the gangue by melting the metal, or by adding a solvent that will remove the metal in solution. Because gold, silver, and copper are so easily obtained from their ores, these were the first metals used by primitive man.

Other metals occur in their ores as compounds, of which the most important types are oxides (Fe_2O_3, SnO_2), sulfides (MoS_2, ZnS), and

Electrolytic refining of aluminum at a plant in Massena, New York. The device at the right is used to add fresh ore to the bath.

carbonates ($MnCO_3$, $PbCO_3$). From these the free metals are obtained by various chemical processes involving reduction. How difficult the reduction is for a given metal is suggested with fair accuracy by the date of its discovery: the metals most easily reduced from their ores were known to primitive man, a few were discovered in the ancient Mediterranean civilizations, a few more were added by the alchemists, whereas those reduced with most difficulty were discovered only in recent times.

copper and tin

Copper, available to primitive man only in limited quantities as the free element, was the first metal which he learned to reduce from its compounds. Probably he discovered the process accidentally, by dropping a chunk of rock in the hot charcoal of his campfire and observing that a few shining drops of liquid metal were formed. In modern language we should say that a copper compound in the rock had been changed by heating to copper oxide (CuO) and that this oxide had been reduced by hot carbon in the charcoal:

$$2CuO + C \longrightarrow 2Cu + CO_2$$

bronze

Much later, about 1500 B.C., primitive man learned that another sort of rock heated with charcoal produced the metal tin, which could be mixed with copper to give the alloy *bronze*. Bronze was valuable to early civilizations because it is stronger and harder than copper or tin, does not corrode readily, and is easier to produce than iron or steel. We would write the equation for the reduction of tin ore as

$$SnO_2 + C \longrightarrow Sn + CO_2$$

Lead and mercury, produced by similar crude processes, were known to the ancient Egyptians and Babylonians.

iron

in refining iron, a flux is added to combine with the gangue to form easily removed liquid slag

Iron presents a more difficult metallurgical problem. In its principal ores, iron occurs as ferric oxide—the simple oxide in the ore mineral *hematite*, the hydrated oxide in the mineral *limonite*. Like the oxides of copper and tin, ferric oxide can be reduced with hot charcoal, but the reaction requires a higher temperature. Separation of the molten metal from gangue requires a further improvement in technique: some substance must be added as a *flux*, to combine with the gangue and make it liquid; the liquid, a light, glassy substance which we call *slag*, then collects on top of the molten iron and can be easily removed.

Although a little iron found in meteorites was used even in primitive times, the trick of reducing it from its ores was not learned until about 600 B.C. Large-scale production of iron began only about A.D. 1600, when coal was substituted for charcoal in the reduction process. In modern blast furnaces, which are huge steel towers (40 to 100 ft high) lined with firebrick, iron is reduced by a continuous process (Fig. 27.4): a mixture of ore with coke and flux is fed in at the top, heated air is forced in near the bottom to burn the coke and so provide heat for the reaction, and molten iron and slag are drawn off at intervals from the bottom. The flux used depends on the nature of the ore; in most iron ores silicon dioxide is the chief impurity, and for these limestone is the best flux. The reactions that take place in the blast furnace may be summarized in the following equations, although the reduction of the oxide and the formation of slag actually take place in steps:

$$2Fe_2O_3 + 3C \longrightarrow 4Fe + 3CO_2$$
$$CaCO_3 + SiO_2 \longrightarrow CO_2 + CaSiO_3 \text{ (slag)}$$

blast-furnace reactions

Blast-furnace slag has a composition similar to that of ordinary glass.

The alchemist added five metals to those known in ancient times: antimony, bismuth, zinc, arsenic, cobalt. Several others were discovered before the end of the eighteenth century, including nickel, manganese, tungsten, molybdenum, chromium—rare metals which have become

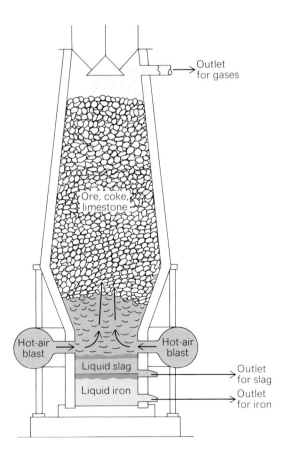

27.4 A blast furnace.

familiar today because of their use in steel alloys. Some of these metals, like the earlier metals, were obtained from their ores by reduction with carbon, but others required more elaborate treatment.

aluminum

In the nineteenth century, metallurgy gained a powerful new tool, electrolysis. The processes of electrolysis are oxidation and reduction reactions reduced to bare essentials: at the cathode, electrons are given directly to one substance, and at the anode electrons are removed from another substance (Chap. 24). To reduce a metal from a compound by electrolysis, the compound must be liquefied or dissolved so that free metal ions may be liberated; then the passage of a current attracts the positive metal ions to the cathode, where electrons are supplied to reduce them. By means of electrolysis we can today reduce even the most active metals, like sodium, potassium, and calcium. Electrolysis also furnishes a convenient method for the final purification of less active metals, such as copper, tin, zinc, and silver.

aluminum is reduced from its ore by electrolysis

One of the great triumphs of electrolysis in metallurgy was the cheap production of the light metal aluminum. Aluminum was first prepared early in the nineteenth century by reducing its chloride with potassium, but the process was so expensive that the metal's only commercial use for some years was in jewelry. Aluminum is an extremely active element, and its compounds resisted all attempts to decompose them by any reducing agents less powerful and less expensive than the alkali metals. Attempts to produce the metal commercially by electrolysis were balked at first by the difficulty of liquefying or dissolving aluminum ores. The problem was finally solved in 1886 by a young American, Charles Martin Hall, at that time just out of college. His great discovery was that cryolite, a mineral obtained in large quantities from Greenland, when melted would dissolve the chief aluminum ore, Al_2O_3. When a current is passed through the solution between graphite electrodes, aluminum is liberated at the cathode and oxygen at the anode. This process makes possible the manufacture of aluminum in quantities sufficient to satisfy the demands of modern industry.

Metallurgy today includes a great variety of chemical processes, but the two principal ones remain the ancient process of reducing ores with hot carbon and the recent one of reducing metallic ions by electrolysis.

GLOSSARY

An element is *oxidized* when its atoms lose electrons and *reduced* when its atoms gain electrons.

A *displacement reaction* is an oxidation-reduction reaction in which one metal (or nonmetal) displaces another metal (or nonmetal) from solution.

OXIDATION AND REDUCTION

In an *electrochemical cell*, electric current is produced by an oxidation-reduction reaction whose two half-reactions take place at different locations.

The *oxidation number* of an atom in a specific situation is assigned on the basis of certain general rules. The oxidation number of an atom increases (becomes more positive) when it is oxidized, and decreases (becomes more negative) when it is reduced.

The *ore* of a metal is any naturally occurring material from which it can be profitably extracted.

EXERCISES

1. Which of the following equations represent oxidation-reduction reactions?
   a. $Zn + S \longrightarrow ZnS$
   b. $H^+ + OH^- \longrightarrow H_2O$
   c. $MnO_2 + 2Cl^- + 4H^+ \longrightarrow Cl_2 + Mn^{++} + 2H_2O$
   d. $CaO + H_2O \longrightarrow Ca^{++} + 2OH^-$

2. In each of the following reactions, identify (*a*) the element that is oxidized, (*b*) the element that is reduced, (*c*) the element whose atoms gain electrons, (*d*) the element whose atoms lose electrons:

   $Mg + 2H^+ \longrightarrow H_2 + Mg^{++}$

   $Ca + S \longrightarrow CaS$

   $2Na + 2H_2O \longrightarrow 2Na^+ + 2OH^- + H_2$

   $F_2 + 2Br^- \longrightarrow Br_2 + 2F^-$

   $2Fe^{3+} + 3H_2S \longrightarrow 2FeS + S + 6H^+$

3. Which loses electrons more easily, Na or Fe? Al or Ag? I^- or Cl^-? Which gains electrons more easily, Cl or Br? Hg^{++} or Mg^{++}?

4. In what part of the periodic table are the elements that are most easily reduced? In what part are those that are most easily oxidized?

5. What would you expect to happen when a knife blade is held in a solution of silver nitrate? Write an equation to show the reaction. (The iron forms ferrous ions.)

6. Which of the halogens will displace bromine from solution? Which will bromine displace from solution? Write an equation for one of these reactions.

7. How could you demonstrate that magnesium is a better reducing agent (that is, more easily oxidized) than hydrogen?

8 Why must water be added periodically to a lead-acid storage battery when it is in normal operation?

9 A charging current is passed through a fully charged lead-acid storage battery. What happens?

10 Determine the changes in oxidation number of the nitrogen, hydrogen, and oxygen in the following reaction:

$$4NH_3 + 3O_2 \longrightarrow 2N_2 + 6H_2O$$

How many electrons does each atom gain or lose in the reaction? Which elements are oxidized and which reduced?

11 Determine the changes in oxidation number of the hydrogen and oxygen in the following reaction:

$$H_2O_2 + H_2O_2 \longrightarrow 2H_2O + O_2$$

How many electrons does each atom gain or lose in the reaction? Which elements are oxidized and which reduced?

12 A common gangue material in ore deposits is the brassy yellow mineral *pyrite*, or "fool's gold" (FeS_2). Account for the yellow-brown and red-brown colors often found in rocks and soil near such deposits.

13 Account for the fact that the color of soil in marshes is nearly always black.

14 Would you expect the danger from spontaneous combustion to be greater in a coal pile containing principally large chunks or in one containing finely pulverized coal? Why?

15 One of the compounds whose oxidation provides our bodies with energy is a simple sugar with the formula $C_6H_{12}O_6$. Write an equation showing the oxidation of this compound. Where does the necessary oxygen come from? What becomes of the products?

16 Mercury is recovered from its principal ore (cinnabar, HgS) by simply heating the ore in air without any reducing agent. The following reaction takes place:

$$HgS + O_2 \longrightarrow Hg + SO_2$$

Which elements change in oxidation number during this process, and in what way? Which ones are reduced and which oxidized?

17 Explain how the reduction of sodium is accomplished by the electrolysis of molten sodium chloride.

18 In a hydrogen-oxygen fuel cell, hollow, porous electrodes are used that are made of an inert, conducting material. These electrodes enable the gases to interact with the electrolyte in a gradual way. A typical electrolyte in such a cell is a solution of potassium hydroxide. At one electrode, hydrogen molecules combine with hydroxide ions to form water, and at the other electrode, oxygen molecules combine with water molecules to form hydroxide ions. Which electrode is negative and which is positive when the cell is in operation? What are the half-reactions that occur at each electrode? Does each half-reaction occur the same number of times as the other?

chapter 28
Organic Chemistry

INTRODUCTION
    carbon compounds
    paraffin hydrocarbons

STRUCTURES OF ORGANIC MOLECULES
    structural formulas
    isomers
    determining structures
    molecular models

CARBON BONDS
    the methane molecule
    carbon-carbon bonds
    the benzene ring

HYDROCARBON DERIVATIVES
    hydrocarbon atom groups
    halogen derivatives
    alcohols
    esters
    organic acids

INDUSTRIAL ORGANIC CHEMISTRY
    plastics
    elastomers
    fibers
    raw materials

In many ways the most remarkable element, carbon is a constituent of hundreds of thousands of compounds. There are dozens of times as many known carbon compounds as there are all other compounds. Further, carbon compounds are the chief constituents of living things—hence the name *organic chemistry* to describe the chemistry of carbon and the name *inorganic chemistry* to describe the chemistry of all the other elements.

organic chemistry is the chemistry of carbon compounds

At one time it was thought that carbon compounds, with the exception of the oxides, the carbonates, and a few others, could be produced only by plants and animals or from other compounds produced by plants and animals. Carbon was supposed to unite with other elements only under the influence of a mysterious *vital force* possessed by living things. This ancient idea was exploded in 1828 by the German chemist Friedrich Wöhler, who prepared the organic compound urea by heating the inorganic compound ammonium cyanate. Since Wöhler's time a great number of organic compounds have been made in the laboratory from inorganic materials, but the general distinction between the chemistry of carbon compounds and inorganic chemistry nevertheless remains useful.

INTRODUCTION

carbon compounds

What properties of carbon enable it to be so prolific in forming compounds?

Let us first examine the periodic table. Carbon is at the head of the middle group of elements, which means that it is a small atom with four valence electrons. Carbon atoms are unable readily either to lose all these electrons in their outer shells or to gain four more required to complete the shells, and so they form compounds almost exclusively by sharing electron pairs. Thus the bonds between carbon atoms and other atoms are covalent, and because of the small size of the carbon atom and the consequent strong attraction of its nucleus for electrons, these bonds are especially strong.

carbon atoms form covalent bonds with each other as well as with other atoms

A carbon atom can form firm attachments not only to many different metallic and nonmetallic atoms but to *other carbon atoms* as well. The strength of the attraction between carbon atoms is demonstrated by the hardness of diamond, a crystalline form of carbon in which each atom is linked to four others by electron-pair bonds. It is this unique capacity of carbon atoms to join together that makes possible the immense number and variety of carbon compounds. A few elements near carbon in the periodic table—for instance, boron, silicon, nitrogen—have this same ability to a small extent, but their chains of atoms are short and unstable.

Because the bonds formed by carbon atoms are covalent, carbon compounds are mostly nonelectrolytes, and their reaction rates are usually slow. The great attraction of carbon atoms and hydrogen atoms for oxygen makes many organic compounds subject to slow oxidation in air and to rapid oxidation if heated. Even in the absence of air, organic compounds are, in general, stable only at ordinary temperatures, and few of them resist decomposition at temperatures over a few hundred degrees centigrade.

organic compounds decompose at high temperatures

paraffin hydrocarbons

The simplest organic compounds are the *hydrocarbons*, compounds that contain only the two elements carbon and hydrogen. Even these relatively simple compounds exist in tremendous variety.

hydrocarbons contain carbon and hydrogen only

ORGANIC CHEMISTRY

One group of hydrocarbons, called the *paraffin* (or methane) hydrocarbons, include substances like CH_4 (methane), C_2H_6 (ethane), C_3H_8 (propane), C_4H_{10} (butane), and so on up to compounds with 30 or more carbon atoms per molecule. We can express the composition of these compounds by the general formula C_nH_{2n+2}, indicating that the number of hydrogen atoms is always two more than twice the number of carbon atoms. The names, formulas, and a few physical properties of some members of this group are given in Table 28.1. Density, boiling point, and freezing point all increase regularly as the molecular mass increases. Many other series of organic compounds exhibit similar changes as the number of carbon atoms per molecule grows larger.

the properties of the paraffin hydrocarbons vary regularly with molecular mass

TABLE 28.1 The paraffin series of hydrocarbons.*

Formula	Name	Freezing point, °C	Boiling point, °C	Commercial name
CH_4	Methane	−182	−161	Fuel gases
C_2H_6	Ethane	−183	−89	
C_3H_8	Propane	−190	−45	
C_4H_{10}	Butane	−138	−1	
C_5H_{12}	Pentane	−130	36	Petroleum ether (naphtha)
C_6H_{14}	Hexane	−95	68	
C_7H_{16}	Heptane	−91	98	Gasoline
C_8H_{18}	Octane	−57	125	
C_9H_{20}	Nonane	−51	151	Kerosene
$C_{10}H_{22}$	Decane	−30	174	
$C_{11}H_{24}$	Undecane	−25	195	
. . .				
$C_{16}H_{34}$	Hexadecane	18	287	

$C_{17}H_{36}$ to $C_{22}H_{46}$, semisolids, constituents of petroleum jelly and lubricating oil
$C_{23}H_{48}$ to $C_{29}H_{60}$, constituents of paraffin

* The data above refer to the *normal*, or straight-chain, compounds. Isomers of these hydrocarbons (p. 541) have somewhat different properties.

The paraffin hydrocarbons occur as constituents of natural gas and petroleum. The lighter ones, methane to butane, make up natural gas; the heavier ones are found in liquid oil. These hydrocarbons are also among the products obtained by heating soft coal in the absence of air. Methane itself has a more widespread occurrence: as a constituent of volcanic gases, for example, and as a minor product of organic decay. The gas that bubbles up from the black ooze at the bottom of stagnant pools is largely methane, commonly called *marsh gas*.

The separation of petroleum into its constituent compounds is a difficult problem, because the properties of the compounds are so similar. We might, for example, try to separate a mixture of pentane and hexane by boiling. Seemingly, pentane should boil off first, since its boiling point is lower (36°C—only a little over room temperature); thus we could leave hexane behind and then condense the vapor to recover the pentane. The difficulty is that hexane evaporates readily at 36°C, since its own boiling point is only 32°C higher, and so the condensed vapor would contain considerable hexane as well as pentane. By this procedure we should, therefore, obtain a vapor relatively rich in the low-boiling compound and a residue relatively rich in the high-boiling compound, but not a complete separation.

Fortunately, for commercial purposes a complete separation is not necessary. The refining of petroleum involves a process of partial separation called *fractional distillation,* in which the oil is heated and its vapors are led off and condensed at progressively higher temperatures. The material coming off at the lowest temperature is largely pentane and hexane, with minor amounts of both lighter and heavier hydrocarbons; it forms a colorless, volatile liquid used as a solvent and cleaning agent (petroleum ether or naphtha). The next fraction, consisting largely of hydrocarbons from hexane to decane, is gasoline. A still heavier and less volatile fraction is kerosene. At higher temperatures lubricating oils are formed and at still higher temperatures the solid hydrocarbon mixtures called petroleum jelly and paraffin.

<small>fractional distillation is a method of separating the different compounds in a mixture by virtue of their different boiling points</small>

Of all these products the most valuable, of course, is gasoline. Unfortunately, the constituents of gasoline make up only a minor fraction of most petroleums. In order to increase the yield of gasoline, it is common practice to supplement the refining process in two ways: by *cracking* the heavier hydrocarbons, which means heating them under pressure in the presence of a catalyst so that they break down into the simpler molecules of gasoline; and by *polymerizing* the lighter hydrocarbons, which means joining together small molecules into larger ones, again under the influence of heat and catalysts.

<small>cracking and polymerization</small>

Molecules of the paraffin hydrocarbons have chains of carbon atoms linked together, forming symmetrical or nearly symmetrical structures. Such structures, as we might expect, give nonpolar molecules, that is, molecules with neither end appreciably more electrically positive or negative than the other. Because of this nonpolar character, the paraffin hydrocarbons are insoluble in water. Chemically they are fairly unreactive, and neither concentrated acids and bases nor most oxidizing agents will affect them at ordinary temperatures. When ignited, as we know, they burn readily in air or oxygen.

<small>the paraffin hydrocarbons have nonpolar molecules</small>

STRUCTURES OF ORGANIC MOLECULES

structural formulas

So far the chemistry of the hydrocarbons seems simple enough. The only complexity we have introduced is the ability of carbon atoms to join together in chains, which makes possible an unusually large number of compounds of the same two elements. But further study of the hydrocarbons shows that this is only one respect in which organic compounds differ from the inorganic compounds that we have discussed. Another important difference is that two or more organic compounds may have the same formula; thus, two different gases share the formula of butane (C_4H_{10}) and three different liquids share the formula of pentane (C_5H_{12}). Among inorganic compounds we seldom think of a formula like NaOH or $FeCl_3$ as representing more than a single substance.

ORGANIC CHEMISTRY

Catalytic cracking unit at an oil refinery breaks down complex hydrocarbons into simpler ones.

The formula of an organic compound is determined in the same manner as other formulas: the molecular weight is found, perhaps by measuring the density of its vapor, and the proportions by weight of the different elements are found by analysis. Like inorganic formulas, the formula of a carbon compound tells us how many atoms of each kind are present in a molecule. Evidently, however, this information is not sufficient to describe an organic compound accurately, since more than one compound can show the same molecular composition. The additional information we need is the *arrangement* of atoms in the molecule.

two or more organic compounds may have the same composition but different molecular structures

To illustrate, we may write the formulas of methane (CH_4) and ethane (C_2H_6) in the pictorial form

$$\begin{array}{c} H \\ \cdot\cdot \\ H : C : H \\ \cdot\cdot \\ H \end{array} \quad \text{Methane}$$

$$\begin{array}{cc} H & H \\ \cdot\cdot & \cdot\cdot \\ H : C : C : H \\ \cdot\cdot & \cdot\cdot \\ H & H \end{array} \quad \text{Ethane}$$

where the dots represent valence electrons (four from each C atom, one from each H atom) arranged in pairs to form covalent bonds. Dashes are often used instead of electron pairs:

$$\begin{array}{c} H \\ | \\ H-C-H \\ | \\ H \end{array} \quad \text{Methane}$$

structural formulas

$$\begin{array}{cc} H & H \\ | & | \\ H-C-C-H \\ | & | \\ H & H \end{array} \quad \text{Ethane}$$

Diagrammatic formulas of this sort are called *structural formulas*; the simpler CH_4 and C_2H_6 are called *molecular formulas*. Besides the information that the molecular formulas give, these structural formulas show that, in both methane and ethane molecules, each hydrogen atom is attached to a carbon atom and that in ethane the two carbon atoms are linked together.

Structural formulas are written according to the ordinary rules of valence. Each carbon atom, with its valence of 4, must be connected with other atoms by four bonds, or dashes. Each hydrogen or chlorine atom can have only one dash; each oxygen atom has two. Since the bonds are all covalent, the positive or negative character of the valence is immaterial. The following formulas illustrate the valence rules:

$$\begin{array}{ccc} H & H & H \\ | & | & | \\ H-C-C-C-Cl \\ | & | & | \\ H & H & H \end{array} \quad \text{Propyl chloride}$$

$$\begin{array}{c} \text{H} \\ | \\ \text{H}-\text{C}-\text{O}-\text{H} \\ | \\ \text{H} \end{array} \quad \text{Methyl alcohol}$$

$$\begin{array}{c} \text{Cl} \\ | \\ \text{Cl}-\text{C}=\text{O} \end{array} \quad \text{Phosgene}$$

isomers

For methane, ethane, and propane the structural formulas given above are the only possible arrangements of carbon and hydrogen atoms that will satisfy the valence rules. Butane, on the other hand, may have its four C atoms and ten H atoms arranged in two *different* ways:

$$\begin{array}{c} \text{H H H H} \\ | \ | \ | \ | \\ \text{H}-\text{C}-\text{C}-\text{C}-\text{C}-\text{H} \\ | \ | \ | \ | \\ \text{H H H H} \end{array} \quad \text{Normal butane}$$

isomers of butane

$$\begin{array}{c} \text{H H H} \\ | \ | \ | \\ \text{H}-\text{C}-\text{C}-\text{C}-\text{H} \\ | \ | \ | \\ \text{H C H} \\ \diagup | \diagdown \\ \text{H H H} \end{array} \quad \text{Isobutane}$$

These formulas show that there are two different compounds with the molecular formula C_4H_{10}. They differ in that one of the carbon atoms in isobutane is linked to three others, while in normal butane the carbon atoms are linked to only one or two others. The physical properties of isobutane are somewhat different from those of normal butane because of this difference in molecular structure; the boiling point of isobutane, for instance, is $-10°C$, whereas that of normal butane, as listed in Table 28.1, is $-1°C$.

Pentane, C_5H_{12}, provides another example of this phenomenon. The five C atoms can be arranged in three different patterns, corresponding to the three known kinds of pentane:

$$\begin{array}{c} \text{H H H H H} \\ | \ | \ | \ | \ | \\ \text{H}-\text{C}-\text{C}-\text{C}-\text{C}-\text{C}-\text{H} \\ | \ | \ | \ | \ | \\ \text{H H H H H} \end{array} \quad \text{Normal pentane}$$

isomers of pentane

$$\begin{array}{c} \text{H H H H} \\ | \ | \ | \ | \\ \text{H}-\text{C}-\text{C}-\text{C}-\text{C}-\text{H} \\ | \ | \ | \ | \\ \text{H H C H} \\ \diagup | \diagdown \\ \text{H H H} \end{array} \quad \text{Isopentane}$$

```
    H H  H
     \|/
      C
H    /|\    H
 \  H | H  /
  C—C—C
 /    |    \
H     C     H
     /|\
    H | H
      H
```
                              Neopentane

isomers have the same molecular formulas but different structural formulas

Compounds that have the same molecular formulas but different structural formulas are called *isomers*. The number of possible isomers increases rapidly with the number of carbon atoms in the molecule; $C_{13}H_{28}$ has 813 theoretically possible isomers and $C_{20}H_{42}$ has 366,319. Only a few of the possible isomers have actually been prepared.

determining structures

Figuring out the possible structural arrangements corresponding to a given molecular formula is an interesting, though not too difficult, game. The big question is: Which structural formula goes with which isomer? This is seldom an easy question to answer and often requires extensive laboratory tests.

To show how the problem is attacked, let us use, instead of hydrocarbons, some more reactive substances that contain oxygen. We choose two that have the molecular formula C_2H_6O. One is the familiar liquid ethyl alcohol, or grain alcohol; the other is a gas named dimethyl ether (not the anesthetic called "ether," but a related compound). The valence rules permit two possible structures for C_2H_6O:

```
  H H                    H     H
  | |                    |     |
H—C—C—O—H          H—C—O—C—H
  | |                    |     |
  H H                    H     H
```

Our problem is to match these structures with the properties of the two substances. Tests such as the following make a choice possible:

1 Sodium reacts with alcohol, liberating hydrogen and forming the compound C_2H_5ONa. No further reaction with the other five H's of the alcohol molecule takes place, so one H must be attached in a different manner from the others.
2 Alcohol reacts with HCl to give water and the gas ethyl chloride, C_2H_5Cl. An O and an H have been replaced by a Cl atom, which suggests that the O and H were together in the original molecule.

We need go no further to assign the first of the above formulas to ethyl alcohol; so the second must represent dimethyl ether.

Many simple examples of this sort give the organic chemist a background of experiences which enables him to handle more difficult problems. He learns, for example, that evolution of hydrogen on addition of

Na to any organic compound indicates an H attached to an O somewhere in the molecule and that reaction with HCl to form water indicates the presence of an OH group. From other reactions he learns that organic compounds that dissolve in water to form acids contain the more complicated *carboxyl* group

$$-C\begin{matrix}\nearrow O \\ \searrow O-H\end{matrix}$$

the carboxyl group

A compound with the *aldehyde* group

$$-C\begin{matrix}\nearrow O \\ \searrow H\end{matrix}$$

the aldehyde group

is not an acid itself but is easily oxidized to an acid.

The organic chemist thus learns to recognize reactions characteristic of certain atom groups, much as the inorganic chemist learns to recognize typical reactions of compounds containing the nitrate group NO_3 or the sulfate group SO_4. The properties of an organic compound are the sum of the properties of its constituent atom groups, each one modified by the presence of the others. By investigating the properties of a substance, an organic chemist can often deduce what groups are present and, by piecing them together, can make a good guess at the formula. Organic chemistry is a study in molecular architecture, dealing with arrangements of atoms in larger units and the fitting together of these units to make molecules.

Physical as well as chemical methods can be used to investigate molecular structures. An example is molecular spectroscopy. The atoms in a molecule can vibrate back and forth about their equilibrium positions, usually in a variety of ways (Fig. 28.1). If a certain group of atoms in the molecule vibrates with the frequency f, it is able to absorb electromagnetic radiation of this frequency. When it does so, the energy of the vibrations increases by the quantum energy hf. Typical frequencies of molecular vibration are 10^{13} to 10^{14} Hz, which corresponds to the infrared part of the spectrum. If we pass infrared light through the liquid or vapor of a particular compound, then, the frequencies that are absorbed correspond to the vibrational frequencies of its molecules. From these frequencies it is often possible to infer details of the structures of the molecules.

molecular spectra help in determining molecular structures

28.1 Three modes of vibration of the H_2O molecule. *Each mode has a different characteristic frequency.*

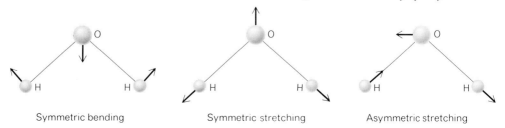

many atom groups have characteristic frequencies of vibration

A number of atom groups have characteristic frequencies of vibration that permit them to be identified in any molecule of which they are a part. Thus the —OH group has a vibrational frequency of 1.1×10^{14} Hz and the —NH_2 group has a frequency of 1.0×10^{14} Hz; the absorption of infrared light of either of these frequencies by a compound means that the corresponding group is present in its molecules. The nature of carbon-carbon bonds can also be established by this technique: the $\equiv$C—C$\equiv$, $>$C=C$<$, and —C$\equiv$C— groups all have different frequencies of vibration.

molecular models

The primary purpose of a structural formula is to summarize the relationships between the atoms in a molecule, not to be an accurate picture of the molecule. Three-dimensional models are useful for the latter purpose. In one type of model, atoms are represented by spheres and the bonds between them by sticks, as in Fig. 28.2a; this method has the virtue of making clear the geometrical configuration of the molecule. In another, more realistic type of model (Fig. 28.2b) the spheres representing the atoms are in contact, so that the result more nearly resembles the actual molecule. The molecule pictured in Fig. 28.2 is that of methane; the symmetrical arrangement of the four H atoms around the C atom is called *tetrahedral* because the H atoms may be thought of as being located at the corners of a tetrahedron.

Models of the two isomers of butane are shown in Fig. 28.3. The zigzag chain of the carbon atoms in normal butane is understandable by reference to the tetrahedral arrangement in space of the covalent bonds of carbon atoms. The symmetry of isobutane is more evident from its model than from its structural formula.

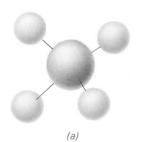

(a)

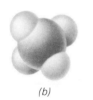

(b)

28.2 (a) Sphere-and-stick and (b) scale models of the methane molecule, CH_4.

CARBON BONDS

A carbon atom has four valence electrons, and we have been taking for granted that all of these electrons can participate to an equivalent extent in covalent bonds. This assumption is true, but a closer look shows that the story of carbon bonding is a good deal more subtle than is at first apparent. A review of Figs. 20.8 and 20.9 may be helpful at this point.

the methane molecule

A carbon atom has two electrons in its $2s$ orbital and one each in its $2p_x$ and $2p_y$ orbitals. The two $2s$ electrons have opposite spins and so should not be able to participate in bonds. Thus we would expect the hydride of carbon to be CH_2, with two σ bonding orbitals and a bond angle of $90°$ or a little more as in the case of H_2O. Yet CH_4 exists and, furthermore, it is perfectly symmetrical in structure with tetrahedral molecules whose C—H bonds are exactly equivalent to one another (Fig. 28.2).

ORGANIC CHEMISTRY

The easiest way to explain the existence of CH_4 is to assume that one of the two $2s$ electrons in C is "promoted" to the vacant $2p_z$ orbital, so that there is now one electron each in the $2s$, $2p_x$, $2p_y$, and $2p_z$ orbitals and four bonds can be formed. To raise an electron from a $2s$ to a $2p$ state means increasing the energy of the C atom, but it is reasonable to suppose that the formation of four bonds (to yield CH_4) in place of two (to yield CH_2) lowers the energy of the resulting molecule more than enough to compensate for this. The foregoing picture suggests that three of the bonds in CH_4 are $s + p$ σ bonds and one of them is an $s + s$ σ bond involving the $1s$ orbital of H and the now singly occupied $2s$ orbital of C; experimentally, however, all four bonds are found to be identical.

The complete explanation for CH_4 is based on a phenomenon called *hybridization* which can occur when the $2s$ and $2p$ states of an atom in a molecule are very close together in energy. In this case the atom can contribute a linear combination of *both* its $2s$ and $2p$ atomic orbitals to *each* molecular orbital if in this way the resulting bonds are more stable than otherwise. The four orbitals which participate in bonding a carbon atom to four hydrogen atoms in CH_4 are hybrids of one $2s$ and three $2p$ orbitals, and we may consider each one as a combination of $\frac{1}{4}s$ and $\frac{3}{4}p$. This particular combination is therefore called a sp^3 hybrid. Its configuration can be visualized as shown in Fig. 28.4. Evidently a sp^3

Normal butane

Isobutane

28.3 The two isomers of butane, C_4H_{10}.

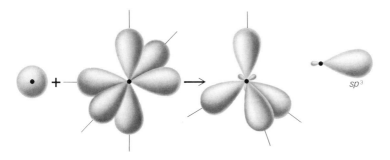

28.4 In sp^3 hybridization, an s orbital and three p orbitals in the same atom combine to form four sp^3 hybrid orbitals.

hybrid orbital is strongly concentrated in a single direction, unlike pure s and p orbitals, which accounts for its ability to produce an exceptionally strong bond—strong enough to compensate for the need to promote a $2s$ electron to a $2p$ state. Figure 28.5 is a representation of the CH_4 molecule in terms of sp^3 hybrid orbitals.

hybrid orbitals produce unusually strong bonds

It must be kept in mind that hybrid orbitals do not exist in an isolated atom, even when it is in an excited state, but arise when that atom is interacting with others to form a molecule.

carbon-carbon bonds

Two other types of hybrid orbital in addition to sp^3 can occur in carbon atoms. In sp^2 hybridization, one valence electron is in a pure p orbital and the other three are in hybrid orbitals that are $\frac{1}{3}s$ and $\frac{2}{3}p$ in character. In sp hybridization, two valence electrons are in pure p orbitals and the other two are in hybrid orbitals that are $\frac{1}{2}s$ and $\frac{1}{2}p$ in character.

Ethylene, C_2H_4, is an example of sp^2 hybridization in which the two carbon atoms are joined by two bonds. Figure 28.6 shows the three sp^2 hybrid orbitals, which are 120° apart in the plane of the paper, and the pure p_x orbital in each C atom. Two of the sp^2 orbitals in each C atom overlap s orbitals in H atoms to form σ bonding orbitals, and the third sp^2 orbital in each C atom forms a σ bonding orbital with the same orbital in the other C atom. The p_x orbitals of the C atoms form a π bond with each other, so that one of the bonds between the carbon atoms is a σ bond and the other a π bond. The conventional structural formula of ethylene is accordingly

the carbon atoms in ethylene are held together by a double bond

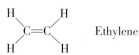 Ethylene

Acetylene, C_2H_2, is an example of sp hybridization in which the two carbon atoms are joined by three bonds. One sp hybrid orbital in each C atom forms a σ bond with an H atom, and the second forms a σ bond with the other atom. The $2p_x$ and $2p_y$ orbitals in each C atom form π bonds, so that one of the three bonds between the carbon atoms is a σ_z bond and the others are π_x and π_y bonds (Fig. 28.7). The conventional structural formula of acetylene is

28.5 The methane molecule. *The overlapping sp^3 hybrid orbitals of the C atom and the s orbitals of the four H atoms form bonding molecular orbitals.*

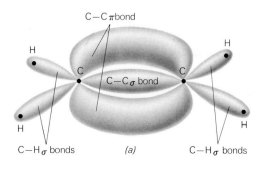

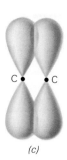

28.6 (a) **The ethylene (C_2H_4) molecule.** *All the atoms lie in a plane perpendicular to the plane of the paper.* (b) *Top view, showing the sp^2 hybrid orbitals that form σ bonds between the C atoms and between each C atom and two H atoms.* (c) *Side view, showing the pure p_x orbitals that form a π bond between the C atoms.*

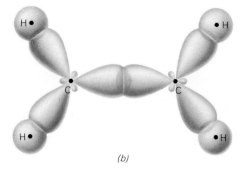

ORGANIC CHEMISTRY

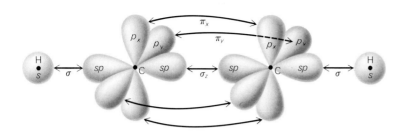

28.7 The acetylene (C_2H_2) molecule. There are three bonds between the C atoms, one σ bond between sp hybrid orbitals and two π bonds between pure p_x and p_y orbitals.

H—C≡C—H   Acetylene

the carbon atoms in acetylene are held together by a triple bond

In both ethylene and acetylene the electrons in the π orbitals are "exposed" on the outside of the molecules. These compounds are much more reactive chemically than molecules with only single σ bonds between carbon atoms, such as ethane,

H H
| |
H—C—C—H   Ethane
| |
H H

in which all the bonds are formed from sp^3 hybrid orbitals in the carbon atoms. Carbon compounds with double and triple bonds are said to be *unsaturated* because they can add other atoms to their molecules in such reactions as

the unsaturated compounds have double or triple carbon-carbon bonds

H   H   H H
 \   / | |
  C=C   + HCl ⟶ H—C—C—H
 /   \ | |
H   H   H Cl

H   H   H H
 \   / | |
  C=C   + Cl₂ ⟶ H—C—C—H
 /   \ | |
H   H   Cl Cl

saturated compounds have only single carbon-carbon bonds and are less reactive than the unsaturated compounds

In a *saturated* compound such as methane or ethane only single bonds are present.

the benzene ring

Benzene, C_6H_6, is an extremely interesting hydrocarbon because its six C atoms are arranged in a flat hexagonal ring (Fig. 28.8). If we think in terms of conventional electron-pair bonds, there are three single and three double bonds between the C atoms in benzene, so that its structure can be described in the following way:

conventional representation of the benzene ring

Benzene

(structural formula of benzene with H and C atoms)

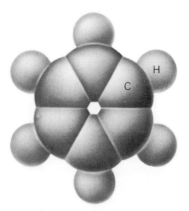

28.8 *Scale model of benzene molecule, C_6H_6.*

There are a number of major objections to this simple picture of benzene, of which we shall mention three. The first is that double bonds are not particularly stable, so that benzene, like ethylene, should form addition compounds readily. What actually happens, on the contrary, is that benzene forms substitution compounds in which one or more of the H atoms are replaced by other atoms or atom groups, with the ring structure remaining intact. In other words, the benzene ring is stronger than the simple structural formula would indicate. Substantiating this observation is the heat of formation of C_6H_6 which turns out to be about 40 kcal/mole—1.7 ev per molecule—*more* than would be expected if there were indeed six C—H, three C—C, and three C=C bonds in each molecule. (The heat of formation of a molecule is the energy given off when it is formed from isolated atoms.)

The second objection arises because C—C bonds are 1.54Å long on the average and C=C bonds are 1.35Å long; the shorter length of the C=C bond is consistent with the participation of four electrons in a double bond, instead of two as in C—C. But benzene molecules are regular hexagons in shape, with all bonds having the same length and all bond angles being 120°. (We recall that 1Å = 10^{-10} m.)

Finally, if C_6H_6 occurred with C—C and C=C bonds alternating, compounds such as dibromobenzene, $C_6H_4Br_2$, ought to occur in four forms, as shown below.

1 Br atoms on opposite atoms:

Paradibromobenzene

2 Br atoms on alternate C atoms:

Metadibromobenzene

3 Br atoms on adjacent double-bonded C atoms:

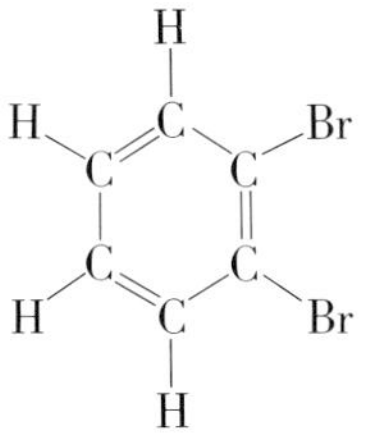

Orthodibromobenzene

4 Br atoms on adjacent single-bonded C atoms:

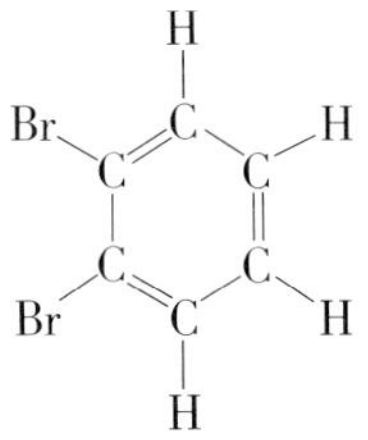

Orthodibromobenzene

Only three forms of $C_6H_4Br_2$ are actually found, however—there is only one kind of orthodibromobenzene molecule, not two. Hence in nature there is no distinction between the double and single bonds we have used to represent the benzene molecule.

A picture of the nature of the benzene molecule that accounts for its observed properties is provided by a consideration of the molecular orbitals involved. Because the carbon-carbon bonds in the benzene ring are 120° apart, we conclude that the basic structure of the molecule is the result of bonding by sp^2 hybrid orbitals. Of the three sp^2 orbitals per C atom, one forms a σ bonding orbital with the $1s$ orbital of an H atom and the other two form σ bonding orbitals with the corresponding sp^2 orbitals of the C atoms on either side (Fig. 28.9). This leaves one $2p_x$ orbital per C atom, which has lobes above and below the plane of the ring. The total of six $2p_x$ orbitals in the molecule combine to produce bonding π orbitals which take the form of a continuous electron probability distribution above and below the plane of the ring. The six electrons belong to the molecule as a whole and not to any particular pair of atoms; these electrons are *delocalized*.

the benzene molecule contains six delocalized electrons that contribute to the bonding

With six delocalized electrons in π bonding orbitals and twelve electrons in the σ bonds between the carbon atoms, each carbon-carbon bond consists of $1\frac{1}{2}$ net bonds, one σ bond and half of a π bond. Confirming this result, each C═C bond in benzene turns out to be 1.40Å long, intermediate between the 1.54Å average length of a C─C bond and the 1.35Å average length of a C═C bond. The extra stability of the benzene molecule is due to the delocalization of the six $2p_x$ electrons.

carbon-carbon bonds in benzene are each equivalent to 1½ net bonds

HYDROCARBON DERIVATIVES

The variety and complexity of organic compounds containing only two elements are amply demonstrated by the hydrocarbons. With one or two

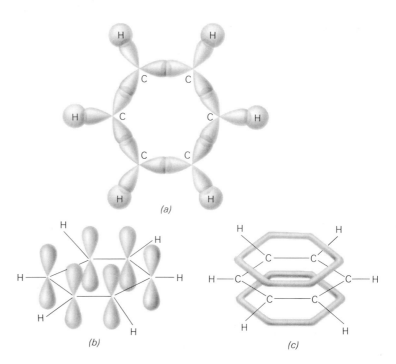

28.9 The benzene molecule. (a) The overlaps between the sp^2 hybrid orbitals in the C atoms with each other and with the s orbitals of the H atoms lead to σ bonds. (b) Each C atom has a pure p_x orbital occupied by one electron. (c) The bonding π molecular orbitals formed by the six p_x atomic orbitals constitute a continuous electron probability distribution around the molecule that contains six delocalized electrons.

more elements added, the number of possible compounds becomes immense. To simplify the problem of classifying these compounds, they are often regarded as *derivatives* of hydrocarbons—that is, as compounds obtained by substituting other atoms or atom groups for some of the H atoms in hydrocarbon molecules. Ordinarily carbon compounds are not prepared in this manner, but their structural formulas suggest that they might be. For example, ethyl alcohol and acetic acid may be regarded as derivatives of ethane and methane, respectively:

The formula of alcohol is derived from that of ethane by substituting an OH group for an H atom, and that of acetic acid is derived from CH_4 by substituting a COOH group for an H atom.

ORGANIC CHEMISTRY

hydrocarbon atom groups

The carbon-hydrogen atom groups that appear in hydrocarbon derivatives are named from the hydrocarbons. Groups corresponding to the hydrocarbons methane, ethane, and propane are

$$\begin{array}{c} H \\ | \\ H-C- \\ | \\ H \end{array}$$ Methyl group

$$\begin{array}{cc} H & H \\ | & | \\ H-C-C- \\ | & | \\ H & H \end{array}$$ Ethyl group

$$\begin{array}{ccc} H & H & H \\ | & | & | \\ H-C-C-C- \\ | & | & | \\ H & H & H \end{array}$$ Propyl group

Polyethylene film forms part of a water desalination system. Light rays from the sun pass through the film to the sea water underneath, but the longer wavelength infrared rays from the heated water cannot escape. The net result is inexpensive preheating of the water, which is distilled in the round tower at upper right.

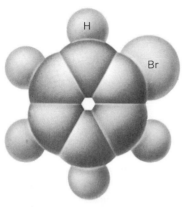

28.10 Scale model of the bromobenzene molecule, C_6H_5Br.

Thus the compound CH_3Cl is methyl chloride, C_3H_7I is propyl iodide, C_2H_5OH is ethyl alcohol, and $CH_3C_2H_5SO_4$ is methyl ethyl sulfate.

Let us examine briefly several important classes of hydrocarbon derivatives.

halogen derivatives

One or more of the H atoms in a hydrocarbon molecule may be replaced by halogen atoms, giving compounds like CH_3Br, CH_2I_2, C_2H_5Cl, and C_6H_5Br (Fig. 28.10). The simpler compounds of this sort are gases and volatile liquids, and, as in the methane series, their boiling points and melting points rise with increasing molecular weight. They can be prepared by the addition of halogens and halogen acids to unsaturated hydrocarbons, as we have seen, but are more conveniently made indirectly from alcohols.

The halogen derivatives are particularly important to the organic chemist because the halogen atoms are easily replaced by other groups in building up complex molecules. A few of the simpler of them are useful for other purposes: $CHCl_3$ is the anesthetic chloroform, CCl_4 is carbon tetrachloride, an important cleaning fluid, and CCl_2F_2 (dichlorodifluoromethane) is one of the gases used in refrigeration under the name Freon.

alcohols

These form a group of hydrocarbon derivatives in which one or more H atoms in the molecule have been replaced by OH groups. The two commonest members of the group are ethyl alcohol (grain alcohol), C_2H_5OH, and methyl alcohol (wood alcohol), CH_3OH. The OH group makes alcohol molecules somewhat polar, so the simpler alcohols are soluble in water. The polarity is not great enough, however, to prevent alcohols from mixing also with a great variety of less polar organic substances. These properties make alcohols, especially ethyl and methyl alcohol, valuable as solvents.

A familiar alcohol with more than one OH group in its molecule is the sweetish, viscous liquid *glycerin*,

$$\begin{array}{c} \text{H} \quad \text{H} \quad \text{H} \\ | \quad | \quad | \\ \text{H}-\text{C}-\text{C}-\text{C}-\text{H} \\ | \quad | \quad | \\ \text{O} \quad \text{O} \quad \text{O} \\ | \quad | \quad | \\ \text{H} \quad \text{H} \quad \text{H} \end{array} \qquad \text{Glycerin}$$

The substitution of an OH group in the molecule of a benzene hydrocarbon produces a compound with properties somewhat different from ordinary alcohols. The simplest is the antiseptic *phenol* (carbolic acid), C_6H_5OH, a very weak but highly poisonous acid.

esters

Alcohols are, so to speak, organic hydroxides, but, unlike inorganic hydroxides, they do not dissociate appreciably in water. They react slowly with acids to form compounds called *esters*. These reactions are superficially similar to the neutralization of an acid by a base but, in contrast to neutralization, are slow and incomplete.

alcohols react with acids to form esters

Esters are analogous to the salts of inorganic chemistry but are nonelectrolytes. They can be formed from either organic acids or inorganic acids. For example, the ester methyl sulfate can be prepared from methyl alcohol and sulfuric acid:

$$CH_3OH + H_2SO_4 \longrightarrow 2H_2O + (CH_3)_2SO_4$$
Methyl alcohol — Sulfuric acid — Water — Methyl sulfate

When the organic acid acetic acid (which we write here "backward" as $HOOCCH_3$, to show how it combines structurally) is added to ethyl alcohol, the ester ethyl acetate results:

$$C_2H_5OH + HOOCCH_3 \longrightarrow H_2O + C_2H_5OOCCH_3$$
Ethyl alcohol — Acetic acid — Water — Ethyl acetate

Many esters have pleasant flowerlike or fruitlike odors and find extensive use in perfumes and flavors. The explosive *nitroglycerin* is an ester formed by the reaction of nitric acid with glycerin:

$$C_3H_5(OH)_3 + 3HNO_3 \longrightarrow C_3H_5(NO_3)_3 + 3H_2O$$

organic acids

The partial oxidation of ethyl alcohol gives CH_3COOH (acetic acid). We have sometimes written this $HC_2H_3O_2$; either formula indicates that only one of the four H atoms is capable of ionizing, but the former shows a little more about the molecular structure. Acetic acid itself is a colorless liquid, freezing at 17°C and boiling at 118°C, miscible with water in all proportions. It is used industrially as a solvent and in the manufacture of dyes, drugs, flavors, and plastics.

acetic acid

In general, any organic compound whose molecule contains the carboxyl group

$$-C\begin{matrix}\nearrow O \\ \searrow O-H\end{matrix}$$

organic acids contain the carboxyl group

which we have abbreviated —COOH, is an acid, since the H atom of this group is capable of ionizing. Most organic acids are very weak. Acids corresponding to the methane hydrocarbons form a series, as do the halogen derivatives and the alcohols, in which the boiling points and freezing points increase steadily with increasing molecular weights. The simpler members of the series, like acetic acid, are soluble in water; the more complex are insoluble. Three members of the series are butyric acid (C_3H_7COOH), which is the acid of rancid butter; capric acid ($C_9H_{19}COOH$), which has an odor like that of goats; and stearic acid ($C_{17}H_{35}COOH$), whose sodium salt is a constituent of soap. More complex organic acids are the citric acid of citrus fruits, the tartaric acid of grapes, and the lactic acid of sour milk.

Organic acids are often produced as intermediate steps in the decay of organic matter. Black soils with abundant decaying plant material, for instance, may be too acidic for the successful growing of crops unless lime or some other basic substance is added to neutralize the acidity. The acids of decay often aid in the slow disintegration of rocks.

INDUSTRIAL ORGANIC CHEMISTRY

The industrial applications of organic chemistry are so many in number and so wide-ranging in scope that it is hopeless even to try to summarize them in a short space. What we shall do instead is briefly examine three important classes of synthetic products created by the chemist: plastics, elastomers, and fibers.

plastics

Earlier in this chapter the unsaturated hydrocarbon ethylene,

ethylene
$$H-\underset{\underset{H}{|}}{\overset{\overset{H}{|}}{C}}=\underset{\underset{H}{|}}{\overset{\overset{H}{|}}{C}}-H$$

was mentioned. Because of the double bond, ethylene molecules can, under the proper conditions, polymerize to form long chains whose formula we might write as

polyethylene
$$\cdots-\overset{H}{\underset{H}{|}}\overset{|}{C}-\overset{H}{\underset{H}{|}}\overset{|}{C}-\overset{H}{\underset{H}{|}}\overset{|}{C}-\overset{H}{\underset{H}{|}}\overset{|}{C}-\overset{H}{\underset{H}{|}}\overset{|}{C}-\overset{H}{\underset{H}{|}}\overset{|}{C}-\cdots$$

monomer and polymer

This material is *polyethylene*, which is widely used as a packaging material because of its inertness and pliability. Ethylene is called the *monomer* in this process, and polyethylene the *polymer*.

Ethylene is the simplest member of a class of hydrocarbons called the *olefins* which, because they possess unsaturated bonds, undergo polymerization fairly readily. Thus the olefin derivative vinyl chloride,

ORGANIC CHEMISTRY

$$H_2C=\underset{\underset{Cl}{|}}{\overset{\overset{H}{|}}{C}}$$

is the monomer in preparing the polymer polyvinyl chloride, sometimes abbreviated PVC, which is used, among other applications, in plastic tubing and in sheet form (Koroseal). Another olefin derivative, methyl methacrylate,

$$H_2C=\underset{\underset{COOCH_3}{|}}{\overset{\overset{CH_3}{|}}{C}}$$

polymerizes to form the transparent plastics Lucite and Plexiglas.

elastomers

Those olefins that contain two double bonds in each molecule are known as *dienes*. The polymers of dienes are, in general, flexible and elastic and are accordingly called *elastomers*. Rubber is a natural elastomer. A widely used synthetic elastomer is Neoprene (actually polychloroprene), a polymer of the monomer chloroprene,

$$H_2C=\overset{\overset{Cl}{|}}{C}-\overset{\overset{H}{|}}{C}=CH_2$$

A valuable property of Neoprene is that liquid hydrocarbons such as gasoline do not affect it, whereas they interact with natural rubber.

Nylon fibers emerging from a die called a spinneret.

fibers

Synthetic fibers are of various kinds. Orlon (actually polyacrylonitrile) is a polymer of the olefin derivative acrylonitrile,

$$H_2C=\underset{H}{\overset{H}{C}}-C\equiv N$$

Saran, which can be manufactured in sheet form also, is a *copolymer* formed by the mutual polymerization of the different monomers vinyl chloride and vinylidene chloride,

$$H_2C=CCl_2$$

Usually the proportions are 90 percent vinylidene chloride and 10 percent vinyl chloride.

Nylon and Dacron are composed of chains of structural elements, just like polymers, but they are produced by chemical reactions rather than by the polymerization of monomer molecules. Thus one type of nylon is prepared by reacting hexamethylenediamine,

$$\underset{H}{\overset{H}{N}}-\underset{H}{\overset{H}{C}}-\underset{H}{\overset{H}{C}}-\underset{H}{\overset{H}{C}}-\underset{H}{\overset{H}{C}}-\underset{H}{\overset{H}{C}}-\underset{H}{\overset{H}{C}}-\overset{H}{\underset{H}{N}}$$

with adipic acid,

$$\underset{O}{\overset{HO}{C}}-\underset{H}{\overset{H}{C}}-\underset{H}{\overset{H}{C}}-\underset{H}{\overset{H}{C}}-\underset{H}{\overset{H}{C}}-\overset{OH}{\underset{O}{C}}$$

The result is a chain whose elements can be written

the structure of nylon

$$-\underset{}{\overset{H}{N}}-\underset{H}{\overset{H}{C}}-\underset{H}{\overset{H}{C}}-\underset{H}{\overset{H}{C}}-\underset{H}{\overset{H}{C}}-\underset{H}{\overset{H}{C}}-\underset{}{\overset{H}{N}}-\overset{O}{\underset{}{C}}-\underset{H}{\overset{H}{C}}-\underset{H}{\overset{H}{C}}-\underset{H}{\overset{H}{C}}-\underset{H}{\overset{H}{C}}-\overset{O}{\underset{}{C}}-$$

The atom group

$$-\underset{}{\overset{H}{N}}-\overset{O}{\underset{}{C}}-$$

is known as an amide linkage, so nylon is called a polyamide. In a similar way Dacron, whose structural elements are different from those of nylon, is a polyester because its elements are linked together by OH groups. The N—H and C=O groups in nylon are polar, and their mutual attraction is what holds adjacent chains of molecules firmly together.

raw materials

The raw materials of the organic chemical industry are very numerous, but three are particularly important: *coal tar*, a black, sticky, seemingly unpromising liquid obtained when coal is heated to make coke; *cellulose*, obtained chiefly from wood and cotton; and *petroleum*.

The distillation of coal tar gives a variety of organic compounds, most of them derivatives of the ring hydrocarbon benzene; among others are benzene itself, toluene, xylene, naphthalene, anthracene, phenol. These relatively simple compounds are the building blocks for an incredible array of synthetic dyes, drugs, perfumes, and other products. Indigo, alizarin, turkey red, Mercurochrome are a few of the dyes. Oil of wintergreen, oil of bitter almonds, synthetic vanilla, the compounds responsible for the fragrance of roses, violets, heliotrope, and narcissus are among the flavors and perfumes. Representative drugs are procaine, aspirin, epinephrine, and sulfanilamide. The explosive trinitrotoluene, TNT, is prepared by the action of nitric and sulfuric acids on toluene. The earliest plastic, Bakelite, was made by polymerizing a mixture of phenol and formaldehyde.

Petroleum and natural gas furnish simple hydrocarbons—methane, ethane, ethylene, propane—which can be used as starting points in building up complex molecules. Some kinds of petroleum also contain ring hydrocarbons like benzene, and these ring hydrocarbons can be built from the simpler ones if not initially present. Thus petroleum and coal tar can both be used as sources of many kinds of hydrocarbons. Which one is used in a given industry depends on availability, cost, and ease of handling; the current trend is strongly toward petroleum and natural gas.

Cellulose treated with nitric acid gives cellulose nitrate, or pyroxylin; further addition of nitric acid gives the explosive guncotton. Pyroxylin is an important plastic and is also an intermediate step in the manufacture of Celluloid, photographic film, and lacquers. Cellulose and its esters may be dissolved in various solvents and squirted through small openings into other liquids, which precipitate the cellulose; in this manner are produced threads of rayon and sheets of cellophane. Today such synthetics as nylon, Dacron, and polyethylene are largely replacing cellulose derivatives.

GLOSSARY

Organic chemistry refers to the chemistry of carbon compounds; *inorganic chemistry* refers to the chemistry of the other elements.

A *hydrocarbon* is an organic compound containing only carbon and hydrogen.

Fractional distillation is a process for separating a mixture of hydrocarbons in the order of their boiling points.

Cracking a hydrocarbon is the process of heating it under pressure in the presence of a catalyst in order to break its molecules down into simpler hydrocarbons.

Polymerization is the process of causing simple hydrocarbon molecules to join together into more complicated ones under the influence of heat and catalysts.

The *structural formula* of a compound is a diagram that shows the arrangement of the atoms in its molecules.

Isomers are compounds that have the same molecular formulas but different structural formulas.

The $2s$ and $2p$ orbitals of an atom can combine to form *hybrid orbitals* which contribute to the molecular orbitals the atom's valence electrons participate in.

An *unsaturated hydrocarbon* is one whose molecules contain more than one covalent bond (that is, shared electron pair) between adjacent carbon atoms. In a *saturated hydrocarbon* there is only one bond between adjacent carbon atoms.

Benzene hydrocarbons contain closed rings of six carbon atoms in their structural formulas; the carbon-carbon bonds in benzene are each equivalent to $1\frac{1}{2}$ net bonds.

A *hydrocarbon derivative* is a compound obtained by substituting other atoms or atom groups for some of the hydrogen atoms in hydrocarbon molecules.

A *halogen derivative* is a hydrocarbon derivative in which one or more hydrogen atoms have been replaced by halogen atoms.

An *alcohol* is a hydrocarbon derivative in which one or more hydrogen atoms have been replaced by OH groups.

An *ester* is an organic compound formed when an alcohol reacts with an acid.

An *organic acid* is any organic compound whose molecules contain the group COOH, because the hydrogen atom of this group is capable of ionizing.

EXERCISES

1 Compare the properties of a simple ester, for instance methyl chloride, with those of a salt, for instance sodium chloride.

2 In each of the following pairs, which substance would you expect to have (*a*) the higher melting point, (*b*) the lower density?
C_6H_{14} and $C_{11}H_{24}$
$C_6H_{13}Br$ and $C_{11}H_{23}Br$
$C_5H_{11}COOH$ and $C_{10}H_{21}COOH$

ORGANIC CHEMISTRY

3 Which of the following would (*a*) dissolve in water, (*b*) turn a litmus solution red, (*c*) be gaseous at ordinary temperatures, (*d*) react with Na to liberate hydrogen, (*e*) react with ethyl alcohol to give esters, (*f*) react with acetic acid to give esters?

C_2H_5COOH C_3H_8
C_2H_4 C_2H_5OH
HCl $C_3H_5(OH)_3$

4 Name one property by which you could distinguish
 a. C_2H_4 from CH_4
 b. CH_3COOH from CH_3OH
 c. C_3H_7OH from $C_5H_{11}OH$
 d. C_2H_5OH from H_2O
 e. CH_4 from O_2

5 Name an unsaturated hydrocarbon, an ester, an organic acid, an alcohol, a sugar, and a derivative of methane.

6 Explain why structural formulas are so much more useful in organic chemistry than in inorganic chemistry.

7 Why are there more carbon compounds than compounds of any other element?

8 Distinguish between saturated and unsaturated hydrocarbons, giving examples of each. What is a hydrocarbon derivative?

9 In what ways do organic compounds, as a class, differ from inorganic compounds?

10 Why are the bonds formed by sp^3 hybrid orbitals stronger than those formed by pure s and pure p orbitals?

11 What are the natures of the double and triple bonds between carbon atoms? Why are compounds that contain such bonds more reactive than those with only single bonds between carbon atoms?

12 What is wrong with the conventional model of the benzene molecule as having alternate single and double bonds?

13 Write structural formulas for three of the isomers of hexane (C_6H_{14}).

14 Show by means of structural formulas the reaction between methyl alcohol and acetic acid to produce methyl acetate.

15 Give the structural formulas of the two isomeric propyl alcohols which share the molecular formula C_3H_7OH.

16 Xylene molecules consist of benzene molecules in which two of the hydrogen atoms have been replaced by CH_3 groups. Give structural formulas for the three isomers of xylene.

chapter 29

The Chemistry of Life

CARBOHYDRATES AND LIPIDS
carbohydrates
lipids

PROTEINS
peptides
protein structures

NUCLEIC ACIDS
nucleotides
DNA
RNA
the origin of life

The chemistry of carbon that was considered in the previous chapter, though interesting and important industrially, does not indicate why the name *organic* chemistry has been given to this subject. The reason we did not encounter living matter in exploring the basic aspects of organic chemistry is that the carbon compounds characteristic of plants and animals are usually extremely complex, often containing hundreds of thousands of atoms arranged in intricate three-dimensional patterns. Until recently only a few of the simpler molecules of living matter had been analyzed and understood, but with the development of advanced techniques it has been possible to elucidate the structure and behavior of a number of the compounds directly involved in the essential phenomena of life. There are four chief classes of organic compounds found in living matter: carbohydrates, lipids, proteins, and nucleic acids. Each of these will be briefly discussed in this chapter.

CARBOHYDRATES AND LIPIDS

carbohydrates

Carbohydrates are compounds of carbon, hydrogen, and oxygen whose molecules contain two atoms of hydrogen for every one of oxygen. They

the carbohydrates are manufactured in green plants by photosynthesis

are manufactured in the leaves of green plants from carbon dioxide and water by the process of photosynthesis, with energy for the reaction being absorbed from sunlight by the catalyst chlorophyll. Sugars, starches, and cellulose are all carbohydrates.

An important group of sugars consist of isomers having the common formula $C_6H_{12}O_6$. Three of these isomers are glucose, fructose, and galactose, whose structural formulas are as follows:

```
         H                    H                    H
         |                    |                    |
         C=O              H—C—OH                  C=O
         |                    |                    |
      H—C—OH                 C=O                H—C—OH
         |                    |                    |
     HO—C—H               HO—C—H               HO—C—H
         |                    |                    |
      H—C—OH               H—C—OH               HO—C—H
         |                    |                    |
      H—C—OH               H—C—OH               H—C—OH
         |                    |                    |
      H—C—OH               H—C—OH               H—C—OH
         |                    |                    |
         H                    H                    H
      Glucose              Fructose             Galactose
```

monosaccharides and disaccharides

Simple sugars like those above are called *monosaccharides*. Two monosaccharides can link together to form a *disaccharide*. Thus sucrose (ordinary table sugar) consists of one unit of glucose and one of fructose; lactose (milk sugar) consists of one unit of glucose and one of galactose; and maltose (malt sugar) consists of two units of glucose. The molecular formulas of all three of these disaccharides is $C_{12}H_{22}O_{11}$, but of course their structures are different.

Polysaccharides are complex sugars that consist of chains of more than two simple sugars. In living things the polysaccharides serve both as structural components and as a medium of energy storage. Thus in plants *cellulose*, which consists of a chain of about 1,500 glucose units, is common as the chief constituent of cell walls; wood is mostly cellulose. *Starch*, whose 300 to 1,000 glucose units are joined together in a slightly different way from that characteristic of cellulose, is the form in which energy is stored in plants for later use. Starch occurs in grains that have an insoluble outer layer, and so remains in the cell in which it is formed until, when needed as fuel, it is broken down into soluble glucose molecules.

cellulose and starch are plant carbohydrates

chitin and glycogen are animal carbohydrates

Two polysaccharides found in animals are *chitin*, which forms the outer shells of insects and crustaceans such as lobsters and crabs, and *glycogen*, which is present in the liver and muscles and is released when energy is required. Glycogen, the animal equivalent of starch, is soluble but its molecules are so large that they cannot pass readily through cell walls. When glucose is needed by an animal, its stored glycogen is split into the much smaller glucose molecules.

The term *respiration* is used to describe the process by which living things obtain the energy they need by the oxidation of nutrient molecules. Generally the nutrient molecule most directly involved is glucose, and

THE CHEMISTRY OF LIFE

its oxidation is an exothermic reaction that yields carbon dioxide and water as products:

$$\underset{\text{Glucose}}{C_6H_{12}O_6} + \underset{\text{Oxygen}}{6O_2} \longrightarrow \underset{\substack{\text{Carbon} \\ \text{Dioxide}}}{6CO_2} + \underset{\text{Water}}{6H_2O} + \text{energy}$$

the oxidation of glucose provides energy to plants and animals; it is the reverse of photosynthesis

This reaction does not take place all at once, as this equation would indicate, but in a complex series of steps that involve a number of other substances; however, the net effect is the oxidation of glucose. The oxidation of glucose is evidently the reverse of photosynthesis, and is the final process in the transformation of the energy in sunlight into the energy used by living things.

Usually the carbohydrate intake of an animal is in the form of disaccharides and polysaccharides. In digestion, these are *hydrolyzed* with the help of water into monosaccharides. For example, the hydrolysis of sucrose into glucose and fructose may be represented by the molecular equation

carbohydrates are converted to simple sugars during digestion

$$\underset{\text{Sucrose}}{C_{12}H_{22}O_{11}} + \underset{\text{Water}}{H_2O} \longrightarrow \underset{\text{Glucose}}{C_6H_{12}O_6} + \underset{\text{Fructose}}{C_6H_{12}O_6}$$

Hydrolysis is promoted by appropriate *enzymes*, which are specialized protein molecules that act as catalysts in most biochemical processes. Although many animals are able to hydrolyze starch into glucose, very few can hydrolyze cellulose. Some plant-eating animals, for instance cattle, are host to microorganisms such as yeasts, protozoa, and bacteria in their digestive tracts whose own enzymes can achieve the hydrolysis of cellulose, and the resulting glucose can then be utilized by the animal itself.

enzymes are protein molecules that catalyze biochemical processes

After digestion, the glucose passes into the bloodstream for circulation throughout the body. Glucose not immediately needed by the cells is converted into glycogen in the liver and elsewhere; if there is too much glucose present for storage as glycogen, it is synthesized into fats.

lipids

Fats and such fatlike substances as oils, waxes, and sterols are collectively known as *lipids*. Like carbohydrates, lipids contain only the elements C, H, and O, which is natural since lipids are synthesized in plants and animals from carbohydrates. The proportions of these elements are different in lipids, though.

A fat molecule consists of a glycerin molecule with three *fatty acid* molecules attached to it. Since glycerin is an alcohol, fats are esters. In a fatty acid molecule a chain of carbon and hydrogen atoms is terminated at one end by a carboxyl group, —COOH. As we know, the possession of a carboxyl group is common to all organic acids; the dissociation of such an acid may be represented as

structure of a fat molecule

$$-COOH \longrightarrow -COO^- + H^+$$

A typical fatty acid is *butyric acid*, C_3H_7COOH, which is found in butter. Its structural formula is

$$\begin{array}{c} \text{H} \;\; \text{H} \;\; \text{H} \;\; \text{O} \\ |\;\;\;\; |\;\;\;\; |\;\;\;\; \| \\ \text{H}-\text{C}-\text{C}-\text{C}-\text{C}-\text{OH} \\ |\;\;\;\; |\;\;\;\; | \\ \text{H} \;\; \text{H} \;\; \text{H} \end{array} \quad \text{Butyric acid}$$

Some other fatty acids are formic acid (HCOOH), acetic acid (CH_3COOH), palmitic acid ($C_{15}H_{31}COOH$), and stearic acid ($C_{17}H_{35}COOH$). Except for formic acid, the compositions of the various fatty acids can be represented by the formula $CH_3(CH_2)_nCOOH$, where $n = 0, 1, 2, \ldots$. Thus acetic acid corresponds to $n = 0$, butyric acid to $n = 2$, and stearic acid to $n = 16$.

The structural formula of glycerin is

$$\begin{array}{c} \text{H} \\ | \\ \text{H}-\text{C}-\text{OH} \\ | \\ \text{H}-\text{C}-\text{OH} \\ | \\ \text{H}-\text{C}-\text{OH} \\ | \\ \text{H} \end{array} \quad \text{Glycerin}$$

When glycerin combines with three fatty acids to form a fat molecule, an H^+ ion from the glycerin molecule joins with an OH^- ion from each of the fatty acids to yield three water molecules. The fatty acids may be the same or different. If we write the structural formula of the fatty acids "backwards" as

$$\begin{array}{c} \text{O} \\ \| \\ \text{HO}-\text{C}-\text{CH}_3(\text{CH}_2)_n \end{array} \quad \text{Fatty acid}$$

we can express the formation of a fat in the manner shown in Fig. 29.1. The atom group

$$\begin{array}{c} \text{O} \\ \| \\ -\text{O}-\text{C}- \end{array} \quad \text{Ester group}$$

between the glycerin and the fatty acid parts of a fat molecule is called an ester group. When the fatty acid molecules are relatively small, the result is a liquid fat; when the molecules are large, the result is a solid fat.

fats contain more biologically usable energy than an equivalent amount of carbohydrates

Fats are used for energy storage and other purposes in most living things. The digestion of a fat molecule involves the breaking of the ester links between its glycerin and fatty acid parts. The enzyme *lipase* is the catalyst for this hydrolysis reaction, during which water molecules are added in the reverse of the synthesis pictured in Fig. 29.1. The oxidation

THE CHEMISTRY OF LIFE

$$
\begin{array}{l}
\text{H} \\
| \\
\text{H—C—OH} \quad + \quad \text{HO—C—CH}_3(\text{CH}_2)_n \\
| \quad\quad\quad\quad\quad\quad\;\; \| \\
\quad\quad\quad\quad\quad\quad\;\; \text{O}
\end{array}
$$

(Glycerin + Three fatty acids → Fat + $3H_2O$)

29.1 The formation of a fat from glycerin and three fatty acids. The H and O atoms that become water molecules in the process are shown in color. The hydrolysis of a fat molecule proceeds in exactly the opposite way.

of the glycerin and fatty acids then proceeds in a fairly complicated way, and is accompanied by the release of considerably more energy than in the case of an equivalent amount of carbohydrate.

PROTEINS

The proteins, which are the principal constituent of living cells, are compounds of carbon, hydrogen, oxygen, nitrogen, and often sulfur and phosphorus; some proteins contain still other elements. The basic chemical units of which the protein molecules are composed are called *amino acids*, of which 23 are known. Typical protein molecules consist of several hundred amino acids joined together in chains, and their structures are accordingly quite complex. The formula of one of the proteins found in milk is $C_{1864}H_{3012}O_{576}N_{468}S_{21}$, which gives an idea of the size some protein molecules have.

Plant and animal tissues contain proteins both in solution, as part of the fluid present in cells and in other fluids such as blood, and in insoluble form, as the skin, muscles, hair, nails, horns, and so forth of animals. Silk is an almost pure protein. The human body contains about 100,000 different proteins, all of which it must make from the amino acids it obtains from the digestion of the food proteins it takes in. One of the great successes of modern biochemistry is the discovery of how living

proteins are the chief constituent of living matter and are composed of amino acids

cells duplicate the complex arrangements of amino acids that are found in the proteins they are composed of.

peptides

The structures of all amino acids follow the pattern

$$\begin{array}{c} H \\ \diagdown \\ H \end{array} N - \underset{R}{\overset{H}{\underset{|}{C}}} - \overset{O}{\underset{}{\overset{\|}{C}}} - OH \qquad \text{Generalized amino acid}$$

amino acid molecules have an amino group at one end and a carboxyl group at the other

where R represents an atomic group that is different in each variety of amino acid. The name "amino acid" arises because the atom group

$$\begin{array}{c} H \\ \diagdown \\ H \end{array} N- \qquad \text{Amino group}$$

is called an *amino group*, and the presence of a carboxyl group,

$$-\overset{O}{\underset{}{\overset{\|}{C}}}-OH \qquad \text{Carboxyl group}$$

is characteristic of an organic acid.

Some amino acids, such as *glycine* and *alanine*, have simple R groups, others have quite complicated ones. The structural formulas of glycine and alanine are as follows:

$$\begin{array}{c} H \\ \diagdown \\ H \end{array} N - \underset{H}{\overset{H}{\underset{|}{C}}} - \overset{O}{\underset{}{\overset{\|}{C}}} - OH \qquad \text{Glycine}$$

$$\begin{array}{c} H \\ \diagdown \\ H \end{array} N - \underset{\underset{H}{\overset{|}{H-C-H}}}{\overset{H}{\underset{|}{C}}} - \overset{O}{\underset{}{\overset{\|}{C}}} - OH \qquad \text{Alanine}$$

peptide bonds join amino acid molecules together end to end

Amino acid molecules are linked together by means of a characteristic *peptide bond:*

$$-\overset{O}{\underset{}{\overset{\|}{C}}} - \underset{H}{\overset{}{\underset{|}{N}}} - \qquad \text{Peptide bond}$$

THE CHEMISTRY OF LIFE

In the formation of such a bond, the amino end of one molecule joins the carboxyl end of the other (Fig. 29.2), a process accompanied by the loss of a water molecule. The result is a unit called a *dipeptide* that also has an amino group at one end and a carboxyl group at the other. A dipeptide can therefore link up with further amino acid molecules at both ends to form more complex peptides.

29.2 Amino acid molecules join together by means of peptide bonds. *The H and O atoms that become water molecules in the process are shown in color.*

protein structures

A protein molecule may contain hundreds of amino acid units in one or more *polypeptide* chains. These chains are usually coiled or folded in intricate patterns. An important aspect of these patterns is the cross-linking that occurs between different chains and between different parts of the same chain. The cross-links in protein molecules are of two chief kinds:

proteins are composed of polypeptides, which are long chains of amino acids

1 *Hydrogen bonds* between the N—H group of one amino acid unit and the C=O group of another. (Hydrogen bonding is described on page 409.) Individual hydrogen bonds are fairly weak, but so many of them may be present in a protein molecule that the resulting structure is stable.
2 *Disulfide bonds* between cysteine units. Cysteine is an amino acid whose molecules have exposed —S—H groups. The —S—H groups of two cysteine units may react to form —S—S— (disulfide) bonds with the loss of two H atoms. Disulfide bonds are covalent and hence are stronger than hydrogen bonds.

Many proteins occur in the form of an *alpha helix* whose successive turns are held in position mainly by hydrogen bonds. An alpha helix may be visualized as a ribbon (representing the polypeptide chain) wrapped in spiral fashion around an imaginary cylinder (Fig. 29.3). Each complete

29.3 The alpha helix form of a protein molecule. *Each amino acid unit in the helix is linked by hydrogen bonds to other units above and below it.*

denaturation

some of the protein structures resemble a pleated ribbon

the sequence of amino acids in a protein molecule together with the form of the molecule gives it a unique character that is reflected in its biological activity

29.4 The pleated sheet form of a protein molecule. *Two or more chains of amino acid units are linked side-to-side along the sheet by hydrogen bonds.*

turn contains 3.7 amino acid units, and hydrogen bonds occur between each amino acid unit and the ones directly above and below it in the helix, which are the third units away from it on either side. Helical protein molecules are usually folded and twisted into compact globules with further bonds to stabilize the assembly.

Heating a protein beyond a certain point ruptures the weak hydrogen bonds, although the member amino acid units of the helix may nevertheless remain together because of the greater strength of the covalent peptide bonds between them. The polypeptide chain then assumes a different structure; the protein has been *denatured*, which is an irreversible process. The coagulation of the white of a boiled egg is an example of denaturation. Proteins can be denatured in a variety of ways other than heating, for instance by exposure to acids and bases.

The polypeptide chains of fibrous proteins such as those found in hair, horn, and cartilage have the form of a pleated ribbon (Fig. 29.4) rather than that of a helix. The pleating occurs because of the angles between the various bonds in each chain. Two or more adjacent chains can bond together side by side with the help of hydrogen bonds.

The sequence of the amino acids in a protein is just as important as which ones they are. There is an astronomical number of possible arrangements of the amino acid units in even a small protein molecule such as insulin, which has 51 units, but only a single arrangement leads to the biological behavior associated with insulin. A close parallel is with the formation of a word from the 26 letters of the alphabet: *run* and *urn* contain the same letters but mean different things because the order of the letters is different. The alphabet of the proteins has only 23 letters, corresponding to the various amino acids, but the words may contain a thousand or more letters whose relative positions in three dimensions are significant. The extraordinary number of different proteins, each serving a specific biological need in an organism of a specific species, is not surprising in view of this picture of protein structure.

The carbohydrates and lipids do not share the specificity of the proteins. Glucose, for instance, is a carbohydrate found in all plants and animals, but there is no protein similarly widespread. Even individuals of the same species may have proteins that are not quite identical, so that tissues cannot be transplanted with the danger of "rejection" of the graft. The matching of blood types before a transfusion is to ensure that the proteins in the blood of the donor are the same as those in the blood of the recipient.

NUCLEIC ACIDS

The nucleic acids are very minor constituents of living matter from a quantitative viewpoint, but, because they control the processes of heredity by which cells and organisms reproduce their proteins and themselves, they are extremely important. If anything may be said to be the key to the distinction between living and nonliving matter, it is the nucleic acids.

THE CHEMISTRY OF LIFE

nucleotides

Nucleic acid molecules consist of long chains of units called *nucleotides*. As in the case of the amino acids in a polypeptide chain, both the kinds of nucleotide present and their arrangement govern the biological behavior of a nucleic acid.

nucleic acid molecules consist of chains of nucleotides

Each nucleotide has three parts, a *phosphate group* (PO_4), a *pentose sugar*, and a *nitrogen base*. A pentose sugar is one that contains five carbon atoms. In *ribonucleic acid* (RNA) the sugar is *ribose*, $C_5H_{10}O_5$, and in *deoxyribonucleic acid* (DNA) the sugar is *deoxyribose*, $C_5H_{10}O_4$, which has one O atom less than ribose. The five nitrogen bases usually found in nucleic acid are adenine, guanine, and cytosine, which occur in both RNA and DNA, and thymine and uracil, which occur respectively in DNA and RNA only. The structures of these nitrogen bases are given in Fig. 29.5.

a nucleotide consists of a phosphate group, a sugar, and a nitrogen base

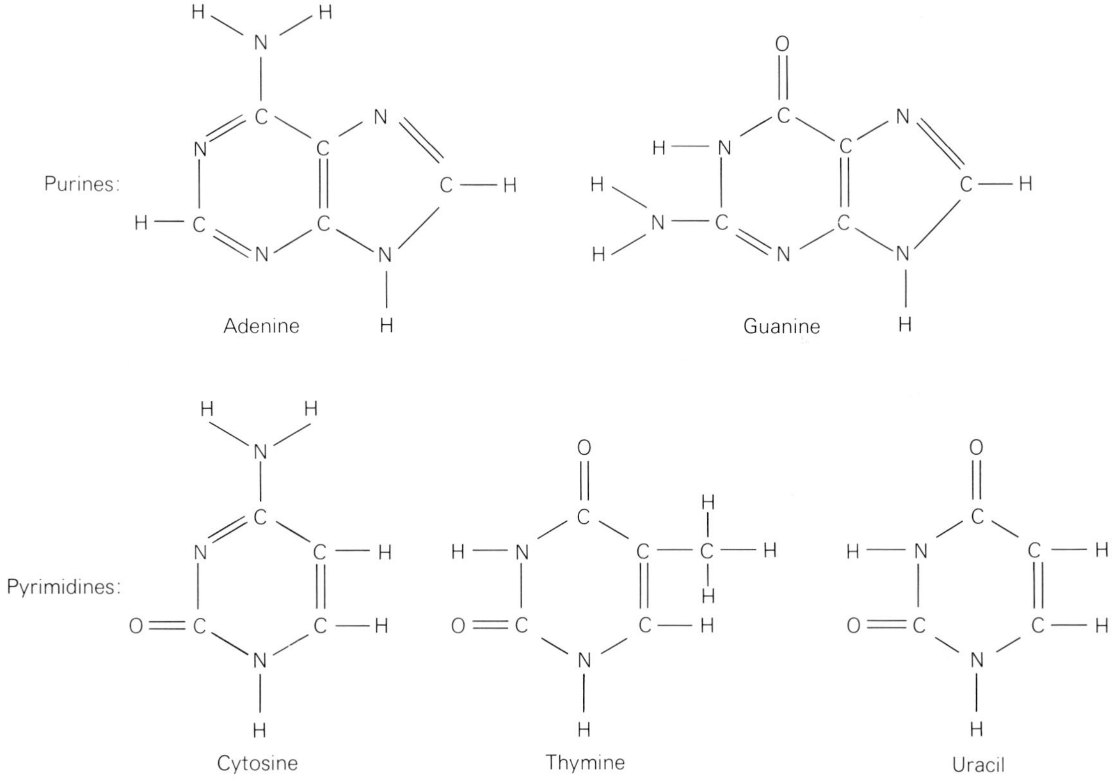

29.5 The structures of the five principal nitrogen bases found in nucleic acids. *Those with single rings are classed as* pyrimidines, *those with double rings as* purines.

DNA

The structure of a DNA molecule is shown in Fig. 29.6. Pairs of nitrogen bases form the links between a double chain of alternate phosphate and deoxyribose groups. Adenine and thymine are always coupled together,

DNA has the form of a double helix linked by nitrogen bases; the sequence of these bases is the genetic code

as are cytosine and guanine. The chains are not flat but spiral around each other in a double helix, as in Fig. 29.6b. Figure 29.6c shows the four "letters" of the genetic code. Each one occurs hundreds of thousands of times in a DNA molecule, and their precise sequence governs the properties of the cell in which the molecule is located. DNA molecules thus represent a kind of biological code which is translated into the processes of life.

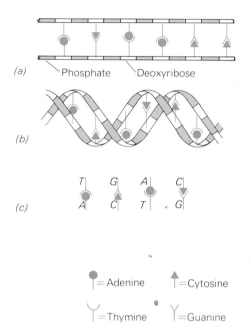

29.6 The structure of DNA. (a) The nitrogen bases link a double chain of alternate phosphate and deoxyribose groups. Adenine and thymine are always paired and cytosine and guanine are always paired. (b) The chains are not flat but form a double helix. (c) The four "letters" of the genetic code.

The reason why only adenine-thymine and cytosine-guanine base pairs are formed is evident from Fig. 29.7, which shows the two hydrogen bonds that occur between the former two bases and the three hydrogen bonds that occur between the latter two bases.

DNA governs protein synthesis in cells and makes possible reproduction and evolution

DNA controls the development and functioning of a cell by determining the character of the proteins it manufactures. This is only one aspect of its role in the life process. Another follows from the ability of DNA molecules to reproduce themselves, so that when a cell divides, all the new cells have the same characteristics (that is, the same *heredity*) as the original cell (Fig. 29.8). Finally, changes in the sequence of bases in a DNA molecule can occur under certain circumstances, for example during exposure to X-rays. These changes will be reflected in alterations in the properties of the cell containing the molecule, and such a *mutation* may result in the descendants of the original organism being different in some way from their ancestor. Thus three fundamental attributes of life can be traced to DNA: the structure of every organism; its ability to reproduce; and its ability to evolve into different forms in subsequent generations.

RNA

The other type of nucleic acid, RNA, differs from DNA in a number of respects. RNA molecules are much smaller than DNA molecules, for example, and usually consist of only single strands of nucleotides. Furthermore, as mentioned earlier, ribose instead of deoxyribose sugar is incorporated in each RNA nucleotide, and uracil replaces thymine as one of the pyrimidine bases.

The function of RNA in a cell is to carry instructions for the synthesis of specific proteins from the DNA in the cell's nucleus to the place where the actual synthesis occurs. The instructions are in the form of a code in which each successive group of three nucleotides determines the particular amino acid to be added next to the protein polypeptide chain being formed. For example, the group GCA (guanine-cytosine-adenine) corresponds to the amino acid alanine, and GGA corresponds to glycine. More than one *codon* (three-nucleotide group) corresponds to each amino acid; thus GCA, GCC, GCG, and GCU all refer to alanine.

Molecular biology is still a young and active discipline. The double-helix form of DNA was only discovered in 1953 (by James D. Watson and Francis H. C. Crick), and the details of protein synthesis and the operation of the genetic code in general are today being explored at a rapid rate. For all that remains to be understood, it is nevertheless no longer appropriate to speak of the "mystery" of life.

the origin of life

Whatever the earth's beginnings may have been, it is a safe assumption that, at some time in the remote past, the surface was considerably warmer than it is at present. The atmosphere of the young earth almost certainly contained compounds of the elements hydrogen, oxygen, carbon, and nitrogen, of which the most likely were methane, ammonia, carbon dioxide, and hydrogen cyanide (HCN) in addition, of course, to water. Eventually the outer part of the earth cooled, and torrents of rain began to fall. The rain carried down with it some of the other atmospheric gases, so that the infant oceans had a certain proportion of these gases dissolved in them. The weathering and erosion of surface rocks began at this time also, and the oceans acquired their salt and mineral content early since these processes must have been exceptionally rapid at first. Life had its origin in these oceans. Today the nature of life no longer seems as impenetrable as it once did, and the transition from lifeless matter to living matter, though still hardly an open book, nevertheless seems more to be an inevitable sequel to the physical and chemical conditions that prevailed on the earth some billions of years ago than a supernatural event.

Chemical reactions occur most readily in liquids, and furthermore water is the best solvent. Hence the early oceans must have been fertile media for chemical processes of all sorts, with ample energy available in sunlight and lightning discharges. Of the great many compounds that must have been formed, five classes have particular biological significance: the sugars, glycerin, the fatty acids, the amino acids, and the nitrogen bases. It is

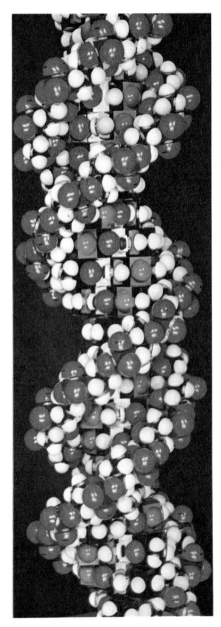

A model of the DNA molecule which enables living organisms to replicate themselves.

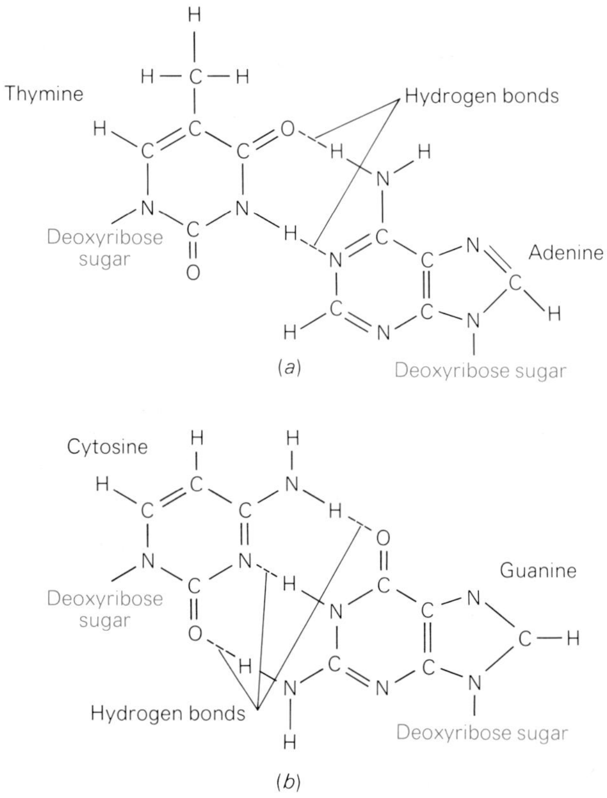

29.7 There are two hydrogen bonds between adenine and thymine and three hydrogen bonds between cytosine and guanine. *In these diagrams some of the base structures are shown in different orientations from those of Fig. 29.5.*

the major types of biochemical compounds are likely to have been formed in the early history of the earth in processes that have been duplicated in the laboratory

naturally not proper to assume that, because a certain compound contains certain elements, merely bringing together these elements in the ocean will yield the compounds. However, the reaction sequences that are necessary to go from the primitive ingredients of the oceans to the specific compounds listed above seem straightforward and likely to have occurred—or, rather, it is hard to see any reason why they should *not* have occurred. And this is one of the rare hypotheses about the early earth that can be directly verified in the laboratory: When an electric discharge simulating lightning is passed through a mixture of water, methane, and ammonia, it is observed that amino acids, fatty acids, and various other compounds of biological importance are created. Even the basic structural parts of the chlorophyll molecule have been produced in this way.

Given molecules of the five classes of significant organic compounds, further reactions of equal plausibility, though all of them have not yet been verified in the laboratory, lead to more complex compounds such as the fats, proteins, and nucleic acids that are directly involved in living matter. And given these latter compounds, most notably the nucleic acids which govern protein synthesis and are able to replicate themselves, the emergence of primitive cells, the basic biological units, becomes inevitable. The progression from an ocean with dissolved gases, salts, and minerals to living organisms certainly did not occur with the neatness and dispatch with which, say, a baker combines certain ingredients, inserts the mixture in an oven, and removes a cake an hour later. But although pure chance must have dictated which molecules came together and reacted to form a more complex one, just when the reaction occurred, and where it did, the ultimate outcome seems not to have been a matter of chance at all. It is estimated that about 2 billion years elapsed between the formation of the earth and the formation of the first cells, and the reaction sequence proposed by biologists to account for the latter seems reasonable in view of this vast span of time.

GLOSSARY

A *carbohydrate* is a compound of carbon, hydrogen, and oxygen whose molecules contain two atoms of hydrogen for each one of oxygen; carbohydrates are manufactured in green plants from water and carbon dioxide in the process of photosynthesis. Sugars, starches, chitin, and cellulose are carbohydrates.

Fats, oils, waxes, and sterols are *lipids*, which are synthesized in plants and animals from carbohydrates. A fat molecule consists of a glycerin molecule with three fatty acid molecules joined to it.

Proteins are the chief constituent of living matter and consist of long chains of amino acid molecules called *polypeptides*. The sequence of amino acids in a protein molecule together with the form of the molecule determines its biological role.

Nucleic acid molecules consist of long chains of *nucleotides* whose precise sequence governs the structure and functioning of cells and organisms. *DNA* has the form of a double helix and carries the genetic code; the simpler *RNA* acts as a messenger in protein synthesis.

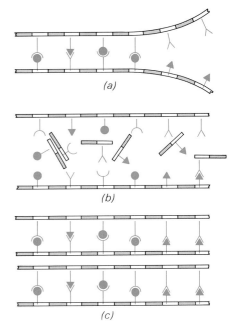

29.8 DNA replication. (a) *When a cell reproduces, each double DNA chain it contains breaks into two single ones, much like a zipper opening.* (b) *The single chains then pick up from the cell material the nucleotides needed to complete their structures.* (c) *The result is two identical DNA chains.*

EXERCISES

1 Trace the energy you use in lifting a book back through the various transformations it undergoes to its ultimate source.

2 How does a plant obtain its carbohydrates and fats? An animal?

3 What are the R groups in the amino acids glycine and alanine?

4 Can you think of any function other than energy storage which body fat might have?

5 What kind of bond is chiefly responsible for holding polypeptide chains in a helical form in proteins?

6 How many bonds are respectively associated with the C, N, O, and H atoms in the various nitrogen bases? What kinds of bonds are these?

7 There are six possible ways in which the nitrogen bases adenine, thymine, cytosine, and guanine can be grouped in pairs, yet only adenine-thymine and cytosine-guanine pairs ever occur in DNA. Why?

part eight
MATERIALS OF THE EARTH

With the next chapter we begin to shift our attention from physics and chemistry to geology and astronomy, a shift that involves a change in point of view as well as in subject matter. In physics and chemistry each aspect of a phenomenon can be isolated and studied in the laboratory, and theories can be verified or disproved by direct experiment. In geology and astronomy, observation rather than experiment is the primary source of data: volcanoes and stars are on too grand a scale to be dissected or duplicated in any terrestrial laboratory.

Another significant difference is that geology and astronomy are concerned not only with natural processes taking place at present but also with the history of natural processes that operated in the past. In order to understand phenomena that occurred long ago and events not subject to laboratory check, new techniques of observation and reasoning are required. The basic pattern of the scientific method—defining a problem, careful acquisition of data, making tentative generalizations, checking predictions based on these generalizations, seeking ever simpler and more inclusive generalizations—is not altered in geology and astronomy, but the details of its application are necessarily different.

We begin our study of geology with a discussion of the materials that constitute the outer part of the earth: rocks, soil, air, water. Rocks are seldom uniform in composition or structure, usually being mixtures of homogeneous substances called minerals. The chief classes of minerals are considered together with how they occur in the three principal types of rock: igneous rock, formed by cooling from the molten state; sedimentary rock, which consolidates from deposits left by water, wind, or ice; and metamorphic rock, produced from buried rocks of any kind by the immense heat and pressure found beneath the earth's surface. Finally we examine the atmosphere and the oceans, both of which participate in processes of geological change as well as being of interest in themselves.

chapter 30

Earth Materials

SILICATES

    composition of the crust

    silicate chemistry

    silicate structures

    glass

MINERALS

    what is a mineral?

    mineral properties

    six common minerals

    clay

When we examine the solid earth beneath us, almost all the material we find is rock. Soil, vegetation, and fragmental substances such as sand and gravel form a thin surface layer in many places, but bedrock is invariably found to underlie this cover. The deepest mines, whose shafts may descend as far as 2 miles, and the deepest wells, which may penetrate 5 miles down, encounter rock material similar to that at the surface. Some of the rock now exposed to our view was once buried several miles or more inside the earth, and the material that makes up some volcanic rock was probably brought up in molten form from still greater depths, perhaps as much as 20 miles down; yet these samples of rock from well below the surface are still very similar to rock formed at higher levels.

Direct observation tells us only that the upper 20 miles or so of the earth's solid body is composed of ordinary rock. This shell of rock, called the *crust* of the earth, is only $\frac{1}{2}$ percent of the earth's 4,000-mile radius, and we have no direct information about the remaining material. In Chap. 36, however, we shall learn of the various ingenious methods that have been developed to probe the interior of our planet and of their results.

A study of the earth requires first of all an acquaintance with the solid materials of the crust. We have gained considerable knowledge about the ultimate constituents of these materials: we know that compounds in the crust contain 90-odd different elements and that the atoms of each element are made up of electrons, protons, and neutrons. We need now some information of a more practical sort, information about the actual sub-

the earth's crust is its outer shell of rock

stances into which the ultimate constituents are combined in nature. In this chapter and the next we seek an acquaintance with the raw rock materials that we find beneath our feet and that we can pick up and examine in the field.

SILICATES

composition of the crust

The average composition of the crust, as determined by numerous chemical analyses of rocks, is given in Fig. 30.1. The numbers are percentages of the various elements by mass. If the compositions of the oceans and atmosphere are included, the percentages are changed only slightly.

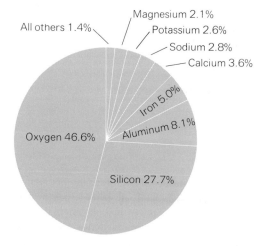

30.1 Average chemical composition of the earth's crust. *Percentages by mass are given.*

oxygen is the most abundant element in the earth's crust, most of it combined with silicon

The figure shows one striking fact about the composition of the crust: A few elements are abundant, but most are exceedingly scarce. Oxygen alone makes up nearly half the mass of the crust; some of it is free in the air, some is combined with hydrogen in water, but the greater part is combined with silicon in silica and the silicates. Silicon and the two metals iron and aluminum account for three-fourths of the rest of the crust's mass. Such familiar metals as copper, tin, lead, and silver are too scarce to be shown. Nitrogen and the halogens are likewise lumped in the 1.4 percent that includes "all others." Carbon and hydrogen, present in all living things, together make up less than 0.2 percent of the total.

silicate chemistry

the majority of crustal rocks are largely composed of silicon compounds

As might be guessed from Fig. 30.1, the chief components in the rocks of the crust are silicon dioxide and various silicates of the six metals Fe, Al, Ca, Mg, Na, and K. The only common rock in which these compounds are not dominant is limestone, which is chiefly calcium carbonate.

It might seem that we could study the chemistry of the silicates in terms of the atom group SiO_3, just as we studied the chemistry of the carbonates in terms of the atom group CO_3. However, there are a number of significant differences between the properties of the silicates and those of the carbonates that make such an analogy invalid. For example, very few of the important silicates are appreciably soluble, and the ion $SiO_3^=$ is seldom encountered. Furthermore, the silicates exhibit a peculiar complexity of structure that makes a classification of their reactions in terms of simple atom groups like SiO_3 all but impossible. Thus we find not one sodium silicate, but several:

Na_2SiO_3 $Na_4Si_3O_8$
Na_4SiO_4 $Na_2Si_2O_5$
$Na_4Si_2O_6$ $Na_2Si_3O_7$, etc.

To study these in the usual manner would mean gathering a separate set of facts about each silicate group—and then we should find that these groups often do not remain intact during reactions.

The above formulas become more intelligible if we write:

Instead of Na_2SiO_3, $Na_2O \cdot SiO_2$
Instead of Na_4SiO_4, $2Na_2O \cdot SiO_2$
Instead of $Na_4Si_2O_6$, $2Na_2O \cdot 2SiO_2$
Instead of $Na_4Si_3O_8$, $2Na_2O \cdot 3SiO_2$, and so on.

That is, the differences between the various sodium silicates may be considered simply as differences in the ratios of the two oxides Na_2O and SiO_2. Other complex silicate formulas may be broken down similarly into combinations of SiO_2 and metallic oxides: thus $MgSiO_3$ and Mg_2SiO_4, two magnesium silicates that occur widely in nature, may be written $MgO \cdot SiO_2$ and $2MgO \cdot SiO_2$, and the more complex silicate $CaAl_2Si_2O_8$ may be written $CaO \cdot Al_2O_3 \cdot 2SiO_2$. Some justification for this procedure besides mere convenience is the fact that many silicates may be prepared artificially by heating oxides together; the two magnesium silicates just mentioned can be made by heating mixtures of MgO and silica in the right proportions, and calcium silicate ($CaSiO_3$) can be made by heating CaO with silica. Reactions between silicates, moreover, are often best interpreted as interchanges of different oxides.

the various silicates are combinations of SiO_2 and the corresponding metal oxides

It is the singular ability of its oxide to form different compounds by combining with various amounts of metallic oxides that distinguishes silicon from other elements and makes possible the great variety and complexity of the silicates.

We might, of course, have studied other compounds in terms of their component oxides: Na_2SO_4 might be written $Na_2O \cdot SO_3$, $CaCO_3$ might be written $CaO \cdot CO_2$, and so on. The reason that we do not use such formulas is simply that in the reactions of these compounds the atom groups that remain together, recur in compound after compound, and appear as ions in solution, are not the oxides but the groups SO_4 and CO_3. In silicate reactions, on the other hand, the groups of elements that seem to cling together are the oxides, so we shall frequently use them rather than larger groups as chemical units.

at ordinary temperatures, the silicates are crystalline solids with widely varying properties

30.2 *Schematic structures of CO_3^- and CO_2.*

30.3 *Schematic structure of SiO_2 (quartz). This is actually a three-dimensional structure with each SiO_4 group being a tetrahedron.*

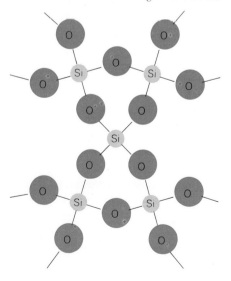

As a class, the silicates are crystalline solids, melting at high temperatures to give viscous liquids. Their variations in composition and structure are reflected in a variety of colors, hardnesses, and crystal forms. The softness of talc, the hardness of zircon and beryl, the transparency of topaz and the deep color of garnet, the platy crystals of mica and the fibrous crystals of asbestos give some idea of the range of silicate properties.

silicate structures

Like so many other solids, crystalline silicates are assemblies of individual atoms bonded together in a continuous lattice, rather than assemblies of molecules. The simple formula of a silicate, even when written out in terms of oxides, tells us only the ratio of the numbers of different atoms present, nothing about how these atoms are put together. Actual structures of many silicates have been worked out with the help of X-rays (see Chap. 21). No convenient method like the structural formulas of organic chemistry has been proposed for diagraming the complex silicate structures, but the general patterns on which silicate crystals are built have been deciphered.

The chemical differences between silicon and carbon hinge on the difference in the sizes of their atoms. Around the tiny carbon atom only two (in carbon dioxide) or three (in carbonates) oxygen atoms can find room; so strong is the covalent bond between C and O that the oxygen atoms are warped out of their usual approximately spherical shape to fit snugly about each carbon atom (Fig. 30.2). In CO_2 each oxygen atom is tightly bound to carbon by two shared electron pairs; since valences are satisfied within each CO_2 molecule, the molecules have little attraction for one another, and they join to form crystals of solid CO_2 only at low temperatures.

The larger size of the silicon atom and its consequent smaller attraction for the shared electrons enable four oxygen atoms to cluster around it without warping. Between each Si and O is a single covalent bond, so that every oxygen atom has one valence electron free to form a bond with another atom. In SiO_2 this bond is with another Si atom; Si and O atoms are linked in a continuous network, each Si joined to four O's and each O joined to two Si's (Fig. 30.3). Thus there are no separate SiO_2 molecules: the atoms are tightly bound in a continuous lattice, and SiO_2 is accordingly a solid with a high melting point.

In silicates as well as in silica, each silicon atom is linked to four oxygen atoms. The O's are symmetrically placed around Si, at the corners of a tetrahedron (Fig. 30.4). In the simplest silicates (for example Mg_2SiO_4) the SiO_4 tetrahedra form separate negative ions, linked together by positive metallic ions—in a general way like the ionic structure of NaCl, with SiO_4 groups taking the place of Cl ions. In other silicates two, three, or more SiO_4 tetrahedra may join together into larger ions, like $Si_2O_7^{6-}$ and $Si_3O_9^{6-}$, some oxygen atoms lying between two silicon atoms.

In more complex structures the tetrahedra are linked together in continuous chains (Fig. 30.5) or sheets (Fig. 30.6), with metal ions lying

EARTH MATERIALS

between. Still more intricate structures involve three-dimensional arrangements of tetrahedra, with some of the Si atoms replaced by Al atoms.

Thus some of the oxygen atoms in a silicate structure link silicon atoms together, while others are held jointly by silicon atoms and metal ions. Just as each silicon atom surrounds itself with four oxygen atoms, so each metal ion tries to surround itself with a certain number determined by its size and valence. The silicon-oxygen structure in any particular case adapts itself to the needs of the metal ions present, leaving enough oxygen atoms with free valences to satisfy the metal ions.

Now we can see why there are so many different silicates: silicon forms a variety of stable structures with oxygen, from simple negative ions to a continuous lattice, and the structure which appears in a particular compound depends on the metal ions present. If silicon were more strongly nonmetallic, like carbon, its atoms would be more closely united to oxygen and it would form simple ions ($SiO_3^=$ or SiO_4^{4-}) exclusively. If silicon were more metallic, it would form positive ions like other metals and silicates would have a simple ionic type of lattice. But silicon is neither metallic nor strongly nonmetallic, and its intermediate character makes possible numerous different silicate structures. This is the explanation of the fact that silica combines with metallic oxides in many different ratios.

glass

Molten silicates as a rule are highly viscous liquids. If cooled very slowly they solidify, like other liquids, into crystalline solids. But if the cooling is fairly rapid, some silicates, instead of crystallizing, simply grow more and more viscous until they become amorphous solids or *glasses*.

The formation of a glass depends on the slowness with which particles move about in a viscous liquid. Rapid cooling does not allow the atom groups sufficient time to find their places in the normal crystal lattice but freezes them in the random arrangement of the liquid state. Because of the random arrangement of its particles, a glass has no definite melting point but softens gradually to a viscous liquid when heated. Some glasses slowly crystallize, or *devitrify*, on standing at room temperature, their particles apparently moving sufficiently even in the solid to arrange themselves in lattice structures. All glasses can be made to devitrify if maintained for a long time at a temperature just below the softening point, where molecular motion is more rapid (see the discussion on p. 398).

Glasses are usually prepared not by melting silicates but by heating together certain mixtures of silica with other oxides. Silica alone forms

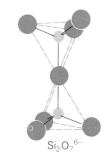
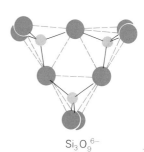

30.4 Silicon-oxygen structures in simple silicates. *These structures alternate with positive metallic ions in the crystal lattices. (Gray spheres are Si atoms, colored spheres are O atoms.) The tetrahedral form of SiO_4^{4-} is shown by dashed lines, solid lines represent bonds between atoms.*

glass is an amorphous solid

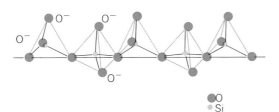

30.5 Chain of SiO_4 tetrahedra. *These chains are linked sidewise by metal ions. The solid lines represent bonds between atoms.*

glass is a mixture of silicates

an excellent glass, but it is too difficult to make and too expensive for ordinary use. Since a glass in general is a complex mixture of silicates made by melting together various oxides, its composition cannot be expressed by a simple chemical formula. Its properties, such as transparency, color, hardness, tendency to devitrify, depend on the amounts and kinds of oxides used, and it is no small part of the glassmaker's art to know just how different oxides and different proportions will affect the final product.

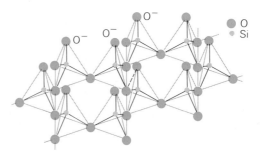

30.6 Sheet of SiO_4 tetrahedra. *The O^- ions on the top of the sheet form ionic bonds with metal ions.*

the preparation of ordinary glass

Ordinary glass is a mixture of complex silicates formed from the oxides Na_2O, CaO, and SiO_2. It is made by heating together clean quartz sand (SiO_2), limestone ($CaCO_3$), and sodium carbonate (Na_2CO_3). Heat drives off CO_2 from the carbonates, leaving Na_2O and CaO:

$$Na_2CO_3 \longrightarrow Na_2O + CO_2$$

$$CaCO_3 \longrightarrow CaO + CO_2$$

The two metallic oxides then combine with silica to form the silicates of the glass. Amounts of the three ingredients are used which will give a final ratio of the three oxides of about $Na_2O:CaO:5SiO_2$. Any considerable deviation from this ratio gives a glass which is partially opaque or which devitrifies easily. Many glass articles made carelessly several decades or several centuries ago have become cracked and opaque through slow devitrification.

Small amounts of other substances added to the glass often markedly change its color and other properties. The green glass used in making cheap bottles has in it a small amount of iron oxide, originally present as an impurity in the sand or limestone. A little cobalt gives the glass a blue color, a little manganese oxide a violet color. Use of potassium oxide instead of sodium oxide gives a glass that softens at a higher temperature than ordinary glass. Pyrex glass, more resistant to temperature changes than ordinary glass, contains a relatively high percentage of silica, considerable B_2O_3, a little K_2O, and only small amounts of Na_2O and CaO. Lead glass, a soft, brilliant glass extensively used in optical instruments, is made up of the three oxides PbO, K_2O, and SiO_2.

A newly cast mirror blank. After cooling, the glass surface will be ground and polished to its final shape.

MINERALS

rocks are composed of homogeneous solids called minerals

Most rocks are heterogeneous solids. The different kinds of material in a coarse-grained rock like granite are apparent to the eye; in a fine-grained rock the separate constituents can be made visible with a microscope. The separate homogeneous substances of which rocks are composed are called *minerals*.

what is a mineral?

The word mineral is used in a variety of ways. Obviously, a doctor who prescribes certain minerals for your diet or an advertiser who tries to sell you water from a mineral spring is not thinking of the constituents of rocks. Even geologists do not always agree on the exact application of the term. Most geological discussions restrict the word to solid, homogeneous, inorganic substances, found in nature either as parts of rocks or in separate deposits. As a rule, minerals are crystalline substances with fairly definite chemical composition.

A consideration of the stabilities of chemical compounds makes possible predictions about the kind of material that can occur as minerals. We should expect to find the more inactive elements in the free state, active elements in compounds. Soluble compounds should be common minerals only in arid regions; easily oxidized compounds should occur only well beneath the surface. So we find free gold, platinum, sulfur, and carbon (both graphite and diamond) as minerals, but elements like sodium, chlorine, and calcium always occur in compounds. Sodium chloride, sodium carbonate, and potassium nitrate form deposits in deserts but are seldom found elsewhere. Such reactive compounds as calcium oxide or phosphorus pentoxide never occur in nature.

silicates are the most abundant minerals

Silicates are by far the most abundant minerals; mica, feldspar, garnet, and topaz are familiar examples. *Carbonates* are another important class, its most abundant representative being the carbonate of calcium, called calcite. *Oxides* and *hydrated oxides* include such common materials as quartz (oxide of silicon); hematite (ferric oxide), the chief ore of iron; and bauxite (hydrated aluminum oxide), the chief ore of aluminum. Several important metals are obtained from deposits of *sulfide* minerals, such as galena (lead sulfide) and sphalerite (zinc sulfide). *Elements* that occur free, or *native*, were mentioned in the last paragraph. Less common as minerals are sulfates, phosphates, and chlorides.

Unfortunately the study of minerals requires the learning of a new list of formidable names, some of them apparently needless duplicates of other names. As an example, the common mineral with the formula $CaCO_3$ is given the name *calcite* instead of the chemical name *calcium carbonate*. For this seeming duplication there are two good reasons:

1 The formula $CaCO_3$ expresses not only the composition of calcite, but also that of aragonite, a less common mineral with different crystal form, hardness, density, and so on; the chemical name calcium carbonate alone does not distinguish between calcite and aragonite.

2 Calcite often contains small quantities of $MgCO_3$ and $FeCO_3$, and its composition is not precisely represented by the formula $CaCO_3$ because the iron and magnesium carbonates form an integral part of the calcite crystals; Fe and Mg atoms simply replace a few of the Ca atoms in the lattice structure.

Many other mineral formulas besides that of calcite apply to two or more distinct substances, and most minerals show a similar slight variability in composition. Hence, orthodox chemical names are not strictly applicable, and the student of minerals finds necessary a new nomenclature.

Luckily, for present purposes we need only a few additions to our vocabulary. More than 2,000 different minerals are known, but most of these are rare. Even among the commoner minerals, the greater number occur abundantly only in occasional veins, pockets, and layers. The number of minerals that are important constituents of ordinary rocks is surprisingly small, so small that acquaintance with less than a dozen will be ample for our future geological discussions.

mineral properties

Common minerals are not only limited in number but are also easily recognizable with some experience, often by appearance alone. To distinguish the rarer minerals elaborate microscopic and chemical tests may be necessary, but for the minerals that compose ordinary rocks such simple physical properties as density, color, hardness, and crystal form make identification relatively straightforward.

In the descriptions given later in this chapter of the important rock-forming minerals, two terms need special attention: *crystal form* and *cleavage*. Most minerals are crystalline solids, which means that their tiny particles (atoms, ions, or atom groups) are arranged in lattice structures with definite geometric patterns. When a mineral grain develops in a position where its growth is not hindered by neighboring crystals, as in an open cavity, its inner structure expresses itself by the formation of perfect crystals, with smooth faces meeting each other at sharp angles. Each mineral has crystals of a most distinctive shape so that well-formed crystals make recognition of a mineral easy; but, unfortunately, good crystals are rare, since mineral grains usually interfere with one another's growth.

crystal form

Even when well-developed crystals are not present, however, the characteristic lattice structure of a mineral may reveal itself in the property called *cleavage*. This is the tendency of a substance to split along certain planes, which are determined by the arrangement of particles in its lattice. When a mineral grain is struck with a hammer, its cleavage planes are revealed as the preferred directions of breaking; even without actual breaking, the existence of cleavage in a mineral is usually shown by flat, parallel faces and minute parallel cracks. The flat surfaces of mica flakes, for instance, and the ability of mica to peel off in thin sheets show that this mineral has almost perfect cleavage. Some minerals (for example,

cleavage

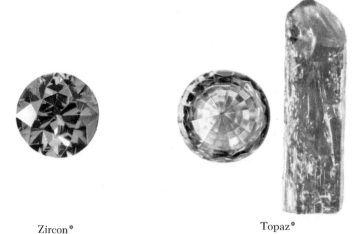

Examples of silicate minerals. Zircon is zirconium silicate, topaz is an aluminum silicate containing fluorine, asbestos is a group of fibrous silicate minerals which contain magnesium and sometimes calcium and iron as well, garnet is a group of silicate minerals which may contain iron, manganese, magnesium, calcium, aluminum, and chromium in various proportions, and beryl is a silicate of beryllium and aluminum.

Zircon* Topaz*

Asbestos* Garnet*

Beryl*

Some minerals and models of their crystal structures. All are found in nature except the potassium dihydrogen phosphate, which is artificially prepared. Left to right, top to bottom: (1) pyrite, (2) copper, (3) alpha quartz, (4) graphite, (5) potassium dihydrogen phosphate, (6) calcite, (7) spinel, (8) perovskite, (9) zinc, and (10) halite (sodium chloride).

EARTH MATERIALS

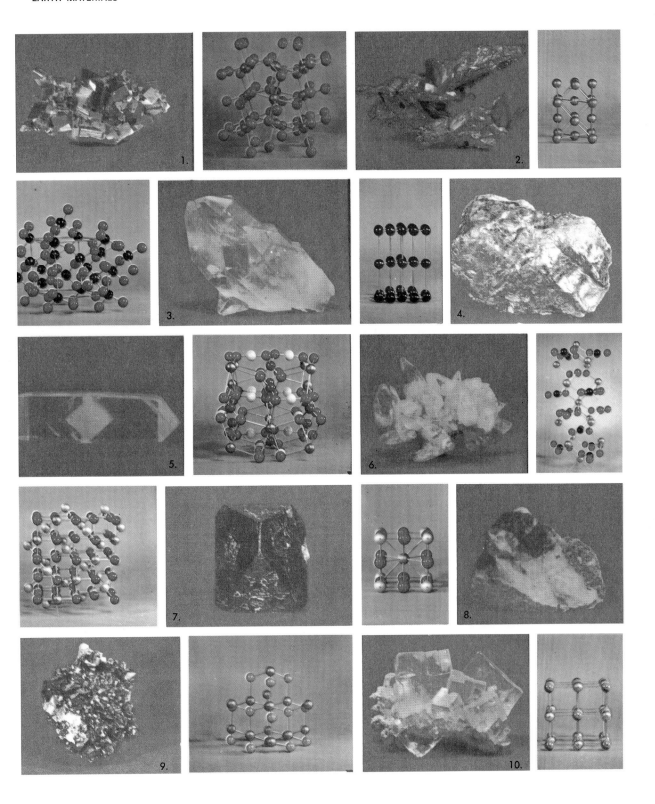

quartz) have practically no cleavage; when struck they shatter, like glass, along random curved surfaces. The ability to recognize different kinds and degrees of cleavage is an important aid in identifying and distinguishing minerals.

six common minerals

Quartz. SiO_2. Quartz crystals when well formed are six-sided prisms and pyramids. They show no cleavage. They are colorless or milky, often gray, pink, or violet because of impurities, with a glassy luster, and are hard enough to scratch glass and feldspar. They occur commonly in many kinds of rocks, abundantly in veins, and often in aggregates of well-formed crystals on the sides of cavities. Clear quartz (rock crystal) is used in jewelry and in optical instruments; smoky quartz, rose quartz, and amethyst are colored varieties used in jewelry.

Feldspar. This is the name of a group of silicate minerals with very similar properties. Two common kinds of feldspar are a silicate of K and Al called orthoclase and a series of silicates of Na, Ca, and Al collectively called plagioclase. The crystals are rectangular, with blunt-pointed ends; they show good cleavage in two directions approximately at right angles. The color is white or light shades of gray and pink, sometimes clear. Feldspar is slightly harder than glass, not so hard as quartz. It is the most abundant single constituent of rocks, making up about 60 percent of the total weight of the earth's crust. Pure feldspar is used in the making of porcelain and as a mild abrasive.

feldspar is the most common type of mineral

Mica. The two chief varieties of this familiar mineral are white mica, a silicate of H, K, and Al, and black mica, a silicate of H, K, Al, Mg, and Fe. Mica is easily recognized by its perfect and conspicuous cleavage in one plane; it is a very soft mineral, only a trifle harder than the fingernail. Large sheets of white mica free from impurities are used as insulators in electrical equipment.

mica is readily split into flat sheets

Ferromagnesian minerals. This name refers to a large group of minerals with diverse properties, all of them silicates of iron and magnesium, nearly all having a dark green to black color. Olivine is a common example. Most of these minerals contain other elements besides iron and magnesium, one of the commonest additional elements being calcium. Black mica belongs to this group, for its composition includes H, K, and Al in addition to Mg and Fe. No general properties of the group besides color and composition can be set down, since the various minerals differ greatly from one another. For our purposes it will be sufficient to remember that the most abundant dark-colored constituents of common rocks belong to this group.

the ferromagnesian minerals are dark green to black in color

Clay minerals. This is a group of closely related minerals that are the chief constituents of clay; they are silicates of H and Al, some with a little Mg, Fe, and K. They are aggregates of microscopic crystals, white or light-colored when pure, often discolored with iron compounds. They have a dull luster. They are very soft, forming a smooth powder when rubbed between the fingers. Their density is low. They are distinguished from chalk by softness and lack of effervescence in acids. Kaolin, one of the clay minerals, is an important ingredient in the manufacture of ceramics, paper, paint, and certain plastics.

Calcite. $CaCO_3$. Calcite crystals are hexagonal, in general appearance resembling those of quartz. They show perfect cleavage in three directions at angles of about 75°, so that fragments of calcite have a characteristic rhombic shape. They are colorless or any light shade, with a glassy luster. They are hard enough to scratch mica or the fingernail, but can be scratched by glass or by a knife blade. They dissolve readily in dilute acid with effervescence. Like quartz, calcite is a common mineral of veins and crystal aggregates in cavities. It is the chief constituent of the common rocks limestone and marble. It is important commercially for many purposes, especially as a source of lime for glass, mortar, and cement.

Of these six kinds of rock-forming minerals, we note that five are compounds of silicon and four are silicates. The light-colored silicates all contain aluminum in addition to silicon and oxygen; two of them (feldspar and mica) contain an alkali metal, two of them (mica and clay) contain hydrogen. The dark-colored silicates contain iron and magnesium.

clay

Clay is worth a further look because of its ability to absorb relatively large amounts of water. The result is a plastic mass that can be molded to any desired form. On heating, or "firing," the wet clay loses both the water it has absorbed and also water from its original structure; what is left has a hard, rocklike character. The transformation of clay from a soft, easily shaped material into a permanent material is the basis for the industrial use of clay in such ceramic products as bricks, pottery, and porcelain.

Ordinary bricks are made from clays containing large amounts of iron oxide and sand. For pottery somewhat purer clays are used and the firing temperatures are higher. The hardening of bricks and pottery in the firing process is due at least partly to the formation of silicates which melt and bind together the clay particles. The red and brown colors of the products are due to varying amounts of iron oxide. Porcelain is made from a mixture of pure white kaolin ("china clay") with finely ground silica and feldspar which is fired at high temperatures (1300 to 1500°F). The mass partially melts and gives a dense, somewhat translucent product.

Pottery and porcelain are usually *glazed* during the firing process. For the cheaper varieties of earthenware this is often accomplished by simply throwing common salt into the kiln; at the high temperature the salt is vaporized and a thin film of sodium silicate glass is formed on the surface of the articles. Tableware is usually given a lead glaze by dipping the

Quartz*

Amazonite

Muscovite

Calcite (on amethyst)

Six common minerals or mineral families are quartz, the feldspars (shown is amazonite), mica (shown is muscovite mica), calcite (on amethyst), the ferromagnesian minerals (shown is olivine), and the clay minerals (shown is kaolin).

Olivine*

Kaolin (china clay)*

EARTH MATERIALS

articles into a paste of lead oxide, clay, and silica; the resulting glaze is a surface layer of lead silicate glass. Other types of glazes are produced by the use of various metallic compounds which form glasses with silica either added or present in the material being fired.

The simplest clay mineral is *kaolinite*, which is the chief constituent of kaolin. Kaolinite has a two-sheet layer structure that essentially consists of a sheet of SiO_4 tetrahedra (Fig. 30.6) and a sheet of $Al(OH)_3$ in which each Al^{+++} ion has six OH^- ions as nearest neighbors (Fig. 30.7). In the SiO_4 sheet three of the O atoms in each tetrahedron are shared by adjacent tetrahedra, leaving one O^- ion per tetrahedron exposed on one side of the sheet. Each exposed O^- ion replaces one OH^- ion in the $Al(OH)_3$ sheet, thereby tying the two sheets together into a firmly bonded flat crystal (Fig. 30.8). Adjacent layers are held together by van der Waals forces to form clay, whose plasticity is the result of the feeble bonds between layers.

The van der Waals forces in clay are actually stronger than otherwise because each layer in kaolinite is polarized, with the oxygen atoms on the silicon side exhibiting a slight negative charge whereas the OH groups on the aluminum side exhibit a slight positive charge. For this reason clay absorbs water readily, with the polar water molecules fitting between adjacent clay layers (Fig. 30.9); this picture explains both the ability of clay to absorb large amounts of water into its structure and the extreme plasticity of wet clay. Talc, another layered silicate mineral, has a more symmetric structure that consists of an $Mg(OH)_2$ sheet with an SiO_4 sheet on both sides. The nonpolar talc layers accordingly show less tendency to stick together than clay layers do, and absorb water much less readily.

the clay minerals have layered structures with weak bonding between the layers

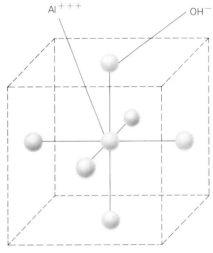

30.7 Structure of $Al(OH)_3$.

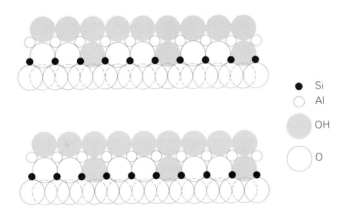

30.8 Side view of the structure of the clay mineral kaolinite. *Successive layers are bonded together by weak van der Waals forces.*

GLOSSARY

The *crust* of the earth is its outer shell of rock which averages about 20 miles in thickness.

Glass is an amorphous solid formed by the rapid cooling of a molten silicate; it is in a sense a *supercooled liquid* rather than a true solid.

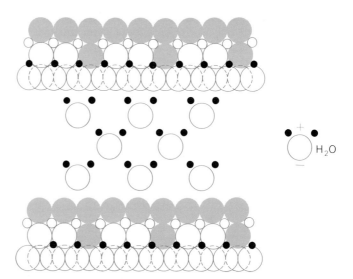

30.9 *Water molecules are readily absorbed into the structure of kaolinite because the O atoms exhibit slight negative charges and the OH groups exhibit slight positive charges.*

Minerals are the separate homogeneous substances of which rocks are composed.

Cleavage is the tendency of a substance to split along certain planes determined by the arrangement of particles in its crystal lattice.

Quartz consists of crystals of silicon dioxide, SiO_2.

Feldspar is the name of a group of light-colored silicate minerals with similar properties.

Mica is a soft mineral with conspicuous cleavage in one plane.

The *ferromagnesian minerals* are a large group of diverse minerals, all silicates of iron and magnesium and all dark green to black in color.

The *clay minerals* are a group of light, soft silicates of hydrogen and aluminum, sometimes with a little magnesium, iron, or potassium, which occur as aggregates of microscopic crystals.

Calcite consists of crystals of calcium carbonate, $CaCO_3$.

EXERCISES

1 Write the formulas of the following silicates in terms of oxides:

Zn_2SiO_4 $\qquad$ $Ca_3Al_2Si_3O_{12}$
$CaMg_3Si_4O_{12}$ $\qquad$ $H_4Mg_3Si_2O_9$
Al_2SiO_5 $\qquad$ $KAlSi_3O_8$

EARTH MATERIALS

2 Name as many minerals as you can in the following categories:
 a. Minerals that contain carbon
 b. Minerals that contain both aluminum and silicon
 c. Minerals that contain sodium and calcium
 d. Minerals harder than glass
 e. Minerals that can be scratched with a knife
 f. Minerals with cleavage

3 Contrast the chemical behavior of silicon with that of carbon. To what differences in their atomic structure is the contrast due?

4 In terms of molecular structure, why is SiO_2 a solid while CO_2 is a gas?

5 Water and carbon dioxide from the air slowly convert many silicates in rocks into clay. A typical reaction is

$$2KAlSi_3O_8 + 2H_2O + CO_2 \longrightarrow H_4Al_2Si_2O_9 + K_2CO_3 + 4SiO_2$$

Rewrite this equation by expressing all formulas as combinations of oxides.

6 Which of the following substances are used in the manufacture of (a) ordinary glass, (b) porcelain, (c) bricks? Clay, sodium carbonate, feldspar, calcium carbonate, silica.

7 What is the difference in composition between glass and blast-furnace slag?

8 Which of the following naturally occurring substances are minerals? Diamond, calcite, petroleum, ice, soil, wood, salt, coal.

9 Which of the following substances would you expect to occur as minerals? Ag, NaOH, Ca, KNO_3, Fe_2O_3, ZnS, $BaCO_3$, P, Fe.

chapter 31

Rocks

ROCKS

rock classification
igneous rocks
sedimentary rocks
metamorphic rocks

SOIL

types of soil

There is hardly any limit to the variety of rocks on the earth's surface. We find coarse-grained rocks and fine-grained rocks, light rocks and heavy rocks, soft rocks and hard rocks, rocks of all sizes, shapes, and colors. But close study reveals that there is order in this diversity, and a straightforward scheme for classifying rocks has been developed which simplifies the problem of understanding their origins and behavior. This scheme is the subject of the first part of this chapter. Later we go on to briefly examine *soil*, a large part of which consists of rock fragments.

ROCKS

The problem of organizing and classifying rocks is rather different from similar problems in physics and chemistry. When Mendeleev sought to classify the elements, he had 60 or so definite substances to work with; each was a distinct material, its properties sharply different from those of the others. In the study of rocks we find few such sharp boundaries between different kinds. We may decide that a light-colored, coarse-grained rock like granite should belong in a different class from a dark, fine-grained volcanic rock, but we can find a whole series of rocks with properties transitional between the two, and so we cannot say with any certainty just where one class ends and the other begins. As in all branches of science dealing with materials that occur in nature, we face the problem of making divisions where natural divisions may not exist. We must fit rocks into pigeonholes as best we can, but we must expect to run across some types with properties intermediate between those of different classes.

rock classification

How then shall we proceed to set up a usable classification of rocks? Since rocks are composed of minerals, we might guess first that they could be classified on the basis of the kinds and amounts of minerals they contain. But we find that rocks of widely different structures and origins have nearly the same mineral composition, and so our classification would group together rocks of obviously different types. A classification based on chemical composition encounters the same difficulty, since it places in the same pigeonhole rocks that are patently unlike; it has the further disadvantage that chemical compositions are not evident from examination of rocks in the field but require costly analyses. We might disregard composition and classify rocks according to their origin. This would be an excellent method if it could be applied to all rocks, but the sad fact is that we simply do not know how some rocks were formed; the origin of many others can be deduced only by detailed study in the laboratory.

Evidently we expect a classification of rocks to fulfill several different purposes. We should like it to summarize something about the origins of different rocks, about their compositions and about their structures, and at the same time we should like to be able to apply it to rocks as we find them in the field. These objectives cannot all be satisfied at the same time. Our recourse is to adopt a compromise, a classification that will accomplish each purpose as well as possible without slighting the others. The particular compromise we shall use is not the only possible one, but it is justified by its simplicity and convenience.

A fundamental division of rocks into three main groups according to differences of origin is agreed on by nearly all geologists:

the three main types of rock are igneous, sedimentary, and metamorphic

1 *Igneous rocks* are those that have cooled from a molten state. Some of these can be observed in process of formation, for instance when molten lava cools on the side of a volcano. For others an igneous origin is inferred from their composition and structure.
2 *Sedimentary rocks* consist of materials derived from other rocks, deposited by water, wind, or glacial ice. Some consist of separate rock fragments cemented together; others contain material precipitated from solution in water.
3 *Metamorphic rocks* are rocks that have been changed, or metamorphosed, by heat and pressure deep under the earth's surface. The changes produced may involve the formation of new minerals or simply the recrystallization of minerals already present.

In the following sections we shall describe the important rock types in each major group. Emphasis here will be on the characteristics by which different rocks can be recognized; in future chapters we shall deal at greater length with processes of rock formation.

igneous rocks

The structure of igneous rocks is characterized by random arrangement of grains, by ragged crystal borders, by intertwinings and embayments such as one might expect in a mass of crystals growing together and interfering with one another's development. In coarse-grained rocks like granite, this structure is visible to the naked eye; in fine-grained rocks it is revealed by the microscope. The principal constituents of these rocks are always minerals containing silicon: quartz, feldspar, mica, and the ferromagnesian group.

igneous rocks have solidified from an originally molten state

The siliceous liquids from which igneous rocks form are thick, viscous materials resembling melted glass both in properties and in composition. Sometimes, in fact, molten lava has the right composition and cools rapidly enough to form a natural glass—the black, shiny rock called *obsidian*. Usually, however, cooling is slow enough to allow crystalline minerals to form. If cooling is fairly rapid and if the molten material is highly viscous, the resulting rock may consist of minute crystals or partly of crystals and partly of glass. If cooling is extremely slow, mineral grains have an opportunity to grow large and a coarse-grained rock is formed. The grain size of an igneous rock, therefore, reveals something about its history and gives us one logical basis for classification.

obsidian is a glassy rock of volcanic origin

Mineral composition provides a convenient means of further classification. Nearly all igneous rocks contain feldspar and one or more of the ferromagnesian minerals; many contain quartz as well. A division of igneous rocks based on relative amounts of these three mineral types is shown in Table 31.1. Thus, a coarse-grained rock containing quartz,

TABLE 31.1 Some igneous rocks. These have solidified from a molten state.

Mineral composition	*Coarse-grained rocks (intrusive)*	*Fine-grained rocks (extrusive)*
Quartz Feldspar Ferromagnesian minerals	Granite	Rhyolite
No quartz Feldspar predominant Ferromagnesian minerals	Diorite	Andesite
No quartz Feldspar Ferromagnesian minerals predominant	Gabbro	Basalt

feldspar, and black mica is granite; a fine-grained rock with no quartz and with feldspar in excess of the dark constituents is andesite, and so on. Not all igneous rocks by any means are shown in the table, but these six are the most important.

This classification is convenient for several reasons:

1 Grain size and usually mineral composition can be determined from inspection in the field. Except for a few fine-grained types, a rock can

Granite is a coarse-grained rock whose constituent minerals are readily discernible.

QUARTZ FELDSPAR HORNBLENDE MICA

The Obsidian Cliff in Yellowstone National Park is composed of black glass of volcanic origin. In the past, Indians in the vicinity used slivers from the cliff for knives and arrowheads.

be named without detailed laboratory study.

2 Even if a rock is too fine for its mineral content to be easily determined, its color often shows its place in the table. Granite and rhyolite, containing only a little ferromagnesian material, are nearly always light-colored; gabbro and basalt, with abundant ferromagnesian minerals, are characteristically dark; diorite and andesite usually have intermediate shades.

3 Grain size usually gives an indication not only of the rate of cooling but also of the environment in which a rock was cooled. Sufficiently rapid cooling to give fine-grained rocks occurs most commonly when molten lava reaches the earth's surface from a volcano and spreads out in a thin flow exposed to the atmosphere. Since fine grain size usually betrays volcanic origin, rhyolite, andesite, and basalt are often called *volcanic* or *extrusive* rocks.

Coarse-grained rocks, on the other hand, have cooled sufficiently slowly for large crystals to have formed, which must have occurred well beneath the earth's surface. Such rocks are now exposed to view only because erosion has carried away the material that once covered them. Since these rocks do not reach the surface as liquids but are intruded into spaces occupied by other rocks, they are often called *intrusive* rocks.

4 The change in mineral composition from top to bottom in Table 31.1 roughly parallels a steady change in chemical composition. The chemistry of rocks is most simply considered in terms of the oxides which make up their minerals, and rock analyses are usually given as percentages of these oxides. Thus granite and rhyolite, rocks which contain abundant quartz and little ferromagnesian material, are spoken of as "rocks with a high content of silica," the silica referring not only to SiO_2 which is free as quartz, but also to the combined SiO_2 in the other minerals. Gabbro and basalt, with no quartz and abundant ferromagnesian constituents, show analyses low in silica and high in the oxides FeO and MgO.

TABLE 31.2 Chemical analyses of igneous rocks. In percent by weight.

Rock	Granite	Andesite	Basalt
SiO_2	70.18	59.59	49.06
Al_2O_3	14.47	17.31	15.70
Fe_2O_3	1.57	3.33	5.38
FeO	1.78	3.13	6.37
MgO	0.88	2.75	6.17
CaO	1.99	5.80	8.95
Na_2O	3.48	3.58	3.11
K_2O	4.11	2.04	1.52
Others	1.54	2.47	3.74

Andesite porphyry found in Nevada. An igneous rock in which large crystals are embedded in a fine-grained mass is called a porphyry.

The decrease in SiO_2 and the increase in metallic oxides down the table are shown by the average analyses of granites, andesites, and basalts in Table 31.2. Since SiO_2 is the oxide of a nonmetal, it is considered an "acidic" oxide (Chap. 25), and rocks like granite and rhyolite, whose analyses show much of this oxide, are commonly called *acidic* rocks. A better term, also commonly used, is *siliceous* rocks—better since "acidic" has so different a meaning in chemistry. Gabbro and basalt, which contain an abundance of the basic oxides FeO, MgO, CaO, are often called *basic* rocks. These chemical differences are sometimes expressed by designating granite and rhyolite *sialic* rocks, and gabbro and basalt *simatic* rocks. (The terms sialic and simatic come from the chemical symbols for silicon and aluminum and for silicon and magnesium respectively.)

From several angles, then, this rather sketchy classification is satisfactory. It groups rocks at least roughly according to their origins; it summarizes information about their mineral and chemical compositions; and, finally, it makes possible their identification from such obvious characteristics as grain size, color, and mineral content.

An outcrop of conglomerate in New York State that contains many fragments of limestone. The large chunk is about 1 ft in length.

sedimentary rocks

Sediments laid down by water, wind, or ice are consolidated into rock by the weight of overlying deposits and by the gradual cementing of their grains with material deposited from underground water. As a class, the resulting rocks are characterized by the usual presence of distinct, somewhat rounded grains, not intergrown like the crystals of igneous rocks. A few sedimentary rocks, however, do consist entirely of intergrowing mineral grains formed by precipitation from solution in water. Since sediments are normally deposited in layers, the majority of sedimentary rocks have a banded appearance owing to slight differences in color or grain size from one layer to the next. Sedimentary rocks may often be recognized at a glance by the presence of fossils—remains of plants or animals interred with the sediments as they were laid down.

Sedimentary rocks may be divided into two groups according to the nature of their original sediments.

This shale in eastern New York State was mud 350 million years ago in the Devonian period.

some sedimentary rocks are formed by the precipitation of material dissolved in water

1 *Fragmental rocks*, made up of the fragments and decomposition products of other rocks
2 *Precipitates*, formed from material once dissolved in water and deposited either as a chemical precipitate or as the shells and bone fragments of dead organisms

The more abundant rock varieties in each group are listed in Table 31.3.

TABLE 31.3 Some sedimentary rocks. These are compacted sediments.

Group	Type	Constituents
Fragmental rocks	Conglomerate	Rock fragments
	Sandstone	Quartz usually most abundant
	Shale	Clay minerals
Chemical and biochemical precipitates	Chert	Microcrystalline quartz
	Limestone	Calcite

The three fragmental rocks are distinguished by their grain size. *Conglomerate* is cemented gravel; its fragments may have any composition and any size from that of small pebbles to large boulders. With decreasing size of fragment, conglomerate grades into *sandstone*. Sand grains may consist of many different minerals, but quartz is generally the most abundant. The hardness of sandstone and conglomerate depends in large measure on how well their grains are cemented together: some varieties crumble easily; others, especially those with silica as the cementing material, are among the toughest of rocks. *Shale* is consolidated mud or silt, a soft rock usually in thin layers. Its chief constituents are usually one or more of the clay minerals.

Limestone, a fine-grained rock composed chiefly of calcite, may be formed either as a chemical precipitate or by the consolidation of shell fragments. Like calcite in larger crystals, limestone is only moderately hard and effervesces readily in acid. Small amounts of impurities may give the rock almost any color. *Chalk* is a loosely consolidated variety of limestone, often made up largely of the shells of tiny, one-celled animals.

Most Indian arrowheads are made either of the igneous rock obsidian or the sedimentary rock *chert*, both much prized among primitive peoples for their hardness and the sharpness of their edges when broken. The chief mineral constituent of chert is microcrystalline quartz. Two familiar varieties of this rock are *flint* and *jasper*. Fragments of chert show the same sharp edges and smooth, concave surfaces as broken quartz or obsidian, but the surfaces have a characteristically duller luster resembling that of wax. Impurities may give the rock almost any color; often a single specimen shows bands and pockets of several different colors. Not nearly as abundant as the other sedimentary rocks just described, chert is nevertheless a common rock in pebble beds and gravel deposits, because its great hardness and resistance to chemical decay enable it to survive much rough treatment from streams, waves, and glaciers.

metamorphic rocks

The terrific pressures and high temperatures a few miles below the earth's surface effect profound changes in sedimentary and igneous rocks that become deeply buried. Minerals stable at the surface are often unstable under the new conditions and may react to form different substances. Other minerals remain stable, but their crystals increase in size. Hot liquids permeating the rocks may add some new materials and dissolve out others. So many kinds of change are possible that no satisfactory general rules can be set down for readily distinguishing metamorphic rocks from others.

Many metamorphic rocks are characterized by a property called *foliation*, which means the arrangement of flat or elongated mineral grains in parallel layers. This arrangement is caused by extreme pressure in one direction which causes crushing and intimate movement all through the rock, the new minerals growing in the planes of movement as the rock is squeezed. Foliated rocks always contain a mineral (like mica) that occurs in thin flakes or a mineral (like some of the ferromagnesian group) that occurs in long needles. Rocks consisting only of minerals like quartz, feldspar, or calcite cannot show foliation, since these minerals have little tendency to grow larger in one direction than in another, even under pressure. Foliation gives a rock a banded or layered appearance, and, when it is broken, the rock tends to split along the bands. Layering is also characteristic of sedimentary rocks, but in them the layering is caused by slight variations in color or grain size; layering in metamorphic rocks is due to the lining up of mineral grains.

An unfoliated metamorphic rock is produced when the chemical composition prevents the formation of minerals with flat or elongated crystals or when a rock is metamorphosed simply by heat without excess pressure in one direction. The random growth of individual mineral grains during metamorphism produces in such a rock a structure much like that typical of igneous rocks, each grain embaying and interlocking with its neighbors. A metamorphic origin for an unfoliated rock can often be recognized from the fact that its mineral composition is unlike that of typical igneous rocks or from the presence of other metamorphic rocks adjacent to it in the field.

metamorphic rocks are formed from sedimentary and igneous rocks under the influence of heat and pressure

many metamorphic rocks are foliated, which gives them a banded or layered appearance

TABLE 31.4 Some metamorphic rocks. These have changed due to heat and/or pressure since originally formed.

Group	Type	Constituents	Origin
Foliated rocks	Slate	Mica and usually quartz, both in microscopic grains	Shale
	Schist	Mica and/or a ferromagnesian mineral, usually quartz also	Shale or fine-grained igneous rock
Foliated and banded	Gneiss	Quartz, feldspar, mica	Various
Unfoliated rocks	Marble	Chiefly calcite	Limestone
	Quartzite	Chiefly quartz	Sandstone

MATERIALS OF THE EARTH

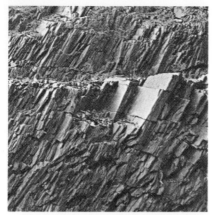

Slate results from the metamorphism of shale under pressure. This scene is in southern New York State.

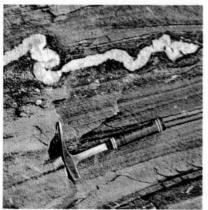

An outcrop of schist in Australia which is cut by a vein of quartz.

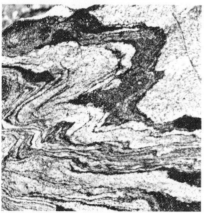

The metamorphic rock gneiss shows foliation and bands of different material.

The commoner metamorphic rocks may be classified according to the presence or absence of foliation, as in Table 31.4.

Slate is produced by the low-temperature metamorphism of shale, the unstable clay minerals changing to tiny flakes of mica. Although the individual flakes are too small to be seen, parallel sheets of mica crystals are responsible for the shiny surfaces produced whenever slate is split along its foliation. Slate is harder than shale, finely foliated, usually black or dark gray but sometimes lighter colored.

Schist is produced from shale by higher temperatures or from fine-grained igneous rocks. In it the mineral grains responsible for the foliation are large enough to be visible, giving the foliation surfaces a characteristic spangled appearance. Schist does not split so easily with the foliation as slate does, and its surfaces are rougher.

Gneiss is a coarse-grained rock, produced under conditions of high temperature and pressure from almost any other rock except pure limestone and pure quartz sandstone. Its composition naturally depends on the nature of the original rock, but quartz, feldspar, and mica are the commonest minerals. In appearance gneiss resembles granite, except for its banding.

The metamorphism of pure limestone and pure quartz sandstone is a relatively simple process. Since each consists of a single mineral of simple composition, heat and pressure can produce no new susbstances, but simply cause the growth and interlocking of new crystals of calcite and quartz. Thus limestone becomes *marble*, a rock composed of calcite in crystals large enough to be easily visible, whereas sandstone becomes the hard rock *quartzite*. The recognition of marble is a simple matter of applying the usual tests for calcite. Quartzite may resemble sandstone in appearance, but its grains are so firmly intergrown that it splits across separate grains when it is broken, giving smooth fracture surfaces in contrast with the rough surfaces of sandstone.

In passing, we should remind ourselves once more that this is an artificial classification of rocks, that in nature we often find no sharp boundaries between different rock types. A medium-grained igneous rock with no quartz and an abundance of feldspar might be hard to classify either as a diorite or an andesite; some limestones contain so much clay that they may be called either shaly limestones or limy shales; there is no exact point in the metamorphic process when a fine-grained rock ceases to be a shale and becomes a slate. In spite of these occasional deficiencies, the classification is valuable because it *organizes* our knowledge about various rocks; henceforth when we label a rock schist, we know at once something about its origin and its general properties. The classification gives us a summary of elementary information on which we can build the generalizations of geology.

SOIL

Though the bulk of the earth's crust is solid rock, what we see on that part of the surface not covered by water is chiefly soil, with only occasional

Cross section of a typical soil. The darkening toward the top is due to the presence of humus.

outcrops of bedrock. Soil originates in the weathering of rock, a complex disintegration process whose result is a coat of rock fragments and clay minerals mixed with varying amounts of organic matter.

Any type of rock—igneous, sedimentary, or metamorphic—may form the parent material of a soil. Typically the particles of rock vary in size down to microscopic fineness and are intimately mixed with dark, partly decomposed plant debris called *humus*. The humus content decreases with depth, and it is customary to call the uppermost layer of soil, which is richest in humus, the *topsoil*, and the underlying layer of accumulated rock fragments the *subsoil*. A great many factors are involved in the

humus is partly decomposed plant material

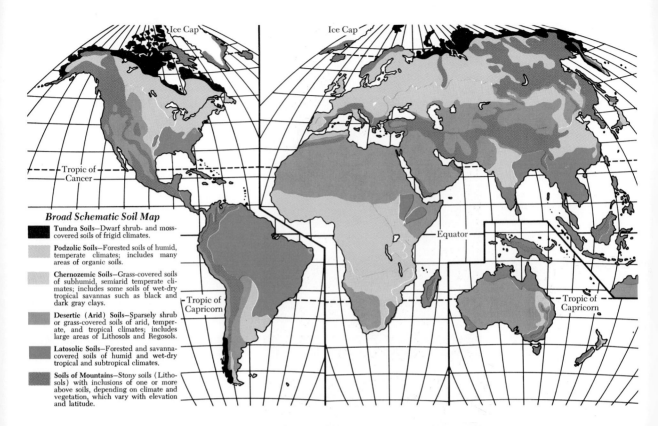

Principal soils around the world. Localized bodies of soil, for instance, the alluvial soils along such major rivers as the Mississippi, Amazon, Ganges, and Yangtze, are not shown.

production of soil, including microorganisms such as bacteria and fungi which are responsible for the decay of plant and animal residues and are important in maintaining the nitrogen content of soil. A significant fraction of the organic material in soil, in fact, consists of the bodies, living and dead, of these microorganisms. Even so lowly a creature as the worm plays a vital role in mixing together the various soil constituents.

types of soil

A wide variety of soil types have been identified, most of which fall into the four broad classes of podzol, latosol, chernozem, and desert soils.

the four chief classes of soil are podzol, latosol, chernozem, and desert

1 *Podzol soils* are mainly found in cool, moist climates under coniferous or partly coniferous forests, as in most of northern Europe and Canada. They are gray in color because most of the iron and other soluble minerals have been washed away ("leached out"), and are acid, which discourages the work of earthworms and other organisms to the extent that there is a sharp line of demarcation between a thin upper layer of partly decayed plant matter and a mineral layer underneath with little organic content. Much of central Europe and the eastern United States is covered with gray-brown, red, and yellow podzolic soils which owe their color and agricultural productivity in part to a smaller degree of acidity, which inhibits leaching and is more favorable for the flourishing of soil organisms.

2 *Latosol soils* are typical of rain forests in hot, humid climates, and they are prevalent in Brazil, west and central Africa, and southeast Asia. Latosols are rich in iron and aluminum oxides; they are very porous and have been largely leached of plant nutrients required for cultivation. Latosol soils are red or yellow in color.

3 *Chernozem* is a Russian word meaning "black earth," and the chernozem soils are indeed black or dark brown in color. They are found in temperate, subhumid climates and were formed under vegetation of prairie grasses rather than forests. Southern Russia, a north-south belt in the central United States, and parts of South America, India, Canada, China, and Australia have soils of this kind, which are extremely fertile.

4 *Desert soils* are of various kinds but, being formed in arid regions with little vegetation, are all light in color due to the lack of organic content. There is an abundance of soluble minerals, and sometimes a crust of alkaline and salt materials is present on the surface. The richness in soluble minerals partly compensates for the absence of humus, and many desert soils can be cultivated with proper irrigation.

GLOSSARY

Igneous rocks are rocks that have been formed from a molten state by cooling.

Sedimentary rocks consist of materials derived from other rocks that have been decomposed by water, wind, or glacial ice; they may be fragments cemented together or material precipitated from water solution.

Metamorphic rocks are rocks that have been altered by heat and pressure deep under the earth's surface.

Volcanic rocks originated when molten lava of volcanic origin cooled rapidly at the earth's surface.

Intrusive rocks are igneous rocks that flowed into regions below the surface already occupied by other rocks and gradually hardened there.

Foliation refers to the alignment of flat or elongated mineral grains characteristic of many metamorphic rocks.

Soil is a mixture of rock fragments and organic matter. The latter is largely partly decomposed plant debris called *humus*.

The Sahara is largely desert today, although at one time vegetation flourished there. Note patterns of ridges caused by the wind.

The Belem-Brazilia highway across Brazil, shown here under construction, cuts through a rain forest growing on latosol soils.

EXERCISES

1. Specify characteristics by which you can distinguish
   a. granite from gabbro
   b. basalt from limestone
   c. schist from diorite
   d. chert from obsidian
   e. conglomerate from gneiss
   f. quartz from calcite

2. Name the following rocks:
   a. A fine-grained, unfoliated rock with intergrowing crystals of quartz, feldspar, and black mica
   b. A finely foliated rock with microscopic crystals of quartz and white mica
   c. A fine-grained rock consisting principally of kaolin
   d. A rock consisting of intergrown crystals of quartz
   e. The rock resulting from metamorphism of limestone
   f. An intrusive igneous rock with the same composition as andesite

3. Arrange the following rocks in three general groups as (*a*) hard, (*b*) moderately hard, and (*c*) soft. Indicate which are igneous, which sedimentary, and which metamorphic.

gneiss	obsidian	andesite
limestone	shale	chalk
quartzite	chert	marble

chapter 32

The Atmosphere

THE ATMOSPHERE
- composition
- escape of planetary atmospheres
- regions of the atmosphere
- stratosphere and mesosphere
- the ionosphere

WEATHER
- meteorology
- atmospheric energy
- atmospheric moisture
- winds
- air masses
- atmospheric circulation

CLIMATE
- tropical climates
- middle-latitude climates
- climatic change

The earth is surrounded by its *atmosphere*, a gaseous envelope whose significance is seldom appreciated. The atmosphere is responsible for many things: its oxygen, nitrogen, and carbon dioxide are indispensable to life; it screens lethal ultraviolet and X-rays from the sun; it transports water over the face of the earth; it weathers rocks and thereby contributes to the erosion processes that continually change the earth's surface; it contains layers of ions that reflect radio waves and permit worldwide radio communication; it makes possible the blue of the sky, the colors of sunrise and sunset, and the dramatic, mysterious aurora. As we shall see in this chapter, such familiar manifestations of the atmosphere as wind and rain are only part of the story of this remarkable portion of our planet.

TABLE 32.1 The composition of dry air near ground level.

Gas	Percentage by volume
Nitrogen	78.08
Oxygen	20.95
Argon	0.93
Carbon dioxide	0.03
Neon	0.0018
Helium	0.00052
Methane	0.00015
Krypton	0.00011
Hydrogen, carbon monoxide, xenon, ozone, radon	<0.0001

THE ATMOSPHERE

composition

Table 32.1 is a list of the principal constituents of the atmosphere and their average abundances. In any given location, of course, the proportions of the various gases may be somewhat different, but only rarely do such differences lead to significant effects at sea level and low altitudes. Water vapor is omitted from Table 32.1 because of its extreme variability, ranging from almost none to as high as 4 percent; we shall discuss atmospheric water later in this chapter in some detail.

Oxygen and nitrogen are important biologically, and each has a characteristic cycle of interaction with living things. The nitrogen cycle was discussed on page 439. Oxygen is required by living things for the oxidation of their foods, which provides energy. A by-product of this process is carbon dioxide, as we learned in Chap. 29. Plants combine carbon dioxide and water to make organic compounds with the help of sunlight and the catalyst chlorophyll in photosynthesis, with oxygen as a by-product. Thus the oxygen–carbon dioxide cycle is an essential aspect of all plant and animal life (Fig. 32.1). Approximately 70 billion tons of carbon dioxide are removed from the atmosphere each year by plants and returned by them as oxygen, and a corresponding amount of oxygen is simultaneously converted by plants and animals into carbon dioxide. There is also a continual exchange of carbon dioxide between the atmosphere and the oceans.

Both the biological and oceanic cycles are, on the average, balanced in their consumption and production of carbon dioxide. But there are also sources of carbon dioxide that have no absorption processes to counter their effects. The most significant of these sources is the burning of coal and oil by man to produce heat for dwellings and mechanical energy for industry and transportation. At present our chimneys and exhaust pipes pour about 12 billion tons of carbon dioxide each year into the atmosphere,

the oxygen–carbon dioxide cycle

THE ATMOSPHERE

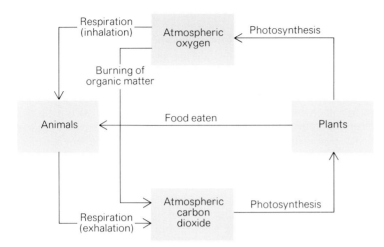

32.1 The oxygen-carbon dioxide cycle in the atmosphere.

and this rate is rapidly increasing. Since 1880 the carbon dioxide content of the atmosphere has gone up by 12 percent (Fig. 32.2). Despite the relatively small proportion of carbon dioxide in the atmosphere—only 330 parts per million—it is a most significant constituent because of its ability to absorb solar energy reradiated by the earth. There is a distinct—and ominous—likelihood that the increase in the average temperature of the atmosphere by about $0.01°C$ per year since 1880 is due to the increase in its carbon dioxide content over this period. We shall return to this topic later in this chapter.

the carbon dioxide content of the atmosphere is steadily increasing

The other major constituent of the atmosphere is argon, which, being chemically inert, escaped detection until the end of the nineteenth century. Argon is more abundant than the other inert gases because the isotope K^{40} of the common element potassium beta-decays into the argon isotope A^{40}.

argon

escape of planetary atmospheres

Since many radioactive substances emit alpha particles (which are helium nuclei), we might suspect that helium should be more prominent a con-

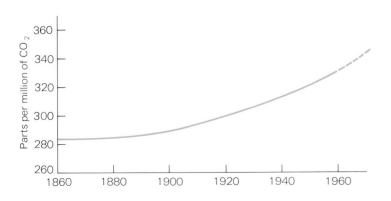

32.2 The carbon dioxide content of the atmosphere since 1860. The increase is believed to be largely due to the combustion of coal and oil.

stituent in the atmosphere than it is. This suspicion is verified by detailed calculations. Upon reflection, we can see that helium is not the only "missing" gas in the atmosphere: hydrogen, so enormously abundant in the universe generally, is strangely sparse also. This problem is evidently related to such similar peculiarities in the solar system as the complete lack of an atmosphere around the moon.

The kinetic theory of gases, which we considered in Chap. 8, provides us with the solution. A planet holds a moving molecule, as it holds any other object, by gravitational attraction. To escape from a planet's attraction a molecule need only acquire sufficient speed in an outward direction. The necessary speed, called the *escape velocity* as we know, is high for a massive planet like Jupiter, low for a light body like the moon (see Table 3.1). The earth's escape velocity is not quite 7 mi/sec: This is the speed which an interplanetary spacecraft must possess in order to leave the earth behind and, similarly, the speed which a molecule at the top of the atmosphere must attain before it can wander off into space. The moon's escape velocity is much smaller, about 1.5 mi/sec. Now air molecules at ordinary temperatures move with average speeds of less than $\frac{1}{2}$ mi/sec, so their chances of escaping from either earth or moon may not seem very good. But this $\frac{1}{2}$ mi/sec is an *average speed*; many air molecules at any given instant are moving more slowly, many others considerably faster. And temperatures on the moon's surface at midday approach 100°C, a temperature at which the average molecular speed is somewhat higher than $\frac{1}{2}$ mi/sec. If a sample of air could be placed on the moon today, a few of its faster molecules would attain speeds over 1.5 mi/sec and drift off into space. As these escaped, collisions in the remaining gas would presently restore the original distribution of molecular speeds and thus give a few more molecules the necessary speed. The air would slowly vanish, molecule by molecule. Probably the moon has been for ages without an atmosphere because the original gas surrounding it escaped by just this process.

To leave the earth, an air molecule would have to reach a speed of 7 mi/sec in the upper atmosphere. Attaining this speed at lower elevations would, of course, do no good, for collisions would cause it to lose energy before it could escape. The number of molecules with speeds so far above the average in the upper atmosphere is negligible, so that our planet is in no danger of losing its valuable air blanket. Even molecules of hydrogen, lightest and hence fastest of all gas molecules at a given temperature, probably cannot now escape the earth's attraction, although if the earth was hotter in the remote past, they may have then escaped in large numbers.

Thus a planet's ability to hold gas molecules depends primarily on its gravitational pull and to a lesser extent on its temperature. Mercury, small and hot, is as barren as the moon. Venus, nearly the size of the earth, has an atmosphere as dense as ours. Mars is heavy enough to hold a thin atmosphere, but the swift molecules of hydrogen and helium must have escaped from it long since. The four giant planets of Jupiter, Saturn, Uranus, and Neptune have enormously thick atmospheres containing abundant hydrogen.

---

escape velocity is the initial speed an object must have to permanently leave a celestial body

the more massive a planet is, the greater its ability to hold an atmosphere

THE ATMOSPHERE

regions of the atmosphere

Those of us who have been in mountainous terrain know that the higher up we go, the thinner and colder the air becomes. This observation is verified by instruments carried in airplanes and balloons, whose average measurements of pressure and temperature are plotted versus altitude in Figs. 32.3 and 32.4. At an elevation of only 5 km the pressure is down to half what it is at sea level, and the temperature is about $-20°C$. At about 11 km the pressure is only one-fourth its sea-level value, which means that 75 percent of the atmosphere lies below, and the temperature has dropped to $-55°C$. Beyond 11 km the pressure continues to fall, but the temperature remains at $-55°C$—which is cold but not so cold as it sometimes is at ground level during the winter in Siberia and northern Canada—for 14 km more.

A passenger in an airplane would notice a marked change in the atmosphere as he passes an elevation of 11 km: above this point there are practically no clouds and no storms, and dust is almost completely absent. Since the character of the atmosphere changes rather abruptly at the 11 km level, this is taken as the boundary between two layers of the atmosphere: the clear, cold upper part, or *stratosphere*, and the denser lower part, or *troposphere*. Such common atmospheric features as clouds and storms, fog and haze, are confined to the troposphere. The boundary between stratosphere and troposphere is higher near the equator, where it is about 16 km above sea level, and lower near the poles, where it is about 6 km; the 11-km figure is an average.

in the lower atmosphere, air temperatures fall an average of 6.5°C per km of altitude (3.6°F per 1,000 ft)

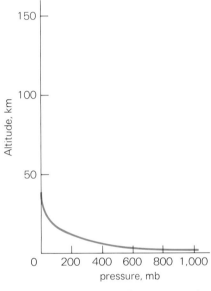

32.3 *The variation of pressure with altitude in the atmosphere.* The average pressure at sea level is 1,013 millibars (mb)

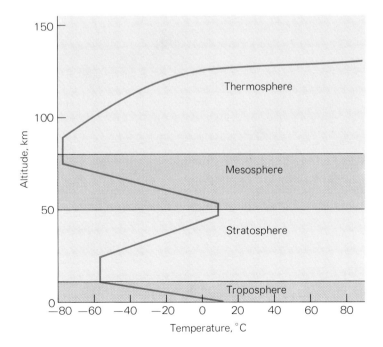

32.4 *The variation of temperature with altitude in the atmosphere.*

At a height of about 25 km the temperature begins to rise, reaches a maximum of 10°C or so in the vicinity of 50 km, and then falls once more to another minimum of about −76°C at 80 km. The portion of the atmosphere between 50 and 80 km is known as the *mesosphere*. Above 80 km the properties of the atmosphere change radically, for now ions become abundant. The *thermosphere* extends upward to a height of about 600 km, with the temperature increasing to about 2,000°C where it levels off. (We must keep in mind that the density of the thermosphere is extremely low, so that despite the high temperatures a slowly moving object there would not get hot if shielded from sunlight.)

mesosphere and thermosphere

The uppermost part of the atmosphere is called the *exosphere*, and it shades off imperceptibly into interplanetary space. These five regions —troposphere, stratosphere, mesosphere, thermosphere, and exosphere— are remarkably independent of one another in many of their properties.

exosphere

stratosphere and mesosphere

Many scientists prefer to regard the stratosphere and mesosphere as being different parts of the same layer, because apart from temperature their properties are similar. Sounding balloons filled with hydrogen or helium routinely penetrate the stratosphere with instruments of various kinds and send down data by radio to ground stations; some aircraft even are capable of exploring the stratosphere. The still higher elevations of the mesosphere require rocket-borne apparatus if direct measurements are to be made, but a number of experimental methods have been devised that enable observatories on the ground to determine some of the physical properties of this part of the atmosphere.

The most striking feature of the stratosphere and mesosphere is the presence of *ozone*, an allotropic form of oxygen whose molecules contain three oxygen atoms. The symbol for ozone is, accordingly, O_3. Ozone is an excellent absorber of ultraviolet radiation, so excellent, in fact, that the relatively small amount of ozone in the upper atmosphere is able to filter out completely the dangerous short-wavelength ultraviolet radiation reaching the earth from the sun. The ozone layer lies between 15 and 35 km. Its maximum density occurs at 22 km, where less than one molecule in 4 million is O_3—hardly an impressive concentration for so efficient a filter. At sea-level temperature and pressure, all the ozone of the atmosphere would form a layer only an inch thick. The elevated temperatures characteristic of the mesosphere are due to the heating effect of the solar ultraviolet energy absorbed there.

ozone in the upper atmosphere absorbs ultraviolet radiation from the sun

the ionosphere

In the year 1901 Marconi was able to send radio signals across the Atlantic Ocean for the first time. Radio waves, like light waves, tend to travel in straight lines, and the curvature of the earth therefore apparently presents an insuperable obstacle to long-distance radio communication. For this reason Marconi's achievement came as a great surprise.

THE ATMOSPHERE

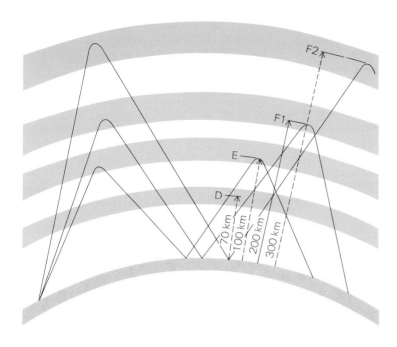

32.5 *The D, E, F1, and F2 layers in the ionosphere. Several possible paths of radio waves are indicated.*

In a short time, however, Oliver Heaviside in England and Arthur Edwin Kennelly in the United States suggested that the effect could be caused by a reflecting layer high up in the atmosphere. Such a layer, together with the sea, could channel radio waves from one side of the Atlantic to the other (Fig. 32.5). Electromagnetic theory was able to predict the mechanism of the reflection: If some of the atoms and molecules in the upper atmosphere are ionized by the action of high-energy solar radiation, the resulting layer of charged particles will behave precisely like a mirror to radio waves (though not to the shorter-wavelength light waves).

Direct experimental confirmation of the presence of ionized layers high up in the atmosphere followed, and today the region that contains substantial numbers of ions is called the *ionosphere*. It extends from 70 km to several hundred km above the earth's surface. The reason that ions are not distributed throughout the entire atmosphere is straightforward: At very great altitudes there is not enough gas present for the solar ultraviolet and X-rays to interact with, and at low altitudes all the solar radiation energetic enough to cause ionization has already been absorbed.

the ionosphere is caused by solar ultraviolet and X-rays

Over 30 years ago a very ingenious method was developed for measuring the properties of the ionosphere: A pulse of radio waves is sent upward from a ground station, and the time required for its echo to return from the ionosphere is measured. Since the velocity of radio waves in the atmosphere is known, the echo return time makes possible a determination of the height of the reflecting layer. Furthermore, by varying the frequency of the radio waves, much valuable information on the nature of the reflecting layers can be obtained.

Four fairly distinct layers have been shown to exist in the ionosphere during the day. The lowest is the D layer, centered at an altitude of about

70 km; next is the E layer at about 100 km; highest are the F1 layer at about 200 km and the F2 layer at about 300 km. At night, much of the ionization disappears to leave only the E and F1 layers in attenuated form. The D layer, unlike the others, tends to absorb rather than reflect radio waves and when it is intensified during solar flares (Chap. 42), it disrupts long-distance radio communication. Normally all it does is reduce the intensity of radio waves that pass through it on their way to the E, F1, and F2 layers, where they are reflected down to the earth again. Several possible paths by which radio waves are propagated over long distances with the help of the ionosphere are shown in Fig. 32.5; for really great distances reflection from the E layer, which is relatively low, is of no significance, since waves reflected by it must undergo a large number of up-and-down passages through the atmosphere, which reduce their intensity.

ionospheric layers reflect radio waves and so permit long-distance communication

The ionosphere is also the part of the atmosphere in which the eerie and magnificent *aurora* occurs, a topic we shall discuss in Chap. 42.

WEATHER

meteorology

The branch of physical science called *meteorology* is devoted to the study of weather and climate. Meteorology is concerned with what may be thought of as a vast, automatic air-conditioning system. Our spinning planet is heated strongly at the equator, feebly at the poles, and its moisture is concentrated in the great ocean basins. It is the task of the atmosphere, from our point of view, to redistribute this heat and moisture so that large areas of the land surface will be habitable. Air conditioning by the atmosphere is far from perfect; it fails miserably in desert regions, on mountain summits, in far northern and southern latitudes. On sultry nights in midsummer or on bitter January mornings we may question its efficiency even in our favored part of the world. But the atmosphere does succeed in making a surprisingly large amount of the earth's surface fit for human habitation.

weather and climate

The two chief functions of any air-conditioning system are the regulation of air temperature and humidity. In addition to these, we expect the atmosphere to perform a third function: it must provide us at intervals with rain or snow. The weather and climate of a given locality describe how effectively these functions are performed. *Weather* refers to the temperature, humidity, pressure, cloudiness, and rainfall at a certain time; *climate* is a summary of weather conditions over a period of years. Important in a description of climate is the variability of temperature and rainfall with the seasons; an outstanding feature of the climate of North Dakota is its extreme warmth in summer and cold in winter, whereas the climate of southern California is characterized by equable year-round temperatures and by a concentration of rainfall in the winter months. Local barometric pressures and the intensity and direction of wind may be important in descriptions of weather and climate.

atmospheric energy

The energy that powers meteorological phenomena comes to us from the sun in the form of electromagnetic radiation of short wavelength—chiefly visible light, together with some ultraviolet and infrared radiation. The energy arriving at the outer atmosphere is called *insolation* (for *in*coming *sol*ar radi*ation*) and amounts to 20 kcal/m² of area perpendicular to the direction of the radiation.

insolation

About 34 percent of the insolation is directly reflected back into space, mainly by clouds. The atmosphere absorbs perhaps 19 percent of the insolation, with ozone, water vapor, and water droplets in clouds taking up most of this amount. Thus 47 percent of the total insolation reaches the earth's surface, where it is absorbed and converted into heat. The warm earth then reradiates its excess energy back into the atmosphere, but the energy now is in the form of long-wavelength infrared radiation. These long waves are readily absorbed by atmospheric carbon dioxide and water vapor. The molecules of these gases, speeded up by absorption of heat energy, give some of this energy to other air molecules during collisions. Thus the chief source of atmospheric heat is radiation from the earth, not the energy of direct sunlight.

atmospheric phenomena are powered by solar energy reradiated from the earth

If the earth had no atmosphere, its heated surface would quickly radiate back into space all the energy that reaches it from the sun. Like the moon, the earth would grow intensely hot during the day, unbearably cold at night. An atmosphere effectively prevents these extremes of temperature; its continual movement makes impossible undue heating of any one region by day, and its ability to absorb and hold the earth's radiation prevents the rapid escape of heat by night. Our atmosphere acts as an efficient trap, admitting the energy of sunlight relatively freely but hindering its escape.

How hot the atmosphere becomes over any particular region depends on a number of factors. Air near the equator is on the average much warmer than air near the poles, because the sun's vertical rays are more effective in heating the surface than the slanting rays of polar regions (Fig. 32.6). Air over a mountain summit may become warm at midday but cools quickly because it is thinner and contains less carbon dioxide and water vapor than air at lower elevations. A region covered with clouds usually has lower air temperatures than an adjacent region in sunlight. Because the temperature of water is changed more slowly than that of rocks and soil by absorption or loss of radiation, the atmosphere near large bodies of water is usually cooler by day and warmer by night than the atmosphere over regions far from water. Desert regions commonly show abrupt changes in air temperature between day and night because so little water vapor is present to absorb heat radiation. The atmospheric temperatures of some regions are influenced profoundly by winds and by ocean currents.

Because the earth's average temperature does not change by very much with time, there must be a balance between incoming and outgoing energy. That such a balance does indeed occur can be seen with the help of Fig. 32.7, which shows how the rates at which radiant energy enters and leaves the earth vary with latitude.

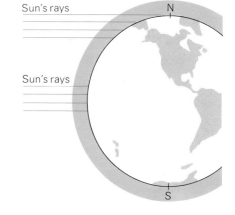

32.6 Air near the equator is on the average much warmer than air near the poles because the sun's vertical rays at the equator are more effective in heating the surface than the slanting rays of polar regions.

32.7 The annual balance between incoming solar radiation and outgoing radiation from the earth. More energy is gained than lost in the tropical regions, and more energy is lost than gained in the polar regions. The latitude scale is spaced so that equal horizontal distances on the graph correspond to equal areas of the earth's surface.

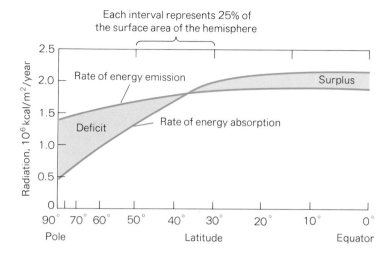

winds and ocean currents carry energy from the tropics to the high latitudes

More energy arrives at the tropical regions than is lost there, and the opposite is true at the polar regions. Why then do not the tropics grow warmer and warmer while the poles grow colder and colder? The answer is to be found in the motions of air and water that shift energy from the regions of surplus to the regions of deficit. About 80 percent of the energy transport around the earth is carried by winds in the atmosphere, and the remainder is carried by ocean currents. We shall examine both of these mechanisms in the remainder of this chapter and in the next chapter.

atmospheric moisture

The moisture content, or *humidity*, of air refers to the amount of water vapor that it contains. Most of the atmosphere's water vapor comes from evaporation of sea water, but a little comes from evaporation of water in lakes, rivers, moist soil, and vegetation. Since water vapor is continually being added to air by evaporation and since it is periodically removed by condensation as clouds, rain, snow, and fog, the humidity of the atmosphere is extremely variable from day to day and from one region to another (Fig. 32.8).

Air is said to be *saturated* with water vapor when it contains the maximum amount that will evaporate at a given temperature. Air is unsaturated when its amount of water vapor is less than this limiting value, since it is capable of holding more water vapor. In effect, air may be regarded as a sort of sponge, filled more or less completely with water vapor. Actually, of course, the air has nothing to do with evaporation; if no air existed, vapor would still escape from bodies of water. But, since air is the agent that transports water vapor from one region to another and since air is the medium in which water vapor condenses as clouds, fog, rain, or snow, we shall find it convenient to think of the air as "taking up" and "holding" different amounts of vapor.

Altocumulus lenticularis clouds appear as semitransparent patches at several levels.

High-altitude cirrus clouds above medium-altitude altocumulus.

Cirrus clouds consist of ice crystals and usually occur at high altitudes (16,000 to 45,000 feet). Shown is cirrus uncinus, commonly called "mares' tails."

Low-altitude stratocumulus clouds formed below 7,000 feet. They are sometimes formed by the horizontal spreading out of cumulus.

Stratus clouds consist of layers of water droplets at low altitudes (here below 3,000 feet) that reveal a horizontal flow of air.

Parallel bands of altocumulus produce a "mackerel sky."

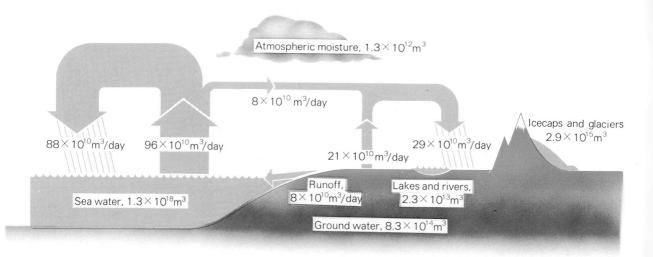

32.8 The world's water content and its daily cycle. *Upward arrows indicate evaporation, downward arrows indicate precipitation.*

relative humidity refers to the extent to which air is saturated with water vapor

We usually describe air as humid if it is saturated or nearly saturated, as dry if it is highly unsaturated. Humid weather is oppressive because little moisture can evaporate from the skin into saturated air, and so perspiration does not produce its usual cooling effect. Very dry air is harmful to the skin because its moisture evaporates too rapidly. Meteorologists express the moisture content of air in terms of *relative humidity*, a number indicating the degree to which air is saturated with water vapor. Usually relative humidity is expressed as a percentage: a humidity of 100 percent means that the air is completely saturated with water, 50 percent means that the air contains half of the maximum it could hold, and 0 percent means perfectly dry air.

The amount of moisture that air can hold increases with temperature. If air saturated at 20°C is heated to 40°C, it can take up more water vapor and so is no longer saturated (in other words, its relative humidity decreases, although the amount of water vapor does not change). If, on the other hand, air saturated at 20°C is cooled to 0°C, some of its water vapor must condense out as liquid water, since at the lower temperature the air can hold only about one-fourth as much vapor as it contained originally. Further, if air at 40°C containing water vapor corresponding to 100 percent relative humidity at 20°C is cooled to 0°C, it grows steadily more saturated until a temperature of 20°C is reached, after which it remains saturated down to 0°C and some of its vapor condenses out. Thus, any sample of ordinary air on heating grows less saturated and on cooling grows more saturated. If the cooling is continued past the saturation point, some liquid water (or ice) must condense out.

The condensation of water from cooled air is an everyday occurrence. Dew forms because the ground surface is cooled by rapid radiation at night, so that air near the ground has its temperature lowered below the saturation point. Fogs are produced when larger masses of air are cooled by contact with cold earth or water bodies. Clouds form when air cools by expansion on rising. Whether the vapor condenses as droplets of liquid water or as tiny particles of ice depends on the temperature.

The growth of tiny water droplets into the larger drops of rain or the growth of minute ice crystals into the lacework patterns of snowflakes can take place only when large masses of air are cooled rapidly. Ordinarily the only method by which cooling on so grand a scale can take place is through the chilling of air by expansion on rising.

snow and rain occur when moisture-laden air is cooled by expansion on rising

winds

Heat and moisture are transported over the earth's surface by movements in the atmosphere called *winds*. If we were somehow able to view the entire pattern of winds over the earth from a great distance, we should find that these air motions fall into three broad categories:

1. On the smallest scale we would discern irregular eddies, some miles across, associated with weather phenomena such as local showers and thunderstorms.
2. Much larger are the enormous eddies, several hundred to a thousand or more miles across, that constitute the *weather systems* so graphically shown on weather maps (Fig. 32.9). These weather systems may center about regions of low pressure, in which case they are called *cyclones* and are characterized by counterclockwise winds in the Northern Hemisphere (clockwise in the Southern), or about regions of high pressure, in which case they are called *anticyclones* and are characterized by clockwise winds in the Northern Hemisphere (counterclockwise in the Southern).
3. And, if we could somehow average out the winds of the great and small whirling air masses, we should discover that there are still vaster systematic movements of air which make up the *general circulation of the atmosphere*.

cyclones and anticyclones

There are three regions whose characteristic winds are conspicuous parts of the general circulation of the atmosphere. As shown in Fig. 32.10, between approximately 60°N and 30°N there are westerly winds, from 30°N to 30°S the winds are predominantly easterly, and between 30°S and 60°S they are again westerly. (Wind direction is conventionally described as the direction from which the wind is blowing.) Of course, the picture has many complications. The easterly winds are relatively weak near the equator and tend to come from the northeast in the Northern Hemisphere and from the southeast in the Southern, for example. But, if we go up to an altitude of a mile or so above the ground, the flow of air is much more regular and either directly east or west. Still higher up *all* the winds tend to assume a westerly character, so that, at an altitude of 7 miles, the flow of air is all from west to east around the earth.

the general circulation of the atmosphere

The various wind zones were important to shipping in the days of sail, as their names indicate. Thus the steady easterlies on either side of the equator became known as the *trade winds*, while the region of light, erratic wind along the equator itself, where the principal movement of air is upward, constitutes the *doldrums*. The *horse latitudes* that separate the trade winds in either hemisphere from the *prevailing westerlies* of

the doldrums and the horse latitudes are belts of calm; the trade winds between them and the prevailing westerlies north and south of the doldrums are relatively steady

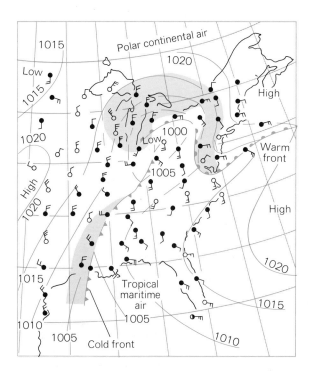 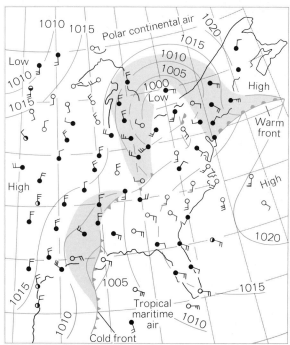

32.9 Weather maps show pressure patterns, winds, and precipitation. *Above, the weather map of the eastern United States one April morning. A cold air mass on the west and north (polar continental air) is separated from a warm air mass (maritime tropical air) by a cold front extending from Louisiana to Michigan and by a warm front from Michigan to Virginia. Where the north end of the warm air mass lies between the two fronts a cyclone has formed, bringing rain (shaded area) to the Great Lakes region. Right, the weather map of the same region 6 hr later. The cyclone has moved northward and eastward, and a new low-pressure area is developing as a kink in a cold front farther south. (The unit of pressure in these maps is the millibar. The small circles indicate clear skies, solid dots indicate cloudy skies. The small lines show wind direction, which is toward the circle or dot, and wind strength; the greater the number of tails, the faster the wind.)*

the middle latitudes are also regions of light winds and are supposed to have been given their name because of the practice of disposing of the horses on board sailing vessels becalmed there and running short of water.

air masses

Day-to-day weather is more variable in the middle latitudes than anywhere else on earth. If we visit central Mexico or Hawaii, in the belt of the northeast trades, we find that one day follows another with hardly any change in temperature, moisture, or wind direction, whereas in nearly all parts of the continental United States abrupt changes in weather are commonplace. The reason for this variability lies in the movement of warm and cold air masses and of storms derived from them through the belts of the westerlies.

In the northern part of the westerly belt an irregular boundary separates air moving generally northward from the horse latitudes and air moving southward from the polar regions. Great tongues of cold air at times sweep down over North America, and at other times warm air from the tropics extends itself far northward. The cold air is ultimately warmed and the warm air cooled, but a large body of air can maintain nearly its original temperature and humidity for days or weeks. These huge tongues of air, or isolated bodies of air detached from them, are the *air masses* of meteorology. The kind of air in an air mass depends on its source: a mass formed over northern Canada is cold and dry, one from the North Atlantic or North Pacific is cold and humid, one from the Gulf of Mexico

A photograph of clouds over the Philippines taken from the Mercury capsule Faith 7. Air over an island is heated to a greater extent than air over the surrounding ocean. As it rises, the moisture-laden air cools, and its content of water vapor condenses into droplets that appear as clouds.

warm and humid, and so on. Weather prediction in this country depends largely on following the movements of air masses from these various source areas (Fig. 32.11).

The contact zone between a warm and a cold air mass is inevitably a zone of disturbance. In general, the lighter air of the warm mass moves

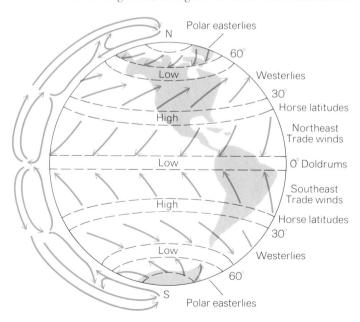

32.10 *Simplified pattern of horizontal and vertical circulation of the atmosphere.* Regions of high and low pressure are indicated.

32.11 The air masses that affect weather in North America. The importance of the various air masses depends upon the season. In winter, for instance, the continental tropical air mass disappears, whereas the continental polar air mass exerts its greatest influence on the weather then.

cold and warm fronts

up over the heavy air of the cold mass, so the contact surface is inclined. This surface is called a *frontal surface*, and the line where the surface meets the ground is called a *front*. As the air masses move, the frontal surfaces at their margins move also, generally eastward with the drift of the westerlies. A front with a cold mass to the west and a warm mass to the east is a *cold front*; a front separating warm air on the west from cold air on the east is a *warm front* (Fig. 32.12). The movement of a cold front brings cold air in place of warm, and the movement of a warm front brings warm air in place of cold; the front is named for the air mass that is bringing about the change.

As warm air rises along an inclined frontal surface it is cooled and part of its moisture condenses out. Clouds and rain, therefore, are commonly associated with both kinds of fronts. A cold frontal surface is generally steeper, since cold air is actively burrowing under warm air, and the temperature difference is greater, so rainfall on a cold front is heavier and of shorter duration than on a warm front. A cold front with a large temperature difference is often marked by violent thundersqualls and tornadoes.

atmospheric circulation

Let us look into how the pattern of the general circulation of the atmosphere comes about. We begin by examining the origin of an important kind of atmospheric movement. Suppose that a small area of the land

THE ATMOSPHERE

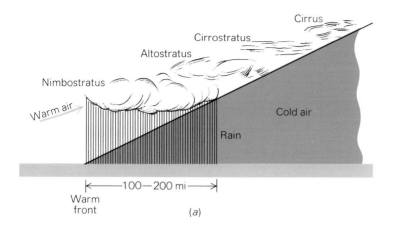

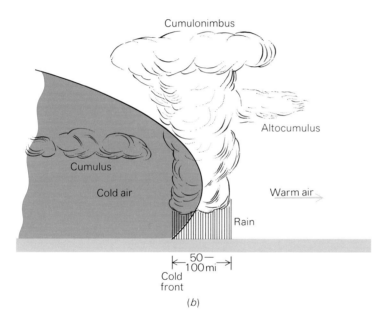

32.12 *Cross-section diagrams of a warm front (a) and a cold front (b).*

surface is heated more strongly than adjacent regions (Fig. 32.13). Air above the heated surface becomes warm, expands, and begins to rise. As it rises, a column of air above it is pushed upward, and air from the upper part of the column spills over into adjacent regions. Because the hot air is light and because air above it moves to each side, pressure over the heated spot is relatively low. Hence cold air from the surroundings moves toward the heated area, is in turn heated, and moves upward. Air currents produced in this manner, as a direct response to unequal heating of the land surface, are called *convection currents*.

The earth as a whole is heated strongly in the equatorial belt and less strongly on either side, so we might expect to find convection currents as part of the general atmospheric circulation. Suppose for the moment

convection currents are produced by uneven heating of the earth's surface

A squall line marks the arrival of a cold front.

32.13 Convection currents produced by unequal heating.

32.14 The convectional circulation that would occur if the earth did not rotate and were heated uniformly at the equator. *The arrows in the center of the diagram indicate surface winds.*

that our planet did not rotate, that it was heated strongly near the equator, and that its surface was made up entirely of either land or water. On such an earth air circulation would depend exclusively on the difference in temperature between equator and poles. Air would rise along the heated equator, overflow at high altitudes toward the poles, and at low altitudes move continually from the poles back toward the equator (Fig. 32.14). We in the Northern Hemisphere would experience a steady north wind. Around the equator would be a belt of relatively low pressure, near each pole a region of high pressure.

A huge convectional circulation of this sort exists, but it is profoundly altered by the earth's rotation: Everywhere except along the equator, winds are deflected from straight lines into curved paths, the deflection being toward the right in the Northern Hemisphere and toward the left in the Southern Hemisphere. Air moving southward from the high-pressure zone in the Northern Hemisphere instead of moving straight south is deflected toward the southwest; air moving northward from this zone is turned toward the northeast and east. This is known as the *Coriolis effect.*

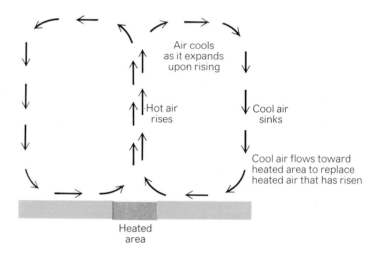

According to current ideas on the subject, the convectional north-south motion of air is broken up by the earth's rotation (acting through the Coriolis force) into the large-scale eddies that constitute the cyclones and anticyclones mentioned earlier (Fig. 32.15). These eddies add their motions to produce a generally eastward drift of air in the middle latitudes and a westward drift in the tropics. The resulting pattern of airflow is further affected by seasonal low- and high-pressure cells caused by unequal heating due to the irregular distribution of land and sea areas. This general concept is strongly supported by calculations based on general principles of fluid flow and from experiments on movements in liquids produced by combined rotation and unequal heating, in addition to being in accord with what measurements can be made of atmospheric circulation directly.

One part of the circulation that is well explained by this hypothesis consists of the *jet streams.* These are narrow currents of air near the

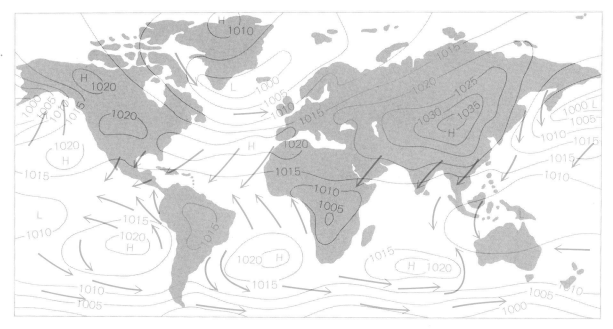

32.15 *Average January sea-level pressures (in millibars) and winds.* High- and low-pressure systems are indicated.

base of the stratosphere in the middle latitudes which have been found to be moving from west to east at speeds on the order of 100 mi/hr faster than the air around them. The jet streams are fairly continuous but shift their positions northward and southward. The shifts can often be correlated with weather patterns, particularly with the paths along which cyclones and anticyclones move.

CLIMATE

Now that we are acquainted with the principal features of the earth's weather, we can go on to consider the climates of various regions.

Precipitation—rain and snow—is one of the most important factors in climate. As mentioned in an earlier section, the production of rain or snow requires that large masses of humid air be cooled to a temperature below the point at which the air is saturated. The necessary water vapor, obtained mostly from evaporation of sea water, may be carried long distances inland by the atmospheric circulation. Sufficient cooling to produce rain is accomplished only by the expansion of air on rising. Three causes for upward movement of large masses of air may be distinguished:

1 Convection currents, either as part of the general circulation or as a result of local heating
2 Wind forced upward by mountain barriers
3 Movements of air masses and associated cyclonic storms

conditions for precipitation to occur

We expect abundant rain or snow, then, wherever air well supplied with moisture moves over a strongly heated area, where mountain ranges lie across the path of prevailing winds, or where cyclonic storms are frequent.

These facts about precipitation, together with those we learned earlier about atmospheric temperatures and the earth's wind system, make possible a plausible guess regarding the climate of almost any point on the globe.

tropical climates

The equatorial belt of calms, with its rapid evaporation and strong rising air currents, provides an ideal situation for superabundant rain. Throughout the year the weather is hot, sultry, with almost daily rains and light, changeable winds. The steaming rain forests of Africa, South America, and the East Indies occur where this belt crosses land.

The horse latitudes, roughly 30° north and south of the equator, are also belts of calm, but their climate is anything but humid. Air in these belts moves chiefly earthward, becoming less compressed and heated and hence less and less saturated with water vapor. Thus the climate is perennially dry, with clouds and rain only at long intervals. In a few regions, however, like the Gulf Coast of the United States, the prevailing aridity is modified by moisture-laden winds of local origin.

Air returning to the warm equatorial belt from the horse latitudes on either side has little reason to lose the small amount of water vapor it possesses, except locally where a mountain range or strong convection currents force it sharply upward. Hence the trade-wind belts, like the horse latitudes, are in general regions of considerable dryness. Seasonal movement of the wind belts to the north and south gives rainfall during part of the year to the equatorial margins of the trade-wind belts. The outer portions of these belts, together with the adjacent horse latitudes, are the regions of the world's great deserts—the Sahara, the deserts of South Africa, the arid districts of Mexico and northern Chile, the dry interior of Australia.

middle-latitude climates

The belts of prevailing westerlies in general have moderate average temperatures. Continental interiors show great seasonal variations in temperature, while oceanic islands and the west coasts of continents have equable temperatures throughout the year. Further generalization about these zones is difficult, because their weather and climate are so largely determined by storms of local origin. Winds vary greatly in strength and direction, and conditions of moisture and temperature vary with them. In the Northern Hemisphere the huge land masses of North America and Eurasia introduce further complications.

climatic conditions in the united states

The complexities of the northern belt of westerlies are well illustrated by a brief survey of climates in the United States. Winds from the Pacific Ocean are forced abruptly upward by a succession of mountain ranges along the West Coast, the western sides of the mountains therefore receiving abundant rainfall. Once across the mountain barriers the westerlies have little remaining moisture, so that the region east to the Great Plains is largely arid. If the westerlies maintained their direction as steadily as do the trade winds, arid conditions would continue across the continent to the East Coast; but wind directions are continually changed by the cyclonic storms characteristic of this belt, so that moisture-laden air is frequently brought from the Gulf of Mexico and the Atlantic Ocean into the Mississippi Valley and the Eastern states. Rainfall increases eastward across the country becoming very large along the Gulf of Mexico. Temperatures on the West Coast, conditioned by the prevailing wind from the ocean, change relatively little from season to season, but in most other parts of the country the difference between summer and winter is very marked. The fine climates of Florida and southern California owe their mildness to nearby warm oceans and to a position approximately at the junction of the belt of westerlies and the horse latitudes.

In the bleak arctic and antarctic regions, summers are short, winters long and cold. Moderate winds are the rule, although violent gales occur at times. The total amount of snow during the year is small simply because the low temperatures prevent the accumulation of much water vapor in the air.

climatic change

Weather we expect to vary, both from day to day and from season to season. Nor are we surprised when one year has a colder winter or a drier summer than the one before. Less familiar are changes in climate. Even though climate represents averages in weather conditions over periods of, say, 20 or 30 years, there is abundant evidence that it, too, is not constant but instead undergoes quite marked fluctuations over long spans of time. The most dramatic such fluctuations were the *ice ages* of the distant past, which we shall examine in Chap. 39.

The last ice age reached its peak about 20,000 years ago when huge ice sheets as much as 4 km thick covered much of Europe and North America. Then the ice began to retreat and climates became progressively less severe; in a period of 12,000 years the average annual temperature of central Europe rose from $-4°C$ to $+9°C$. By about 4000 B.C. average temperatures were a few degrees higher than those of today. A time of declining temperatures then set in, reaching a minimum in Europe between 900 and 500 B.C. A gradual warming-up followed that came to a peak between A.D. 800 and 1200; so generally fine were climatic conditions then that the Vikings established flourishing colonies in Iceland and Greenland from which they went on to visit North America. The subsequent deterioration led to cool summers, exceptionally cold winters, and extensive freezing of the Arctic Sea from 1300 to 1700. So extreme was the weather in the first half of the seventeenth century that it has been called the "Little Ice Age." Greenland became a much less attractive place than formerly and the colony there disappeared, the coast of Iceland was surrounded by ice for several months per year (in contrast to a few weeks per year today), and glaciers advanced farther across alpine landscapes than ever before or since in recorded history.

A cyclonic storm system 1200 miles north of Hawaii, photographed during the Apollo 9 space flight.

world temperatures have been generally increasing for the past 150 years

By 1800 or so a trend toward higher temperatures became evident which has led to a marked shrinkage of the world's glaciers. In the first half of this century especially pronounced temperature increases took place whose most noticeable consequences were milder winters in the higher latitudes. In Spitzbergen, for instance, January temperatures averaged from 1920 to 1940 were nearly 8°C (14°F) higher than those averaged from 1900 to 1920, and Greenland became less inhospitable than before.

As mentioned earlier in this chapter, the recent sharp increase in worldwide temperatures parallels the increase in the carbon dioxide content of the atmosphere that is probably due to the burning of coal and oil. Since carbon dioxide absorbs long-wavelength electromagnetic energy radiated by the earth, there is a plausible link between the two phenomena. But it is not possible to be sure about such a cause-effect relationship, partly because a great many other factors also influence climate—for example, wind patterns, solar emissions of fast ions (Chap. 42), the moisture and ozone contents of the atmosphere, the presence of dust from volcanic eruptions, and ocean currents that involve motions of cold and warm water around the globe. And further doubt concerning the effect of the increasing carbon dioxide on climate follows from the observation that, in the past 20 years, temperatures have been on the *decrease* in the far north. Clearly more time is needed to establish whether the decrease represents a temporary fluctuation in a continuing long-term upward trend or the start of a new long-term downward trend. The only certainty on the question of whether man is inadvertently producing a cold future for himself is uncertainty—an excellent argument for trying to moderate the amount of carbon dioxide so casually discarded into the atmosphere.

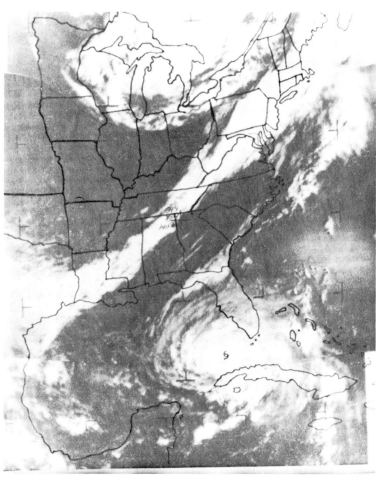

This photograph of Hurricane Inez was taken on October 5, 1966 by a U.S. weather satellite. The northward progress of the storm is being blocked by a high-pressure system (whose boundary appears as a diagonal line of clouds), and it went on to move west-southwest toward Mexico. The land outlines were drawn after the photo was radioed back to earth.

GLOSSARY

The earth's *atmosphere* is its gaseous envelope.

The five regions of the atmosphere are, from the earth's surface upward, the *troposphere*, the *stratosphere*, the *mesosphere*, the *thermosphere*, and the *exosphere*. The *ionosphere* is part of the thermosphere and is characterized by the presence of a high concentration of ions.

Ozone is a form of oxygen whose molecules consist of three oxygen atoms each. It is an efficient absorber of ultraviolet radiation.

Meteorology is the science of atmospheric phenomena such as temperature, pressure, humidity, winds, and precipitation.

Weather refers to the state of the atmosphere in a particular place at a particular time, while *climate* refers to the weather trends in the same region through the years.

Relative humidity is the ratio between the amount of moisture in a volume of air and the maximum amount of moisture that volume of air can hold when completely saturated. It is usually expressed as a percentage.

Cyclones are weather systems centered about regions of low pressure; in the Northern Hemisphere cyclones are characterized by counterclockwise winds, and in the Southern Hemisphere by clockwise winds.

Anticyclones are weather systems centered about regions of high pressure; the characteristic winds of anticyclones are opposite in direction to those of cyclones.

Convection currents result from the uneven heating of a fluid; the warmer parts of the fluid expand and rise because of their buoyancy, while the cooler parts sink.

The *Coriolis effect* is the deflection of winds to the right in the Northern Hemisphere, and to the left in the Southern; this effect is a consequence of the earth's rotation.

The *doldrums* at the equator and the *horse latitudes* at about latitudes 30°N and 30°S are regions of protracted calms separated by the steady *trade winds*. In the Northern Hemisphere the trade winds blow from northeast to southwest, and in the Southern Hemisphere from southeast to northwest. In the middle latitudes of either hemisphere are the *prevailing westerlies*.

Jet streams are swift currents of air near the base of the stratosphere which move from west to east.

An *air mass* is a huge tongue of air of a particular temperature and humidity (or an isolated body of air detached from one) projecting into a region whose air has different temperature and humidity.

A *frontal surface* is the surface separating a warm and a cold air mass; a *front* is where this surface touches the ground. A *cold front* generally involves a cold air mass moving approximately eastward; a *warm front* generally involves a warm air mass moving approximately eastward.

EXERCISES

1. Account for the abrupt changes in temperature between day and night in desert regions.

2. Why does the skin feel cooler in a breeze than in quiet air at the same temperature?

3. The California Current along the California coast is cooler than the ocean to the west. How does this fact explain the numerous fogs on this coast?

4. The island of Oahu (one of the Hawaiian Islands) is at latitude 21°N and is crossed by a mountain range trending roughly northwest-southeast. Account for the more abundant rainfall on the northeastern side of the range.

5. What are the characteristics of the atmosphere that have led to its division into a series of layers?

6. Suppose that you are traveling upward in an airplane not equipped with an altimeter. How can you tell when you are approaching the top of the troposphere?

7. Discuss the relationship between relative humidity and saturation.

8. Why are frontal surfaces inclined at an angle to the ground?

9. Why is the E layer of the ionosphere unimportant in long-distance radio communication?

10. If you were a yachtsman planning to sail from the United States to Europe, what route would you take? If you were planning to sail back to the United States, what route would you take?

11. What would be the consequences of the disappearance of ozone from the upper atmosphere?

12. Considering the speed of radio waves in the atmosphere to be the same as the speed of light in vacuum, how long would a pulse of radio waves emitted from a ground station take to reach the F2 layer and be reflected back to the ground?

chapter 33

The Oceans

PROPERTIES OF THE OCEANS

ocean basins

composition

life in the ocean

MOVEMENTS OF THE OCEANS

waves

ocean currents

the tides

Almost 71 percent of the earth's surface is covered by the oceans and the shallower seas that join with them, and the oceans accordingly are a major factor in shaping the environment of life on this planet. The oceans themselves are home to abundant plant and animal life—indeed, life probably began in the primordial ocean—and in addition influence continental life in a variety of indirect ways. For one thing, the oceans provide the reservoir from which water is evaporated into the atmosphere, later to fall as rain and snow on the land. The oceans participate in the oxygen–carbon dioxide cycle both through the life they support and through the vast quantities of these gases dissolved in them. And the oceans help determine climates by their ability to absorb solar energy and transport it around the world. The list of chemical and physical cycles on the earth in which the oceans play significant roles is a long one.

PROPERTIES OF THE OCEANS

ocean basins

Each of the world's oceans lies in a vast basin bounded by continental land masses. Typically an ocean bottom slopes gradually downward from the shore to a depth of 130 m or so before starting to drop more rapidly (Fig. 33.1). The average width of this *continental shelf* is 65 km, but it ranges from less than a kilometer off such mountainous coasts as the western coast of South America to over 1,000 km off the low arctic coasts of the Eurasian land mass. The North, Irish, and Baltic Seas are

<small>continental shelf</small>

part of the European continental shelf, while the Grand Banks off Newfoundland are part of the North American shelf. A sharp change in gradient marks the transition from the continental shelf to the steeper *continental slope*, which after a fall of perhaps 2 km joins the *abyssal plain* of the ocean floor via the gentle *continental rise*.

the steep continental slope marks the true margin of a continent

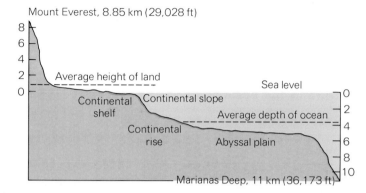

33.1 *Profile of the earth's surface. The vertical scale is greatly exaggerated. Heights and depths are in km.*

The ocean basins average 3.7 km in depth, while the continents average only about 0.8 km in height above sea level. The deepest known point of the oceans, 11 km below the surface, is found in the Marianas Trench southwest of Guam in the Pacific; by contrast, Mt. Everest is only 80 percent as high above sea level. If the earth were smooth, it would be covered with a layer of water perhaps 2.4 km thick, but it seems likely that the oceans have always been confined to more or less distinct basins and presumably will continue to be.

the antarctic icecap contains most of the world's permanent ice

A considerable amount of water is stored as ice in the form of the glaciers and ice caps which cover one-tenth the land area of the earth. About 90 percent of this ice is located in the Antarctic ice cap, about 9 percent in the Greenland ice cap, and the remaining 1 percent in the various glaciers of the world. If suddenly melted, the ice would raise sea level by perhaps 75 m. (By comparison, if all the water vapor in the atmosphere were condensed, sea level would go up by only about 3 cm.) Over a long period of time the rise in sea level would be reduced by about one-third by changes in the levels of the continents and the ocean floors brought about by the changes in the weights they have to bear (Chap. 38). During the most recent ice age sea level was 130 m lower than at present owing to the water locked up in the huge continental ice sheets.

topography of the ocean floor

The topography of the ocean floor, like that of the continents, is marked by mountain ranges and valleys, isolated volcanic peaks and vast plains, many of them rivaling or exceeding in size their terrestrial counterparts. The Hawaiian Islands, for instance, are volcanoes that rise as much as 30,000 ft above the ocean floor, about half of their altitude being above sea level. Less conspicuous from the surface is the Mid-Atlantic Ridge, an immense submarine mountain range that extends from Iceland past the tip of South America before swinging into the Indian Ocean. Such islands as the Azores, Ascension Island, and Tristan da Cunha are all that protrude from the ocean of this ridge.

composition

Sea water has a salt content (or *salinity*) that averages 3.5 percent. The composition of sea water is shown in Fig. 33.2. Evidently the ions of ordinary table salt (NA^+ and Cl^-) account for over 85 percent of the total salinity. The salinity of sea water varies around the world, but the proportions of the various ions are virtually the same everywhere because of mixing by currents both on and below the sea surface. Figure 33.3 illustrates the origins of the ions found in sea water.

the salinity of sea water averages 3.5 percent; most of the ions present are Na^+ and Cl^-

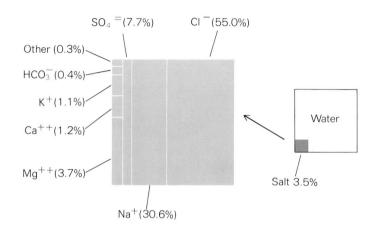

33.2 The composition of sea water. *In the open ocean the total salt content varies about an average of 3.5% but the relative proportions of the various ions are quite constant.*

Sea water contains dissolved gases as well as salt ions. Because of the constant exchange of gases at the sea-air boundary, the uppermost layer of the oceans is *saturated* with atmospheric gases: there is an equilibrium in which as much gas of any kind leaves the water as enters it. Just below the surface layer, though, the photosynthetic activities of plant life lead to a disproportionately greater oxygen concentration. Deep in the ocean,

atmospheric gases are present in solution in sea water

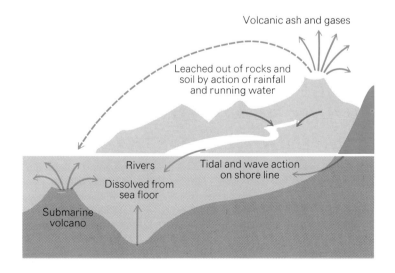

33.3 Origins of seawater salts.

The underwater exploration vessel Deepstar is self-propelled and can operate at depths of 4,000 feet. It carries a crew of three.

there is relatively little oxygen because animal life there consumes oxygen and there is not enough sunlight for plants to produce more in photosynthesis.

Only a fraction of the carbon dioxide in the ocean remains in the form of CO_2 molecules. Most of it reacts with water to form hydrogen ions and bicarbonate (HCO_3^-) ions:

$$CO_2 + H_2O \rightleftharpoons H^+ + HCO_3^-$$

Some bicarbonate ions further dissociate into hydrogen ions and carbonate ($CO_3^=$) ions:

$$HCO_3^- \rightleftharpoons H^+ + CO_3^=$$

Because water can hold much more carbon dioxide in the form of HCO_3^- and $CO_3^=$ ions than in the form of CO_2 molecules, the total carbon dioxide content of the oceans is far greater than its rather low (0.03 percent by volume) concentration in the atmosphere would suggest. It is estimated that there are 1.3×10^{14} tons of carbon dioxide in the oceans of the world, and the continual exchange of CO_2 between the atmosphere and the oceanic reservoir probably involves more CO_2 than the exchange between the atmosphere and the biological reservoir of plant and animal life.

life in the ocean

As on land, all life in the ocean ultimately depends upon photosynthesis. This vital process is carried out in the ocean largely by minute plants collectively called *phytoplankton*. Most phytoplankton are algae. Floating in water that contains the right ions and dissolved gases and irradiated with sunlight, phytoplankton flourish, and indeed they constitute a larger food reservoir than land vegetation.

The phytoplankton in the upper hundred meters of the oceans, where sunlight can penetrate, form the first link in the chain of marine life. Upon them feed the *zooplankton*, small animals such as protozoa and tiny shrimp. In turn the zooplankton (and, to some extent, phytoplankton also) form the diets of most fish and aquatic mammals such as whales and porpoises.

On land the tropics are regions of abundant plant and animal life, yet tropical waters are barren of marine life as compared with the colder waters nearer the poles. One reason for this apparent paradox lies in the floating of the warmer and hence less dense surface water on the colder water beneath it. A typical temperature profile of the warm regions of the ocean is shown in Fig. 33.4. The *thermocline* between the surface layer at more than 20°C and deeper water at less than 5°C is in effect a boundary that impedes the vertical mixing of ocean water. The algae in the surface layer consume nutrient elements there, but upon death

in much of the ocean a layer of warm water floats on cold water beneath it; the boundary between them is called the thermocline

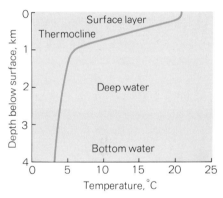

33.4 Typical variation of water temperature with ocean depth.

Sea water samples from various depths are taken during oceanographic surveys in order to determine their composition and biological content.

their remains sink into the deeper water where they are unavailable for reuse in the cycle of life. The result is a depletion in the nutrient elements of the surface layer, a decrease in the phytoplankton there, and a relative scarcity of animal life.

In the seas of the far north and far south, however, there is no layering and vertical circulation is relatively rapid because *all* the water is cold. Algae are especially abundant in these seas, which not surprisingly turn out to be the location of the chief fishing grounds of the world.

Marine life is one of the factors that determines the color of sea water. The deep blue of much of the ocean has the same origin as the blue of the sky. When we look at the sea or the sky, what enters our eyes is scattered sunlight, and it happens that the shorter the wavelength of the light, the more efficient the scattering. Since blue light has the shortest wavelengths in the visible spectrum, it is preferentially scattered. Skylight as well as direct sunlight illuminates the ocean, which enhances its blue appearance, as does the fact that water absorbs blue light to a smaller extent than light of longer wavelengths.

the sea is blue chiefly because water scatters blue light more efficiently than light of other colors

The green color of parts of the ocean can usually be traced to the presence of plant and animal life there. The suspended particles of matter common where aquatic life occurs scatter green light more readily than pure sea water does, and in addition various pigments associated with phytoplankton are greenish in color. Some marine organisms lead to brown and even red colors in sea water, as in the "red tides" that occasionally appear in coastal areas.

MOVEMENTS OF THE OCEANS

The oceans are not static bodies of water. Winds ruffle their surfaces into waves and drive great currents that profoundly influence climates around the world. And the twice-daily rise and fall of the tides is familiar to every coastal dweller.

waves

Waves are produced by winds blowing over the surface of a body of water. When a wind starts to blow over a smooth water surface, ripples (the "cat's paws" of the sailor) begin to form whose crests are rounded and glassy in appearance. An increase in wind speed brings true waves with sharper crests; the stronger the wind, the longer it blows, and the greater the distance over which the wind has been in contact with the water, the higher the waves. These three factors govern the amount of energy transferred to the water, and thus they govern the violence of the disturbance that results.

The relationship between winds at sea and the waves they cause is given in Table 33.1. In this table it is assumed that the wind has been blowing for enough time and over enough distance for the waves to attain their maximum height. For example, a 30-knot offshore wind produces waves averaging about $13\frac{1}{2}$ ft high after blowing for 10 hr or more over

TABLE 33.1 The relationship between wind speed and the state of the sea surface in open water. The *Beaufort scale* is often used by mariners to describe wind speed. (1 knot = 1 nautical mi/hr = 1.15 mi/hr = 0.515 m/sec)

Beaufort number	Wind description	Wind speed, knots	Sea condition	Wave height, ft
0	Calm	Less than 1	Sea like a mirror	—
1	Light air	1–3	Ripples with scaly appearance, no foam crests	$\frac{1}{4}$
2	Light breeze	4–6	Small wavelets whose crests have a glassy appearance and do not break	$\frac{1}{2}$
3	Gentle breeze	7–10	Large wavelets with crests beginning to break, perhaps scattered whitecaps	2
4	Moderate breeze	11–16	Small waves becoming longer, frequent whitecaps	$3\frac{1}{2}$
5	Fresh breeze	17–21	Moderate waves of greater length, many whitecaps and some spray	6
6	Strong breeze	22–27	Large waves begin to form, whitecaps everywhere, more spray	$9\frac{1}{2}$
7	Moderate gale	28–33	Sea heaps up and streaks of foam are blown from breaking waves	$13\frac{1}{2}$
8	Fresh gale	34–40	Moderately high waves of greater length, edges of crests break into spindrift, well-marked streaks of foam	18
9	Strong gale	41–47	High waves whose crests begin to topple over, dense streaks of foam, sea begins to roll	23
10	Whole gale	48–55	Very high waves with long overhanging crests, sea is white with foam and rolls heavily, visibility reduced	29
11	Storm	56–63	Exceptionally high waves, sea completely covered with foam, small- and medium-sized ships lost to view behind waves, visibility further reduced	37
12	Hurricane	More than 64	Sea completely white with driven spray, air filled with foam, visibility very poor	45

100 mi or more of open sea, but after 3 hr the average wave height is only 8 ft at this distance and 10 mi offshore it does not exceed 4 ft no matter how long a period of time is involved.

As we saw in Fig. 9.1, the particles of water at the surface that participate in the passage of a wave move in circular orbits whose diame-

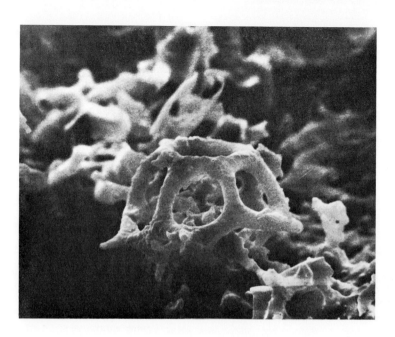

Scanning electron micrograph of diatoms, a type of phytoplankton.

deep water is unaffected by surface waves

ters are equal to the wave height. At the crest of the wave, the surface water moves in the same direction as the wave; in the trough, the surface water moves in the opposite direction. This orbital motion also occurs below the water surface, but the diameters of the orbits decrease rapidly (Fig. 33.5). At a depth equal to $\lambda/4$, where λ is the length of the waves, the orbital diameter is 21 percent of the wave height, at a depth of $\lambda/2$ the orbital diameter is 4 percent of the wave height, and at a depth of λ the orbital diameter is 0.2 percent of the wave height. Below $\lambda/2$, then, there is hardly any disturbance of the water due to waves on the surface. This is the reason why submarines always submerge in storms.

In water less than about $\lambda/2$ deep, the orbital motions that accompany the passage of a wave are affected by the presence of the sea bottom. The waves slow down, which causes their lengths to shorten and their heights to increase. The crests become steep and narrow while the troughs flatten out, so that in shallow water the waves appear as isolated crests relatively far from each other. As the water depth decreases near a beach, the waves get steeper and steeper, until finally there is not enough water in front of the advancing crests to support their continuation toward the shore. When the water depth is about 1.3 times the wave height, the oncoming waves collapse with the forward orbital motion of the water in the crests sending it spilling down the wave fronts. Breaking waves are a common phenomenon on gently shoaling coasts.

the oceans act as heat reservoirs that moderate the climates of adjoining land areas

ocean currents

The oceans affect climate in two ways. First, they act as reservoirs of heat which moderate the temperature extremes of the seasons. In spring

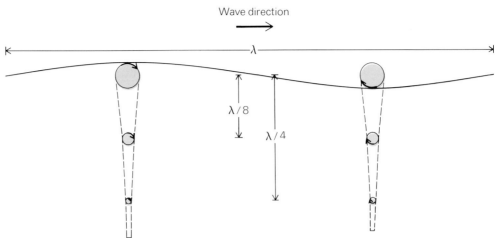

33.5 The orbits of water particles decrease rapidly in diameter with increasing depth as a wave passes by in deep water. Below a depth of about half a wavelength orbital motion is negligible.

and summer the oceans are cooler than the regions bordering them, since the heat they absorb is dissipated in a greater volume than in the case of solid, opaque land. The heat retained in the ocean depths means that in fall and winter the oceans are warmer than the regions bordering them. Heat flows readily between moving air and water; with a sufficient temperature difference, the rate of energy transfer from warm water to cold air (or from warm air to cold water) can exceed the rate at which solar energy arrives at the top of the atmosphere. With no such heat reservoir nearby, continental interiors experience lower winter temperatures and higher summer temperatures than those of coastal districts. In Canada, for instance, temperatures in the city of Victoria on the Pacific coast range from an average January minimum of $36°F$ to an average July maximum of $68°F$, whereas in Winnipeg, in the interior, the corresponding figures are $-8°F$ and $80°F$.

The stronger the wind, the longer it blows, and the greater the distance over which the wind has been in contact with the water, the higher the waves that are produced.

surface ocean currents are caused by winds

Also important in influencing climate are surface drifts in the oceans produced by the friction of wind on water. Such drifts are much slower than movements in the atmosphere, with the fastest normal surface currents having speeds of about 7 mi/hr.

The wind-impelled surface currents parallel to a large extent the major wind systems. The northeast and southeast trade winds drive water before them westward along the equator, forming the *equatorial current*. In the Atlantic Ocean this current runs head on into South America, in the Pacific into the East Indies. At each of these points the current divides into two parts, one flowing south and the other north. Moving away from the equator along the continental margins, these currents at length come under the influence of the westerlies, which drive them eastward across the oceans. Thus gigantic whirlpools are set up in both Atlantic and Pacific Oceans on either side of the equator (Fig. 33.6). Many minor complexities are produced in the four great whirls by islands, continental projections, and undersea mountains and valleys.

the gulf stream

The western side of the North Atlantic whirl, a warm current moving partly into the Gulf of Mexico, partly straight north along our southeastern coast, is the familiar Gulf Stream. Forced away from the coast in the latitude of New Jersey by the westerlies, this current moves northeastward across the Atlantic, splitting on the European side into one part which moves south to complete the whirl, another part which continues northeastward past Great Britain and Norway into the Arctic Ocean. To compensate for the addition of water into the polar sea, a cold current moves southward along the east coast of North America as far as New York; this is the Labrador Current. Down the west coast of North America moves the Japan Current, the southward-flowing eastern part of the North Pacific whirl.

Since ocean currents retain for a long time the temperatures of the latitudes from which they come, they exert a direct influence on the temperatures of neighboring lands. The influence is greatest, of course, where the prevailing winds blow shoreward from the sea. Thus the warm

33.6 *Principal ocean currents of the world.* Warm currents are shown by the colored arrows, cool currents by the black ones.

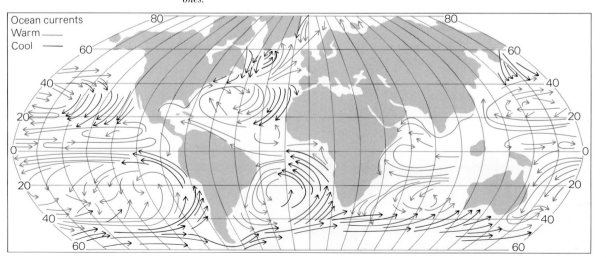

Gulf Stream has a much greater effect in tempering the climate of northwest Europe than that of eastern United States, since prevailing winds in these latitudes are from the west. Cyclonic storms bring east winds to the Atlantic Seaboard frequently enough, however, so that the Gulf Stream helps to raise temperatures in the South Atlantic states, while the Labrador Current is in part responsible for the rigorous climate of New England and eastern Canada.

Currents also occur deep in the ocean, though their speeds are usually slower than those of surface currents. In the polar regions of both hemispheres cold water sinks because of its greater density and flows toward the equator several miles below the surface. These cold currents keep tropical waters cooler than they otherwise would be, and they also bring oxygen to the lower depths of the ocean which enables plant and animal life to occur there.

Thus the oceans, besides acting as water reservoirs for the earth's atmosphere, play a direct part in temperature control—both by preventing abrupt temperature changes in lands along their borders and by aiding the winds, through the motion of ocean currents, in their distribution of heat and cold over the surface of the earth.

33.7 *Explanation of the tides.* The moon's attraction is greatest at A, least at B.

the tides

That there is a relationship of some kind between the moon and the rhythm of the tides was known even in ancient times. Dwellers beside the ocean have been aware for thousands of years of such aspects of tidal behavior as that at new moon and full moon the difference between high water and low water is larger than it is when only half the moon's disk is visible. Kepler, who discovered the laws of planetary motion, also tried to establish the nature of the connection between the moon and the tides. This was scoffed at by Galileo, who ridiculed Kepler for having "given his ear and assent to the moon's predominancy over the water, and to occult properties and suchlike trifles."

Not until Newton was the cause of the tides finally ascribed to the gravitational attraction of the moon and, to a smaller extent, of the sun as well. At first glance the mechanism seems obvious: the moon pulls the waters of the earth toward it, and as the earth rotates the positions of high and low water stay the same with respect to the moon but shift with respect to the earth. This picture is not correct. For one thing, almost everywhere there are two, not one, tidal cycles per day of about $12\frac{1}{2}$ hours each. For another, the sun's gravitational force on the earth greatly exceeds that of the moon—but the tides follow the moon. A more subtle analysis is required.

The key to the origin of the tides is that the moon attracts different parts of the earth with slightly different forces. For matter at A in Fig. 33.7 the attraction is strongest, since the distance from the moon is least; at B, the point farthest from the moon, the attraction is weakest. Because of these unequal forces, the moon pulls matter at A away from the rest of the earth, which explains the tidal bulge there. The moon also pulls the entire earth away from matter at B, which tends to be left behind

origin of the tides

In the Bay of Fundy between Nova Scotia and New Brunswick, the tidal range is fifty to sixty feet, the greatest anywhere in the world. A large volume of water flows in the wide mouth of the bay, causing this huge change in level at the bay's narrow head.

in consequence. Solid rock resists the bulging effect to a large extent (though there *are* detectable tides in the earth's crust), but the fluid ocean responds easily. Water is heaped up on the sides of the earth facing and directly opposite the moon, and is drawn away from other parts of the earth. As the earth rotates on its axis, the water bulges are held in position by the moon. The earth, so to speak, moves under the bulges, and a given point on its surface therefore experiences two high tides and two low tides per day.

A cubic meter of water has a mass of 10^3 kg, and the earth attracts it with a force of $mg = 9.8 \times 10^3$ newtons. At the center of the earth, the gravitational force of the moon on 1 m^3 of water would be 0.033 newton. At A in Fig. 33.7, the point on the earth's surface directly facing the moon, the lunar force is 0.034 newton, and at B, the point farthest from the moon, the lunar force is 0.032 newton. Hence the differential tide-producing force on 1 m^3 of water at both A and B is only 0.001 newton—only 10^{-7} of the force the earth exerts on such a parcel of water. Clearly so trifling a force is not capable of bodily lifting the oceans by several feet. However, although the moon cannot raise tides directly, it is able to achieve this result by sliding the waters of the seas and oceans horizontally. Anyone can manage to budge a boat weighing several tons in still water, even though lifting it is out of the question. It is because water flows so readily that tides occur. Figure 33.8 shows the components parallel to the earth's surface of the differential forces on a given mass of matter at various locations; these components are called *tractive forces*, and they are what cause the tides.

tractive force

Although the direct gravitational force exerted by a body r away varies as $1/r^2$, the tractive force varies as $1/r^3$. For this reason the sun, despite its vastly greater mass than that of the moon, exerts only 46 percent as much tractive force on the earth as the much closer moon does. Thus the solar tides appear as modifications of the dominant lunar tides. Roughly twice a month, when sun, moon, and earth are in a straight line, solar tides are added to lunar tides to form the unusually high (and low) *spring* tides; when the line between moon and earth is perpendicular

spring and neap tides

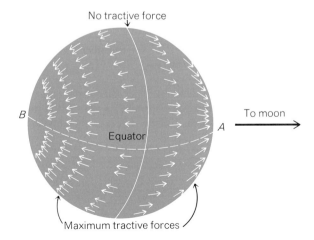

33.8 Tractive forces around the earth. *These forces are responsible for the tides.*

to that between sun and earth, the tide-raising forces oppose each other and tides are the *neap* tides whose range is small (Fig. 33.9).

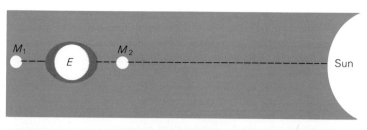

33.9 Variation of the tides. Spring tides are produced when the moon is at M_1 or M_2, neap tides when the moon is at M_3 or M_4.

GLOSSARY

An ocean bottom at first descends gradually from the shore in a region called the *continental shelf,* then steeply in the *continental slope,* and again gradually in a *continental rise* until the *abyssal plain* is reached.

The *salinity* of sea water refers to its salt content, which averages 3.5 percent.

Tiny, primitive plants floating in the ocean are called *phytoplankton;* their animal counterparts are called *zooplankton.*

The *thermocline* is the boundary between the cold bottom water of the ocean and its warm upper layer.

Breaking waves occur in shallow water when incoming waves become too steep and their crests collapse.

The *tides* are twice-daily rises and falls of the ocean surface. They are produced by *tractive forces* parallel to the earth's surface that in turn are due to the different attractive forces exerted by the moon on different parts of the earth.

Spring tides have a large range between high and low water; they occur when the sun and moon are in line with the earth and thus add together their tide-producing actions. *Neap tides* have small ranges and occur when the sun and moon are 90° apart relative to the earth and thus their tide-producing actions tend to cancel each other out.

THE OCEANS

EXERCISES

1. With the help of Chap. 7 calculate the pressure at the deepest known point in the oceans, which is 11,000 m below the surface.

2. Would sea level change if the ice floating in the Arctic and Antarctic oceans were to melt?

3. The salinity of the Red Sea is about 4 percent, whereas that of the Baltic Sea is less than 1 percent. Can you account for the difference?

4. The Gulf Stream can be identified by its brilliant blue color, which contrasts with the green color of the colder waters on either side of it. What does this observation indicate about the abundance of plankton in the Gulf Stream?

5. The earth takes almost exactly 24 hr to make a complete rotation about its axis, so one might expect successive high tides to occur 12 hr apart. In reality, the interval between high tides is 12 hr 25 min. Can you account for the difference?

part nine
BASIC GEOLOGY

Now that we know a little about the rocks and minerals, atmosphere and oceans that clothe our planet, we can turn to the physical and chemical processes whose action has produced the wrinkled face of the earth we are familiar with. The wearing away of rocks by water, wind, and ice and the subsequent deposition of the debris elsewhere are influential factors in shaping landscapes everywhere. The endless cycles of erosion and sedimentation are not the only large-scale influences that affect the contours of the earth; also of major significance are the processes of vulcanism, which involve movements of liquid rock, and of diastrophism, which involve movements of solid crustal rock. Finally we descend to the earth's interior, at whose center is a core of molten iron and nickel. Between the core and the thin outer crust is an intermediate layer called the mantle, which is probably composed of iron and magnesium silicates. The earth's magnetic field almost certainly originates in the core, where immense electric currents flow.

chapter 34

Erosion and Sedimentation

EROSION

- weathering
- stream erosion
- evolution of landscape
- glaciers
- wind
- waves and currents

SEDIMENTATION

- processes of sedimentation
- sedimentary rocks

GROUNDWATER

- movement of groundwater
- deposition by groundwater

It takes some effort even today to get used to the idea that this solid earth around us, composed largely of strong, hard rock, is in a state of constant change. We might stand on a great granite cliff on an ocean shore and watch the ceaseless battering of the oncoming waves—and note that the rock stands apparently unaffected, the same now as it was in the time of our grandfathers and the same as it will be in the time of our grandchildren.

Yet can we be so certain that rocks do not change—slowly perhaps, but nevertheless steadily? Watch the granite shore again during a storm, when huge waves lift pebbles and boulders from the beach to grind against the cliff; such powerful rasping must surely dig scratches and grooves into the rock. Note the widened cracks in the granite along the shore, the caves and arches, the rounded granite boulders on the beach; how else could these form if not by the slow wearing of waves and wave-driven pebbles? At Niagara Falls, records show that the rocky cliff is yielding slowly to the river's constant gnawing, so that in places the brink of the falls has retreated as much as 200 ft in 50 years.

Even where waves and rivers are not active, careful examination of rock surfaces shows evidence of their slow disintegration and wearing away. Rocks, hills, and mountains are permanent only by comparison with the brief span of human life, and the long history of the earth extends back not scores of years but millions of years. In these immense stretches of time the slow processes by which solid rock is decayed and worn down have wrought great changes in the earth's appearance. Shorelines have been pushed back. Waterfalls have disappeared; even mountain ranges have been leveled and broad seas filled with their debris. We can read such events of the past in the rocks of the present day, and we find at least a partial explanation for them in the processes by which these same rocks are now being destroyed.

EROSION

erosion and gradation

All the processes by which rocks are worn down and by which the debris is carried away are included in the general term *erosion*. The underlying cause of erosion is gravity. Such agents of erosion as running water and glaciers derive their destructive energy from gravity, and gravity is responsible for the transport of removed material to lower and lower elevations. The leveling of landscape by erosion is often referred to as *gradation*.

weathering

weathering refers to the disintegration of rock surfaces in the open air

We have all seen the obliteration of ancient inscriptions and the rough, pitted surfaces of old stone buildings. This sort of disintegration, brought about simply by rain water and the gases in the air, is called *weathering*.

Weathering is in part a chemical process, in part a mechanical process. It participates in erosion in an important way by preparing rock material for easy removal by the more active erosional agents. Among the active agents, those whose work is most obvious are streams, glaciers, wind, and waves. Less apparent is the erosional work of groundwater, water in crevices and channels beneath the surface. All these agents are capable of cutting slowly into solid, unweathered rock, but their work is greatly speeded by the disintegration of rocks into the softer material of the weathered layer.

chemical weathering

Some of the minerals in igneous and metamorphic rocks are especially susceptible to *chemical weathering*, since they were formed under conditions very different from those at the earth's surface. Ferromagnesian minerals are readily attacked by atmospheric oxygen, aided by carbonic acid (formed by solution of carbon dioxide in water) and by organic acids from decaying vegetation. Their ferrous iron is oxidized to a hydrate of ferric oxide, whose red and brown colors commonly appear as stains on the surface of rocks containing these minerals. Feldspars and other silicates containing aluminum are in large part altered to clay minerals. Among common sedimentary rocks limestone is most readily attacked by chemical weathering because of the solubility of calcite in carbonic acid. Exposures of this rock can often be identified simply from the pitted surfaces and enlarged cracks that solution produces. Quartz and white

EROSION AND SEDIMENTATION

mica are extremely resistant to chemical attack and usually remain as loose grains when the rest of a rock is thoroughly decayed. Rocks consisting wholly of silica, like chert and most quartzites, are practically immune to chemical weathering.

Mechanical weathering is often aided by chemical decay; not only is the structure of a rock weakened by the decomposition of its minerals, but fragments are actively wedged apart because the decay of a mineral grain usually results in increased volume. The most effective process of mechanical disintegration that does not require chemical action is the freezing of water in crevices. Just as water freezing in an automobile radiator on a cold night may burst the radiator, so water freezing in tiny cracks is an effective wedge for disrupting rocks. Plant roots aid in rock disintegration by growing and enlarging themselves in cracks. Expansion and contraction of a rock's surface caused by temperature changes from day to night may gradually weaken it and aid in its destruction, but the importance of this effect is significant only in a few rocks.

Weathering processes clothe the naked rock of the earth's crust with a layer of debris made up largely of clay mixed with rock and mineral fragments. The upper part of the weathered layer, in which rock debris is mixed with decaying vegetable matter, is the soil. From a human point of view weathering is significant chiefly because it contributes to the formation of soil.

stream erosion

By far the most important agent of erosion is the running water of streams. The work of glaciers, wind, and waves is impressive locally but, by comparison with running water, they play only minor roles in the shaping of the earth's landscapes. Even in deserts, mountain sides are carved with the unmistakable forms of stream-made valleys.

A stream performs two functions in erosion:

1 Active cutting at the sides and bottom of its channel
2 Transportation of debris supplied by weathering and by its own cutting

Its effectiveness in carrying debris depends on its slope and on its volume of water. Its effectiveness in cutting its channel depends on these two factors and also on the amount and kind of debris with which it is supplied. Sand grains, pebbles, and boulders are the tools that a stream uses to dig into its bed. Scraping them along its bottom, ramming them against its banks, a stream can cut its way through the hardest rocks; the rounded forms of stream channels in hard rock and the smoothly rounded surfaces of the pebbles in stream gravel are testimony to the effectiveness of this grinding and pounding mechanism.

An additional factor of prime importance in determining how rapidly a stream will erode its valley is the frequency of violent storms in its neighborhood. Often during a few hours of a heavy rain a stream accomplishes more than in months or years of normal flow. One reason that running water is the dominant erosional agent in deserts is that desert

This topography on the Utah-Arizona border was produced by stream erosion of a sandstone plateau. The natural bridge, the world's largest, is 309 ft high and 378 ft across.

A swift stream undercutting its bank.

34.1 Successive stages in the development of a river valley. *The general appearance of the valley at various times is shown at the right and the corresponding cross sections at the left.*

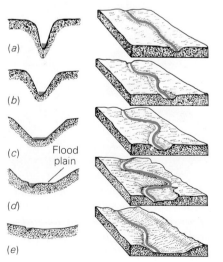

storms, when they do occur, are violent enough to send raging torrents down the normally dry valleys.

If we could watch the development of an idealized river, cutting downward through rock of uniform hardness, we would see repeated in slow motion the events that mark the growth of a gully cut in the soft material of a hillside. By the temporary stream formed during each successive rain, the gully is deepened, lengthened, and widened. Deepening is accomplished by the downcutting of the stream. Lengthening takes place at the head of the gully, where the stream eats farther and farther into the hill. Widening is a direct result not of the stream's activity but of rainwash and slumping of material on the gully's sides. Thus the stream itself cuts like a blunt knife downward and backward into the hill, while secondary processes widen the gash. The combination of deepening and widening gives the gully its characteristic V-shaped cross section; the V is steep when downcutting is rapid compared with the work of rainwash and slumping, broader when downcutting is slow. As a gully grows older the rate of downcutting slackens, and the processes of widening make its cross section a broader and broader V.

In the same manner, at first the river would dig a deep gorge, its cross section steeply V-shaped; a profile drawn to represent the slope of the river would show a steep channel with numerous rapids and waterfalls (Fig. 34.1). Then the part of the gorge near the river's mouth would be deepened to the level of an adjacent valley or perhaps nearly to sea level. Below this the river cannot cut, and downcutting would gradually slacken along its entire course. Its long profile would grow less steep and its cross section more broadly V-shaped. Later, when the slope of the stream becomes a smooth curve, downcutting would practically cease. From this point the stream would devote its energy to cutting into the sides of its channel, giving its valley a flat floor, or *flood plain.* In dry weather the stream would wander over its plain in a sinuous, meandering channel; in very wet weather it would rise out of its channel and overflow the plain. The flood plain would grow wider and wider, the stream would become more and more sluggish, the sides of its valley would become lower and lower.

evolution of landscape

During this development of the major valley, tributaries would extend their smaller valleys on either side. Soon the characteristic treelike (or *dendritic*) pattern would be fully developed, separated from the patterns of adjacent streams by sharp divides (Fig. 34.2). As flood plains develop along the main streams, divides would slowly be lowered by attack from the streams on either side. In the final stages of valley growth, when flood plains are wide and rivers broadly meandering, most of the divides would be obliterated and those remaining would be low and rounded (Fig. 34.3).

So a landscape is altered progressively by the erosional work of streams. Whatever its original form, streams begin their attack by cutting a few deep canyons; valleys grow and tributaries develop until the entire region is dissected into V-shaped valleys and sharp ridges; flood plains form along the main streams, and divides are lowered; finally the rivers widen their

EROSION AND SEDIMENTATION

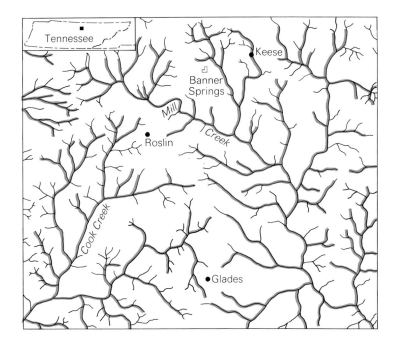

34.2 *The typical treelike pattern of a fully developed drainage system.*

plains until the entire region is reduced to a nearly flat surface, with a few low hills representing the higher of the old divides.

For convenience we may speak of a landscape as *young* when streams are beginning their work and valleys are few and steep, *mature* when tributaries are well developed and most of the area is cut into ridges and valleys, *old* when flood plains are broad and the work of erosion is nearly complete. These are terms without precise definitions, but are often useful in describing the general appearance of a landscape.

young, mature, and old landscapes

The plateau of northern Arizona, which is cut into by the Grand Canyon and a few smaller gorges, is a good example of a young landscape. Mature topography is common in mountain regions and is especially well shown by the Cumberland Plateau region west of the Appalachians. Simple landscapes in old age are not common; parts of the lower Mississippi Valley show the characteristic broad flood plains, the meandering streams, the low divides, but the topography here is complicated by other geologic processes besides stream erosion.

Actual landscapes seldom show precisely the simple valley shapes and patterns just described. Perhaps the commonest reason is the presence of rocks of different hardness: hard rocks usually remain as cliffs and high ridges, while the more easily eroded soft rocks wear away. In the Grand Canyon, the typical V shape of a young stream valley is modified by steps because the stream has cut through horizontal layers of hard and soft rocks; the hard layers form the cliffs, the soft layers the more gentle slopes between. In the Appalachian Mountains treelike patterns of tributaries are not developed, because here alternate hard and soft layers are tilted on end and streams preferentially erode the soft layers (Fig. 34.4). Many of the odd shapes produced by erosion are due simply to differences in resistance from one rock layer to the next.

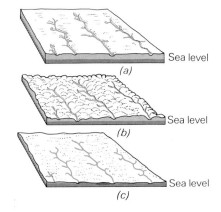

34.3 *The development of a landscape by stream erosion, from youth (a) through maturity (b) and old age (c).*

The flood plain of the Christian River, Alaska. The meandering, shifting channel is typical of the old age of a river.

the ultimate result of stream erosion would be a flat plain nearly at sea level if no other processes act

a glacier is a moving mass of ice

34.4 Parallel ridges and valleys produced by stream erosion in tilted layers of hard and soft rocks. *Soft layers underlie the valleys, hard layers the ridges. Landscapes and rock structures of this sort are typical of the Appalachian Mountains.*

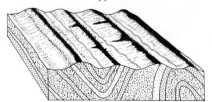

Whatever the valley shapes produced in various stages of landscape development, whatever different kinds of rock may be present, the ultimate goal of stream erosion is to reduce the land surface to a nearly flat plain approximately at sea level. This goal is never attained, however, because, as we shall learn, other geologic processes continually counteract the effects of running water.

glaciers

In a cold climate with abundant snowfall, the snow of winter may not completely melt during the following summer, and so a deposit of snow accumulates from year to year. Partial melting and continual increase in pressure cause the lower part of a snow deposit to change gradually into ice. If the ice is sufficiently thick, gravity forces it to move slowly downhill. A moving mass of ice formed in this manner is called a *glacier*.

Existing glaciers are of two principal types:

1 Easily accessible glaciers—in the Alps, on the Alaskan coast, in the western United States—are patches and tongues of dirty ice lying in mountain valleys that are called *valley glaciers*. These move slowly down their valleys, melting copiously at their lower ends; the combination of downward movement and melting keeps their ends in approximately the same position from year to year. Movement in the faster valley glaciers (a few feet per day) is sufficient to keep their lower ends well below timber line.

2 Glaciers of another type cover most of Greenland and Antarctica: huge masses of ice thousands of square miles in area, engulfing hills as well as valleys, and appropriately called *icecaps*. These, too, move downhill,

but the "hill" is the slope of their upper surfaces. An icecap has the shape of a broad dome, its surface sloping outward from a thick central portion of greatest snow accumulation; its motion is radially outward in all directions from its center (Fig. 34.5).

Just how a glacier moves is not altogether clear, but it is in part by sliding, in part by internal fracture and healing in the crystals of solid ice. Like a stream, a glacier picks up rock fragments that act as tools in cutting its bed. Some fragments are the debris of weathering that drop on the glacier from its sides; others are torn from its bed when melted water freezes in rock crevices. Fragments at the bottom surface of the glacier, held firmly in the grip of the ice and dragged slowly along its bed, gouge and polish the bedrock and are themselves flattened and scratched. Smoothed and striated rock surfaces and deposits of debris containing boulders with flattened sides are common near the ends of valley glaciers. Because such evidence of the grinding and polishing of ice erosion is found in regions far from present-day glaciers, geologists have good reason to infer that glaciation was more extensive in the past.

Valley glaciers form in valleys carved originally by streams, and by long erosion produce characteristic changes in the valley shapes. A mountain stream cuts like a knife vertically downward, letting slope wash, slumping, and minor tributaries shape its valley walls; by contrast, a glacier is a blunt erosional instrument which grinds down simultaneously all parts

34.5 Schematic cross section of an icecap. *The arrows show the direction of ice movement.*

The drainage pattern of the Wadi Hadramaut in Aden is conspicuous from the air. The Gulf of Aden is at the top.

of its valley floor and far up the sides as well. Effects of this erosion are best seen in valleys that have been glaciated in the past but in which glaciers have dwindled greatly or disappeared. Typically such valleys have U-shaped cross sections with very steep sides, instead of the V shapes produced by stream erosion. Their heads are round, steep-walled amphitheaters called *cirques*, in contrast to the small gullies at the heads of stream valleys. Tributaries often drop into glaciated valleys over high cliffs because glaciers cut their valleys much more actively than do their tributaries; such tributary valleys, left stranded high above their main valleys, are called *hanging valleys*.

<small>glacial erosion produces U-shaped valleys</small>

Lakes and swamps are numerous in glaciated valleys, formed where the glaciers gouged basins in their channels or left piles of debris as dams. Divides between cirques and between adjacent U-shaped valleys are often extremely sharp ridges because of the steepness of the valley walls. In general, since valley glaciers produce deep gorges, steep slopes, and knifelike ridges, their effect is to make mountain topography extremely rugged. The earth's most spectacular mountain scenery is in regions (the Alps, the Rockies, the Himalayas) where valley glaciers were large and numerous several thousand years ago.

The influence of icecaps on landscapes is very different from that of valley glaciers. We cannot, of course, observe directly the effect of existing icecaps on the buried landscapes of Greenland and Antarctica, but larger icecaps that once covered much of Northern Europe and North America have left clear records of their erosional activity. We shall discuss the evidence for the existence of these glaciers in more detail later; here we need only note the rounded hills and valleys, the abundant lakes and swamps so characteristic of these regions. Like a gigantic piece of sandpaper, an icecap rounds off sharp corners, wears down hills, and fills depressions with debris, leaving innumerable shallow basins which form lakes when the ice recedes.

<small>northern europe and north america were once covered by icecaps</small>

Glacial erosion is locally very impressive, particularly in high mountains. The amount of debris and the size of the boulders which a glacier can carry or push ahead of itself are often startling. But by and large, the world over, the erosional work accomplished by glaciers is small. Only rarely have they eroded rock surfaces deeply, and the amount of material transported long distances is insignificant compared with that carried by streams. Most glaciers of today are but feeble descendants of mighty ancestors, but even these ancestors succeeded only in modifying somewhat landscapes already shaped by running water.

wind

The erosional activity of wind is limited to places where fine material is abundant and unprotected by vegetation—exposed beaches, arid and subarid lands too intensively farmed, and especially deserts. Fine dust particles can be carried long distances by wind, as the dust storms which plague Kansas and New Mexico in drought years demonstrate all too forcefully. Sand grains are swept along close to the ground, locally grooving and polishing rock surfaces. Fragments larger than sand grains cannot be carried by wind, except rarely in violent local storms. Just how much erosion wind can accomplish is not certain, but compared with streams its effect is slight, even in favorable localities. In the desert strong winds blow frequently and rain falls but seldom, yet the sides of desert mountains are carved with the characteristic patterns of stream-eroded valleys.

Wind is a highly efficient sorting agent, winnowing fine silt and dust cleanly from sand and gravel. Much of the fine material is carried to great heights and for long distances, spreading so thinly over land and sea that it loses its identity. Where wind-blown dust accumulates in sheltered places, it forms a soft, fine, claylike material called *loess*. Great deposits of loess are found in northern China, representing the long accumulation of dust swept from the desert basins of Central Asia. Wind piles sand grains on exposed beaches and in deserts into the low, shifting hills called *dunes*. Dune sand is typically free from dust or mud particles and has well-rounded grains.

accumulations of wind-blown dust form loess

waves and currents

Waves are produced by the friction of wind on open water. The breaking of waves on a beach gives rise to two kinds of currents: *rip currents*, which are narrow streams of water that move seaward amidst the incoming surf; and *long-shore currents*, movements of water parallel to the shore set up by waves which approach the shore obliquely. Cutting of the shore by waves and removal of the debris by currents are responsible for the sometimes spectacular coastline scenery of the world.

Direct erosion by waves on a lake shore is limited to a narrow vertical range approximately at lake level, on the ocean shore to a somewhat wider range determined by high and low tides. Thus wave erosion acts like a horizontal saw, cutting always farther inland along the same level. Where mountainous country is adjacent to the coastline this erosion results in steep cliffs, their steepness maintained and their faces continually freshened by the undermining action of the waves. Wave erosion cuts a broad platform underwater, the seaward margin of the platform often being extended by material deposited by currents (Fig. 34.6a). Portions of the

wave erosion

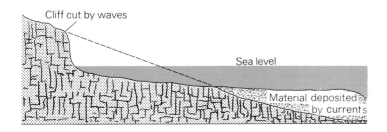

34.6a Cross section of a steep, rocky shoreline.

shoreline formed of hard rock, less easily eroded than adjacent softer rocks, may be left as projections or as small rocky islands. On gently sloping coastlines a similar cliff-and-platform profile is produced by wave

The Johns Hopkins Glacier in Alaska is a valley glacier.

erosion, but its curves are more gentle; the "cliff" may be simply the seaward slope of a bar or beach (Fig. 34.6b).

34.6b Cross section of a low, sandy shoreline.

Like streams and glaciers, waves erode solid rock effectively only when supplied with rock debris to use as cutting tools. The rounded pebbles of beach gravel show the constant wear to which these tools are subjected. Just as streams do their most effective cutting during severe rainstorms, so waves accomplish the greater part of their erosional work when strong gales drive them against the shore with higher energies than usual.

SEDIMENTATION

Most of the material transported by the agents of erosion is eventually deposited to form *sediments* of various sorts. Only substances in solution can escape such deposition; ions of various salts carried by streams to lakes and oceans may remain dissolved indefinitely. The salt of the sea is an accumulation of material dissolved out of rocks by rain, rivers, and groundwater through long geologic ages. Under some conditions part of the dissolved material may even form sediments; slightly soluble salts like calcium carbonate precipitate readily, and others appear when evaporation concentrates the water of a salt lake or an arm of the sea.

The ultimate destination of erosional debris is the ocean, and the most widespread sediments accumulate in shallow parts of the ocean near continental margins. But much sedimentary material is carried to the sea in stages, deposited first in thick layers elsewhere—in lakes, in desert basins, in stream valleys. Each of the various erosional agents has its own peculiar methods of depositing its load, and these methods leave their stamp on the character of the deposits formed. Since sediments laid down ages ago often retain many of their original characteristics, an acquaintance with the processes of deposition enables us to infer the probable origin of older deposits. In this way we can reconstruct past conditions of erosion and sedimentation and so gain insight into many periods of earth history.

The Greenland Icecap covers an area of over 500,000 sq miles and is as much as 2 miles thick in places. The icebergs that menace shipping in the North Atlantic are fragments that have broken off from the western edge of this icecap.

An important characteristic of sedimentary deposits is their *degree of sorting*, which refers to the extent of separation of fine material from coarse—whether boulders, clay, and sand are mixed up together or segregated in different layers. Another feature that often helps to betray a sediment's origin is the kind of *stratification* (layering) it shows. Some sediments have practically no layering, but retain the same color and texture through great thicknesses; in others each bed or stratum is sharply marked off from those above and below by differences in color and grain size. The layers of stratified deposits are sometimes uniform and parallel over long distances, sometimes show abrupt variations in thickness.

sorting and stratification

Occasionally sediments show a type of layering called *cross-bedding*, in which thin, curved beds lie at moderate angles to the general trend of stratification. Sand dunes often exhibit large-scale cross-bedding.

cross-bedding

processes of sedimentation

Stream Deposits. Streams, the chief agents of erosion, deposit some of the abundant debris they carry whenever their speeds drop or their volumes of water decrease. Four sites of deposition are common:

sites of sediment deposition

1 Debris carried in time of flood is deposited in gravel banks and sandbars when the swiftly flowing waters begin to recede.
2 The flood plain of a meandering river is a site of deposition in occasional floods, when the river overflows its banks and loses speed as it spreads over the plain. In Egypt, for example, the fertility of the soil has been maintained for centuries by the deposit of black silt left each year when the Nile is in flood.

3 A common site of deposition, especially in the western United States, is the point where a stream emerges from a steep mountain valley and slows down as it flows onto a plain. Such a deposit, usually taking the form of a low cone pointing upstream, is called an *alluvial fan*.

<small>origin of a delta</small>

4 A similar deposit is formed when a stream's flow is stopped abruptly as it flows into a lake or sea. This kind of deposit, built largely under water and with a surface usually much flatter than that of an alluvial fan, is called a *delta*.

Streams can transport fragments of all sizes, from fine clay particles to large boulders. Turbulent stream currents are only fair as sorting agents, so that individual layers in their deposits commonly contain mixtures of sand and gravel or of sand and clay. Pebbles and boulders, angular when they start out, become well rounded when a stream has rolled them along its channel. The constant shifting of stream courses, the filling of old channels and cutting of new ones, the building and tearing down of gravel banks and sandbars, lead to conspicuously uneven bedding. This irregular bedding—thin layers of gravel, sand, and silt which thicken or pinch out abruptly, often cross-bedded—is the outstanding characteristic of stream deposits.

Fragments carried by a swift stream are subjected to continual hammering against one another and against the stream bed, and are constantly exposed to the agents of weathering. In such an environment only the toughest and most resistant minerals can survive. Stream gravels that have not traveled far from their source may contain pebbles of many different rock types, but gravels that a stream has carried for long distances contain hard rocks almost exclusively. Since quartz is the hardest and most resistant of common minerals, material that has been battered down to sand-grain size usually contains an abundance of quartz. Fine material in stream sediments consists chiefly of clay minerals.

Glacial Deposits. The material scraped from its channel by a glacier is in part heaped up at its lower end and pushed forward as a low ridge by the slow ice movement, in part dumped into depressions and spread as a layer of irregular thickness beneath the ice. The pile of debris around the end of the glacier, called a *moraine*, is left as a low ridge of hummocky topography when the glacier melts back. Moraines in mountain valleys and in the North Central states are part of the evidence for a former wide extent of glaciation.

<small>till is the material deposited by a glacier</small>

All the material deposited directly by ice goes by the name of *till*, characteristically an indiscriminate mixture of fine and coarse material. Huge boulders are often embedded in the abundant, fine, claylike material that a glacier produces by its polishing action. Typically, most of the boulders are angular; a few are rounded and show the flat scratched faces produced as they were dragged along the bed of the glacier.

Stratified till is sometimes associated with stream deposits, for the melting of glacial ice furnishes abundant water which sorts and stratifies some of the glacially eroded material.

Undersea Deposits. Most important of the agents of deposition, since they handle by far the largest amount of sediment, are the currents of the seas and oceans. Currents deposit not only the materials eroded from coastlines by wave action, but also abundant debris brought to the ocean by streams, wind, and glaciers. Visible deposits of waves and currents include beaches and sandbars, but the great bulk of the sediments brought to the ocean are laid down under water.

Deposition beneath the sea takes place in several ways. Sand, gravel, and clay are dragged outward along the bottom by offshore currents, to drop where the current becomes too weak to carry them. Some salts, notably calcium carbonate, are deposited as chemical precipitates when sea water becomes locally oversaturated. In places living organisms are so abundant that their shells, when the organisms die, become an important part of the material deposited.

These depositional processes operate chiefly in the shallow parts of the ocean bordering the continents, out to depths of 2,000 to 3,000 ft. In shallow arms of the sea, like Hudson Bay and the Baltic Sea, active deposition may take place over the entire bottom. Little material from the land is directly deposited in the deeper parts of the ocean, but some of the mud and sand deposited on the shallow continental shelves may move to greater depths in the form of *turbidity currents*. These are masses of water and sediment, with the sediment held in suspension by extreme turbulence, which occasionally roll down the relatively steep slope beyond the shelf when triggered by earthquakes or oversteepening.

> turbidity currents shift sediments deposited in shallow water to greater depths offshore

Coarse fragments in marine sediments are well rounded, since before deposition they are used by waves in their battering of the shore. The constant back-and-forth shifting of loose material on the bottom by waves and currents provides a good sorting mechanism, so that the grain size in any one layer of sediment is fairly uniform. Often, however, there is a very gradual change in grain size away from shore, since outgoing currents become progressively feebler; if fragments of all sizes are available, coarse gravel is dropped near shore where currents are strong, sand farther out, and clay in still deeper water. Since the sea floor is nearly flat, and since marine deposition as a rule takes place uninterruptedly for long periods of time, marine sediments are characterized by even, parallel beds often of considerable thickness. Cross-bedding is of minor importance and is always on a small scale; it suggests deposition on beaches and sandbars. Fossils are more abundant in marine deposits than in any other type.

The feebler currents of an inland lake also sort out the debris supplied by streams and by wave erosion and spread it in even layers over the lake bottom. Commonly the currents are too weak to transport any but the finer particles out to deep water, so that lake deposits are characterized by thin, uniform beds of clay. In arid regions evaporation of lake water causes the precipitation of various salts (calcium carbonate, calcium sulfate, sodium carbonate), which are mixed with the clay or in separate layers. Deposits of shallow lakes and swamps in humid regions often contain partially decayed plant material in great abundance.

Crawford Notch in New Hampshire has the typical U-shaped cross section of a valley cut by a glacier.

sedimentary rocks

Sediments buried beneath later deposits are gradually hardened into rock, a process called *lithification*. Lithification is often a complex process, completed only after slow changes have gone on for thousands or even millions of years. One important change in the sediment is compaction, the squeezing together of its grains under the pressure of overlying deposits. Some recrystallization may accompany compaction; the calcite crystals of limy sediments, in particular, grow larger and interlock with one another.

Chemical changes brought about by circulating groundwater are largely responsible for the hardening of many sediments—the grains of coarse sediments are cemented by material precipitated from solution in groundwater, and some sediments have much of their original material dissolved away and replaced by other substances. The profound changes groundwater

groundwater promotes the hardening of many sediments into rock

can accomplish are strikingly illustrated by petrified wood, in which the original organic compounds have been removed, molecule by molecule, and replaced by silica—the whole process taking place so gradually that the finest details of wood structure may be preserved.

The most common cementing materials with which groundwater binds together the grains of sediments are silica, calcium carbonate, and hydrated ferric oxide. Ferric oxide betrays its presence by the red, yellow, and brown colors of many sandstones and conglomerates. The hardening of sediments by cementation, compaction, and recrystallization produces the ordinary varieties of sedimentary rock. Gravel beds are cemented into tough conglomerates, sand deposits into sandstones. Layers of clay become shale, and precipitates of calcium carbonate become limestone. The origin of chert is more obscure, perhaps because it may come into being in more than one way.

The structures of a sediment—such as stratification, cross-bedding, fossils—are usually preserved through the hardening process, although delicate structures may be partly obliterated. Hence examination of a sedimentary rock often reveals a good deal about the conditions under which the original sediment was laid down. For example, a hard sandstone in thick, even layers that contains impressions of fish skeletons or clamshells is almost certainly a marine deposit. A sandstone in thin beds, interstratified with shale and conglomerate and showing strong cross-bedding, was probably laid down by a stream. A sandstone free from clay and gravel, made up of well-sorted, rounded grains, and cross-bedded

The incessant battering of waves has cut deeply into the shores of Buzzards Bay, Massachusetts.

The action of winds is probably responsible for this striking example of cross-bedding in Utah.

springs are channels where groundwater emerges from beneath the surface

in large, sweeping curves, represents dune material which has become hardened.

Sedimentary rocks are peculiarly important in geology because they contain material that was deposited at or near the earth's surface. In the nature and structure of this material is preserved a record of the changing surface conditions of past time. If we read the record with sufficient insight and imagination, we find spread out before us a panorama of earth history: seas that once spread widely over the land, the advance and retreat of ancient glaciers, the winds and torrential streams of long-vanished deserts. Dimly we can see even the living creatures that inhabited lands and seas of the past, for in sedimentary rocks are entombed abundant fossil remains of plants and animals. Whereas igneous and metamorphic rocks reveal in their structures something about conditions in the earth's interior, sedimentary rocks tell us the more varied and interesting history of surface landscapes.

GROUNDWATER

Much water that falls as rain does not run off immediately in streams but soaks into the ground. All water that thus penetrates the surface is called *groundwater.*

The soil, the weathered layer, and any porous rocks beneath act together as a huge sponge, taking up great quantities of water into their pore spaces. During and immediately after a heavy rain all available pores in the sponge may be filled, and the ground is then said to be *saturated* with water. When the rain has stopped, water slowly drains away from hills into the adjacent valleys. Thus a few days after a rain porous material in the upper part of a hill has little moisture, while that in the lower part may still be saturated. Another rain would raise the upper level of the saturated zone, prolonged drought would lower it. The fluctuating upper surface of the saturated zone is called the *water table.*

movement of groundwater

Beneath valleys the water table is commonly nearer the surface than under adjacent hills, since water from the saturated zone continually moves outward into valleys. These general relations are shown in Fig. 34.7. The movement of groundwater in the saturated zone is principally a slow seepage downward and sideward into streams, lakes, and swamps. The motion is rapid through coarse material like sand or gravel, slow through fine material like clay. It is this flow of groundwater that maintains streams when rain is not falling; a stream goes dry only when the water table drops below the level of its bed. A *spring* is formed where groundwater comes to the surface in a more or less definite channel.

Groundwater is the source of water used by plants and of much water used for drinking and irrigation. Because it is so vitally important to human welfare and because its movements are hidden beneath the ground surface, groundwater has long been a subject of extravagant superstitions. Actually, groundwater moves according to a few simple rules: its source

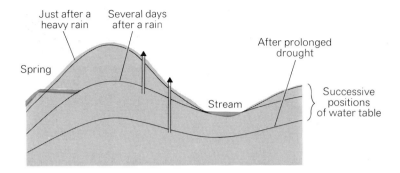

34.7 Schematic cross section through a landscape underlain by porous material. *The position of the water table is shown just after a heavy rain, several days later, and after a prolonged drought. The spring and upper well would be dry during the drought.*

is rain; it moves in a general downward direction; its motion consists of a slow seepage, faster in sand or gravel than in clay, faster in large cracks than in small ones; definite underground streams or pools are rare except in limestone caverns. From these simple facts, together with adequate knowledge of the topography and rock characteristics of a given region, a geologist can usually make accurate predictions as to the whereabouts and motions of available groundwater.

Although groundwater movement is slow, its erosional activity is by no means negligible. It can accomplish little mechanical wear, but its intimate contact with rocks and soil enables it to dissolve much soluble material. The dissolved substances are in part transported to neighboring streams, in part redeposited at other points in the weathered layer or bedrock. Dissolved material is responsible for the *hardness* of water from many wells. In regions underlain by limestone, most easily soluble of ordinary rocks, the dissolving action of groundwater makes itself conspicuous by the formation of caves. A cave is produced when water moving through tiny cracks in limestone gradually enlarges the cracks by dissolving and removing adjacent rock material.

limestone caves

The activity of groundwater extends downward for some hundreds or thousands of feet, the depth in a given region depending on the kinds of rock present. Nearly all rocks within a short distance of the surface have sufficient pore space or are sufficiently cracked to permit some circulation, although in massive igneous and metamorphic rocks the amount of groundwater may be very small. At lower levels cracks become too small and too scarce to permit free movement. Thus, deep mines often have abundant water in their upper parts but so little at lower levels that dust from drilling and blasting becomes a problem.

deposition by groundwater

Deposition by groundwater is almost entirely chemical precipitation from solution. The precipitation may be brought about by evaporation of the water, by mixing of groundwater from different sources, by a lowering of temperature, or by the escape of a gas when pressure is reduced. One of the most important geologic functions of groundwater is the depositing of material in the pore spaces of sediments, which helps to convert the sediments into rock. Much dissolved material is deposited along open

Luray Caverns, Virginia, were produced by the solvent action of groundwater on limestone.

cracks to form *veins*, common in all kinds of rocks. The commonest vein minerals are quartz and calcite, but some veins contain commercially valuable minerals. More spectacular examples of groundwater deposition are the *stalactites* that hang from the roofs of limestone caverns and the colorful deposits often found around hot springs and geysers.

GLOSSARY

Erosion refers to all processes by which rock is disintegrated and worn away and its debris removed.

Weathering is the surface disintegration of rock by chemical decay and mechanical processes such as the freezing of water in crevices.

Stream erosion is erosion caused by the action of running water.

A *flood plain* is the wide, flat floor of a river valley produced by the sidewise cutting of the stream when its slope has become too gradual for further downcutting.

A *glacier* is a large mass of ice that gradually moves downhill.

Sediments consist of material that has been transported by erosional agents and then deposited.

An *alluvial fan* is a deposit of sediments where a stream emerges from a steep mountain valley and flows onto a plain.

A *delta* is a deposit of sediments where a stream flows into a lake or the sea.

A *moraine* is the pile of debris around the end of a glacier left as a low, hummocky ridge when the glacier melts back. The deposited material is called *till*.

Rip currents are narrow streams of water that move seaward amidst incoming surf.

Veins consist of dissolved material deposited along open cracks.

Groundwater is rainwater that has soaked into the ground. The *water table* is the upper surface of that part of the ground whose pore spaces are saturated with water. A *spring* is a channel through which groundwater comes to the surface.

EXERCISES

1 Which would you expect to be more effective in the tropics, mechanical weathering or chemical weathering? Which on a high mountain peak?

2 Why are streams the principal agent of erosion on the earth's surface?

3 After long erosion in a temperate region underlain with marble, quartzite, basalt, shale, and granite, which types of rock would you expect to find just beneath the surface of
   a. hills?
   b. valleys?

4 In general, would you expect the debris deposited by a glacier to show more or less chemical decay than stream sediments?

5 The effects of past glaciation in the Sierra Nevada of California extend to much lower elevations in valleys on the west side of the range than in valleys on the east. Suggest a possible reason.

6 In sand derived from the attack of waves on granite, what mineral or minerals would you expect to be most abundant?

7 What characteristics of a sedimentary rock might indicate an arid climate at the time the original sediment was deposited?

8 Imagine two small hills separated by a stream valley, one consisting chiefly of sand and gravel, the other chiefly of clay. Show by a diagram the position of the water table in each (*a*) immediately after a heavy rain, (*b*) several days after a heavy rain, (*c*) after prolonged drought. Indicate by vertical lines the depth to which a well must penetrate from the top of each hill to maintain a supply of water during drought.

9 What are the chief characteristics of (*a*) the material in an alluvial fan at the base of a short, steep slope; (*b*) sediments laid down a short distance from shore along a steep, rocky seacoast; (*c*) the material in a sand dune; (*d*) the material in a moraine?

10 What is the probable origin of the following sedimentary rocks?
  a. A thick, evenly bedded limestone
  b. A conglomerate with well-rounded boulders and numerous thin beds of sandy and clayey material, showing conspicuous cross-bedding
  c. A sandstone consisting of well-sorted, well-rounded grains of quartz with conspicuous large-scale cross-bedding

11 Why are hot-spring deposits thicker than the deposits found around ordinary springs?

chapter 35

Vulcanism and Diastrophism

VULCANISM

> volcanic eruptions
> intrusive rocks
> intrusive bodies
> problems of vulcanism

DIASTROPHISM

> recent diastrophic movement
> faults
> folds
> regional movement
> dynamothermal metamorphism

Erosion and sedimentation are basically leveling processes through which the higher parts of the earth's surface are worn down and the lower parts filled with sediment. If their work could be carried to completion, these processes would ultimately reduce all the continents to plains beneath the level of the sea. The simple fact that continents still exist, with high mountains and deep valleys, is good evidence that the work of erosion and sedimentation is opposed by other agencies capable of creating irregularities in the earth's crust. Such agencies may be grouped under two heads:

1 Processes of *vulcanism*, which involve the movement of liquid rock
2 Processes of *diastrophism*, which include all movements of the solid materials of the crust

Vulcanism and diastrophism are not wholly independent, for movement of liquid rock often causes considerable distortion of adjacent rock layers, and major diastrophic movements are often accompanied by volcanic activity.

The crater of Kitsimbanyi Volcano in the Congo is split in the center by lava flow.

VULCANISM

a volcano is an opening through which liquid rock emerges

By definition, a *volcano* is an opening in the crust through which molten rock, usually called *magma* while underground and *lava* aboveground, pours forth. Because much of the material that escapes from a volcano accumulates as solid rock near the orifice, most volcanoes in the course of time build up mountains of characteristic shape—roughly conical, steepening toward the top, with a small depression or crater at the very

volcanic eruptions

A volcanic eruption is one of the most awesome spectacles in all nature. Usually a few hours or a few days beforehand there is a warning in the form of earthquakes—minor shocks probably caused by the movement of gases and liquids underground. An explosion or a series of explosions begins the eruption, sending a great cloud billowing upward from the crater. In the cloud are gases from the volcano, water droplets (since water vapor is a prominent volcanic gas), fragments of solid material blown from the crater and the upper part of the volcano's orifice, dust, and larger solid fragments representing molten rock blown to bits and hurled upward by the violence of the explosions. Gas continues to issue in great quantities, and explosions recur at intervals. The cloud may persist for days or weeks, its lower part glowing red at night. Activity gradually slackens, and a tongue of white-hot lava may spill over the edge of the crater or pour out of a fissure on the mountain slope. Other flows may follow the first, and explosive activity may continue with diminished intensity. Slowly the volcano becomes quiescent, until only a small steam cloud above the crater suggests its activity.

Not all eruptions by any means follow this particular pattern. Volcanoes are notoriously individualistic, each one having some quirks of behavior not shared by others. In one group of volcanoes the explosive type of activity is dominant, little or no fluid lava appearing during eruptions. Cones of these volcanoes, built entirely of fragmental material ejected

explosive volcanoes emit solid material that forms steep cones

Pahoehoe is a type of rock that is formed by the slow cooling of lava.

in a solid or nearly solid state, are very steep sided. Good examples are found in the West Indies, in Japan, and in the Philippines. Other volcanoes, like those of Hawaii, have eruptions characterized by quiet lava flows with little explosive activity. Mountains built by these volcanoes are broad and gently sloping, quite different from the usual volcanic structure. The most common kind of volcano is neither wholly of the "explosive" type nor wholly of the "quiet" type, but has eruptions in which both lava flows and gas explosions figure prominently.

quiet volcanoes emit lava that solidifies into gently sloping mountains

The chief factors that determine whether an eruption will be a largely quiet lava flow or primarily explosive are the viscosity of the magma and the amount of dissolved gas it contains. A magma is a complex mixture of the oxides of various metals with silicon dioxide, usually containing an abundance of gas dissolved under pressure. Like most silicate melts, it is exceedingly viscous; with rare exceptions, molten lavas creep downhill slowly, like thick syrup or taffy. The viscosity depends upon chemical composition; magmas with high percentages of silica are in general more viscous than those with large amounts of metallic oxides. Gas content also affects viscosity; magmas with little gas are the most viscous. If the magma feeding a volcano has a high gas content and is still siliceous enough to be fairly viscous, eruptions will be explosive; basic magmas with relatively small amounts of gas are required for quiet eruptions.

the properties of the magma (underground liquid rock) involved determine the nature of a volcanic eruption

The gaseous products of volcanic activity include carbon dioxide, nitrogen, water vapor, hydrogen, compounds of sulfur, and minor amounts of halogens and metallic compounds. Most of these mix at once with the atmosphere. Much of the water vapor condenses, giving rise to the torrential rains that often accompany eruptions.

Molten lava solidifies to form one or another of the various volcanic rocks. All these are characterized by their fine grain size, since lava flows cool rapidly. Basalt is by far the commonest volcanic rock and forms the largest flows; coming from a basic magma, it is fluid enough to flow long distances. Rhyolite, the most siliceous of ordinary lavas, forms small, thick flows. Rhyolitic lava is sometimes so viscous and cools so rapidly that crystallization does not take place; then the natural glass obsidian is formed. All kinds of volcanic rocks frequently show rounded holes left by expanding gas that was trapped during the final stages of solidification; viscous lavas are sometimes so filled with gas cavities that the light, porous rock *pumice* is formed.

tuff and breccia are formed by the consolidation of fine and coarse rock fragments, respectively, emitted during a volcanic eruption

During explosive eruptions much of the liquid rock is blown into fragments which solidify immediately or during their flight through the air. The fragments range in size from fine dust particles to large "volcanic bombs" several feet in diameter. Most of this material settles on the slopes of the volcano, though the finer dust may be carried long distances by wind. Deposits of the finer material may be cemented to form the rock *tuff*, and deposits of the coarser material may form a kind of conglomerate called *volcanic breccia* (Fig. 35.1).

Lava flows and thick falls of volcanic dust may do considerable property damage during an eruption but are seldom responsible for much loss of life. The really destructive feature of volcanic activity is a phenomenon often associated with explosive eruptions: part of the cloud above the crater becomes so choked with solid debris, and both debris and gas are in such an extreme state of turbulence, that the mixture is capable of moving down the mountainside like a liquid. Within this rapidly moving cloud are hot, poisonous gases and solid fragments of all sizes in a maelstrom of violent movement. No living thing can survive in its path. Such a cloud from the volcano Mt. Pelée in 1902 completely destroyed the city of St. Pierre, taking the lives of 25,000 people in seconds.

35.1 Diagrammatic cross section of a volcano. Lava flows (color) alternate with beds of tuff and volcanic breccia.

Active volcanoes of the present are found around the borders of the Pacific Ocean, on some of the islands in the Pacific and South Atlantic, in Iceland, in the Mediterranean region, in the West Indies, and in East Africa (Fig. 35.2). In many other parts of the world volcanoes have been

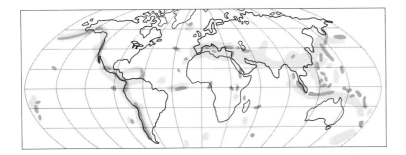

35.2 The principal earthquake (light color) and volcanic (dark color) regions of the world.

active in the past. Where volcanoes have become extinct in the recent geologic past, we find evidence of their former activity in isolated, cone-shaped mountains, in solidified lava flows, and in hot springs, geysers, and steam vents. Some of the great mountains in the western United States are old volcanoes, some of their lava flows so recent that vegetation has not yet gained a foothold on them. In regions where volcanoes have been dead for a longer period, erosion may have removed all evidence of the original mountains and left only patches of volcanic rocks to indicate former igneous activity.

intrusive rocks

Molten rock that rises through the earth's crust but does not reach the surface solidifies to form intrusive bodies (often called *plutons*) of various kinds. Because cooling in these bodies is slower than at the surface, intrusive igneous rocks are in general coarser-grained than volcanic rocks. We find such coarse-grained rocks exposed at the surface only when deep erosion has uncovered them long after their solidification.

intrusive rocks have coarser grains than volcanic rocks because their slower cooling has permitted larger mineral crystals to form

The igneous origin of volcanic rocks is evident enough, for we can actually watch fluid lava harden to solid rock. But no one has ever seen a rock like granite in its original liquid state; the evidence that it is an igneous rock must be entirely indirect. The kind of evidence that supports the statement that granite was once molten is suggested by the following observations:

1 Under the microscope granite shows the same relations among its minerals that volcanic rock shows: the separate grains are intergrown,

why granite is believed to have hardened from a molten state

Nearly all the islands in the Atlantic Ocean are the tops of volcanoes that have risen from the ocean bottom. Helping confirm this hypothesis was the emergence of this volcanic island in 1957 in the Azores.

and those with higher melting points show by their better crystal forms that they crystallized a little earlier.

2 In some small intrusive formations every gradation can be found between coarse granite and a rock indistinguishable from the volcanic rock rhyolite, whose igneous origin is established by direct observation.

3 Granite is found in masses that cut across layers of sedimentary rock and from which small irregular branches and stringers penetrate into the surrounding rocks; sometimes blocks of the sedimentary rocks are found completely engulfed by the granite.

4 That granite was at a high enough temperature to be molten is shown by the baking and recrystallization of the rocks that it intrudes.

These four types of evidence apply equally well to the other intrusive rocks.

intrusive bodies

A *dike* is a wall-like mass of igneous rock intruded along a fissure that cuts across existing rock layers. The largest dikes have thicknesses in

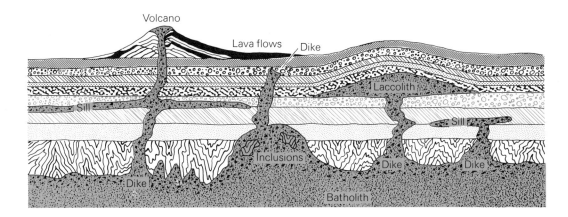

thousands of feet, but the commoner ones range from a few inches to a few tens of feet thick. The distinction between dikes and veins is that a dike is molten rock that has filled a fissure and solidified, whereas a vein consists of material deposited along a fissure from solution in water. Any kind of igneous rock may occur in a dike; rapid cooling in small dikes may give rocks similar to those of volcanic origin, and slow cooling in larger dikes often gives coarse-grained rocks. Dikes may cut any other kind of rock. They are frequently associated with volcanoes; apparently some of the magma forces its way into cracks instead of ascending through the central orifice. In regions of intrusive rocks, dikes are often found as offshoots of larger masses, as in Fig. 35.3. Also shown in the figure are *sills* and *laccoliths*, intrusive bodies that lie parallel to the strata they are found in.

Batholiths are very large bodies of intrusive rock that extend downward as much as 10 km. Visible exposures of batholiths cover hundreds of thousands of square miles; the great batholith that forms the central part of the Sierra Nevada in California, for example, is some 500 miles long and, in places, over 100 miles wide. Granite is the principal rock in batholiths, although many have local patches of diorite and gabbro. Batholiths are always associated with mountain ranges, either mountains of the present or regions whose rock structure shows evidence of mountains in the distant past. This association, however, does not mean that batholiths are the cause of mountain ranges; the intrusion of a batholith is merely one incident in mountain building.

Batholiths are so large and remain hot for so long a time that they have a profound effect on rocks along their borders. Mineral grains in the intruded rocks grow larger and lock together; minerals stable only at low temperatures disappear and new minerals are formed from them. Soft sedimentary rocks become hard and change their appearance completely; limestone changes to marble, sandstone to quartzite, and shale to a hard rock called *hornfels*. These metamorphic changes, produced by heat at the border of an igneous intrusion, are included in the term *contact metamorphism* or *thermal metamorphism* (Fig. 35.4). The metamorphic rocks produced are typically unfoliated, since they have not been subjected to movement or directed pressure which would orient the newly formed minerals.

35.3 A batholith and associated dikes and sills; a laccolith and a volcano are also shown.

The Devil's Tower in Wyoming is an ancient mass of intrusive rock that has survived the erosion of surrounding rocks.

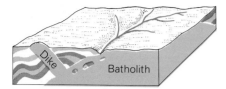

35.4 A batholith and a dike intruded into a series of sedimentary beds. *Since it solidified, the igneous rock has been uncovered and partly removed by erosion. The sedimentary beds near the contacts of the batholith show the effects of thermal metamorphism.*

The magma that forms a batholith, like the magmas that come to the surface in volcanoes, contains a large quantity of dissolved gases. As the magma solidifies its gases are in part driven out into the surrounding rocks, in part concentrated under ever-increasing pressure in the residual, uncrystallized liquid. In the last stages of crystallization this liquid is largely hot water, containing in solution besides gaseous material many soluble, low-melting substances concentrated from the original magma. At length pressures become great enough to force the liquid out into cracks, not only in the nearly frozen batholith but in nearby rocks as well. Like ordinary groundwater, the hot liquid deposits some of its dissolved material as it goes, forming veins. (We recall the distinction between dikes and veins: a *dike* is molten rock which has filled a fissure and solidified, whereas a *vein* consists of material deposited along a fissure from solution in water.) The commonest material of these veins is quartz, but the hot water associated with batholiths sometimes has concentrates of commercially valuable metals and their compounds. Many important deposits of gold, silver, and copper have been formed in this way.

problems of vulcanism

Though a great deal is understood about vulcanism, a number of major questions remain without definitive answers.

First of all, where does liquid rock come from? In the days when the earth was believed to be a molten sphere with a thin, solid crust an answer to this question was easy: volcanoes and batholiths were formed where material from the fluid interior had worked its way into the crust. But today we have reliable measurements of the earth's rigidity and the speed of earthquake waves which indicate that the earth is not liquid but largely solid, and that no really large mass of liquid can exist anywhere in the interior except near the earth's center. The magma that forms a batholith or feeds a volcano must be a relatively small mass of liquid generated locally. So the problem is to discover some process in the earth's crust which can produce bodies of liquid rock in isolated regions.

magma may be produced from solid rock by a local reduction in pressure below the earth's surface

A plausible answer is local release of pressure. We know that both pressure and temperature increase downward in the crust; it may be that rock material at lower levels is at a temperature above its normal melting point at the surface but is kept from melting by the enormous pressures. So if some sort of diastrophic movement, some kind of readjustment in the upper part of the crust, could reduce the pressure locally, a part of the rock at greater depths might liquefy. This hypothesis is especially attractive because it accounts nicely for the common association of batholiths and volcanoes with large-scale diastrophic disturbances, but it remains far from proved.

Another major problem is the manner by which liquid rock works its way upward through the crust. All kinds of igneous phenomena give evidence that magma is formed at lower levels than those where we now find it and that its motion in the liquid state is dominantly upward. For the magma which forms dikes or that which escapes from volcanoes the motion is not hard to understand: the liquid rock moves toward a zone of lower pressure along fissures, probably widening the fissures as it moves. But the migration of the magma that forms batholiths is not so readily explained.

Certainly the magma does not simply move into open cavities, for large cavities at such depths in the crust can neither form nor be maintained. Sedimentary layers are often cut off sharply at a batholithic contact, suggesting that emplacement of the magma is accomplished at least in part by removal of the preexisting rock. Perhaps the removal is accomplished by melting the rock, perhaps by the detachment and sinking of huge blocks through the liquid. At other batholithic contacts layers in the adjoining rocks have been squeezed and contorted into tight folds, which apparently indicate that the batholith made room for itself in part by compressing and pushing aside the intruded rocks. Another possible way for a batholith to work its way upward would be to lift or dome up the rocks above it; this would be difficult to demonstrate, since we do not know that a batholith exists until erosion has worn away most of its cover. Just what part is played by these several processes in the movement of large bodies of magma remains unclear.

Three dikes of igneous rock intruded in sedimentary beds. The beds on either side of the large dike do not match, which suggests that the dike was intruded along a fault.

DIASTROPHISM

Terra firma, the solid earth, has come to be a symbol for stability and strength. On foundations of rock, man confidently anchors his buildings, his dams, his bridges. The massive rock of mountain ranges seems strong enough to withstand any conceivable force that might be exerted upon it. Yet even casual observation shows at once how naïve these simple notions of the earth's stability are. Entombed in the strata of high mountains we find shells of marine animals, shells that can be there only if solid rock formed beneath the sea has been lifted high above sea level. Sedimentary rocks, which must have been deposited originally in horizontal layers, are found tilted at steep angles or folded into arches and basins. Other layers have broken along cracks, and the fractured ends have moved apart. Despite superficial appearances, there must be gigantic forces in the crust capable of lifting, bending, crumbling, and breaking even the strongest rocks.

recent diastrophic movement

Nor can we conclude that these forces have racked the earth's crust only in some remote prehistoric age and that our planet since then has acquired its apparent stability. Careful observation shows clearly that today as in the past diastrophic movements are changing the earth's landscapes. Major earthquakes, for example, sometimes give rise to permanent displacements in the crust; cracks open in the ground, and the material on one side of a crack may shift up, down, or sidewise with respect to the other side. At Yakutat Bay, Alaska, the shoreline was raised 40 ft during an earthquake in 1899; in 1964 another earthquake raised 12,000 square miles of southeastern Alaska by from 3 to 8 ft, while an adjacent area of similar

diastrophic movement occurs today as well as in the past

The Alaska earthquake of 1964 caused substantial damage to the city of Anchorage.

size subsided. In the San Francisco quake of 1906 fences and roads were displaced sidewise along the line of fracture, some by as much as 20 ft. These are relatively small movements, of course, but a succession of earthquakes in the same region may in time produce large displacements.

Shoreline features often show evidence of recent shifts of sea level. One of the most striking is a ruined market building in Pozzuoli, near Naples, built in Roman times. When the base of the structure was excavated a century or so ago, the bottoms of the three columns left standing were found riddled with holes made by a boring clam that lives in the nearby Mediterranean Sea. The building was obviously constructed on dry land and stands today on dry land, but the clam borings show clearly that at some time during the past 2,000 years this region has been several feet below sea level. Another conspicuous change in elevation has occurred in Scandinavia, where accurate records of the shoreline kept for over a century show that the land is steadily rising out of the Baltic Sea, in places by as much as a centimeter per year.

Elevation of a coastline in the not too distant past is often shown by a wave-cut cliff and terrace high above the present shore. Such a feature does not necessarily mean movement within historic time; but geologically the rising of the land or the sinking of the sea must be fairly recent events,

VULCANISM AND DIASTROPHISM

This coastal scene in Oregon shows wave-cut cliffs and terraces above a beach, which indicates an uplift of the land relative to the sea surface.

for stream erosion will quickly obliterate cliff and terrace from the landscape. Recent sinking of the land with respect to sea level is most clearly shown by long, narrow bays filling the mouths of large stream valleys. A body of water like Chesapeake Bay (Fig. 35.5), for example, could not be formed by wave erosion, since wave attack normally straightens a coastline rather than deeply indenting it; the shape of the bay suggests that it lies in a stream-carved valley whose lower part has been submerged beneath the sea.

origin of chesapeake bay

From evidence of this sort, evidence that could be multiplied a thousand times, we must conclude that the earth's solid outer shell is not as substantial as it seems, but that it has yielded and is yielding to powerful forces of deformation.

There are three principal kinds of diastrophic movement, which we shall discuss in turn:

1. The slippage of rocks along a fracture, or *fault*
2. Movement resulting in the formation of *folds*
3. Broader, regional movements of uplift, subsidence, or tilting

types of diastrophic movement

faults

A fault is any rock surface along which movement has taken place. All kinds of rock are cut by numerous fractures, but only those fractures showing definite evidence that one side has moved with respect to the

a fault is a rock fracture one of whose sides has moved relative to the other

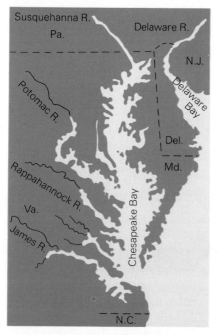

35.5 Drowned valleys on the Atlantic Coast of the United States.

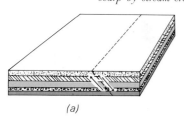

35.6 The development of a normal fault. (a) Strata before faulting; (b) after faulting, erosion assumed negligible; (c) modification of the fault scarp by stream erosion.

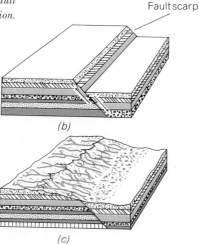

other are called faults. In an outcrop a fault is commonly recognized as a fairly straight line against which sedimentary layers and other structures end abruptly. Near the fault, layers may be bent or crumpled, and along the fault itself streaks of finely powdered material may have developed from friction during movement.

Faults are distinguished by the direction of relative movement along them. Three important kinds are illustrated in Figs. 35.6, 35.7, and 35.8.

1 A *normal fault* is an inclined surface along which the rocks above the surface have slipped down with respect to those below the surface. Note that this kind of movement requires that the strata fill a greater horizontal area after faulting than before; in other words, normal faults would be expected in parts of the crust where the chief forces are of a *tensional*, or "stretching apart," nature.

2 A *thrust fault* is an inclined surface along which the rocks above the surface have moved up with respect to those below. Thrust faulting decreases horizontal area, hence presumably is the result of *compressional* forces.

3 A *strike slip*, or *lateral, fault* is a vertical or nearly vertical surface along which one side has moved approximately horizontally with respect to the other side.

Movement along faults usually takes place in a series of small sudden displacements, with intervals of years or centuries between successive jerks. The immediate topographic effect of displacement along a thrust fault or normal fault is the production of a small cliff. Erosion attacks the cliff at once, and may obliterate it before the next movement. If successive movements follow one another fast enough, erosion may not be able to keep pace with diastrophism, and a high cliff may be formed only slightly modified by erosion. Cliffs of this sort are called *fault scarps*. Good examples of scarps produced by normal faults are the steep mountain fronts of many of the desert ranges in Utah, Nevada, and eastern California. A more deeply eroded scarp produced by thrust faulting is the eastern front of the Rocky Mountains in Glacier National Park.

The most famous strike-slip fault in the United States is the San Andreas fault of California, movement along which caused the San Francisco earthquake. Along much of its course the line of this fault is represented by a straight valley. The valley is not a direct result of faulting but is due to rapid stream erosion in rock material fractured and pulverized by the fault movement.

35.7 A thrust fault. *The dashed lines show the material removed by erosion.*

folds

Folds always result in a shortening of the crust (Fig. 35.9), hence are in general produced by compressional forces. Brittle rocks yield to compression by thrust faulting; less brittle and deeply buried rocks yield by folding. Not uncommonly major thrust faults are associated with intense folding. Two names often used to designate types of folds are *anticline*, referring to an arch or a fold convex upward, and *syncline*, referring to a trough or a fold convex downward. In regions of intense folding, anticlines and synclines follow one another in long series.

Folding apparently takes place by slow, continuous movement, in contrast to the sudden displacements along faults. Sometimes folding produces hills and depressions in the landscape directly, but more commonly erosion keeps pace with folding and obliterates its direct topographic effects. Indirectly folds affect topography by exposing tilted beds of varying degrees of resistance to the action of streams, so that characteristic long, parallel ridges and valleys develop, like those of the Appalachian Mountains. In these mountains as in many others, the actual folding is very ancient; the present ridges are due entirely to deep erosion after successive uplifts of the stumps of the old folds.

anticline and syncline

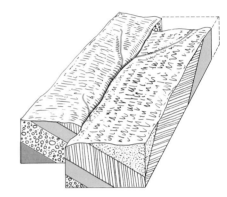

35.8 Diagram of a strike-slip fault. *Note stream flowing along fault.*

regional movement

Regional uplift and subsidence may involve whole continents or large parts of them. It may be a vertical movement; it may involve some tilting; it may mean broad, large-scale bending or warping. Just how the movement takes place is not clear, but as a rule it is imperceptibly slow.

Subsidence creates a basin of deposition, in which marine sediments accumulate if the land drops sufficiently low, otherwise river and lake sediments. Uplift increases the vigor of stream erosion: an aged landscape raised a few hundred feet is cut into by youthful streams, and the long history of valley and landscape development begins anew. Along shorelines recent uplift and subsidence are recorded in the raised beaches and drowned valleys mentioned earlier.

In the landscapes and sedimentary rocks of the United States a multitude of regional movements are recorded. For example, across the Mississippi Valley lie thick, horizontal beds of marine sandstones and limestones, recording a time when this part of the continent lay beneath the sea and a later time of uplift when the sea retreated. Shapes and patterns of river

35.9 Cross sections showing effects of folding in horizontal strata.

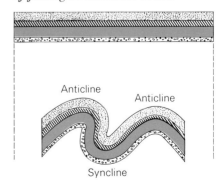

valleys in the Appalachians and the Rockies show that these great mountain systems have been uplifted repeatedly since the original mountains were formed. The thick accumulations of sediment on and near the Mississippi Delta and in the central valley of California suggest that these are regions where subsidence is in progress. These are only a few of the more obvious of such movements.

Theories of how the immense forces involved in diastrophic movement originate are the subject of Chapter 38. In that chapter the most remarkable example of such movement, the "drift" of the continents over the earth's surface, is also considered.

dynamothermal metamorphism

Pressures resulting from diastrophic movement, especially at depths where temperatures are high, may produce conspicuous changes in the rocks undergoing deformation. The pressures are in general *directed* pressures, more intense in one direction than in others. In response to such pressures new minerals develop, particularly those that can grow in long needlelike forms or those like mica that grow in flat, platy crystals. The needles and flat plates grow in directions determined in part by the maximum pressure, in part by structures within the rock, and hence arrange themselves in roughly parallel layers to give foliated rocks like slates, schists, and gneisses. Because these rocks show the effect of both heat and

A fault is a weakness in the earth's crust and is especially susceptible to stream erosion. This mosaic of aerial photographs shows the San Andreas fault rift zone from Hughes Lake to Lake Palmdale, California.

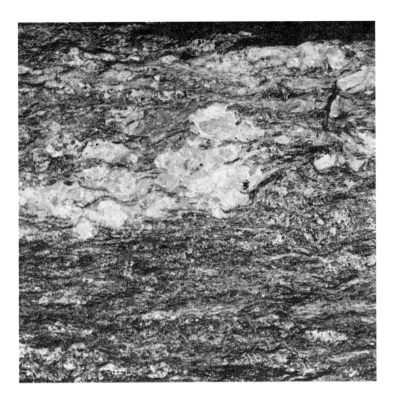

Mica schist produced by heat and pressure.

pressure, the process of alteration is called *dynamothermal metamorphism*, in contrast to *thermal* metamorphism produced by heat alone at the borders of igneous intrusives. Foliated rocks produced by dynamothermal metamorphism are abundant in any region which has undergone intense folding followed by deep erosion; parts of New England and much of northeastern Canada are good examples.

Shale is particularly susceptible to dynamothermal metamorphism, since its chief constituents, the clay minerals, alter readily to mica even at relatively low temperatures and pressures. The first step in the alteration of shale is slate, whose shiny cleavage surfaces betray the parallel arrangement of myriads of tiny mica flakes. More intense heat and pressure convert slate into coarse-grained mica schist; under extreme conditions part of the mica is changed to feldspar and the rock becomes a gneiss. Pure limestone and pure quartz sandstone subjected to dynamothermal metamorphism do not give foliated rocks, since their composition does not permit the formation of minerals with platy or needlelike crystals; as in thermal metamorphism, these rocks change to marble and quartzite, respectively. Volcanic rocks commonly become schists and under extreme conditions gneisses. Granite may change directly to gneiss. These various changes are often clearly evident in the field, a layer of shale, for example, altering within a few thousand feet to slate and then to schist as the region of greater deformation is approached.

Rocks that have undergone dynamothermal metamorphism often show by their crumpling and complex folding that there has been much move-

foliated rocks are produced by dynamothermal metamorphism

the metamorphism of shale

ment not only between separate layers but also within the solid rock itself. Original structures such as sedimentary beds and fossils have small chance of preservation. Parts of the rock may show streaks of finely granulated material produced by the crushing of one mineral grain against another; by bringing particles of different substances into intimate contact, such crushing doubtless speeds up the chemical reactions involved in the production of new minerals. Despite the intense deformation, however, these rocks ordinarily *show no evidence of melting* during metamorphism. The processes involved are chemical reactions between solid particles, aided by crushing and probably by small amounts of water, but not, except locally, by actual fusion of the rock material.

This is an important point. Dynamothermal metamorphism is in general a process taking place several miles below the surface, and its products accordingly give us valuable information about conditions at those depths. We find that rocks here are hot enough and under sufficient pressure to be readily deformed; they will crumple and flow like cold pitch squeezed slowly between the jaws of a vise, and their constitution may be profoundly altered, but they remain for the most part solid.

GLOSSARY

Vulcanism refers to geological processes that involve the movement of molten rock.

Diastrophism refers to geological processes that involve movements of the solid materials of the crust.

A *volcano* is an opening in the earth's crust through which molten rock emerges.

Molten rock is called *magma* when underground and *lava* when aboveground.

A *dike* is a wall-like mass of igneous rock intruded along a fissure that cuts across existing rock layers. A *sill* is a sheetlike intrusive body lying parallel to the rock layers; a *laccolith* is a larger body of this kind.

A *batholith* is a very large body of intrusive rock that extends downward as much as 10 km.

A *fault* is a rock fracture along which movement has occurred.

An *anticline* is a fold that is convex upward, forming an arch; a *syncline* is a fold that is concave upward, forming a trough.

Dynamothermal metamorphism refers to the changes that occur in rocks due to intense heat and pressure.

EXERCISES

1. Which of the following rocks might you expect to find (*a*) in lava flows, (*b*) in dikes, (*c*) in batholiths?

   rhyolite   andesite
   diorite   granite
   marble   obsidian
   basalt   conglomerate

2. What kind of rock would you expect to find as the chief constituent of lava flows from a volcano whose eruptions are dominantly of the explosive type? What kind in flows from a volcano of quiet type? Why?

3. What characteristic topographic features do active volcanoes produce? From what topographic features could you conclude that volcanoes had once been active in a region where actual eruptions have long since ceased?

4. List all the evidence you can for each of the following statements:
   a. Granite is an igneous rock.
   b. Mica schist is a rock that has been subjected to nonuniform pressure.
   c. Compressional forces exist in the earth's crust.
   d. Diastrophic movement is going on at present.

5. Suppose that you are studying a series of horizontal layers of shale and sandstone at some distance from the border of a batholith. If you should follow the layers toward the batholith, what changes in the rocks would you expect to observe? What other evidences of igneous activity might you find before you reached the actual contact?

6. Suppose that you find a nearly vertical contact between granite and sedimentary rocks, the sedimentary beds ending abruptly against the granite. How could you tell whether the granite had intruded the sedimentary rocks or had moved up against them after its solidification by diastrophic movement along a fault?

7. When stream erosion has been active for a long time in a region underlain by folded strata, what determines the position of the ridges and valleys?

8. Where would you expect to find the wider zone of thermal metamorphism, at the contact of a dike or a batholith? Why?

9. Indicate whether each of the following rocks is formed by thermal or dynamothermal metamorphism (or neither). Indicate also from what kind of igneous or sedimentary rock each might be produced.

mica schist   gneiss   slate
marble   quartzite hornfels

10 Distinguish between the foliation of a metamorphic rock and the stratification of a sedimentary rock.

11 Figure 35.10 is a schematic diagram of the rock cycle. Indicate which of the arrows represent processes of erosion and sedimentation, which of diastrophism, and which of vulcanism.

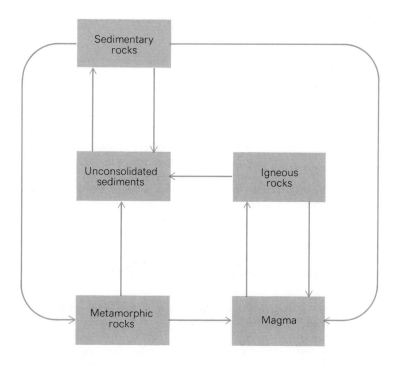

35.10 The rock cycle.

12 An experiment is performed to determine the lowest temperature at which a sialic magma can exist within the earth by melting a granite sample in a furnace. How meaningful are the results of this experiment?

chapter 36

Within the Earth

PROPERTIES OF THE EARTH
- size
- shape
- mass
- density

EARTHQUAKE WAVES
- earthquakes
- earthquake waves

INTERIOR STRUCTURE
- core and mantle
- the crust
- composition
- temperature

TERRESTRIAL MAGNETISM
- the external magnetic field
- fossil magnetism
- origin of the field

The earth's crust, containing the continents and oceans, and its atmosphere, the gaseous envelope so necessary for life, are directly accessible to our instruments, and so we may legitimately hope one day to understand their structure and behavior. The interior of the earth, however, is beyond our direct reach; with the exception of shafts which ultimately may be sunk deep enough to barely get through the crust, it is likely to remain untouched. Yet we know a great deal about this mysterious, fabled region, and we are continually learning more.

BASIC GEOLOGY

PROPERTIES OF THE EARTH

We shall begin our discussion of the earth's interior by describing how its size, shape, and mass are determined, since this will tell us the average density of the materials of which it is composed and thereby provide a clue to their nature. Then we shall proceed to consider the composition, structure, and temperature of the interior, and finally we shall examine the perplexing problem of the earth's magnetism whose source is deep in the core.

size

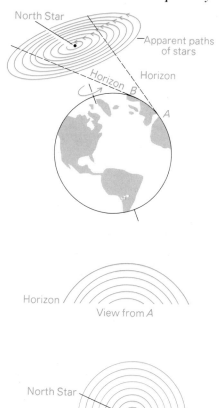

36.1 At high latitudes more stars remain above the horizon all night than at low latitudes, evidence of the earth's sphericity.

As early as the fifth century B.C., Parmenides of Elea declared that the earth was a sphere. It is likely that he was led to this conclusion from an analysis of the tales of travelers, who discovered that, when they went north, a greater number of stars remained above the horizon all night (Fig. 36.1) and that, when they went south, additional stars became visible (for instance, Canopus, not visible in Greece). The early travelers also reported that the length of the day changed with what we call latitude, a rather difficult fact to explain in terms of a flat earth.

Another major Greek figure sometimes credited with teaching that the earth was a globe was Pythagoras. There is no doubt that the school he founded, which flourished for 200 years, ultimately adopted this view, but, inasmuch as his successors tended to attribute all discoveries to their master, it is by no means certain that Pythagoras himself held to the sphericity of the earth. The Pythagoreans, for a time, were about the only men who had faith in this idea, and they went on to develop an impressively ingenious though quite incorrect theory of the universe.

In time, as the notion that the earth was round became accepted in ancient Greece, attempts were made to estimate its size. Aristotle quotes 400,000 stadia for the circumference—much too big—without stating the source of this figure, which was probably obtained from the work of the earlier and highly talented mathematician and astronomer Eudoxus. Archimedes, also omitting references, later gave the circumference as 300,000 stadia—which is better, though still 20 percent in error.

The best of the early measurements of the earth's circumference was made by Eratosthenes (276–194 B.C.), who spent the latter half of his life in charge of the great library at Alexandria. Eratosthenes knew that, at Syene, which was due south of Alexandria, the sun was directly overhead at noon on the first day of summer. On the first day of summer in 250 B.C. he carefully measured the extent to which the sun's rays slanted away from the vertical at noon in Alexandria, an angle he found to be $\frac{1}{50}$ of a complete circle, or a little over 7°. Since the distance from Syene to Alexandria was 5,000 stadia, the circumference of the earth, corresponding to a full circle of 360°, must be 50 times 5,000, or 250,000 stadia. How long is a stadium? There were several different stadia in use in the ancient world, but it seems likely that Eratosthenes used the stadium of 517 ft that was employed by the professional pacers of the time in surveying distances. This means a circumference of 24,500 miles, not far from the 24,860-mile meridional circumference of more recent determinations. In metric units, the earth's mean radius is 6.3712×10^3 km.

the earth's radius is approximately 6,400 km, or 4,000 miles

shape

The earth, we are usually told, is a sphere. Actually, it is closer to being an *oblate spheroid*, meaning that it is somewhat flattened at the poles and bulges slightly at the equator. Why does it have this particular shape? Why is it not shaped like an egg, like a pyramid, or like a corkscrew?

the earth is an oblate spheroid

We can answer the above question by considering the pressures beneath the earth's surface. Pressures in water are familiar enough: swimmers can descend only a few dozen feet, submarines only a few hundred, before the pressure becomes dangerously high. These crushing forces are due simply to the weight of overlying liquid, to the earth's gravitational attraction for the upper layers of water. Less familiar are pressures in solid rocks. Most ordinary rocks are between two and three times as heavy as water—which means that pressures in rocks, owing simply to their weight, should be more than twice as great as pressures at similar depths in the ocean. With increasing distance beneath the earth's surface, rock pressures quickly become enormous. Below a depth of about 20 km, the pressure is so great that solid rock will flow in response to it. We have no direct information about what kinds of material exist 20 km down, but we can be sure that they behave somewhat like very thick liquids in response to pressure changes. This means that one part of the earth cannot project out very much farther than other parts; if it did, pressures beneath it would be greater than under surrounding regions, and the rock beneath it would flow out to the sides until pressures were equalized.

gravity is the cause of the earth's spherical shape

In other words, gravity alone tends to give the earth a spherical shape, to keep all parts of its surface at the same distance from the center. Such minor irregularities as mountains and ocean basins do not greatly disturb the pressure balance, but no large protuberances can exist.

The cause of the bulge at the equator is inertia. The force of gravity on an object on the rotating earth is reduced by the centripetal force $F_c = mv^2/R$ necessary to keep the object moving in a circle despite its tendency to proceed along a straight path. Here m is the object's mass, v is its speed as it moves with the rotating earth, and R is its distance from the axis of rotation of the earth (which is a line joining the North and South Poles). The *effective* weight of the object is therefore

$$W = mg - \frac{mv^2}{R}$$

where the first term represents the gravitational pull of the earth and the second represents the force needed to keep the object traveling in a circle. At the equator R is large but v^2 is extremely large, whereas in the polar regions R is small but v^2 is extremely small; hence the magnitude of the second term is greatest at the equator, where it amounts to about $1/300$ of the gravitational force mg there. As Fig. 36.2 indicates,

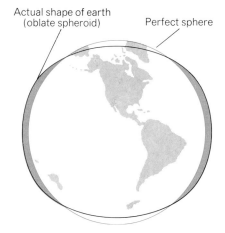

36.2 The influence of its rotation distorts the earth into an oblate spheroid. *The effect is greatly exaggerated in the figure; the equatorial diameter of the earth is actually only 27 miles more than its polar diameter.*

centrifugal distortion

the greater effective downward force at the poles than at the equator leads to a flattening of the earth at the poles and a bulging at the equator, with the resulting shape being that of an oblate spheroid. The total distortion is not great, for the earth is only 43 km wider than it is high. The effect is customarily called *centrifugal distortion*, since the equatorial regions are farthest from the center.

The earth is about 0.34 percent away from being a perfect sphere. Venus, which rotates very slowly, has negligible centrifugal distortion; Jupiter, Saturn, and Uranus, all of which turn rapidly on their axes, exhibit distortions of 6.2, 9.6, and 6 percent, respectively, and are conspicuously flattened at their poles.

The overall flattening of the earth is not its only deviation in shape from a perfect sphere. It is somewhat pear-shaped: the equatorial bulge is about 8 m south of the exact middle of the earth. Our planet is also warped and dented, and on top of this its skin is wrinkled into mountains and valleys both above and below sea level. Still, although the earth is a pretty irregular object referred to the size of man, viewed from outside it would seem smoother than the finest billiard ball. The total range from the Pacific depths to the summit of Everest is a trifle less than 20 km, which is less than $\frac{1}{3}$ of 1 percent of the earth's radius (Fig. 36.3).

There are three widely used models of the earth's shape. These are illustrated in Fig. 36.4. The first is the sphere, but for a great many scientific purposes this form is not adequate. Next is the oblate spheroid, which incorporates the equatorial bulge. This is the form the earth would have if there were no irregularities in its internal structure. The "standard earth" was computed in terms of an oblate spheroid by J. F. Hayford in 1910, and the dimensions of the *Hayford spheroid*, as it has become known, are those of the regular geometric form that is closest to the contours of the actual earth.

The next approximation to the configuration of the earth is the *geoid*. The geoid mirrors the bulges and indentations that are averaged out in the Hayford spheroid, but it ignores such small-scale irregularities as mountains. In essence the geoid is the sea-level equivalent earth; a plumb bob suspended anywhere on the geoid would always hang perpendicular to its surface, which is not true of the idealized Hayford spheroid or, for that matter, of the actual earth.

Over a hundred years ago Sir George Stokes pointed out that the geoid could be determined on the basis of gravity measurements made all over the world. Today the art of evaluating the force of gravity in terms of the acceleration g that it imparts to free-falling bodies is highly developed, and exceedingly precise gravimeters have been built that make use of either a vibrating spring or an oscillating pendulum. How can gravity measurements yield the shape of the geoid? The answer is that, if we are far from the center of the earth on the crest of a bulge, the force of attraction is weak, whereas, if we are in a depression, the attractive force is greater than at the "normal" surface. (Of course, if we descend into the earth's interior in a deep hole, the gravitational force will diminish until it is zero at the center. However, the force is actually larger than usual if we are on the earth's surface in a dent, so to speak, since then we are closer to the bulk of the mass of the earth without having any above us.) Means have been devised to take into account the effects on

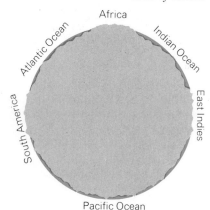

36.3 Cross section of the earth at the equator. (*The vertical scale is exaggerated.*) Note relatively large extent of oceans.

gravity of local anomalies, such as mineral deposits, which would otherwise make it impossible to evaluate the geoid.

mass

It is possible to turn the pages of a reference handbook and find there the statement that the earth has a mass equivalent to 5,983,000,000,000,000,000,000,000 (or 5.983×10^{24}) kg, give or take a thousand billion billion kg or so. To someone not acquainted with the indirect methods of science this must seem a preposterous assertion, particularly since it implies that the density of the earth as a whole is twice that of any of the materials of which its surface is composed. Yet there is no dispute about this figure in the scientific community, and in fact it is one of the more secure aspects of our knowledge of our planet.

Isaac Newton was the first to suggest a method for finding the mass of the earth. His idea was this: when we suspend an object from a string, *ideally* it hangs "straight down"—meaning that it points toward the center of the earth. However, if there is a mountain nearby, the gravitational pull of the mountain causes the plumb bob to be deflected from the vertical (Fig. 36.5). The precise amount of the deflection is a measure of the ratio between the mountain's mass and the mass of the rest of the earth. Finding the mass of the mountain is, in theory, a simple matter: first determine its volume, an elementary problem in surveying, and then multiply by the density of the rock of which it is made. Since density is mass per unit volume, this gives the required number, and from the angle made by the plumb bob the earth's mass can then be computed.

In 1738, eleven years after Newton's death, Pierre Bouguer carried out an experiment of this kind. It is evident that the larger the mountain, the more the deflection and, hence, the greater the accuracy. Bouguer went to Mt. Chimborazo, in what is now Ecuador, whose 6,295 m (20,640-ft) elevation is indicative of its enormous size. His data were not especially good, however, partly because of uncertainty in the mass of the mountain and partly because of the crude nature of his instruments, but at least he was able to establish that the earth's interior was neither empty nor filled with water, as certain of his contemporaries believed.

The next attempt was made in 1776 by Nevil Maskelyne, then Astronomer Royal, in Scotland. His estimate of the terrestrial mass was 4.9×10^{24} kg, which was later raised to 5.4×10^{24} kg when the mass of the mountain (Schiehallion) was ascertained more precisely. The latter figure is not too far away from the current value of 5.983×10^{24} kg, which is remarkable in view of the approximate nature of the entire experiment.

A much better method was proposed by John Mitchell. Let us focus our attention on a 1-kg mass on the earth's surface. Its weight W, which is the force with which the earth attracts it, is given by

$W = mg$

$= 1 \text{ kg} \times 9.8 \text{ m/sec}^2$

$= 9.8 \text{ newtons}$

Sphere

Spheroid

Geoid

36.4 Three approximations to the shape of the earth. *The deviations from a perfect sphere are greatly exaggerated in the sketches.*

weight of 1 kg on the earth's surface

There is another way of finding the force of attraction on this 1-kg mass. Newton's law of gravitation states that, in general,

$$F = \frac{Gm_1 m_2}{r^2}$$

Here, if the force is that exerted by the earth on the 1-kg mass, we can let m_1 be 1 kg and m_2 be the unknown mass of the earth m_{earth}. R is the distance between the center of the earth and the surface of the earth, which is an average of 6.37×10^6 m. The constant G is 6.67×10^{-11} newton-m^2/kg^2. Thus,

gravitational force of earth on 1 kg calculated from the law of gravitation

$$F = \frac{6.67 \times 10^{-11} \times 1 \times m_{earth}}{(6.37 \times 10^6)^2} \quad \text{newtons/kg}$$

$$= 1.64 \times 10^{-24} \, m_{earth} \quad \text{newtons/kg}$$

But this force of attraction is just the weight of the 1-kg mass, which we found to be 9.8 newtons. Therefore

$$W = F$$

$$9.8 \text{ newtons} = 1.64 \times 10^{-24} \, m_{earth} \quad \text{newtons/kg}$$

$$m_{earth} = \frac{9.8}{1.64 \times 10^{-24}} \text{ kg}$$

$$= 5.98 \times 10^{24} \text{ kg}$$

The mass of the earth is 5.98×10^{24} kg, which, in more familiar units, turns out to be 6.6×10^{21} tons.

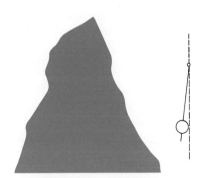

36.5 *The earth's mass can be determined from the amount a plumb bob is deflected from the vertical near a mountain of known mass.*

density

The average radius of the earth is 6.37×10^6 m and its mass is 5.98×10^{24} kg. From these figures we can compute the earth's mean density, which is, since the volume of a sphere of radius r is $\frac{4}{3}\pi r^3$,

$$d = \frac{\text{mass}}{\text{volume}} = \frac{m}{\frac{4}{3}\pi r^3} = \frac{5.98 \times 10^{24} \text{ kg}}{\frac{4}{3}\pi \times (6.37 \times 10^6 \text{ m})^3}$$

$$= 5.52 \times 10^3 \frac{\text{kg}}{\text{m}^3} = 5.52 \frac{\text{g}}{\text{cm}^3}$$

the earth's average density is double that of surface rocks

Since the mean density of surface rocks is about 2.7 g/cm^3, the earth as a whole has more than twice the density of the crust. Evidently the earth's interior must consist of extremely dense materials. What are they likely to be? What is their physical state? Is the interior a homogeneous mass or does it have a definite structure? In the remainder of this chapter we shall examine some plausible answers to these questions and the indirect methods by which they have been obtained.

EARTHQUAKE WAVES

earthquakes

Probably the most valuable assistants we have in exploring the depths of the earth are earthquakes. An earthquake, the most destructive of natural phenomena, consists of rapid vibratory motions of rock near the earth's surface. A single shock usually lasts no more than a few seconds, but in that time may do immense damage to life and property. The rapidity of the vibrations rather than the actual amount of motion is responsible for the damage: Rigid, man-made structures are shaken to pieces because they are unable to follow the fast back-and-forth motions of the underlying rock.

Unlike most volcanic eruptions, earthquakes come without warning. Usually the first shock is the most severe, with disturbances of lessening intensity following at frequent intervals for days or months afterward. A major earthquake may be felt over an area of many thousands of square miles, but its destructiveness is limited to a much smaller area. Small quakes shake some part of the crust daily, but destructive shocks are commonly separated by intervals of months or years.

Earthquakes are produced in a number of ways—by landslides, artificial explosions, movement of magma before and during volcanic eruptions—but most are due to the sudden dislocation of solid rock along faults. Such faults, as we know, are the scars of earlier fractures, which occurred when the stresses developed within the earth became too great for the rock to support. An additional stress, if large enough, may cause a further slippage, and this slippage in turn sends out shock waves that can be felt over thousands of square miles in the case of a major earthquake. As far as the earth's interior is concerned, an earthquake is like a vast explosion that sends out vibrations everywhere. The great majority of earthquakes originate within a few miles of the surface, but some have been recorded which originated at depths as great as 700 km.

most earthquakes are due to rock movement along faults

A number of extremely sensitive instruments, called seismographs, have been devised which respond to the vibrations of even distant earthquakes (Figs. 36.6 and 36.7). Several hundred seismological stations are in operation around the earth, and the data they obtain are routinely compared and correlated to extract from them the maximum of information. It is possible to infer from seismological data the precise location (or *focus*) of an earthquake and something about the energy it has released.

seismographs detect vibrations due to earthquakes

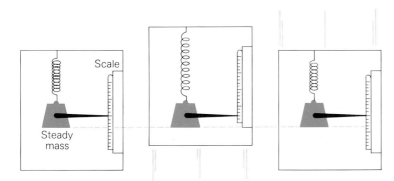

36.6 The principle of the vertical seismograph. *The suspended mass has a very long period of oscillation, hence it remains very nearly stationary in space as the box and scale move up and down when earthquake waves arrive. Only vertical movements of the earth's surface are recorded by this instrument.*

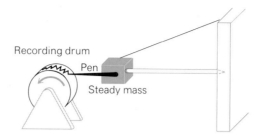

36.7 A horizontal pendulum seismograph. *This instrument responds to horizontal movements of the earth's surface.*

the richter scale of earthquake magnitude

Earthquake magnitudes are usually expressed on the *Richter scale*, which is based upon the maximum amplitude of an earthquake's vibrations. Each step of 1 on this scale represents a change in vibration amplitude of a factor of 10 and a change in energy release of a factor of about 30; thus an earthquake of magnitude 5 produces vibrations 10 times larger than one of magnitude 4 and evolves 30 times more energy. An earthquake of magnitude 0 is one barely capable of being detected nearby, and is equivalent to the explosion of perhaps a pound of TNT, whereas the very greatest earthquakes have magnitudes of about 8.6, equivalent to the explosion of perhaps 7 billion tons of TNT. The energy released in a magnitude 0 earthquake is just about sufficient to blow up a tree stump, whereas that released in a magnitude 8.6 earthquake is approximately double the energy content of the coal and oil produced each year in the entire world.

major earthquakes and active volcanoes both occur in the same belts around the earth

Of the million or so earthquakes per year strong enough to be experienced as such (that is, of magnitude 2 or more), no more than a small proportion liberate enough energy to do serious damage to man-made structures. Only about fifteen really violent earthquakes (magnitude 7 or more) occur each year on the average. Regions in which severe earthquakes are comparatively frequent include the mountain chains fringing the Pacific and a broad belt extending from China across southern Asia into the Mediterranean basin. Major earthquakes have occurred sporadically elsewhere, but the greater number have been concentrated in these zones. In or close to the earthquake belts lie most of the world's active volcanoes—a fact that suggests a connection between the origins of the two phenomena (see Fig. 35.2).

36.8 Two kinds of earthquake waves. *The P waves are longitudinal pressure waves, condensations and rarefactions like those in sound. The S waves are transverse waves, like waves in a taut string.*

earthquake waves

The vibrations caused by an earthquake spread out by means of three kinds of waves, P (for primary), S (for secondary), and L (for long). P waves are pressure waves that involve back-and-forth motions essentially the same as those of sound waves (Fig. 36.8). S waves, on the other hand, are transverse motions in a rigid material in which the vibrations take place at right angles to the direction in which the wave travels. They might be compared with the vibrations in a taut string that move down the string when one end is given a quick jerk. An easy way to distinguish between P and S waves is to think of P waves as "push-pull" vibrations

and S waves as "shakes." Earthquake P waves usually occur 2 or 3 sec apart, whereas S waves are 10 to 15 sec apart.

P and S waves are both body waves that occur within a volume of matter and hence can travel through the earth's interior. L waves, however, are restricted to the earth's surface; they are analogous to water waves in that particles of matter in their paths oscillate in complex orbits, not just back and forth or up and down. (See Fig. 9.1). Earthquake L waves usually have periods of from 10 sec to 1 min.

A simple seismogram is shown in Fig. 36.9. P waves are the fastest, and they therefore arrive first at a seismograph station when an earthquake occurs somewhere. The S waves, which are slower, arrive next, with the time lag between them depending upon the distance from the earthquake focus to the seismograph. The surface L waves, which have to travel *around* the earth instead of *through* it, arrive last, but their vibrations may be stronger than the others, particularly when the distance is no more than a few thousand miles. Figure 36.10 shows the relationships between travel time and distance (measured on the earth's surface) for P, S, and L waves. The farther P and S waves have gone, the greater their average speed, which signifies that these waves travel faster the deeper they penetrate into the earth. The speeds of P and S waves range from $5\frac{1}{2}$ to 14 km/sec and from 3 to 7 km/sec respectively. L waves travel along the surface at the more or less uniform speed of 4 km/sec.

The time-distance curves of Fig. 36.10 make it easy to find the location of an earthquake from seismograms recorded at several stations. The observed time interval between the arrivals of the P and S waves at each station is compared with the curves, and the distance for which $t_{S\ waves} - t_{P\ waves}$ equals the observed time interval is the surface distance from that station to the earthquake. Then a circle is drawn on a globe around each station, with the radius of the circle equal to the computed distance. The intersection of the circles is the location of the earthquake's *epicenter*, the point on the earth's surface directly above its focus. (Fig. 36.11). The depth of the focus can be inferred from such factors as the relative intensities of the various waves (the surface L waves of a deep-focus earthquake will be weak or absent, for instance) and discrepancies in the epicenter location as obtained by stations at different distances away.

INTERIOR STRUCTURE

Earthquake P and S waves do not travel in straight lines within the earth. There are two reasons for this. The first is that, as we have already noted, the speeds of both kinds of waves increase with depth, so that their paths are normally curved owing to refraction. The second reason is more

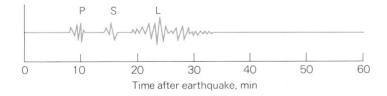

Seismic surveys are valuable in searching for oil and gas deposits. In the background a small ship detonates explosive charges underwater in the North Sea while seismographs in the ship in the foreground record waves reflected from various layers in the sea floor.

36.9 A simple seismogram of waves from an earthquake that occurred about 5,000 km from the recording station.

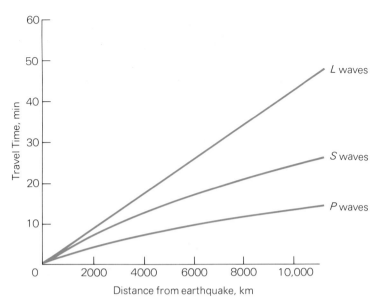

36.10 Time-distance relationships for earthquake waves. P and S waves have higher speeds the deeper they penetrate into the earth; L waves, which travel on the surface, have a constant speed.

spectacular: there are layers of materials having different properties within the earth.

core and mantle

When an earthquake wave traveling in one layer reaches the boundary, or *discontinuity*, that separates it from another layer in which its speed is different, both refraction and reflection occur (Fig. 36.12). The refracted wave shows an abrupt change in direction, unlike the more gradual

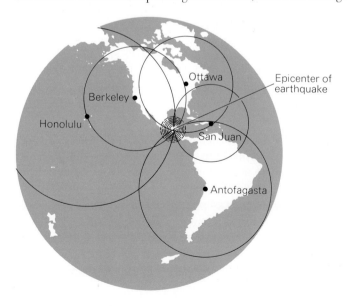

36.11 The time of arrival of earthquake waves at a seismograph depends upon the distance of the quake and the nature of the waves. If the times of arrival of P and S waves at several seismographic observatories are known, the location of the earthquake can be determined graphically.

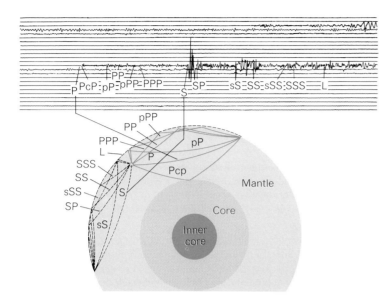

36.12 A typical seismograph record of an earthquake showing the paths taken by the various waves. The earthquake here occurred at a greater depth than usual.

36.13 Earthquake waves spreading through the earth. The existence of a shadow zone for each earthquake is evidence of a central core.

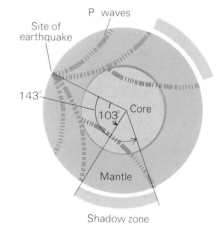

change due to speed variations within each layer. The wave reflected from a discontinuity may be of the same kind as the incident wave, in which case it is called a PP or SS wave if one reflection has occurred, or a PPP or SSS wave if two have occurred. It is also possible for the incident wave to be transformed upon reflection into a wave of the other kind, so that a PS wave is one that started out as a P wave but became an S wave on reflection and an SP wave has had the opposite history; PPS, SPS, and similar multiply-transformed waves have been observed.

Now let us suppose that an earthquake of large magnitude occurs somewhere. We consult the various seismological observatories and find that most though not all of them have recorded P waves from this event. Curiously, the stations that did not detect any P waves all lie along a band from 103° to 143° distant from the earthquake (Fig. 36.13), and we would find, if we consulted the records of other earthquakes, that no matter where they took place, similar *shadow zones* existed. This is the clue that confirmed an early suspicion that the earth's interior is made up of concentric layers.

Figure 36.13 shows why this conclusion is necessary. In the picture the earth is divided into a central *core* and a surrounding *mantle*. P waves leaving the earthquake are able to go directly through the mantle only to a limited region of the surface slightly larger than a hemisphere. Those P waves that impinge upon the core are bent sharply toward the center of the earth, and, when they emerge, they are 4,000 km or more away from those P waves that just barely cleared the core. From an accurate analysis of the available data, it was found that the mantle is 2,900 km thick, which means that the core has a radius of nearly 3,500 km, over half the earth's total radius. However, the core constitutes less than 20 percent of the earth's volume.

Supporting the above finding and giving further important information about the nature of the core is the behavior of the S waves. These, it

shadow zones

the earth has a central liquid core and an outer solid mantle

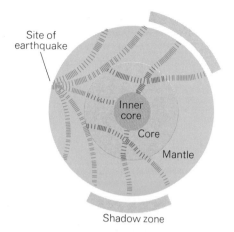

36.14 The inner core. *Faint earthquake waves in the shadow zone suggest the presence of a smaller solid inner core within the liquid core.*

is found, cannot get through the core at all. The only explanation is that the core is in liquid form; this would account not only for the absence of S waves in the core but also for the marked changes in the velocity of P waves when they enter and leave the core. The molten state of the core has other important consequences, most notably with respect to the earth's magnetism, which we shall return to shortly.

As time went on, the division of the interior of the earth into a liquid core and a solid mantle began to be less certain. Especially sensitive seismographs detected faint traces of P waves in the shadow zone, which should not have been able to get there at all. It seemed impossible to account for the appearance of such P waves without attributing it, more or less vaguely, to diffraction, by analogy with the bending of light waves around the edge of an obstacle. Unfortunately this stratagem, although qualitatively plausible, did not give the correct results when worked out in detail.

In 1936 it was proposed that within the liquid core there was a small solid inner core that could bend certain of the P waves reaching it so that they could reach the shadow zone. This effect is shown schematically in Fig. 36.14, and subsequent research has indicated that this is substantially correct. The radius of the inner core is 1,360 km.

the crust

the crust of the earth is relatively thin

From observations made on a 1909 earthquake it became clear that there was a distinct difference between the surface regions of the earth and the underlying mantle. In fact, the line of demarcation between the mantle and the crust above it is quite sharp and is known as the Mohorovicic discontinuity, after its discoverer. Under the oceans it is seldom much more than 5 km thick; under the continents it averages about 33 km, and it may reach 60 km under some mountain ranges (Fig. 36.15). For this reason the once projected hole through the crust to the mantle—the Mohole—was to be drilled through the ocean floor.

36.15 *General structure of the earth's crust.*

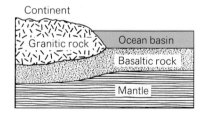

composition

the mantle may consist of olivine

We have found four different layers in the earth: a central body that is probably solid, a liquid outer core surrounding it, a solid mantle, and a thin surface crust. Of what are they composed?

Pending the boring of a hole 6,371 km deep, anything said about the earth's interior is essentially conjecture, but there is a good deal of indirect evidence to support the conjectures that have been made. Most of the evidence in hand concerning the mantle suggests that it consists largely of the mineral olivine, whose several forms are variations on the basic ferromagnesium silicate $(Mg,Fe)_2SiO_4$. Low down in the mantle, where immense pressures are encountered (over a million pounds per square inch) the usual crystal structure of olivine has probably changed to another form. In the inner 2,000 km of the mantle, olivine no longer exists as such, but its components are thought to exist in separate phases such as silica, iron oxide, and magnesia.

Now we come to the liquid core. Since the material constituting the mantle is not much denser than surface rocks, the core must be very heavy in order to account for the large average density of the earth. There are several clues that point to iron as the logical candidate. It has the right density, it is in the liquid state at the pressure and temperature of the core, it is a rather abundant element in the universe generally, and it is a good conductor of electricity, which is a necessary qualification in order that the earth's magnetism may be explained. Because meteorites—fossil fragments left over from the creation of the solar system—contain iron only as an alloy with a little nickel, a good guess is that there is nickel in the core also.

the core probably consists largely of molten iron

Between the outer and inner cores is another transition region, whose presence is known from seismic data but whose properties remain mysterious. Then we finally arrive at the inner core, the kernel of the earth, which many geophysicists believe to be mostly crystalline iron.

temperature

Whether or not the fires of Hell lie beneath the surface of the earth, there is no doubt that it is hot enough there to satisfy most critical requirements for this region. Many lines of evidence point to the existence of such high temperatures. The most direct evidence comes from simply taking a thermometer down into the deepest mines and wells; we find that, on the average, the temperature goes up by about 1°C for every 30 m of depth. If we were to extrapolate this rate of increase all the way down to the earth's center, the temperature would have to rise to 200,000°C—which is absurd, of course, but nevertheless an indication that considerable temperatures may be expected there.

One clue to the temperatures inside the earth and their cause is the flow of heat outward. Measurements of this flow have been made in various locations, and it has been found to be virtually the same all over the earth. The total amount of heat evolved per year is immense, about 10^{21} joules, which is perhaps 100 times greater than the energy involved in such geological events as volcanoes and earthquakes (but still small compared with the energy received from the sun and reradiated into the atmosphere). There is plenty of heat to spare to account for mountain building and other deformations that occur in the crust. In fact, the geological history of the earth is predominantly a consequence of the steady heat streaming through its outer layers.

Part of the earth's heat is a relic of its formation, but most of it comes from radioactive materials. The four isotopes involved in the radioactive heating are potassium 40, uranium 235, uranium 238, and thorium 232. (Ordinary potassium contains only a minute fraction of its radioactive isotope, but there is enough potassium in the earth for it to be significant.) We can estimate the amounts of these isotopes that are present and their distribution. By combining these guesses with the conjectured original temperatures, we can trace back what has been happening to the earth since its formation. The results show that in the interior, below about 800 km from the surface, the earth has been growing warmer and warmer since it solidified. At the core-mantle boundary, the temperature has gone

radioactivity is responsible for heating the interior

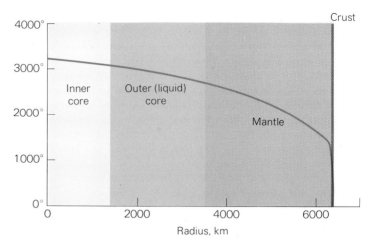

36.16 *Estimated temperatures within the earth, in °C.*

up about 300°C. The very top layers heated up at first but, after an initial billion years or so, began to cool gradually.

A strong point in favor of this theory of thermal history of the earth is that it suggests that the crust may have been so strongly heated in the billion years after the earth was formed that it remelted. This would account for the puzzling fact that, although the earth is at least $4\frac{1}{2}$ billion years old, rocks older than about $3\frac{1}{2}$ billion years have never been found.

The present temperature distribution within the earth is believed to increase fairly rapidly in the mantle from less than 1,000°C at its top to perhaps 2,700°C at the core boundary. The rise is slower in the core, and the temperature at the center of the earth is estimated to be in the neighborhood of 3,200°C—although some geologists think it may be as much as 6,000°C (Fig. 36.16). These figures are not to be taken too literally, of course, but they are indicative of the temperatures we might expect to find in the earth's interior.

TERRESTRIAL MAGNETISM

Although the earliest description of the compass and its use in navigation that we have was published by Alexander Neckham in 1180, there is

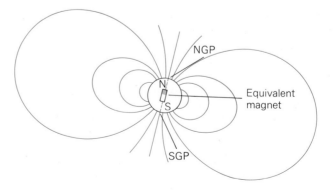

36.17 *The earth's magnetic field. The configuration of the field is approximately that which would be produced by an imaginary bar magnet located within the earth. The positions of the geomagnetic poles (NGP and SGP) are indicated.*

little doubt that knowledge of the compass was widespread even further back in antiquity. Until 1600, however, it was believed that this phenomenon had its origin in an attractive force exerted by Polaris, the North Star, on magnetized needles. In that year, Sir William Gilbert, physician to Elizabeth I, wrote of experiments he had performed with spherical pieces of lodestone, a naturally magnetized mineral. By comparing the direction of the magnetic force on a test iron needle at various positions near the lodestone sphere with similar measurements made over the earth's surface by explorers, Gilbert concluded that the earth behaves like a giant magnet—"magnus magnes ipse est globus terrestris."

the external magnetic field

Today a great deal is known about geomagnetism, and an equally substantial amount is still beyond our grasp. Essentially the information we have is confined to the magnetic field at the earth's surface and above it. From this information we can deduce that the magnetic field originates within the earth and that the field must be very strong in the interior. But the big problem of how the field is generated has had only a partial solution so far.

Nearly all of the magnetic field at the earth's surface is equivalent to the field that would be produced by an ordinary bar magnet (or *dipole*) of enormous power located a few hundred miles from the center of the earth and tilted by 11° from the direction of the earth's axis (Fig. 36.17). No such magnet can possibly exist, since iron loses its magnetic properties above about 1,400°C and temperatures exceeding this figure are present in all but the upper 25 km of the earth. However, this dipole is a useful fiction for making computations and serves as a definite target for theories of the origin of the earth's magnetism to aim at. The points at which the magnetic axis intersects the earth's surface are called the *geomagnetic poles* and are indicated in Fig. 36.17.

the external geomagnetic field is like that of a bar magnet

Besides this dipole magnetic field there are irregularities of various kinds also present in the magnetic field of the earth. Some of them are due to local deposits of iron ore, and magnetic prospecting is a well-developed tool for investigating geological structures as well as for simply looking for iron. Other anomalies come from within the earth and, in a sense, represent malfunctioning in the operation of the geomagnetic dynamo. The net result is that compasses point to slightly different "magnetic poles" in different regions around the world, and these variations must be indicated on navigation charts. A point called a *dip pole* has been located in each hemisphere at which a compass needle, if suspended freely, would point straight down, and these dip poles are popularly called the *magnetic poles*.

Apart from its very existence, the most intriguing aspect of the earth's magnetism lies in the changes it has experienced in the past. That such changes occur was discovered quite early in the game, when Henry Gellibrand noticed in 1634 that the direction to which a compass needle pointed in London seemed to vary from year to year. From 1576, the earliest date for which appropriate records exist, a compass in London

the geomagnetic field varies both in strength and in direction

The first comprehensive magnetic surveys of the earth were made by the wooden ship Carnegie. Modern practice is to tow a magnetometer some distance behind a ship to be clear of the effects of its steel hull and electrical machinery.

would have gone from 8° east of true north all the way over to 24° west in 1823, whereupon it would have begun meandering back until now it is about 8° west. Magnetic records for Paris, starting in 1617, exhibit similar wandering. From these data it would seem that the geomagnetic axis is moving around within the earth so that it makes a complete circle in about 500 years. But, when records for other parts of the globe are examined, this conclusion is less attractive; these compass variations are probably local affairs not shared by the entire planet.

The strength of the geomagnetic field itself has been measured for over a century and in that period has dropped by about 6 percent. This decrease is definitely not a local anomaly, since the measurements from which it is derived were obtained all over the world. The immediate conclusion that we can draw is that the field must originate in the liquid core, where physical changes can take place rapidly (a century is a very brief interval on a geological time scale) and not in the mantle, whose solid nature inhibits any but long-term phenomena. Although this does no more than reinforce our assumption that the earth's magnetism arises in the core owing to the possibility of electric currents there, it is comforting to have such corroboration.

fossil magnetism

Quite apart from these small but precisely known magnetic variations of the recent past, evidence has been found for changes of a spectacular nature that occurred much further back in the history of the earth.

From the earliest times it has been known that certain rocks are naturally magnetized, but it has been possible only recently to interpret the data so as to yield information about the magnetic field that existed when and where the rocks were formed. In studying fossil magnetism the procedure is to cut a specimen from a geological formation of known age, marking its orientation, and then, in the laboratory, to determine the direction in which it is magnetized. By comparing this direction with the orientation the specimen had in its parent rock, the local direction of the geomagnetic field (that is, the direction in which a compass needle would point) at the time the rock was formed can be found. And these directions often differ considerably from the direction of the present magnetic field.

fossil magnetism

Primary in the interpretation of fossil magnetism is the hypothesis that the geomagnetic and geographic poles have never been far apart. From a theoretical standpoint, as we shall see, this seems to be a necessary consequence of the generation of the magnetic field in the liquid core; experimentally, there are indications that the present $11°$ angle between the geomagnetic axis and the rotational axis is unusually large and that this angle has averaged half this amount or less in the past.

Going on the assumption that the geomagnetic and geographic poles have always roughly coincided, the fossil-magnetism studies show that the continents have changed their positions drastically both with respect to one another and with respect to the earth's interior. This is evident from Fig. 36.18, which shows the apparent tracks made by the North Pole as determined from data obtained in Europe, North America, Australia, India, and Japan. But of course there is only one North Pole, so the continents must have been arranged differently in the past. Other evidence for continental drift is considered in Chap. 38.

polar wandering and continental drift

Another remarkable discovery in fossil magnetism is that the earth's magnetic field has reversed itself a great many times. In numerous locations rock specimens of different ages exhibit opposite magnetic polarities, and it seems that the only explanation in most cases is that the earth's field has reversed itself periodically during the period of formation of the various rocks. In the past 76 million years 171 field reversals are believed to have occurred (Fig. 36.19). These reversals do not represent an incessant sliding of the earth's crust over the mantle. Instead, each time, the geomagnetic field probably dropped to zero in a period of several thousand years and shortly was resurrected with the opposite polarity. Then, after a longer interval of from fifty thousand to a few million years, the field reversed itself once more. Apparently such flip-flops are a regular feature of the dynamo within the earth's core and

the geomagnetic field has undergone many reversals of direction

36.18 *The "wandering of the North Pole in the past few hundred million years" according to fossil magnetic data from various parts of the world.* The different tracks are evidence that the continents have shifted their positions around the globe.

36.19 *Changes in the direction of the earth's magnetic field in the past 76 million years.* The color bands represent intervals when the direction was the same as it is now; white bands represent intervals of reversed direction.

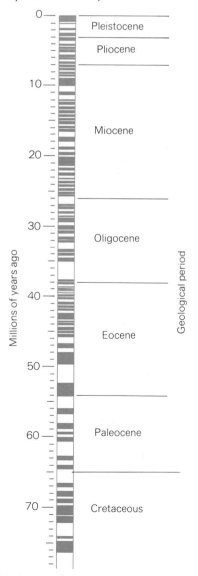

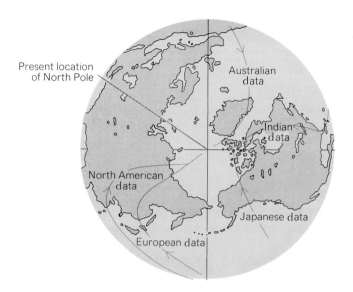

represent one more problem for those who have the temerity to try to explain such a dynamo.

origin of the field

The most successful approach to the origin of the earth's magnetism is based upon the quite new science of hydromagnetism. (New as it is, it has already undergone a change of name from the more cumbersome magnetohydrodynamics.) We must not be fooled into thinking that magnetic water is involved; rather, hydromagnetism treats of the interactions between magnetic and fluid dynamic effects, interactions that take place only in gases and liquids that are good conductors of electricity. Accordingly, hydromagnetism is concerned principally with such astrophysical phenomena as sunspots and galactic structure, but finds application also in the liquid iron core of the earth. (We might remark, as an aside, that hydromagnetic principles are being applied to the generation of thermonuclear power, which, when perfected, will represent almost as great a jump in technology as the first utilization of fire.)

One of the basic principles of physics is that all magnetic fields arise from electric currents. Conversely, all currents are surrounded by magnetic fields. When a coil of wire that is free to rotate is placed in a magnetic field, passing the proper current through it will cause its resultant magnetic field and the external magnetic field to exert forces on each other so as to turn the coil. This is the mechanism behind the electric motor, which operates through the agency of magnetic forces.

There is also a reciprocal effect: when a wire or other conductor of electricity is subjected to a changing magnetic field, a current is induced in it. If we rotate a coil of wire in a magnetic field, then, we have a generator of electric current.

Now let us connect an electric motor and a generator together, electrically and mechanically. At first glance it appears that the combination will go on forever, once we have given it an initial push, with the generator supplying power to the motor, which in turn rotates the generator to produce more power. Unfortunately this cannot actually occur, owing to the inevitable presence of friction, resistance in the connecting wires, and similar agencies of power loss. However, given a small external energy source sufficient to make up the power dissipated in friction, there is no reason why the motor-generator should not continue indefinitely. And, in the vicinity of the combination, we should encounter a magnetic field resulting from the various currents.

In essence most theories of the earth's magnetism invoke a mechanism of this kind, in which there is a coupling of mechanical, electrical, and magnetic phenomena. The required auxiliary energy is presumably supplied by the solid central core in the form of heat, which then produces convective motions in the liquid iron much as a hot radiator produces convective motions in the surrounding air. There is no agreement on the manner in which the heat itself is produced: some authorities feel that it comes from radioactive materials there, while others argue for chemical processes and crystallization as the answer. In any event the required heat is not large, and no one doubts that enough is present.

The chain of events in the core that lead to the observed geomagnetic field may possibly be something like the following. Magnetic lines of force that exist initially are dragged around by the fluid motion so that they form closed loops, like parallels of latitude (Fig. 36.20). (What are actually involved are bodies of fluid in which electric currents are flowing, but it is convenient to speak in terms of lines of force as a kind of shorthand summary of the situation.) These loops cannot be detected outside the core and are believed to be very numerous there. It is in the formation of the loops that energy is fed into the magnetic field, since work must be done by the fluid motions in the core in stretching the original lines of force into their new shapes. Portions of the loops are then twisted by the combination of convection and the effect of rotation into smaller loops that lie in meridional planes (a meridional plane corresponds to a thin orange segment). These loops then coalesce into the dipole—or bar-magnet—field we perceive at the earth's surface, and the cycle starts over.

Of course, the true picture is surely much more involved, but at least the mechanism described can take care of the main features of the observed field. There is no difficulty in explaining the alignment of the magnetic and rotational axes, since the symmetry of motion imposed by the spinning earth must be reflected in the field generated as a result of the spin. Further, the reversals in the external dipole field that took place in the past could have occurred as results of rather minor changes in the pattern of fluid currents in the core. And, most important, the feedback between the two systems of lines of force, the dipole and the internal closed loops, is self-regulating, so that the external field remains fairly constant in the intervals between reversals.

the geomagnetic field originates in the liquid core through a coupling between fluid motions and electric currents there

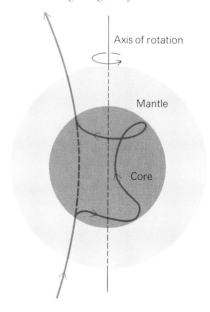

36.20 Formation of loops of magnetic flux within the earth's core, according to one theory of the origin of the geomagnetic field. *Energy is fed into the field through the formation of these loops, which in turn regenerate the external geomagnetic field.*

GLOSSARY

The *centrifugal distortion* of a rotating body refers to its bulging at the equator and flattening at the poles caused by the rotation.

The *geoid* is a theoretical model of the earth's shape such that a plumb bob suspended anywhere on its surface would point directly to its center.

An *earthquake* consists of rapid vibratory motions of rock near the earth's surface; it is usually caused by the sudden movement of rock along a fault.

Earthquake P (for primary) waves are longitudinal oscillations in the solid earth like those in sound waves; earthquake S (for secondary) waves are transverse oscillations in the earth like those in a stretched string; earthquake L (for long) waves are oscillations of the earth's surface like those in water waves.

A *seismograph* is a sensitive instrument designed to detect earthquake waves.

The earth's *core* is a spherical region around the center of the earth whose radius is about 3,500 km; it consists of a solid *inner core*, probably mainly solid iron, surrounded by a liquid *outer core*, probably mainly liquid iron.

The earth's *mantle* is the solid part of the earth from the core to the crust; it is about 2,900 km thick.

The *geomagnetic poles* are the points on the earth's surface through which the magnetic axis of the earth passes.

Hydromagnetism, or *magnetohydrodynamics*, is the science of the behavior of electrically conducting liquids and gases, such as those believed responsible for sunspots and the earth's magnetism.

EXERCISES

1 Distinguish among the L, P, and S waves of earthquakes. Which are the fastest? Which are the slowest?

2 Why is the earth's mantle thought to be solid?

3 Why is the earth's core thought to be composed principally of iron? What is the evidence for the core being liquid?

4 What is the evidence for the existence of a solid inner core?

5 State the various hypotheses involved in the conclusion from magnetic data that the continents have shifted relative to the earth's interior.

6 What provides the energy required to power the dynamo mechanism that causes the earth's magnetic field?

7 Justify the statement made in the text that the geological history of the earth is predominantly a consequence of the steady heat streaming through its outer layers. What is the origin of this heat?

8 What are the intervals (to the nearest minute) between the arrivals of the P and S waves of earthquakes that occur 3,000, 6,000, and 9,000 km from a seismograph?

9 The S waves of an earthquake arrive at a seismograph 8 min after the P waves. How far away did the earthquake occur? When will the L waves arrive?

part ten
THE EVOLVING EARTH

Geological process are, for the most part, very gradual, so that widespread and radical changes in world landscapes do not occur within human lifetimes, but in the course of millions of years they have profound consequences. Tracing the origins of the earth's surface features, like many of the classic problems of science, resembles the solving of a complex jigsaw puzzle; all or almost all of the pieces are in hand, and the trick is to put them together into a coherent pattern. Today, though some pieces are lacking or do not quite fit the rest of the picture, the outlines of that picture exist, and the development of the earth's topography is no longer the impenetrable mystery it once seemed.

Perhaps the most exciting discovery of the geologist in his role of historian is that the continents have been—and still are—drifting around the globe as if on huge conveyor belts. The rock shell, or lithosphere, of the earth is broken up into a number of rigid plates, which are moving apart along fissures in the ocean floors where new molten rock wells up from below. Where the plates come together they buckle, throwing up mountain ranges at the margins of continents and chains of islands in the open ocean. After examining the theory of continental drift, we pass on to an outline of the history of the earth in the $4\frac{1}{2}$ billion years since its formation. Naturally the recent past is easier to trace in the rock record, but a surprising amount of information about the remote past is also available for the curious geologist.

chapter 37

Interpreting the Rock Record

THE PRINCIPLE OF UNIFORM CHANGE
the catastrophic hypothesis
hutton and lyell
the principle of uniform change

METHODS OF HISTORICAL GEOLOGY
dating rock formations
unconformities
the grand canyon

GEOLOGIC TIME
radioactive dating
fossils
geochronology

Geology, the earth science, is in large part an application of physics, chemistry, and biology to the earth. To pursue further our study of rocks would require knowledge of the chemistry of silicates and the physics of crystal structure. To learn more about the earth's interior would require detailed study of wave motion in solids and the behavior of materials under high temperatures and stresses. To delve into the lore of fossils would necessitate an extensive acquaintance with some branches of biology. In all its many aspects geology leans heavily on other sciences.

But in one important sense geology is more than an application of other fields of knowledge: it deals, more than any other science, with the problem of past time. A geologist is concerned not alone with relationships among processes and substances in the present-day earth, but also with the remote origins of earth materials and the changes they have undergone. Concerning a mountain, a chemist might inquire: "What are its rocks made of? How is their composition changing on exposure to the atmosphere?" A physicist would be more curious about the strength of its constituent rocks or about changes in the force of gravity near the mountain. To a biologist the mountain's chief attraction might be the plant and animal assemblages he would find at different elevations. But a geologist would regard all these matters as secondary to the questions: "Why is the mountain here? How can its past history account for its present shape? When was the mountain formed, and what sort of landscape preceded it?"

THE PRINCIPLE OF UNIFORM CHANGE

Already in several discussions we have had occasion to mention the earth's great age and to explain landscapes and rock structures by slow processes acting through long ages. These ideas about past time, so fundamental in modern geology, have been generally accepted among scientists for hardly more than a century. In the first part of this chapter we shall look back into history to see how concepts of the earth's past have changed and how modern ideas have been justified.

the understanding of fundamental geological processes is only about a century old

the catastrophic hypothesis

The emphasis of geology on events in the earth's past hindered early development of the science largely because scientific thinking along such lines quickly ran afoul of firmly established ideas. The book of Genesis, for instance, tells the story of the earth's beginning with beauty and simplicity and an uncomfortable preciseness as to dates. So carefully are events dated in the Biblical account that a learned seventeenth-century theologian, Bishop Ussher, found it possible to compute that the creation of our planet from formless void took place at 9 o'clock in the morning of October 12, 4004 B.C. Now the most casual dabbling in geology shows that changes in climate and sea level clearly recorded in the present landscape cannot possibly have occurred in the space of a few thousand years; yet the best efforts of many early investigators were nevertheless spent in trying to make their findings fit into a literal interpretation of Genesis.

Even apart from Biblical preconceptions, many of the notions current among educated people two centuries ago regarding the earth's history seem to us fantastic. Today it is commonly accepted that most valleys are formed by stream erosion; in the eighteenth century they were often ascribed to the downward slipping of fragments of the earth's crust. Mountains were believed to have risen by gigantic upheavals in some dim, chaotic stage of the earth's beginning. References to ages of universal flood and universal fire are frequent in literary works.

Newton's spectacular success, near the close of the seventeenth century, in reducing the complex motions of the solar system to order and simplicity proved a mighty spur to eighteenth-century scientists to find some similar order and simplicity in the phenomena of the earth's surface. Lacking Newton's genius and faced with a far more complex problem, these lesser philosophers too often let imagination supplant patient observation. One who stood out was a German, Abraham Gottlob Werner. Near his home in Freiberg, Werner found granite overlain by folded, somewhat metamorphosed rocks, these in turn overlain by flat sedimentary beds. Untraveled and deaf to the reports of others, Werner considered this sequence worldwide. Each of the three types of rock, he imagined, was deposited by a universal ocean, granite precipitating first and the flat upper beds last. Thus all rocks, in Werner's system, were sedimentary rocks; and the history of the earth consisted of three sudden precipitations from a primeval ocean, followed by the disappearance of most of the water.

Another man who greatly influenced geology at the beginning of the nineteenth century was the French biologist Georges Cuvier. Primarily an anatomist, Cuvier extended his work from animals of the present to the creatures whose fossilized shells and bones he found in great abundance in the rocks near Paris. This was the first careful, critical study of fossils, the first attempt to catalog the precise similarities and differences between living forms of the past and the present.

In successive rock layers of the Paris region Cuvier found distinct assemblages of animals, different from one another and from the present animals of that region; he concluded that each assemblage appeared on the earth as the result of a special creation and that each was destroyed by a universal cataclysm before the next creation. Thus Cuvier regarded the earth's history as a succession of catastrophes, separated by intervals of more stable conditions.

The theories of Cuvier and Werner have this much in common: both suppose that the history of our planet has been punctuated by tremendous events which have no counterpart in the natural processes we observe today. No modern ocean has ever suddenly precipitated great masses of rock over its entire basin, as did Werner's universal ocean, nor has any modern animal assemblage been suddenly created or destroyed. Because both theories involved events, or "catastrophes," which apparently violated the natural laws of the present world, their central idea came to be known as *catastrophism*. A similar idea, of course, pervades all the other notions of worldwide fire and flood, of the formation of mountains and valleys in a time of chaotic movement in the earth's crust.

according to the catastrophic hypothesis, sudden, spectacular events are responsible for geological changes

hutton and lyell

The first to combat the catastrophic ideas was a Scotsman, James Hutton, greatest of the eighteenth-century scientists who sought to parallel Newton by establishing a complete and coherent "theory of the earth." More clearly than his predecessors, Hutton recognized the erosive work of streams in carrying material from the land into the sea. Opposed to the wearing down of the land, he believed, was the formation of rock from ocean deposits by heat and pressure, and the gradual reelevation of this new rock above sea level. Thus the cycle was continuous: rocks were formed from deposits in the sea, raised up to make land, worn away by streams, and again deposited in the ocean. For every step of this process Hutton thought he could find evidence in the present-day world. The biography of the earth needed no catastrophes, but, as far back as the rocks gave evidence, consisted of repetitions of the same cycle. "In the phenomena of the earth," Hutton said in concluding his book, "I see no vestige of a beginning, no prospect of an end."

hutton saw a cycle of continuous change in the rock record

In detail, Hutton's views are often as erroneous as Werner's. He went especially astray in that part of his cycle which deals with the formation of rocks. More traveled than Werner, Hutton paid more heed to volcanic phenomena, hence to the part that heat and pressure play in rock formation. He recognized granite and basalt as igneous rocks, formed by solidification from the liquid state rather than as precipitates from solution in sea water. But he went too far in asserting that all rocks, even such sedimentary rocks as limestones and sandstone, owed their hardness to pressure, partial melting, and infiltration of hot liquids. He made no clear distinction between igneous and metamorphic rocks, simply regarding heat and pressure beneath the earth's surface as essential for the formation of all rocks.

observational evidence is central in geology, as in every other science

Perhaps the greatest contribution of Hutton and his contemporary supporters to geology was their emphasis on the importance of direct observational evidence. After Hutton came no more philosophers content to fabricate grandiose speculations about the earth's past; instead there came to the fore, especially in England, a group of geologists who understood that what their science needed most was a careful gathering of facts.

A notable exponent of fact gathering in the early nineteenth century was William ("Strata") Smith, a quarryman and surveyor, whose lack of education was partly compensated by immense patience and acuteness of observation. In the wandering over England which his business made necessary, Smith studied carefully the fossils of different localities. Like Cuvier, he observed that fossil assemblages differed from one rock stratum to another; and he presently discovered, as Cuvier had not, that the sequence of fossil assemblages is the same in different localities, so that each stratum can be followed from place to place, even if it is partly covered with soil and vegetation, by means of the fossils it contains. Eventually Smith succeeded in following several different strata over the whole of England and Wales. Sometimes the layers were flat, sometimes arched or tilted, but always they followed one another in regular order, and always they could be recognized by their fossils.

present processes of landscape change are slow, but there is enough time in the earth's history for such processes to have created the observed geological record

Hutton's most important champion was Sir Charles Lyell (1797–1875), who made it the business of his life "to explain the former changes of the earth's surface by forces now in operation." Lyell reemphasized Hutton's point that processes which seem incapable of altering the landscape appreciably may produce great changes if given sufficient time. If the earth's history is confined to a few thousand years, then catastrophism is indeed necessary; but if geologic time extends back for millions of years, the slow processes that we find at work today are sufficient to account for the earth's eventful past.

In extensive travels through Europe and America Lyell sought always for positive data regarding geologic changes of the present—the shifting of stream courses, the building of deltas, advances and retreats of shorelines, the outpouring of lava and ash from volcanoes. Then from rocks and landscapes he read the slow accumulation of these changes through the long past—rivers now cutting far below their former channels, coasts with remnants of old beaches high above the present shore, immense lava flows where no volcanoes exist today. Lyell was the first to distinguish metamorphic rocks as a group, and so to understand clearly the cycle of changes by which rocks are formed, destroyed, and re-formed. He continued the work of Cuvier and his English followers in studying the succession of animals whose remains are entombed in the rocks of France and England. Nowhere did he find evidence of past changes brought about by agencies other than those in the world about him. Lyell's immense accumulation of facts and observations was overwhelming; within a few years after his *Principles of Geology* was published, most of his fellow geologists, at first skeptical, had accepted his views.

One last step was supplied in 1859, when Darwin introduced the theory of evolution into biology. Not only changes in the inorganic world of rocks, but, according to Darwin, changes in living things as well could be explained in terms of processes operating in the present world. Thus the distinct assemblages found by Cuvier in the Paris basin were not special creations but stages in a continuous line of development. Lyell understood at once the importance of Darwin's work and became one of his earliest and most active supporters. The theory of organic evolution knocked the last prop from under the idea of catastrophism.

theory of evolution

the principle of uniform change

The principle of uniform change is simply Lyell's thesis that past changes of the earth's surface are adequately explained by processes now in operation. This concept is as fundamental to geology as the law of conservation of energy is to physics or the periodic law to chemistry.

the principle of uniform change is fundamental to geology

Proposed at first as an antidote to extreme catastrophism, Lyell's earliest concept of uniform change was extreme in the opposite direction, but in later years he modified his position considerably. Modern geologists interpret the law very broadly. They hold with Lyell's basic tenet that processes now in operation are sufficient to account for changes in the

An unconformity separates the Moenkopi formation (lower right) from the overlying Chinle formation in the Uinta Mountains of Utah.

past, but they find good evidence that these processes have not always operated with their present intensity. For example, climates of the world have been colder at some periods, so that glaciers were more widespread; volcanic activity has at times been much greater than at present; mountains have been fewer and lower, so that stream action has been less effective; widespread desert conditions and periods of intense cold have at times wiped out whole races of animals and hastened the development of other forms. Thus our modern concept is a partial compromise with catastrophism: although present-day processes can adequately explain the past, certain combinations of these processes have at times made the face of the earth very different from its present aspect.

The acceptance of uniform change should not obscure the fact that this law, like all scientific laws, is no more than a generalization based on long and careful observation. Just as the law of conservation of energy rests on the circumstance that all known physical processes conform to it, so the principle of uniform change depends on the failure of long searches to find any good evidence for an opposing view.

Many people still prefer to believe in some form of catastrophism. Such a belief cannot be proved wrong; clearly, a sufficient number of local catastrophes and special creations would be capable of accounting for the present form of the earth's surface. But if advocates of this view were to consider *all available evidence* (as ordinarily they do not), they would find necessary an enormous number of assumptions. The idea of uniform change is accepted because it involves the fewest arbitrary assumptions and hence gives the *simplest* explanation for all available facts.

How far back toward the earth's beginning does the principle of uniform change hold? Somewhere it must break down, because the earth in its primitive state must have been shaped by processes different from those now in evidence. All geology can say for certain is that the oldest rocks now exposed show a clear record of the action of processes very similar to those of the present. We shall see in Chap. 45 how astronomy is able to shed some light on the way the earth came into being and the events of its earliest history.

METHODS OF HISTORICAL GEOLOGY

The geologic events of the recent past have been recorded and preserved in the rocks and landscapes of the present, and one task of the geologist is to try to reconstruct these events from his knowledge of the processes currently at work which are once again reshaping the face of the earth. Thus, from moraines, lakes, and U-shaped valleys we learn of the spread and retreat of ancient glaciers; wave-cut cliffs and terraces above the sea suggest recent elevation of the land; hot springs and isolated, cone-shaped mountains show past volcanic activity. Earlier episodes are recorded more dimly in the rocks. A geologist finds a bed of salt or gypsum buried beneath other strata, and he knows that the region must once have had a desert climate in which a lake or an arm of the sea evaporated; from a layer of coal he reconstructs an ancient swamp in which partly decayed vegetation accumulated; a limestone bed with numerous fossils suggests a clear, shallow sea in which grew clams, snails, and other hard-shelled organisms. As the long history is carried further and further back, the evidence becomes always more fragmentary and the geologist's reconstruction of the earth's surface correspondingly vague.

dating rock formations

Historical geology poses two fundamental problems: to arrange in order the events recorded in a single outcrop or in the rocks of a single small region, and to correlate events in this region with events in other parts of the world so as to give a connected history of the earth as a whole. We begin with the first of these problems.

Some of the principles used in reading the history of a small area are not hard to arrive at:

1 In a sequence of sedimentary rocks, the uppermost bed is the youngest and the lowermost the oldest. Thus, in Fig. 37.1, bed *A* must have been deposited before the others and bed *E* after those below it. A not very common exception to this rule is a sequence of strata overturned by intense folding.
2 Diastrophic movement resulting in folding or tilting is later than the youngest bed affected by the deformation. Thus the strata of Fig. 37.1 were obviously not folded until after the deposition of bed *E*.
3 Diastrophic movement along a fault is later than the youngest bed cut by the fault. Faulting in Fig. 37.1 could not have occurred before the deposition of bed *C*.
4 An intrusive igneous rock is younger than the youngest bed that it intrudes. The granite shown in Fig. 37.1 is younger than bed *D*. (This assumes that the *age* of an igneous rock refers to the time at which it solidifies; as magma the rock material may have existed long before the intruded sediments were laid down.)

Obvious as these statements sound, their application in regions of intricately folded and faulted strata requires much ingenuity. The problem is especially difficult in regions where much of the rock structure is hidden by later sediments or vegetation.

unconformities

A structure like that shown in Fig. 37.2 requires further attention. Here the lower, tilted sedimentary beds are cut off abruptly by an uneven surface on which rest the upper horizontal beds. An irregular surface of this sort, separating two series of rocks, is called an *unconformity*.

At first glance the surface looks as if it might be a fault. But unconformities do not show the characteristic minor features of faults—no distortion of beds, no finely granulated material, no grooved and polished surfaces indicating the grinding of one rock against another. An unconformity, moreover, can usually be traced from outcrop to outcrop over an area of many square miles, with the same bed always occurring just above it. Evidently some explanation other than faulting is called for.

elementary principles of historical geology

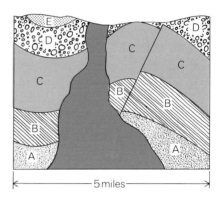

37.1 *Schematic cross section showing folded sedimentary rocks intruded by granite and displaced along a fault.*

37.2 *An unconformity is an irregular surface that separates tilted lower rock strata from horizontal ones.*

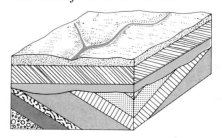

The Grand Canyon in Arizona, 1 mile deep and 13 miles wide, represents about 7 million years of erosion by the Colorado River of uplifted sediments laid down at intervals from 300 million to over 1 billion years ago.

Presumably the lower tilted or folded beds were deposited first, in order from the lowest to the highest. After deposition diastrophic movement deformed them, and subsequently the higher part of the series was removed. The only conceivable agent besides faulting that might have accomplished this removal is erosion. After erosion, conditions changed so that sedimentation began anew, and the horizontal strata were deposited on the old eroded surface.

Thus an unconformity is interpreted as a buried surface of erosion. It always records at least three geologic events:

an unconformity is a buried surface of erosion

1 Diastrophic movement resulting in uplift and exposure of the older rocks
2 A period of erosion when the land surface was above sea level
3 A change of conditions resulting in deposition of sediment on the eroded landscape

Usually the third event involves subsidence, lowering the eroded surface either beneath the sea or to a level where stream deposition could occur. An unconformity is most easily recognized when the lower beds are tilted or folded, as shown in Fig. 37.2, but if the original diastrophic movement is simply a vertical uplift, the lower beds may be horizontal and therefore parallel to those above. It is not necessary that the lower series should consist of sedimentary layers, for the eroded surface may equally well be carved on igneous or metamorphic rocks. The one essential feature of an unconformity is that it represents a surface formed by the processes of erosion and buried beneath later deposits.

Unconformities are important in historical geology for several reasons. First of all, they make possible the approximate dating of past diastrophic movements; obviously the movement responsible for an unconformity must have occurred *after* the latest rocks of the older series and *before* the oldest of the upper layers. Secondly, unconformities tell us something about the distribution of land and sea at different periods of the earth's past, for an unconformity always means that dry land must have existed during the period of erosion that formed it. And unconformities are also important in a negative sense, for they indicate gaps in the geologic record, times when no deposits were forming in particular regions. An unconformity tells us that a region was above sea level, but all details of the region's history for that period are lost.

the grand canyon

For an example of the piecing together of geologic history in a particular region, let us turn to the Grand Canyon of northern Arizona, where the Colorado River has cut a mile-deep gash into the earth's crust.

The essential features that an observant visitor would see in the rocks of the canyon are shown diagrammatically in Fig. 37.3. In the upper part are the massive, nearly horizontal sedimentary layers responsible for the sculptured cliffs and the brilliant hues which have made the canyon so famous. Near the top of the steep inner gorge where the river is now

INTERPRETING THE ROCK RECORD

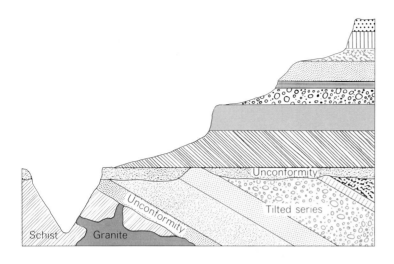

37.3 *Diagrammatic cross section of the Grand Canyon*

cutting, the lowermost of these beds rests on an uneven surface which bevels a series of tilted sedimentary strata. The tilted beds in turn are separated by an irregular surface from a still lower series of dark-colored schists and gneisses, complexly folded and intruded by dikes and irregular masses of gray granite.

The oldest rocks in the canyon are evidently the schists and gneisses of the inner gorge. The history of these early rocks is obscure, for metamorphism has all but obliterated their original structures. Probably they were once sedimentary layers interbedded with lava flows, lying in an approximately horizontal sequence. In a period of diastrophism they were folded and metamorphosed; at the same time or later they were intruded by granite. These are events that accompany mountain building, so we may picture a range of mountains here at some distant time in the earth's past.

a mountain range once stood in the region of the grand canyon

The uneven surface that planes off schist and granite alike is an unconformity, representing an immensely long period of erosion when the ancient mountains were reduced to a nearly level plain. Sinking of the land or the rising of other mountains nearby at length made this plain a basin of deposition, and beds which now form the tilted series were laid down—originally, of course, in a more nearly horizontal position. These beds contain no fossils, so we find it difficult to reconstruct accurately the conditions under which they were formed. Because the rocks are mostly fine-grained sandstone and shale, because each stratum is relatively thin and often shows irregularities of bedding, we may picture as a probable site of deposition the flood plain of a large river. After the deposition of these layers they were tilted in another period of diastrophism, and then long erosion reduced the land to the nearly level surface of the second unconformity, on which the horizontal strata rest.

From here on the record becomes clearer. Many of the horizontal beds contain fossils, and from these or other structures the conditions of their deposition can be inferred with confidence. The thick limestone at the top of the canyon and some of the massive, well-sorted sandstones are

the region of the grand canyon was subsequently below sea level

Sampling the ocean floor with a hollow steel corer off the California coast.

eventually the grand canyon region became raised and was eroded by the colorado river

typical marine deposits. Some thin shale beds show tracks of land reptiles, and so probably represent ancient mudflats along a river or a beach. A sandstone layer shows the rounded grains, good sorting, and large-scale cross-bedding of windblown sand in an arid climate. Transitions between some layers are gradual, suggesting that deposition was continuous in spite of changing conditions. Other layers are separated by distinct unconformities, showing that at times the land was elevated sufficiently for erosion to take place. We need not follow the history in detail, but careful study of the different layers would make possible an accurate reconstruction of changing conditions through a long part of earth history.

The last geologic events recorded at the canyon are elevation of the land high above the sea and erosion by the Colorado River, two events which probably took place simultaneously.

In few places are rocks of many different ages exposed as magnificently as at the Grand Canyon, but patient investigation makes possible the working out of similar histories even where exposures are poor. The histories differ widely in detail, but often the general pattern is much like the one just described: fragmentary glimpses of episodes in the far distant past, separated by enormously long periods of erosion whose separate events have left no trace; a fairly connected story of seas and rivers and deserts recorded in more recent rocks; and finally, events of the immediate past clearly suggested in the present landscape.

INTERPRETING THE ROCK RECORD

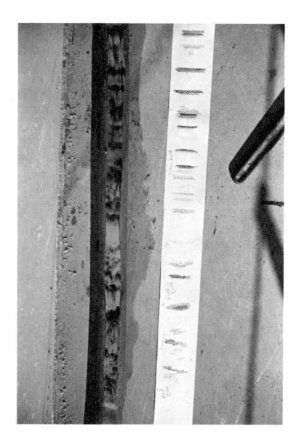

Cores of sediment, such as this post-glacial clay obtained by driving a hollow tube into the floor of Lake Michigan, provide information on marine life and geological events over spans of several million years. White strip is logging paper to record information before changes occur. Gray is ship's deck.

GEOLOGIC TIME

The second major problem in historical geology is the correlation of processes such as those we have described that occurred in different regions. We might trace the sequence of geologic events in two widely separated places on the globe, but we cannot begin to understand the worldwide picture of the evolution of the crust unless we can determine which events at each place happened at the same time. This would present no problem if various rock layers could be followed continuously from one region to the other. In general this is impracticable, however, since a given layer when followed for a long distance is either cut off by erosion or concealed by later deposits.

radioactive dating

Only one means is available for determining geologic time in years, and this method is limited to a few kinds of rock. It involves the measurement of the amount of various radioactive elements present in a rock and depends on the fact that radioactive decay goes on at a constant rate regardless of external conditions. Thus, if a certain amount of a radioactive element is present in a rock at the time of its formation, after a million

the presence of long-lived radioisotopes and their decay products permits the age of a rock to be determined

years a known fraction of the element will have decayed, regardless of what other elements it is combined with or what high temperatures and enormous pressures the rock undergoes.

The procedure may be illustrated with a rock that contains uranium. The chief isotope in this element is U^{238}, which disintegrates in a series of 14 steps into other radioactive elements such as radium and polonium, a series ending with the stable lead isotope Pb^{206} (see Table 17.1). Each transformation in the series takes place by either beta decay or alpha decay so that the end products are the two isotopes Pb^{206} and He^4. Since the rate of decay of U^{238} itself is extremely slow compared with those of the intermediate radioactive elements, for approximate calculations the end products only need to be considered.

The necessary data are the amounts of Pb^{206} and undecayed U^{238} or the amounts of helium and undecayed U^{238}. The amount of either lead or helium gives the quantity of uranium that must have decayed, and this figure added to the amount still there gives the original quantity that must have been present. Comparison of the initial amount with the amount that has decayed tells us the age of the rock, since the rate of decay is accurately known (1.33×10^{-8} percent per year). Because helium is a gas and some may escape from the rock, the most reliable calculations of age are based on lead.

Time measurements with the "radioactive clock" tell us that manlike creatures appeared on the earth about 2 million years ago, that rocks with the first fossil remains of mammals are about 200 million years old, that sea animals with hard shells first became abundant about 600 million years ago. The oldest rocks whose ages have been determined are intrusive rocks from South Africa, roughly 3,500 million years old. These intrude metamorphosed sedimentary rocks which must be still older, but how much older is unknown.

Figures like these are extremely valuable, for they give us an accurate idea of the immense reaches of time involved in geologic processes. Unfortunately, rocks with sufficient uranium to make the measurements possible are scarce. Other radioactive elements can be used, notably Rb^{87} (which beta-decays to Sr^{87} with a half-life of 4.7×10^{10} years), but most of them are not significantly more abundant than uranium. One exception is potassium, a common and widespread element whose K^{40} isotope exhibits feeble natural radioactivity. A product of the radioactive decay of K^{40} is the argon isotope Ar^{40} that can be detected in very minute quantities; the half-life of K^{40} is 1.3×10^9 years. The potassium-argon and rubidium-strontium methods have been extensively used in recent years.

fossils

fossils are the remains or traces in rocks of living things of the past

Perhaps the most fascinating and ingenious technique the geologist has at his command for establishing relationships among rocks of different regions and for arranging beds in sequence makes use of *fossils*. Fossils are the remains or traces of organisms preserved in rocks. The most common, of course, are the hard parts of animals, such as shells, bones, and teeth. On rare occasions an entire animal may be preserved; ancient insects have been preserved in amber, and immense woolly mammoths have been found frozen in the arctic. Plant fossils are relatively scarce, since plants do not contain easily preserved hard parts. The structure of tree trunks is sometimes beautifully shown in petrified wood, which is wood whose original organic materials have been replaced by silica deposited from solution in groundwater. Incomplete decay of buried leaves and wood fragments produces black, carbonaceous material which may preserve the original organic structures—*coal* is a thick deposit of such material. Occasionally fine sediments preserve impressions of delicate structures like leaves, feathers, and skin fragments, even when all trace of the original materials has vanished. Some fossils are merely trails or footprints left in soft mud and covered by later sediments.

Conditions necessary for preservation have been much the same throughout geologic history. Chemical decay, bacteria, and scavengers have quickly disposed of most of the organisms that have lived on the earth, and only special conditions of burial occasionally permit the survival of fossil groups. In general, these conditions are best realized on the floors of shallow seas, where life is abundant and deposition of sediment is sometimes rapid. Our picture of marine life in the past is, accordingly, far more complete than our picture of the organisms that lived on land, but even the marine record is fragmentary. Thick marine strata often contain no fossils at all, and the fossils that do occur are frequently broken and poorly preserved.

Although the fossil record is far from complete, a careful study reveals a great deal about living things of the past. One important fact that emerges from such a study is that groups of organisms show a progressive change in form from those entombed in ancient rocks to those of the most recent strata. In general, the change is from simple forms to more complex forms; in general, also, the change is from forms very different from those in the present world to creatures much like those we find today. These observations are a part of the factual basis for Darwin's theory that life has evolved by a continuous development from simple forms to the complex organisms of the present.

living things have changed gradually from previous simple forms to present complex ones

Because plants and animals have changed continuously through long ages, rock layers from different periods can be recognized by the kinds of fossils they contain. This fact makes possible the arrangement of beds in a time sequence, even when their geologic relationships are not directly visible, and also provides a means of correlating the strata of different localities. If, for example, fossil snail shells and clamshells are found in a rock layer in New York that are exactly similar to fossil shells from a layer in the Grand Canyon, the two layers must be approximately the same age. Suppose that above the layer in New York is an unconformity and that in the Grand Canyon continuous deposition is recorded into a higher layer with a different group of marine fossils; then we can infer that, in this later time, the New York region was a land area undergoing erosion while northern Arizona was still covered by the sea. Thus fossils are an all-important tool in historical geology for linking together the events of distant regions.

Fossils are useful not only in tracing the development of life and in correlating strata but also in helping us to reconstruct the environment in which the organisms lived. Some creatures, like barnacles and scallops,

Removing fossil specimens is usually a delicate task.

live only in the sea, and it is probable that their close relatives in the past were similarly restricted to salt water. Other animals can exist only in fresh water. On land some organisms prefer desert climates, others cold climates, others warm and humid climates. Evidently many details about the environment in which a rock was formed are revealed by its fossil organisms.

geochronology

Fossils make possible a chronological arrangement of geologic events over the entire earth. Enough of these events can be dated accurately by measurements of radioactive decay so that good estimates can be made for the dates of the others.

cenozoic, mesozoic, and paleozoic eras

The most recent 600 million years of geologic history have been divided into three major divisions called *eras*. The era in which we are now living, the *Cenozoic* ("recent life") *era*, began about 65 million years ago. Before that came the *Mesozoic* ("intermediate life") *era*, which lasted 160 million years, and the *Paleozoic* ("ancient life") *era*, which lasted about 375 million years. The geologic record of events before the Paleozoic era is so dim that geologists are not agreed about the proper division of this early time into eras. Although time before the Paleozoic makes up three-fourths of all geologic history, we shall do best to consider it as a single long division, *Precambrian time*—just as in human history we might lump together the longest but least known part of man's development in a chapter called prehistoric man.

precambrian time

The Cenozoic, Mesozoic, and Paleozoic eras have been subdivided into shorter time intervals called *periods*, and the periods themselves into *epochs*. Table 37.1 shows the accepted divisions of geologic time together with information on the living things in each division; only the Cenozoic epochs are included.

INTERPRETING THE ROCK RECORD

The divisions of geologic time shown in Table 37.1 were originally made on the basis of evidence of dramatic changes that seemed to have occurred at intervals in the earth's history—changes in landscapes, in climates, in types of organisms. For example, the disappearance of dinosaurs marks the division between the Mesozoic and Cenozoic eras. However, the boundaries between epochs, periods, and eras are not nearly so clear-cut as they once seemed to be. In one region there may have been at some time in the past widespread diastrophic movement, as revealed to us by marked unconformities, while elsewhere at the same time a more gradual transition occurred between earlier and later rocks. But it is still useful to think of eras, periods, and epochs, and we must accept the difficulty of separating adjacent ones as the price of their convenience.

GLOSSARY

An *unconformity* is an uneven surface separating two series of rocks. It is a buried surface of erosion involving at least three geologic events: diastrophic uplift and exposure of older rocks, erosion, and the deposit of sediments on the eroded landscape.

TABLE 37.1 Geologic time. The earth was formed 4,500 million years ago, and the oldest known surface rocks were formed 3,500 million years ago.

Millions of years before the present	Era	Period	Epoch	Duration in millions of years	The biological record	
65	Cenozoic	Quaternary	Recent	0.01		
225			Pleistocene	2	Rise of man; large mammals abundant	Age of Mammals
600		Tertiary	Pliocene	5	Flowering plants abundant	
			Miocene	19	Grasses abundant; rapid spread of grazing mammals	
			Oligocene	12	Apes and elephants appear	
			Eocene	16	Primitive horses, camels, rhinoceroses	
			Paleocene	11	First primates	
	Mesozoic	Cretaceous		71	First flowering plants; dinosaurs die out	
		Jurassic		44	First birds; dinosaurs at their peak	Age of Reptiles
		Triassic		45	Dinosaurs and first mammals appear	
	Paleozoic	Permian		45	Rise of reptiles; large insects abundant	
		Pennsylvanian		35	Large nonflowering plants in enormous swamps	
		Mississippian		45	Large amphibians; extensive forests; sharks abundant	
		Devonian		50	First forests and amphibians; fish abundant	
3,500		Silurian		40	First land plants and air-breathing animals (scorpions)	
		Ordovician		60	First vertebrates (fish) appear	
		Cambrian		100	Marine shelled invertebrates (earliest abundant fossils)	
	Precambrian time	Late Precambrian		1,900?	Marine invertebrates, mainly without shells	
4,500		Early Precambrian		2,000?	Marine algae (primitive one-celled plants)	

Fossils are the remains or traces of organisms preserved in rocks.

The past 600 million years are divided into three *eras* according to the following approximate chronology: *Cenozoic era,* which began 65 million years ago; *Mesozoic era,* which began 225 million years ago; *Paleozoic era,* which began 600 million years ago.

Precambrian time is the name given to the part of the earth's history before the Paleozoic era.

EXERCISES

1. Figure 37.4 represents a cross section of a region about 10 miles wide. On the figure label the following:
   a. The youngest rock
   b. The oldest rock
   c. An unconformity
   d. An intrusive contact
   e. A syncline
   f. A fault on which movement occurred before the intrusion
   g. A fault on which no movement has occurred since the basalt was extruded

2. In order from the earliest to the latest, list as far as possible the geologic events recorded in the cross section of Fig. 37.4. Indicate the events that cannot be dated accurately.

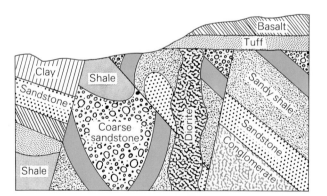

37.4

3. In parts of Colorado and Wyoming, a long period during which marine sediments were deposited in a subsiding basin ended with intense diastrophic movement at the close of the Mesozoic era. A mountain range was formed, which in the early part of the Cenozoic era was worn down by erosion to a nearly level plain. On parts of this plain stream and lake sediments were deposited. Describe the rocks and rock structures that you would expect to find in this region.

4 What are the chief assumptions involved in the determination of the age of a rock by measurements of its uranium and lead content?

5 Why are fossils more useful than measurements of radioactive decay in correlating rocks from one region with another?

6 Why is our knowledge of the evolution of marine life more complete than our knowledge of the evolution of land organisms?

chapter 38

Continental Drift

THE FLOATING CRUST

lithosphere and asthenosphere

isostasy

mountain building

CONTINENTAL DRIFT

wegener's theory

laurasia and gondwanaland

SEA-FLOOR SPREADING

the ocean floors

hypothesis of sea-floor spreading

the magnetic evidence

mechanism of spreading

During the past several years a major advance has occurred in our understanding of the large-scale forces that are at work shaping and reshaping the earth's crust. The notion that the continents are slowly drifting relative to one another across the globe—a notion over half a century old, but largely derided until very recently—has turned out to be the only way to explain a variety of striking observations, and these same observations also provide clues to the physical processes that cause the continents to drift. So far-reaching are the implications of the new picture of continental drift, and so suddenly have they come to light, that it is legitimate to speak of a revolution in geological thought. Before we consider the new concepts, though, it is appropriate to sketch some of the background ideas from which they crystallized.

THE FLOATING CRUST

How do the immense forces originate that are capable of wrinkling and breaking the strongest rocks? Why do the continents, an average of 800 m above sea level, not sink down so that they are on the same level as the ocean floor, an average of 3,700 m below sea level? These are

lithosphere and asthenosphere

At the heart of all current explanations for major crustal diastrophic changes is the idea that the mantle of the earth is not entirely a rigid rock structure, like the crust, but contains an outer layer capable of plastic flow. The existence of enormous temperatures and pressures within the earth supports this view: the outer mantle becomes plastic because it is very hot but not under too much pressure, whereas lower down the increasing pressure counteracts the heat to give a rigid material. Insofar as brief, suddenly applied forces are concerned, the mantle behaves like a solid and can transmit the transverse vibrations of earthquakes, for example, for thousands of miles. But when forces act on the mantle over long periods of time, some of it behaves like a thick, viscous fluid and flows gradually in order to conform to the applied forces.

The preceding picture is not quite the whole story, because the extreme upper part of the mantle is still fairly stiff. From the point of view of mechanical strength both the crust and the outer mantle are often grouped together into the *lithosphere*, a rigid rock shell perhaps 50 km deep around the earth. The lithosphere has no sharp boundary, as the crust does, but gradually turns into the underlying plastic *asthenosphere*, a region 100 to 400 km thick. The crust is distinguished from the mantle on the basis of their different compositions and different seismic-wave speeds; the lithosphere is distinguished from the asthenosphere on the basis of their different degrees of rigidity.

the lithosphere is a rigid rock shell that consists of the earth's crust and a little of its mantle

the asthenosphere is a layer of plastic rock just under the lithosphere

isostasy

Let us look first into the problems of the support of the continents and the origin of their vertical motions. A clue is that the crust is much thicker beneath the continents than beneath the oceans (Fig. 36.15). At first glance this seems only to magnify the difficulty of explaining the support of the continental masses, but, when we recall that the density of the earth as a whole is double that of the crust, everything fits together.

Suppose that we place several blocks of wood of different sizes in a pool of water. The larger blocks float higher than the smaller while simultaneously extending down farther into the water (Fig. 38.1). Thus, if the lithosphere is imagined as floating in equilibrium on a denser asthenosphere capable of plastic deformation, we have the analog of wooden blocks of different sizes floating on water. This implies, if carried to its logical end, that exceptionally elevated regions—mountain ranges and plateaus—have corresponding roots extending an exceptional distance downward. Such is actually the case; in fact, its discovery led the British scientist Sir George Biddle Airy a century ago to propose the floating of the entire lithosphere.

38.1 Isostasy. *Large blocks of wood float higher and extend farther downward than smaller blocks of wood. The application of this fact in order to explain various properties of the earth's crust is known as isostasy.*

The idea that irregularities in the crust are supported because of their buoyancy is known as *isostasy*. But irregularities in the earth's crust, as

isostasy

we know, are not at all stationary. Every mountain range is continually undergoing erosion, and the eroded material is deposited largely in adjacent valleys or oceans. The mountain block accordingly becomes lighter and the valley block or ocean block becomes heavier as the weight of sediment increases. Eventually the difference in pressure beneath the mountain range and the adjacent blocks becomes too great for the material at depth to withstand, so the mountain range rises and the basins of deposition sink until the isostatic equilibrium is restored (Fig. 38.2). The situation may be duplicated on a small scale by a large cork and a small cork floating in water: if material is removed from the top of the large cork and placed on the small one, the large cork will rise a little out of the water and the small one will sink.

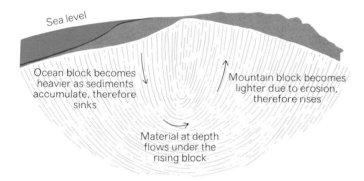

38.2 *The evolution of the crust is in part a consequence of erosion and isostasy.*

mountain building

Mountains can form in a number of ways. Some are accumulations of lava and fragmental material ejected by a volcano. Some are small blocks of the earth's crust elevated along faults. But the great mountain ranges of the earth, like the Appalachians, the Rockies, the Alps, and the Himalayas, have a much longer and more complex history involving sedimentation, folding, faulting, igneous activity, repeated uplifts, and deep erosion. The formation of a mountain range is a major event of earth history, and it leaves an indelible record in the rocks which can be read long after the range itself has vanished.

A careful observer would note one conspicuous difference between the sedimentary rocks exposed in most mountain ranges and the corresponding rocks under adjacent plains: as each layer is followed from the plains toward the mountains, its thickness increases enormously (Fig. 38.3). Now, in general, the sedimentary rocks in both plains and mountains are formed from deposits that accumulated in shallow seas or on low-lying parts of the land—that is, on surfaces not very far above or below sea level. For successive beds to be laid down the land must have been slowly sinking, the accumulating deposits keeping the surface at roughly the same level. Since the thickness of sediments in the present mountain region is so much greater, this part of the surface during

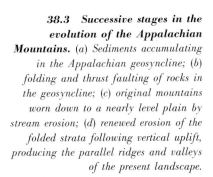

38.3 Successive stages in the evolution of the Appalachian Mountains. (a) Sediments accumulating in the Appalachian geosyncline; (b) folding and thrust faulting of rocks in the geosyncline; (c) original mountains worn down to a nearly level plain by stream erosion; (d) renewed erosion of the folded strata following vertical uplift, producing the parallel ridges and valleys of the present landscape.

<small>a geosyncline is a subsiding basin in which sediments accumulate</small>

<small>diastrophic action transforms a geosyncline into a mountain range</small>

sedimentation must have been sinking more rapidly than adjacent areas. In other words, the thickness of sedimentary layers in a mountain range suggests that the region where the range now stands was once a broad subsiding basin in which sediments were accumulating rapidly.

A huge sinking basin of this sort is called a *geosyncline.* A good example of a small-scale geosyncline in the present world is the central valley of California. This is a flat-bottomed valley about 500 miles long and close to 100 miles wide with most of its surface just above sea level; at present it receives sediments brought down by rivers from the mountains on either side. Oil wells drilled in the rocks beneath the valley show that the strata are almost entirely soft sedimentary rocks formed from shallow-water marine and lake deposits. The deepest well has penetrated over 15,000 ft of such material, and calculations indicate that parts of the valley must have at least 10,000 ft more. Thus the floor of the valley has been slowly sinking for millions of years while these sediments piled up, and the sedimentation has been approximately keeping pace with subsidence. Sinking troughs like this but on a grander scale must once have existed where the Appalachians and the Rockies now stand. The Appalachian geosyncline reached from the mouth of the St. Lawrence River to the Gulf of Mexico, and the still larger Cordilleran geosyncline extended from Mexico to the Arctic Ocean.

Another conspicuous feature of the sedimentary rocks in mountain ranges is their complex structure. They are widely folded and often cut by huge thrust faults and minor normal faults. Thus, the long sinking of a geosyncline must be followed by a period of diastrophism in which the thick pile of sediments is subjected to intense compressional forces. The compression raises part of the folded layers above the sea, and erosion begins to wear down the exposed beds as folding continues. In this manner the geosyncline is transformed into a mountain range (Fig. 38.3).

Where erosion has cut deeply into the rocks of a mountain range, much of the old sedimentary material is found converted into such metamorphic rocks as slate, schist, and marble. Since the earlier rocks of the geosyncline were deeply buried beneath later sediments, it seems reasonable that intense diastrophic forces should have induced metamorphism on a large scale. Also brought to light by long erosion in a mountain range are intrusive igneous rocks—dikes of various kinds and huge batholiths of granite, often forming the highest central part of the present range because of their resistance to erosion. Since these igneous masses intrude the folded sedimentary and metamorphic rocks and since the igneous rocks usually show evidence of only minor deformation, we must infer that they were intruded in the last stages of mountain building, when diastrophic movement had nearly ceased. Sometimes lava flows and other volcanic products suggest that the internal intrusive activity was accompanied by volcanic eruptions at the surface.

During and after the diastrophic movement that forms a mountain range, erosion shapes its surface features. As more and more material is removed from the range, the isostatic balance between it and adjacent segments of the crust is disturbed, until at length the balance is restored by an uplifting of the mountain block. This leads to an increase in the rate of erosion, further removal of material from the mountains, and eventually another upward movement. The later history of a mountain range is punctuated by these successive vertical uplifts. Thus the features of most present-day mountain landscapes are due not primarily to the

Sedimentary strata were uplifted and tilted in the formation of the Front Range of the Rocky Mountains in Colorado about 60 million years ago.

compressional forces that originally folded and faulted their rocks but instead to long erosion in regions subjected to periodic uplifts.

When a mountain system has finally succumbed to the processes of erosion and is worn down to a region of low hills or a plain, evidence of its former existence is still preserved in the rocks. All the original folded and faulted sedimentary layers have disappeared, leaving exposed only metamorphic rocks intruded by igneous masses; but these show clearly by their intense folding and crushing that they once formed the roots of a mountain range. Whenever a geologist comes across numerous outcrops of contorted schists intruded by dikes and batholithic masses, he knows that mountains once existed in that region.

isostasy cannot account for the horizontal forces involved in mountain building

The concept of isostasy accounts fairly well for the long life of mountain ranges and to some extent for the repeated uplifts they undergo, but it offers no explanation for the vigorous folding and faulting that occur. All horizontal movement in the lithosphere and probably much vertical movement as well must have some other cause.

CONTINENTAL DRIFT

A casual glance at a map of the world suggests the possibility that at some time in the past the continents were joined together in one or two giant supercontinents. If the margins of the continents are taken to be on their continental slopes (see Fig. 33.1) at a depth of 3,000 ft, instead of their present sea-level boundaries, the fit between North and South America, Africa, Greenland, and western Europe is remarkably exact, as Fig. 38.4 shows. But merely matching up outlines of continents is not by itself sufficient evidence that the continents have migrated around the globe. The first really comprehensive theory of continental drift was proposed early in this century by the German meteorologist Alfred Wegener, who based his argument on biological and geological evidence.

wegener's theory

Wegener was troubled by the parallel evolution of living things. Going back through the ages, the fossil record shows that, until about 200 million years ago, whenever a new species appeared it did so in many now-distant regions where suitable habitats existed. Until sometime in the Mesozoic era evolution proceeded at the same rate and in the same way in continents and oceans that today are widely separated. Only afterward did plants and animals in the different continents develop in different ways.

At one time the standard explanation for the similarity of patterns of early life around the world was a series of land bridges linking the continents together. But this meant that the oceans were then separated from one another, so a series of channels had to be devised to permit aquatic plants and animals to pass between the oceans. No really believable scheme of bridges and channels could be devised, and even if one had been, it would still be necessary to account for the disappearance of all traces of them. Wegener was on firm ground when he searched for an alternative to this notion.

CONTINENTAL DRIFT

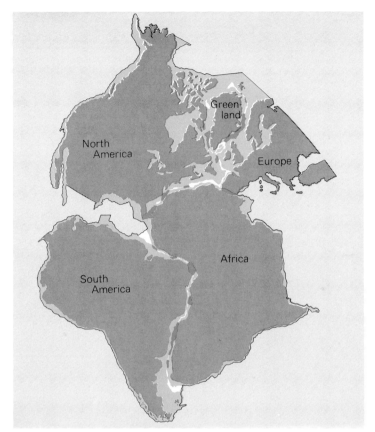

38.4 How some of the continents fit together. *The boundary of each continent is taken at a depth of 3,000 ft on its continental slope; the gray regions represent land above sea level at present, and the light colored regions represent submerged land on the continental shelf and slope. Overlaps are shown in dark color and gaps in white.*

What Wegener suggested instead was that originally the continents were all part of a huge land mass he called Pangaea. A few hundred million years ago Pangaea was supposed to have begun to break up and the continents to slowly drift to their present locations. This model found additional support in geological data regarding prehistoric climates. A little over 200 million years ago South Africa, India, Australia, and part of South America were burdened with great ice sheets, while at the same time a tropical rain forest covered North America, Europe, and China. At various other times, there was sufficient vegetation in Alaska and Antarctica for coal deposits to have resulted, and so currently frigid a place as Baffin Bay was a desert.

Wegener and his followers examined what was known about the climates of the distant past, and tried to arrange the continents in each geological period so that the glaciers were near the poles and the hot regions were near the equator. The results, in general, were quite convincing, and in some cases startlingly so: deposits of glacial debris and fossil remains of certain distinctive plant species follow each other in the same succession in Argentina, Brazil, South Africa, Antarctica, India, and Australia, for example. A recent discovery of this kind was the identification of a skull of the reptile Lystrosaurus in a sandstone layer in the Alexandra mountain range of Antarctica. This creature, which was about three feet long, flourished long ago in Africa. It is as unlikely that

wegener believed all the continents were once part of a single land mass

there is both geological and biological evidence for continental drift

The Himalayas, which include the 29,028-ft Mt. Everest, are a young mountain range with deep roots. The Himalayas were once part of the Tethys Sea and were later thrust upward by the collision of the Indian and Eurasian lithosphere plates.

Lystrosaurus swam the 2,700 miles between Africa and Antarctica as it is that a land bridge this long connected them, only to vanish completely later on.

laurasia and gondwanaland

laurasia and gondwanaland are thought to have been the sources of all the continents

Today it is widely thought that the continents originally came from two, not one, primeval land masses. The northern supercontinent, *Laurasia*, consisted of what is now North America, Greenland, Europe, and most of Asia. The southern supercontinent, *Gondwanaland*, consisted of what is now South America, Africa, Antarctica, India, and Australia (Fig. 38.5).

38.5 The land masses of the earth as they may have appeared 200 million years ago and as they are today.

Laurasia and Gondwanaland were almost equal in size. The notion of two original supercontinents is supported by detailed geological and biological evidence, for instance certain differences between Laurasian and Gondwana fossils of the same age. It seems probable that there was a land bridge between Laurasia and Gondwanaland during the Mesozoic era.

Laurasia and Gondwanaland were separated by a body of water called the *Tethys Sea*. Today a little of the Tethys Sea survives as the Mediterranean, Caspian, and Black Seas, but its original extent can be gauged from the sediments that were subsequently uplifted to form the mountain ranges that stretch from Gibraltar eastward to the Pacific. The Pyrenees, Alps, and Caucasus of Europe, the Atlas Mountains of North Africa, and the Himalayas of Asia all were once part of the Tethys Sea.

the tethys sea separated laurasia and gondwanaland

Continental drift, then, has some very attractive aspects. Why was it not widely accepted until very recently? Wegener, who lacked a knowledge of the mechanical properties of the various parts of the earth's crust, envisioned the continents as floating freely over the mantle and having no trouble in moving *through* the ocean floors. If this were the case only relatively weak forces would be needed to move the continents over the face of the earth, and Wegener was able to cite several such forces. But the ocean floors are in fact extremely hard and strong, and if enough force could somehow be applied, it seems likely that a continent would buckle rather than pass through the ocean floor.

the ocean floor is hard and rigid

An entirely different mechanism has proved to be involved, and until its discovery in the middle 1960s continent drift, for all its allure, remained discredited by most geologists.

SEA-FLOOR SPREADING

The mountains and valleys, plains and plateaus of the continents have been known for a long time, and few surprises are in store for future explorers. But the continents occupy less than 30 percent of the area of the earth's crust, while the rest lies hidden in perpetual darkness thousands of meters below the seas and oceans. Only in the past two decades have the floors of the oceans been mapped and their physical characters elucidated. It is largely these findings that have clarified the evolution of the crust.

the ocean floors

The methods used to investigate the ocean floors are not particularly subtle—the real problem has been the vastness of the area to be covered.

These days depths are charted by means of echo sounders: such an instrument sends out a pulse of high-frequency sound waves, and the time needed for it to reach the sea floor, be reflected there, and then to return to the surface is a measure of how deep the water is (Fig. 38.6). A variant

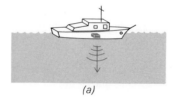 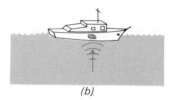

38.6 The principle of echo sounding. (a) A pulse of high-frequency sound waves is sent out by a suitable device on a ship. (b) The time at which the pulse returns to the ship is a measure of the sea depth.

of this method reveals something of the structure of the sea floor itself. What is done is to detonate an explosive charge in the water and study the returning echoes—one echo will come from the top of the sediment layer, and a later one from the hard rock underneath. Samples of the sea floor can be obtained by dropping a hollow tube to the bottom on a long cable and then pulling it up filled with a core of the sediments into which it sank. These sediments can be examined later in the laboratory for their composition, their age (by radioactive dating), the fossils they contain, their magnetization, and so forth. Another important technique is to tow a magnetometer behind a survey ship to obtain an idea of the direction and intensity of the magnetization of the rocks of the ocean floor over wide areas.

Four findings about ocean floors have proved of crucial importance:

the ocean floors are relatively young

1 The ocean floors are, geologically speaking, very young. The sediment deposits go back no more than about 135 million years, in contrast

to the generally much older continental rocks, some of which came into being 3,500 million years ago.

2. There is a worldwide system of ridges that runs across the oceans (Fig. 38.7). In a few places—Iceland, the Azores, Ascension Island, and Tristan da Cunha in the Atlantic are examples—the ridges poke through the water, but for the most part they are submerged. These ridges are offset at intervals by fracture zones that indicate transverse shifts of the ocean floors.

oceanic ridges

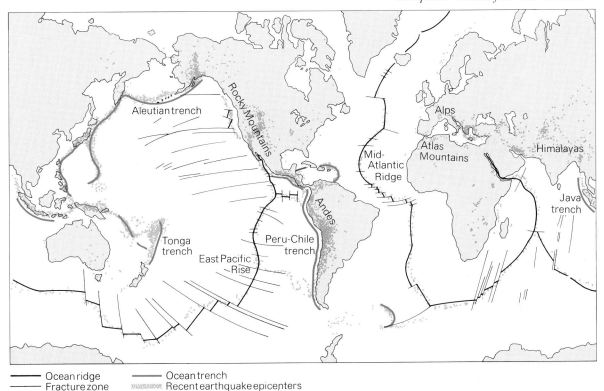

38.7 The world-wide system of oceanic ridges and trenches. *The ridges are offset by transverse fracture zones. Color dots represent epicenters of earthquakes recorded from 1957 to 1967.*

3. There is also a worldwide system of trenches, giant troughs that rim the Pacific Ocean. These trenches are in the same regions where the majority of earthquakes and volcanoes are found (compare Figs. 38.7 and 35.2), except for the belt of active diastrophic and volcanic activity that extends across southern Asia to the Mediterranean. As we shall find, it is significant that this latter belt coincides with the former location of the Tethys Sea.

oceanic trenches

4. Ocean-floor rocks reverse their directions of magnetization at regular intervals perpendicular to the mid-ocean ridges and these reversals are symmetric on both sides of each ridge. Parallel to a ridge at a given distance away, the direction of magnetization is the same. Hence strips of opposite magnetization lie along the ridges (Fig. 38.8).

magnetization of the sea floor

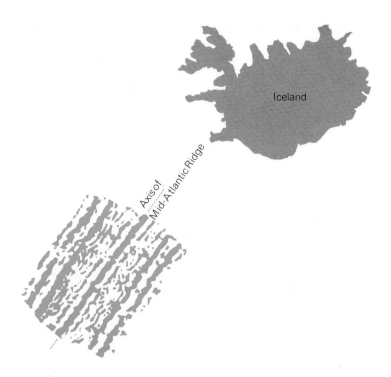

38.8 Pattern of magnetization along the Mid-Atlantic Ridge southeast of Iceland. *Sea-floor rocks whose directions of magnetization are the same as that of today's geomagnetic field are shown in color; the intervening white spaces represent rocks whose magnetization is in the opposite direction.*

hypothesis of sea-floor spreading

A single hypothesis is able not only to correlate these observations but also to link them to continental drift. This is the hypothesis of sea-floor spreading, independently proposed in the early 1960s by the American geologists Harry H. Hess and Robert S. Dietz. (A very similar theory was put forward in 1928 by Arthur Holmes in England, but it remained practically unnoticed because supporting data were lacking.)

sea-floor spreading

The basic idea of sea-floor spreading is that molten rock is continually rising up along the mid-ocean ridges and flowing outward from them. The parts of the lithosphere on either side of a ridge are pushed apart at speeds of 1 to 10 cm per year, with the new material taking their place as it hardens.

the lithosphere consists of a number of huge, moving plates

The lithosphere is cracked not only along the ridges but also along the trenches of the ocean floor. These cracks break the lithosphere into perhaps six huge plates and a number of smaller ones (Fig. 38.9). As a plate is pushed away from a mid-ocean ridge, its opposite edge is forced over or under the plate that adjoins it, as in Fig. 38.10, or, if the motion is slow, the plate edge may simply buckle, as is believed to occur in the Himalayas. Thus the plates are shifted about despite their rigidity. This rigidity suggests that where one plate overrides another the crust will be violently deformed—and indeed such zones are the sites of most of the world's severe earthquakes and of its youngest mountain ranges.

collisions between plates lead to the formation of mountain ranges

A striking example is the collision of the eastern Pacific plate (which is drifting eastward) with the western Atlantic plate (which is drifting

Egypt, Saudi Arabia, the Middle East, and the Red Sea, photographed from Gemini spacecraft. Baja California and the Gulf of California.

westward) along the western margin of South America. The result is the formation of a trench along the coasts of Peru and Chile where the Pacific plate sinks into the asthenosphere and melts, and also the formation of a mountain range, the Andes, where the Atlantic plate is thrust upward.

origin of the andes

The western Pacific plate, in its westward movement, has produced the island arcs that border the Asiatic side of the Pacific—the Aleutians, Japan, the Philippines, Indonesia, the Marianas. It seems possible that a part of the rock in these and similar island arcs elsewhere (the West

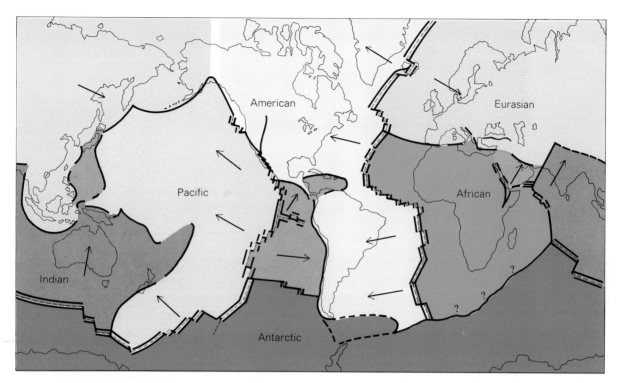

38.9 The six chief lithosphere plates whose motion results in continental drift. *The plates are bounded by ridges or trenches. The arrows show the direction of motion of the plates except for the African plate, which is thought to be stationary. There are several smaller plates as well.*

origin of the himalayas

Indies, for instance) consists of the lighter materials of the lithosphere plates carried downward to melt in the hot asthenosphere. After melting, the buoyancy of these lighter materials would carry them upward again to penetrate fissures in the lithosphere overhead, forming volcanoes at the surface and bodies of intrusive rock below the surface. According to one point of view, all the continents had their origins in processes of this kind. As a plate is forced under and melts, the sialic rocks rise to create island arcs that eventually grow into continents. Traces of what may have been former island arcs have been detected in the interiors of such present-day continents as North America.

The moving apart of lithosphere plates and their collisions are not confined to ocean basins. Thus the Red Sea and the Gulf of California, both extensions of ocean ridges, are currently widening. The massive Himalayas that divide India from the rest of Asia apparently owe their existence to the enhanced upthrust due to forcing of the Indian subcontinent against the original Asian land mass. As we can see from Fig. 38.7, the Himalayas lie along an extension of the Java Trench off southeast Asia.

The lithosphere plates do not all travel at the same speed. Where the plates are moving apart rapidly, more than say 3 cm per year, the upwelling molten rock spreads out over a wide region and forms a broad ridge whose sides slope down gradually. An example is the East Pacific Rise. On the other hand, a slow separation of adjacent plates gives time for the new material to accumulate, which leads to steep crests such as those that characterize the Mid-Atlantic Ridge.

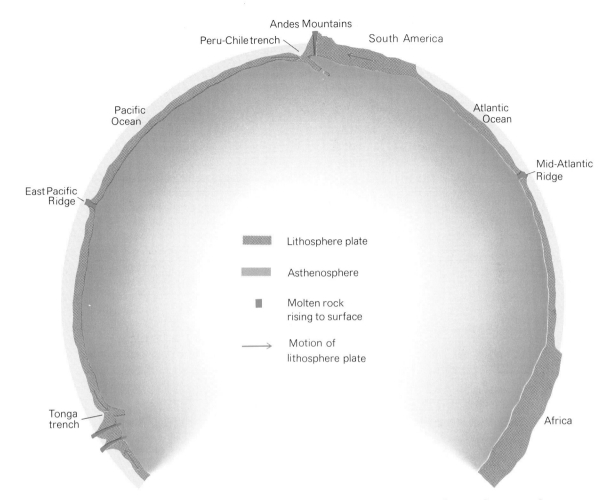

38.10 The sea floor spreads apart at mid-ocean ridges where molten rock rises to the surface of the lithosphere. At a trench, one lithosphere plate is forced under another into the asthenosphere, where it melts. Mountain ranges, volcanoes, and island arcs are found where plates collide. The vertical scale is greatly exaggerated.

The oceanic ridges are not continuous but are interrupted every few hundred km by transverse faults. The ridge segments are displaced out of line at these faults (Fig. 38.7). As the sea floor spreads outward from each ridge segment, then, solid rock must move past solid rock along the faults, giving rise to frequent earthquakes.

As we might expect, undersea volcanoes are common near the spreading margin of a moving plate. Most of them are active for no more than 20 or 30 million years, during which period they have been carried several hundred km from the constantly renewed plate edge and presumably out of reach of the magma body that fed them. The Pacific floor contains a great many submerged ancient volcanoes, called *guyots*, that once were above the ocean surface but later sank below it as the underlying crust moved away from the zone of spreading and subsided. Some oceanic volcanoes, however, have persisted in their activity for 100 million years or more—the Hawaiian Islands in the Pacific and the Canary Islands in the Atlantic are notable examples.

guyots

the magnetic evidence

The fourth key observation mentioned earlier, which concerns the magnetization of rocks on either side of an ocean ridge, confirms the hypothesis of sea-floor spreading in a convincing way. As Fig. 38.8 shows in the case of a portion of the Mid-Atlantic Ridge southeast of Iceland, successive strips of rock lying parallel to the ridge are magnetized in alternate directions. To interpret this pattern we draw upon the fact that the earth's magnetic field has periodically reversed itself many times in the past (Fig. 36.19). What must have been happening is obvious. As molten rock, unmagnetized in its liquid state, comes to the surface of the crust at a ridge, it hardens and the iron content of its minerals becomes magnetized in the same direction as that of the prevailing geomagnetic field. With the passage of time the direction of the geomagnetic field reverses, and the molten rock that cools thenceforth becomes magnetized in the opposite direction. Thus strips of alternate magnetization follow one another going away on both sides from a ridge (Fig. 38.11).

The reversals of the earth's magnetic field have been dated by measurements made on magnetized lava flows on land using the potassium-argon method. This information can be used to ascertain the ages of the magnetized strips that make up the sea floor, and thereby to establish the speeds with which the lithosphere plates have been moving. Some of the results are shown in Fig. 38.12.

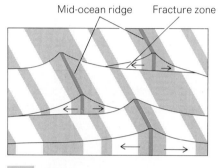

38.11 *As magma from the asthenosphere hardens into new sea-floor rock, it is magnetized in the direction of the geomagnetic field at the time.* Successive reversals of the field therefore result in strips of alternate magnetization on the sea floor parallel to the ridge between spreading lithosphere plates.

mechanism of spreading

What is the mechanism that drives the lithosphere plates as though on huge conveyor belts from the ridges where they are formed to the trenches where they disappear back into the asthenosphere? Although the motion of the plates, and the drift of the continents that are embedded within them, seems well-established, the details of how the motion occurs are still in doubt.

The picture many geologists accept starts from the observations that the asthenosphere is plastic enough to permit the lithosphere plates to slide over it and that its temperature increases from bottom to top. Not much imagination is needed to go on from these facts to speculate whether convection currents might not occur in the asthenosphere. We have encountered such currents before both in the atmosphere and in the oceans, and they are familiar to anyone who has placed a pan of cold soup on a stove. As the soup is heated, liquid from underneath rises to the surface in some places while surface liquid, which is cooler and therefore denser, sinks downward in other places; the motion in between is horizontal.

A convective circulation of some kind, on a huge scale and very slow, might well be dragging the lithosphere plates across the face of the earth. Figure 38.13 shows an idealized flow pattern in the asthenosphere that fits the motion of the plates. A virtue of this pattern is that it is in accord with the forcing upward of the lithosphere where the asthenosphere material is rising and with its being sucked downward where the asthenosphere material is sinking.

CONTINENTAL DRIFT

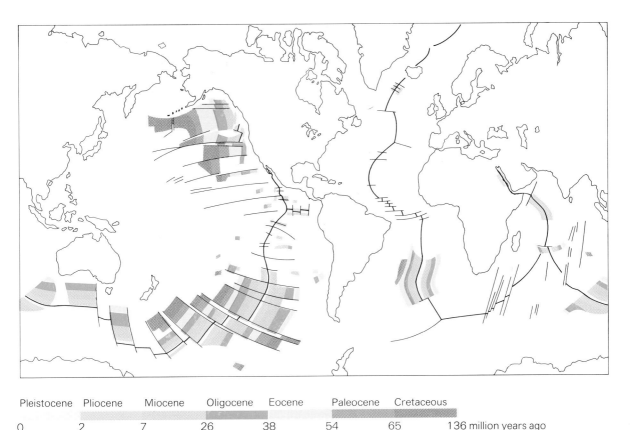

Pleistocene	Pliocene	Miocene	Oligocene	Eocene	Paleocene	Cretaceous	
0	2	7	26	38	54	65	136 million years ago

Time scale

38.12 Sea-floor ages. *The various bands represent regions that hardened in the indicated geologic periods or epochs. Evidently the Pacific floor is spreading more rapidly than the Atlantic floor. However, since the leading edges of the Pacific lithosphere plates are being forced under the continents that rim the Pacific, that ocean may be gradually shrinking in area.*

But for all its elegance, there are serious objections to the theory of convection currents. The chief difficulty is that the dense, viscous asthenosphere is supposed to flow in a manner that does not seem to be consistent with what is known about convection from theory and laboratory experiment. Another source of conflict is the zigzag pattern of the oceanic ridges, which are divided into displaced segments by transverse faults. In some cases adjacent segments are offset by distances greater than the lengths of the segments themselves (Fig. 38.7), which is hard to reconcile with a regular pattern of convection.

Another hypothesis for the origin of the plate movement that avoids such objections has received some support. This is the idea that the elevated ridges push sideways on the adjacent plates simply by virtue of their weight. Forced upward by its buoyancy through the crack in the lithosphere between two plates, molten rock from the asthenosphere piles up and, as in the case of a man whose feet are in two adjacent rowboats, the plates inevitably spread apart. Calculations show that the speeds at which the plates are moving, several centimeters per year, is about what would be expected on this basis. But a final answer to the problem seems a long way off.

Regardless of the details of how their motions occur, the continents are definitely drifting across the globe, and new land masses are coming

the hypothesis of gravity spreading

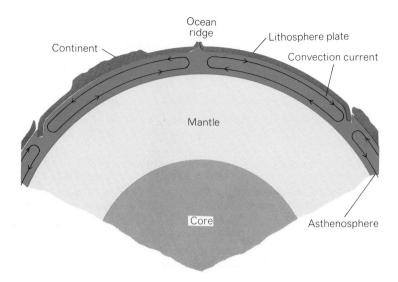

38.13 One proposed system of convection currents in the asthenosphere. *There are serious objections to this picture of the origin of plate movement. The vertical scale is greatly exaggerated.*

into being at the various island arcs. A projection of present trends 30 million years into the future yields a picture like that shown in Fig. 38.14. The Atlantic Ocean has grown wider, the Pacific narrower. California has been detached from the rest of North America, and the Arabian peninsula has been forced around to become an integral part of Asia. The West Indies island arc has grown into a land bridge between the Americas, and the western Pacific island arcs have also increased markedly in extent.

And after that? All that can be said is that the face of the earth will probably continue to change in the future, just as it has been changing as far back in the past as we have any evidence.

GLOSSARY

The *lithosphere* is the earth's outer shell of rigid rock. It consists of the crust and the outermost part of the mantle.

The *asthenosphere* is a layer of rock, neither wholly rigid nor wholly fluid but capable of plastic deformation, that is just below the lithosphere in the mantle.

Isostasy refers to the idea that irregularities in the lithosphere are supported by their buoyancy.

A *geosyncline* is a broad basin that sinks relatively rapidly as sediments accumulate upon it.

According to the theory of *continental drift,* today's continents were once part of two primeval supercontinents called *Laurasia* (North America, Greenland, Europe, and most of Asia) and *Gondwanaland* (South America, Africa, Antarctica, India, and Australia), which were separated by the *Tethys Sea.*

CONTINENTAL DRIFT

38.14 How the earth may appear in 30 million years from now if present trends continue.

Sea-floor spreading pictures continental drift as originating in the moving apart of huge lithosphere *plates* due to upwelling of molten rock at oceanic ridges.

EXERCISES

1 How is it possible to reconcile the notion of a plastic asthenosphere with the transmission of transverse seismic waves through it?

2 Distinguish between the earth's crust and its lithosphere.

3 If the entire mantle were to become as rigid as the lithosphere, so that vertical motions due to isostasy and to continental drift no longer would occur, would erosion eventually lead to a perfectly smooth earth?

4 Can you suggest the kind of evidence that might be found in the Alps and Himalayas to confirm that these regions were once part of the floor of the Tethys Sea?

5 Which of today's continents are thought to have been part of Laurasia? Of Gondwanaland?

6 How do the ages of the ocean floors compare with the ages of continental rocks?

7 Can the continents be regarded as floating on the asthenosphere and drifting through soft ocean floors?

8 Why is there a mountain range on the western edge of South America but not one on its eastern edge?

9 What are two possible mechanisms for pushing the lithospheric plates across the earth's surface?

chapter 39

Earth History

PRECAMBRIAN TIME
precambrian rocks

THE PALEOZOIC ERA
early paleozoic
late paleozoic
coal and oil

THE MESOZOIC ERA
mesozoic north america
dinosaurs and mammals

THE CENOZOIC ERA
geologic activity
animal life
the ice age

Two kinds of events in the history of the earth's crust are significant. In one category are physical changes, such as the drift of continents, the rise and subsidence of mountains, the spread and retreat of seas, the advance and recession of glaciers and ice sheets. In the other are changes in the living things that populate the earth, from primitive one-celled creatures to the complex plants and animals of today.

In the past biological changes have been brought about largely by changes in the physical environment, so the biological narrative follows the physical one through the ages. The modern world seems to be witnessing a reversal of this sequence. Man is not only able by virtue of his ingenuity to flourish in the environment that nature has provided, but he is able to alter the environment in a number of ways. These alterations, small but numerous, have permitted a vast number of people to live relatively free (by historical standards) of starvation and disease. Apparently no irreversible worldwide damage to the interplay between organism and environment has as yet occurred. But continued into the future and swollen many times over in scale by population growth, man's present patterns of industry, agriculture, and warfare seem certain to menace this interplay. The indiscriminate use of insecticides and artificial fertilizers has already destroyed the ecological balance in many areas; numerous inland waters have been poisoned by industrial waste; noxious gases fill the air in most densely populated regions; deposits of radioactive byproducts lie waiting for chance catastrophe to disperse them; and so on. Perhaps most ominous of all is that, whatever the character of man's terrestrial activity, its multiplication is inevitable: from perhaps 500 million in 1650, 1 billion in 1850, 2 billion in the 1920s, and 3 billion in the 1960s, the world's population will climb to an estimated 6 billion by the year 2000.

PRECAMBRIAN TIME

the earth is 4½ billion years old

The earth came into existence about 4.5 billion years ago. The evidence for this age, and theories of how the earth and the other planets were formed, are considered in Chap. 45. With the oldest surface rocks, which date back 3.5 billion years according to radioisotope measurements, the story of the earth leaves the realm of speculation. These ancient rocks are schists and gneisses with interbedded layers of marble and quartzite, metamorphic rocks evidently derived at least in part from sediments. So the visible record of the earth's history begins at a time when the cycle of erosion and diastrophism was already well established.

precambrian rocks

PRECAMBRIAN TIME (4,500–600)

early precambrian (4,500–2,500) marine algae (primitive one-celled plants)

late precambrian (2,500–600) marine invertebrates, mainly without shells

[dates are millions of years ago]

precambrian rocks are at the surface in eastern canada

Nothing is known about the locations of the continents in Precambrian time. Possibly they were then all united in a single mass like Wegener's Pangaea; possibly they were divided into the two supercontinents Laurasia and Gondwanaland. Both of these hypotheses imply that continental drift is a relatively new phenomenon, since the splitting of Laurasia and Gondwanaland into today's continents began less than 200 million years ago, a mere 4 percent of the earth's age and 6 percent of the age of the oldest known rocks. If sea-floor spreading had been going on farther in the past, the Precambrian continents no doubt were differently arranged in the midst of a quite different set of oceans. This point of view would suggest that all mountain ranges originated in the collision of lithosphere plates, not just the Andes and other youthful ones which can be correlated with present-day plate movement. Hence ancient mountain ranges, or what is left of them after erosion did its work, are located where former oceans shrank and closed, as the Himalayas represent the scar where the Tethys Sea was squeezed out of existence by the northward migration of India.

In any case, the geological history of the continents themselves in Precambrian time can be traced in a number of regions. For instance, Precambrian rocks are exposed at the surface over a broad area covering most of eastern Canada and adjacent parts of the United States. This immense region of ancient rocks, one of the largest in the world, has stood above sea level for most of the 600 million years since the beginning of the Paleozoic era. Smaller areas of Precambrian rocks are found in many parts of the United States, particularly in the cores of mountain ranges where repeated uplifts and deep erosion have combined to expose them. In the Grand Canyon the Colorado River has cut through more than 5,000 ft of Paleozoic strata to reveal the older rocks at their base.

Although in places the later Precambrian strata are practically unaltered sedimentary and volcanic rocks, these old rocks in general show considerable metamorphism and often intense deformation. This is hardly surprising in view of their great age and the deep erosion they have undergone. Careful study of even the highly metamorphosed rocks usually reveals something about their origin, and the less altered strata often give surprisingly detailed records of their history. Among Precambrian rocks we find varieties formed in nearly all the environments recorded in later geologic ages. The sedimentary beds are in part stream deposits, in part marine. In late Precambrian strata we even find rocks of glacial origin—coarse, angular conglomerates containing smoothed and striated boulders, resting on grooved rock surfaces—indicating at least two distinct periods of glaciation in this early stage of earth history. The volcanic rocks include all types, with basalt flows then as now the most common. Intrusive rocks are represented in great abundance and variety. Evidently, geologic processes a billion years ago were not very different from those in the modern world.

Unlike rocks of later ages, Precambrian rocks are almost devoid of fossils. This does not mean that life was absent, for occasional finds of structures produced by algae, by sponges, and probably by worms prove that primitive organisms did exist. The scarcity of fossils is probably due to lack of animals with preservable shells rather than to any actual scarcity of living things. Indirect evidence for the presence of abundant life in Precambrian seas comes from thick beds of limestone, which presumably require the action of organisms for their deposition, and from occasional layers of graphite, most reasonably explained by the metamorphism of organic debris. Precisely where and how life began on the earth we do not know, but marble and graphite in very old Precambrian rocks suggest that primitive sea-dwelling creatures existed nearly 3 billion years ago.

The scarcity of fossils coupled with widespread metamorphism and deformation makes the correlation of Precambrian events very difficult. Some progress has been made by comparing rock compositions, degrees of metamorphism and deformation, and extent of intrusive activity, and in recent years by radiometric age determinations, but these make possible only a very general outline of Precambrian history. Like the later history these early chapters are largely a record of successive periods of sedimentation, separated by intervals of erosion and mountain building. In the later Precambrian lava flows were especially numerous and widespread, and glaciation was extensive in two distinct periods. Precambrian time ended with considerable diastrophic movement in many regions 600 million years ago. Continents were uplifted and then eroded, producing the unconformity which often separates Precambrian rocks from Paleozoic strata.

A sample of Precambrian rock that shows ripples and cracks characteristic of fine sediments deposited in shallow water and occasionally dried by exposure to the sun. Similar conditions occur along the shores of present-day lakes and seas.

precambrian life was too primitive to leave many fossils

THE PALEOZOIC ERA

The Paleozoic ("ancient life") era begins in mystery, but its history is remarkably complete. No longer are there the doubts and the vagueness that characterize Precambrian correlations, for Paleozoic strata are widely exposed, and their wealth of fossils makes possible correlation of rocks and events from one side of the earth to the other.

PALEOZOIC ERA

cambrian (600–500)
marine shelled invertebrates (earliest abundant fossils)

ordovician (500–440)
first vertebrates (fish) appear

silurian (440–400)
first land plants and air-breathing animals (scorpions)

devonian (400–350)
first forests and amphibians; fish abundant

mississippian (350–305)
large amphibians; extensive forests; sharks abundant

pennsylvanian (305–270)
large nonflowering plants in enormous swamps

permian (270–225)
rise of reptiles; large insects abundant

[dates are millions of years ago]

early paleozoic

North America in the early Paleozoic bore little resemblance to the present continent. To the northeast was a highland of Precambrian rocks, perhaps with erosional remnants of late Precambrian mountains still standing. At either side of the continent was a broad subsiding trough, or geosyncline (Fig. 39.1); during most of this time this was covered with a shallow sea. Between the geosynclines stretched a low plain, parts of which were submerged at intervals by spreading of the shallow seas. The Appalachian geosyncline was continuous with a geosyncline in northwestern Europe, which during at least most of the Paleozoic formed part of a single continent with North America.

The chief physical events of the early Paleozoic were advances and retreats of the shallow sea which covered the geosynclines most of the time and parts of the continental interior occasionally. At one time nearly 65 percent of the continent was under water (Fig. 39.2). In the intervals between periods the continent was uplifted so that even the geosynclines were partly or wholly dry. In general this was a time when the earth's crust in North America was stable, its movements consisting simply of minor ups and downs of the continent as a whole. The only important exception was a mountain-building disturbance in New England toward the end of the Ordovician period.

The nature and distribution of fossil animals suggest that the early Paleozoic climates were mild over the entire continent, without the marked zoning into hot and cold regions that we find today. One exception was an interval of desert conditions recorded by beds of salt and gypsum in the Silurian strata of New York.

The animals whose fossils appear in the earliest Paleozoic beds were *invertebrates*, creatures without internal skeletons but with external shells of calcium carbonate, silica, or chitin. All the major groups of the invertebrates are represented, some by organisms that must have had complex internal structures. It seems impossible that such varied and complex forms could have evolved during the erosion interval at the end of Precambrian time, but the odd fact remains that fossils of these creatures are not found in late Precambrian rocks. A possible way out of this dilemma is the hypothesis that fairly complex invertebrates were indeed present in the seas of late Precambrian times but that they were not yet able to grow hard shells that could be preserved. This idea would allow a long period of evolutionary development during Precambrian times, the only radical advance just before the Cambrian being the acquisition of shells.

What the lands of early Paleozoic times looked like we can only surmise. Probably primitive plants had gained a foothold even in the Precambrian, but the oldest definite fossils of land plants (primitive fungi and mosses) are Silurian. The first animal to succeed in adapting itself to an air-breathing existence, so far as available records show, was a relative of the modern scorpion that came out of the sea in the late Silurian.

the first air-breathing animal was a type of scorpion

late paleozoic

Rocks of the later periods reveal, in contrast with the wide seas and the minor amount of diastrophic movement in the early Paleozoic, restricted seas and great diastrophic activity. Sedimentation, largely in shallow marine waters, continued through most of this time in parts of the two major geosynclines. Occasionally seas spread over the continental interior, but never widely as in earlier periods. A peculiar type of sedimentation in some regions characterized part of the Mississippian and Pennsylvanian periods: marine deposition alternated with nonmarine, so that thin beds of sandstone, shale, and limestone follow one another in a regular sequence repeated time after time. The nonmarine parts of these cycles, especially in the Pennsylvanian, often contain layers of coal, indicating times of widespread, low-lying swamps with abundant vegetation. These are the great coal deposits of the central Mississippi Valley and the Appalachian region. So abundantly was coal formed at this time both in this country and in Europe that the Mississippian and Pennsylvanian are frequently lumped together as the *Carboniferous period*.

39.1 Map of North America in the early Paleozoic era. *The Cordilleran geosyncline is to the left, the Appalachian geosyncline to the right. The colored areas are under water.*

carboniferous period

The mild climate of early Paleozoic periods persisted over most of this continent until mid-Permian times. But with continental uplift and the rise of mountain ranges at the end of the era came radical climatic changes. Broad deserts formed in the lee of mountains; icecaps and valley glaciers developed in many parts of the world. Extremely arid climates during the Devonian in western Canada and during the Permian in Texas and New Mexico are indicated by thick deposits of salts, formed by evaporation of restricted arms of the sea—not only gypsum and sodium chloride, but far more soluble salts of potassium and magnesium as well.

Volcanic activity in the Devonian period is shown by lava flows in the extreme eastern part of Canada, and flows of later periods are found in British Columbia and the northwestern part of the United States. Minor mountain-building disturbances occurred in New England during the late Devonian, in the Appalachian region and Oklahoma at the end of the Mississippian, in western Texas toward the close of the Pennsylvanian. But these and earlier Paleozoic disturbances were dwarfed by the disturbances which ended the Permian period. At this time of intense diastrophic activity, affecting many other parts of the world besides the North American continent, the sediments that had accumulated for more than 300 million years in the Appalachian geosyncline were crumpled, fractured, and uplifted into a mountain chain which must have rivaled any modern range in height and grandeur. At this time also the entire continent was uplifted, and shallow seas disappeared except for small areas.

39.2 Map of North America during part of the Ordovician period of the Paleozoic. *Shallow seas are shown in light color and present-day outcrops of Ordovician rocks are shown in dark color.*

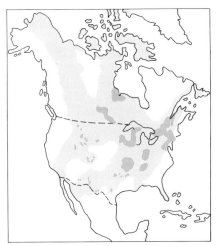

Marine life in the late Paleozoic shows many changes from that in earlier seas, but was still far different from marine life of the present. Clams and snails increased in numbers and show considerable evolutionary development. Corals built widespread reefs in the middle Devonian, but thereafter were not conspicuous. Starfish and sea urchins were not common, but some of their distant relatives that today are extinct or rare were extremely numerous. Fishes were more abundant than in earlier periods and showed a greater variety of form.

In the late Paleozoic rocks we find for the first time abundant evidence of land-dwelling organisms. In the coal swamps of Pennsylvanian times grew dense forests of primitive plants—huge fernlike trees, enormous horsetails, primitive conifers. A modern man wandering through such a forest would find no bright-colored flowers, no grasses, few plants at all familiar except possibly some of the ferns and mosses. In and near these primeval forests lived a great variety of animals: scorpions, land snails, primitive insects of many kinds.

Of most interest to us as humans, since they are early members of our own family tree, are the land-living vertebrates of the late Paleozoic. Fossil *amphibians,* oldest of the land vertebrates, appear first in Devonian rocks. These are relatives of modern frogs and salamanders, sluggish creatures that laid their eggs and spent the early part of their lives in water. Their body structure and their dependence on water suggest that they were descended from fishes, but the complete line of development cannot be traced. In Pennsylvanian rocks appear fossils of reptiles, animals that looked at first much like their amphibian ancestors but had the great advantage of being able to lay their eggs on dry land. The dry climate at the end of this era wrought havoc upon the amphibians, but the reptiles, no longer dependent on water for hatching their eggs, multiplied rapidly and developed a great variety of species. During the Permian the reptiles became the dominant creatures of the land.

The end of the Paleozoic saw harsh climates, dwindling water supplies, and rapid changes in environment, and it was a time of profound change in the organic world. Many kinds of plants and animals were unable to cope with the new conditions, and others survived only because rapid evolutionary developments made them better adapted to a changing world. So, in the earliest Mesozoic rocks, we find that many of the common Paleozoic organisms are missing and that both land and water are inhabited by a host of new forms.

Fossils of trilobites that lived about 450 million years ago. The larger one is 3 in. long.

coal and oil

coal beds are the sites of ancient swamps

Coal and oil, so necessary to our technological civilization, are abundant in Paleozoic rocks. Coal was formed from plant material that accumulated under conditions where complete decay was prevented. The only possible situation where large amounts could accumulate under such conditions was in swamps; so a bed of coal nearly always implies an ancient swamp. Coal has been formed in swamps from the Devonian to the present, but seldom have conditions been so favorable for large-scale accumulation as in the Pennsylvanian period. Apparently these special conditions were broad swamps almost at sea level, periodically submerged so that partly decayed vegetation was covered with thin layers of marine sediments.

The formation of coal begins with slow bacterial decay, chiefly of the cellulosic material of plants. Taking place largely under water and in the absence of air, this decay results in a gradual removal of oxygen and hydrogen from the cellulose and a concentration of carbon in complex compounds of unknown composition. These are the compounds that break down when heated to give the hydrocarbons with ring structures in their molecules, like benzene and naphthalene, that are so valuable in chemical industry. Aiding decay in coal formation is the action of heat and pressure resulting from burial beneath later sediments.

The origin of petroleum is more obscure, for two reasons: fossils cannot be preserved in a fluid, and oil often migrates long distances from the place where it forms. Because petroleum hydrocarbons can be detected in modern marine sediments, because oils resembling petroleum can be prepared artificially from organic material, and because petroleum is commonly found in rocks formed from sediments deposited in shallow seas, most geologists consider that the origin of petroleum from organic material is established. Probably both plant and animal matter contribute to its formation, the substances involved being largely proteins, fats, and waxes rather than cellulose. From these materials slow bacterial decay in the absence of air produces the characteristic hydrocarbons of petroleum, those with straight-chain molecules like butane and octane. Natural gas, usually associated with petroleum and probably formed by the same type of decay, consists of the lighter hydrocarbons.

Both gas and oil, like groundwater, can migrate freely through such porous rocks as loosely cemented sandstones and conglomerates. Wherever they may be formed, they often find their way into porous beds, and it is from these beds that they are obtained by drilling. Migration along a porous bed may be induced by a number of factors—gravity, pressures due to compaction or to diastrophic movement, gas pressure, movement of groundwater. Since both oil and gas are lighter than water, they may be displaced by groundwater and so move upward to the surface to form oil seeps. In general, oil becomes available in large quantities only when it is trapped underground by impervious material. Two common kinds of traps are illustrated in Fig. 39.3, in one of which oil is trapped under an anticline and in the other against a fault.

THE MESOZOIC ERA

The earliest Mesozoic ("intermediate life") sediments were laid down about 225 million years ago, a long time by ordinary reckoning. But the earth was already very old. Some 375 million years had elapsed since the beginning of the Paleozoic and $3\frac{1}{4}$ billion years since the oldest known rocks in the Precambrian. All the time that we include in the Mesozoic and Cenozoic eras is only one-sixteenth of the history recorded in rocks of the earth's crust.

Fossil ferns of the Pennsylvanian period found in Illinois.

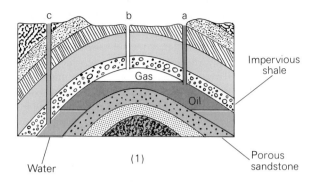

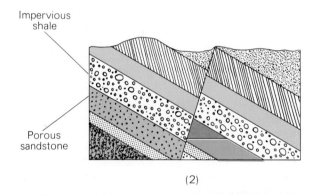

39.3 Diagrammatic cross sections showing two common types of structural traps in which oil accumulates. (1) *A trap formed by an anticline;* (2) *a trap formed by a fault. In both cases oil in a porous layer is prevented from moving upward by an impervious layer. Note that a well drilled at (a) would strike oil, one drilled at (b) would strike gas, one at (c) only water.*

mesozoic north america

laurasia and gondwanaland broke up in the mesozoic era

MESOZOIC ERA

triassic (225–180) dinosaurs and first mammals appear

jurassic (180–136) first birds; dinosaurs at their peak

cretaceous (136–65) first flowering plants; dinosaurs die out

[dates are millions of years ago]

The Mesozoic era saw the breakup of the supercontinents of Laurasia and Gondwanaland, whose appearance in the late Paleozoic was something like that sketched in Fig. 38.5. Early in the Mesozoic era North America began to part from Europe, and somewhat later, perhaps 120 million years ago, South America and Africa began to drift apart. By the end of the era Gondwanaland no longer existed: Australia, New Zealand, and India had all left Africa, though Arabia still remained attached. Africa itself was in the process of a shift northeastward, thus closing the western end of the Tethys Sea, while India, well on its way toward Asia, was moving into the eastern end. The Mid-Atlantic Ridge was already a prominent feature of the floor of the infant Atlantic Ocean.

For the North American continent—first as part of Laurasia, then as a separate continent—the history of the era can be reconstructed from the distribution of Mesozoic rocks. At first almost the entire landmass stood above sea level, its surface a broad plain except for the new-formed Appalachians on the east and a few low ranges elsewhere. Streams from the eastern Appalachians carried sediments into narrow valleys along the mountain front. Arid conditions in the Southwest made possible the heaping up of wind-blown sand. Then, toward the middle of the era,

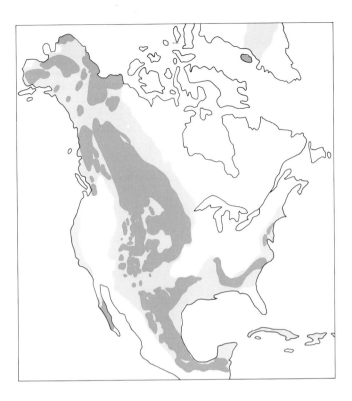

39.4 Map of North America during the Cretaceous period of the Mesozoic era. *Shallow seas are shown in light color and present-day outcrops of Cretaceous rocks are shown in dark color.*

Two oil wells being drilled simultaneously from an offshore platform in the Gulf of Mexico.

shallow seas along the Pacific Coast became more prominent. From the Arctic Ocean a sea invaded the Cordilleran geosyncline, at the time of its widest extent spreading out over the plains states and connecting with the Gulf of Mexico (Fig. 39.4). Fluctuations of this sea led to the formation of inland basins where river and lake deposits were formed. Finally, at the end of the era, the sea withdrew and left the continent once more above water.

The Mesozoic, like the early Paleozoic, was for the most part a time of crustal stability interrupted only by minor uplifts and subsidences. One local disturbance broke into this peaceful picture, perhaps coinciding with the beginning of the westward drift of North America away from Europe: Toward the end of the Jurassic, folding and the intrusion of granite batholiths formed a mountain range on the site of the present Sierra Nevada of California and lesser ranges to the north and south. Volcanic activity during the early Mesozoic is recorded in basalt flows and intrusive sheets in the red Triassic sediments of the Appalachians. Volcanic materials are also abundant in Triassic beds in British Columbia and in the northwestern states. Eruptions in California and adjacent regions accompanied the mountain building at the end of the Jurassic.

Climates of the Mesozoic were in general mild, although arid conditions in the southwestern states are recorded for at least part of this era. When the late Mesozoic sea was widespread, equable climates extended to far northern latitudes, for fossil palms and breadfruit trees are found in the Cretaceous rocks of Greenland.

Like the Paleozoic, the Mesozoic era in North America closed with a time of intense diastrophic activity, this time centering in the Cordilleran geosyncline. The sediments of this great trough, which had accumulated intermittently since earliest Cambrian times, were folded, thrust eastward, and uplifted high above sea level. At the same time other parts of the continent were raised vertically; in particular the old Appalachian range, worn down by the end of the Cretaceous to a nearly level plain, was warped upward so that erosion could begin anew. This period of disturbance, like earlier ones, was accompanied by extremes of climate and by rapid shifts in the position of land and sea. Many groups of Mesozoic animals succumbed during this period, and others underwent rapid evolutionary development to keep pace with the changing environment.

dinosaurs and mammals

the mesozoic was the age of reptiles, of which the largest were the dinosaurs

On the Mesozoic lands there developed a group of reptiles that included some of the largest animals the earth has seen—the race of *dinosaurs*, descendants of the few primitive reptiles that had survived from the Paleozoic. Some were carnivores, their bodies designed for pursuing and eating other animals. Some were herbivores, with jaws and digestive organs adapted for a vegetarian diet. Active forms were adapted for life in open plains, more sluggish forms for life in swamps. Some developed bony armor for protection; others depended on speed to escape their enemies. Not all the dinosaurs by any means grew to large size, but the biggest ones were enormous beasts over 80 ft in length and weighing more than 40 tons.

mammals first appeared in the mesozoic, but were not prominent until much later

Meanwhile other land organisms were undergoing important developments. Flowering plants appeared in mid-Mesozoic and with them a host of modern-looking insects adapted for helping in the pollination of flowers. The first true birds, with feathered wings rather than membranes, developed from reptilian ancestors in the Jurassic. Sometime in the Triassic appeared the first *mammals*, tiny creatures probably descended from a group of small Permian reptiles. All during the Mesozoic the mammals remained small and inconspicuous, but in several respects they represented an evolutionary advance over the reptiles: They were warmblooded, hence better able to cope with changes of temperature; they had bigger brains relative to their body size; and they cared for their young after birth, so some of the experience of one generation could be passed on to the next. These traits enabled the mammals to survive the drastic climatic changes at the end of the Cretaceous period, as the highly specialized reptiles could not.

THE CENOZOIC ERA

In many ways the Cenozoic ("recent life") era, in which we exist today, has been markedly different from preceding eras. During the Cenozoic the continents have stood for the most part well above sea level. No longer do shallow seas spread widely; in North America, marine beds are found only in narrow strips along the Pacific Coast and on the Atlantic Coast from New Jersey south to Yucatan. The locally thick Tertiary beds east and west of the Rocky Mountains are river, lake, and wind deposits made in continental basins. And climates during much of the Cenozoic have had a diversity like those of the present: The distribution of plants and animals shows that, instead of having widespread moderate climates like those of other eras, Cenozoic continents have had zones of distinct hot, cold, humid, and dry climates.

geologic activity

A characteristic of Cenozoic times has been widespread volcanic activity. From the Rockies to the Pacific Coast lava flows and tuff beds testify to the former presence of volcanoes, some of which have only recently become extinct. In the mid-Tertiary, immense flows of basalt inundated an area of nearly 200,000 square miles in Oregon, Idaho, and Washington; some of these flows today form the somber cliffs of the Columbia River gorge.

Finally, the Cenozoic has been a time of almost continuous diastrophic disturbance, in contrast with the long periods of crustal stability in previous eras. Movements associated with the mountain-building episodes that divide the Cenozoic from the Mesozoic lasted well into the Tertiary. In mid-Tertiary the Alps and Carpathians of Europe and the Himalayas of Asia were folded and uplifted, marking the end of the Tethys Sea. Toward the end of the Tertiary the Cascade range of Washington and Oregon was formed, and other mountain-building movements began around the border of the Pacific which have continued to the present day. Mountain ranges that had been folded earlier—the Appalachians, the Rockies, the Sierra Nevada—were repeatedly uplifted during the Cenozoic, and erosion following these uplifts has shaped their present topography.

It is not hard to associate the above reshaping of continental landscapes with the spreading of sea floors and the grinding together of lithospheric plates that are still in action today. In the Cenozoic the continents continued their earlier drifts, and in addition Greenland parted from Norway, Australia parted from Antarctica (New Zealand had done so earlier), and the Bay of Biscay opened up. More recently the Arabian peninsula broke off from Africa, the Gulf of California opened to separate Baja California from mainland Mexico, and Iceland rose above the surface of the Atlantic Ocean.

animal life

Like the reptiles at the start of the Mesozoic, the mammals in the early Tertiary evolved quickly from a few primitive forms into species adapted for many different modes of life. Carnivores like cats and wolves, herbivores like horses and cattle, sluggish armored beasts like rhinoceroses, agile creatures built for speed like deer and rabbits—ancestors of all these

---

CENOZOIC ERA

tertiary
- paleocene (65–54) *first primates*
- eocene (54–38) *primitive horses, camels, rhinoceroses*
- oligocene (38–26) *apes and elephants appear*
- miocene (26–7) *grasses abundant; rapid spread of grazing mammals*
- pliocene (7–2) *flowering plants abundant*

quaternary
- pleistocene (2–0.01) *rise of man; large mammals abundant*
- recent (0.01–0) *man is most complex and successful form of life*

[dates are millions of years ago]

volcanic and diastrophic activity have been prominent in the cenozoic

the cenozoic is the age of mammals

A dinosaur being assembled despite missing parts.

modern forms roamed the early Tertiary landscape. A few mammals, like the whales and porpoises, became adapted to life in the sea; another line, the bats, developed wings. By the middle Tertiary, mammals dominated the earth as the reptiles had before them.

Some of the mammalian lines have left fossils in sufficient abundance so that their development can be traced in detail from Mesozoic forms. In general, evolutionary changes in the Tertiary involve an increase in size, an adaptation of tooth structure for special diets, a specialization of limb structure for various postures and modes of life, and an increase in brain size. The modern horse, for example, is descended from a tiny rabbitlike creature of the early Tertiary; succeeding generations increased in size and brain capacity, their original five toes decreased to one powerful hoof, their legs became strengthened for speed, their teeth developed into efficient grinding instruments for a diet of grass. Side by side with the mammals developed modern birds, modern insects, and the

The fossil-rich Badlands of South Dakota were formed from sediments laid down 40 million years ago when the region was covered by a shallow sea.

deciduous trees of modern forests. As the end of the Tertiary approached, both the physical and the organic worlds assumed more and more closely their present aspect.

Until well into the Mesozoic era primitive mammals were apparently able to move more or less freely between many of today's continents despite the fragmentation of Laurasia and Gondwanaland. Land bridges joined North America with both Europe and Asia, for instance, so even though an inland sea from the Tethys Sea to the Arctic Ocean cut Europe off from Asia directly, a roundabout route between them existed. But in the Cenozoic the biological isolation of the continents became increasingly effective—there was not even a land link between North and South America until only a few million years ago—and separate lines of mammalian development appeared. In place of the division of the world's land area into two self-contained biological units in the Mesozoic, when reptiles were the dominant form of animal life, the Cenozoic lands seem to have been effectively split into at least eight biological units.

The Quaternary period of the Cenozoic has lasted for about 2 million years. This is roughly 3 percent of the entire era. The 65 million years that make up the Cenozoic represent in turn about 2 percent of the total time since the earliest dated Precambrian rocks were formed. If the geologic record encompassed one year instead of 3.5 billion years, the Quaternary would cover a few hours. About this brief period we have a great deal of information, for its events are very close to our own time.

the ice age

During the Pleistocene epoch, which constitutes the first part of the Quaternary, great icecaps formed in Canada and northern Europe, and valley glaciers advanced in high mountains elsewhere. About a quarter of the earth's land area was covered by ice. This was but the latest in a series of glacial periods that have punctuated earth history, but, since these particular glaciers have taken part so directly in shaping present landscapes, the Pleistocene is often referred to simply as the Ice Age.

the origin of the ice age is uncertain

Why the climate of this particular time became cold enough for glaciers to form remains an unsolved problem. There are many theories, some more plausible than others but none really accepted. In one it is assumed that the Ice Age began as a result of a reduction in the amount of solar energy reaching the earth, possibly due to the screening effect of large amounts of volcanic dust in the atmosphere, or conceivably to a transient diminution in the sun's energy output. In another it is assumed that the Ice Age began as a result of an *increase* in the solar energy reaching the earth. This caused the earth to heat up, evaporating huge quantities of water which formed clouds that provided an effective shield against further solar radiation. Then the earth, under the concealing clouds, cooled down, and the Ice Age came into being.

Glacial deposits in North America show that ice spread outward from three centers of accumulation in Canada, the ice front in its farthest advance reaching the Missouri River on the west and the Ohio River to the east (Fig. 39.5). Throughout this area the numerous lakes and

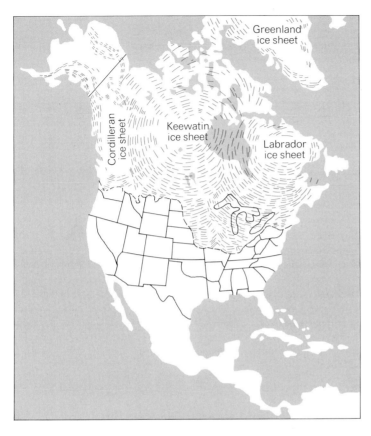

39.5 Map of North America showing the maximum extent of Pleistocene glaciers.

swamps, rounded hills, and abundant boulders of Precambrian rocks that spread over the surface of eroded Paleozoic strata would suggest the work of glaciers even to a casual observer. Closer study reveals fresh, almost unweathered deposits overlying older, deeply weathered glacial material which sometimes contains plant fragments. Thus the ice must have spread not once but several times, melting back between the periods of advance sufficiently for vegetation to flourish. Four different times of ice advance can be distinguished, with long interglacial periods between; during at least one of these interglacial times the ice disappeared completely, and the climate became warmer than it is today.

The changing climates of the Pleistocene proved a severe ordeal for the mammals. In the present world mammals are still dominant, but in numbers and diversity of species they have declined markedly since the later Tertiary. Against this background of shifting ice fronts and a declining mammalian population was played the drama of early human history. Probably sometime in the late Tertiary man's evolutionary branch split off from that of the apes and monkeys, for man-like fossils and crude stone implements have been found in uppermost Pliocene beds in east Africa. By early in the Ice Age human creatures had spread widely over the earth; their remains have been found in Java and China as well as in Africa. The later record is most complete in Europe, where stone

early human fossils appear in african tertiary strata

implements, burial sites, drawings on caves, and skeletal fragments give a fairly connected history. Unfortunately the history does not show a continuous development but rather a succession of races, each flourishing for a time and then being supplanted by another. The immediate ancestors of present-day Europeans did not appear until the retreat of the last ice sheet, some 30,000 years ago.

Because the Cenozoic era is so brief and because it is a time of mountain building, of continental elevation, and of climatic change, there is some question whether it should be considered an era at all—at least in the sense in which we regard the Mesozoic and Paleozoic as eras. Events in the Cenozoic seem important and the times involved very long simply because the events are so close to us compared with those of earlier times. From a longer perspective the whole Cenozoic may perhaps be better regarded as the terminal throes of the Mesozoic era.

Whatever the answer to this question, it is certain that we live in a time that is, geologically speaking, highly unusual. The widespreading shallow seas of earlier ages are nearly absent; mountains stand high above sea level; large parts of the earth are desert; temperatures vary widely from one part of the earth to another. Similar conditions are found in the past only during times of major geologic change. If we are living in a time of such change, the future may hold more diastrophic activity, continued volcanic eruptions, perhaps another advance of glaciers. Or perhaps we live just at the start of a new era, so that the future should bring a time of milder climates, of gradually lowering mountains, and of seas spreading far inland.

GLOSSARY

Invertebrates are animals without internal skeletons but with external shells.

Coal was formed chiefly from cellulosic plant material that accumulated under conditions preventing complete decay. *Petroleum* is believed to have originated from the protein, fat, and wax content of plant and animal matter.

The *dinosaurs* were reptiles of various kinds, many huge in size, which flourished in the Mesozoic era.

Mammals are warm-blooded animals that bear live young. They have relatively large brains and care for their young after birth.

EXERCISES

1 What possible explanation can you suggest for the fact that Mesozoic dinosaurs did not survive into the Cenozoic era whereas mammals did?

2 In rocks of what era or eras would you expect to find fossils of the following:
   a. Horses
   b. Ferns
   c. Clams
   d. Insects
   e. Apes

3 Describe the landscape in the Appalachian Mountain region of eastern Pennsylvania as it appeared (*a*) in the early Paleozoic, (*b*) during the Pennsylvanian period, (*c*) at the beginning of the Mesozoic.

4 At what time in geologic history
   a. was most of the Mississippi Valley covered by a shallow sea?
   b. were extensive mountains first formed in the Rocky Mountain region?
   c. were the North Central states covered by an icecap?

5 Discuss the differences in origin of coal and oil.

6 What kinds of geologic processes are exhibited in Precambrian strata?

part eleven

THE SUN AND ITS FAMILY

The earth is an inconspicuous member of the solar system, the group of nine planets that circle the sun. In the first of the chapters that follow, the properties of the various planets are outlined, with special attention to the possibility—by no means small—that life of some kind exists on some of them. Then the moon is honored with a chapter of its own, chiefly because it has been the destination of the first manned space flight to another celestial body. This transcendent adventure represents another milestone on the endless road of science: not so much because it is a triumph of advanced technology as because of the quantum jump it represents in astronomical research.

Finally we take up the sun itself, the bright orb at the heart of the solar system whose steady shine furnishes the energy needed by living things wherever they exist in the solar system. Although we cannot perform experiments on the sun (or other stars) to find out how they function, we can examine the light they emit with the spectroscope and infer a great deal. Among the characteristics of a star revealed by its spectrum are its structure, temperature, composition, condition of matter, and state of motion—a remarkable list. The source of solar energy and the mystery of sunspots are among the aspects of solar behavior that are examined.

chapter 40

The Solar System

THE FAMILY OF THE SUN

the planets

bode's law

meteors

comets

THE PLANETS

mercury

venus

mars

jupiter

saturn

uranus, neptune, pluto

The greatest modern telescope provides no more direct information about the stars than the naked eye does—to both a star is simply a tiny point of light. On the other hand, most of the planets are magnified to clear disks by telescopes of even modest power. This does not mean that the stars are smaller than the planets, of course, but only that they are much farther away. It is about 4×10^9 miles from the sun to the outermost planet, 2.5×10^{13} miles from the sun to the nearest star. To give these distances meaning, let us try a drastic reduction of scale. If we take a golf ball to represent the sun, we must put a small sand grain a dozen feet away to represent the earth. The farthest planet, Pluto, will be another sand grain 500 ft from the golf ball. Within the 1,000-ft-wide orbit of Pluto are all the other planets. But to place the nearest star in our model, we must take another golf ball 600 miles away!

The earth and the sun and the other eight planets are isolated in space. This set of nine spheres that circle the bright sun is poised in emptiness and separated by unimaginable distances from everything else in the universe. Because the sun is its central figure, the family of bodies that accompanies it is called the *solar system*, and in this chapter we shall survey briefly what is known about it.

THE FAMILY OF THE SUN

Until the seventeenth century the solar system was thought to consist of only five planets besides the earth and moon. In 1609, soon after having heard of the invention of the telescope in Holland, Galileo built one of his own and was able to add four new bodies to the system: the brighter of the moons (or *satellites*) that revolve around Jupiter. Since Galileo's time telescopic improvements have made possible the discovery of many more members of the sun's family. The list of planets now includes nine; in order from the sun they are Mercury, Venus, Earth, Mars, Jupiter, Saturn, Uranus, Neptune, and Pluto (Fig. 40.1). All except Mercury, Venus, and Pluto have satellites, smaller bodies revolving around them as the moon revolves about the earth. Thousands of small objects called *asteroids*, all less than 500 miles in diameter, follow separate orbits about the sun in the region between Mars and Jupiter. Comets and meteors, in Galileo's time thought to be atmospheric phenomena, are now recognized as still smaller members of the solar system.

there are nine planets in orbits around the sun

asteroids, comets, and meteors are also members of the solar system

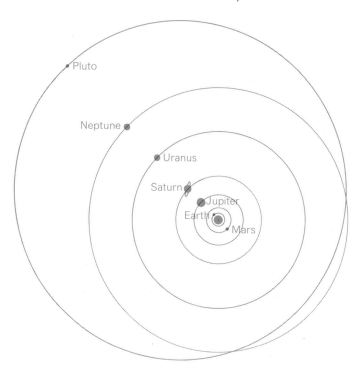

40.1 The solar system. *The orbits of Mercury and Venus are too small to be shown on this scale. Pluto's orbit is by far the most elliptical. Diameters of sun and planets are exaggerated.*

the planets

Not only is the entire solar system isolated in space, but each of its principal members is separated from the others by vast distances. From the earth to our nearest neighbor, the moon, is about 238,000 miles; from the earth to the sun is about 93 million miles. It took the Apollo 11 spacecraft 3 days to reach the moon, and at the same rate of progress more than three years would be needed to reach the sun.

THE SOLAR SYSTEM

Let us return for a moment to the model mentioned at the start of this chapter in which a golf ball represented the sun and a grain of sand 12 ft away the earth. On this scale the moon would be scarcely more than a dust speck, about $\frac{1}{2}$ in. from the sand grain. The largest planet, Jupiter, would be the size of a small pebble, 60 ft from the golf ball. With three smaller pebbles, three more sand grains, and a few more dust specks, all within the 1,000-ft-wide orbit of Pluto, the model is complete (Fig. 40.2). An extremely empty structure, this solar system, with its members separated by distances enormous compared with their size.

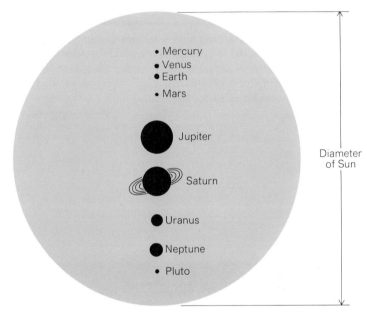

40.2 *Relative sizes of planets and sun.*

Planets *revolve* about the sun and *rotate* on their axes. Their motion around the sun follows Kepler's three laws: Each planet describes an ellipse having the sun at one focus; the motion is fastest when the planet is nearest the sun, slowest when the planet is farthest away; planets with small orbits move rapidly, planets with large orbits slowly. Asteroids follow similar elliptical orbits, though not in the same plane as those of the planets. Satellites move in small ellipses around their planets.

Two facts about the motions of the solar system are notable:

1 Nearly all the motions—revolutions of planets, asteroids, and satellites, axial rotations of sun and planets—are in the same direction; only the rotation of Venus and the revolutions of a few minor satellites run contrary to the general motion. (Uranus is an exception of a different kind, since it rotates about an axis only 8° from the plane of its orbit.)
2 All the orbits except those of comets lie nearly in the same plane (Fig. 40.3).

As a consequence of these facts, the solar system has a great deal of angular momentum, nearly all of which resides in the motions of the

most of the angular momentum of the solar system resides in the motions of the planets and satellites

40.3 The orbits of the planets seen edgewise, showing that they lie nearly in the same plane.

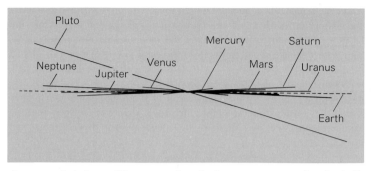

planets and their satellites even though the sun accounts for the bulk of the entire system's mass. The strange distribution of angular momentum in the solar system has important implications as regards both the origin of the system and the possibility of planetary systems about other stars, topics we shall consider in Chap. 45.

THE SOLAR SYSTEM

Planets, asteroids, and satellites shine only by reflected sunlight, and observation of any of these objects is limited to the half that is directly exposed to the sun. Planets with orbits larger than the earth's never come between us and the sun, so we can always see nearly the whole of their illuminated sides. Mercury and Venus, however, have orbits smaller than the earth's and are between us and the sun for a good part of each revolution. In this position their dark sides are turned toward us, and we see them either not at all or as thin crescents.

planets shine by reflected sunlight

bode's law

Perhaps the most remarkable regularity in the solar system is the "law" that relates the distances of the planets from the sun. Kepler thought he had found such a law when he noticed that, if a cube were placed inside a sphere with the radius of the orbit of Saturn, a sphere inscribed

This crater near Winslow, Arizona, was caused by the impact of a huge meteorite.

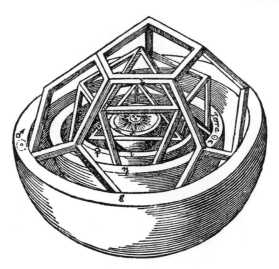

Kepler's model of the regularity of the solar system.

within the cube had almost the identical radius as Jupiter's orbit. A tetrahedron inscribed in Jupiter's sphere will just enclose a sphere whose radius is that of the orbit of Mars; and so on, with the five regular solids interposed between spheres representing the orbits of the six planets known at the time. This celestial model gave Kepler more joy than any of his other discoveries, but its imprecision and the finding of additional planets which cannot be fitted into the model make it impossible for us to share his pleasure.

In 1772 an equally unusual relationship turned up which has lost neither validity nor inexplicability in the two centuries since then. It is so well established that some astrophysicists have designed theories of the origin of the solar system primarily to account for it, whereas others prefer to pretend it is really not there at all, in order to avoid its explanation. This rule, known as Bode's law after the man who first published it, states that the distance from the sun to any planet, counting the earth's distance from the sun as 1.0, may be found by taking the number 0.4 and adding to it 0 for Mercury, 0.3 for Venus, 0.6 for the Earth, 1.2 for Mars, and so on, doubling the number to be added to the constant 0.4 for each successive planet. The result is a table of planetary distances which compares with the actual distances as follows:

bode's law

Planet	Predicted distance	Actual distance
Mercury	0.4 + 0.0 = 0.4	0.39
Venus	0.4 + 0.3 = 0.7	0.72
Earth	0.4 + 0.6 = 1.0	1.00
Mars	0.4 + 1.2 = 1.6	1.52
Asteroids	0.4 + 2.4 = 2.8	2.9 (average)
Jupiter	0.4 + 4.8 = 5.2	5.20
Saturn	0.4 + 9.6 = 10.0	9.54
Uranus	0.4 + 19.2 = 19.6	19.18
Neptune	0.4 + 38.4 = 38.8	30.06
Pluto	0.4 + 76.8 = 77.2	39.44

Evidently Bode's law is quite accurate out to Uranus, is rather poor for Neptune, and is hopeless for Pluto—although Neptune's predicted distance is fairly close to Pluto's actual one, and there is evidence (which we shall examine later in this chapter) that a violent event of some sort involving Neptune and Pluto took place in the past.

In Bode's time the asteroids, Uranus, Neptune, and Pluto were still unknown. Bode had no reason to suppose that planets beyond Saturn existed, but he felt strongly about the apparent gap between Mars and Jupiter—"Can we believe that the Creator of the world has left this space empty? Certainly not!" The first confirmation of Bode's law came unexpectedly with Herschel's discovery of Uranus, which fitted almost perfectly into the table after Saturn. Evidently this was no accidental relationship, but one to be taken seriously. The search for the missing planet began in earnest, and soon Piazzi, a Sicilian astronomer, found a tiny object in almost exactly the right place—Ceres, about 500 miles across. Shortly afterward a number of other minor planets were found whose orbits lie between those of Mars and Jupiter. Their average distance from the sun is very close to the prediction of Bode's law. Of the several thousand asteroids whose properties are known (there are many more) the majority are between 10 and 25 miles across; their total mass is a good bit less than that of the moon.

the asteroids form a belt between the orbits of mars and jupiter; most are very small

A further reason for taking Bode's law seriously is the fact that the satellites of Jupiter, Saturn, and Uranus have orbits whose radii are given by formulas mathematically identical with Bode's law, though of course with different numbers in each case. Newton was able to discern the law of gravitation in Kepler's laws of planetary motion; perhaps a new Newton will be required to interpret Bode's law.

meteors

Meteoroids are small fragments of matter that the earth meets as it travels through space. Most meteoroids have masses of less than a gram. Moving swiftly through the atmosphere, meteoroids are heated rapidly by friction. Usually they burn up completely about 60 miles above the earth, appearing as bright streaks in the sky (*meteors,* or "shooting stars"). Sometimes, though, they are so large to begin with that a substantial portion may get through the atmosphere to the earth's surface. The largest known fallen meteoroids, called *meteorites,* weigh several tons. The smallest meteoroids are so light that they float through the atmosphere without burning up; many tons of these fine, dustlike *micrometeorites* reach our planet daily.

meteoroids are chunks of matter moving through space; meteors are the flashes of light they produce in the atmosphere; meteorites are what reach the ground

A keen observer on an average clear night can spot as many as 10 meteors an hour. Most of these meteors are random in occurrence, following no particular pattern either in the time or place in the sky that they appear. At several specific times of year, however, great meteor showers occur, with 50 to 100 or more meteors visible per hour. The showers come about when the earth moves through a swarm of meteoroids all following the same orbit about the sun. The most conspicuous meteor showers can be seen about May 4–6, August 10–14, and October 19–23 every year.

there are two chief kinds of meteorites

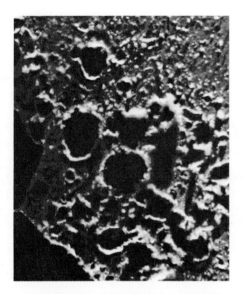

Pellets fired from a 30-ft hydrogen-powered gun at about 23,000 ft/sec simulate meteorite impacts on alloyed aluminum and titanium panels such as might be used in spacecraft. Some of the struck material is vaporized by the sudden transfer of energy, and the accompanying transfer of momentum produces pressure waves that cause internal damage.

Nearly all meteorites that have been examined fall into two classes: (1) stony meteorites, whose compositions are similar to those of ordinary silicate rocks, and (2) iron meteorites, which consist largely of iron with a small percentage of nickel alloyed with it. Both types of meteorites are sufficiently different from terrestrial rocks to permit their unambiguous identification. All lines of evidence indicate that meteoroids, random as well as shower, are members of the solar system and pursue regular courses about the sun until they collide with the earth or some other planet. Recent studies strongly suggest that meteoroids are tiny remnants of the primordial matter from which the larger bodies of the solar system—the planets and their satellites—formed.

comets

Comets appear as small, hazy patches of light, often accompanied by long, filmy tails. Most comets are visible only telescopically, but occasionally one becomes conspicuous to the unaided eye. Watched for a few weeks or months, a comet at first grows larger, its tail longer and more brilliant; then it fades slowly, losing its tail and at length disappearing altogether.

Paths followed by comets are quite different from the nearly circular planetary orbits. Usually a comet moves toward the sun from far out in space beyond the orbit of Pluto, approaches the sun closely, swings around it, then disappears toward the outer parts of the solar system. Sometimes the orbit is a long, narrow ellipse, so that the comet returns at long intervals to the sun and is definitely a part of the solar system. Sometimes the orbit appears to be an open curve (parabola or hyperbola), along which the comet would move indefinitely outward into space and never return to the sun. Probably all comets belonged originally to the solar system, and one is deflected into an open-curve orbit only if it chances to pass too close to a planet on its journey around the sun.

A five-inch meteorite.

| April 26 | April 27 | April 29 | April 30 | May 1 |

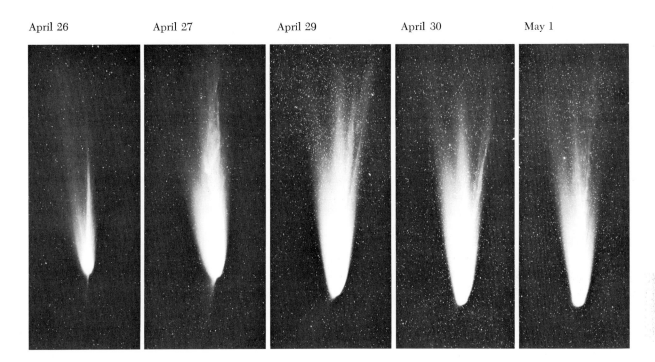

Comets are visible only when they are close to the sun. This is partly because of the reflection of sunlight by cometary material and partly because of an excitation of the cometary gases similar to the excitation that makes the gas neon glow in an illuminated sign. Also, comet tails appear only in the solar vicinity, and they have the striking characteristic of always pointing away from the sun, even when the parent comet is receding (Fig. 40.4). The explanation for this curious behavior is again twofold: to some extent it is a result of the pressure of the sun's radiation pushing gases from the head of the comet, and to some extent it is a result of the streams of protons and electrons constantly emitted by the sun (the "solar wind") sweeping these gases outward. In many comets, apparently, one or the other of these causes is predominant; in others both are effective.

The modern picture of a comet has it as an aggregate of ice, frozen ammonia, and frozen methane, with some other materials probably also present. Out in the far reaches of the solar system comets range from about 1 to 50 miles across. Near the sun, though, the ice, ammonia, and methane melt and vaporize, forming the visible comet head tens of thousands of miles across that we are able to see with our naked eyes. Once past the sun, the comet begins to freeze again into a relatively small body. Some comets have disintegrated into swarms of meteors, which the earth meets whenever it crosses the original paths of the comets. In the past possible collisions between the earth and comets have often caused much anxiety, but in the light of present knowledge it seems unlikely that even a direct hit by the head of a comet would be noticed as anything more than an unusually brilliant shower of meteors.

These photographs of the Arend-Roland comet were taken on successive nights from April 26 to May 1, 1957.

comets glow and have tails only when near the sun

40.4 *The tail of a comet always points away from the sun because of pressure from the sun's radiation and from solar wind of ions.* The tail is longest near the sun, and is probably absent far away from the sun.

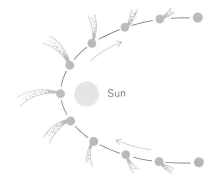

THE PLANETS

Table 40.1 summarizes the principal data about the planets and their orbits; it is a repetition of Table 3.1. Evidently the planets fall into two categories. The *inner planets* of Mercury, Venus, Earth, and Mars are relatively small, have densities 4.0 to 5.5 times that of water, and rotate fairly slowly on their axes. The *outer planets*, except for Pluto, are large, have densities only 0.7 to 2.0 times that of water, and rotate fairly rapidly on their axes. Pluto's properties resemble those of the inner planets more closely than those of the outer ones, which suggests that the details of its origin may have been different from those of the latter.

mercury

Mercury, smallest and swiftest of the planets, is too close to the sun for detailed observation of its surface. Radar measurements indicate that it rotates once every 59 days. Mercury has either no atmosphere at all or an exceedingly thin one, which is in accord with its low escape velocity. Surface temperatures on the sunlit side are higher than the melting point of lead, and those on the dark side are extremely cold, at most $-123°C$.

venus

venus rotates "backwards" very slowly

In size and mass the planet Venus resembles the earth more closely than any other member of the sun's family. Apart from the sun and the moon, Venus is the brightest object in the sky, and is even visible in daylight. Venus has the distinction of spinning "backward" on its axis; that is,

TABLE 40.1 The solar system

Body	Mean distance from sun, millions of miles	Mean diameter, thousands of miles	Relative mass, $m_{earth} = 1$	Period of revolution[a]	Period of rotation[b]	Mean density, g/cm^{3c}	Acceleration of gravity, g^d	Escape velocity, mi/sec^e	Known satellites
Sun		865	333,000		25–30 days	1.4	28	383	(9 planets)
Moon	[f]	2.16	0.01	27⅓ days	27⅓ days	3.4	0.2	1.2	—
Mercury	36	3	0.05	88 days	59 days	5	0.4	2.7	0
Venus	67	7.8	0.82	225 days	243 days[g]	5.0	0.9	6.5	0
Earth	93	7.9	1.00	365 days	24 hr	5.5	1.0	7.0	1
Mars	142	4.2	0.11	687 days	24.6 hr	4.0	0.4	3.2	2
Jupiter	483	89	318	12 years	9.8 hr	1.3	2.7	38	12
Saturn	887	75	95	29 years	10.3 hr	0.7	1.2	23	10
Uranus	1,784	31	15	84 years	10.8 hr[h]	1.5	1.0	14	5
Neptune	2,795	28	17	165 years	15.7 hr	2.0	1.4	16	2
Pluto	3,672	4?	0.2?	248 years	6 days(?)	6?	?	?	0

[a] Orbital period.
[b] Spin period.
[c] Density of water = 1.0 g/cm^3.
[d] $1 \text{ g} = 32 \text{ ft/sec}^2 = 9.8 \text{ m/sec}^2$.
[e] Speed necessary for permanent escape from gravitational field of the body.
[f] The mean distance of the moon from the earth is 238,000 miles.
[g] The direction of rotation of Venus is opposite to that of the other planets.
[h] The axis of rotation of Uranus is only 8° from the plane of its orbit.

THE SOLAR SYSTEM

Viewed through a telescope, Venus appears as a bright disk or crescent whose size varies with the distance from the Earth. Because of the dense cloud cover on Venus, no surface detail is visible.

looking downward on its north pole, Venus rotates clockwise, whereas the earth and the other planets rotate counterclockwise. An observer on Venus would find the sun rising in the west and setting in the east. The rotation of Venus is extremely slow, so that a "day" on that planet represents 243 of our days.

The surface of Venus is obscured by what seem to be thick layers of clouds. The atmosphere above the clouds, whose own composition is unknown, is mainly carbon dioxide, with a little nitrogen and still less water vapor also present. On the earth carbon dioxide is an important absorber of radiation from the earth that prevents the rapid loss of heat from the ground after sunset. Venus, blanketed more effectively by far than the earth, retains more heat; estimates based on data radioed back by spacecraft suggest an average surface temperature of about $400°C$, about the melting point of zinc. Since the temperature is so high and since no oxygen and relatively little water vapor can be detected in the atmosphere, the existence of life on Venus seems impossible.

venus is probably too hot to support life of any kind

mars

The reddish planet Mars has long fascinated astronomers and laymen alike, for it is the only other known body on which surface conditions seemed suitable for life of some kind. Yet Martian climates are exceedingly severe by our standards, and the thin atmosphere does little to screen solar ultraviolet radiation. If life exists on Mars, it is adapted to an environment that would soon destroy most earthly organisms.

Mars rotates on its axis in a little over 24 hr; its revolution about the sun requires nearly 2 years; and its axis is inclined to the plane of its orbit at nearly the same angle as the earth's. These facts mean that the Martian day and night have about the same lengths as ours and that Martian seasons are 6 months long and at least as pronounced as ours. Farther from the sun than the earth, Mars receives considerably less light and heat. Its atmosphere, largely carbon dioxide, is extremely thin, so little of the sun's heat is retained after nightfall. Daytime temperatures

mars has a thin atmosphere of carbon dioxide

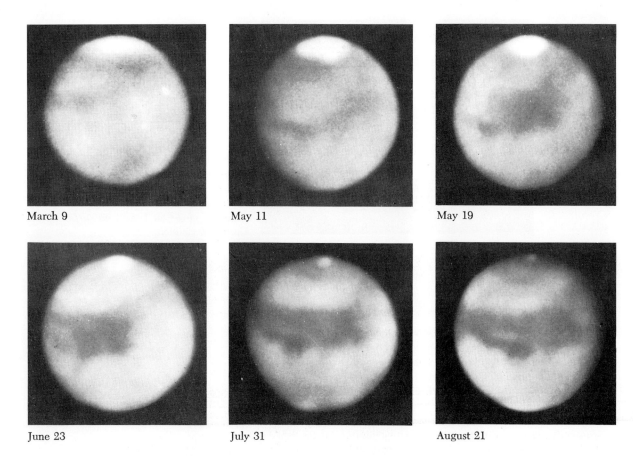

Photographs of Mars taken several weeks apart. The shrinking of the white polar cap and the change in appearance of the darker areas are evident.

in summer rise to perhaps 30°C, but at night fall to perhaps −75°C. Another major difficulty that life must face on Mars is the scarcity of water. A trifle is there, as water vapor in the atmosphere and possibly in the white polar caps as well, but only enough to be barely detectable. The polar caps, which increase in area in winter and decrease in area in summer, are believed to be almost entirely frozen carbon dioxide ("Dry Ice").

Despite these inhospitable conditions, one line of evidence suggests that life of some kind may nevertheless exist on Mars. This is based on the observation that certain areas, apart from the poles, show conspicuous seasonal color changes, from red-brown in winter to dark green in spring and summer. Other explanations are possible, but it is conceivable that these changes are due to the growth of vegetation during the warm season. As to what this vegetation might be, we cannot even guess, because no nitrogen or nitrogen compounds such as ammonia (NH_3) have been detected in the Martian atmosphere—and all forms of life on earth employ nitrogen in important roles. If there is life on Mars, it would have to be very different from what we are acquainted with.

Early in this century the Italian astronomer Giovanni Schiaparelli and the American Percival Lowell reported that the surface of Mars was covered with networks of fine dark lines, popularly called canals (a poor English translation of the Italian *canali*, meaning "channels"). The apparent straightness and geometric patterns of these canals were considered evidence of the handiwork of intelligent beings. But the canals cannot be photographed, and other observers have failed to find them. Such fine details, at the limit of visibility, are difficult to observe, because of blurring due to irregularities and currents in the earth's atmosphere. The pictures radioed back by the three spacecraft to pass near Mars show no signs of canals, though there do seem to be several regions where a number of craters are approximately in line. Probably the canals are optical illusions; certainly the existence of Martian creatures advanced enough to be capable of digging actual canals is highly unlikely.

In 1965 the American spacecraft Mariner 4 passed within a few thousand miles of Mars and radioed back 21 photographs of the Martian surface (Fig. 40.5). Four years later, on July 31 and August 5, 1969 respectively, Mariners 6 and 7 flew past Mars again and with improved equipment transmitted a total of 1,379 photographs from distances ranging from 5,700 to 2,200 miles. Terrain of three distinctive kinds was found:

1 Heavily cratered regions. Some craters are large (several km to several hundred km across) with flat bottoms, others are small with round bottoms. Only a modest amount of erosion seems to have occurred.
2 Regions broken up into irregular short ridges and depressions, with no craters. This chaotic topography may have been caused by the slumping of the ground due, possibly, to the sudden melting of a thick

the "canals" of mars

The American spacecraft Mariner 4 passed near Mars on July 14, 1965, and radioed back data from which photographic representations of the Martian surface were made. This picture was taken when Mariner 4 was 7,800 miles from Mars and covers a region 150 by 170 miles. The surface features visible are reminiscent of those of the moon.

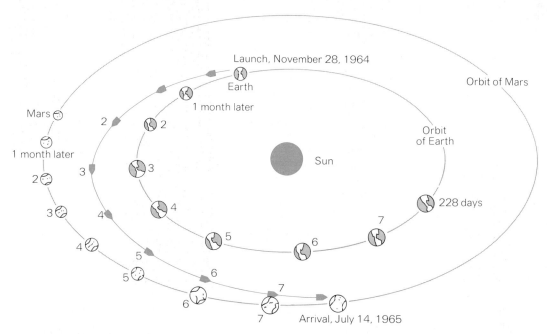

40.5 The flight of the American spacecraft Mariner 4 to Mars. *In 1969 Mariners 6 and 7 took only two-thirds as long to travel to Mars.*

layer of permafrost of some kind underneath. (On the earth, permafrost is water-saturated soil and rock that is permanently frozen below the surface in the arctic regions.)

3 Featureless regions, some of them larger and smoother than any on the moon.

The presence of many craters was expected, since the asteroid belt between Mars and Jupiter contains numerous chunks of matter whose orbits overlap that of Mars and eventually collide with it. Erosion of some kind—not the result of running water or glaciers, since so little water is present on Mars—must have obliterated the craters in regions 2 and 3. Because there are no large-scale Martian features such as mountains, continents, or ocean basins, the internal stresses that wrack the earth's lithosphere could not have occurred on Mars for a long time, if indeed they had ever been active there.

jupiter

The giant planet Jupiter, like Venus, is so shrouded in clouds that its surface cannot be seen. But Jupiter's clouds are conspicuously banded and show other semipermanent markings, such as the Red Spot some tens of thousands of miles across, that make possible a determination of the planet's period of rotation. This turns out to be less than 10 hr, which means that points on Jupiter's equator travel at the enormous speed of 28,000 mi/hr; the earth's equatorial speed is only 1,040 mi/hr. Because of its rapid rotation, Jupiter bulges more conspicuously at the equator than the earth does.

THE SOLAR SYSTEM

The four satellites of Jupiter that Galileo discovered over three centuries ago are conspicuous objects in a small telescope. The largest is as big as Mercury, and the smallest is about the size of the moon. The other eight satellites are very small (15 to 150 miles in diameter), and one of them escaped detection until 1951.

Jupiter's volume is about 1,300 times that of the earth, but its mass is only 300 times as great. The resulting low density—only a third more than that of water—means that Jupiter cannot be composed of a mixture of rock, iron, and nickel as is the earth. Like the other giant planets (Saturn, Uranus, and Neptune), Jupiter must consist chiefly of hydrogen and helium, the two lightest elements. Jupiter's interior is believed to be very hot, about 500,000°C according to some estimates, but not hot enough for nuclear reactions to occur in its hydrogen content whose release of energy would turn Jupiter into a star (these reactions are described in Chap. 42). But if Jupiter's mass were 30 times greater, the increased internal pressure would push the temperature to 20 million °C, and the result would be a miniature star.

jupiter, saturn, uranus, and neptune are composed chiefly of hydrogen and helium

The outermost layer of the clouds that mask the Jovian surface consists largely of crystals of frozen ammonia at about $-120°$C. Underneath are thought to be warmer layers respectively of ammonia vapor, ice crystals, and water vapor. The atmosphere in which these clouds float consists of hydrogen, helium, and methane. Probably Jupiter does not have an actual surface; instead, its atmosphere gradually becomes thicker and thicker with increasing depth until it becomes a liquid, and farther down the liquid turns into a solid. A terrestrial analogy might be the slushy surface of a snowbank on a warm winter day.

the atmosphere of jupiter

It seems entirely possible—some biologists think probable—that the combination of moderate temperatures and the presence of ammonia, methane, and water vapor in Jupiter's lower atmosphere has led to the evolution of life of some kind there. As mentioned in Chapter 29, laboratory experiments show that when a mixture of these gases is exposed to energy sources such as are usually present in a planetary atmosphere (for instance lightning, ultraviolet light, streams of fast ions), various amino acids, nitrogen bases, and other compounds characteristic of life are formed.

saturn

In its setting of brilliant rings, Saturn is the most beautiful of the earth's kindred. The planet itself is much like Jupiter: similarly flattened at the poles by rapid rotation, similarly possessing a dense atmosphere, its surface similarly hidden by banded clouds. Farther from the sun than Jupiter, Saturn is considerably colder; ammonia is largely frozen out of its atmosphere, and its clouds consist mostly of methane.

The famous rings, two bright ones and a fainter inner one, surround the planet in the plane of its equator. This plane is somewhat inclined to Saturn's orbit. Hence, as Saturn moves in its leisurely 29-year journey around the sun, we see the rings from different angles. Twice in the 29-year period the rings are edgewise to the earth; in this position they

saturn's rings

Jupiter.

are practically invisible, which suggests that their thickness is small, in fact only 10 or 20 miles as compared with the 170,000-mile diameter of the outer ring.

The rings are not the solid sheets they appear to be, but instead consist of myriad small bodies each of which revolves about Saturn like a miniature satellite. No satellite of substantial size can exist close to its parent planet because of the disruptive effect of tide-producing forces, which are proportionately less the farther distant the satellite. The *Roche limit* is the minimum radius that a satellite orbit must have if the satellite is to remain intact; the limit is named in honor of E. A. Roche, who investigated the origin of Saturn's rings a century ago. For Saturn the Roche limit is calculated to be 2.4 times the planet's radius, and in fact the outer rim of the outer ring is 2.3 radii from the center of Saturn and the closest satellite never approaches closer than 3.1 radii from the center. Saturn has ten ordinary satellites outside the rings; the innermost of these was discovered in 1966.

roche limit

uranus, neptune, pluto

The three outermost planets, Uranus, Neptune, and Pluto, owe their discovery to the telescope. Uranus was found quite by accident in 1781, during a systematic search of the sky by the great English astronomer William Herschel. It is just barely visible to the naked eye, and in fact had been identified as a faint star on a number of sky maps prepared during the preceding hundred years. Herschel suspected Uranus to be a planet because, through the telescope, it appeared as a disk rather than as a point of light. Observations made over a period of time showed its position to be changing relative to the stars, and its orbit was determined from these data. The discoveries of Neptune in 1846 and of Pluto in 1930 were made as the result of predictions based on their gravitational effects on other planets, as described in Chap. 3.

Uranus and Neptune are large bodies, each with a diameter about $3\frac{1}{2}$ times that of the earth. Pluto is somewhat smaller than Mars, and may once have been a satellite of Neptune that somehow was pulled away to pursue its own orbit around the sun. In most of their properties Uranus and Neptune resemble Jupiter and Saturn. Their atmospheres are largely methane, which accounts for their greenish color, with some hydrogen present as well. Because these planets are so far from the sun, their surface temperatures are below $-200°C$, and any ammonia present would be frozen out of their atmospheres. Pluto is so small, so far away, and so feebly illuminated that reliable information about it is difficult to obtain.

A unique opportunity for investigating the outer planets is presented by their positions late in the 1970s. If a spacecraft is launched in 1977 in just the right direction at just the right speed, it will first pass near Jupiter, whose gravitational attraction will swing it into a new path that will carry it to Uranus. Uranus in turn will swing the spacecraft into a path leading to Neptune. Thus a single spacecraft could make close-up observations of these three planets, which would be radioed back to earth, in a grand tour lasting 11 years (Fig. 40.6)—a tour that sheer coincidence makes possible at exactly the time when our technology is adequate for the purpose. Another feasible trajectory can take a spacecraft launched in 1979 to Jupiter, Saturn, and Pluto. Such grand tours will not become possible again for nearly 200 years.

grand tour of the outer planets

Saturn.

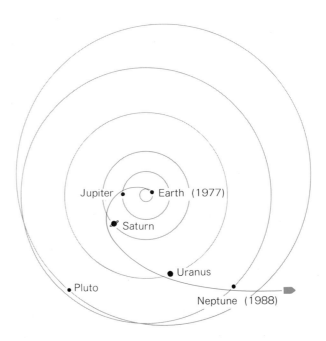

40.6 *A spacecraft with the right launching speed and direction will pass close to three of the largest planets in a period of 11 years.* The positions of the outer planets are shown approximately where they will be at the times of encounter with the spacecraft; the orbits are not to scale.

GLOSSARY

A *satellite* is, in general, an astronomical body that revolves around some other body. Usually the term refers to the satellites of the planets; thus the moon is a satellite of the earth.

Bode's law is a simple relationship that gives the distances of the planets from the sun.

An *asteroid* is one of a large number of relatively small bodies (less than 500 miles across) that revolve around the sun in orbits lying between those of Mars and Jupiter.

Meteoroids are pieces of matter moving through space; *meteors* are the flashes of light they produce when entering the earth's atmosphere; and *meteorites* are the remains of meteoroids that reach the ground.

Comets are aggregates of matter pursuing regular orbits in the solar system and appearing as small, hazy patches of light when near the sun, often accompanied by long, filmy tails.

EXERCISES

1 About how much time elapses between sunrise and sunset on Mars? On Jupiter?

2 Would you expect the danger of being struck by a meteoroid to be greater or less on Mars than on the earth?

3 Why is the average temperature on Mars lower than that on the earth? Why is there a greater variation in Martian temperatures?

4 Suppose you were on Mars, observing the earth through a telescope. Describe the changes in the earth's appearance as it moves around in its orbit.

5 How does the idea of Roche's limit bear on theories of the past history of the moon?

6 If the moon circled Jupiter in an orbit the same size as its present one, would its period of revolution be the same, shorter, or longer? Why?

7 The escape velocities of Mars and Mercury are nearly the same, yet Mars has an atmosphere and Mercury does not. Why?

chapter 41
The Moon

MOTIONS OF THE MOON
phases
eclipses
tidal friction

PROPERTIES OF THE MOON
mass
flight of apollo 11
the lunar environment
the lunar landscape
lunar rocks

ORIGIN OF THE MOON
the moon as a fragment
the moon as a captured planet
the moon as our twin planet

To look at the moon is to wonder. What is it made of? What is its origin? What is the nature of its landscape? What geological processes occur on its surface and in its interior? Is there life on the moon? What is the ultimate destiny of the earth-moon system? Until July 20, 1969 the study of the moon was more notable for the questions asked than for the answers available. On that day Neil Armstrong set foot on the moon, the first man ever to do so, after a four-day voyage aboard the spacecraft Apollo 11 with two companions. Four days after that they returned to earth, bringing with them samples of the lunar surface. Since that historic expedition questions about the moon still outnumber answers, but now the questions are more precise and detailed, reflecting the great advances in our knowledge about our nearest neighbor in space.

MOTIONS OF THE MOON

Although closer to the earth than any other celestial body, the moon is nevertheless an average of 384,400 km (almost 240,000 miles) away. Its diameter of 3,476 km (2,160 miles)—a little more than a quarter of the earth's diameter—places it among the largest satellites in the solar system. The moon circles the earth every $27\frac{1}{3}$ days and, like the earth, turns on its axis as it revolves. In the case of the moon the rotation keeps pace exactly with the revolution, so the moon turns completely around only once during each circuit of the earth. This means that the same face of our satellite is always turned toward us and that the other side remains hidden from the earth, though not from spacecraft.

the same hemisphere of the moon always faces the earth

phases

Each month the moon completes its familiar cycle of *phases*. First there is a thin crescent in the western sky at sunset which grows and moves eastward (relative to the stars); then a half-moon, until after 2 weeks the full moon rises in the east at sunset; next the moon wanes, becoming a thin crescent that rises just before the sun; finally it disappears altogether a few days before its next appearance as a crescent.

These different aspects represent the amounts of the moon's illuminated surface visible to us in different parts of its orbit. When the moon is full, it is on the opposite side of the earth from the sun; so the side facing

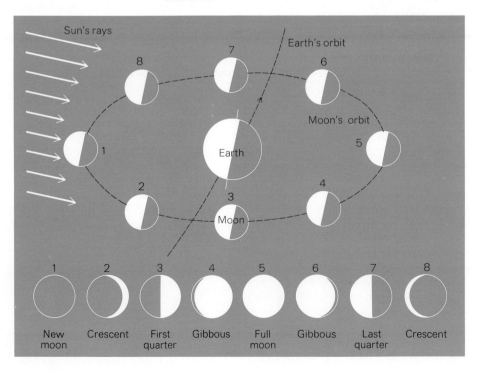

41.1 *The origin of the moon's phases.* As the moon revolves around the earth we see it from different angles; when it is between us and the sun we see only the dark side (new moon), and when it is on the opposite side of us from the sun we see only the illuminated side (full moon). At other times we see parts of both sides.

THE MOON

us is fully illuminated. In the "dark of the moon" (or new moon), it is moving approximately between us and the sun, so the side toward the earth is in shadow. Figure 41.1 illustrates the origin of the moon's phases.

eclipses

When the moon is behind the earth (full moon), how can the sun illuminate it at all? Why doesn't the earth's shadow hide it completely? Again, when the moon passes between sun and earth, why isn't the sun hidden from view?

The answers to these questions follow from the fact that the moon's orbit is tilted at an angle to the earth's orbit, so that ordinarily the moon passes either slightly above or slightly below the direct line between sun and earth (Fig. 41.2). On the rare occasions when the moon does pass more or less directly before or behind the earth, an *eclipse* occurs—an eclipse of the moon when the earth's shadow obscures the moon, an eclipse of the sun when the moon's shadow touches the earth.

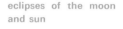

eclipses of the moon and sun

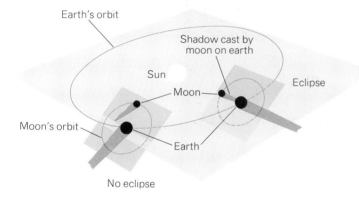

41.2 The orbit of the moon is tilted with respect to that of the earth. *For this reason the moon normally passes above or below the direct line from the sun to the earth. Eclipses occur only on the rare occasions when the moon passes exactly between the earth and the sun (a solar eclipse, with the moon blocking out light from the sun) or exactly behind the earth (a lunar eclipse, with the earth blocking out light from the sun).*

Total eclipses of sun and moon, with either body completely obscured, occur because, though the sun's diameter is about 400 times greater than that of the moon, it is also about 400 times farther away. Hence the apparent diameters of both sun and moon are the same as seen from the earth, and total eclipses are possible.

tidal friction

In Chap. 33 the phenomenon of the tides was traced to the influence of the moon's gravitational field. Water is heaped up on the parts of the earth nearest and farthest from the moon, and is drawn away from other parts. As the earth rotates, the water bulges are held in position by the moon, so a given point on its surface experiences two high and two low tides per day.

The earth does not rotate smoothly beneath the tidal bulges, but tries to carry them around with it. The moon's attraction prevents them from

This photograph of the moon was taken from the Apollo 11 spacecraft when it was 10,000 nautical miles away, homeward bound.

41.3 The tidal bulges are dragged ahead of the moon by the earth's rotation. *The friction between the bulges and the earth slows the earth's rotation on its axis, while the gravitational pull of the bulges speeds up the moon's revolution in its orbit. The effect is considerably exaggerated in the figure.*

being dragged very far, but the line between the bulges is somewhat inclined to the line between earth and moon (Fig. 41.3). In this position the moon holds the bulges firmly, and the upraised water drags back on the earth as it rotates. Friction between water and rotating earth is not very great in the open ocean, but along irregular coasts it may be considerable. The effect of the friction is, of course, to slow the earth's rotation; the tidal bulges act like huge but inefficient brake bands, clamped on opposite sides of the spinning planet.

In other words, because of tidal friction the day is slowly growing longer. Verification of this effect has come from records of ancient eclipses among other sources. Using the day's present length, astronomers can calculate precisely when and where past eclipses should have occurred. These calculations do not agree with the observations recorded by ancient Egyptian and Babylonian astronomers, the discrepancies being greatest for the oldest eclipses. Calculated and observed positions, however, agree well if the slow increase in the day's length is considered. The rate of increase is very small: the time between sunrise and sunrise is longer today by $\frac{1}{1,000}$ sec than it was 100 years ago, longer by $\frac{1}{50}$ sec than in the days of Julius Caesar.

The decrease in the earth's rotational speed is accompanied by an increase in the speed of the moon's revolution. Not only does the moon

The Apollo 11 spacecraft and its Saturn V launch vehicle before takeoff.

Edwin Aldrin installing seismographic equipment on lunar surface. Nearby is a reflector for laser beams aimed from the earth. The lunar module is at right.

pull back on the nearer tidal bulge (Fig. 41.3), but with an equal force the bulge pulls the moon ahead in its orbit. Work is done on the moon by the force exerted by the tidal bulge, and in consequence the moon's orbit steadily increases in diameter. That is, our satellite is gradually becoming more and more distant.

the moon is receding from the earth

The rate at which the moon is receding, like the rate of the earth's slowing, seems unimportantly small—about 5 ft each century. At this rate the moon will take 100,000 years to increase its distance by 1 mile. Yet over longer periods the cumulative effect of such small changes becomes impressive. As the moon spirals outward, it needs more time to circle the earth. The slowing down of the earth's rotation occurs at a more rapid rate, so eventually—if everything else stays the same—the day and the month will become equal, nearly seven of our present weeks in length. The earth and the moon would then be stationary with respect to each other, so the earth would not move under the lunar tidal bulges any more. The smaller tides due to the sun would continue, however, and their friction would act to slow the earth's rotation. Now the tidal bulges will be slightly ahead of the moon, and this would tend both to speed up the earth's rotation and to slow down the moon's revolution. Ultimately the moon would spiral in closer to the earth than the Roche limit, and then break up into fragments like those in the rings of Saturn.

The above predictions are based on a continuation of the present states of the sun, earth, and moon. But billions of years would be needed before the moon's ultimate breakup, and it seems unlikely that the waters of the earth will remain even approximately as they are today over so long a period. As we shall learn in Chap. 45, it is probable that the sun's radiation will eventually increase, leading to an earth surrounded by steam instead of liquid water. Still later the sun will cool, and after another interval of surface water there will be a progressive freezing which will leave a permanent coat of ice. Clearly we cannot count on present tidal effects to continue indefinitely.

PROPERTIES OF THE MOON

The moon was hardly a mystery even before the voyages of Apollo 11 and of the spacecraft that preceded it there. Even a small telescope reveals the chief features of the lunar landscape: wide plains, jagged mountain ranges, and innumerable craters of all sizes. Each mountain stands out in vivid clarity, with no clouds or haze to hide the smallest detail. Mountain shadows are black and sharp-edged. When the moon passes before a star, the star remains bright and clear up to the moon's very edge. From these simple observations we conclude that the moon has little or no atmosphere. Water is likewise absent, as indicated by the complete lack of lakes, oceans, and rivers.

the moon has neither an atmosphere nor surface water

But there is still no substitute for direct observation and laboratory analysis, and each spacecraft that has landed on the moon and returned to earth, manned or unmanned, has brought back information and samples of the greatest value. The lack of a protective atmosphere and of running water to erode away surface features means that there is much to be

THE MOON

learned on the moon about our common environment in space, both past and present. And from the composition and internal structure of the moon hints can be gleaned of its origin and past history, which may well have a lot in common with those of the earth. Thus the study of the moon is also a part of the study of the earth, doubly justifying the effort of its exploration.

mass

The mass of the earth can be calculated from its radius, the acceleration of gravity g at its surface, and the universal constant of gravitation G, as we saw in Chap. 36. Given the earth's mass, the moon's mass can also be determined, even without a measurement of the acceleration of gravity on the lunar surface.

The procedure is based upon the fact that, as the moon revolves around the earth, the earth also revolves around the moon. More precisely, both objects are circling a common center of mass (Fig. 41.4). This center of mass is the balance point of the earth-moon system, and because of the much greater mass of the earth, it lies within the earth about 4,700 km (2,900 mi) from its center. The center of mass of the earth-moon system is what follows a smooth, elliptical orbit around the sun; the earth itself oscillates on either side of this orbit in its monthly circuit of the center of mass. The latter oscillation shows up in apparent irregularities in the motions of the other planets, from which the location of the center of mass can be inferred.

the earth and the moon both revolve around the center of mass of the system

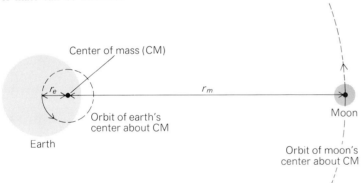

41.4 *The earth and moon both revolve about their common center of mass, which is located about 1,000 mi below the earth's surface.* Each revolution requires 1 lunar month of 27.3 days.

If r_e and r_m are the distances of the earth and moon from their common center of mass and m_e and m_m are their respective masses, then from the definition of the center of mass as the balance point of the system

$$r_m m_m = r_e m_e$$

$$m_m = m_e \frac{r_e}{r_m}$$

Since $m_e = 5.98 \times 10^{24}$ kg and r_e/r_m is found to be $1/81.3$, the moon's mass is

The moon presents a bleak landscape. The earth is in the background, 238,000 miles away.

The Mare Serenitatis is surrounded by highlands and mountains.

Crater Lake, in Oregon, is shown here looking north along its west rim. The crater is believed to have been caused by the collapse of a volcanic cone. A similar origin for some lunar craters is possible.

$$m_m = \frac{5.98 \times 10^{24} \text{ kg}}{81.3}$$
$$= 7.36 \times 10^{22} \text{ kg}$$

This figure has been confirmed by spacecraft measurements of the acceleration of gravity near the moon.

The average density of the moon is only 3.3 g/cm^3, which is comparable with the 2.7 g/cm^3 density of surface rocks on the earth but much smaller than the 5.5 g/cm^3 average density of the entire earth. The low density of the moon is partly due to its small mass, only 1.2 percent that of the earth. The greater the mass of an astronomical body, the greater the pressure in its interior from the weight of the overlying material. Because the moon's interior is compressed to a smaller extent than that of the earth, the density there ought to be less as well. However, the difference in density is too large to arise from this factor alone. It seems probable that the moon lacks the high proportion of iron found in the earth, and in addition may possibly contain significant quantities of such low-density substances as graphite.

the moon's average density is much smaller than that of the earth, and hence must have a different composition

flight of apollo 11

The Apollo 11 spacecraft and the Saturn V launch vehicle that sent it on its journey to the moon are shown in Fig. 41.5. The entire assembly at the moment of ignition weighed 3,240 tons; the command module, the only part that returned to earth, weighed a little over 6 tons. The launch vehicle consisted of three rocket stages, the reason for which was discussed in Chap. 4. The spacecraft itself had four principal parts:

1 *Launch escape system* (LES). If the Saturn V caught fire or seemed likely to explode on the ground or just after launching, the rocket motor of the escape system would have carried the command module containing the astronauts to safety. The parachute system that was part of the command module would then have brought it to the ground some distance away. The LES was jettisoned after the first stage had separated and the second one ignited.

2 *Command module* (CM). The CM contained a compartment for the crew, instruments, a number of control systems, and the various components of the earth landing system such as parachutes and inflatable buoyancy bags.

3 *Service module* (SM). The SM contained the rocket engine that slowed the spacecraft when it was to go into orbit around the moon and that later propelled the spacecraft into the path that it followed back to the earth. The SM also contained fuel cells for electric power, communications antennas, an environmental control system for regulating the pressure, temperature, and composition of the atmosphere in the command module, and systems for guidance, stabilization, attitude control, and so forth. Until just before reentry into the earth's atmosphere the CM and SM remained coupled together as a unit called the CSM.

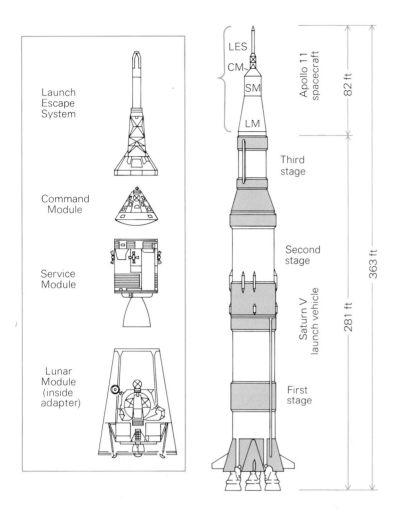

Figure 41.5

4. *Lunar module* (LM). The LM is a miniature spacecraft that was used for travel from the orbiting CSM to the lunar surface and back. In the first part of the flight the LM was protected by the "adapter," a conical cover later jettisoned.

The flight of Apollo 11 to the moon and back is outlined in Fig. 41.6. The chief events were these, keyed to the numbers in the diagram:

1 Launch on July 16, 1969. The first-stage rocket engines were ignited and the entire assembly left the ground. After 2 min 41 sec the first-stage engines were cut off and the stage separated from the rest of the vehicle. At this point the spacecraft was 217,655 ft above the ground and had a speed of 6,141 mi/hr. The second-stage engines were then ignited and the LES jettisoned. A few minutes later the second-stage engines were shut down and the stage separated; the spacecraft was then about 115 mi from the earth and traveling at

launch of apollo 11

15,468 mi/hr. The third stage was ignited for a short time, and about 12 min after the launch the spacecraft was in a nearly circular "parking" orbit around the earth. At this point the combined weight of the Apollo 11 spacecraft and the third-stage rocket was only 4 percent of the original weight on the launching pad.

injection into trans-lunar path

2. After twice circling the earth, the spacecraft was placed in a trajectory leading to the moon by the reignition of the third-stage engine. The speed was increased to 24,182 mi/hr.

rearrangement of the spacecraft

3. The CSM separated from the rest of the assembly, the LM adapter was cast off, and the CSM turned around and was attached to the LM with its rocket engine facing outward. The third stage was then jettisoned and set into orbit around the sun.

injection into lunar orbit

4. A little over three days after the launch, Apollo 11 neared the moon and its rocket engine was fired to slow it down. The first lunar orbit attained was highly elliptical; a brief later firing of the engine led to a more nearly circular orbit 86,000 mi long and 71,000 mi wide in which the spacecraft speed was 3700 mi/hr.

separation of lunar module

5. On the thirteenth circuit of the moon the LM, with Neil Armstrong and Edwin Aldrin aboard, separated from the CSM. Michael Collins, the third astronaut, remained in the orbiting CSM.

landing on moon

6. Two hours later the LM approached the moon, braked with the help of its descent engine, and landed on the Sea of Tranquility. (There are no actual seas on the moon; the name was given by early astronomers because of the appearance of this region through the telescope.)

Lunar highlands are thrown into bold relief by glancing sunlight.

Armstrong and Aldrin descended to the lunar surface, explored the immediate vicinity on foot, and collected samples of the rocks there. They also set up a device for measuring the solar wind (fast protons and electrons continually streaming from the sun), a detector for moonquakes, and a reflector to enable a laser beam from the earth to measure the earth-moon distance accurately. After $2\frac{1}{2}$ hr on the lunar surface, the astronauts returned to the LM.

7. Twenty-two hours after landing on the moon, the ascent stage of the LM took off, leaving the lower descent stage, with its braking engine, behind. *(ascent from moon)*

8. Less than four hours later the LM had reached the orbiting CSM, docked with it, and the LM crew transferred to the CSM. The LM was then cast adrift. *(return to spacecraft)*

9. The CSM engine was fired to increase the speed past the 5,300 mi/hr escape velocity from the moon. As the spacecraft approached the earth, its speed increased owing to the pull of the earth's gravity. *(injection into trans-earth path)*

10. The SM was detached from the CM near the earth, and the CM oriented so its heat shield would be facing ahead as it entered the atmosphere. The CM was moving at 24,641 mi/hr when it entered the atmosphere, where friction reduced its speed to only about 60 mi/hr. *(command module re-enters atmosphere)*

11. Splashdown on July 24, 1969. Near the earth the parachutes were deployed, and the CM landed in the Pacific exactly where planned. *(splashdown)*

the lunar environment

The visits of Apollo 11 and of subsequent manned spacecraft to the moon have confirmed the inhospitability of the lunar environment. There is no atmosphere and no surface water. In the two weeks of lunar day at a given place, the surface temperature may reach 117°C; in the two weeks of lunar night, the surface temperature drops as low as $-173°$C. Cosmic rays (see Chap. 44) and meteoroids of all sizes rain down unimpeded on the moon, and its sunlit half is bathed by solar X- and ultraviolet radiation and by the fast protons and electrons of the solar wind. Fortunately a space suit offers adequate protection against the X- and ultraviolet radiation, the solar wind, and the dustlike micrometeoroids which far outnumber their larger, more dangerous brethren.

Cosmic rays are more of a problem. The earth is partially shielded from these high-energy atomic nuclei which fill our galaxy by its magnetic field and by its atmosphere. Those cosmic-ray particles that do reach the earth offer no direct hazard to living things, though they are responsible for at least some of the random mutations that are involved in biological evolution. On the unprotected moon, the cosmic-ray intensity at the surface is 10 or 20 times that on the earth's surface, which is still not past the danger level and, though such intensity is undesirable genetically, nobody is likely to spend enough time there during his reproductive years to cause anxiety on this score. But at irregular intervals events called *flares* occur in the sun's atmosphere which send out energetic X-rays and cosmic-ray particles in profusion. Flares occur with little or no warning, *(cosmic rays)* *(solar flares)*

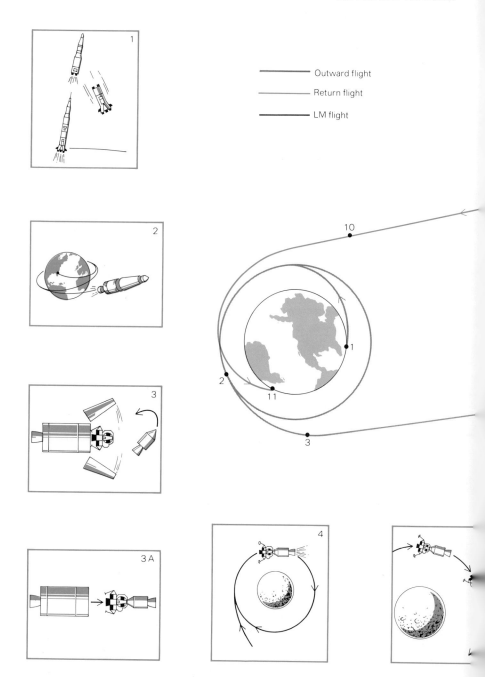

41.6 *The flight of Apollo 11 to the moon and back.*

and, even though flares violent enough to affect space-suited astronauts only happen once every few years on the average, their unpredictability makes them a factor to be taken into account in plans for colonizing the moon.

THE MOON

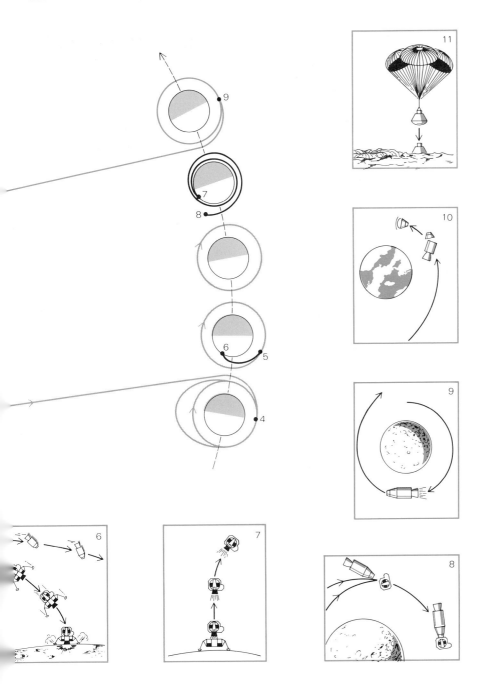

the lunar landscape

With the help of no more than binoculars it is easy to distinguish the two main kinds of lunar landscape, the dark, relatively smooth *maria* and the lighter, ruggedly mountainous highlands.

This rille is named Ariadaeus Rille.

maria are large, dark, smooth regions on the lunar surface

mascons are dense bodies of matter under some maria

Mare means "sea" in Latin, but the term is still used even though it has been known for a long time that these regions are not covered with water. The largest of the maria is Mare Imbrium, the Sea of Showers, which is over 1,000 km across. The maria are circular depressions covered with dark, loosely packed material—not solid rock. They are not perfectly smooth, but are marked by small craters, ridges, and cliffs. The latter features are only a few hundred meters high, but may be hundreds of kilometers long. It is curious and perhaps significant that nearly all the maria are on the lunar hemisphere that faces the earth.

A remarkable and wholly unexpected finding is that underneath some—but not all—of the maria are large, dense concentrations of mass—*mascons* for short. The presence of mascons was deduced from irregularities in the motions of artificial satellites set in orbit around the moon. When a satellite approached one of these maria, its speed increased, suggesting the presence of denser material whose gravitational pull is greater than that of the rest of the moon. To give an idea of the kind of object that could be responsible, calculations show that the gravitational anomaly near Mare Imbrium could be produced by a nickel-iron sphere 70 km in diameter whose center lies 50 km below the surface.

The first thought that comes to mind is that mascons are giant meteorites, the remains perhaps of asteroids that struck the moon long ago. The impact of such an asteroid would have produced a huge crater, which would later have filled in with debris and lava resulting from the collision to produce the maria of today. But this explanation does not seem altogether satisfactory. For one thing, it is hard to see why an asteroid colliding with the moon would not have shattered, even vaporized, at once. Another objection is that even if a nickel-iron meteorite of the required size were placed gently on the lunar surface, it would soon have migrated to the center of the moon by virtue of the pressure it would have exerted on the underlying material; even solid granite would give way to such an object.

There is no lack of alternative possibilities, which are also tied to theories of the origin of the maria. Perhaps a mascon results from lava that flowed into a large crater or into a low-density, porous part of the lunar crust, with its surface later broken up by meteoroid impacts; perhaps it results from dust that settled in a crater and became compacted by its own weight; perhaps it results from sediments deposited in a mare by running water that may once have existed on the moon. A successful theory of mascons must account not only for their presence beneath certain maria, but also for their absence from beneath others.

The lunar highlands are scarred by innumerable craters ranging up to 236 km in diameter. Most of the craters are circular with raised rims that are steeper on the inside than on the outside, and some have mountain peaks at their centers. Certain craters, such as Tycho and Copernicus, have conspicuous streaks of light-colored matter radiating outward. These *rays* may extend for hundreds or thousands of km, and probably consist of lunar material sprayed outward by whatever events caused the craters.

the craters of the moon

The lunar craters strongly resemble in form known meteorite craters on the earth (see the photograph on page 778), and may well have had a similar origin. This view is supported by their random locations on the moon. An objection is that the craters on the earth are fairly small—the

meteoritic theory of lunar craters

Lunar rock sample No. 12036 collected during the Apollo 12 mission is composed of olivine (yellow-green), feldspar (white), and a dust covering (gray).

spectacular one in Arizona is only a mile across—whereas there are tens of thousands of lunar craters larger in size. However, traces have been found in Canada, Africa, and Europe of what seem to have been meteorite craters as much as 65 km in diameter. Many lunar craters are even larger than this, but geological change is rapid on the earth and the scarcity of meteorite craters is to be expected here.

volcanic theory of lunar craters

A volcanic origin for the lunar craters also has its supporters. Huge craters called *calderas* result from the collapse of volcanic cones on the earth; the basin of Crater Lake in Oregon is a good example (see page 803). If the maria are indeed filled with lava or volcanic ash as many think, volcanic activity may well have existed elsewhere on the moon. Furthermore, rows of small craters have been found in various locations on the moon that occur along what seem to be faults, which indicates a volcanic origin for them at the least. On balance, a meteoritic origin for most of the lunar craters seems probable, with a minority the result of volcanic activity.

The mountains of the moon are thousands of meters high, which means that the moon's surface is about as irregular as the earth's. In order to support these irregularities, the moon must be quite rigid, which in view of the moon's size suggests that the lunar interior is nowhere near as hot as that of the earth. The existence of mascons is another argument for the rigidity of the moon. Even if they are thin layers of dense material rather than spherical lumps of iron, they cannot be resting on plastic rock like the asthenosphere of the earth or they would long ago have sunk far below where they are now.

rilles resemble river beds

The *rilles* of the highlands are especially intriguing. These are narrow channels up to 250 km long that look like nothing else than dried-up river beds. One of them, called Rima Hadley, is 1.5 km across and 300 to 500 m deep, and meanders down from the edge of a mountain range to the adjoining mare. No really plausible hypothesis of the origin of rilles exists, but some suggestions have been advanced. For instance, although there is no water today on the moon's surface, and no trace of hydrated minerals in the Apollo 11 and 12 samples, subsurface water may be conceivably present, and could be released by an event such as a meteoroid impact. The water gushing forth could carve out a rille before finally evaporating into space perhaps a century later. Another theory of rilles, equally speculative, holds that they were created by the collapse of subsurface lava-filled tubes.

lunar rocks

The samples brought back by the Apollo 11 and 12 and the automated Russian Luna 16 spacecraft are similar in character to their terrestrial counterparts but different in many details. Minerals such as plagioclase, pyroxene, ilmenite, and olivine are apparently common on the moon, as they are on the earth, but titanium is markedly more abundant in lunar minerals and oxygen less abundant. The relative scarcity of oxygen is underlined by the finding of particles of pure iron of nonmeteoritic origin in lunar samples, whereas terrestrial iron almost always occurs in compounds, usually in combination with oxygen. A distinctive feature of the lunar terrain is the presence of glass everywhere: glassy patches on top of the soil, glassy beads in the soil, and glassy crusts and glass-lined pits in the rocks.

rocks from the sea of tranquility

Rocks found at the Sea of Tranquility landing place of Apollo 11 were found by radioactive dating techniques to have solidified about 3.5 billion years ago. Fine granules from the soil there apparently came into being much earlier, about 4.6 billion years ago, which is when the solar system originated. The inference is that a billion years after the moon's formation the material in the Sea of Tranquility hardened from a molten state, which is supported by the igneous nature of the rocks found there. The melting of the surface may have been due to volcanic activity or, somewhat less likely, to meteoroid bombardment. The highlands may have escaped this melting and thus represent the original material of the lunar surface.

rocks from the sea of storms

Apollo 12 landed on the Sea of Storms about 1,500 km away from the Apollo 11 site. Most rocks from this region were also formed by solidification from a molten state about 3.5 billion years ago. One difference between samples from the two landing places is that those from the Sea of Storms are poorer in titanium and richer in iron and nickel than those from the Sea of Tranquility. The oldest rock thus far found on the moon came from the Sea of Storms: rich in uranium, thorium, and potassium, it is 4.6 billion years old, the same age as that estimated for the moon.

The lunar surface has been disturbed in a variety of small-scale ways since it was formed. Many rocks are rounded on top, suggesting some kind of erosion—perhaps due to being struck by small particles ejected upon meteoroid impacts. Traces of cosmic rays should decrease rapidly with depth, yet some soil samples from 10 cm down show as many such traces as samples from the uppermost layer, which implies that some sort of redistribution of the rock fragments in the soil has occurred. And there is direct evidence of meteoroid bombardment in the form of a glass fragment, presumably formed when surface material melted under the impact of a meteoroid, whose age is at most 35 million years.

ORIGIN OF THE MOON

Theories of the origin of the moon fall into three categories:

1 The moon was initially part of the earth and split off from it to become an independent body.
2 The moon was formed elsewhere in the solar system and was later captured by the earth's gravitational field.
3 The moon and the earth came into being together as a double-planet system.

Each of these approaches was very fashionable at one time or another, and each has strong arguments both for and against it. Further exploration of the moon, and in particular seismic and magnetic studies that will reveal its inner structure, will undoubtedly narrow the choice.

the moon as a fragment

The first hypothesis assumes that the original earth was spinning so fast that it became unstable and broke in two. If we add together the earth's angular momentum of rotation on its axis and the moon's angular momentum of revolution around the earth, a straightforward calculation shows that the original earth would have rotated every $5\frac{1}{2}$ hr or so on this basis. This is quite fast, but not nearly fast enough for such a body to break up. Furthermore, even if it had done so, the mass ratio of the fragments ought to be in the neighborhood of $8:1$, not the $81:1$ mass ratio of the earth and moon. And as a clincher, if the moon had originated in this way, it would have escaped from the earth's gravitational field altogether instead of going into orbit around it.

But it is not hard to imagine a modification of the above hypothesis that can be taken more seriously. Suppose the original earth had been spinning fast enough to be unstable and that two pieces had split off, not just one. The larger of the two would have had roughly $\frac{1}{8}$ the mass of the remaining earth, and it could have carried off with it the excess angular momentum now missing from the earth-moon system. The moon could perfectly well have remained behind as a satellite of the earth in such a situation. The larger fragment would escape the earth's gravitational field but not that of the sun, and it would go into an orbit around the sun. Is there an object of the required kind in the solar system? Yes—the planet Mars, whose mass is $\frac{1}{9}$ that of the earth and whose density is intermediate between those of the earth and moon. This bold concept evidently has its attractive aspects, but there are many details that do not quite fit, such as why the orbit of Mars suits Bode's law so neatly and why the composition of moon rocks is different from those on the earth.

the moon as a captured planet

The second hypothesis must overcome the problem that a body approaching the isolated earth from somewhere far away will, if no collision occurs, simply swing past the earth and move off again. However, the earth is not isolated in space; the sun, with its great mass, is close enough to influence matters. It seems possible that the moon could have been an independent planet early in the history of the solar system which, through chance, got caught in an orbit that sent it around the earth for a number of circuits.

If nothing else had happened, the moon would eventually have moved away to circle the sun once more. But if the moon lost energy in some way near the earth, it could have been permanently trapped. Apparently only a small energy change would have been enough; three suggestions are friction in a very extensive early atmosphere of the earth, friction in a dust cloud filling the early solar system, and the impact of a large meteoroid. The chief difficulty with this hypothesis is perhaps that chance plays more than one role in it. On the other hand, only the earth has a satellite as relatively massive as the moon, so the origin of the moon ought to be different from the origins of the other planetary satellites.

the moon as our twin planet

The third hypothesis regards the earth and moon as twin planets that developed in the primeval solar system instead of a single one. Since the solar system probably originated in the gradual fusing together of small particles of matter in a cloud around the sun (Chap. 45), it seems possible that there could have been two centers of accumulation rather than one. However, nobody has yet proposed a definite mechanism whereby a double planet instead of a single one could have evolved. And if the moon and earth were formed from the same material, why is the moon's density so much less than that of the earth? The origin of the moon is virtually an open question at this time—only the breaking away of the moon from the earth as a single fragment seems definitely excluded, and none of the other ideas is overwhelmingly convincing.

GLOSSARY

The *phases of the moon* occur because the amount of the illuminated side of the moon visible to us varies with the position of the moon in its orbit.

A *lunar eclipse* occurs when the earth's shadow obscures the moon; a *solar eclipse* occurs when the moon obscures the sun.

The *maria* of the moon are dark, smooth regions on its surface once thought to be seas. Some maria have dense bodies of matter called *mascons* under them.

EXERCISES

1 In what phase must the moon be at the time of a solar eclipse? At the time of a lunar eclipse?

2 Explain with the help of a diagram why Mercury and Venus are the only planets to exhibit phases to the earth.

3 Explain with the help of a diagram how you would go about calculating the height of a mountain on the moon from a photograph showing the shadow cast by the mountain.

4 A radar pulse is aimed at the moon from a station on the earth. How long does it take for the radar echo to return?

5 The mascon under Mare Imbrium is equivalent in its gravitational effects on lunar satellites to a nickel-iron sphere about 70 km in diameter. (*a*) What is the mass of such a sphere, assuming that it is 90 percent iron and 10 percent nickel (which is typical of nickel-iron meteorites)? (*b*) If the mascon is a nickel-iron disk 5 km thick formed by the

flattening out on impact of an original spherical asteroid, what is its diameter? (Note: The volume of a sphere of radius r is $\frac{4}{3}\pi r^3$; the area of a circle of radius r is πr^2.)

6 Why would you not expect the moon to have either an atmosphere or surface water?

7 Would you expect spring or neap tides when an eclipse of the moon occurs? An eclipse of the sun?

chapter 42

The Sun

TOOLS OF ASTRONOMY

the telescope
the spectroscope
spectrum analysis
doppler effect

THE SUN

properties of the sun
solar energy
sunspots
the aurora
the solar atmosphere

The sun is the glorious body that dominates the solar system, and the origin and destiny of the earth as well as our daily lives are closely connected with solar phenomena. The astronomer has another reason for studying the sun closely, for it is in many ways a typical star, a rather ordinary member of the family of perhaps 10^{20} stars that constitutes the known universe. Although we are the enormous distance of 93 million miles from the sun, the next nearest star (Proxima Centauri) is more than 270,000 times farther away. Light takes 8 min to reach us from the sun but more than four years from Proxima Centauri—and billions of years from the most distant observed stars. These great distances mean that all stars save the sun appear simply as points of light in even the most powerful telescopes. The properties of the sun that we can observe by virtue of its relative closeness, then, are interesting not only in themselves but also because they provide information about stars in general that would otherwise be inaccessible.

TOOLS OF ASTRONOMY

the telescope

Modern stellar astronomy began toward the end of the eighteenth century, with the work of Sir William Herschel. For nearly a century astronomers

The mounting of the 200-in.-diameter reflecting telescope at Mt. Palomar, California, permits it to examine all parts of the sky.

had been concerned chiefly with refining and elaborating Newton's great synthesis of the solar system. Herschel sought to find in the stars something of the same regularity of structure and orderliness of motion that Newton and his predecessors had found in the sun's family of planets. Like a pioneer in any branch of science, Herschel began with observation and classification; he spent many years in cataloging, counting, and observing the apparent motions of stars in all accessible parts of the sky. From this study he was able to deduce a structure for the universe which in all essentials is the same as the one that modern astronomers believe to be correct.

In Herschel's time, as in ours, the telescope was the basic astronomical instrument, and his success rested largely on the improvements he introduced in telescope construction. Herschel was the first to build and use a large *reflecting* telescope, an instrument invented by Newton in which light is reflected from a concave mirror instead of being refracted through a lens. All the larger modern telescopes used in studying the stars are of the reflecting type. The largest, located on Mt. Palomar in California, has a mirror 200 in. (16 ft 9 in.) in diameter.

modern astronomical telescopes are all of the reflecting type

In stellar astronomy the chief purpose of a big telescope is not to secure great magnification, for no obtainable magnification can make the stars appear larger than points of light. The great advantage of big mirrors and big lenses is their light-gathering power; more light from a given object can be brought to a focus by a large surface than by a small one. Thus, faint objects that would otherwise be invisible are revealed by a large telescope, and more light from other objects is available for study.

large telescopes are useful primarily because of their light-gathering ability, not because of their magnification

Within a century after Herschel's time another extremely valuable instrument was introduced into stellar astronomy, the camera. The camera is used with a telescope, so that light collected by the latter's lens or mirror falls on a photographic plate rather than on the astronomer's eye. Photographic film and plates have a great advantage over the eye in that they respond to the *total amount* of light falling on them over the period of time during which they are exposed, whereas the eye responds only to the *brightness* of the light reaching it. The longer a plate is exposed to faint light, the more distinct the resulting image is. A telescope with

A spectrograph mounted on the 24-in.-diameter telescope at Lowell Observatory, Flagstaff, Arizona.

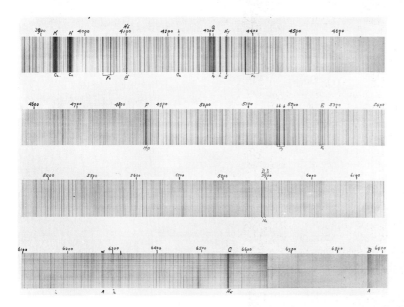

A portion of the solar spectrum. Wavelengths are in units of 10^{-10} m.

camera attached can be trained on the same star for hours or, if necessary, for several nights, so the image of the star can be made as conspicuous as desired. Objects much too faint for the eye to detect are faithfully recorded by the camera, and with its aid we can see more objects and more distant objects than would otherwise be possible. Photographic plates have the further advantage that their records are permanent and so positions of stars as we see them can be compared with those photographed years ago.

the spectroscope

The most valuable item in the modern astronomer's equipment is an instrument whose extraordinary usefulness has become apparent only during the twentieth century. This is the *spectroscope*, the same instrument that has contributed so much to our knowledge of atomic structure. The spectroscope is more important than the telescope and camera, even though it is nearly always used in combination with these instruments, because to the spectroscope we owe a vast amount of information concerning the stars that telescope and camera alone could not possibly give.

the spectroscope is the most important tool of the astronomer

A spectroscope is designed to break up light into its separate wavelengths, as we learned earlier. The resulting band of colors, with each wavelength separate from the others, is the *spectrum*, which is recorded on a photographic plate. A remarkable amount of information can be deduced from the spectra of sunlight and starlight.

The spectrum of a star is not at all impressive. If photographed in natural colors, it generally consists of a rainbow band crossed by a multitude of fine dark lines. Ordinarily color film is not used, so the spectrum shows simply black lines on a light gray background. Usually a photograph of the spectrum of a reference element is taken on the same plate, to make possible a direct comparison of lines in the star's spectrum with the known lines of that element's spectrum.

At first glance it does not seem that a few black lines on a photographic plate can get us very far in understanding the stars. But each of those lines has its own story to tell about the conditions that produced it, and an expert can piece together the information from different lines into a comprehensive picture of an entire star. Some types of information directly obtainable from spectra are outlined below.

spectrum analysis

Structure. A spectrum of dark lines on a continuous colored background is the type that we have called in Chap. 18 an *absorption spectrum*. It is produced when light from a hot object passes through a cooler gas; atoms and molecules of the gas absorb light of certain wavelengths— wavelengths that they would emit if they themselves were hot—and so leave narrow gaps in the band of color. Thus a star that has this kind of spectrum (and nearly all of them do) reveals at once something of its structure: It must have a hot, incandescent interior surrounded by a relatively cool gaseous atmosphere.

most stars have hot interiors surrounded by a cooler atmosphere

The continuous background of the spectrum, however, tells us little about the star's interior. Continuous spectra are produced by very hot solids, liquids, or fairly compressed gases—any form of matter in which the atoms are close enough to interfere with one another's motions. A heated gas at low or moderate pressures, in which each atom can emit its own characteristic radiation without interference from its neighbors, gives a discontinuous *emission spectrum* consisting of isolated bright lines. Thus we know that the interior of an average star cannot be a gas at low pressure, but whether it consists of solid, liquid, or compressed gas the continuous spectrum does not tell us.

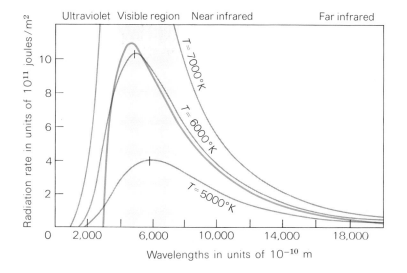

42.1 *Relative intensity of the wavelengths of light emitted by bodies with the temperatures indicated.* The wavelength whose intensity is greatest is shorter for hot bodies than for cooler ones. The curve in color represents measurements of the sun's photosphere.

Temperature. One valuable datum obtained from the continuous background of a star's spectrum is the temperature of its surface—that is, the part of the star from which we receive light. In principle, the measurement is very simple. One need only determine where in the spectrum the star's radiation is most intense; since the wavelength of maximum intensity decreases as the temperature rises, the point of maximum intensity in the spectrum is a direct measure of temperature. This relation holds for incandescent materials on the earth as well as for the stars, so stellar temperatures can be correlated with temperatures measured in the laboratory (Fig. 42.1). For quantitative work a correction is needed for the difference between the surface nature of a star and luminous objects on the earth, but qualitatively we may generalize that the hottest stars are blue (maximum intensity at the short wavelength end of the spectrum), stars of intermediate temperature white, and the coolest visible stars red.

blue stars are hottest and red stars are coolest

Composition. Since each element has a spectrum consisting of lines with characteristic wavelengths, the elements present in a star's atmosphere can be identified from the dark lines in its spectrum. One need only determine the wavelength of each line in the spectrum and compare these wavelengths with those produced by various elements in the laboratory.

Condition of matter. In practice, the identification of lines in a star's spectrum is not so easy as it sounds. The wavelengths and intensities of the lines characteristic of a given element depend not only on the element but also on such conditions as temperature, pressure, and degree of ionization. These difficulties prove to be blessings in disguise, however, for once the lines are identified they reveal not only what elements are present in a star's atmosphere but something about the physical conditions in which the elements exist.

Chemical compounds also have spectral lines of recognizable wavelengths, so spectra provide a means of determining how much of the matter in a star's atmosphere is in the form of compounds rather than elements.

A serious limitation on the use of spectral lines for determining the composition and physical conditions of a star's outer layers is imposed by the absorption of light in the earth's atmosphere. A small amount of ozone in the stratosphere effectively blocks out all ultraviolet radiation except for a small range of wavelengths just below the violet—a fortunate circumstance for living things on earth but a calamity to astronomers, since the lines of several elements occur only in the ultraviolet region. Isolated wavelengths elsewhere in the spectrum are obscured by other atmospheric gases. Spectroscopes mounted in sounding rockets, high-altitude balloons, artificial satellites, and in spacecraft are necessary to investigate those parts of solar and stellar spectra that cannot penetrate to terrestrial observatories.

the earth's atmosphere absorbs light of certain wavelengths, mainly in the ultraviolet

THE SUN

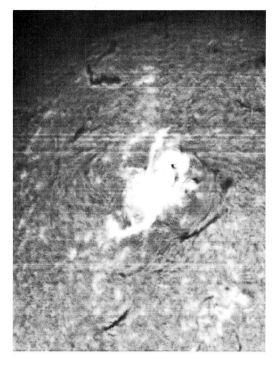

The granulated structure of the solar surface is due to small-scale convective motions of gases heated from below.

Magnetic fields. The presence of a magnetic field causes individual energy levels within atoms to split into several sublevels. When such atoms are excited, their characteristic spectral lines are accordingly split each into a number of lines close to the original lines. This phenomenon is called the Zeeman effect after its discoverer, the Dutch physicist Pieter Zeeman (Chap. 18). With the help of the Zeeman effect the magnetic nature of sunspots has been established, and a large number of stars have been discovered that appear to be strongly magnetized—information of the greatest significance to the science of astrophysics.

the zeeman effect permits stellar magnetism to be studied

doppler effect

We all know that sounds produced by vehicles moving toward us seem higher pitched than usual, whereas sounds produced by vehicles receding from us seem lower pitched than usual. Anybody who has listened to the whistle of a train as it approaches and then leaves a station or to the siren of a fire engine as it passes by at high speed is aware of these apparent changes in frequency, called the Doppler effect (Chap. 9). Similarly, a star moving toward the earth has a spectrum whose lines are shifted toward the violet (high-frequency) end, and a star moving away from the earth has a spectrum in which each line is shifted toward the red (low-frequency) end, as in Fig. 42.2. From the amount of the shift we can calculate the speed with which the star is approaching or receding.

42.2 The Doppler effect in stellar spectra. Star A is stationary with respect to the earth. Star B is receding from the earth; it moves the distance b during the emission of a single light wave, whose wavelength is therefore increased by b. Star C is approaching the earth; it moves the distance c during the emission of a single light wave, whose wavelength is therefore decreased by c. Hence stars receding from the earth have spectra whose lines are shifted toward the red (long wavelength) end, while stars approaching the earth have spectra whose lines are shifted toward the blue (short wavelength) end.

the motion of a star toward or away from the earth is revealed by doppler shifts in its spectral lines

Let us see how the Doppler effect in spectra comes about. Imagine a star emitting light of only one wavelength. If the star is receding, it moves a short distance away from us between the emission of each wave and the next. Thus each wave starts from a point a little more distant than the last, and the distance between waves appears to be a little greater than it would be if the star were motionless. Now a greater distance between waves, or a longer wavelength, means a slight change in color toward the red. So, if the star were at first motionless and then started to recede, the single line in its spectrum would shift toward the red end, the amount of shift depending on the rate of motion. Similar reasoning applies to an approaching star: The wavelength of its light would be shortened because the star moves a short distance toward us between each wave and the next.

In the radiation from a star all wavelengths are shifted by the same amount in one direction or the other. The dark lines shift as well as the continuous background, since wavelengths absorbed are affected just as much as those which are not. Note that this spectral shift records only motion of *approach* or *recession*; motion of a star across the line of sight causes no change in its spectrum.

THE SUN

Now we turn to the sun and examine what the telescope and spectroscope reveal about the one star available for close scrutiny.

properties of the sun

The gross physical properties of the sun are listed in Table 42.1. To convey some idea of what these figures imply, we might note that the sun's mass

THE SUN

TABLE 42.1   Properties of the sun.

Radius	6.96×10^8 m = 4.32×10^5 miles
Mass	1.99×10^{30} kg = 2.2×10^{27} tons
Acceleration of gravity at solar surface	274 m/sec^2 = 28 g
Emitted radiation	3.86×10^{26} joules/sec
Apparent surface temperature	6,000°K
Period of rotation	Approx. 25 days near equator
	Approx. 30 days near poles

is over 300,000 times greater than that of the earth and that its volume is so immense that 1,300,000 earths would fit into it.

The surface temperature of the sun, which is 6,000°K, is determined both from the variation of brightness with wavelength in its spectrum and from the amount of energy it emits, as described above. At this temperature all known substances are gaseous, which means that the surface of the sun is a glowing gas envelope. The thickness of this envelope, which is known as the *photosphere*, is about 200 miles; below it the solar gases contain a small amount of *negative* hydrogen ions (hydrogen atoms with an extra electron) which are so effective in absorbing light of all colors that this region is almost totally opaque; above it the gases are thinner, emit little radiation, and are therefore practically transparent. There is no sharp break between the photosphere and the outer atmosphere of the sun on one side and the solar interior on the other but rather gradual transitions between them. The density of the photosphere is only about 1 percent of that of the earth's atmosphere at sea level.

the photosphere is the glowing gas envelope around the sun

Of the thousands of dark lines in the sun's spectrum, many can be identified with those of elements known on the earth. The remaining lines are produced not by unknown elements but by familiar elements under extreme conditions of temperature and pressure. In all, about 70 of the known elements have been detected in the sun's atmosphere, and the others would probably be found if it were possible to examine the far ultraviolet part of the spectrum. Some of the elements are in their normal states and others are ionized. Lines of only a very few extremely stable compounds are recognizable; temperatures in the parts of the solar atmosphere where most of the light is absorbed are high enough to decompose nearly all molecules into atoms.

Although conditions on the sun are so different from those on the earth, the elementary substances that make up the two bodies appear to be the same. Even the relative amounts of different elements are similar, except for a much greater abundance of the light elements hydrogen and helium on the sun. At the low temperatures prevailing on the earth, most of the elements have combined to form compounds; in the sun the elements are present mostly as individual atoms, many of them ionized owing to the great heat. The general similarity in composition between earth and sun is good support for the idea that the earth's material was once a part of the sun or that both earth and sun were formed from the same nebular material.

hydrogen and helium are extremely abundant in the sun

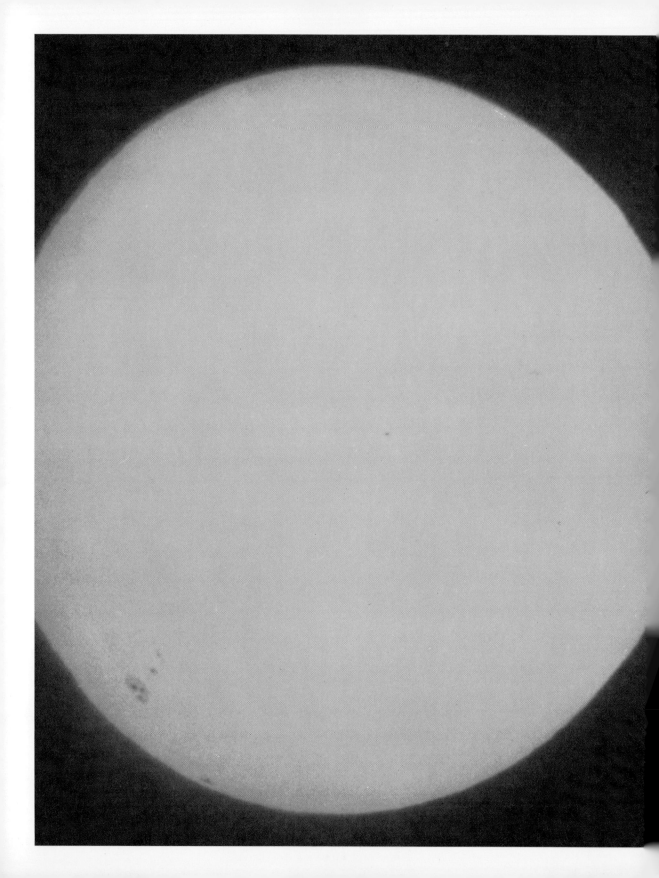

solar energy

Here on the earth, 93 million miles from the sun, a surface 1 m² in area exposed to the vertical rays of the sun receives an average of nearly 20 kcal of energy per minute. Adding up all the energy received over the earth's surface gives a staggering total, although this is but a tiny fraction of the sun's total radiation. And the sun has been emitting energy at this rate for billions of years. Where does it all come from?

The answer that first suggests itself is combustion, for fire is the only familiar source of energy that seems at all comparable to the sun. But a moment's reflection shows how impossible any kind of combustion theory is. The sun is *too* hot to burn: Burning implies the combination of other elements with oxygen to form compounds, but in the sun all compounds are decomposed by the terrific heat. Even if burning were chemically possible, the heat obtainable from the best fuels known would be hopelessly inadequate to maintain the sun's temperature.

Solar energy must somehow be produced by processes taking place in the sun's interior. Although the interior cannot be observed directly, theoretical considerations make possible reliable estimates of conditions that exist there. Pressures must be high even at moderate depths, simply because of the weight of overlying material. Temperatures must increase rapidly toward the interior, since a continuous flow of energy is supplied to the outermost layer to make good the prodigious losses by radiation. Mathematical analysis and a few reasonable assumptions lead to an estimate of 14 million °C for the temperature and 1 billion atm for the pressure near the sun's center; the density of the matter there is about 8 times that of lead on the earth's surface.

It is possible to infer the properties of matter under these conditions from laboratory experiments with very small numbers of atoms. In the sun's interior, these experiments indicate, atoms of the lighter elements would have lost all their electrons, and atoms of the heavier elements would retain only their inmost electron shells. Thus matter in the sun's interior probably consists of atomic debris—free electrons in great numbers and positive nuclei surrounded by a few electrons or none at all.

the interior of the sun contains positive ions and electrons under conditions of high temperature and pressure

These atomic fragments are in extremely rapid motion, traveling far more rapidly than gas molecules at ordinary temperatures. Such speeds mean that two atomic nuclei that collide may get close enough to each other—despite the repulsive electrical force arising from their positive charges—to react, forming a single larger nucleus. When this occurs among the light elements, the new nucleus usually weighs a little bit *less* than the combined weights of the reacting nuclei. The missing mass is converted into energy in the process according to Einstein's formula $E = mc^2$. So huge an amount of energy is evolved in nuclear reactions of this kind that it is not difficult to regard them as the source of solar energy.

The basic energy-producing reaction in the sun is the conversion of hydrogen into helium. This takes place both directly by collisions of hydrogen nuclei (protons) and indirectly by a series of steps in which

solar energy is derived from the conversion of hydrogen to helium in nuclear reactions

A sunspot group is at the lower left in this photograph of the sun.

carbon nuclei absorb a succession of hydrogen nuclei (Figs. 42.3 and 42.4). Each step can be duplicated on a small scale in the laboratory, and for each the energy required and the energy given out can be measured. In the sun's interior conditions are ideal for such energy-producing collisions—not as events affecting rare, isolated atoms, like those in our laboratories, but as commonplace events occurring many times a second in almost every cubic inch of the sun's material. For the entire process by either mechanism, the energy available per helium atom corresponds to the difference in mass between four hydrogen nuclei ($4 \times 1.0073 = 4.0292$ amu) and one helium nucleus (4.0015 amu), or 0.0277 amu. Hence every 4 kg of helium that is formed in the sun means the liberation of

$$E = mc^2$$
$$= 0.0277 \text{ kg} \times (3 \times 10^8 \text{ m/sec})^2$$
$$= 2.5 \times 10^{15} \text{ joules}$$

The relative probabilities of the carbon and proton-proton cycles depend upon temperature. In the sun and stars like it, which have interior temperatures in the vicinity of 14 million °C, the proton-proton cycle predominates. Most of the energy of hotter stars comes from the carbon cycle, while in cooler stars the proton-proton cycle is the chief energy source.

Every second the sun converts over 4 million tons of matter into energy, and its hydrogen content is such that it should be able to continue releasing energy at this rate for many billions of years more. In fact, the amount of matter lost in all of geologic history is not enough to have changed the sun's radiation appreciably—which confirms other evidence

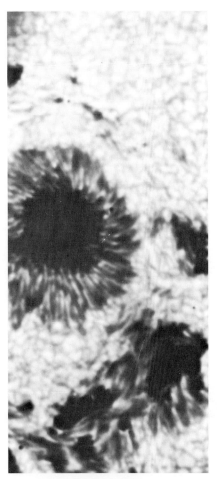

An active sunspot group photographed from a balloon-borne telescope at an altitude of 80,000 ft to avoid the distorting effects of atmospheric irregularities. The spots consist of relatively cool cores surrounded by filaments of outward-moving gases. The cellular background pattern arises from small-scale convection currents in the hot gases of the solar surface. Particle streams from this spot group produced magnetic disturbances and auroras on the earth.

42.3 The proton-proton cycle. *This is one of the two nuclear reaction sequences that take place in the sun with the evolution of energy.*

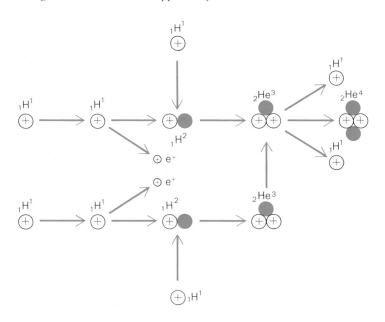

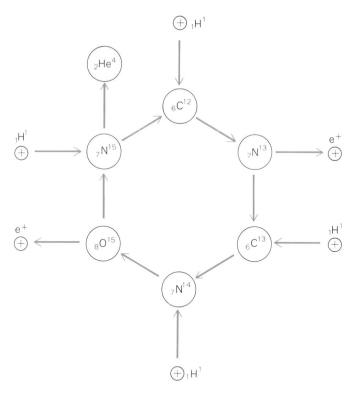

42.4 The carbon cycle also involves the combination of four hydrogen nuclei to form a helium nucleus with the evolution of energy. *The $_6C^{12}$ nucleus is unchanged by the series of reactions.*

that the earth's surface temperature has remained approximately constant during this period.

sunspots

Marring the intense luminosity of the sun's surface are dark markings called *sunspots*. Sunspots change continually in form, each one growing rapidly and then shrinking, with lifetimes of from 2 or 3 days to more than a month. The largest sunspots are many thousands of miles across, large enough to engulf several earths. Galileo, one of the first to study sunspots, noted that they moved across the sun's disk, evidence which he interpreted, as we do today, as indicating that the sun rotates on its axis. Sunspots appear black only because we see them against a brighter background; the blackest spots have temperatures of about $4,500°$K, sufficiently hot to glow brilliantly but nevertheless $1,000°$K cooler than the rest of the solar surface.

sunspots appear dark only by comparison with the brighter solar surface around them

Sunspots generally appear in groups, each with a single large spot together with a number of smaller ones. Some groups contain as many as 80 separate spots. They tend to occur in two zones on either side of the solar equator and are rarely seen either near the equator or at latitudes on the sun higher than $35°$. Spectroscopic examination reveals that sunspots consist of gas in rapid, spiral motion. The gas seems to move upward from the sun's interior, expanding and cooling as it spirals out.

THE SUN AND ITS FAMILY

This auroral form is known as a "drapery."

A double-arc aurora (both photographs taken in northern Canada).

The spectra of sunspots have a most remarkable feature: Spectral lines split into several components close to one another, a phenomenon characteristic of the Zeeman effect which indicates the presence of a magnetic field. The larger the spot, it is found, the more intense its magnetic field. The fields are so strong that there is little doubt that they are an important factor in the development and maintenance of sunspots. Some current theories of sunspots link them to a hypothetical vast magnetic field within the sun, with the spots representing the intersections with the solar surface of lines of magnetic force that have leaked outward somehow from the interior; others view the magnetic nature of the spots as being a more local phenomenon originating in the outer layers of the sun.

The number of spots visible on the sun changes with time. Part of the variation is due merely to the sun's rotation, which carries spotted regions out of our sight, but there is also a real variation in the number of spots from year to year. Approximately every 11 years the number of visible sunspots reaches a maximum, diminishing afterward so that 6 or 7 years later there are virtually no spots at all. Then the number increases to another maximum, and the *sunspot cycle*, as this periodic fluctuation is called, repeats itself.

The sunspot cycle has aroused much interest because a number of effects observable on the earth—such as disturbances in the terrestrial magnetic field (called *magnetic storms*), shortwave radio fadeouts, changes in cosmic-ray intensity, and unusual auroral activity—follow this cycle, and there is evidence suggesting that some aspects of the weather do so also. There seem to be two different mechanisms responsible for events on the earth that are synchronized with sunspots:

1 Intense bursts of ultraviolet light and perhaps X-rays as well are emitted from the sun more frequently during sunspot maximum, which cause radio fadeouts by interacting with the ionosphere.
2 Vast streams of energetic protons and electrons shoot out of the sun from the vicinity of sunspot groups, and these streams lead directly to magnetic storms and spectacular auroras and indirectly to variations in cosmic rays.

the aurora

The aurora (or *northern lights*) is one of nature's most awesome spectacles. In a typical auroral display, colored streamers seem to race across the sky, and glowing curtains of light appear, pulsating as they change their shapes into weird forms and images. In the climax of the display the heavens seem on fire, with silent green and red flames dancing everywhere. Then, after a while, the drama fades away, and only a faint reddish arc remains. Auroras are most common in the far north and far south; *aurora borealis* is the name given to this phenomenon in the Northern Hemisphere, and *aurora australis* in the Southern.

The reason that we discuss the aurora in a chapter devoted to the sun is that it is apparently caused by the ionized solar streams we mentioned above. These particles require about a day to reach the earth from the

sun, and when they enter the atmosphere they interact with the gaseous nitrogen and oxygen there so that light is given off. The process is similar to what occurs in a neon-filled glass tube when electricity is passed through it: The gas molecules are given energy by the passage near them of charged particles, and this energy is then radiated as light in the characteristic wavelengths of the particular element. The green hues of an auroral display come from oxygen, and the reds originate in both oxygen and nitrogen.

The incoming streams of solar protons and electrons are affected by the earth's magnetic field in a complicated way and, as a result, the greatest number of auroral displays occur in doughnutlike zones about 1,400 miles in diameter centered about the geomagnetic north and south poles. Sometimes, though, the cloud of particles from the sun is so immense that auroras are produced much closer to the equator than usual.

Although normally visible only near the polar regions, auroras sometimes occur in the middle latitudes. This photograph was taken on February 10, 1959, in Parkersburg, West Virginia. The 30-min exposure does not reveal the delicate, flickering patterns characteristic of most auroras.

the solar atmosphere

Above what we have loosely been speaking of as the solar surface is a rapidly thinning atmosphere consisting principally of hydrogen, helium, and calcium. From this atmosphere great, flamelike *prominences* sometimes extend out into space, much like sheets of gas standing on their

solar prominences

sides. Prominences occur in a wide variety of forms; a typical prominence is about 120,000 miles long, 6,000 miles wide, and 30,000 miles high. Prominences are often associated with sunspots and, like sunspots, exhibit behavior that strongly suggests the presence of substantial magnetic fields. In subsequent chapters, too, we shall find magnetism playing an important part in astronomy—in fact, the realization of the importance of magnetism in understanding the workings of the astronomical universe is one of the most significant developments in this science in recent years.

the corona

During a total eclipse of the sun, when the moon obscures the sun's disk completely, a wide halo of pearly light can be seen around the dark moon. This halo, or *corona*, may extend out as much as a solar diameter and seems to have a great number of fine lines extending outward from the sun immersed in its general luminosity. During sunspot maximum the corona is roughly circular in outline, but when the sunspot activity is at its minimum the corona becomes markedly smaller at the poles. Much of the coronal gas consists of ionized atoms and electrons, and its temperature, as revealed by spectroscopic measurements, is about 1 million °C. Although the corona that we can see is relatively near the sun, indirect evidence indicates that, in very diffuse form, it also pervades much of the region between earth and sun. Most authorities even regard the sun's atmosphere as extending well beyond the earth's orbit—a radical change indeed from the older idea that interplanetary space is an all but total vacuum. The outward flow of protons and electrons in this atmosphere constitutes the *solar wind* which has been detected by rocket-borne instruments.

the solar wind

GLOSSARY

The *photosphere* is the relatively thin gaseous layer on the sun that emits nearly all the light the sun radiates into space.

There are two basic energy-producing nuclear reactions in stars that both involve the conversion of hydrogen to helium: the *proton-proton cycle* in which a series of collisions of protons results in the formation of helium nuclei, and the *carbon cycle* in which carbon nuclei absorb a succession of protons, ultimately resulting in the formation of helium nuclei and the reemergence of the original carbon nuclei.

A *sunspot* is a dark marking on the solar surface; sunspots range up to some thousands of miles across, last from several days to over a month, and have temperatures as much as 1000°C cooler than the rest of the solar surface.

The *sunspot cycle* is a regular variation in the number and size of sunspots whose period is about 11 years.

An *aurora* is a display of colored, changing patterns of light that appear in the sky particularly at high latitudes. The aurora arises from the excitation of atmospheric gases at high altitudes by streams of energetic ions of solar origin.

A solar *prominence* is a large sheet of luminous gas projecting from the solar surface.

The sun's *corona* is a vast cloud of extremely hot, rarefied gas which surrounds the sun. It is visible during solar eclipses.

EXERCISES

1. A photograph of a star cluster (such clusters are described in Chap. 44) shows many more stars than can be seen by direct visual observation of the cluster with the same telescope. Explain why this difference occurs.

2. Outline the procedure you would follow to establish whether or not gold is present in the sun's atmosphere.

3. Arrange the following types of stars in order of decreasing surface temperature: yellow stars, blue stars, white stars, red stars.

4. What part of a star produces the continuous background of its spectrum? What happens to the radiation produced underneath this part of the star? What part produces the dark absorption lines? For what part of a star can the composition be determined?

5. What evidence suggests that the sun is almost wholly gaseous?

6. Suggest one piece of evidence for sunspots being cooler than their surroundings.

7. Suppose that you examine the spectra of two stars and find that lines in one are displaced slightly toward the red end in comparison with those in the other. What conclusion can you draw?

8. If the earth were moving toward a star instead of the star toward the earth, would lines in the star's spectrum appear to be shifted? If so, toward which end of the spectrum?

9. Can you think of any evidence in favor of the hypothesis that the sun's radiation rate has not changed appreciably in at least 2 billion years?

10. Give two different methods for determining the sun's rotation rate.

11. Calculate the velocity of the streams of solar particles responsible for the aurora on the assumption that they require one day to reach the earth and that they travel in straight lines. If they do not travel in

A multiple-exposure photograph of the solar eclipse of June 30, 1964. The glowing gaseous envelope around the sun's disk is the corona.

straight lines but are deflected by interplanetary magnetic fields, what kind of limit does the above velocity represent?

12. What would be the effect on the aurora of the disappearance of the earth's magnetic field?

13. Why do we conclude that the sun's energy originates in its interior rather than in its surface layers?

14. Stars whose temperatures are 4500°C—about the temperature of the darkest parts of sunspots—nevertheless shine brightly. Explain why.

part twelve

THE UNIVERSE

Stars vary widely in most of their properties (except, significantly, in mass), but most of them seem to fall into a so-called main sequence that runs from hot, bright stars in a consistent order down to cool, dim ones. In addition, there are a number of giant red stars and dwarf white ones which form two groups with properties different from those of main-sequence stars. Modern astronomy has been able to explain these observations in terms of the evolution of stars. Even such remarkable objects as the flickering pulsars, as regular in their bursts of energy as the very best clocks, apparently fit into one of the grand life cycles of stars.

Stars do not occur at random in the universe, but are found concentrated in vast clusters and nebulae. The sun itself is part of an immense disklike galaxy of stars; we see the other stars in the galaxy as the Milky Way. Many other galaxies are scattered through the sky, all sharing a number of features with one another and with our own galaxy. The origin of cosmic rays lies in the depths of the galaxy and turns out to be intimately connected with the magnetic fields that govern the structure of the galaxy. A remarkable feature of the astronomical universe is an apparent recession of the galaxies, leading to the conclusion that the universe is undergoing a general expansion. The implications of the expanding universe and several hypotheses of its origin are discussed in the final chapter together with the origin of the elements and the probable history of the planet earth.

chapter 43

The Stars

THE STARS IN SPACE
- stellar distances
- intrinsic and apparent brightness
- variable stars
- stellar motions

STELLAR PROPERTIES
- mass
- temperature
- size
- spectra

STELLAR EVOLUTION
- hertzsprung-russell diagram
- white dwarfs
- stellar evolution
- old stars
- supernovae
- pulsars

Of the unimaginably large number of stars in the universe, none appears as more than a point of light to even the most powerful telescope. As recently as the last century, before the techniques of spectroscopic analysis were discovered, most scientists despaired of ever knowing the physical nature of the stars. Today, however, we not only have a great deal of detailed information on thousands of stars but are also able to trace the general course of stellar evolution from the birth of a star through its maturity to its last agonies and eventual "death." We owe nearly all this knowledge to the spectroscope. Before proceeding to the properties of the stars themselves, let us first consider stellar distances and motions and how they are measured.

THE STARS IN SPACE

stellar distances

Aristotle pointed out long ago that, if the earth revolves around the sun, the stars should appear to shift in position, just as trees and buildings shift in position when we ride past them. Since he could detect no such shift, Aristotle concluded that the earth must be stationary. Another interpretation of the lack of apparent movement among the stars, suggested by some of the Greeks and later by Copernicus, is that the stars are simply too far away for the movement to be detected. When the copernican theory of the solar system had become well established and astronomical instruments had been improved, many observers tried in vain to find the small shift in position that should result from the earth's motion.

An undoubted shift for one star was finally discovered in 1838 by the German astronomer Friedrich Wilhelm Bessel, and in the following years several others were found. These shifts are so exceedingly small that the long failure to detect them is not strange.

Bessel's discovery made possible the direct measurement of distances to the nearer stars. The method is simple, similar to that used in the range finder of a camera (Fig. 43.1). The position of a star is determined twice, at times 6 months apart. From the measured change in the angle of the telescope, together with the fact that the telescope was displaced by 186 million miles during the 6 months, the unknown distance to the star can be found by simple trigonometry.

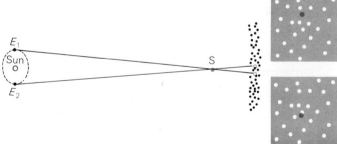

43.1 Measuring stellar distances. *The apparent shift in the position of a star in relation to more distant ones as the earth moves in its orbit may be used to determine the distance to the star.*

a light-year is the distance light travels in a year

The *parallax*, or apparent shift in position, is large enough to be measurable for only a few thousand of the nearer stars. Even for these the measurement is difficult; measuring the parallax of the closest star is equivalent to measuring the diameter of a dime seen from a distance of nearly 4 miles. The distance to Proxima Centauri, the star other than the sun that is nearest to us, is 4×10^{16} m (2.5×10^{13} miles). To help make such enormous distances mean something to us, they are often expressed in terms of the distance that light travels in a year. This distance is called a *light-year;* since light has a speed of 3×10^8 m/sec (186,000 mi/sec), a light-year is 9.5×10^{15} m (about 6×10^{12} miles). Thus Proxima Centauri is 4 light-years away—which means that we see the star not as it is today but as it appeared 4 years ago. Only about 40 stars are within 16 light-years of the solar system. Such distances between stars are typical for much of the visible universe. This means that space is all but completely empty, far more empty even than the solar system with its tiny isolated planets.

intrinsic and apparent brightness

Direct measurements of parallax are possible only for distances up to about 300 light-years. Several indirect methods are available for finding distances to stars farther away than this; the most useful depends on the measurement of the intrinsic brightnesses of stars by means of a spectroscope.

The *apparent* brightness of a star is its brightness as we see it from the earth; this measurement expresses simply the amount of light that reaches us from the star. The *intrinsic* brightness, however, is the real brightness of a star, a number expressing the total amount of light that it radiates into space. The apparent brightness of a star depends on two things: its intrinsic brightness, and its distance from us. (Apparent brightness varies inversely with the square of the distance, so a star twice as far from the earth as another, identical star seems only one-fourth as bright as the latter.) A star that is actually very bright may appear faint because it is far away; a star that is actually faint may have a high apparent brightness because it is close.

If the apparent brightness of a star is accurately measured and if its distance is known, the intrinsic brightness can be calculated by figuring out how bright an object would have to be at the known distance to send us the observed amount of light. Since the apparent brightness is easily determined for any star, intrinsic brightnesses are known for all stars whose distances can be measured.

Now let us reverse the problem: If both the apparent brightness and the real brightness of a star are known, it should be possible to calculate its distance. The calculation now involves finding out how far away an object of known brightness must be placed in order to send us the amount of light we observe. This calculation is not hard, and so we can find the distance to a star beyond the 300-light-year limit, provided some method is available for determining the star's intrinsic brightness.

Such a method for finding intrinsic brightness, involving the use of the spectroscope, was discovered by the American astronomer W. S. Adams. Studying the spectra of the nearer stars, for which intrinsic brightnesses are known, Adams observed that the relative intensities of certain spectral lines (for stars of any one type) depended on the star's intrinsic brightness. That is, the spectrum of a bright star showed certain relationships among the intensities of its lines, and the spectrum of a faint star showed somewhat different relationships. So definite was the connection between relative intensities of the lines and intrinsic brightnesses of the stars that Adams found it possible to predict and check the brightness of a star simply by examining its spectrum. Assuming that the relationship would hold for more distant stars, Adams could then use their spectra to find their intrinsic brightnesses and, therefore, the distances to them.

the apparent brightness of a star depends upon its intrinsic brightness and upon its distance

Superimposed photographs taken at different times of the region of the sky in which the variable star WW Cygni appears. Only the brightness of this star has changed.

This method is a characteristic example of a scientist at work. It began with accurate direct measurements of distance for a great many nearer stars. Intrinsic brightnesses for these stars were calculated, and a relationship was found between the brightnesses and certain characteristics of their spectra. Finally, the relationship was assumed to hold for stars so distant that a direct check is no longer possible.

The spectroscopic method is applicable to any star bright enough to give a good spectrum, except for comparatively uncommon stars of certain spectral types for which Adams's relation does not hold. Through its use stellar distances have been determined up to several thousand light-years.

variable stars

An extension of the above method for finding stellar distances is based upon the properties of a certain type of *variable star*. A variable star is one whose brightness varies continually; some show wholly irregular fluctuations, but the greater number repeat a fairly definite cycle of change. A typical variable grows brighter for a time, then fainter, then brighter once more, with irregular minor fluctuations during the cycle. Periods separating times of maximum brightness range all the way from a few hours to several years. Maximum brightness for some variables is only slightly greater than minimum brightness, but for others it is several hundred times as great. Since the sun's radiation changes slightly during the sunspot cycle, we may consider it a variable star with an extremely small range in brightness (a few percent at most) and a long period (about 11 years).

The light changes in a few variable stars are simply explained; the stars are actually double stars whose orbits we see edgewise, so that one component periodically eclipses the other (Fig. 43.2). But the fluctuations in most variables cannot be accounted for so easily. In some the appearance of numerous spots at regular intervals may dim their light; others might be pulsating, expanding and contracting so that their surface areas change periodically. Perhaps the irregular variables are passing through or behind clouds of gas and fine particles, which would explain their strange behavior.

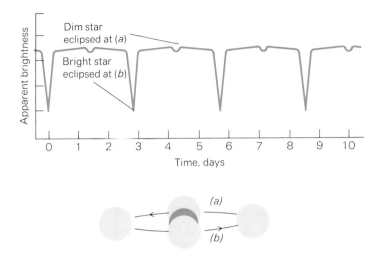

43.2 *The apparent brightness of the double star Algol varies as the bodies, which have different sizes and absolute magnitudes, revolve around their center of gravity.*

A special class of variable stars, called *Cepheid variables*, have turned out to be exceedingly useful to astronomers in determining exactly how far from us certain distant star groups are. The Cepheid variables all have relatively short periods, and take their name from a typical example discovered in the constellation Cepheus.

The usefulness of Cepheid variables in measuring distances was discovered by Henrietta Leavitt in a study of the Cepheids in a single large cluster of stars called the Small Magellanic Cloud, which is visible in the Southern Hemisphere. The Cepheids of this cluster show a curious relationship between their apparent brightnesses and the periods during which their light fluctuates: A bright Cepheid requires a long time (2 to 3 months) to complete its cycle of changes, but a faint Cepheid requires only a few days (Fig. 43.3). Now the cluster is so very far away that all its stars may be considered to be approximately the same distance from the earth, and so differences in apparent brightness must be due solely to differences in intrinsic brightness. Therefore the observed rela-

the intrinsic brightness and the period of a cepheid variable star are related

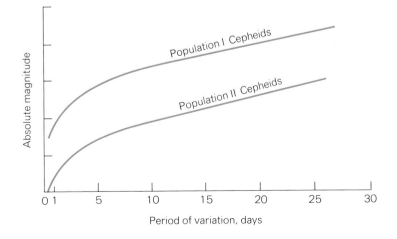

43.3 *The variation of the luminosity of a Cepheid variable star with its period.* (*The two stellar populations are discussed in Chap. 44*).

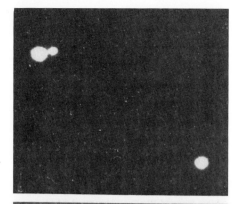
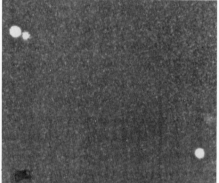
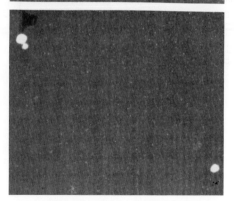

The double star at upper left was photographed in 1908, 1915, and 1920. About 50 years are required for a complete rotation.

tionship among the Cepheid variables must be due to some connection between their periods of variation and their actual brightnesses.

Values for the intrinsic brightnesses of stars in the cluster could not be determined directly, because of their immense distance. For other Cepheids closer to the earth, however, an estimate of intrinsic brightness was possible, and, by combining figures for these closer Cepheids with data from the cluster, it was possible to establish a definite numerical relationship between brightness and period of variation. If we assume that this relationship holds for Cepheids elsewhere in the universe, then the intrinsic brightness of any Cepheid can be found simply by measuring its period of variation. A comparison of intrinsic brightness with apparent brightness then gives the approximate distance, as we know. This method can be extended to much greater distances than determinations with the spectroscope, since the period of a Cepheid can be measured regardless of its faintness, but the method is handicapped by its limitation to a single unusual type of star.

stellar motions

The speeds of stars moving toward or away from the earth can be found from the Doppler shifts in their spectral lines that were mentioned in the previous chapter. Motion across the line of sight can be followed by direct observation. The great distances of the stars make their apparent movements exceedingly slow, so slow that we commonly refer to the stars as "fixed stars." Yet the motion is sufficient to have caused perceptible changes in the shapes of some constellations during the few thousand years since accurate observations began (Fig. 43.4).

Most stars, of course, are moving neither directly along nor directly across our line of sight but at an angle oblique to it. The spectrum of such a star reveals a certain speed of approach or recession, and direct observation also shows motion in a certain direction across the line of sight. If the distance of the star is known, the actual speed of this latter motion can be calculated. Then the star's total apparent motion can be found by adding vectors corresponding to the two observed velocities. Such calculations show that most stars are moving with speeds of 10 or more km per second relative to the sun.

Since other stars are moving, we might surmise that the sun is moving also. Such a motion should reveal itself in regularities in the apparent movements of the stars: If the sun is moving toward a certain part of the sky, stars in that direction, on an average, should appear to be approaching us and to be radiating out from a point, much as trees in a forest seem to approach and spread out when we drive toward them. Average stellar motions of this sort are observed in the neighborhood of the constellation Cygnus, and in the opposite part of the sky stars are apparently receding and coming closer together. Careful study of these motions indicates that the sun and its family of planets are moving toward Cygnus at a speed of 200–300 km/sec.

THE STARS

STELLAR PROPERTIES

We now turn to the properties of the stars themselves. There are many different kinds of star, most of which fit into a pattern that can be understood in terms of a regular evolutionary sequence and some that are still puzzles for the astronomer.

mass

The points of light that appear to the eye as single stars are frequently *double stars*, two stars close together in space. The components of most double stars are close enough for there to be a strong gravitational attraction between them, which means that each star moves in an elliptical orbit about the center of gravity of the pair.

Often a star that appears single even in a large telescope turns out to be double when examined spectroscopically. Only one spectrum can be obtained for the star, of course, but photographs of the spectrum made on successive days show small periodic shifts in some lines, periodic doubling of others. The periodic shifts and the doubling suggest that two separate stars are responsible for the lines. On certain days one of the stars is moving toward the earth, so that its lines are shifted (by the Doppler effect) slightly toward the violet end of the spectrum, and the other star is moving away from the earth, so that its lines are shifted toward the red end. On other days neither star is moving directly toward or away from the earth, and so all lines are single. These effects occur when the stars revolve in orbits that we see nearly edgewise (Fig. 43.5).

Thus the spectroscope makes possible the detection of double stars whose components are so close together as to be almost touching. These stars move about each other rapidly, completing a revolution in a few

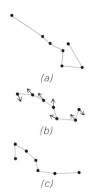

43.4 *The constellation Big Dipper (a) as it was 200,000 years ago, (b) as it is today (arrows show directions of motion of the stars), and (c) as it will be 200,000 years from now.*

some double stars can be detected spectroscopically

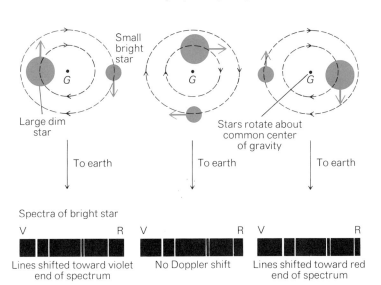

43.5 *How the motion of a double star causes periodic shifts in its spectral lines.*

double stars make possible mass determinations

days or hours. Pairs that the telescope shows as separate stars are many millions of miles apart and require years to complete each revolution.

One of the reasons that double stars are so interesting to astronomers is that a direct determination of mass is possible only for stars noticeably affected by gravitational forces. Of course every star is attracted to some extent by all other stars in the universe and its motion is controlled by these combined forces, but most stellar distances are so great that the attraction between any two stars is not measurable. Only in double stars do we find stars sufficiently close together for gravitational effects to be perceptible, and only for these can accurate measurements of mass be made. There are so many double stars, though, that this is less of a limitation than it might seem.

stellar masses are limited in range

The oddest fact that emerges from measurements of star masses is that all stars seem to contain roughly the same amount of matter. No star is known with a mass smaller than one-tenth that of the sun, and only a few have masses greater than ten times that of the sun. In view of the enormous variations among the stars in brightness, in diameter, and in density, it seems extraordinary that the range of masses should be so small.

temperature

The temperature of a star is determined from its spectrum by finding the part of its spectrum in which the star's radiation is most intense (see Fig. 42.1). This measurement gives the temperature of the star's *sur-*

The principal types of stellar spectra.

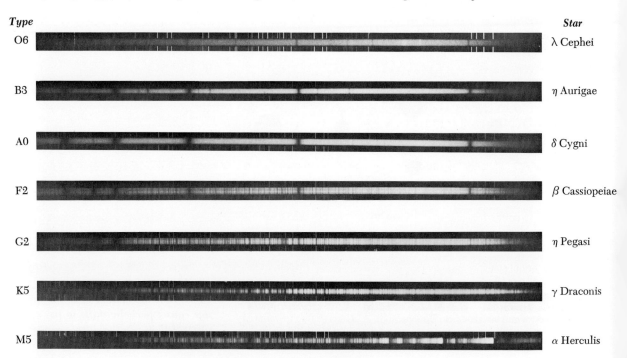

THE STARS

The Ring Nebula in Lyra is probably a shell of gas moving outward from the star in the center.

face—the part from which radiation is emitted. Measured surface temperatures of a few very hot stars range up to 30,000°K, but the great majority have temperatures between 3,000 and 12,000°K. Probably many stars have temperatures below 3,000°K (which is about the boiling point of iron), but, unless they are relatively close, their radiation is then too feeble for us to detect. Like the sun, stars must have enormously high internal temperatures to maintain their surface radiation.

The surface temperature of a star, like that of a poker held in a fire, is intimately related to its color. In general, as we have said, the hottest stars are blue-white, those of intermediate temperature are white or yellow, and the coolest are red.

the temperature of a star may be found from its color

size

Measurements of a star's temperature and its intrinsic brightness make possible an estimate of its diameter. Since the temperature determines the intensity of radiation from the star's surface, a measurement of temperature (plus a few reasonable assumptions about the nature of the surface) gives a value for the amount of radiation emitted from every square meter of the star's area. The intrinsic brightness is a measure of the total radiation from the star's entire surface. We need only divide

the size of a star can be determined from its temperature and intrinsic brightness

43.6 The range of stellar sizes, from Antares (at the bottom) through the sun (color dot) to a large white dwarf (black dot).

the total radiation by the radiation per square meter to find the number of square meters in the star's surface area, and from the area the diameter and volume are easily computed. (There is also a more direct method of measuring stellar diameters based on the interference of light that can be used on the larger stars. Results obtained in this way agree with estimates from temperatures and intrinsic brightness.)

The diameters of stars, unlike their masses, have an enormous range (Fig. 43.6). The smallest stars, like the faint companion of the bright star Sirius, are little bigger than the earth. The largest, like the giant red star Antares in the constellation Scorpio, have diameters of more than 300 million miles. Antares is so huge that, if the sun were placed at its center, the four inner planets could pursue their normal orbits inside the star with plenty of room to spare.

stellar sizes and densities vary widely

If the mass and volume of a star are known, finding the average density means simply dividing one by the other. Like the volumes, densities vary greatly from star to star. Giant stars like Antares have densities less than one-thousandth that of ordinary air—densities corresponding to a fairly good vacuum here on earth. At the other extreme are the incredible densities of some small stars, for instance the companion of Sirius, densities so great that a cubic inch of their substance would weigh more than a ton.

spectra

The analysis of starlight with the spectroscope has provided us with information on the distances, temperatures, sizes, and motions of the stars. What about the spectra themselves? Are they identical for all stars, are they totally different, or do they exhibit regularities of some kind?

Even a superficial examination of stellar spectra reveals two important facts:

1 Almost all stars have absorption (dark-line) spectra, like that of the sun, which imply a hot interior surrounded by a relatively cool atmosphere.
2 The various spectral lines found can be identified with lines found in the spectra of elements known on the earth.

the matter of the visible universe is like that on the earth

Thus we know that the matter of the visible universe has the same basic nature as the matter we are familiar with on earth and, further, that most of the large aggregates of matter in the visible universe follow the same general pattern. This uniformity of material and structure is perhaps the most striking discovery of stellar astronomy.

When examined carefully, stellar spectra show considerable variety. Some have relatively few lines, others have many; some have only sharp lines, others have diffuse bands; some have prominent lines of hydrogen, others have prominent lines of certain metals. Comparison of large numbers of spectra shows that nearly all can be arranged in a single sequence, depending on the intensities of different lines. Lines that are prominent in spectra at one end of this sequence grow less intense in successive

stellar spectra fall into a natural sequence

THE STARS

spectra while other lines become prominent, and then these become faint and still others grow conspicuous. Thus, between any two distinct types of spectra, there are gradual changes, and the changes follow one another in a single regular order.

Each spectrum shows the composition and the physical state of a certain layer in a star's atmosphere. Differences in spectra, therefore, represent differences in the makeup of this layer. These differences may be due to slight variations in temperature and pressure in various stars rather than to any fundamental difference in composition, but they nevertheless make possible a convenient classification of the stars. We need not go into the technical details of this classification, but, generally speaking, the spectral sequence distinguishes a number of groups, with smooth gradations from stars of one group to stars of the next group (Table 43.1).

STELLAR EVOLUTION

hertzsprung-russell diagram

Two astronomers, Ejnar Hertzsprung in Denmark and Henry Norris Russell in America, independently discovered a definite relationship between the position of a star in the spectral sequence and its intrinsic brightness. This relationship is shown by the graph in Fig. 43.7 (called the *Hertzsprung-Russell diagram*), on which each point represents the intrinsic brightness and the spectral type of a single star. Intrinsic brightnesses are plotted along the vertical axis and spectral types on the horizontal axis. Obviously, most stars belong to the *main sequence*, but there are also a considerable number in the *red giant* class at the upper right and a few in the *white dwarf* class at the lower left. (The names *giant*

the hertzsprung-russell diagram is a plot of intrinsic brightness versus temperature for stars

TABLE 43.1 Spectral classes of the stars.

Class	Examples	Color	Surface temperature, °K	Spectral characteristics
O	10 Lacertae	Blue	Over 25,000	Lines of ionized helium and other ionized elements, hydrogen lines weak
B	Rigel, Spica	Blue-white	11,000–25,000	Hydrogen and helium prominent
A	Sirius, Vega	White	7,500–11,000	Hydrogen lines very strong
F	Canopus, Procyon	Yellow-white	6,000–7,500	Hydrogen lines weaker, lines of ionized metals becoming prominent
G	The sun, Capella	Yellow	5,000–6,000	Lines of ionized and neutral metals, especially calcium, prominent
K	Arcturus, Aldebaran	Reddish	3,500–5,000	Lines of neutral metals and band spectra of simple compounds present
M	Betelgeuse, Antares	Red	2,000–3,500	Band spectra of many compounds prominent

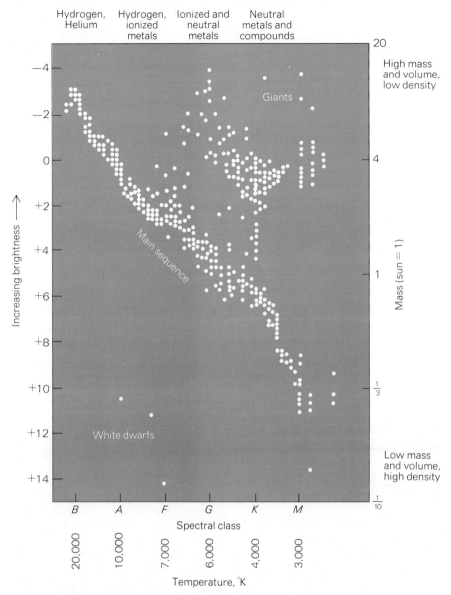

43.7 The Hertzsprung-Russell diagram for Population I stars, showing the distribution of stars according to spectral type and intrinsic brightness. *The numbers at the left express absolute magnitude (the astronomical measure of intrinsic brightness) with low numbers indicating bright stars and high numbers, faint stars. The letters at the bottom designate spectral classes. The temperatures of the giant stars are somewhat lower than those indicated, which hold for the main sequence stars.*

and *dwarf* refer, as we might expect, to very large and very small stars, respectively.)

There are a number of interesting correlations between the position of a star on the Hertzsprung-Russell diagram and its physical properties. Temperature varies with spectral type, so that hot stars are at the left of the diagram and cool stars at the right. In general, the brighter stars are those with both greater mass and greater volume, so that large, massive stars are found at the top of the diagram and smaller, less massive stars at the bottom. Densities are commonly greater for the small stars, so that densities at the bottom of the diagram are greater than those at the top. These correlations are very general, to be sure. They do not mean, for example, that all stars with the same absolute magnitude have the same mass and density or that all stars of the same spectral type have precisely the same temperature.

Stars at the upper end of the main sequence are large, hot, massive bodies, with prominent lines of hydrogen and helium in their spectra. Stars at the lower end are small, dense, and reddish, with low enough temperatures so that chemical compounds form a considerable part of their atmosphere. In the middle part of the main sequence are average stars like our sun, with moderate temperatures, densities, and masses, rather small diameters, and spectra in which lines of metallic elements are prominent. The majority of stars show these definite combinations of principal characteristics.

main-sequence stars

To the *giant* class belong the huge, diffuse stars like Antares, with low densities and diameters up to several hundred million miles. Many of these stars have low surface temperatures, as their reddish color indicates, but their enormous surfaces make them very bright.

red giants

white dwarfs

The position of the *white dwarfs* in the Hertzsprung-Russell diagram suggests their peculiar combination of properties: intensely hot surfaces but small total radiation. These properties imply that their surface areas and, hence, their volumes must be exceedingly small. Most white dwarfs are somewhat larger than the earth, although a few are smaller, but their masses are not correspondingly small. For those white dwarfs that are

The Veil Nebula in Cygnus consists of high-speed gas filaments ejected from an exploding star over 50,000 years ago.

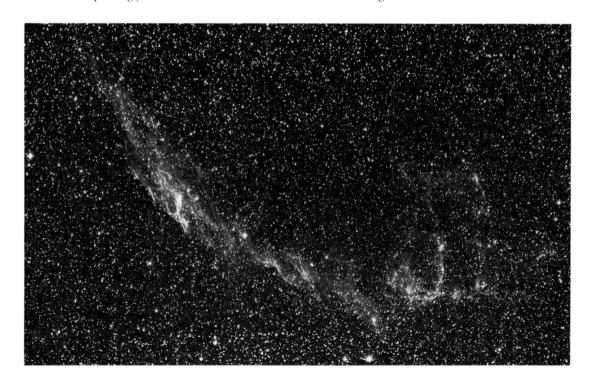

white dwarfs have the size of a planet but the mass of a star and are very hot

components of double stars calculation shows masses comparable with that of the sun. Now, if the sun's mass were contained in a star with the volume of a planet, its density would be about 10^6 g/cm^3! A pinhead of such matter would weigh nearly a pound here on earth, and a cupful would weigh several tons.

Densities like this seem incredible, but they have been checked by enough methods to leave little doubt of their correctness. (See, for instance, the discussion of the gravitational red shift in white dwarfs on page 266.) The only possible explanation is that atoms in these stars have partially collapsed: Instead of ordinary atoms with electrons following wide orbits around their nuclei, white dwarfs must have electrons and nuclei packed closely together. Matter in this state has never been found or prepared on earth, but the theories of physics permit us to calculate its properties. These calculations indicate that white dwarfs cannot have masses greater than 1.2 times the sun's mass; but of course a more massive star can lose matter in a variety of ways to become a white dwarf.

Only a few hundred white dwarfs are known. Their scarcity may be more apparent than real, since they are so faint that only the nearer ones can be seen even in large telescopes. Enough of them have been found in recent years to suggest that the universe probably contains great numbers of these peculiar stars.

stellar evolution

So striking a series of relationships as that revealed by the Hertzsprung-Russell diagram cannot have occurred through blind chance. Are the stars in different parts of the diagram perhaps in various stages of development? Does the mass or the density of a star perhaps exert some controlling influence over its temperature and the composition of its atmosphere? Astronomers have not reached full agreement about the answers to these questions, but in general terms the diagram finds a satisfactory explanation in modern theories of the life history of a star.

a body of matter with the mass and composition of a star *must* shine

A star shines because it is a large, compact aggregate of matter that contains abundant hydrogen. A body of this sort *cannot avoid* being luminous because of the energy liberated in the conversion of its hydrogen into helium. If it were not radiating, the mutual gravitation of its particles would cause it to contract, and energy liberated by the contraction would then heat the interior sufficiently to start the hydrogen-helium transformation; thereafter, the radiation would maintain itself as long as any hydrogen remained. A star does not shine because some occult force has started it shining; it shines because it has a certain mass and a certain composition. If we could somehow build a star by heaping together sufficient matter of the right composition, it would start to shine of its own accord.

early history of a star

We may imagine as the most reasonable starting point in a star's history a stage when its matter was an irregular mass of cool, diffuse gas and small, solid particles. Gravitation in such a mass would ultimately concentrate it into a smaller space. The gradual contraction would heat the gas, much as the gas in a tire pump is heated by compression. At length the temperature would grow high enough for hydrogen to be converted into helium, and the mass would begin to glow brightly. From this time on the tendency to contract would be counterbalanced by the pressure of radiation from the hot interior, so shrinking would stop and the star would maintain a nearly constant size. The diameter of a star is thus determined by an equilibrium between gravitational forces pulling its material inward and forces due to radiation pushing its material outward.

The temperature a star attains is determined by its mass. Gravitation in a large mass is stronger than in a small mass and, hence, requires more intense radiation to counterbalance it. Calculations show that the relationship between mass and temperature, for stars with abundant hydrogen, should be exactly that shown by stars in the main sequence of the Hertzsprung-Russell diagram. The large, heavy stars at the upper end of the main sequence have high temperatures and high intensities of radiation; the small stars at the lower end are relatively cool and only faintly luminous. Stars in the main sequence, therefore, may be thought of as *normal* stars in which the hydrogen-helium reaction is pouring out energy and in which the rate of energy production is adjusted to the masses of the stars.

temperatures of heavy stars are higher than those of light stars

Hydrogen is so abundant in the universe that any star beginning its career as a contracting mass of gas and small particles would have sufficient fuel to keep on shining for a good many billion years. When at last the hydrogen supply begins to run out, the character of the star must change radically. It is interesting to speculate on what changes would take place—even though such speculations cannot be based on actual observations, because human life is too short for us to see more than a tiny part of a star's life history. The best we can do is to observe the various kinds of stars existing at the present moment, to assume that they represent stars in different stages of development, and then on the basis of our knowledge of nuclear reactions to try to arrange them in a reasonable sequence. It is the sort of problem that might be faced by an intelligent butterfly having a life span of a single day, gifted with a rudimentary knowledge of physiology, and trying to arrange in order of development the various kinds of human beings it might encounter during that day.

old stars

A star consumes its hydrogen rapidly if it is large, slowly if it is small. A fairly small star like our sun makes its supply of hydrogen last for a period on the order of 10 billion years; probably the sun is now about halfway through this part of its career. When the hydrogen supply at last begins to run low in a star like the sun, the life of the star is by no means ended but enters its most spectacular phase. Further gravitational contraction makes the interior still hotter, and other nuclear reactions become possible—particularly reactions in which atoms of heavier elements are made by a combination of helium atoms. These reactions, once started, give out so much energy that the star expands to become a giant. Energy is now being poured out at a prodigious rate, so the star's life as a giant is much shorter than the earlier part of its existence.

stars grow larger as they grow old

This series of photographs made in 1937, 1938, and 1942 records a supernova in the constellation Virgo.

novae

at the end of its life cycle a star becomes a white dwarf

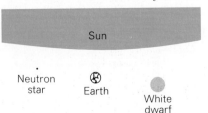

43.8 A comparison of a white dwarf and a neutron star with the sun and the earth. *Both white dwarfs and neutron stars are thought to have masses similar to that of the sun.*

Eventually the new energy-producing reactions run out of fuel, and again the star shrinks—although probably not without a few last brief flare-ups, which we see from the earth as *novae* ("new stars"), shining brilliantly for a week or two and then subsiding into insignificance. The shrinking ultimately reduces the star to the white dwarf state, in which the atoms themselves have partly collapsed. As a slowly contracting dwarf the star may remain luminous for more billions of years, its energy now coming from the contraction, from nuclear reactions involving elements heavier than helium, and from proton-proton reactions in a very thin outer atmosphere of hydrogen.

Thus the giant stars to the right of the main sequence on the Hertzsprung-Russell diagram are stars of small or average mass that are passing through the middle part of their life histories, when much of their hydrogen has been used up and some of their energy is coming from reactions involving helium nuclei. The dwarfs are the old stars, stars that have exhausted their hydrogen and helium and have shrunk until their matter has enormous densities. If we should plot the life history of a star on the Hertzsprung-Russell diagram, it could be represented by a line moving from the main sequence up into the area of the giants, then downward and to the left across the main sequence into the area of the white dwarfs.

The distribution of stars on the Hertzsprung-Russell diagram can evidently be interpreted fairly satisfactorily by a hypothesis of evolutionary development. For the sun, the star in which we have the most personal interest, the evolutionary hypothesis predicts that several billion more years will pass with no marked change from present conditions. Then the sun's temperature will gradually increase, and life on earth will ultimately become impossible—not, as was once thought, because the sun will cool off but rather because it will grow too hot. Eventually the sun will expand into a giant, as large around, perhaps, as the orbit of Venus, and still later it will collapse into a white dwarf. A dismal end, but one that our most remote progeny are not likely to survive to witness.

supernovae

Heavy stars at the upper end of the main sequence, in fact all stars whose masses are more than about 1.5 times that of the sun, have a somewhat different history. They consume their hydrogen more rapidly than smaller stars and, hence, spend less time in the main sequence; helium reactions begin in their interiors long before their hydrogen is exhausted; and in the middle part of their careers they may expand only slightly. Eventually they become unstable and explode violently, emitting enormous amounts of material. Such explosions we note on the earth as *supernovae*, flare-ups on the order of 10,000 times as luminous as ordinary novae. Having drastically reduced their masses by explosion, these stars can then subside like their smaller brethren into the condition of dwarf stars.

Many astronomers believe that the residual dwarfs of supernovae are different from ordinary white dwarfs because of the unusually large mass of their parent stars and the equally unusual violence attending their birth. These hypothetical dwarfs are calculated to have densities far in excess of ordinary dwarfs, with masses comparable to that of the sun packed into spheres perhaps 15 km (9 miles) in diameter (Fig. 43.8). The matter of such a star would weigh billions of tons per cubic inch. (If the earth were this dense, it would fit into a large apartment house.) Under the pressures that would be present the most stable form of matter is the neutron. Until recently the notion of neutron stars was purely speculative, but the discovery of pulsars has made their existence quite plausible.

The Crab Nebula is the remnant of a supernova that appeared in A.D. 1054.

pulsars

In 1967 unusual radio signals were detected that came from a source in the constellation Vulpecula. The signals were found to fluctuate with an extremely regular period, exactly 1.33730113 sec. Since then over two dozen more *pulsars* have been discovered, all with periods between $\frac{1}{30}$ and 3 sec. At first only radio emissions from pulsars could be observed, but in 1969 flashes of visible light were found to be given off by a pulsar in the center of the Crab nebula that are exactly synchronized with the radio signals.

The power output of a pulsar is about 10^{26} watts, which is comparable with the total power output of the sun. But it is inconceivable that so strong a source of energy can be switched on and off in as little time as $\frac{1}{30}$ sec, or even that such a source can be the size of the sun: if the sun were to suddenly stop radiating, it would take 2.3 sec before light stopped reaching us, because all parts of the solar surface that we see are not at the same distance away. Hence we can conclude at once that a pulsar must have roughly the mass of a star, in order to be capable of emitting so much energy, and that it must be very much smaller than a star, in order that its signals fluctuate so rapidly.

From the above and other considerations it seems likely that pulsars are neutron stars which are spinning rapidly. Conceivably a pulsar has a strong magnetic field whose axis is at an angle to the axis of rotation, and this field traps tails of ionized gases that do the actual radiating. Whatever the actual mechanism, though, a pulsar is apparently like a lighthouse whose flashes are due to a rotating beam of light. The identification of pulsars with neutron stars is supported by evidence that the periods of pulsars are very gradually decreasing, which would be expected as they continue to evolve energy.

The fast pulsar in the Crab nebula is of exceptional interest. This nebula is the remnant of a supernova that was seen in A.D. 1054, and it has been expanding rapidly and glowing brightly ever since. Until now it was hard to understand exactly where all the energy of the nebula is coming from, since the explosion itself occurred so long ago. However, the light and radio flashes from the Crab pulsar are apparently powerful enough to furnish the entire nebula with its energy. This pulsar is both the fastest pulsar (1 flash per $\frac{1}{30}$ sec) and is slowing down at the most rapid rate known—1 part in 2,400 per year. Both these observations are in accord with its supposed formation 900 years ago. As time goes on, presumably the Crab pulsar will rotate more slowly and at a steadier rate, as other pulsars do.

GLOSSARY

The *parallax* of a star is the apparent shift in its position relative to more distant stars as the earth revolves in its orbit.

A *light-year* is the distance light travels in one year; it is 9.5×10^{15} m, or about 6×10^{12} miles.

The *apparent brightness* of a star is its brightness as seen from the earth; the *intrinsic brightness* of a star is its actual brightness, which depends upon the amount of light it radiates into space.

A *variable star* is one whose brightness continually varies.

A *Cepheid variable* is a variable star of a particular type whose intrinsic brightness and period of variation are related.

A *double star* is actually a pair of stars both revolving about their mutual center of gravity.

The *spectral sequence* of the stars is an arrangement of different types of stars according to their spectra; at one end of the sequence are stars in whose spectra lines of hydrogen and helium are most prominent; at the other end are stars in whose spectra lines of neutral metals and simple compounds are most prominent. The former are hot, bluish-white stars; the latter are relatively cool, red stars.

The *Hertzsprung-Russell diagram* is a graph on which are plotted the intrinsic brightness and spectral type of the different stars. Most stars belong to the *main sequence*, which follows a diagonal line on the diagram, but there are also small, hot *white dwarf* stars and large, cool *red giant* stars occupying other positions there.

A *nova* is a star that suddenly flares up and shines brilliantly for a week or two before fading back to a less conspicuous luminosity.

A *supernova* is a star that explodes with spectacular brightness so that it may be visible even in daylight and ejects vast amounts of material into space.

A *neutron star* is an extremely small star, smaller than a white dwarf, which is composed almost entirely of neutrons. *Pulsars* are stars that emit extremely regular flashes of light and radio waves; they are believed to be neutron stars that are spinning rapidly.

EXERCISES

1. Give all the evidence you can in support of each of the following statements:
    a. The sun is a star.
    b. Most stars have relatively cool atmospheres surrounding hot interiors.
    c. Distances to the stars are very great.
    d. Some stars that appear to be single even in a telescope are actually double stars.
    e. The sun is moving with reference to the other stars.

2 What methods can be used for determining the intrinsic brightness of a star? What assumption is involved in the spectroscopic method?

3 How is a star's diameter estimated from measurements of temperature and intrinsic brightness?

4 For what stars is a direct determination of mass possible? On what characteristic of these stars do you suppose the determination depends?

5 Would it be possible for a star to have any shape other than that of a sphere or spheroid? Could a star, for instance, be shaped like a cube or like a corkscrew? Explain your answer.

6 What data are needed for the determination of a star's average density? How would you expect the density to change from the surface layers to the interior of a star?

7 What are the chief characteristics of an average star
 a. in the upper left-hand corner of the Hertzsprung-Russell diagram?
 b. in the lower left-hand corner?
 c. in the upper right-hand corner?
 d. in the middle of the main sequence?

8 What is the evidence for the enormous average densities ascribed to white dwarfs?

9 Sirius, the Dog Star, is a bluish-white star with a very great intrinsic brightness. From these facts what, if anything, can you conclude about each of the following?
 a. Its temperature
 b. Its average density
 c. The principal lines in its spectrum
 d. Its position in the Hertzsprung-Russell diagram

10 Explain the steps you would take to determine the distance to a star cluster that contains Cepheid variables.

11 Why must a star radiate energy?

12 It is a common misconception that stars begin by being very bright and gradually grow dimmer. What reasons, based on physical principles, can you give to refute this idea?

13 Why do small stars have life histories different from those of large stars?

14 Justify with a diagram and a calculation the statement in the text that if the sun were to suddenly stop radiating, 2.3 sec would elapse before all light from it ceased to reach the earth.

chapter 44

Structure of the Universe

OUR GALAXY

    the milky way
    star clusters
    galactic nebulae
    radio astronomy

ISLAND UNIVERSES

    spiral galaxies
    "island universes"

COSMIC RAYS

    primary cosmic rays
    secondary cosmic rays
    radiocarbon dating
    cosmic-ray origin
    galactic magnetic fields

The great band of misty light we call the Milky Way forms a continuous ring around the heavens. When it is examined with a telescope, the Milky Way is an unforgettable sight. Instead of a dim glow we see countless individual stars, stars as thick as the sand grains on a beach although so faint and far away that the naked eye cannot distinguish them. In other parts of the sky the telescope reveals many stars not visible to the eye, but nowhere else in such incredible numbers. Clearly, the stars are not uniformly distributed in space—a simple observation which, as we shall see, has profound implications for the problem of the structure of the universe.

OUR GALAXY

If we could travel through space wherever we wished, regardless of direction or distance, would we find ourselves surrounded by stars no

A mosaic of several photographs of the Milky Way between the constellations Sagittarius and Cassiopeia.

matter how far we went, or would we at last come to a part of space in which stars were absent?

To answer this question without actually making such a journey, all we need is to possess a powerful telescope and leave a series of photographic plates exposed to the same part of the sky for longer and longer times. By increasing the exposure time, we are able to detect stars farther away. If stars are present everywhere in space in about the same numbers as we find them near the sun, then each increase in the distance our instruments can penetrate should lead to an increase in the number of stars visible. Furthermore, this increase in number should follow a regular law: When we penetrate farther into space, the *volume* we reach goes up as the cube of the distance. Thus, when the distance at which we can barely see stars of a certain brightness is increased 10 times, the volume of space we are looking at is 10^3, or 1,000 times, greater, and the total number of stars visible should increase 1,000 times; when the distance is increased 20 times, the number of stars should be 8,000 times as great, and so on.

Counting stars in this manner shows the expected rate of increase for relatively small distances, but a rapid falling off in the rate of increase for large distances. This means that a rocket traveler moving away from the sun would, after a while, find the stars thinning out, and eventually would come to a region where they are exceedingly scarce. The stars do not occur with the same frequency everywhere in space; our sun is part of an aggregate of stars with fairly definite limits. We shall find later that there are other similar aggregates far beyond the limits of this one, each separated from the others by vast reaches of empty space. Such aggregates are called *galaxies*, and the particular one to which our sun belongs is termed *our* galaxy.

a galaxy is an island universe of stars

the milky way

The falling off in the number of stars at increasing distances is least rapid for sections of the sky near the Milky Way, most rapid for sections at right angles to it. In the direction of the Milky Way we seem to be looking out through a far greater thickness of stars than we find in other parts of the sky. These facts tell us something about the arrangement of the nearby stars in our galaxy: They must form a relatively thin, flat structure, shaped like a thin pocket watch (Fig. 44.1). From our vantage point deep within the galaxy we look along the plane of the watch toward its edge and see the thick mass of stars in the Milky Way, and when we look out through the face or the back of the watch, we see the relatively small number of stars at right angles to the Milky Way.

Since the earth appears to be nearly in the plane of the Milky Way, the sun must be close to the central plane of the galaxy. This conclusion is borne out by accurate counts of stars in opposite directions at right angles to the Milky Way: In these directions the increase in the number of stars begins to fall off at about the same distance from us, indicating that we should have to travel about as far to reach the front of the watch as to reach its back. Our position relative to the edges of the watch cannot be so readily determined from star counts because of the great distances involved and because parts of the Milky Way are hidden by dark obscuring masses. Studies of the motion of the sun and nearer stars indicate, however, that we are probably about two-thirds of the distance out from the center of the galaxy toward one edge. The center most probably lies in the direction of the densest part of the Milky Way, in the constellation Sagittarius.

The stars of the galaxy are almost all revolving about its center, which is what they must be doing if the galaxy is not to gradually collapse because of the gravitational attraction of its parts. (The planets do not fall into the sun because of their orbital motions, too.) The orbital speed of the sun and the stars in its vicinity around the galactic center is about 140

the galaxy of which the sun is a member appears in the sky as the milky way

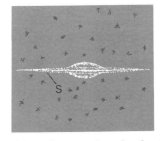

44.1 Cross-section sketch of our galaxy. *Globular clusters are shown as color patches, and the sun is located at S. The diameter of the galaxy is about 100,000 light-years.*

Globular cluster in Hercules.

(Left) Andromeda Nebula photographed in blue light shows giant and super-giant stars of Population I in the spiral arms. The hazy patch at the upper left is composed of unresolved Population II stars. (Right) NGC 205, companion of the Andromeda Nebula, photographed in yellow light shows stars of Population II. The brightest stars are red and 100 times fainter than the blue giants of Population I. (The very bright, uniformly distributed stars in both pictures are foreground stars belonging in our own Milky Way galaxy.)

our galaxy is about 100,000 light-years across

mi/sec. At this rate the sun makes a complete revolution once every 220 million years.

Superimposed on the galactic rotation is a smaller random motion, like that of molecules in a gas, so each star seems to be moving with respect to its neighbors in addition to partaking with them of a common circumnavigation of the center of the galaxy. The sun's speed relative to its neighboring stars seems to be about 12 mi/sec. The comparison with molecular motion in a gas is actually quite accurate, except that the gas must be exceedingly tenuous to resemble the emptiness of the galaxy. We must imagine a gas more rarefied than the best vacuum we can produce on earth, so rarefied that collisions between its molecules are extremely uncommon.

The size of the galaxy is difficult to determine, for stars near its borders are much too faint for distances to be found by spectroscopic means. Estimates made from star counts, from studies of Cepheid variables (which we discussed in the last chapter), and in other indirect ways indicate that the galaxy is roughly 100,000 light-years across and between 10,000 and 15,000 light-years thick. Included in this enormous system are 100 billion or so stars, among which our sun is inconspicuous indeed. Surrounded by countless similar bodies, the sun is not notable for its position, for its size, for its temperature, or for its motion. It is probably not even unusual in possessing a family of planets; there may well be over a million inhabited planets in our galaxy alone, according to the estimates of some astronomers.

star clusters

Stars are not uniformly distributed even within our galaxy. About one in a thousand seem to be segregated into groups called *galactic clusters*, each of which consists of a great many stars relatively close to one another that travel together about the center of the galaxy. The Pleiades make up such a cluster; the naked eye can make out six of its stars, and a telescope reveals several hundred more. The stars contained in galactic clusters are much like the more isolated stars that constitute the vast celestial merry-go-round that is the Milky Way. These relatively slow-velocity stars are termed *Population I* stars.

<small>population I stars</small>

But there are a number of other stars associated with our galaxy whose quite rapid motions are not regular at all and whose properties differ in other ways as well from those of Population I. These *Population II* stars are often found in *globular clusters*, great assemblies of stars which, although definitely part of our galaxy, lie outside the main watch-shaped aggregate we see as the Milky Way. To the naked eye the largest globular clusters are barely visible on clear evenings as faint patches of light. Through a telescope they are spectacular aggregates of stars, roughly spherical in form, bright and dense near the center and thinning out toward the edges. About 100 of these objects have been discovered. In

<small>population II stars</small>

The Great Nebula in the constellation Orion is a gas cloud excited to incandescence by hot stars in its center.

photographs of one of the largest, the great cluster in Hercules, more than 50,000 stars have been counted. These are only the very brightest stars, since the cluster is so far away that faint ones cannot be seen; estimates place the probable total number of stars at close to a million. The nearest clusters appear to be about 20,000 light-years away from us and the farthest more than 100,000 light-years away. Light from the great Hercules cluster travels 33,000 years before reaching our eyes; we see the cluster not as it would look today but as it appeared toward the end of the Ice Age. The average distance separating the stars in a globular cluster is about 1 light-year; so stars in a cluster, particularly toward its center, are considerably more closely packed than those near the sun. Even so, collisions between stars in a cluster are infrequent.

The globular clusters and the other Population II stars travel in randomly oriented elliptical orbits at very high speeds about the galactic center, and their motions have nothing to do with the orderly procession of Population I stars in the flat disk of the galaxy. This is not the only difference between stellar Populations I and II. Apparently Population II stars, approximately 5 billion years in age, are all much older than Population I stars. In order to have survived for so lengthy a period, the Population II stars are relatively cool and faint; by burning up their nuclear energy reserves slowly, they last longer. Population I stars, though, are often hot and bright, with spectacular but brief careers.

structure of the galaxy   Actually, there seems to be good reason to suspect that our galaxy is itself a vast globular cluster of Population II stars, spherical in shape with a thin disk of younger Population I stars at its center. In other words, perhaps the Milky Way is just a part of our galaxy, containing 10 percent or so of its total number of stars, with most of the contents of the galaxy scattered through a great volume of space over 100,000 light-years across.

galactic nebulae

The telescope reveals two kinds of luminous objects in the sky besides stars and members of the solar system, *galactic nebulae* and *spiral nebulae*. Both appear as faint patches of diffuse light, and this appearance led to their early name of nebula, Latin for "cloud." It is unfortunate that the same name is still applied to both, for they are altogether different in nature.

The galactic nebulae are irregular masses of diffuse material within our galaxy. Some appear as small luminous rings or disks surrounding stars, some take the form of lacy filaments, and many are wholly irregular in outline. The brightest, the great nebula in Orion, is barely visible with the naked eye, but most of them are so faint that long exposure of a photographic plate is necessary to bring out details of structure. Probably these nebulae are masses of rarefied gas and tiny solid particles, shining only because they reflect light from nearby stars or because electrically charged particles, X-rays, and ultraviolet radiation from stars excite their matter to luminescence.

Masses similar to galactic nebulae but without any luminosity sometimes reveal their presence as dark patches obscuring the light of stars beyond them. Such *dark nebulae,* rarefied clouds of gas and small solid particles, may be fairly abundant; they are difficult to find by telescopic exploration, because bright stars shine through them except where they are especially dense. In fact, much of our galaxy is filled with exceedingly rarefied nebular material, with a density somewhere near one atom per cubic centimeter, and the dark nebulae are only local concentrations of this interstellar matter. Empty space is not nearly so empty as it was once thought to be, but in most places the concentration of interstellar material is so small that starlight meets little interference in passing through it. The luminous galactic nebulae are probably no different in composition from the dark nebulae or the interstellar matter in general, but they happen to be in places where their particles are made luminous by radiation from adjacent stars.

interstellar space is not completely empty

radio astronomy

A new tool for the investigation of dark nebulae has been made available to astronomers by the discovery that radio waves reach the earth from many parts of the universe. Various sources of radio waves have been found, and not all are well understood, but many are known to be masses of cool hydrogen gas that emit electromagnetic radiation with a wavelength of 21 cm. With the help of special *radio telescopes,* giant antennas connected to sensitive radio receivers, the intensity of this radiation can be measured in different parts of the sky. The concentrations of interstellar gas forming dark nebulae can now be mapped with much greater accuracy than would ever be possible using visible light. Studies of these concentrations of gas and small particles have become matters of special interest because, on theoretical grounds, it seems likely that some of the dark nebulae are places where new stars are in process of formation.

a radio telescope is a sensitive radio receiver connected to a directional antenna

Extraterrestrial radio waves other than those from cool hydrogen seem to originate in two different ways. In most cases the properties of the waves indicate that they come from the random thermal motion of ions and electrons in a hot gas, and a number of such sources have been identified as luminous nebulae. In other cases the radio waves are generated by high-speed electrons moving in magnetic fields, and sources of this kind have been observed whose emission of energy as radio waves exceeds that emitted as light waves. The strongest radio sources, called *quasars,* are of the latter type and have power outputs of about 10^{38} watts. Quasars are discussed in Chap. 45.

ISLAND UNIVERSES

spiral galaxies

The other kind of luminous nebulae, the spiral nebulae, are by far the more interesting of the two. To the naked eye only one of these nebulae is visible in the Northern Hemisphere, the spiral in Andromeda, and this appears only as a small, hazy patch of light. But in photographs taken

The North American Nebula in Cygnus is a gas cloud that is excited to radiate by ultraviolet light from stars embedded in it. The obscured areas that simulate the Atlantic Ocean and Gulf of Mexico are due to absorption of nebular light by dust clouds between the nebula and the earth.

with large telescopes and long exposures the spiral nebulae are impressive objects. The photographs differ from those of galactic nebulae in several respects, the most immediately obvious being a greater regularity of shape. The shapes range from fuzzy, spherical objects practically without structure to distinct, flat spirals like Fourth-of-July pin wheels. Members of the most common variety consist of a definite spiral structure with two curving arms radiating from a brighter nucleus. The telescope shows us spiral nebulae from different angles: some full face, some obliquely, and some edgewise, as in the accompanying illustrations. The general shape of the average nebula is that of a thin circular disk somewhat thicker in the center. Some celestial objects of irregular shape may represent spiral nebulae in process of colliding with each other. Such collisions are strong emitters of radio waves.

Radio telescope at Greenbank, West Virginia, picks up radio waves from planets, stars, and nebulae.

Good spectra of the spiral nebulae are difficult to obtain because of their extreme faintness. The brighter ones give spectra closely resembling ordinary star spectra—dark absorption lines on a continuous background, most of the lines identifiable with those of familiar elements. This strongly suggests that the nebulae are aggregates of stars, a suggestion borne out by photographs taken with large telescopes, which show the outer portions of the biggest spirals resolved into separate stars. Possibly the more diffuse spirals and the central parts of others are masses of gas and small particles, but the chief components of the principal nebulae are certainly stars. For this reason spiral nebulae are usually called *spiral galaxies* today.

spiral nebulae are galaxies of stars far away from our own galaxy

The exceeding faintness of the stars that can be made out in the brightest spirals implies that these objects are very remote. Accurate estimates of distances to a few of the nearer ones are made possible by the presence among their stars of recognizable Cepheid variables. Measurements of the apparent brightnesses and periods of fluctuation of these variables in the nearest of the large galaxies, the one in Andromeda, give a distance of about $1\frac{1}{2}$ million light-years. Several others are in the range of a few million light-years. Estimates of distances to the farther spirals cannot be based on so sure a foundation as the periods of Cepheid variables, but the faintest spirals that have been photographed cannot be much nearer the earth than 2 billion light-years. This means that the light from these objects that darkens our photographic plates has been moving through space since far back in Precambrian time.

There is little doubt today that spiral galaxies are aggregates of stars far outside the limits of our own galaxy. Such aggregates are exceedingly numerous: Within a distance of a billion light-years from our galaxy there are at least 100 million spirals large enough to be recorded by our telescopes. Evidently these objects hold the key to any investigation of the structure of the universe as a whole.

"island universes"

William Herschel, the first to study spiral nebulae intensively, suggested that they might be other galaxies of stars which resemble our own, "island universes" in the sea of space. In his time this proposal was hardly more than a brilliant guess, but more recently, particularly because of the work of the American astronomer Edwin Hubble, it has been abundantly justified.

Let us examine a few of the similarities between spiral galaxies and our galaxy.

Shape. Here the resemblance is evident, for star counts show that our galaxy has the same flat, watchlike form so characteristic of the spirals. Recent evidence, both from the visible structure of the Milky Way and from the distribution of dark hydrogen clouds as revealed by radio astronomy, makes it almost certain that our galaxy has the usual two curved arms.

Size. Our galaxy ranks with the larger spirals. Of the 19 known galaxies within a distance of 4 million light-years (the so-called *local group* of galaxies), the largest is the one in the constellation Andromeda and the second in size is our galaxy.

Globular clusters. The discovery of globular clusters near some of the brighter spiral galaxies is further evidence for their similarity in structure to our galaxy. Star clusters would be too small and faint for detection except around the closest ones.

Rotation. The rotation of some of the spirals about their centers, similar to the rotation of our galaxy, has been detected by the displacement of lines in their spectra.

Composition. The resemblance between galactic spectra and stellar spectra and also observations of the presence of stars in the brightest spirals indicate that at least a great many spirals are similar in composition to our galaxy. Although some spiral galaxies may consist largely of gas and tiny particles with few stars, this is more a matter of relative proportion than of fundamental difference; the contents of our own galactic disk are divided approximately evenly between stars and diffuse matter. It is likely that the bright Population I stars originate in interstellar gas and dust clouds, and so spirals devoid of stars are simply in an earlier stage of development than our galaxy.

Thus we are justified in picturing the universe as made up of galaxies of stars, each one isolated in space and separated from its nearest neighbors by distances of a million light-years or more. In all directions, in unbroken succession, these galaxies extend to the farthest parts of the universe that our instruments can penetrate. Not only is the earth an undistinguished planet circling an undistinguished star; even the great galaxy that includes the sun is no different from millions of others.

the universe is filled with spiral galaxies isolated from one another

Through all this vast array of uncounted suns and unimaginable distances runs a startling uniformity of material and structural pattern. The elements of our laboratories are the elements of the spiral nebulae, the sun generates energy by a process repeated in billions of other stars, and the form of our galaxy recurs again and again in the nebulae. Everywhere we find the same ultimate particles of matter, the same kinds of energy, the same patterns of structure. We can study at firsthand no more than a tiny fragment of the universe, yet so ordered and uniform is the whole that from this fragment we can extend our knowledge wherever our instruments enable us to see.

COSMIC RAYS

We now return to our galaxy to study one of its most unusual phenomena—*cosmic rays*.

The story of these once mysterious rays began early in this century, when it was discovered that the ionization in the atmosphere *increased* with altitude. At that time most scientists thought that the small number of ions always present in the air were due to the alpha, beta, and gamma rays emitted by the naturally radioactive substances, such as radium, that are found everywhere on the earth in minute quantities. If this were the correct explanation, then when we go high into the atmosphere away from the earth and its content of radioactive materials, the proportion of ions that we find should drop. Instead it increases, as a number of balloon-borne experimenters learned between 1909 and 1914. Finally Victor Hess, an Austrian physicist, suggested an explanation: From somewhere *outside* the earth ionizing radiation is continually bombarding our atmosphere. This radiation was later called *cosmic radiation* because of its extraterrestrial origin.

primary cosmic rays

Before we discuss where cosmic rays come from, let us look briefly into what we know of their properties.

Primary cosmic rays, which are the rays as they travel through space before reaching the earth, are ordinary atoms that have tremendous speeds. The speeds are so great, in fact, that the atoms move nearly as fast as light, or almost 186,000 mi/sec. Therefore cosmic rays are really not rays (like light rays or X-rays) at all. The fact that they go so fast means

primary cosmic rays are high-energy atomic nuclei that are present in space

The Great Nebula in Andromeda is the spiral galaxy nearest to the earth. Its period of rotation is about 200 million years.

that each atom has a large amount of energy, and this energy is what gives cosmic rays their unique properties. The cosmic-ray atoms reaching the earth contain no electrons; they are just bare nuclei, having lost their orbital electrons at some early stage in their flight through space.

Primary cosmic rays include almost all the known elements of the universe in their usual proportions. The great majority are hydrogen and helium, with heavier atoms, such as carbon, nitrogen, and iron, also present in much smaller amounts.

the earth's magnetic field influences the paths of the slower primary cosmic rays

Much of our knowledge of cosmic rays has been made accessible by the earth's magnetic field. Moving charged particles, as we know, are deflected by magnetic fields. The effect of the earth's field is to direct the less

Spiral nebulae. At the right, the spiral nebula in Canes Venatici; at the left, that in Coma Berenices seen edge-on.

energetic cosmic-ray primaries toward high latitudes but to permit the faster ones to reach the earth everywhere. Some incoming particles are trapped by the field, oscillating back and forth along the lines of force while simultaneously drifting around the earth. These trapped particles, together with the protons from neutron decay mentioned below, constitute the zones of radiation discovered by James van Allen in rocket and artificial-satellite measurements of cosmic rays.

Cosmic-ray energies are not all the same, varying from a few Gev (billion electron volts) for the greater number to more than 1 billion Gev for a few especially energetic ones. (In comparison, the largest particle accelerator currently operating produces protons with energies of 76 Gev—on a par with the feebler cosmic rays but a long way indeed from the strongest.) Over a billion billion primary cosmic rays strike the earth each second, representing a power of about 6 billion watts—equivalent to the output of a dozen of the largest power plants in existence.

secondary cosmic rays

When the primary cosmic rays strike the blanket of air surrounding the earth, they collide with oxygen and nitrogen molecules in their paths, thereby producing showers of secondary particles. It is these secondaries that actually reach the earth's surface, with eight of them passing through each square inch at sea level per minute. There are three chief kinds of cosmic-ray secondaries:

primary cosmic rays interact with the atmosphere to produce the secondary cosmic rays that reach the ground

1 *Neutrons* and *protons* knocked out of the nuclei of atoms struck by the primaries. These protons and neutrons are often fast enough themselves to disrupt other nuclei. Thus a single primary may lead to the destruction of a number of atoms. The liberated protons ultimately pick up electrons and become hydrogen atoms, adding to the minute hydrogen content of the atmosphere. Some of the neutrons go out into space, ultimately beta-decaying into protons which are often captured by the geomagnetic field where they remain trapped for relatively long periods of time. Many of the neutrons that remain in the atmosphere are captured by the nuclei of nitrogen atoms in the atmosphere to form radiocarbon, as described in the next section.

2 *Mesons* and *hyperons*, which are, as we mentioned in Chap. 17, unstable particles all heavier than electrons. Table 17.3 summarizes the proper-

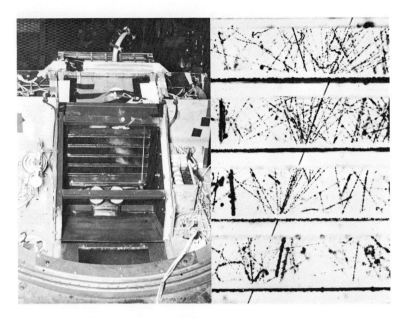

At left, a cloud chamber containing ¼-in. lead plates used in balloon-borne cosmic-ray experiments. At right, a photograph of the tracks of a 2.8-Gev cosmic-ray electron and the shower of secondary electrons it produces as it passes through the cloud chamber. For clarity, the path of the electron is drawn across the fronts of the lead plates in the photograph.

soft and hard showers

ties of these particles, which exist only briefly before decaying into other forms.

3 *Electrons* and *gamma rays*, which arise largely from the decay of mesons and multiply as they go through the atmosphere to form showers of particles and photons. As we know from Chap. 17, a high-energy gamma ray can *materialize* into an electron and a positron. No conservation laws of physics are violated in this process: The total energy of the electron and positron equals the initial energy of the gamma ray (reckoning the energy equivalent of a mass M as Mc^2); the total charge of the electron and positron together is zero—which, of course, was true of the gamma ray; and the gamma-ray momentum becomes the total momentum of the electron and positron. This materialization is called pair production, since each time it happens an electron-positron pair is produced.

Furthermore, a fast electron or positron is able to produce gamma rays when it strikes matter, the process being similar to that involved in the creation of X-rays. Thus we have the picture of a continuous cascade of electrons, positrons, and gamma rays, as shown in Fig. 44.2. Starting from, say, a single gamma ray, we first get an electron-positron pair. Each of these produces several additional gamma rays. The new gamma rays form more electron-positron pairs, which in turn produce further gamma rays, and so on. Each generation of particles and gamma rays is more numerous than the one before it, though the individual energies are smaller. Finally, there are a large number of electrons, positrons, and gamma rays, but now they are too weak to produce more and are soon absorbed. The cascades of electrons, positrons, and gamma rays are called *soft showers*, because they are usually absorbed in the atmosphere tens of thousands of feet above the earth. *Hard showers* of muons, on the other hand, are not easily absorbed and so can penetrate to great depths in the earth. (Muons are the next-to-last stage in the decay of most mesons and hyperons.)

radiocarbon dating

When a cosmic-ray neutron is captured by a nitrogen atom in the atmosphere, the following nuclear reaction occurs:

$$_7N^{14} + {_0}n^1 \longrightarrow {_6}C^{14} + {_1}H^1$$

The newly formed carbon isotope C^{14} (called *radiocarbon*) is radioactive, and beta-decays to N^{14} with a half-life of 5,600 years. Although the radiocarbon on the earth decays steadily, the cosmic-ray bombardment constantly replenishes the supply. It is estimated that there is a total of about 90 tons of radiocarbon distributed throughout the earth at the present time.

Shortly after being produced in the atmosphere, a radiocarbon atom attaches itself to an oxygen molecule to form radioactive carbon dioxide. Green plants take in carbon dioxide in order to live, and so every plant contains radioactive carbon which it absorbed along with its ordinary intake of carbon dioxide. Animals eat plants and in so doing become radioactive themselves. Consequently, every living thing on earth is slightly radioactive because of absorbed radiocarbon. The mixing of radiocarbon is very efficient, hence living plants and animals all have the same proportion of radiocarbon to ordinary carbon (C^{12}).

After death, however, the remains of living things no longer ingest radiocarbon, and the radiocarbon they contain keeps decaying away to nitrogen. After 5,600 years, then, they have only half as much radiocarbon left—relative to their total carbon content—as they had as living matter, after 11,200 years only one-fourth as much, and so on. Accordingly, by determining the proportion of radiocarbon to ordinary carbon it is possible to evaluate the ages of ancient objects and remains of organic origin (Fig. 44.3). This elegant method permits the dating of archeological specimens such as mummies, wooden implements, cloth, leather (for instance, the leather wrappings of the Dead Sea Scrolls), charcoal from campfires, and similar remains of ancient civilizations as much as 40,000 years old.

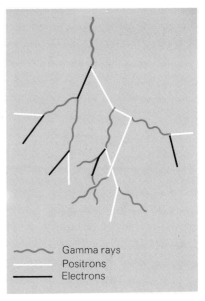

44.2 Schematic diagram of a cosmic-ray cascade of electrons, positrons, and gamma rays.

the proportion of radiocarbon in living matter decreases after death, which permits the remains to be dated

cosmic-ray origin

What have cosmic rays, interesting as they are, to do with the structure of the universe? A hint of their significance is the fact that about as much energy arrives on the earth in the form of cosmic rays as in the form of starlight. Now, cosmic rays do not come from the sun—if they did, cosmic-ray intensity would vary markedly between day and night, but, on the contrary, it is quite constant. Also, nobody has been able to account for a generating process on the sun capable of putting out billion-Gev energies. (To be sure, *some* low-energy particles are sent out by the sun that contribute to cosmic-ray fluctuations, but the bulk of the steady rain

the energy in cosmic rays is comparable with the energy in starlight

Time after death of animal or plant	C^{14} content of sample	C^{12} content of sample
0 years	C^{14} ●	C^{12}
5,600 years ($\frac{1}{2}$ of original C^{14} remains undecayed)	C^{14} ●	C^{12}
11,200 years ($\frac{1}{4}$ of original C^{14} remains undecayed)	C^{14} •	C^{12}
16,800 years ($\frac{1}{8}$ of original C^{14} remains undecayed)	C^{14} •	C^{12}

44.3 The principle of radiocarbon dating. *The radioactive C^{14} content of a sample of dead animal or plant tissue decreases steadily, while its C^{12} content remains constant. Hence the ratio of C^{14} to C^{12} contents indicates the time that has elapsed since the death of the organism.*

of cosmic radiation must come from elsewhere.) Hence cosmic rays must be characteristic of interstellar space, and, since they contain as much energy as that radiated by the stars, their generation must represent a process as basic to the universe as the existence of stars.

Let us narrow down the region in which cosmic rays can originate. Our galaxy rotates, and this rotation gives the solar system a speed through space of 140 mi/sec. If cosmic rays came from outside our galaxy, either from other galaxies or from intergalactic space, then more of them would strike the forward side of the earth than the backward side. (Similarly, if you run through the rain, more drops splash on your face than on the back of your head.) This would give a 24-hr variation—but no such variation exists. The conclusion, then, is that most cosmic rays come from within our galaxy, the Milky Way.

most cosmic rays originate in our galaxy

Many theories of cosmic-ray origin have been advanced, but only one has survived the ultimate test of comparison with experimental data. It is impossible to think of a way of providing primary cosmic rays with energies of 1 billion Gev or more in a single act, the way a bullet is given all its energy in a single explosion. Some kind of gradual process must be involved that accelerates cosmic rays little by little over a long period of time. But what kind of gradual process can function in the virtual void of galactic space?

The answer was provided by Enrico Fermi, the nuclear physicist who also designed the first nuclear reactor. Fermi noted that the galaxy contains immense clouds of dust and ionized hydrogen, many of which neither emit nor absorb light to a sufficient extent for their presence to appear as conspicuous galactic nebulae. Starlight passing through these clouds is partially *polarized,* a fact that astronomers interpret as indicating that the clouds contain weak magnetic fields. (In polarized light, the waves in any direction all lie in one plane, while in ordinary light the waves are randomly oriented. See Chap. 13.) Such magnetic fields would align dust grains in the clouds, and starlight scattered from grains roughly parallel to one another would be partially polarized. Fermi reasoned that cosmic-ray primaries traveling through space will, from time to time, collide with these magnetized clouds, and he showed mathematically that, on the average, the effect of the collisions would be to give additional energy to the primaries. These "collisions" are between the primaries and the magnetic fields, so the heavier nuclei in cosmic rays do not become smashed to pieces by striking particles of matter.

Fermi's hypothesis has been confirmed by studies of radio noise coming from various regions of the galaxy. How are these radio waves generated? The most reasonable explanation seems to be that some high-energy cosmic-ray primaries do, on occasion, collide with particles in the magnetized interstellar clouds. When this occurs, a shower of electrons and positrons results from the decay of the mesons created during the collision. When the electrons and positrons move within the magnetic field of the cloud they lose energy all the time, largely in the form of radio-frequency radiation. When this is worked out quantitatively, the notion that many of the radio signals reaching us from the depths of the galaxy come from interactions between cosmic rays and magnetized clouds seems attractive, though it is by no means proved.

As to the origin of the primary nuclei themselves, they probably receive their initial accelerations during the explosions of supernovae. Such explosions occur an estimated once per hundred years in our galaxy, which is apparently often enough to make up for the loss of primary cosmic rays through collisions with gas particles.

cosmic rays are accelerated by interactions with magnetic fields in the galaxy

galactic magnetic fields

The discovery that our galaxy and, presumably, others as well contain magnetic fields immediately struck the fancy of astronomers everywhere. It became clear that the very structures of spiral galaxies were closely

In order to measure its C^{14} content, the carbon in a sample of organic origin is usually converted into a gas such as carbon dioxide which is then used to fill a special beta-sensitive detector. An early stage in this process is the heating of the sample to remove its volatile constituents, leaving nearly pure carbon which is shown being removed from the tube in which the reduction took place. Next the sample will be further purified in an acid bath.

magnetic fields play a role in determining the structures of spiral galaxies

related to their magnetic fields: This galactic material does not expand outward partly because it is held together by magnetic fields whose lines of force lie along the arms of the galaxies. And, very recently, laboratory experiments with magnetized gaseous material have led to the creation of objects whose resemblance to spiral galaxies is uncanny. These laboratory objects exist for only a fraction of a second, and their method of production is not likely to have been duplicated during the formation of the universe, but it is hard to dismiss their similarities in content and form to astronomical bodies.

Another piece of evidence contributed by cosmic-ray research concerns the dimensions of our galaxy. Particles with energies of many Gev in energy are bent only to a small extent by the weak magnetic fields characteristic of galactic space. For particles to be trapped within the galaxy, as they must have been in order to acquire such fantastic energies, the size of the galaxy must be enormously greater than that of the thin disk of the Milky Way. Calculations indicate that it should be roughly spherical in shape and about 100,000 light-years in diameter—the same conclusion that we reached earlier by considering the distribution of globular clusters. Radio astronomy, which observes radio noise emanating from regions outside the Milky Way, further confirms the picture of our galaxy as a spherical, not flat, collection of stars, nebulae, dust, and gas isolated in space and held together by magnetic as well as gravitational forces.

GLOSSARY

Galaxies are aggregates of stars separated from other such aggregates by much greater distances than those between the member stars.

Globular clusters are roughly spherical assemblies of stars outside the plane of our galaxy but associated with it.

There are two general categories of stars, *Population I* and *Population II*. In our vicinity the stars in the plane of the galaxy that revolve about its center belong to Population I, while the members of the globular clusters belong to Population II.

Galactic nebulae are irregular masses of diffuse material within our galaxy.

Spiral galaxies are immense, spiral-shaped aggregates of stars which closely resemble our galaxy.

A *radio telescope* is an antenna, often located at the center of a metallic "dish," connected to a sensitive radio receiver in order to pick up radio waves from space.

Primary cosmic rays are high-speed atomic nuclei, largely protons, which move through the galaxy and frequently strike the earth's atmosphere.

Secondary cosmic rays consist of the atomic debris that results from the collisions of primary cosmic rays with the oxygen and nitrogen atoms present in the upper atmosphere.

Radiocarbon dating is a procedure for establishing the approximate age of once-living matter on the basis of the relative amount of radioactive carbon it contains.

EXERCISES

1. Compare the motions of (*a*) stars within a galaxy, and (*b*) the galaxies themselves, with the motions of molecules in a gas.

2. What are the chief differences between
   a. a globular cluster and a spiral nebula?
   b. a galactic nebula and a spiral nebula?

3. What is the evidence that the spiral nebulae are not a part of our galactic system?

4. What is the evidence that the spiral nebulae are galaxies of stars?

5. Suppose that you were trying to demonstrate the rotation of a spiral nebula spectroscopically. Which would you choose for your study, a nebula of which you could see the full-face view or one turned edgewise toward you? Why?

6. Give two observations that tend to favor the idea that our galaxy as a whole is spherical in shape. How can this be reconciled with the evidence that the Milky Way has the shape of a thin disk?

7. State as many differences as you can between Population I and Population II stars. Where is each type found?

8. If the earth's magnetic field were to disappear, what effect would this have on the distribution of cosmic rays over the earth?

9. What steps would you take to determine the age of an ancient piece of wood by radiocarbon dating?

10. State four different motions the earth undergoes in space.

11. Why are radio-astronomical studies of the distribution of hydrogen in the universe of greater interest than studies of the distribution of other elements?

12. Can you think of any experiments that would determine whether cosmic-ray primaries are bare nuclei or complete atoms?

13 What ultimately becomes of the protons and neutrons knocked out of oxygen and nitrogen atoms in the atmosphere by cosmic rays?

14 How would you use atomic-particle counters and lead blocks to distinguish hard showers of mesons from soft showers of electrons, positrons, and gamma rays?

15 How can the present theory of cosmic-ray origin account for the scarcity of energetic electrons in the primary cosmic radiation?

16 Experiments are under way to monitor the 21-cm radio waves emitted by neutral hydrogen in the universe to seek signals produced by intelligent beings on planets outside our solar system. Can you think of any reasons why this particular wavelength was chosen for monitoring?

17 Do radio telescopes actually magnify anything? If not, why are larger and larger ones being built? What would you suppose to be the limiting factors in the size of a radio telescope's dish-shaped antenna?

chapter 45

Evolution of the Universe

THE EXPANDING UNIVERSE

    red shifts

    the expanding universe

    quasars

EVOLUTION OF THE UNIVERSE

    theories of origin

    the primeval fireball

    origin of the elements

    origin of the earth

    other solar systems

We have now explored all the major provinces of physical science. In each province we have had always the same goal: to explain, to simplify, to see behind complex events an underlying order and consistency.

We have found, too, the same method of attack: the asking of a question inspired by curiosity or dissatisfaction with existing ideas, followed by careful observation and experiment, the formulation of hypotheses and laws, and eventually the combination of laws into more general laws and broad theories—with continual checks on each new generality by further observation. We have seen one part of the universe after another yield to this attack. Behind the apparent motions of the sun and planets we have found the orderly arrangement of the solar system; the infinitely complex behavior of ordinary matter we have interpreted in terms of elementary particles; from rocks and from the scrutiny of present landscapes we have read much of the earth's history; and, finally, from starlight, cosmic rays, and radio signals from space we have learned something about the earth's relationship to the rest of the universe.

In this final section we shall go further into speculation than we have before in order to provide a glimpse into some current ideas regarding the history of the universe and, in fact, into the very meaning of what we so glibly call *the universe*.

THE EXPANDING UNIVERSE

red shifts

There is a curious feature in the spectra of spiral galaxies that we have not yet mentioned. The lines in all but a few of these spectra are shifted toward the red end, the amount of the shift increasing with the distance of the galaxy from us. This displacement is illustrated in Fig. 45.1, which shows two of the absorption lines of calcium (indicated by arrows) in the spectra of several galaxies. Each galactic spectrum is shown between two comparison spectra, so that the shift of the two lines toward the red (right in this picture) for galaxies at different distances is clearly evident.

galactic red shifts

In Chap. 42 we saw that red shifts in stellar spectra result from motion away from the earth. If we also attribute galactic red shifts to the Doppler effect, we must conclude that all the galaxies in the universe (except the 19 in the local group in our own vicinity) are receding from us. From the extent of the red shifts the recession velocities can be computed, and the results are startling: several hundred kilometers per second for the nearer galaxies, over 200,000 km/sec for the farthest ones. By comparison, the speed of light—the maximum relative velocity anything can have—is 300,000 km/sec.

the farther away a galaxy is, the faster it is receding

If we plot the recession velocities of the galaxies shown in Fig. 45.1 versus their distances from the earth, as in Fig. 45.2, we find that these quantities are proportional. The greater the distance, the faster the galaxies are traveling. The speed increases by about 30 km/sec (nearly 19 mi/sec) per million light-years of distance. When this graph is extended to cover the still faster and more distant galaxies that have been studied, the experimental points fall on the same line. The proportionality between galactic speed and distance was discovered in 1929 by the astronomer Edwin Hubble and is known as *Hubble's law*. In equation form, Hubble's law is

hubble's law

$$V = HR$$

where V is the speed of recession, R is the distance, and H is about 30 (km/sec)/million light-years. Because the distances of the farthest galaxies are hard to measure, there is some uncertainty about the exact magnitude of Hubble's constant H.

No exceptions to Hubble's law have been found, which permits astronomers to use red shifts to establish distances for galaxies that cannot be determined in any other way. The largest red shifts that have been observed place the objects involved about 8 billion light-years away, which is therefore the minimum radius of the universe.

the expanding universe

It might seem at first as though our galaxy had some strange repulsion for all other galaxies, forcing them to move away from us with ever-increasing speed. But it is hardly probable that we occupy so important

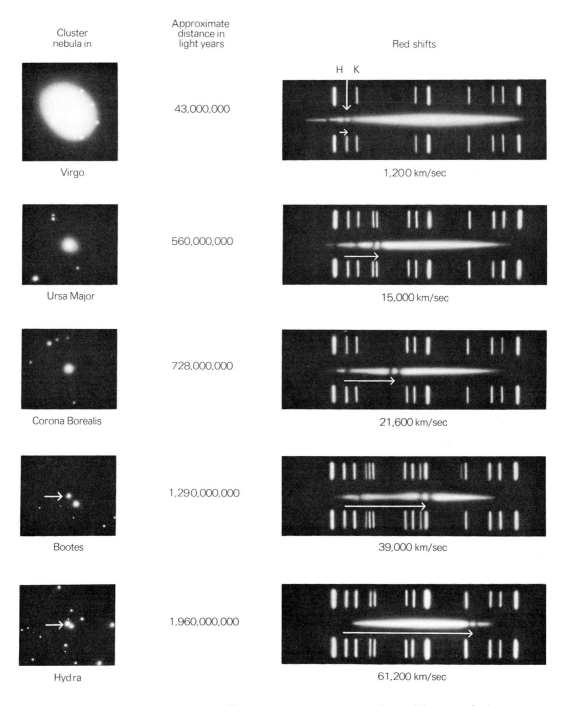

45.1 The red shift in the spectral lines of distant galaxies increases with increasing distance. *The lines marked H and K occur in the spectrum of calcium.*

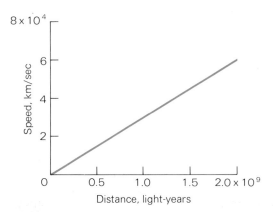

45.2 *Graph of recession speed versus distance for the galaxies of Fig. 45.1. The speed of recession averages 30 km/sec (19 mi/sec) per million light years, which corresponds to an age of 10 billion years for the universe.*

a place in the universe. If we make the more reasonable guess that all the galaxies are drifting away from one another, spreading apart like the fragments of a bursting rocket, an observer on our galaxy *or on any other* would get this illusion of his neighbors fleeing in all directions. The universe, in other words, seems to be rapidly expanding, its component galaxies moving ever farther apart (Fig. 45.3).

the entire universe is expanding

This leads immediately to a striking result. If the red shift means that the galaxies are flying apart, we might infer that at some time in the past they must have been very much closer together. We cannot know, of course, how long the present speeds of recession have been maintained, but, on the assumption that the speeds have not varied greatly, we can calculate backward to a time when the galaxies were close enough for their materials to intermingle.

We know from the value of Hubble's constant H that a galaxy that is now a million light-years away has a velocity of 30 km/sec. Since 1 light-year = 9.5×10^{15} m and 1 km/sec = 10^3 m/sec, we have

$R = 10^6$ light-years = 9.5×10^{21} m

$V = 30$ km/sec = 3×10^4 m/sec

The galaxy must therefore have started out at a time

$$t = \frac{R}{V} = \frac{9.5 \times 10^{21} \text{ m}}{3 \times 10^4 \text{ m/sec}}$$

$$= 3.2 \times 10^{17} \text{ sec}$$

the universe is approximately 10 billion years old

ago. There are 3.2×10^7 sec in a year, which makes the age of the universe 10^{10} years—10 billion years.

The latter age figure—which is only approximate since the value of H is not known with any great confidence—is double what had been accepted by astronomers as recently as two decades ago. The reason for the change is interesting, and is further evidence for the impossibility of ever taking anything for granted in science. In 1952 it became clear that there are actually two types of Cepheid variable stars, one type belonging to Population I and the other to Population II. The relationship

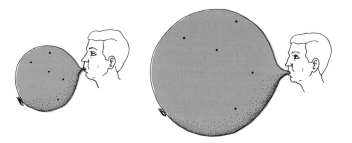

45.3 Two-dimensional analogy of the expanding universe. As the balloon is inflated, the spots on it become farther apart. A bug on the balloon would find that the farther away a spot is from his location, the faster it seems to be moving away from him. This is true no matter where the bug is.

between the intrinsic brightness of a Cepheid and the period of its variation that had been assumed to hold for *all* Cepheids turned out to be valid for Population II Cepheids only, while those in Population I are actually about four times more luminous for a given period (Fig. 43.3). This meant that the previously reckoned galactic distances which had been based on Population I Cepheids were all half as great as they should have been, thereby changing the red-shift calibration based upon these distances and doubling the size of the universe. The reasoning is straightforward: Because light spreads out more and more as it moves away from a source, the apparent brightness of a star drops inversely with the square of the distance from it. The Population I Cepheids each emit four times more light than had been thought, which means that they must be two times farther away to keep the ratio between intrinsic brightness and the square of the distance the same as before—the stars all shine just as they did before the discovery of two kinds of stellar populations, and the apparent brightness of the stars depends upon the latter ratio. Doubling the size of the universe means doubling its age as well, since the recession speeds determined from the red shifts in galactic spectra are not changed.

A few astronomers are still reluctant to believe in the expansion of the universe. There is no question of the reality of the red shifts, nor of the distance scale associated with them, but instead it is the interpretation of the origin of the shifts that has been disputed. For instance, it has been suggested that photons of light may somehow lose energy in their immense journeys through space, and, since the energy of a photon and its frequency are proportional ($E = hf$, where h is Planck's constant), a loss in photon energy means a shift to lower frequencies: a red shift. The trouble with this hypothesis is that no mechanism has ever been discovered by which a photon can lose energy while in flight, and in fact theoretical studies have shown such a mechanism to be impossible. Of course, existing theories may be incorrect and are probably incomplete, but thus far the combined electromagnetic and quantum theories of light have been among the most successful in all of physical science.

Another theory concerning the cause of red shifts has a better foundation, but is unlikely to have much relevance to the problem of distant nebulae. An important conclusion of the general theory of relativity is that photons of light are attracted by gravitational fields to a very small extent. Perhaps the most spectacular example of this attraction is the "bending" of starlight that passes close to the sun, an effect that has been verified by observations made during solar eclipses when the moon blocks out solar radiation. Recently extremely precise laboratory measurements

gravitational red shift

have shown that light is indeed affected by gravity in the predicted manner, further confirming Einstein's work of a half-century ago. Because of the gravitational attraction of the star from which it is emitted, a photon of light must lose some of its initial energy in escaping into space (see page 266). Thus spectral lines of stellar origin are all shifted toward the red, but the magnitude of the shift depends upon the ratio between the mass and radius of the star involved and is so small as to be imperceptible for most stars. Until now it has been detected only in the spectra of white dwarfs, whose immense densities lead to high mass/radius values. However, that this relativistic effect is responsible for the observed nebular red shifts is a far more unlikely assumption than that the shifts are due to the Doppler effect. For instance, all spiral nebulae would have to be composed of stars more dense than the densest white dwarfs known, which means discarding a number of basic laws of physics, and these densities would have to increase regularly with distance from the earth, a strange idea. The interpretation of the observed red shifts as a consequence of a general expansion of the universe still seems the only plausible one.

quasars

The most remarkable astronomical objects whose spectra show red shifts are *quasars*. In even the most powerful telescopes, quasars appear as mere points of light, exactly as stars do. But unlike stars, quasars have a prodigious output of radio waves; hence their name, a contraction of *quasi-stellar radio sources*. More than a hundred quasars have been identified, all with the same combination of properties: a large red shift; powerful emission of radio energy; extremely high intrinsic brightness; and both the light and radio outputs fluctuate markedly, sometimes over periods of only a few weeks.

The big question about quasars is where they are in the universe—far away or nearby. The evidence for their being far away depends upon their red shifts. The enormous quasar red shifts, in some cases corresponding to recession velocities of over 80 percent of the speed of light, suggest great distances from the earth, by analogy with the results for ordinary galaxies in the universe. This analogy is plausible because it interprets the quasar speeds as originating in the expansion of the universe.

But if the quasars are far away, at the very limits of the known universe, how can the radio waves we receive from them be so strong? Only a minute proportion of the sun's energy output is in the form of radio waves, and the sun is a typical star. To compound the puzzle, quasar emissions of light and radio waves vary too rapidly in intensity for them to be large in size. If a quasar were 100,000 light-years across, as our galaxy is, even a simultaneous change in its energy output (if it were possible) over a month's time would appear to us as spaced over 100,000 years. Hence the observed output fluctuations of several weeks' duration point to a quasar diameter of only a millionth that of our galaxy—but the energy output is 100 times greater.

Quasi-stellar radio sources (or "quasars"), such as the one in the center of this photograph, apparently emit as much energy in the form of radio waves as the total energy output of 10^8 to 10^9 suns. How quasars can produce so much energy is not known. The quasar shown is thought to be about 4 billion light-years away.

On the other hand, if quasars were close to us, they would not have to be emitting so much energy in order that strong radio signals arrive here. The same is true of the quasar luminosity, which could be more modest if they were nearby. However, there is now the problem of the red shifts. If they are supposed to arise from the Doppler effect, there are serious objections. First, there is no known process that can account for such high speeds if the quasars are near our galaxy where the recession speed due to the expansion of the universe is relatively low (Fig. 45.2). But nobody today can claim that everything about the universe is fully understood, and we are at liberty to hypothesize some sort of colossal explosion, involving an as-yet unknown mechanism, that caused quasars to fly away from nearby galaxies. The difficulty with this approach is that only *red* shifts are observed, corresponding to motion *away* from us, whereas we would expect on this basis at least some *blue* shifts, corresponding to motion *toward* us. No way around the absence of blue shifts has yet been proposed.

Another possibility is that the red shifts are not due to the Doppler effect at all, but instead to gravity. Though possible in principle, as discussed above, gravitational explanations for quasar red shifts have (thus far, anyway) proved unable to reconcile the small size and great mass required for the observed shifts with the other, equally conspicuous properties of quasars. The best guess today seems to be that quasars are indeed very far away and their red shifts are due to the expansion of the universe.

quasars are the most distant known objects in the universe if their red shifts follow hubble's law

A hint as to the nature of quasars comes from the discovery of other objects that seem to be intermediate between ordinary galaxies and quasars. For instance, there seem to be a large number of so-called *quasi-stellar galaxies*, which are similar to quasars except that their outputs of radio waves are too small to be detected. Less common are the *Seyfert galaxies* which have dense, very hot central regions whose brightness fluctuates over fairly short periods. Some Seyfert galaxies are also powerful radio sources. The main difference between quasars and Seyfert galaxies seems to be that the quasars have greater intrinsic brightnesses and that only their spectra show large red shifts. Quasars are thus important for two reasons: a novel energy-generating mechanism may be responsible for their behavior, and they may represent a link in the evolution of the universe.

EVOLUTION OF THE UNIVERSE

theories of origin

The picture of the origin of the universe that currently has the widest acceptance is the aptly named *big bang* theory. According to this theory, the universe began as a "primeval atom," an exceedingly hot and dense lump of matter that exploded violently 10 or so billion years ago. As the material spread out in space, it condensed by gravitational attraction into individual aggregates that became the galaxies. The evolution of the

the big bang theory

universe since the big bang seems in accord with all known physical principles, a strong point in its favor, and it accounts very simply for the observed expansion of the universe and Hubble's law.

A number of scientists are unhappy with the big bang theory, however, despite the lack of strong evidence against it. For one thing, it leaves unanswered the questions that always arise when a precise date is given for the creation of the universe: Why was all matter concentrated in one tiny region? What was the universe like before the explosion started? Where did the matter come from in the first place? And so on—profound questions that require some audacity even to think seriously about.

the cyclic theory is a modification of the big bang theory

One way out that has its attractive points holds that there did not have to be a single "origin" of the universe. If the cosmic explosion whose consequences we observe as the red shift of distant galaxies was not violent enough, the expansion of the universe may not continue forever. Instead, the universe may ultimately begin to contract as gravitational forces take over. As the contraction proceeds, galaxies and then individual stars will collide, and finally all of them will come together into a single huge mass at extremely high temperature. So high a temperature, in fact, that all the complex atoms will be broken down into hydrogen. Then another great explosion will occur, leading to an expansion of primordial matter that condenses into galaxies and stars as it speeds toward the outer reaches of space. Finally the expansion phase will cease, and the universe will collapse again. Calculations indicate a span of at least 30 billion years for each complete cycle, so we might now be about a third of the way through an expansion.

The cyclic theory of the universe does not require that the universe "began" at some single instant; it says that, given the 10^{50} or more kilograms of the known universe, existing laws of physics can account for a succession of expansions and contractions extending infinitely far into the past and infinitely far into the future. We are left with but a single question: Where did the matter come from?

the steady-state theory

The British astronomers Hoyle, Gold, and Bondi have proposed a rather different hypothesis. Their so-called *steady-state* universe is one in which the galaxies were never any closer together than they are now and never will be farther away. According to this theory, new galaxies are coming into being continually at a rate just enough to counterbalance the thinning effect of expansion. Thus the average density of matter in the universe remains constant despite the spreading out of the existing galaxies; the necessary matter for the new galaxies is supplied, in this view, by the spontaneous and continuous creation of matter in free space. The universe is in a state of equilibrium, according to Hoyle, Gold, and Bondi: The galaxies move apart all the time, and new ones form which draw their substance from newly created matter. The two processes proceed simultaneously, and we cannot speak of either the beginning or the end of the universe.

This bold and unconventional approach to cosmology has a number of features to recommend it, particularly the philosophical conception of the universe as extending unbounded in space and time, always evolving yet always the same. But many scientists find the notion of the continuous creation of matter absurd, and recent evidence definitely points away from it. Much experiment and observation remain before we can draw any really firm conclusions about the history of the universe.

the primeval fireball

According to the big bang theory, the universe started out as a compact, intensely hot aggregate of matter. Such a body of matter must radiate energy in the form of electromagnetic waves—exactly as a hot object such as a poker thrust in a fire does. The higher the temperature of the object, the shorter the wavelength of the most intense part of its spectrum (see Fig. 42.1). In the case of the primeval fireball, the radiation must have been chiefly in the form of X- and gamma rays of short wavelength and hence high energy. As the universe expanded, the radiation that was emitted rapidly decreased in intensity and energy. However, the original radiation should have continued to spread out with the rest of the universe, so that even today remnants of it must be everywhere.

Owing to the expansion of the universe, an observer today would expect to find these remnants to have undergone a red shift to long wavelengths, in the range of radio waves. The radiation would not be easy to find,

Radio waves thought to have originated in the primeval fireball were first detected with a sensitive receiver attached to this 50-ft-long antenna at the Bell Telephone Laboratories research facility at Holmdel, New Jersey.

since it would be very weak, but, according to theory, it ought to have two distinctive characteristics that would permit it to be identified: the radiation should come equally strongly from all directions, and its spectrum should be the same as that which an object at about $3°K$ would radiate.

Remarkably enough, radiation of exactly this kind has been discovered; in fact, it apparently is one of the sources of the "snow" on television screens. The universe is bathed in a sea of radio waves whose ultimate origin seems to have been a primeval fireball. The existence of this radiation is strong support for the big bang theory of the origin of the universe, and it also rules out the steady-state theory, which denies that a primeval fireball could have existed. What is perhaps most exciting about the discovery is that it takes the question of the early history of the universe out of the realm of speculation and into the realm of measurement.

remnants of the radiation from the primeval fireball constitute a sea of radio waves that pervade the universe

origin of the elements

Let us now turn to the possible ways in which the matter of which the universe is composed came into being. By far the most abundant element is hydrogen, whose atoms are the simplest possible, but atoms with hundreds of protons and neutrons in their nuclei do exist and must be accounted for. Is there any simple way in which the various elements can have originated?

Our first guess might be that the complex elements were all built up from hydrogen alone with the help of appropriate nuclear reactions. We saw in Chap. 42 how helium is formed from hydrogen in the sun and other stars. On the basis of other known nuclear reactions we can readily construct a scheme whereby, step by step, primordial hydrogen could have yielded almost all the known elements and their isotopes. In the laboratory we must go to a considerable amount of trouble to cause these reactions to occur, but conceivably the conditions of immense heat and pressure that we might suppose to have existed in the dawn of the universe just as it began expanding would have made the reactions possible. This idea becomes even more attractive when we reflect that, given enough pressure, the electrons of the hydrogen atoms can collapse into the protons to form neutrons. Nuclear reactions involving neutrons occur much more readily than those involving protons, which, being charged, are repelled electrically by other nuclei. George Gamow, who pioneered this theory, felt that all the elements could have been formed in a few minutes in the primeval fireball by this mechanism. Speed, of course, would have been necessary so that the reactions would all have occurred before the expansion progressed too far.

the elements could not all have been formed in the primeval fireball

Unfortunately there are no atoms of mass number 5 or mass number 8 that are stable, so it is hard to see how Gamow's theory can account for elements heavier than helium. At first this objection did not seem too serious, since the theory has so many desirable features—for instance, it predicts the relative abundance of the elements with some accuracy—but time has shown it to be crucial. And recent work has shown that the abundance of the various elements is markedly different in different stars, so different that it is hard to believe that the matter in these stars was the same at the beginning of their histories. We must look elsewhere for the origin of the elements.

We have already discussed the synthesis of helium from hydrogen that is currently going on in most stars. It is hardly a major jump to ask whether it is possible that complex elements may be formed within stellar interiors, with plenty of time available instead of only several minutes. Here we have better luck. Given enough time and the variety of physical states that a star passes through during its evolution, we can conceive of sequences of reactions that could lead to all the elements known. Not, to be sure, so simple and elegant a hypothesis as that based exclusively on successive neutron captures, but one that contains no internal contradictions. For example, we can get around the absence of stable mass numbers 5 and 8 by invoking the direct combination of three helium nuclei to form carbon, $3He^4 \longrightarrow C^{12}$—a most uncommon reaction, but one which, with billions of years available, occurs often enough. To obtain the light elements between helium and carbon is not a serious problem, for they may well be "chips" resulting from the bombardment of heavy atoms with the fast protons of hot stellar interiors. In this manner we can just manage to account for the presence of all the elements.

the elements were probably formed in stellar interiors

Clearly, the origin of the elements does not provide us with a definite yardstick with which to measure the plausibility of theories of the evolution of the universe. However, it is reassuring to know that no physical processes as yet unverified need be invoked in order to explain the existence of the complex atoms found in matter.

origin of the earth

Where does our humble earth fit into the grandiose universe we have been discussing? Insignificant though it is in terms of the cosmos, it is nevertheless our home and therefore of considerable interest to us. Although nothing like a complete picture of the origin of the earth has been devised, some idea of the probable sequence of events in its formation is available.

The story begins with a local eddy in the swirling gas and dust of the primordial galaxy. This eddy in time condensed through the action of gravity into a cloud stable enough to resist disruption. The cloud, the so-called *protosun*, was somewhat farther across than the present solar system. Eventually it began to contract, taking perhaps 80 million years to make the transition from gas cloud to star, and, in the course of the contraction, some of the gas—about 10 percent—remained behind to form a diffuse spherical nebula about the nascent sun.

protosun

In time the nebula cooled and flattened into a thin disk. In doing so it increased in density and ultimately reached the point at which mutual gravitational forces within the disk became as great as the tide-producing forces exerted by the sun. This created an unstable situation, which

protoplanets resolved itself when the flattened nebula broke up into separate smaller clouds. These were the *protoplanets* from which the planets of the present developed. Had the nebular density been less than it was, many tiny planets would have resulted; if it had been much greater one or more small stars would have been formed as companions to the sun.

While the protoplanets were coming into existence the sun was still dark, not yet having completed the shrinkage that, by increasing the pressure and temperature of its interior, would permit it to evolve energy and become luminous. The protoplanets were huge affairs, protoearth, for instance, being 500 times more massive than and 2,000 times as far across as the current earth. With the sun in darkness the heavier constituents of the protoplanets migrated inward, collecting in the center of each protoplanet to form a heavy core surrounded by lighter gases. At this time the satellites got their start. In the case of protoearth, for some reason there was only a single large secondary body (see Chap. 41).

Then the sun began to shine, and the peaceful evolution of the protoplanets reached a more dramatic phase. In addition to the light it emits, the sun puts out an intense flux of fast, ionized particles. Today these particles lead to auroral displays and magnetic disturbances on the earth and the deflection of comet tails away from the sun, among other effects; during the early stages of planetary evolution they may have swept the solar system free of the remnant nebular gas that pervaded interplanetary space. Meanwhile the cold protoplanets grew warm, and their own envelopes of gas and vapor began to boil away. The earth and the other planets nearest the sun suffered an immense diminution in their sizes and weights, since the light hydrogen and helium of which they were largely composed could escape easily. Finally, after the greater part of a billion years, the relatively bare, shrunken planets of today emerged permanently from the mists that enshrouded them.

Meanwhile, the protoplanet cores were proceeding to consolidate themselves. In the case of the earth this led to pronounced heating, due partly to radioactivity. Ultimately the entire earth melted, and the iron and silicate components separated to form the core and mantle, respectively. Then, gravitational energy exhausted, the molten earth began to solidify, and, with the development of the crust and its continents, a recognizable ancestor of our present planet appeared. This occurred $4\frac{1}{2}$ billion years ago, completing a process that had begun a billion or so years earlier.

the future of the solar system What of the future course of the earth? As we mentioned in Chap. 43, it seems quite likely the next few billion years will see the sun swell into a luminous giant, as large around as the orbit of Venus and emitting a hundred times more radiation than it does now. The earth will grow warmer, and ultimately the oceans will reach the boiling point. Steam will fill the atmosphere, and all life will perish except, conceivably, some exceedingly resistant spores. This situation will last perhaps a billion years, until the sun begins to decline in an evolutionary process that will end with it a feeble white dwarf. The steam will condense into new oceans as the earth cools, terminating in a globe completely covered with ice and snow. Doubtless a little volcanic activity will still persist, but the ultimate picture will be one of lifeless desolation as darkness draws near.

other solar systems

The modern view of the origin of the solar system is that it is a natural aspect of the evolution of the sun. Since our sun is by no means an exceptional star in any respect it is reasonable to suspect that other stars are also attended by planetary systems.

there are probably a great many other planetary systems in the universe

The above suspicion is reinforced by the observation that, while the majority of stars for which appropriate Doppler-shift data are available rotate rapidly on their axes, some—about 10 percent—spin rather slowly. The sun is one of the latter, requiring four weeks for each complete rotation. The reason the sun turns so slowly is that almost all the angular momentum (the rotational equivalent of linear momentum) of the solar system is concentrated in the planets, with the sun having less than 1 percent of the total. It is therefore hard to escape the conjecture that the stars that rotate slowly do so because they are accompanied by swarms of planets which share the angular momentum available. If this conjecture is correct, over a billion stars in our galaxy alone should have planetary systems of some kind (one other star has already been found to have planets circling it), and many of them should resemble our own solar system. Of these hundreds of millions of planets similar to the ones that circle the sun, a certain proportion surely meet, in the words of Harlow Shapley, "the happy requirements of suitable distance from the star, near-circular orbit, proper mass, salubrious atmosphere, and reasonable rotation period—all of which are necessary for life as we know it on earth." True, the combination of immense distances between stars and the impossibility of attaining speeds faster than that of light makes it unlikely that the hypothesis of life on other worlds will be directly verified in the near future, but no serious arguments that dispute this hypothesis have been advanced. It seems probable that we are not alone in the universe.

GLOSSARY

The term *expanding universe* refers to the fact that all the galaxies visible to us seem to be rapidly receding from one another. The evidence for the recession is the observed *red shift* in galactic spectra, which is a shift of their spectral lines toward the red end of the spectrum that is interpreted as a Doppler effect.

A *quasar* is a strong source of radio waves whose spectral lines show large red-shifts and whose radiation fluctuates markedly. Quasars appear as points of light in a telescope, and may be tiny, extremely bright objects of unknown nature at the limits of the observable universe.

According to the *big bang* theory, the universe originated in the explosion of a compact mass of matter 10 billion years ago. According to the *cyclic theory*, the universe alternately expands and contracts. According to the *steady-state theory*, matter is continually being created everywhere in the universe.

The *protosun* was the original cloud of dust and gas that evolved into our sun; the *protoplanets* were the smaller clouds of dust and gas that remained after the contraction of the protosun and eventually became the planets.

EXERCISES

1 Outline the chain of reasoning that led to the picture of our universe as currently undergoing an expansion.

2 What are the objections to the theory that the elements all came into being in a short time during the creation of the universe? Can you think of any objections to the theory that the elements were created in stellar interiors exclusively?

3 How do the expansion, expansion-contraction, and continuous-creation theories of the universe explain the red shift?

4 On the basis of the latest theory of the origin of the solar system, why is it reasonable to suppose that many other stars have planetary systems circling them?

5 Is it possible that present estimates of the "age" of the universe may have to be modified in the future? What sort of evidence might prompt such a modification?

6 Why is it extremely unlikely that the Compton effect is responsible for observed nebular red shifts?

Mathematics Refresher

A modest degree of familiarity with basic mathematics is necessary to appreciate much of physical science. This review is included primarily to assist those readers whose mathematical skills have become rusty, but it is sufficiently self-contained to introduce such useful ideas as powers-of-ten notation to those who have not been exposed to them elsewhere.

ALGEBRA

Algebra is the arithmetic of symbols that represent numbers. Instead of being restricted to relationships among specific numbers, algebra can express more general relationships between quantities whose numerical values need not be known. To give an example, in the theory of relativity it is shown that the "rest energy" of any object—that is, the energy it possesses by virtue of its mass alone—is

$$E = mc^2$$

What this formula does is give a prescription for calculating the rest energy E in terms of mass m and speed of light c. The formula is not restricted to a particular object, but can be applied to any object whose mass is known. What is being expressed is the manner in which rest energy E varies with mass m *in general*; if we are told merely that the rest energy E of some object is 5 joules, we do not know upon what factors E depends or precisely how the value of E varies with those factors. (The joule is a unit of energy widely used in physics; it is equal to 1 kg·m^2/sec^2.) The quantities E and m are *variables*, since they have no fixed values. On the other hand, c^2 is a *constant*, since it is the square of the speed of light c and c is a constant of nature whose value—299,792,800 m/sec, or about 186,000 mi/sec—is the same everywhere in the universe. Thus the formula $E = mc^2$ tells us, in an extremely simple and straightforward way, that the rest energy of an object varies only with its mass and also how to go about obtaining a numerical value for E if we are given the mass m of a particular body.

The convenience of the algebraic approach in science is enhanced by the use of standard symbols for most constants of nature: c always represents the speed of light, π always represents the ratio between the circumference and diameter of a circle, e always represents the electric charge of the electron, and so on.

Before going any further, it is worth reviewing how the arithmetical operations of addition, subtraction, multiplication, and division are expressed in algebra. Addition and subtraction are straightforward:

$$x + y = a$$

means that we obtain the sum a by adding the two quantities x and y together, while

$$x - y = b$$

means that we obtain the difference b when quantity y is subtracted from quantity x.

In algebraic multiplication, no special sign is ordinarily used, and the symbols of the quantities to be multiplied are merely written together. Thus

$$xy = c \text{ means } x \times y = c$$

and

$$xyz = d \text{ means } x \times y \times z = d$$

When the quantity x is to be divided by y to yield the quotient e, we would write

$$\frac{x}{y} = e$$

which can also be expressed in the form

$$x/y = e$$

whose meaning is the same.

If several operations are to be performed in a certain order, parentheses and brackets are used to indicate this order. Thus

$$3\left[\frac{(x-y)z}{w}\right] + v = k$$

means that we are to carry out the following sequence of operations with the numbers represented by v, w, x, y, and z to obtain the value of k:

1 Subtract y from x to give $(x - y)$.
2 Multiply $(x - y)$ by z to give $(x - y)z$.
3 Divide $(x - y)z$ by w to give $[(x - y)z/w]$.
4 Multiply $[(x - y)z/w]$ by 3 to give $3[(x - y)z/w]$.
5 Add v to $3[(x - y)z/w]$ to give k.

If we keep in mind that the rules of algebra are the same as those of arithmetic, seemingly complex expressions turn out to be entirely straightforward. Several examples of the products of sums and differences underline this point:

$$(a + b)(c + d) = a(c + d) + b(c + d) = ac + ad + bc + bd$$
$$(a - b)(c + d) = a(c + d) - b(c + d) = ac + ad - bc - bd$$
$$(a + b)^2 = (a + b)(a + b) = a(a + b) + b(a + b)$$
$$= a^2 + ab + ba + b^2 = a^2 + 2ab + b^2$$

exercise set I

Carry out the indicated operations. Answers for all the exercises are given at the end of the *Mathematics Refresher*.

MATHEMATICS REFRESHER

1 $(a + b)c =$
2 $(a - b)c =$
3 $(a + b)/c =$
4 $(a + b)/(a + b) =$
5 $(a + b)^2/(a + b) =$
6 $(a + b)(a - b) =$
7 $(a + b)(a + c)/a =$
8 $(a - b)(a + b^2 - c) =$
9 $(a + b + c)(a - b - c) =$
10 $(a - b)^2 =$

POSITIVE AND NEGATIVE QUANTITIES

In the physical sciences some quantities may have either positive or negative values. It is therefore important to keep in mind the rules for multiplying and dividing positive and negative numbers:

1 A positive number times a positive number yields a positive number. For example, $(+5) \times (+3) = +15$.
2 A negative number times a positive number yields a negative number. For example, $(-5) \times (+3) = -15$.
3 A negative number times a negative number yields a positive number. For example, $(-5) \times (-3) = +15$.

Similar rules hold for division, so that

$$\frac{+28}{+7} = +4$$

$$\frac{+28}{-7} = -4 \quad \text{and} \quad \frac{-28}{+7} = -4$$

$$\frac{-28}{-7} = +4$$

An example of the use of positive and negative quantities occurs in physics, where there are two kinds of electric charge, one of which is designated positive and the other negative. The force F that a charge q_1 exerts on another charge q_2 a distance r away is given by Coulomb's law as

$$F = k\frac{q_1 q_2}{r^2}$$

where k is a universal constant of nature. By convention, a positive value of F signifies a repulsion between the charges—the force tends to push q_1 and q_2 apart. A negative value of F signifies an attraction between the charges—the force tends to pull q_1 and q_2 together. Evidently a positive (= repulsive) force acts when *either* both charges are + *or* both are −: "like charges repel." When one charge is + and the other one −, the force is negative (= attractive): "opposite charges attract." Both of the above experimental observations regarding the types of force that occur together with the way in which the precise strength of F varies with the magnitudes of q_1 and q_2 and with their separation r are included in the simple formula $F = kq_1 q_2/r^2$.

To help clarify the various rules mentioned above, let us find the numerical value of a in the formula

$$a = \frac{x(y - 2z)}{x + y} - 8x$$

when $x = -2$, $y = \frac{1}{2}$, and $z = 4$. We begin by substituting the given values of x, y, and z in the formula, and then proceed to carry out the indicated operations:

$$a = \frac{-2\left(\frac{1}{2} - 2 \times 4\right)}{-2 + \frac{1}{2}} - 8 \times (-2)$$

$$= \frac{-2\left(\frac{1}{2} - 8\right)}{-1\frac{1}{2}} + 16$$

$$= -\frac{15}{1\frac{1}{2}} + 16$$

$$= -10 + 16$$

$$= 6$$

exercise set II

Evaluate the following:

1 When $x = 5$, $y = -2$, $\dfrac{3(x + y)}{2} =$

2 When $x = 3$, $y = 2$, $\dfrac{1}{x - y} - \dfrac{1}{x + y} =$

3 When $x = 1$, $y = -2$, $\dfrac{4xy}{y + 3x} + 5 =$

4 When $x = -2$, $y = 2$, $z = 4$, $\dfrac{x^2 + y^2}{2z} + \dfrac{z}{x - y} =$

5 When $x = 2$, $y = 8$, $z = 10$, $\dfrac{(x + z)^2}{y} + \sqrt{xy} =$

EXPONENTS

There is a convenient shorthand way to express a quantity that is to be multiplied by itself one or more times. In this scheme a superscript number called an *exponent* is used to indicate how many times the multiplication is to be carried out, as follows:

MATHEMATICS REFRESHER

$$a = a^1$$
$$a \times a = a^2$$
$$a \times a \times a = a^3$$
$$a \times a \times a \times a = a^4$$

and so on. The quantity a^2 is read as "a squared" because it is equal to the area of a square whose sides are a long, and a^3 is read as "a cubed" because it is equal to the volume of a cube whose edges are a long. Past an exponent of 3 we read a^n as "a to the nth power," so that a^5 is "a to the fifth power."

Suppose we have a quantity raised to some power, say a^n, that is to be multiplied by the same quantity raised to another power, say a^m. In this event the result is that quantity raised to a power equal to the sum of the original exponents:

$$a^n \times a^m = a^n a^m = a^{n+m}$$

To convince ourselves that this is true, we can work out $a^3 \times a^4$:

$$(a \times a \times a) \times (a \times a \times a \times a) = a \times a \times a \times a \times a \times a \times a$$
$$a^3 a^4 = a^7$$

Because the process of multiplication is basically one of repeated addition,

$$(a^n)^m = a^{nm}$$

where $(a^n)^m$ means that a^n is to be multiplied by itself the number of times indicated by the exponent m. Thus

$$(a^2)^4 = a^{2 \times 4} = a^8$$

because

$$(a^2)^4 = a^2 \times a^2 \times a^2 \times a^2 = a^{2+2+2+2} = a^8$$

Reciprocal quantities are expressed according to the above scheme but with negative exponents:

$$\frac{1}{a} = a^{-1}$$
$$\frac{1}{a^2} = a^{-2}$$
$$\frac{1}{a^3} = a^{-3}$$
$$\frac{1}{a^4} = a^{-4}$$

and so on. Exactly the same rules as before hold for negative exponents and for mixtures of negative and positive exponents. For instance,

$$a^5 a^{-3} = a^{5-3} = a^2$$

which may be easier to understand if we write it out as

$$a^5 a^{-3} = \frac{a^5}{a^3} = \frac{a \times a \times a \times a \times a}{a \times a \times a} = a \times a = a^2$$

A few further examples of the use of negative exponents may be helpful:

$$a^2 a^{-3} = a^{2-3} = a^{-1}$$
$$a^{-1} a^{-4} = a^{-1-4} = a^{-5}$$
$$(a^{-2})^4 = a^{-2 \times 4} = a^{-8}$$
$$(a^{-3})^{-2} = a^{-3 \times (-2)} = a^6$$

A quantity raised to the zeroth power, a^0 for instance, is always equal to 1. To verify this important statement, we note that

$$\frac{a}{a} = 1$$

can also be written $\frac{a}{a} = aa^{-1} = a^{1-1} = a^0$ so that it must be true that $a^0 = 1$.

It is not necessary that an exponent be a whole number. A fractional exponent is used to indicate a "root" of a quantity. The "square root" of a, familiarly written $\sqrt{a}$, is that quantity which, multiplied by itself, is equal to a:

$$\sqrt{a} \times \sqrt{a} = a$$

In terms of exponents we would write the square root of a as

$$\sqrt{a} = a^{1/2}$$

because

$$a^{1/2} \times a^{1/2} = (a^{1/2})^2 = a^{2 \times 1/2} = a^1 = a$$

In a similar way the "cube root" of a, which is $\sqrt[3]{a}$, is indicated by the exponent $\frac{1}{3}$ since

$$a^{1/3} \times a^{1/3} \times a^{1/3} = (a^{1/3})^3 = a^1 = a$$

In general, the nth root of any quantity is indicated by the exponent $1/n$:

$$\sqrt[n]{a} = a^{1/n}$$

A few examples will indicate how fractional exponents fit into the general pattern of exponential notation:

$(a^6)^{1/2} = a^{(1/2)\times 6} = a^3$

$(a^{1/2})^6 = a^{6\times 1/2} = a^3$

$(a^3)^{-1/3} = a^{(-1/3)\times 3} = a^{-1}$

$a^6 a^{1/2} = a^{6+1/2} = a^{6\,1/2}$

exercise set III

Evaluate the following:

1. $a^2 a^5 =$
2. $a^{3/2} a^{1/2} =$
3. $a^{1/2} a^{1/2} a^{1/2} =$
4. $a a^{1/3} =$
5. $a^4 a^{-2} =$
6. $a^6 a^{-1/2} =$
7. $a^{14} a^{-14} =$
8. $\dfrac{a^6}{a^2} =$
9. $\dfrac{a^2}{a^6} =$
10. $\dfrac{a}{a^{1/2}} =$
11. $\dfrac{a^2}{a^{-1/2}} =$
12. $\dfrac{a^5 a^{-2}}{a^8} =$
13. $(a^4)^{1/2} =$
14. $(a^4)^{-1/2} =$
15. $(a^{-4})^{1/3} =$
16. $a^5 + a^{1/2} =$
17. $a^2 + a^2 =$
18. $a^2 - a^2 =$

EQUATIONS

An equation is a statement of equality: whatever is on the left-hand side of any equation is equal to whatever is on the right-hand side. An example of an arithmetical equation is

$3 \times 9 + 8 = 35$

since it contains only numbers, and an example of an algebraic equation is

$5x - 10 = 20$

since it contains a symbol as well as numbers. To "solve" an equation for a quantity means to perform whatever algebraic operations—addition, subtraction, multiplication, etc.—are necessary to isolate that quantity on one side of the equation. If we are to solve $5x - 10 = 20$ for x, we must perform the following operations.

$$5x - 10 = 20$$

Add 10 to both sides: $5x - 10 + 10 = 20 + 10$

$$5x = 30$$

Divide both sides by 5: $\dfrac{5x}{5} = \dfrac{30}{5}$

$$x = 6$$

The solution of the equation is $x = 6$, since only when x has that value is the equation true.

Sometimes it is necessary to solve an equation that contains several different symbols for one of them. Let us solve for v the equation

$$m = \frac{m_0}{\sqrt{1 - v^2/c^2}}$$

which gives the mass m of an object moving at the speed v in terms of its mass measured at rest m_0 and the speed of light c. (This famous equation is one of the results of the theory of relativity and is discussed in Chap. 5.) We proceed as follows, listing only the results of each operation:

$$m = \frac{m_0}{\sqrt{1 - v^2/c^2}}$$

Multiply both sides by $\sqrt{1 - v^2/c^2}$: $\quad m\sqrt{1 - v^2/c^2} = m_0$

Square both sides: $\quad m^2(1 - v^2/c^2) = m_0^2$

Divide both sides by m^2: $\quad 1 - \dfrac{v^2}{c^2} = \dfrac{m_0^2}{m^2}$

Add v^2/c^2 to both sides and reverse the equation:

$$\frac{m_0^2}{m^2} + \frac{v^2}{c^2} = 1$$

Subtract m_0^2/m^2 from both sides: $\quad \dfrac{v^2}{c^2} = 1 - \dfrac{m_0^2}{m^2}$

Multiply both sides by c^2: $\quad v^2 = c^2(1 - m_0^2/m^2)$

Take the square root of both sides: $\quad v = c\sqrt{1 - m_0^2/m^2}$

In working out the above examples, we have been employing the basic rule that any operation performed on one side of an equation must be performed on the other. Thus an equation remains valid when

1. The same quantity is added to both sides.
2. The same quantity is subtracted from both sides.
3. The same quantity multiplies both sides.
4. The same quantity divides both sides.
5. Both sides are raised to the same power (that is, each side is multiplied by itself the same number of times).
6. The same root of each side is taken (for instance, the square root).

exercise set IV

Solve each of the following equations for x:

MATHEMATICS REFRESHER

1 $\dfrac{4x - 35}{3} = 9(1 - x)$ **2** $\dfrac{3x - 42}{9} = 2(7 - x)$

3 $x^3 + 27 = 0$ **4** $2x^4 - 32 = 0$

5 $3x^2 = 6x$ **6** $(x + 3)(x - y) = z^2 + x^2$

7 $y\sqrt{2x} = 12$ **8** $z = \dfrac{x + y}{x - y}$

9 $\dfrac{x}{y} = \dfrac{4}{z}$ **10** $\dfrac{y}{x} = \dfrac{x}{5}$

POWERS OF TEN

There is a convenient and widely used method for expressing very large and very small numbers that makes use of powers of 10. Any number in decimal form can be written as a number between 1 and 10 multiplied by some power of 10, a positive power for numbers larger than 10 and a negative power for numbers smaller than 1. Positive powers of ten follow this pattern:

$10^0 = 1$ = 1 with decimal point moved 0 places
$10^1 = 10$ = 1 with decimal point moved 1 place to the right
$10^2 = 100$ = 1 with decimal point moved 2 places to the right
$10^3 = 1,000$ = 1 with decimal point moved 3 places to the right
$10^4 = 10,000$ = 1 with decimal point moved 4 places to the right
$10^5 = 100,000$ = 1 with decimal point moved 5 places to the right
$10^6 = 1,000,000$ = 1 with decimal point moved 6 places to the right

and so on. The exponent of 10 in each case indicates the number of places through which the decimal point is moved to the right from 1.00000 . . . ; or, equivalently, the exponent gives the number of zeroes that follow the 1.

Negative powers of ten follow a similar pattern:

$10^0 =$ 1 = 1 with decimal point moved 0 places
$10^{-1} =$ 0.1 = 1 with decimal point moved 1 place to the left
$10^{-2} =$ 0.01 = 1 with decimal point moved 2 places to the left
$10^{-3} =$ 0.001 = 1 with decimal point moved 3 places to the left
$10^{-4} =$ $0.000,1$ = 1 with decimal point moved 4 places to the left
$10^{-5} =$ $0.000,01$ = 1 with decimal point moved 5 places to the left
$10^{-6} =$ $0.000,001$ = 1 with decimal point moved 6 places to the left

and so on. Here the exponent of 10 in each case indicates the number of places through which the decimal point is moved to the left from 1; the number of zeros between the decimal point and the 1 is one less than the exponent, that is, $n - 1$.

Some examples of powers-of-ten notation follow:

$$8,000 = 8 \times 1,000 = 8 \times 10^3$$

$$347 = 3.47 \times 100 = 3.47 \times 10^2$$
$$8,700,000 = 8.7 \times 1,000,000 = 8.7 \times 10^6$$
$$0.22 = 2.2 \times 0.1 = 2.2 \times 10^{-1}$$
$$0.000,035 = 3.5 \times 0.000,01 = 3.5 \times 10^{-5}$$

An advantage of powers-of-ten notation is that it makes calculations involving large and small numbers easier to carry out. The rules for manipulating exponents that were reviewed in a previous section hold for exponents of 10, and so we have here

Multiplication: $\qquad 10^n \times 10^m = 10^{n+m}$

Division: $\qquad \dfrac{10^n}{10^m} = 10^{n-m}$

Raising to power: $\qquad (10^n)^m = 10^{nm}$

Taking a root: $\qquad (10^n)^{1/m} = 10^{n/m}$

An example will show how a calculation involving powers of ten is worked out:

$$\frac{460 \times 0.000,03 \times 100,000}{9,000 \times 0.006,2} = \frac{(4.6 \times 10^2) \times (3 \times 10^{-5}) \times (10^5)}{(9 \times 10^3) \times (6.2 \times 10^{-3})}$$
$$= \frac{4.6 \times 3}{9 \times 6.2} \times \frac{10^2 \times 10^{-5} \times 10^5}{10^3 \times 10^{-3}}$$
$$= 0.25 \times \frac{10^{2-5+5}}{10^{3-3}} = 0.25 \times \frac{10^2}{10^0}$$
$$= 25$$

Another virtue of this notation is that it permits the accuracy with which a quantity is known to be expressed unambiguously. The speed of light in free space c is often given as simply 3×10^8 m/sec. If c were written out as 300,000,000 m/sec we might be tempted to think it is precisely equal to this number, right down to the last zero. Actually, the most accurate figure for the speed of light is given by experiment as 299,792,800 m/sec, with the last three digits uncertain: they are somewhere between 750 and 850. For our purposes we do not need this much detail, and by writing just $c = 3 \times 10^8$ we automatically indicate both how large the number is (the 10^8 tells how many decimal places are present) and how precise the quoted figure is (the single digit 3 means that c is closer to 3×10^8 than it is to either 2×10^8 or 4×10^8 m/sec). If we wanted more precision, we could write $c = 2.998 \times 10^8$ m/sec, and again how large c is and how precise the quoted figure is are both obvious at a glance. To be sure, sometimes one or more zeros in a number are meaningful in their own right, and not solely decimal-point indicators. In the case of the speed of light, we can legitimately state that, to three-digit accuracy

$$c = 3.00 \times 10^8 \text{ m/sec}$$

since c is closer to this figure than to 2.99×10^8 or 3.01×10^8 m/sec. In the sample calculation of the preceding paragraph, the quantity $(4.6 \times 3)/(9 \times 6.2)$ actually equals

0.2473118 . . . , but it is rounded off to 0.25 because the result of a calculation may have no more significant digits than those in the least precise of the numbers that went into it.

exercise set V

Express the following numbers in powers-of-ten notation:

1. $720 =$
2. $890{,}000 =$
3. $0.02 =$
4. $0.000{,}062 =$
5. $3.6 =$
6. $0.4 =$
7. $49{,}527 =$
8. $0.002{,}943 =$
9. $0.0014 =$
10. $49{,}000{,}000{,}000 =$
11. $0.000{,}000{,}011 =$
12. $1.4763 =$

Express the following numbers in decimal notation:

13. $3 \times 10^{-4} =$
14. $7.5 \times 10^3 =$
15. $8.126 \times 10^{-5} =$
16. $1.01 \times 10^8 =$
17. $5 \times 10^2 =$
18. $3.2 \times 10^{-2} =$
19. $4.32145 \times 10^3 =$
20. $6 \times 10^6 =$
21. $5.7 \times 10^0 =$
22. $6.9 \times 10^{-5} =$

Evaluate the following in powers-of-ten notation:

23. $\dfrac{30 \times 80{,}000{,}000{,}000}{0.0004} =$
24. $\dfrac{30{,}000 \times 0.000{,}000{,}6}{1{,}000 \times 0.02} =$
25. $\dfrac{0.0001}{60{,}000 \times 200} =$
26. $5{,}000 \times 0.005 =$
27. $\dfrac{5{,}000}{0.005} =$
28. $\dfrac{200 \times 0.000{,}04}{400{,}000} =$
29. $\dfrac{0.002 \times 0.000{,}000{,}05}{0.000{,}004} =$
30. $\dfrac{500{,}000 \times 18{,}000}{9{,}000{,}000} =$

ANSWERS TO EXERCISES

exercise set I

1. $ac + bc$
2. $ac - bc$
3. $a/c + b/c$
4. 1
5. $a + b$
6. $a^2 - b^2$
7. $a + b + c + bc/a$
8. $a^2 + ab^2 - ac - ab - b^3 + bc$
9. $a^2 - 2bc - b^2 - c^2$
10. $a^2 - 2ab + b^2$

exercise set II

1 4.5   **2** 0.8   **3** −3   **4** 0   **5** 22

exercise set III

1 a^7 **2** a^2 **3** $a^{3/2}$ **4** $a^{4/3}$ **5** a^2
6 $a^{11/2}$ **7** 1   **8** a^4 **9** a^{-4} **10** $a^{1/2}$
11 $a^{5/2}$ **12** a^{-5} **13** a^2 **14** a^{-2} **15** $a^{-4/3}$
16 $a^5 + a^{1/2}$ **17** $2a^2$ **18** 0

exercise set IV

1 $x = 2$
2 $x = 8$
3 $x = -3$
4 $x = 2$
5 $x = 2$
6 $x = (z^2 + 3y)/(3 - y)$
7 $x = 72/y^2$
8 $x = -y(1 + z)/(1 - z)$
9 $x = 4y/z$
10 $x = \sqrt{5y}$

exercise set V

1 7.2×10^2 **2** 8.9×10^5 **3** 2×10^{-2}
4 6.2×10^{-5} **5** 3.6×10^0 **6** 4×10^{-1}
7 4.9527×10^4 **8** 2.943×10^{-3} **9** 1.4×10^{-3}
10 4.9×10^{10} **11** 1.1×10^{-8} **12** 1.4763×10^0
13 0.0003   **14** 7,500   **15** 0.000,081,26
16 101,000,000   **17** 500   **18** 0.032
19 4,321.45   **20** 6,000,000   **21** 5.7
22 0.000,069   **23** 6×10^{15} **24** 9×10^{-4}
25 8.3×10^{-11} **26** 2.5×10^1 **27** 10^6
28 2×10^{-8} **29** 2.5×10^{-5} **30** 10^3

Glossary

Absolute zero is that temperature at which the volume of an ideal gas sample would shrink to zero if its pressure were held constant. This temperature is $-273°C$. There would be no random molecular movement in the gas at absolute zero. The *absolute temperature scale* gives temperatures in degrees centigrade above absolute zero, denoted $°K$; thus the freezing point of water is $273°K$.

The **acceleration** of a body is the rate of change of its velocity. A body is accelerated when its speed changes, when its direction of motion changes, or when both change. Acceleration is expressed in such units as feet per second squared (ft/sec^2), meters per second squared (m/sec^2), and so on.

The **acceleration of gravity** is the acceleration with which bodies fall in the absence of friction. The acceleration of gravity is the same for all bodies near the earth's surface, and its value is $32 \ ft/sec^2$, which is equal to $9.8 \ m/sec^2$.

An **acid** is any substance whose molecules contain hydrogen and whose water solution contains hydrogen ions.

The **activation energy** of a reaction is the energy that must be supplied initially for the reaction to start.

An **alcohol** is a hydrocarbon derivative in which one or more hydrogen atoms have been replaced by OH groups.

The **alkali metals** are a family of soft, light, extremely active metals with similar chemical properties. The alkali metals are lithium, sodium, potassium, rubidium, and cesium in order of atomic mass.

An **alluvial fan** is a deposit of sediments where a stream emerges from a steep mountain valley and flows onto a plain.

An **alpha particle** is the nucleus of a helium atom; it consists of two protons and two neutrons, and has a charge of $+2e$.

An **alternating current** is an electric current that reverses its direction periodically.

An **amorphous solid** is one whose atoms or molecules exhibit no regularity of arrangement.

The **ampere** is the unit of electric current; it is equal to a flow of one coulomb of charge per second.

The **amplitude** of a wave is the maximum value attained by whatever quantity is periodically varying.

The **angular momentum** of a rotating body is a measure of its tendency to continue to spin at the same rate about an axis that does not change in orientation. Angular momentum is a vector quantity. The angular momentum of an isolated body or system of bodies is *conserved* (remains unchanged).

Archimedes' principle states that the buoyant force on an object immersed in a fluid is equal to the weight of the fluid displaced by the object.

An **asteroid** is one of a large number of relatively small bodies (less than 500 miles across) revolving around the sun in orbits lying between those of Mars and Jupiter.

The **asthenosphere** is a layer of rock, neither wholly rigid nor wholly fluid but capable of plastic deformation, that is just below the lithosphere in the earth's mantle.

The earth's **atmosphere** is its gaseous envelope. The five regions of the atmosphere are, from the earth's surface upward, the **troposphere,** the **stratosphere,** the **mesosphere,** the **thermosphere,** and the **exosphere.**

Atmospheric pressure is the force with which the atmosphere presses down upon each unit area at the earth's surface. Atmospheric pressure is normally about $1.013 \times 10^5 \ newtons/m^2$ ($14.7 \ lb/in.^2$), corresponding to the pressure of 76 cm (about 30 in.) of mercury.

The **atomic mass** of an element is the average mass of its atoms relative to that of a carbon atom of the most abundant kind, where the latter is assigned the atomic weight of precisely 12.

The **atomic number** of an element is the number of protons in the nuclei of its atoms. Thus an element is a substance all of whose atoms have the same atomic number.

An **aurora** is a display of changing colored patterns of light that appear in the sky, particularly at high latitudes. The aurora arises from the excitation of atmospheric gases at high altitudes by streams of energetic ions of solar origin.

Avogadro's hypothesis states that equal volumes of all gases, under the same conditions of temperature and pressure, contain the same numbers of molecules.

Avogadro's number is the number of atoms in a gram atom of any element; it is also the number of formula units in a gram mole of any compound.

A **base** is any substance whose molecules contain OH groups and whose water solutions contain OH^- ions.

A **batholith** is a very large body of intrusive rock, mainly granite, that extends downward as much as 10 km.

Benzene hydrocarbons contain closed rings of six carbon atoms in their structural formulas; each ring contains the equivalent of three double and three single bonds.

Beta particles are electrons or positrons ejected from atomic nuclei during radioactive decay.

Bode's law is a simple relationship that gives the distances of the planets from the sun. Its significance is unknown.

In the **Bohr model of the atom,** electrons are supposed to move around nuclei in circular orbits of definite size; when an electron jumps from one orbit to another, a photon of light is either emitted or absorbed whose energy corresponds to the difference in the electron's energy in the two orbits.

Boyle's law states that the volume of a gas is inversely proportional to its pressure provided that the temperature is held constant.

A **carbohydrate** is a compound of carbon, hydrogen, and oxygen whose molecules contain two atoms of hydrogen for each one of oxygen; carbohydrates are manufactured in green plants from water and carbon dioxide in the process of photosynthesis. Sugars, starches, chitin, and cellulose are carbohydrates.

The **carbon cycle** is one of two series of energy-producing nuclear reactions that take place in the sun and other stars and involve the conversion of hydrogen to helium. In the carbon cycle, carbon nuclei absorb a succession of protons; this absorption ultimately results in the formation of helium nuclei and the reemergence of carbon nuclei.

A **catalyst** is a substance that can alter the rate of a chemical reaction without itself being permanently changed by the reaction.

Cathode rays are streams of electrons produced in an evacuated tube when opposite charges are placed on electrodes at each end.

The **Cenozoic era** refers to the past 65 million years of the earth's history; the term means "recent life."

The **centrifugal distortion** of a rotating body refers to its bulging at the equator and flattening at the poles caused by the rotation.

Centrifugal force is the apparent outward force experienced by a body moving in a circle; it is a manifestation of inertia, the tendency of moving things to travel in straight lines.

The **centripetal force** on a body moving in a circle is the inward force that must be exerted to produce this motion. It always acts toward the center of the circle in which the body is moving.

A **Cepheid variable** is a variable star of a particular type whose intrinsic brightness and period of variation are related.

A **chain reaction** is a succession of nuclear fissions in which neutrons produced by each fission induce further fissions in other atoms.

Charles's law states that the volume of a gas is proportional to its absolute temperature provided that the pressure is held constant.

A **chemical equilibrium** occurs when a chemical reaction and its reverse reaction both take place at the same rate.

Chlorophyll is the catalyst that makes possible the reaction of water and carbon dioxide in green plants in order to produce carbohydrates. The reaction is called *photosynthesis*.

The **clay minerals** are a group of light, soft silicates of hydrogen and aluminum, sometimes with a little magnesium, iron, or potassium, which occur as aggregates of microscopic crystals.

Cleavage is the tendency of a substance to split along certain planes determined by the arrangement of particles in its crystal lattice.

Coal is a solid fuel, largely carbon, that was formed chiefly from cellulosic plant material that accumulated under conditions preventing complete decay.

Combustion is the rapid oxidation of a substance accompanied by the evolution of noticeable heat and light.

Comets are aggregates of matter pursuing regular orbits in the solar system and appearing as small, hazy patches of light often accompanied by long, filmy tails.

A **compound** is a homogeneous combination of elements in definite proportions. The properties of a compound are generally very different from those of its constituent elements.

The **Compton effect** occurs in the scattering of X-rays by electrons; the scattered X-rays have lower frequencies than they had originally, corresponding to the loss of energy by their photons upon colliding with electrons.

A **conductor** is a substance through which electric current can flow readily.

A **constellation** is a group of stars whose arrangement in the sky suggests a particular pattern.

According to the theory of **continental drift**, today's continents were once part of two primeval supercontinents called Laurasia (North America, Greenland, Europe, and most of Asia) and Gondwanaland (South America, Africa, Antarctica, India, and Australia), which were separated by the Tethys Sea.

Convection currents result from the uneven heating of a fluid; the warmer parts of the fluid expand and rise because of their buoyancy, while the cooler parts sink.

In the **Copernican system** the sun and stars are fixed, the earth and the other planets revolve around the sun, the moon revolves around the earth, and the earth rotates on an axis passing through Polaris.

The earth's **core** is a spherical region around the center of the earth whose radius is about 3,500 km (2,100 miles); it consists of a solid *inner core*, probably mainly iron, surrounded by a liquid *outer core*, probably mainly molten iron.

The **Coriolis effect** is the deflection of winds to the right in the Northern Hemisphere, to the left in the Southern; this effect is a consequence of the earth's rotation.

The sun's **corona** is a vast cloud of extremely hot, rarefied gas which surrounds the sun. It is visible during solar eclipses.

Primary cosmic rays are high-speed atomic nuclei, largely protons, which move through the galaxy and frequently strike the earth's atmosphere. *Secondary cosmic rays* consist of the atomic debris that results from the collisions of primary cosmic rays with the oxygen and nitrogen atoms present in the upper atmosphere.

Coulomb's law states that the force between two electric charges is directly proportional to the product of their charges and inversely proportional to the distance between them; the force is repulsive when the charges have the same sign, attractive when they have different signs.

Covalent compounds are compounds formed by atoms that share pairs of electrons. A *covalent crystal* consists of individual atoms which share electrons with their neighbors.

Cracking a hydrocarbon is the process of heating it under pressure in the presence of

GLOSSARY

a catalyst in order to break its molecules down into simpler hydrocarbons.

The **crust** of the earth is its outer shell of rock which averages 33 km under the continents, less than 5 km under the oceans.

A **crystalline solid** is one whose atoms or molecules are arranged in a definite pattern.

An electric **current** is a flow of charge from one place to another. A direct current is one that always flows in one direction; an alternating current periodically reverses its direction of flow.

Cyclones are weather systems centered about regions of low pressure; in the Northern Hemisphere cyclones are characterized by counterclockwise winds, in the Southern Hemisphere by clockwise winds. *Anticyclones* are weather systems centered about regions of high pressure; the characteristic winds of anticyclones are opposite in direction to those of cyclones.

The law of **definite proportions** states that the elements that make up a chemical compound are always combined in the same definite proportions by weight.

A **delta** is a deposit of sediments where a stream flows into a lake or the sea.

Diastrophism refers to geological processes involving movements of the solid materials of the crust.

A **dike** is a wall-like mass of igneous rock intruded along a fissure that cuts across existing rock layers.

The **dinosaurs** were reptiles of various kinds, many huge in size, which flourished in the Mesozoic era.

A **displacement reaction** is an oxidation-reduction reaction in which one metal (or nonmetal) displaces another metal (or nonmetal) from solution.

The **Doppler effect** is the change in perceived frequency of a wave due to relative motion between its source and a listener.

A **double star** is actually a pair of stars both revolving about their mutual center of gravity.

The **dynamo effect** refers to the production of an electric current in a conductor by electromagnetic induction, which involves either a changing magnetic field or motion relative to a magnetic field.

An **earthquake** consists of rapid vibratory motions of rock near the earth's surface; it is usually caused by the sudden movement of rock along a fault. Earthquake P (for primary) waves are longitudinal oscillations in the solid earth like those in sound waves; earthquake S (for secondary) waves are transverse oscillations in the earth like those in a stretched string; earthquake L (for long) waves are oscillations of the earth's surface like those in water waves.

A **lunar eclipse** occurs when the earth's shadow obscures the moon; a *solar eclipse* occurs when the moon obscures the sun.

A body possessing **electric charge** is capable of producing a spark and of attracting small light objects. *Negative charge* is that produced on a rubber rod rubbed with fur; *positive charge* is that produced on a glass rod rubbed with silk. The *coulomb* is the unit of charge.

In an **electrochemical cell**, electric current is produced by an oxidation-reduction reaction whose two half-reactions take place at different locations.

Electrolysis is the liberation of free elements from a liquid by passing an electric current through it.

An **electrolyte** is a substance that separates into free ions on solution in water.

An **electromagnet** is a coil of wire with an iron core to enhance its magnetic field.

Electromagnetic induction refers to the production of a current in a wire moving through a magnetic field or in a wire loop in the presence of a changing magnetic field.

Electromagnetic waves are coupled periodic electric and magnetic disturbances that spread out from accelerated electric charges. Among the various kinds of electromagnetic waves, distinguished only by their frequencies, are gamma rays, X-rays, ultraviolet radiation, visible light, infrared radiation, millimeter waves, microwaves, and radio waves. Electromagnetic waves all travel in vacuum with the *speed of light*.

An **electron** is a tiny, negatively charged particle found in matter whose charge is -1.6×10^{-19} coul and whose mass is 9.1×10^{-31} kg.

The **electron affinity** of an atom is the amount of energy that it releases when it picks up an extra electron beyond its normal complement.

The **electron volt** is a unit of energy equal to 1.6×10^{-19} joule, which is the amount of energy acquired by an electron accelerated by a potential difference of 1 volt. A **Mev** is 10^6 ev and a **Gev** is 10^9 ev.

The **electronegativity** of an atom refers to its tendency to attract shared electrons when it participates in a chemical bond.

An **elementary particle** is one of the various indivisible particles found in nature.

Elements are substances that can neither be decomposed nor transformed into one another by ordinary chemical or physical means. There is a limited number of elements and all other substances are combinations of them in various proportions.

An **endothermic reaction** is one that must be supplied with energy in order to occur.

Energy is the property something has that enables it to do work. Energy has the same units as work. Some forms of energy are kinetic energy, potential energy, heat energy, chemical energy, electrical energy, magnetic energy, radiant energy, and mass energy. The *law of conservation of energy* states that energy can be neither created nor destroyed, although it may go from one form to another.

Energy bands. The proximity of the atoms in a crystal alters slightly the energy states of their outer electron shells. As a result the crystal as a whole has energy bands of myriad separate levels close together. Gaps between the bands are called *forbidden bands*, and electrons in the crystal cannot have energies that fall in these gaps.

An **equation** is a mathematical statement of the relationship between a certain quantity and others upon which it depends.

Erosion refers to all processes by which rock is disintegrated and worn away and its debris removed.

The **escape velocity** is the speed a body requires if it is to permanently escape from the gravitational pull of a particular astronomical body.

An **ester** is an organic compound formed when an alcohol reacts with an acid.

The Pauli **exclusion principle** states that no more than one electron in an atom can have the same set of quantum numbers.

An **exothermic reaction** is one that liberates energy.

The **expanding universe** refers to the fact that all of the galaxies visible to us seem to be rapidly receding from one another. The evidence for the recession is the observed *red shift* in galactic spectra, which is a shift of their spectral lines toward the red end of the spectrum that is interpreted as a Doppler effect.

A **fault** is a rock fracture along which movement has occurred.

Feldspar is the name of a group of light-colored silicate minerals with similar properties.

The **ferromagnesian minerals** are a large group of diverse minerals, all silicates of iron and magnesium and all dark green to black in color.

A **field of force** is a region of altered space (for example, around a mass, an electric charge, a magnet) that exerts a force on appropriate bodies placed in that region.

A **flood plain** is the wide, flat floor of a river valley produced by the sidewise cutting of the stream when its slope has become too gradual for further downcutting.

Foliation refers to the alignment of flat or elongated mineral grains characteristic of many metamorphic rocks.

A **force** is any influence capable of producing a change in the motion of a body, in other words, capable of accelerating it. In the British system the unit of force is the *pound* (*lb*); in the metric system it is the *newton*.

The **formula mass** of a compound is the sum of the atomic masses of its constituent elements, each multiplied by the number of times it appears in the formula of that compound.

Fossils are the remains or traces of organisms preserved in rocks.

Fractional distillation is a process for separating a mixture of chemical compounds in the order of their boiling points.

The **frequency** of a wave is the number of waves that pass a given point per second. The frequency of a sound wave is its *pitch*.

A **frontal surface** is the surface separating a warm and a cold air mass; a *front* is where this surface touches the ground. A *cold front* generally involves a cold air mass moving approximately eastward; a *warm front* generally involves a warm air mass moving approximately eastward.

The four **fundamental interactions** are the gravitational, weak nuclear, electromagnetic, and strong nuclear interactions, in order of strength. They give rise to all the physical processes in the universe.

Galactic nebulae are irregular masses of diffuse material within our galaxy.

Galaxies are aggregates of stars separated from other such aggregates by much greater distances than those between the member stars.

A **galvanometer** is a device for measuring electric currents with the help of the motor effect.

Gamma rays are very high frequency electromagnetic waves.

The **geomagnetic poles** are the points on the earth's surface through which the magnetic axis of the earth passes.

A **geosyncline** is a broad basin that sinks relatively rapidly as sediments accumulate upon it.

A **glacier** is a large mass of ice that gradually moves downhill.

Glass is an amorphous solid formed by the rapid cooling of a molten silicate; it is in a sense a supercooled liquid rather than a true solid.

Globular clusters are roughly spherical assemblies of stars outside the plane of our galaxy but associated with it.

Gondwanaland was the primeval supercontinent which later split up into what are today South America, Africa, Antarctica, India, and Australia.

A **graph** is a pictorial representation of the relationship between two quantities.

Newton's **law of gravitation** states that every particle in the universe attracts every other particle with a force that is directly proportional to the product of their masses and inversely proportional to the square of the distance between them.

A body is **grounded** when its charge is permitted to flow to the earth.

Groundwater is rain water that has soaked into the ground. The *water table* is the upper surface of that part of the ground whose pore spaces are saturated with water. A *spring* is a channel through which groundwater comes to the surface.

The **half-life** of a radioactive element is the period of time required for one-half of an original sample of the element to decay.

The **halogens** are a family of highly active nonmetals with similar chemical properties. The halogens are fluorine, chlorine, bromine, and iodine in order of atomic mass.

Heat is energy of random molecular motion. The heat that a body possesses depends upon its temperature, its mass, and the kind of material of which it is composed. The unit of heat called the *kilocalorie* is defined as the heat necessary to raise the temperature of 1 kg of water by 1°C.

The **heat of fusion** of a substance is the amount of heat required to change 1 kg of the substance from solid to liquid at its freezing point; it is also the heat that 1 kg of the substance liberates when it changes from liquid to solid at its freezing point.

The **heat of vaporization** of a substance is the amount of heat required to change 1 kg of the substance from liquid to vapor at its boiling point; it is also the heat that 1 kg of the substance liberates when it changes from vapor to liquid at its boiling point.

The **Hertzsprung-Russell diagram** is a graph on which are plotted the intrinsic brightness and spectral type of the different stars. Most stars belong to the *main sequence*, which follows a diagonal line on the diagram, but there are also small, hot *white dwarf* stars and large, cool *red giant* stars occupying other positions not on the main sequence.

A **hydrocarbon** is an organic compound containing only carbon and hydrogen. An *unsaturated hydrocarbon* is one whose molecules contain more than one bond (that is, shared electron pair) between adjacent carbon atoms. In a *saturated hydrocarbon* there is only one bond between adjacent carbon atoms.

A **hydrocarbon derivative** is a compound obtained by substituting other atoms or atom groups for some of the hydrogen atoms in hydrocarbon molecules.

In a **hydrogen bond**, a nearly bare proton on the outside of a molecule exerts an exceptionally strong van der Waals force on a neighboring molecule.

Hydromagnetism, or *magnetohydrodynamics*, is the science of the behavior of electrically conducting liquids and gases, such as those believed responsible for sunspots and the earth's magnetism.

The **hydronium ion** consists of a water molecule with a hydrogen ion attached; its symbol is H_3O^+. (Sometimes more than one water molecule may be attached to a single hydrogen ion; the notion of the hydronium ion is for convenience only.)

A **hypothesis** is a scientific generalization as originally presented; a *law* is a scientific generalization whose correctness has been clearly demonstrated; a *theory* is a logical structure built on several fundamental generalizations that explains a wide variety of phenomena. (These terms are often used in senses slightly different from the ones indicated, but the definitions stated here indicate their usual meanings.)

Igneous rocks are rocks that have been formed from a molten state by cooling.

An **indicator** is a substance that changes color when neutralization occurs.

The **inert gases** are a family of totally inactive elements consisting of helium, neon, argon, krypton, xenon, and radon in order of atomic mass.

Inertia is the apparent resistance a material body offers to any change in its state of motion.

An **insulator** is a substance through which charges can move only with difficulty.

Interference occurs when two or more waves of the same kind pass the same point in space at the same time. If the waves are "in step" with each other, their amplitudes add together to produce a strong wave; this situation is called *constructive interference*. If the waves are "out of step" with each other, their amplitudes tend to cancel out and the resulting wave is weaker; this situation is called *destructive interference*.

Intrusive rocks are igneous rocks that flowed into regions below the surface already occupied by other rocks and gradually hardened there.

Invertebrates are animals without internal skeletons.

An **ion** is an atom or molecule possessing an electric charge. The process of forming ions is called *ionization*.

Ionic compounds are compounds formed by the transfer of electrons from one atom to another.

An **ionic crystal** consists of individual ions in an equilibrium array in which the attractive forces between ions of opposite charge balance the repulsive forces present.

The **ionization energy** of an atom is the amount of work required to remove an electron from it.

The **ionosphere** is a region in the upper atmosphere characterized by the presence of a high concentration of ions.

Isomers are compounds that have the same molecular formulas but different structural formulas.

Isostasy is the idea that irregularities in the earth's crust are supported by their buoyancy.

The **isotopes** of an element are varieties of the element whose atoms have the same atomic number but different atomic masses; that is, their nuclei contain the same number of protons but contain different numbers of neutrons.

The **jet stream** consists of swift currents of air near the base of the stratosphere which move from west to east.

Kepler's laws of planetary motion state that (1) the paths of the planets around the sun are ellipses, (2) the planets move so that their radius vectors sweep out equal areas in equal times, and (3) the ratio between the square of the time required by a planet to make a complete revolution around the sun and the cube of its average distance from the sun is a constant for all the planets.

Kinetic energy is the energy a body has by virtue of its motion. The kinetic energy of a moving body is equal to $\frac{1}{2}mv^2$, one-half the product of its mass and the square of its speed.

According to the **kinetic theory of gases**, the absolute temperature of a gas is directly proportional to the average kinetic energy of its molecules.

The **kinetic theory of matter** states that matter is composed of tiny discrete particles called *molecules* and that these molecules are in constant motion.

A **laser** is a device that produces an intense beam of monochromatic, coherent light from the cooperative radiation of excited atoms. (The waves in a coherent beam are all in step with one another.)

Laurasia was the primeval supercontinent which later split up into what are today North America, Greenland, Europe, and most of Asia.

Lava is molten rock on the earth's surface.

Le Châtelier's principle states that when the conditions of a system in equilibrium are altered, the equilibrium will shift in such a way as to attempt to restore the initial conditions.

The **left-hand rule** states that when a current-carrying wire is grasped so that the thumb of the left hand points in the direction in which the electrons move, the fingers of that hand point in the direction of the magnetic field around the wire.

A **light-year** is the distance light travels in one year; it is about 6×10^{12} miles.

Lines of force are imaginary lines used to describe a field of force; their direction is that of the force which the field would exert on a positive test particle, and their density is proportional to the strength of the field (that is, they are closest together where the field is strongest, farthest apart where the field is weakest).

Fats, oils, waxes, and sterols are **lipids**, which are synthesized in plants and animals from carbohydrates. A fat molecule consists of a glycerin molecule with three fatty acid molecules joined to it.

The **lithosphere** is the earth's outer shell of rigid rock. It consists of the crust and the outermost part of the mantle.

In a **longitudinal wave** the particles constituting the wave oscillate back and forth in the same direction as that in which the wave is moving. Sound waves and compressional waves in a spring are longitudinal waves. Water waves are a combination of longitudinal and transverse waves.

The **Lorentz contraction** is the decrease in the measured length of an object that is moving relative to an observer compared with its length when measured by an observer at rest relative to the object.

Magma is molten rock below the earth's surface.

A **magnet** is a body with the property of attracting iron objects. When freely suspended, the *north pole* of a magnet points north while its *south pole* points south. Like poles repel each other; unlike poles attract.

Mammals are warm-blooded animals that bear live young. They have relatively large brains and care for their young after birth.

The earth's **mantle** is the solid part of the earth from the core to the crust; it is about 2,900 km (1,800 miles) thick.

The **maria** of the moon are dark, smooth regions on its surface once thought to be seas. Some maria have dense bodies of matter called *mascons* under them.

The **mass** of a body is the property of matter that manifests itself as inertia; it may be thought of as the quantity of matter in the body.

The **law of conservation of mass** in chemical reactions states that the total mass of the products of a chemical reaction is always the same as the total mass of the original materials.

A **matter wave** is associated with rapidly moving bodies, whose behavior in certain respects resembles wave behavior. The matter waves associated with a moving body are in the form of a group, or packet, of waves which travel with the same speed as the body.

The **mechanical equivalent of heat** is the amount of energy equivalent to one unit of heat; it is 4,185 joules/kcal.

Pi mesons (or *pions*) are elementary particles involved in holding atomic nuclei together. Their masses are between those of the electron and proton, and they may either have charge or be neutral. Charged pi mesons decay into the light *mu mesons* (or *muons*), which in turn decay into electrons. Uncharged pi mesons decay into a pair of gamma rays.

The **Mesozoic era** is the period in the earth's history from 225 million years ago to 65 million years ago; the term means "intermediate life."

The **metallic bond** has its origin in a "gas" of freely moving electrons that pervades every metal.

Metals possess a characteristic sheen (metallic luster) and are good conductors of heat and electricity. They combine with nonmetals more readily than with one another. The more active metals liberate hydrogen from dilute acids and their oxides react with water to form bases.

Metamorphic rocks are rocks that have been altered by heat and pressure deep under the earth's surface.

Meteoroids are pieces of matter moving through space; *meteors* are the flashes of light they produce when entering the earth's atmosphere; and *meteorites* are the remains of meteoroids that reach the ground.

Mica is a soft mineral with conspicuous cleavage in one plane.

Minerals are the separate homogenous substances of which rocks are composed.

The **molarity** of a solution is the number of moles of a dissolved substance it contains per liter.

A **mole** of any substance is that amount of it which contains Avogadro's number of atoms, molecules, or formula units, depending upon the nature of the substance. Hence a mole of an element has a mass equal to its atomic mass expressed in grams, and a mole of a compound has a mass equal to its formula mass expressed in grams.

A **molecule** is an electrically neutral combination of two or more atoms held together strongly enough to be experimentally observable as a particle.

The linear **momentum** of a body is the product of its mass and its velocity. Momentum is a vector quantity, possessing both a magnitude and a direction; the direction is that of the body's motion. The law of *conservation of momentum* states that, when several bodies interact with one another (for instance, in an explosion or a collision), if outside forces do not act upon the bodies involved, the total momentum of all the bodies before they interact is exactly the same as their total momentum afterward. (See also **angular momentum**.)

A **moraine** is the pile of debris around the end of a glacier left as a low, hummocky ridge when the glacier melts back. The deposited material is called *till*.

Newton's **first law of motion** states that every body continues in its state of rest or of uniform motion in a straight line if no force acts upon it. The *second law of motion* states that the acceleration of a body is directly proportional to the magnitude of the force acting upon it and inversely proportional to its mass; the acceleration is in the direction of the applied force. The *third law*

GLOSSARY

of motion states that for every force there is an equal and opposite force.

The **motor effect** is the sidewise force a magnetic field exerts on a current-carrying wire; the force is perpendicular both to the direction of the current and the direction of the field.

The **neutralization** of a strong acid by a strong base in water solution is a reaction between hydrogen and hydroxide ions to form water.

The **neutrino** is a massless, uncharged particle that is emitted during beta decay and in the decay of various elementary particles.

The **neutron** is an electrically neutral elementary particle whose mass is approximately that of the proton. Atomic nuclei consist of neutrons and protons.

A **neutron star** is an extremely small star, smaller than a white dwarf, which is composed almost entirely of neutrons.

Nonmetals have an extreme range of physical properties; in the solid state they are usually lusterless and brittle and are poor conductors of heat and electricity. Some nonmetals form no compounds whatever; the others combine more readily with active metals than with one another. Soluble nonmetal oxides react with water to form acids.

A **nova** is a star that suddenly flares up and shines brilliantly for a week or two before fading back to a less conspicuous luminosity.

Nuclear fission occurs when a heavy nucleus splits into two or more lighter nuclei. Considerable energy is evolved each time fission occurs.

Nuclear fusion occurs when two light nuclei unite to form a heavier one. Considerable energy is evolved in such processes; this energy is called *thermonuclear energy*.

A **nuclear reactor** is a device in which fissions occur at a controlled rate.

Nucleic acid molecules consist of long chains of *nucleotides* whose precise sequence governs the structure and functioning of cells and organisms. *DNA* has the form of a double helix and carries the genetic code; the simpler *RNA* acts as a messenger in protein synthesis.

The **nucleus** of an atom is its small, heavy core, containing all of the atom's positive charge and most of its mass.

Ohm's law states that the current in a circuit is equal to the potential difference between the ends of the circuit divided by the resistance of the circuit; symbolically, $i = V/R$. Resistance is expressed in units called *ohms*.

An **orbital** is the wave function that describes an electron in an atom or molecule; an orbital can contain two electrons, one of either spin. A *bonding molecular orbital* is one in which there is an increased electron probability density between two atoms, leading to a covalent bond.

Organic chemistry refers to the chemistry of carbon compounds; *inorganic chemistry* refers to the chemistry of the other elements.

Oxidation is the chemical combination of a substance with oxygen. An element is *oxidized* when its atoms lose electrons and *reduced* when its atoms gain electrons.

The **oxidation number** of an atom in a specific situation is assigned on the basis of certain general rules. The oxidation number of an atom increases (becomes more positive) when it is oxidized, and decreases (becomes more negative) when it is reduced.

Ozone is a form of oxygen whose molecules consist of three oxygen atoms each.

Pair production is the materialization of an electron and a positron (or a proton and an antiproton) when a sufficiently energetic gamma ray passes near an atomic nucleus.

The **Paleozoic era** is the period in the earth's history from 600 million years ago to 225 million years ago; the term means "ancient life."

The **parallax** of a star is the apparent shift in its position relative to more distant stars as the earth revolves in its orbit.

Mendeleev's **periodic law** in modern form states that if the elements are listed in the order of their atomic numbers, elements with similar properties recur at definite intervals. A tabular arrangement of the elements showing this recurrence of properties is called a *periodic table*.

Petroleum is a naturally occurring mixture of hydrocarbons believed to have originated from the protein, fat, and wax content of plant and animal matter.

The **pH** of a solution is a method for expressing the exact degree of acidity or basicity of a solution in terms of its hydrogen-ion concentration. A pH of 7 signifies a neutral solution, a smaller pH than 7 signifies an acid solution, and a higher pH than 7 signifies a basic solution.

The **phases of the moon** occur because the amount of the illuminated side of the moon visible to us varies with the position of the moon in its orbit.

The **photoelectric effect** is the emission of electrons from a metal surface when light is shone upon it.

Photons are packets of energy whose motion constitutes the propagation of light. The energy content of electromagnetic waves is quantized into tiny bursts, and these bursts are called photons or, alternatively, quanta.

The **photosphere** is the relatively thin gaseous layer on the sun that emits nearly all the light the sun radiates into space.

Photosynthesis is the reaction between water and carbon dioxide to produce carbohydrates that occurs in green plants with the help of the catalyst chlorophyll.

A **planet** is a satellite of the sun that appears in the sky as a bright object whose position changes relative to the stars.

Plankton are tiny, primitive plants and animals that float in the ocean.

Polar covalent compounds are covalent compounds in which one part of each molecule is relatively negative and another part relatively positive.

A **polar molecule** is one that behaves as if it were negatively charged at one end and positively charged at the other. A *polar liquid* is a liquid whose molecules are polar whereas a *nonpolar liquid* has molecules whose charge is symmetrically arranged.

Polymerization is the process of causing simple hydrocarbon molecules to join together into more complicated ones under the influence of heat and catalysts.

The **potential difference** between two points in an electric circuit is the amount of potential energy lost by 1 coul of charge flowing from one point to the other. The unit of potential difference is the *volt*, equal to a potential energy of 1 joule/coul of charge.

Potential energy is the energy a body has by virtue of its position. The gravitational potential energy of a body is equal to Wh, the product of its weight and its height above some reference level. (In the metric system gravitational potential energy is given as mgh, the product of the mass of the body, the acceleration of gravity, and the height.)

Power is the rate at which work is being done. The unit of power in the metric system is the *watt*, equal to 1 joule/sec.

Precambrian time is the name given to the period in the earth's history previous to the Paleozoic era, which began 600 million years ago.

A **precipitate** is a solid that forms as a result of a chemical reaction in solution.

The **pressure** on a surface is the force acting perpendicular to the surface divided by the area of the surface. A pressure is a force per unit area. Pressure units are pounds per square inch (lb/in.^{2}), pounds per square foot (lb/ft^2), and newtons per square meter (newtons/m^2).

Proteins are the chief constituents of living matter and consist of long chains of amino acid molecules called *polypeptides*. The sequence of amino acids in a protein molecule together with the form of the molecule determines its biological role.

The **proton** is a positively charged elementary particle constituting the nucleus of the hydrogen atom. Its charge is $+1.6 \times 10^{-19}$ coul and its mass is 1.67×10^{-27} kg.

The **proton-proton cycle** is one of two series of energy-producing nuclear reactions that take place in the sun and other stars and involve the conversion of hydrogen to helium. In the proton-proton cycle, a series of collisions of protons results in the formation of helium nuclei.

The **protosun** was the original cloud of dust and gas that evolved into our sun; the *protoplanets* were the smaller clouds of dust and gas that remained after the contraction of the protosun and eventually became the planets.

The **ptolemaic system** is a hypothesis of the astronomical universe in which the earth is at the center with all of the other celestial bodies revolving around it in more or less complex orbits.

A **pulsar** is a star that emits extremely regular flashes of light and radio waves; pulsars are believed to be neutron stars that are spinning rapidly.

Quanta, or *photons*, are tiny packets of energy whose motion constitutes the propagation of light; under certain circumstances light appears as an electromagnetic wave phenomenon, under others as a quantum phenomenon. The energy of each quantum of light is hf, where h is *Planck's constant* and f is the frequency of the light.

Quantum mechanics is a highly mathematical theory of atomic phenomena which treats only experimentally measurable quantities and does not invoke mechanical models that contradict the uncertainty principle.

Four **quantum numbers** are required to specify completely the physical state of an atomic electron. These are the *total quantum number*, n, which governs the energy of the electron; the *orbital quantum number*, l, which determines the magnitude of the electron's angular momentum; the *magnetic quantum number*, m_l, which determines the direction of the electron's angular momentum; and the *spin magnetic quantum number*, m_s, which determines the orientation of the electron's spin.

Quartz consists of crystals of silicon dioxide, SiO_2.

A **quasar** is a strong source of radio waves whose spectral lines show large red shifts and whose radiation fluctuates markedly. Quasars appear as points of light in a telescope, and may be tiny, extremely bright objects of unknown nature at the limits of the observable universe.

A **radio telescope** is an antenna, often located at the center of a metallic "dish," connected to a sensitive radio receiver in order to pick up radio waves from space.

Radioactivity is the property certain substances have of spontaneously radiating charged particles or very high frequency electromagnetic waves, or both.

Radiocarbon dating is a procedure for establishing the approximate age of once-living matter on the basis of the relative amount of radioactive carbon it contains.

Reflection occurs when a wave bounces off a surface; light is reflected by a mirror, sound by a wall.

Refraction occurs when a wave is bent on passing from one medium to another in which its speed is different.

Relative humidity is the ratio between the amount of moisture in a volume of air and the maximum amount of moisture that volume of air can hold when completely saturated. It is usually expressed as a percentage.

The **relativistic length contraction** is the decrease in the measured length of an object that is moving relative to an observer compared with its length when measured by an observer at rest relative to the object.

The **relativistic time dilation** is the increase in the length of time intervals when measured by an observer in motion relative to a clock compared with their length when measured by an observer at rest relative to the clock.

The **special theory of relativity** deals with problems that arise when one frame of reference moves at a constant velocity with respect to another. Its two postulates are that all laws of nature are the same in all frames of reference moving relative to one another at constant velocity and that the speed of light in vacuum is the same in all such frames of reference.

The **general theory of relativity** deals with problems that arise when one frame of reference is accelerated with respect to another.

The **relativity of mass** refers to the greater mass of an object when measured by an observer in relative motion as compared with the mass measured by an observer at rest relative to the object.

The **rest energy** of the object is the product of its mass measured when it is at rest and the square of the speed of light, namely m_0c^2.

The **salinity** of sea water refers to its salt content, which averages 3.5 percent.

A **salt** is one of a class of ionic compounds most of which are crystalline solids at ordinary temperatures and most of which consist of a metal combined with one or more nonmetals. Any salt can be formed by mixing the appropriate acid and base and evaporating the solution to dryness.

A **satellite** is, in general, an astronomical body revolving about some other body. Usually the term refers to the satellites of the planets; thus the moon is a satellite of the earth.

A **saturated solution** is one that contains the maximum amount of solute at a given temperature.

The **scientific method** is a general scheme for attacking scientific problems that may be thought of as consisting of four steps: (1) *formulating a problem* concerning some aspect of the physical world; (2) *observation*, that is, the collection of facts bearing upon the problem; (3) *generalization*, that is, the statement of the pattern to which the observed facts seem to conform, or an explanation of the observations in terms of simpler patterns or processes; (4) *checking the generalization* by performing new experiments or making new observations to verify predictions made on the basis of the generalization.

Sea-floor spreading pictures continental drift as originating in the moving apart of huge lithosphere plates due to upwelling of molten rock at oceanic ridges.

Sedimentary rocks consist of materials derived from other rocks that have been decomposed by water, wind, or glacial ice; they may be fragments cemented together of material precipitated from water solution.

Sediments consist of material that has been transported by erosional agents and then deposited.

A **seismograph** is a sensitive instrument designed to detect earthquake waves.

An electron **shell** in an atom consists of all the electrons having the same principal quantum number n. When a particular shell contains all the electrons possible it is called a *closed shell*. An electron *subshell* in an atom consists of all the electrons having both the same principal quantum number n and the same orbital quantum number l. When a particular subshell contains all the electrons possible it is called a *closed subshell*.

A **sill** is a sheetlike mass of igneous rock lying parallel to preexisting rock layers.

Soil is a mixture of rock fragments, clay minerals, and organic matter. The latter is largely partially decomposed plant debris called humus.

A **solar prominence** is a large sheet of luminous gas projecting from the solar surface.

A **solution** is a homogeneous combination of elements or compounds without any definite proportions. The constituents of a solution retain most of their original properties. In a solution containing two substances, the one present in the larger amount is the *solvent* and the other the *solute*. When solids or gases are dissolved in a liquid, the liquid is always considered the solvent.

The **specific heat** of a substance is the heat required to change the temperature of 1 kg of the substance by 1°C.

The **spectral sequence** of the stars is an arrangement of different types of stars according to their spectra; at one end of the sequence are stars in whose spectra lines of hydrogen and helium are most prominent; at the other end are stars in whose spectra lines of neutral metals and simple compounds are most prominent. The former are hot, bluish-white stars; the latter are relatively cool, red stars.

The lines in the spectrum of an element fall into several **spectral series**, in which the frequencies of the lines are related by simple formulas.

A **spectroscope** is a device for analyzing a beam of light into its component colors.

A **spectrum** is the band of different colors produced when a light beam passes through a glass prism or is diffracted by a device called a *grating*. An *emission spectrum* is one produced by a light source alone; it may be a *continuous spectrum*, with all colors present, or a *bright-line spectrum*, in which only a few specific wavelengths characteristic of the source appear. An *absorption spectrum* is one produced when light from an incandescent source passes through a cool gas; it is also called a *dark-line spectrum* because it appears as a continuous band of colors crossed by dark lines corresponding to characteristic wavelengths absorbed by the gas.

The **speed** of a body is the rate at which it covers distance; more precisely, it is the distance that the body moves in a period of time divided by that period of time. Speed is expressed in such units as feet per second (ft/sec), meters per second (m/sec), miles per hour (mi/hr), and so on.

Spiral nebulae are immense, spiral-shaped aggregates of stars which closely resemble our galaxy.

A **star** is a large, self-luminous body of gas held together gravitationally and obtaining its energy from nuclear fusion reactions in its interior.

Stellar brightness. The *apparent* brightness of a star is its brightness as seen from the earth; the *intrinsic* brightness of a star is a measure of the total amount of light it radiates into space.

The **stratosphere** is the relatively constant-temperature part of the atmosphere from the top of the troposphere to an altitude of about 50 km.

The **structural formula** of a compound is a diagram that shows the arrangement of the atoms in its molecules.

A **sunspot** is a dark marking on the solar surface; sunspots range up to some thousands of miles across, last from several days to over a month, and have temperatures as much as 1000°C cooler than the rest of the solar surface. The *sunspot cycle* is a regular variation in the number and size of sunspots whose period is about 11 years.

A **supernova** is a star that explodes with spectacular brightness so that it may be visible even in daylight and emits vast amounts of material into space.

A **supersaturated solution** is one in which more solute is dissolved than normally possible at that temperature; it is unstable and the excess solute readily crystallizes out.

The **temperature** of a body is a measure of the average energy of random motion of its constituent particles; when two bodies are in contact, heat flows from the body at the higher temperature to the one at the lower temperature.

The first law of **thermodynamics** is the law of conservation of energy; the second law of thermodynamics states that, in every energy transformation, some of the original energy is always changed into heat energy not available for further transformations.

A **thermometer** is a device for measuring temperature. In the *Fahrenheit* temperature scale, the freezing point of water is defined as $32°F$ and the boiling point of water as $212°F$; in the *centigrade* temperature scale, the freezing point of water is defined as $0°C$ and the boiling point of water as $100°C$.

The **tides** are twice-daily rises and falls of the ocean surface. They are produced by *tractive forces* parallel to the earth's surface that in turn are due to the different attractive forces exerted by the moon on different parts of the earth. *Spring tides* have a large range between high and low water; they occur when the sun and moon are in line with the earth and thus add together their tide-producing actions. *Neap tides* have small ranges and occur when the sun and moon are $90°$ apart relative to the earth and thus their tide-producing actions tend to cancel each other out.

Till is the material that is left behind when a glacier recedes.

A **transformer** is a device that transfers electrical energy in the form of alternating current from one coil to another by means of electromagnetic induction.

In a **transverse wave** the particles or fields constituting the wave oscillate at right angles to the direction of motion of the wave. Waves in a stretched string are transverse waves.

The **troposphere** is the lower part of the atmosphere, from sea level to an altitude of about 11 km, in which most weather phenomena take place.

The **uncertainty principle** states that it is impossible to determine simultaneously accurate values for the position and velocity of a very small particle, notably an electron. Hence, in dealing with electrons within atoms, all we can consider are probabilities rather than specific positions and states of motion.

An **unconformity** is an uneven surface separating two series of rocks. It is a buried surface of erosion involving at least three geologic events: diastrophic uplift and exposure of older rocks, erosion, and the deposit of sediments on the eroded landscape.

The **valence** of an element or atomic group refers to its combining ability. An atom that donates n electrons to a chemical bond is said to have a valence of $+n$, and an atom that receives m electrons from a chemical bond is said to have a valence of $-m$. Metals have positive valences, nonmetals negative valences. In covalent compounds the valence of an element or atomic group is the number of electron pairs that are shared.

Van der Waals forces have their origin in the electrostatic attraction between asymmetrical charge distributions in atoms and molecules.

The **vapor pressure** of a substance at a given temperature is the pressure its vapor exerts when confined above a sample of the substance. It is a measure of how readily the molecules of the substance escape from its surface.

A **variable star** is one whose brightness continually changes.

A **vector** is an arrow whose length is proportional to the magnitude of some quantity and whose direction is that of the quantity; a *vector quantity* is a quantity that has both magnitude and direction.

Veins consist of dissolved material deposited along cracks in rocks.

The **velocity** of a body refers both to its speed and the direction of its motion. A car has a speed of 30 mi/hr; its velocity is 30 mi/hr to the northwest.

Viscosity is the resistance of fluids to flowing motion; liquids are more viscous than gases.

Volcanic rocks originated when molten lava of volcanic origin cooled rapidly at the earth's surface.

Vulcanism refers to geological processes involving the movement of molten rock.

The variable quantity that characterizes the matter waves of a moving particle is called its **wave function**, symbol ψ. The probability of finding the particle at a certain place at a certain time is proportional to the value of ψ^2.

In **wave motion** a change or distortion in a medium is propagated. A wave carries energy from one place to another, but there is no net transport of matter.

The **wavelength** of a wave is the distance between adjacent crests, in the case of transverse waves, or between adjacent compressions, in the case of longitudinal waves.

Weather refers to the state of the atmosphere in a particular place at a particular time, whereas *climate* refers to the weather trends in a region through the years.

Weathering is the surface disintegration of rock by chemical decay and mechanical processes such as the freezing of water in crevices.

The **weight** of a body is the force with which gravity pulls it toward the earth.

Work is the product of a force and the distance through which it acts. If the force is not parallel to its displacement, the component of the force parallel to the displacement must be used in calculating the work it does. The unit of work in the British system is the *foot-pound* (*ft-lb*); in the metric system, the *joule*.

X-rays are high-frequency electromagnetic waves produced whenever fast electrons are brought to rest quickly.

Answers to Odd-numbered Exercises

CHAPTER 1

1 *a.* Velocity zero, acceleration to right.
b. Velocity to right, acceleration zero.
c. Velocity zero, acceleration to left.

3 The ball will remain stationary with respect to the barrel, since both are falling with the same acceleration.

5 3×10^5 km/sec = 186,000 mi/sec.

7 4.8 sec.

9 A straight line sloping upward to the right; a straight line sloping downward to the right.

11 y is directly proportional to $x + 3$, with $y = \frac{1}{2}(x + 3)$.

13 Here $v_1 = 50$ mi/hr and $v_2 = 0$, so $a = (v_2 - v_1)/t = (0 - 50 \text{ mi/hr})/3$ sec $= -16.7$ (mi/hr)/sec $= -6 \times 10^4$ mi/hr².

15 Here $v_1 = 20$ ft/sec and $v_2 = 14$ ft/sec, so $a = (v_2 - v_1)/t = (14 \text{ ft/sec} - 20 \text{ ft/sec})/4$ sec $= -1.5$ ft/sec².

17 19.8 m/sec downward; 59 m/sec downward. [$v_2 = v_1 + at = 10$ mi/sec + $(9.8 \text{ m/sec}^2)t$.]

19 3 sec; 6 sec; 96 ft/sec downward. [At the highest point, $v_2 = 0$ and $v_1 = gt$.]

21 12 sec. [$a = (50 \text{ mi/hr})/20$ sec $= 2.5$ (mi/hr)/sec. Use this value of a and $v_1 = 50$ mi/hr in $v_2 = v_1 + at$ to find t.]

23 *a.* 0.61 sec.
b. 0.59 sec.
[The constant speed of the elevator in (*a*) is shared by the stone when it is dropped, hence the acceleration of the stone *relative to the elevator* is 32 ft/sec²; in (*b*) it is 34 ft/sec². In both cases $d = \frac{1}{2}at^2$, so $t = \sqrt{2d/a}$.]

CHAPTER 2

1 The moon's motion around the earth is accelerated because the moon's direction of motion continually changes. Because it is accelerated, a force must be acting upon it. The force is directed toward the earth.

3 A person who jumps onto loose earth is slowed down more gradually than if he jumps onto concrete, and so the force acting on him is less.

5 The forces act on different bodies.

7 When the pole is used, the vertical component of the force goes into pressing the sled against the snow, which increases the frictional resistance to its motion. When the rope is used, the vertical component of the force goes into lifting the sled, which decreases the frictional resistance.

9 15 lb; 0.

11 No, it falls during its entire flight; no; yes, because air resistance varies with size and shape; only to the last question, since without air resistance, the rate of fall is independent of size and shape.

13 30 newtons.

15 The force is his weight plus his mass times his upward acceleration, or 944 newtons.

17 60 pounds.

19 50 miles.

21 A wind from the northwest blows toward the southeast. The wind has southerly and easterly components of $(10 \sin 45°)$ mi/hr = $(10 \cos 45°)$ mi/hr = 7.07 mi/hr. Relative to the boat, the smoke's initial velocity is 25 mi/hr to the east, so its final velocity relative to the boat has an easterly component of 32.07 mi/hr and a southerly component of 7.07 mi/hr. Their vector sum is 33 mi/hr at 168° from the boat's heading.

23 The calculated speeds are 100.5, 111.4, and 140.0 m/sec.

25 27.5 newtons.

27 12.9 lb.

CHAPTER 3

1 Sprinters could not improve their time in the 60-yd dash on the moon because their masses are the same there as on the earth. With the force that their legs can exert also unchanged, their acceleration will be the same, and hence their motion will not differ from that on the earth.

3 At the equator; at the poles. The reason is that the equator is farthest from the earth's axis of rotation, while the poles are on the axis.

5 The sun's gravitational pull on the earth varies during the year since the distance from the earth to the sun varies.

7 The earth must travel faster when it is nearest the sun in order to counteract the greater gravitational force of the sun.

9 12.5 newtons.

11 Here $m = 3{,}500 \text{ lb}/(32 \text{ft/sec}^2) = 109$ slugs, $r = 1{,}200$ ft, and $F_c = 800$ lb. Hence $v = \sqrt{rF_c/m} = 94$ ft/sec = 64 mi/hr.

13 3.05×10^{-7} m/sec^2. [Hint: The tip of the minute hand travels $2\pi r$ in 60 min = 3,600 sec.]

15 The centripetal force mv^2/r must equal the weight mg. Hence $v = \sqrt{rg} = 11.31$ ft/sec. The circumference of the circle is $2\pi r = 8\pi$ ft = 25.13 ft, so $t = 25.13 \text{ ft}/(11.31 \text{ ft/sec}) = 2.2$ sec.

17 4.4 lb.

19 4.17×10^{23} newtons; 1.3×10^4 m/sec. [From Table 3.1, Jupiter's mass is 318 times the earth's mass, which is given in the text as 5.98×10^{24} kg. For the second part of the problem, setting equal the gravitational and centripetal forces yields $v = \sqrt{Gm_{\text{sun}}/r}$.]

21 The force on each mass is 6.67×10^{-10} newton. The acceleration of the 2-kg mass is 3.33×10^{-10} m/sec^2, and that of the 5-kg mass is 1.33×10^{-10} m/sec^2.

CHAPTER 4

1 Yes, because all changes require the performance of work.

3 When it is farthest from the sun; when it is closest to the sun. The work needed to pull a planet away from the sun to a given distance increases with the distance, so the planet's PE is greatest the farthest it is from the sun. The gravitational force of the sun on a planet is greatest when it is closest to the sun, hence its speed is also greatest there in order that gravitational and centripetal forces be in balance.

ANSWERS TO ODD-NUMBERED EXERCISES

5 The resulting water will add to the oceans, and more of the earth's mass will be distant from its axis than before. Conservation of angular momentum requires that the earth in that event spin more slowly to compensate, and the day will be longer.

7 0.102 m; 1 m.

9 5 kg.

11 11.3 m; 10,000 joules.

13 The PE of the hammer is initially $wh = 19{,}000$ ft-lb, and the work done is $F_{av} \times 0.5$ ft. Setting these equal yields $F_{av} = 38{,}000$ lb.

15 900 ft-lb; 800 ft-lb; friction in pulleys.

17 From conservation of momentum, $m_1 v_1 + m_2 v_2 = (m_1 + m_2)V$ in general, where V is the final speed of the composite body. Here $v_2 = 0$ and $m_1 = m_2$, so that $V = v_1/2 = 22$ ft/sec. The difference between $\frac{1}{2} m_1 v_1^2$ and $\frac{1}{2}(m_1 + m_2)V^2$ is 53,000 ft-lb, which is the KE lost in the collision.

19 16 ft/sec.

CHAPTER 5

1 The two postulates are (1) all laws of nature are the same in all systems of reference moving uniformly relative to one another; and (2) the speed of light in vacuum is the same in all systems of reference moving uniformly relative to one another. These postulates had to have experimental verification.

3 The length of the rod is greatest in the frame of reference in which it is at rest.

5 $L = L_0 \sqrt{1 - v^2/c^2}$
$= 6 \text{ ft} \times \sqrt{1 - \dfrac{(1.5 \times 10^8)^2}{(3 \times 10^8)^2}}$
$= 6 \text{ ft} \times \sqrt{1 - \frac{1}{4}}$
$= 6 \text{ ft} \times \sqrt{0.75}$
$= 5.2 \text{ ft.}$

7 It is necessary to solve $\sqrt{1 - v^2/c^2} = 0.99$ for v. The result is
$v = c\sqrt{1 - (0.99)^2}$
$= 3 \times 10^8 \times \sqrt{1 - (0.99)^2} \text{ m/sec}$
$= 4.2 \times 10^7 \text{ m/sec.}$

9 A time interval measured as t_0 in the rocket ship is measured as $t = t_0/\sqrt{1 - v^2/c^2} = 1.15 t_0$ on the earth. Given that $t - t_0 = 60$ sec, then $t_0 = t - 60$ sec, and $t = 1.15(t - 60$ sec). Solving for t yields 460 sec = 7 min 40 sec.

11 1 ev = 1.6×10^{-19} joule = mc^2. Hence $m = 1.6 \times 10^{-19}$ joule/$(3 \times 10^8 \text{ m/sec})^2 = 1.8 \times 10^{-36}$ kg.

13 3.6×10^{26} watts. [4×10^9 kg $\times (3 \times 10^8 \text{ m/sec})^2 = 3.6 \times 10^{26}$ joules and 1 watt = 1 joule/sec.]

CHAPTER 6

1 The highest temperature that water can have while remaining liquid at atmospheric pressure is its boiling point. Increasing the rate of heat supply thus increases the amount of steam produced without changing the temperature of the boiling water.

3 Alcohol has a smaller specific heat than water and hence is less efficient at absorbing heat from the engine and carrying it to the car's radiator where it is dissipated into the atmosphere.

5 A piece of ice at 0°C is more effective in cooling a drink than the same weight of water at 0°C because of the latent heat of fusion which must be added to the ice before it melts. Hence the ice will absorb more heat from the drink than the cold water.

7 $-297°F$.

9 $37.8°C$.

11 3.72 kcal; 3.72 kcal.

13 The heat gained by the punch equals the heat lost by the silver. If the final temperature of the mixture is T, then $m_p c_p (T - 0°C) = m_s c_s (20°C - T)$, and $T = 2.14°C$.

15 620 kcal. [The heat of vaporization of water is 540 kcal/kg, and 80 kcal/kg are lost by water whose temperature falls from 100°C to 20°C since the specific heat of water is 1 kcal/kg·°C.]

17 The stone's energy is $mgh = 9,800$ joules. The mechanical equivalent of heat is 4,185 joules/kcal, so the stone's energy is equal to $(9,800/4,185)$ kcal $= 2.34$ kcal. Since $Q = mc\,\Delta T$, $\Delta T = Q/mc = 2.34$ kcal/$(10^4$ kg $\times$ 1 kcal/kg·°C$) = 2.34 \times 10^{-4}$°C.

19 Since 1 watt = 1 joule/sec, the energy given off by a 600-watt heater in 1 hr is 600 joules/sec $\times$ 3600 sec $= 2.16 \times 10^6$ joules. Since 4,185 joules = 1 kcal, this energy is equivalent to $(2.16 \times 10^6/4,185$ kcal $= 516$ kcal.

21 The heat of fusion of a mass m of ice at 0°C is $80\,m$ kcal $= 80\,m$ kcal $\times$ 4,185 joules/kcal $= 3.348 \times 10^5\,m$ joules. Setting this equal to mgh yields $h = (3.348 \times 10^5/9.8)\,m = 3.4 \times 10^4$ m.

CHAPTER 7

1 The block will stay where it is, since there is no water underneath it to furnish a buoyant force.

3 The pan containing the kilogram of lead goes down because the feathers, whose density is less, experience the greater upward force due to air buoyancy.

5 Lower; lower.

7 1.6 ft.

9 4.5 lb/in.^{2}

11 Atmospheric pressure corresponds to the pressure of about 76 cm (30 in.) of mercury. Since the density of water is only 1/13.6 as great, a water column 13.6×76 cm $= 1,034$ cm $= 10.34$ m high (or 408 in. $= 34$ ft high) would be involved in a water barometer.

13 2,000 lb/(62 lb/ft^3) = 32 ft^3; 2,000 lb/(4.8 $\times$ 10^2 lb/ft^3) = 4.2 ft^3.

15 The volume of water whose mass is 60 kg is 60 kg/(10^3 kg/m^3) = 0.06 m^3. By Archimedes' principle, the raft rises by a height h such that its submerged volume decreases by 0.06 m^3 when the man dives off. Hence $h = 0.06$ m^3/(3 m $\times$ 2 m) $= 0.01$ m $= 1$ cm.

17 The density of gold is 19 g/cm^3, so 50 g of gold has a volume of (50/19) cm^3 = 2.63 cm^3. Hence the bracelet is not pure gold. To find the proportion of gold it contains, we note that $m_{gold} + m_{lead} = 50$ g and $(m_{gold}/d_{gold}) + (m_{lead}/d_{lead}) = 4.0$ cm^3. From these equations we find that $m_{gold} = 14$ g and $m_{lead} = 36$ g, so that the bracelet is 28 percent gold by weight.

19 333 cm^3; 2,000 cm^3.

21 The ratio of absolute pressures equals the ratio of absolute temperatures, which is $300°K/273°K = 1.099$. The absolute pressure of the air in the tire at $0°C$ is $(14.7 + 24)$ lb/in.^{2} = 38.7 lb/in.^{2}, hence its absolute pressure at $27°C$ is 42.5 lb/in.^{2} and the corresponding gauge pressure is 27.8 lb/in.^{2}

CHAPTER 8

1 Bombardment by air molecules does not produce brownian movement in large objects because the mass of the air molecules is so much smaller than that of the objects.

3 The thermal energy of a solid resides in oscillations of molecules about their equilibrium positions.

5 No change, since the average molecular speed depends only upon the gas temperature.

7 By heating it gradually; if it is glass, it will sag slowly, but this will not occur if it is a crystalline solid.

9 The pressure doubles because doubling the number of molecules present doubles the number of molecular collisions per second with the walls of the tank.

11 0.267 mi/sec.

13 5.65×10^{-21} joule; 7.72×10^{-21} joule. [KE = $\frac{3}{2}kT$, where $k = 1.38 \times 10^{-23}$ joule/$°K$ and T is the absolute temperature.]

15 Alcohol; $78°C$; $94°C$; 0.46 cm Hg; increase pressure to 169.95 cm Hg.

CHAPTER 9

1 No. Air could equally well be described as a continuous elastic fluid from the point of view of sound propagation.

3 The wave energy "missing" at locations of destructive interferences supplements the normal wave energy at locations of constructive interference to give a greater amplitude there than if no interference occurred. The total energy remains the same as it would be without interference, but is differently distributed.

5 The pitch change is too small to be detected when the speed of relative motion is negligible relative to the speed of sound.

7 0.12 cycles/sec = 0.12 Hz.

9 The speed of sound is 1,100 ft/sec, so the wavelength of the waves is (1,100 ft/sec)/1,044 sec^{-1} = 1.054 ft. Hence there are 50 ft/1.054 ft = 47 waves in a wavetrain 50 ft long, which means that the string vibrated 47 times while its sound traveled 50 ft.

11 The speed of light is so much greater than that of sound that it may be assumed there is a negligible time interval between the lightning flash and the time it is seen. Hence the distance is $v_s t$ = 1,100 ft/sec × 4 sec = 4,400 ft.

13 The distance of the man from the spike is $v_s t$ = 1,100 ft/sec × 2 sec = 2,200 ft. Hence the speed of sound in the rail is 2,200 ft/0.14 sec = 15,700 ft/sec.

CHAPTER 10

1 The various possible experiments include the use of an electroscope to show that all charges either add to or cancel out charges of vitreous or resinous origin placed on it initially, and the use of

suspended pith balls to show that all electrostatic charges act as though of either vitreous or resinous origin.

3 This is still a useful notion insofar as the transfer of heat and electric current energy is concerned. The experiment evidence against heat as a fluid is that the supposed fluid cannot have mass or, indeed, any other physical properties, whereas the kinetic theory of heat is a result of experimental observations. The evidence against electricity as a fluid is similar, with the discovery of the motion of electrons and ions whenever current flows contributing support.

5 There are many reasons. An obvious one is that all gravitational forces are observed to be attractive, whereas if they were electrostatic in origin some would have to be repulsive.

7 It will not move; it will rotate about its center.

9 *a.* 0.00022 newton.
 b. 0.004 newton.
 c. 0.018 newton.
 d. 0.001 newton.

11 $1 \text{ coul}/(1.6 \times 10^{-19} \text{ coul/electron}) = 6.25 \times 10^{18}$ electrons must be added. Their total mass would be 5.7×10^{-12} kg. This is an almost impossible experiment since the mutual repulsion of the electrons is so great that they would continually fly off the pith ball.

13 $E = F/q_1$, where F is the force on a charge of $q_1 = 1$ coul located at the point in question. Hence $E = Kq_1q_2/q_1r^2 = Kq_2/r^2 = (9 \times 10^9 \text{ newton-m}^2/\text{coul}^2 \times 7 \times 10^{-6} \text{ coul})/(0.4 \text{ m})^2 = 3.94 \times 10^5$ newtons/coul.

15 5×10^{-4} newton; 5×10^{-4} joule. [The work done on the particle by the electric field is force × distance, and the particle's kinetic energy is equal to this amount of work.]

CHAPTER 11

1 Bring up a glass rod that has been stroked with a silk cloth so that it is positively charged. Then, without touching the rod to the electroscope, ground the electroscope. Remove the ground, and then remove the rod; the electroscope will now be negatively charged.

3 The production of electricity by friction involves the removal of electrons from one of the contacting substances, which is then positively charged, and their adherence to the other substance, which is then negatively charged. Since the number of electrons missing from the positively charged substance is equal to the number of excess electrons on the negatively charged substance, the amounts of charge on both are the same.

5 Cathode rays are deflected by electric and magnetic fields, so they possess charge, and from the amount of deflection their charge-to-mass ratio can be determined, so they possess mass.

7 A person taking a bath is grounded through the conducting bath water and the drain piping, and so much larger currents can flow through his body if he is in contact with a source of current than otherwise.

9 Electrons have little inertia, and so are able to go around bends readily.

11 0.625 amp; 192 ohms; 75 watts.

13 3,600 watts; 36 bulbs.

15 60 kcal of heat must be added, which is 60 kcal × 4,185 joules/kcal = 2.51×10^5 joules. Since 1 kw-hr = 10^3 watts × 3,600 sec = 3.6×10^6 joules, 60 kcal = 2.51×10^5 joules/(3.6×10^6 joules/kw-hr) = 0.0697 kw-hr. At 3¢/kw-hr, the cost is 0.2¢.

17 1.55 amp.

ANSWERS TO ODD-NUMBERED EXERCISES

19 KE = $\frac{1}{2}mv^2$ = $\frac{1}{2}$ × 9.1 × 10^{-31} kg × $(10^6$ m/sec$)^2$ = 4.55 × 10^{-19} joule. Since 1 ev = 1.6 × 10^{-19} joule, this energy is equal to 2.84 ev.

21 The potential difference across the entire circuit is $V = iR$, where R is the combined resistance of R_1 and R_2. Hence $iR = iR_1 + iR_2$, and $R = R_1 + R_2$.

CHAPTER 12

1 To prepare such a map, worldwide measurements of the magnitude and direction of the earth's magnetic field are required.

3 *a.* Alternating current.
b. Direct current.
c. Alternating current.

5 With a direct current, the coil rotates in the magnetic field until it is in equilibrium; with an alternating current, the coil oscillates back and forth.

7 The moving electrons constitute electric currents, which give rise to magnetic fields.

9 0° or 180°; 90°.

11 Because the flux through the loop undergoes both an increase and a decrease in each rotation.

13 North; south.

15 The ratio of turns is 15:1. Since there are more turns in the secondary winding, the potential difference across it is 15 × 550 volts = 8,250 volts and the current in it is 10 amp/15 = 0.67 amp.

CHAPTER 13

1 The electric and magnetic fields of an electromagnetic wave are perpendicular to each other and to the direction of propagation.

3 When light is absorbed, the absorbing material is heated.

5 Blue stars have the highest temperature, red stars the lowest.

7 Red; red; black.

11 Irregularities in the atmosphere cause the light from a star to be randomly deviated to a small extent by refraction. These deviations are perceived as twinkling; they are not apparent to an observer above the atmosphere.

13 The wavelengths in visible light are very small relative to the size of a building, whereas those in radio waves are more nearly comparable.

15 The listener 3,000 miles away. Sound takes 100 ft/(1,100 ft/sec) = 0.091 sec to travel 100 ft, but light takes only 3,000 miles/(186,000 mi/sec) = 0.016 sec to travel 3,000 miles.

17 Light takes 14 miles/(186,000 miles/sec) = 7.53 × 10^{-5} sec to make a round trip. In this time the wheel must turn through 1/100 of a complete revolution, so it must make 1/(100 × 7.53 × 10^{-5} sec) = 133 revolutions/sec.

19 The mirror must be half your own height. It does not matter how far away you are.

CHAPTER 14

1 Electrons possess mass, while photons do not. Electrons possess charge, while photons do not. Electrons may be stationary or move with velocities of up to almost the speed of light, while photons always travel with the speed of light. Electrons are constituents of ordinary matter, while photons are not. The energy of a photon depends upon its frequency, while that of an electron depends upon its velocity.

3 Diffraction and interference are much easier to demonstrate than such quantum phenomena as the photoelectric effect.

5 In the photoelectric effect, the entire photon energy is absorbed by an electron, while in the Compton effect, only part of the photon energy is absorbed.

7 Each atom in a solid is restricted to a definite small region of space, since otherwise the assembly of atoms would not be a solid. Because the uncertainty in the position of each atom must therefore be finite, Δmv cannot be 0, and the momentum (and hence energy) of the atom cannot be 0. In the case of an ideal gas, there is no restriction on the position of each molecule, and in a large container Δx can be essentially ∞. Hence $\Delta mv = 0$, which means that it is possible for the molecular energy to be 0 (as predicted by the kinetic theory of gases) without violating the uncertainty principle.

9 1.3×10^{-17} joule; 1.3×10^{-28} joule.

11 3 photons.

13 The frequency corresponding to a wavelength of 5.5×10^{-7} m is $f = c/\lambda = (3 \times 10^8 \text{ m/sec})/5.5 \times 10^{-7} \text{ m} = 5.45 \times 10^{14} \text{ sec}^{-1}$, so $hf = 6.63 \times 10^{-34}$ joule-sec $\times 5.45 \times 10^{14} \text{ sec}^{-1} = 3.61 \times 10^{-19}$ joule. Hence there are $(1{,}400 \text{ joules/m}^2\text{-sec})/(3.61 \times 10^{-19} \text{ joule/photon}) = 3.88 \times 10^{21}$ photons/m²-sec reaching the earth from the sun.

15 The energy of the scattered photon is 5×10^4 ev $\times 1.6 \times 10^{-19}$ joule/ev $= 8 \times 10^{-15}$ joule. Hence its frequency is $f = E/h = 8 \times 10^{-15}$ joule/6.63×10^{-34} joule-sec $= 1.21 \times 10^{19}$ sec$^{-1} = 1.2 \times 10^{19}$ Hz.

17 a. The electron's mass is $m = m_0/\sqrt{1 - v^2/c^2} = 9.1 \times 10^{-31}$ kg$/\sqrt{0.75} = 1.05 \times 10^{-30}$ kg. Hence its wavelength is $\lambda = h/mv = 4.2 \times 10^{-12}$ m.
b. 1.5×10^{-38} m.

19 Since 1 ev $= 1.6 \times 10^{-19}$ joule, the energy of each electron is 6.4×10^{-15} joule and, from KE $= \frac{1}{2}mv^2$, its speed is

$$v = \sqrt{\frac{2 \times 6.4 \times 10^{-15} \text{ joule}}{9.1 \times 10^{-31} \text{ kg}}}$$
$$= 1.19 \times 10^8 \text{ m/sec.}$$

The corresponding de Broglie wavelength is $\lambda = h/mv = 6.63 \times 10^{-34}$ joule-sec$/(9.1 \times 10^{-31}$ kg $\times 1.19 \times 10^8$ m/sec$) = 6.12 \times 10^{-10}$ m.

CHAPTER 15

1 The change from water to ice is a physical change because chemically the substance remains the same; the only differences between ice and water are in their physical properties.

3 Water may be shown to be a compound rather than a mixture by decomposing it into hydrogen and oxygen and showing that the ratio between the amounts of these evolved gases is a constant in all samples.

5 The idea that air is an element may be shown false by demonstrating that it is composed of more than one gas.

7 The law of conservation of mass is based upon experiments involving the precision determination of weights; adequate balances did not exist before the time of Lavoisier.

9 See text. The answer to the last part of the question is no.

11 Homogeneous: carbon dioxide gas, solid carbon dioxide, iron, rust, air, oxygen, salt.

ANSWERS TO ODD-NUMBERED EXERCISES

13 They agree that the positive matter in the atom has much greater mass than the negative matter, but in the Thomson model the positive matter occupies the entire atomic volume whereas it occupies only a tiny region at the center of the atom in the Rutherford model.

15 Two.

17 If the hydrogen molecule contained only one atom, the combination of one volume of hydrogen with any amount of chlorine to give hydrogen chloride would result in only one volume of hydrogen chloride. Since two volumes of hydrogen chloride are actually produced, the hydrogen molecule must contain at least two atoms of hydrogen.

CHAPTER 16

1 *a.* The statement is no longer exact. Experiments involving the bombardment of atomic nuclei with other particles have shown that atoms may be changed into other atoms or even be completely decomposed into their constituent neutrons, protons, and electrons. A correct statement would be that atoms are indivisible and indestructible by ordinary chemical or physical means.
b. This statement is no longer exact, since it does not take into account the conversion of mass into energy and energy into mass. A correct statement would be that the total amount of energy plus mass energy in the universe is constant.
c. This statement is still correct.
d. This statement is no longer exact, since the atoms of which an element consists may be decomposed as in (*a*).
e. This statement is no longer exact, since experiments have shown that isotopes, which are atoms of an element having different atomic masses, exist. A correct statement would be that all atoms of any one element have the same number of protons in their nuclei and the same number of electrons in their electron shells.

3 Oxygen: $8p$, $8n$, $8e$; iron: $28p$, $28n$, $28e$; iodine: $53p$, $74n$, $53e$; bismuth: $83p$, $126n$, $83e$.

5 The chief difference is that in fission heavy nuclei split into lighter ones, while in fusion light nuclei join to form heavier ones. The chief similarity is that in both processes mass is converted into energy.

7 Neutrons are uncharged and so are not repelled by the positively charged nuclei; hence it is easier for them to enter nuclei and react with them.

9 *a.* 9×10^{13} joules.
b. 2.25×10^4 tons.

11 6, 12; carbon.

13 $_1H^3$: 0.0092 amu, 8.6 Mev.
$_2He^3$: 0.0084 amu, 7.8 Mev.

CHAPTER 17

1 Radium is considered an element because its spontaneous decomposition into radon and helium cannot be affected by ordinary chemical or physical means.

3 Electrons, positrons, neutrinos, alpha particles, photons.

5 More, because the neutron mass is greater than the proton mass.

7 The mass number is reduced by 4 and the atomic number by 2, since the mass number of an alpha particle is 4 and its atomic number is 2. The residual nucleus therefore has an atomic number of 86 and a mass number of 222.

9 92; 233; uranium.

11 0.5 kg; 0.125 kg; radon and helium.

CHAPTER 18

1 *a.* An absorption line spectrum, since the continuous spectrum emitted by the sun must pass through the cooler solar atmosphere.
   b. A continuous emission spectrum.
   c. An emission line spectrum.
   d. An absorption line spectrum.

3 The Bohr theory assumes that the position and velocity of each electron in an atom may be definitely known at the same time, which is prohibited by the uncertainty principle.

5 The results of quantum mechanics are in better quantitative agreement with experiment than those of the Bohr theory.

7 The electron's total energy is governed by n, the magnitude of its angular momentum by l, the direction of its angular momentum by m_l, and the direction of its spin by m_s.

9 The centripetal force mv^2/r on the electron is provided by the electrostatic attraction Ke^2/r^2 of the hydrogen nucleus. Hence $mv^2/r = Ke^2/r^2$ and $v = \sqrt{Ke^2/mr}$. In the $n = 1$ orbit, $r = 5.3 \times 10^{-11}$ m, so

$$v = \sqrt{\frac{9 \times 10^9 (1.6 \times 10^{-19})^2}{9.1 \times 10^{-31} \times 5.3 \times 10^{-11}}} \text{ m/sec}$$
$$= 2.19 \times 10^6 \text{ m/sec}$$

This speed is too small for relativistic effects to be appreciable.

11 *a.* $n = 2$ to $n = 1$.
    b. $n = 6$ to $n = 2$.
    c. $n = 1$ to $n = 2$.

13 The average molecular energy at the absolute temperature T is $\tfrac{3}{2}kT$ and the binding energy of the hydrogen atom is 2.2×10^{-18} joule. Hence $\tfrac{3}{2}kT = 2.2 \times 10^{-18}$ joule and

$$T = \frac{2 \times 2.2 \times 10^{-18} \text{ joule}}{3 \times 1.38 \times 10^{-23} \text{ joule}/°K}$$
$$= 1.06 \times 10^5 \text{ °K}$$

CHAPTER 19

1 Sodium is a very active metal, and so it combines readily, whereas platinum is highly inactive, and therefore does not tend to combine at all.

3 *a.* Solid.
   b. 2.
   c. Slightly soluble.
   d. HAt.
   e. KAt, CaAt$_2$.
   f. Less stable.

5 2.

7 2, with opposite spins.

CHAPTER 20

1 Both are ionic compounds.

3 Electrons are liberated from metals illuminated by light more easily than from nonmetals because the outer electrons of metal atoms are less tightly bound, which is also the reason they tend to form positive ions. Electrons are most readily liberated from metals in Group I of the periodic table.

5 The two isotopes of chlorine are identical in atomic structure except for a difference in the number of neutrons in their respective nuclei. Since their electron structures are the same, the chemical behavior of the two isotopes is the same.

7 Any number of Li atoms can join together to form a solid since there are seven vacant $n = 2$ states in an Li atom whereas there is only one vacant $n = 1$ state in an H atom. See page 407 for a further discussion.

9 An orbit is the path followed by a certain particle; an orbital is the wave function that describes the probability of finding a certain particle in a certain region of space.

11 A chlorine ion has a closed outer shell, whereas a chlorine atom lacks an electron of having a closed outer shell.

13 The bonding electrons are, on the average, closest to the atom with the highest electronegativity, so with the help of Table 20.3 the answer to the first part of the question is Cl, S, O, O, O. The larger the difference in electronegativity between the atoms in a compound, the greater the ionic character of the compound, so the answer to the second part of the question is NO, CuS, CO, HgO, and KCl.

CHAPTER 21

1 Ionic, NaCl; covalent, diamond; van der Waals, ice; metallic, copper. Electrostatic attraction.

3 Liquids have in common with gases the ability to flow readily. The densities of liquids and their compressibilities are similar to those of solids, however, and liquids have short-range internal structures similar to those of amorphous solids except that the groupings of liquid molecules are not stable.

5 These forces are too weak to hold inert gas atoms together to form molecules against the forces exerted during collisions in the gas phase.

7 The H^+ ion is simply a proton, whose size is very much smaller than that of an atom or ion of any other kind. Hence the electrical forces it can exert are stronger, since these forces vary as $1/r^2$ and the H^+ ion can be closer to whatever negative charge it interacts with. H_2O.

9 The volume of 1 kg of copper is $V = m/d = 1 \text{ kg}/(8.9 \times 10^3 \text{ kg/m}^3) = 1.12 \times 10^{-4} \text{ m}^3$. The energy stored in this volume of copper is $1.12 \times 10^{-4} \text{ m}^3 \times 10^7 \text{ joules/m}^3 = 1.12 \times 10^3$ joules. In heat units the stored energy equals $1.12 \times 10^3 \text{ joules}/(4,185 \text{ joules/kcal}) = 0.268$ kcal. The rise in temperature of the copper is therefore $\Delta T = Q/mc = 0.268 \text{ kcal}/(1 \text{ kg} \times 0.093 \text{ kcal/kg} - °C) = 2.9°C$. Hence the final temperature is about 203°C. This result is independent of the mass of the copper sample.

CHAPTER 22

1 +2; −3; +4; +3; +2; +2; +4.

3 $CaCO_3 \longrightarrow CaO + CO_2$

5 $2HCl \longrightarrow H_2 + Cl_2$; yes, since the same numbers of molecules are liberated.

7 a, c, d, f, g.

9 a. $2H_2 + O_2 \longrightarrow 2H_2O$
b. $2C + O_2 \longrightarrow 2CO$
c. $SO_3 + H_2O \longrightarrow H_2SO_4$
d. $2K + S \longrightarrow K_2S$
e. $Ba + H_2O \longrightarrow BaO + H_2$

11 1 ton = 2,000 lb × 454 g/lb = 9.08×10^5 g. The atomic mass of lead is 207, so there are $9.08 \times 10^5 \text{ g}/(207 \text{ g/g atom}) = 4.386$ g atoms in a ton of lead. There are N_0 atoms in a g atom, hence 4.386 g atoms × 6.02×10^{23} atoms/g atom = 2.64×10^{27} atoms in a ton of lead.

13 Ag_5SbS_4. The proportion of Ag by mass

in Ag_5SbS_4 is $539.35/789.39 = 0.68 = 68$ percent, whereas in Ag_3AsS_3 it is $323.61/494.74 = 0.65 = 65$ percent.

15 The formula mass of N_2 is $2 \times 14 = 28$, so there are 0.0042 g/$(28$ g/mole$) = 1.5 \times 10^{-4}$ moles of N_2 in the sample and 1.5×10^{-4} moles $\times 6.02 \times 10^{23}$ molecules/mole $= 9.03 \times 10^{19}$ molecules there. The volume of N_2 is 1.5×10^{-4} mole $\times 22.4$ liters/mole $= 3.36 \times 10^{-3}$ liter.

CHAPTER 23

1 Like hydrogen, sodium reduces many metallic oxides to the pure metal and burns brightly in oxygen. Both have valences of $+1$. Hydrogen and the more active metals burn brilliantly in oxygen as well as in chlorine. Chlorine and oxygen have negative valences.

3

$MgCO_3$	Magnesium carbonate
$HgSO_4$	Mercuric sulfate
SiO_2	Silicon dioxide
$AgNO_3$	Silver nitrate
$AgCl$	Silver chloride
Na_3N	Sodium nitride
K_2CO_3	Potassium carbonate
NiS	Nickel sulfide
$Al_2(SO_4)_3$	Aluminum sulfate
$Zn(NO_3)_2$	Zinc nitrate
UF_6	Uranium hexafluoride

5 *a*. Phosphorus trioxide, phosphorus pentoxide.
b. Mercurous chloride, mercuric chloride.
c. Ferrous hydroxide, ferric hydroxide.

7 Chlorine.

9 More metallic, because metallic behavior becomes more pronounced going down a given group of the periodic table; the atoms are larger, and their valence electrons are correspondingly less tightly bound. Germanium will not form compounds with hydrogen analogous to the hydrocarbons. (*a*) GeO_2; (*b*) $GeCl_4$; (*c*) Na_2GeO_3.

CHAPTER 24

1 During electrolysis, electric energy is converted to chemical energy (that is, electron potential energy). This energy change is reversed in batteries.

3 There are two simple ways of determining whether a sugar solution is saturated: (1) add some additional sugar and see if it dissolves; (2) cool the solution and see if any sugar crystallizes out.

5 A solution of an electrolyte conducts electricity, while a solution of a nonelectrolyte does not.

7 *a*. When Ag^+ is added to a solution containing Cl^-, AgCl is precipitated, but nothing happens if Ag^+ is added to a solution of NO_3^- since $AgNO_3$ is soluble.
b. When Cl^- is added to a solution containing Ag^+, AgCl is precipitated, but nothing happens if Cl^- is added to a solution of Na^+ since NaCl is soluble.
c. When $CO_3^=$ is added to a solution containing Ca^{++}, $CaCO_3$ is precipitated, but nothing happens if $CO_3^=$ is added to a solution containing Na^+ since Na_2CO_3 is soluble.
d. Cu^{++} is blue in color, while Ca^{++} is colorless.

9 NaCl will precipitate out since its solubility is less than that of KCl whereas the solubilities of $NaNO_3$ and KNO_3 are greater than that of KCl. (KCl is the less soluble of the two initial compounds.)

11 Here $Q = 50$ amp $\times$ 10 min $\times$ 60 sec/min = 30,000 coul, the atomic mass of Na is 22.99, and its valence is 1. Hence $m = (30,000/96,500) \times 22.99$ g = 7.15 g.

13 The passage of 1 faraday transfers Avogadro's number N_0 of electrons. Since 2 electrons must be transferred to decompose each H_2O molecule, $\frac{1}{2}N_0$ molecules are decomposed. There are N_0 molecules in a mole, hence $\frac{1}{2}$ mole of H_2O is decomposed. The molecular mass of H_2O is 18, so $\frac{1}{2}$ mole is 9 g. Thus 9 g of H_2O is decomposed by the passage of 1 faraday.

CHAPTER 25

1 An indicator can be used: Thus blue litmus turns red in an acid solution, while red litmus turns blue in a basic solution. In a neutral solution neither red nor blue litmus will change color. Another method is taste: Acids are sour, bases are bitter.

3 HBr is a strong acid because, like HCl, it is completely dissociated into ions in solution.

5 Dissolve in water and heat. If (a), ammonia will be evolved, and if (b), carbon dioxide will be evolved.

7 $3Na^+ + BO_3^{3-} + 3H^+ + 3Cl^- \longrightarrow H_3BO_3 + 3Na^+ + 3Cl^-$; basic, because H_3BO_3 dissociates less than NaOH.

9 Add a strong acid, for instance, HCl:

$2Na^+ + S^= + 2H^+ + 2Cl^- \longrightarrow H_2S + 2Na^+ + 2Cl^-$

11 $CO_3^=$ hydrolyzes more because H_2CO_3 (or better, $CO_2 + H_2O$) is less dissociated than $HC_2H_3O_2$. The weaker the acid, the greater the extent to which its ion hydrolyzes.

13 $Ca(C_2H_3O_2)_2 + H_2SO_4 \longrightarrow 2HC_2H_3O_2 + CaSO_4.$

CHAPTER 26

1 Exothermic: a, b, e, f.

3 In an atomic-bomb explosion, the liberated energy comes from rearrangements of particles within atomic nuclei, while in a dynamite explosion, the liberated energy comes from rearrangements within the electron clouds of atoms.

5 a. Increase the temperature.
b. Increase the concentration of the sulfuric acid.
c. Use powdered zinc or zinc filings to increase the exposed surface area.

7 At room temperature few of the molecules will have energies as great as the activation energy, and since only these few molecules can react, the process is a slow one.

9 a. Increased, because the greater the gas pressure, the more of it dissolves.
b. Decreased, because the solubility of gases decreases with increasing temperature.
c. Decreased, because KOH is a strong base.
d. Increased, because removing $S^=$ ions by the precipitation of Ag_2S reduces the rate at which H_2S leaves the solution without affecting the rate at which H_2S enters it.

11 Decrease the yield because the reaction is exothermic; increase the yield because in the reaction 3 molecules combine to form only 2; use a catalyst.

CHAPTER 27

1 *a, c.*

3 Na, Al, I⁻; Cl, Hg⁺⁺.

5 When a knife blade is held in a solution of silver nitrate, the iron is oxidized and the silver reduced; hence the iron goes into solution while metallic silver deposits out. The equation is

$$Fe + 2Ag^+ \longrightarrow Fe^{++} + 2Ag$$

7 To show that magnesium is a better reducing agent than hydrogen, it may be placed in an acid solution. Hydrogen gas is evolved, meaning that the magnesium has reduced hydrogen ions in the solution.

9 Electrolysis of water in the electrolyte.

11 The oxygen atoms in each O_2 molecule changed their oxidation numbers from -1 to 0 in losing an electron each, and were oxidized. The oxygen atoms in the H_2O molecules changed their oxidation numbers from -1 to -2 in gaining an electron each, and were reduced. The oxidation number of the hydrogen does not change since the hydrogen atoms neither gained nor lost electrons.

13 Marshes contain decaying plant matter whose oxidation is accompanied by the reduction of ferric compounds in the soil to black ferrous compounds.

15 $C_6H_{12}O_6 + 6O_2 \longrightarrow 6CO_2 + 6H_2O$. The oxygen comes from air inhaled into the lungs; the carbon dioxide and some of the water are exhaled from the lungs, and the rest of the water is excreted in urine.

17 The flow of current through molten NaCl during its electrolysis transfers electrons from the Cl⁻ ions to the Na⁺ ions, which results in Na atoms and Cl_2 molecules.

CHAPTER 28

1 Esters are nonelectrolytes, while salts in solution are electrolytes. Salts (such as sodium chloride) are crystals in their pure state, while the simpler esters (such as methyl chloride) are liquids or gases.

3 *a.* C_2H_5COOH; C_2H_5OH; HCl; $C_3H_5(OH)_3$.
b. C_2H_5COOH; HCl.
c. C_3H_8; C_2H_4.
d. C_2H_5COOH; HCl; C_2H_5OH; $C_3H_5(OH)_3$.
e. C_2H_5COOH; HCl.
f. C_2H_5OH; $C_3H_5(OH)_3$.

5 *a.* Ethylene and benzene are unsaturated hydrocarbons.
b. Ethyl acetate and nitroglycerin are esters.
c. Acetic acid and citric acid are organic acids.
d. Ethyl alcohol and glycerin are alcohols.
e. Glucose and sucrose are sugars.
f. Acetic acid and methyl chloride are derivatives of methane.

7 There are more carbon compounds than compounds of any other element because of the ability of carbon atoms to form bonds with one another.

9 Organic compounds invariably contain carbon and are usually nonelectrolytes; their reactions are usually slow; their molecular weights may be very great; they are usually stable only at moderate temperatures; these general properties differ from those of inorganic compounds.

11 In a double bond, one is a σ bond and one is a π bond. In a triple bond, one is a σ bond and the others are π bonds.

The π bonds are weaker than σ bonds, and so are more easily broken.

CHAPTER 29

1 The ultimate source is the conversion of hydrogen to helium in thermonuclear reactions in the sun. Solar energy reaches the earth as electromagnetic radiation, which is utilized by plants in photosynthesis to produce carbohydrates. The energy in carbohydrates is liberated in animal respiration.

3 H; CH_3.

5 Hydrogen bonds.

7 The molecular structures of these compounds only permit hydrogen bonding between two of the possible pairs (see Fig. 29.7).

CHAPTER 30

1 $2ZnO \cdot SiO_2$; $CaO \cdot 3MgO \cdot 4SiO_2$; $Al_2O_3 \cdot SiO_2$; $3CaO \cdot Al_2O_3 \cdot 3SiO_2$; $3MgO \cdot H_2O \cdot 2SiO_2$; $K_2O \cdot Al_2O_3 \cdot 6SiO_2$.

3 See discussion in text.

5 $K_2O \cdot Al_2O_3 \cdot 6SiO_2 + 2H_2O + CO_2 \longrightarrow 2H_2O \cdot Al_2O_3 \cdot 2SiO_2 + K_2O \cdot CO_2 + 4SiO_2$.

7 Most glass consists of one part each of sodium and calcium oxides together with five parts of silicon dioxide, while blast-furnace slag consists largely of calcium oxide and silicon dioxide in equal proportions.

9 Ag, Fe_2O_3, ZnS, and $BaCO_3$ because they are relatively insoluble and do not react with water or air appreciably.

CHAPTER 31

1 *a.* Granite, which contains only a small proportion of ferromagnesian material, is light in color, while gabbro, with much ferromagnesian material, is dark.
   b. Limestone effervesces readily in acid, unlike basalt.
   c. Schist is foliated, while diorite is not.
   d. Obsidian is black and shiny, while chert may be any color and is waxy in appearance.
   e. Gneiss is coarse-grained and foliated, while conglomerate is cemented gravel.
   f. Quartz scratches glass and has no cleavage, while calcite is scratched by glass and has perfect cleavage.

3 *a.* Hard: quartzite (metamorphic)
              chert (sedimentary)
   b. Moderately hard: limestone
                    (sedimentary)
              gneiss
                    (metamorphic)
              andesite
                    (igneous)
              marble
                    (metamorphic)
   c. Soft: chalk (sedimentary)
              shale (sedimentary)

CHAPTER 32

1 The air over desert regions contains little water vapor and hence absorbs little heat radiation.

3 When warm moist air from the west blows over the colder California Current, its temperature drops and moisture from the now supersaturated air condenses into tiny droplets to form a fog.

5 The irregular variation of temperature with altitude and the presence of ionized regions in the upper atmosphere.

7 Relative humidity is the ratio between the moisture content of an air sample at a given temperature and the maximum amount of moisture it can hold at that temperature. Saturated air contains the maximum amount of moisture possible, hence has a relative humidity of 100 percent.

9 The E layer is too low to be of help in really long distance communication.

11 Ultraviolet light from the sun could reach the earth's surface, which would destroy almost all life there.

CHAPTER 33

1 The pressure is $p = dgh = 1.03 \times 10^3$ kg/m^3 $\times$ 9.8 m/sec^2 $\times$ 1.1 $\times$ 10^4 m = 1.11 $\times$ 10^8 newtons/m^2. Since atmospheric pressure is 1.013×10^5 newtons/m^3, the pressure 11,000 m below the sea surface is 1,100 times greater than atmospheric pressure.

3 The Red Sea lies in the tropics, where extensive evaporation increases the salt concentration. Also, the adjoining lands are arid and do not contribute much fresh water to dilute the sea water. The Baltic Sea is near the Arctic, where average temperatures are low and hence evaporation is minor; furthermore, the melting of winter snow contributes considerable fresh water to the Baltic in the summer months.

5 The moon revolves around the earth in the same direction as the earth's rotation, so it is in the same place relative to a point on the earth's surface about 50 min later every day.

CHAPTER 34

1 Chemical weathering; mechanical weathering.

3 *a.* Quartzite, granite.
  b. Marble, basalt, shale.

5 Rainfall is greater on the west side of the Sierra Nevada range, so that the glaciers on that side were larger and therefore able to extend to lower elevations before melting completely.

7 The presence of well-sorted, rounded grains, the absence of clay and gravel, and cross-bedding in large, sweeping curves.

9 *a.* An irregularly bedded mixture of sand and gravel or sand and clay together with rounded pebbles and boulders.
  b. Well-sorted sand, gravel, and clay; sometimes salts such as calcium carbonate and the shells of marine organisms.
  c. Dune sand is free of dust and mud particles, has well-rounded grains, and shows large-scale cross-bedding.
  d. A fine, claylike material in which are embedded coarser materials and, often, large boulders which are usually angular but may be rounded and show scratched faces.

11 The solubility of minerals is greater in hot water than in cold water, and hence hot-spring deposits are thicker than those near ordinary springs.

CHAPTER 35

1 *a.* Rhyolite, andesite, obsidian, basalt.
  b. Rhyolite, andesite, diorite, basalt, granite, conglomerate.
  c. Diorite, granite, marble.

3 Volcanic mountains are conical, steepening toward the top, with small craters at the summit. The presence of isolated, conical mountains, lava flows, hot springs, geysers, and steam vents, as well as of volcanic rocks, indicates a region where volcanoes had once been active.

5 The shale becomes hornfels and the sandstone, quartzite. Veins of quartz and other materials may be found near the batholith.

7 The hardness of the exposed rocks.

9 Thermal metamorphism: marble from limestone, quartzite from sandstone, hornfels from shale. Dynamothermal metamorphism: mica schist from slate, gneiss from mica schist, slate from shale.

11

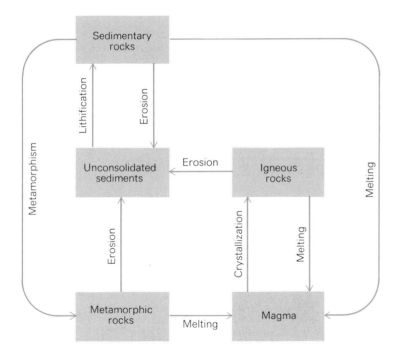

CHAPTER 36

1 P waves are longitudinal, S waves are transverse, and L waves are a combination of both, much like water waves. P waves are the fastest, L waves the slowest.

3 The core is believed to be liquid because earthquake S waves cannot get through it. It is believed to be largely iron because iron has the right density to account for the large average density of the earth, it is in the liquid state at the pressure and temperature of the core, it is rather abundant in the universe generally, and it is a good conductor of electricity.

5 The chief hypotheses are that the geographic and geomagnetic poles have never been far apart and that magnetized rocks have not changed their orientations since solidifying and becoming magnetized.

7 The amount of heat is many times greater than the energy involved in geological events, and, by causing convection and melting subsurface rocks, can account, directly or indirectly, for known geological processes. This heat is partly a relic of the formation of the earth, but most of it comes from radioactive minerals.

9 About 6,500 km; about 10 min after the S waves.

CHAPTER 37

3 The original sedimentary material will have been converted into such metamorphic rocks as slate, schist, and marble, which will be exposed by the subsequent erosion. The outcrops of metamorphic rocks will be intruded by igneous rocks—granite batholiths and dikes of various kinds. Unconformities will occur where subsequent stream and lake sediments are deposited. Thin sandstone beds, interstratified with shale and conglomerate and exhibiting cross-bedding, will predominate in the overlying sedimentary rocks.

5 Radioactive methods have limited use in correlating rocks partly because not all rocks contain the required radioisotopes and partly because small age differences are difficult to determine. Fossils are useful because groups of organisms have changed continuously with time, permitting rocks from the same period to be recognized by the similarity of the fossils they contain.

CHAPTER 38

1 When a large stress is applied over a long period of time, the asthenosphere gradually flows in response to it. When brief, relatively small forces are applied, as is the case with seismic waves, the asthenosphere is rigid enough to transmit them as a solid does.

3 Not necessarily. For example, volcanic activity could still create mountains and local subsidence could occur due to the solvent action of groundwater.

5 Laurasia: North America, Greenland, Europe, Asia (except for India). Gondwanaland: South America, Africa, Antarctica, India, Australia.

7 No; the ocean floors are actually extremely rigid, and continental drift is a more complex process involving the motion of lithospheric plates in which the continents are embedded.

9 Convection currents in the asthenosphere; sidewise forces due to the weight of the elevated mid-ocean ridges.

CHAPTER 39

1 Mammals were able to survive into the Cenozoic era because, being warm-blooded, they could cope with temperature changes; they had relatively larger brains; and they cared for their young after birth, permitting one generation to pass on some of its experience to the next.

3 *a*. Shallow seas.
   b. Heavily vegetated swamps.
   c. High mountains with prominent faulting and folding.

5 Coal forms from plant material that accumulates under conditions where complete decay is prevented. Such conditions occur in swamps that are almost at sea level, so that partially decayed vegetation is periodically covered with thin layers of marine sediments. During the decay oxygen and hydrogen are gradually removed from the plant cellulose, leaving complex compounds consisting largely of carbon. The origin of oil is less understood, but it apparently is produced from the protein, fat, and wax constituents of both plants and animals by slow bacterial decay in the absence of air.

CHAPTER 40

1 Mars, 12.3 hr; Jupiter, 4.9 hr.

3 Mars is farther from the sun than the

earth is; its atmosphere is less dense than that of the earth.

5 No satellite can have been closer to the earth than the distance given by Roche's limit or it would have disintegrated. This sets a lower limit to the size of the moon's orbit in the past.

7 The surface temperature of Mercury is much greater than that of Mars.

CHAPTER 41

1 New moon; full moon.

5 a. The radius of a sphere 90 km in diameter is 35 km = 3.5×10^4 m, so its volume is $V = \frac{4}{3}\pi (3.5 \times 10^4$ m$)^3 = 1.75 \times 10^{14}$ m^3. The volume of iron present is $0.9V = 1.58 \times 10^{14}$ m^3 and the volume of nickel is $0.1V = 1.75 \times 10^{13}$ m^3. From Table 7.1 the densities of iron and nickel are respectively 7.8×10^3 and 8.9×10^3 kg/m^3, so their masses are respectively 1.23×10^{18} kg and 0.16×10^{18} kg for a total mass of 1.4×10^{18} kg.
b. The volume of a disk h high and of radius r is $\pi r^2 h$, so equating this volume with the volume of the above sphere gives $\pi r^2 (5 \times 10^3$ m$) = 1.75 \times 10^{14}$ m^3. Solving for r yields $r = 1.06 \times 10^5$ m = 106 km, so the diameter is 212 km.

7 Spring tides in both cases, since the earth, sun, and moon lie along a straight line in eclipses.

CHAPTER 42

1 More stars appear in the photograph because the exposure time can be long, permitting even faint stars to produce images.

3 Blue, white, yellow, red.

5 The 6000°C temperature of the sun's visible surface suggests that the sun is almost wholly gaseous.

7 The star whose spectral lines are displaced to the red is moving away from the earth faster than the other star.

9 The sun's radiation cannot have changed appreciably in the past 2 billion years since the temperature of the earth's surface, which depends upon the intensity of the radiation, has been approximately constant in that period.

11 If the particles travel in straight lines, their speed is 93×10^6 mi/24 hr = 3.9×10^6 mi/hr (or 1.49×10^{11} m/8.64×10^4 sec = 1.72×10^6 m/sec) since the radius of the earth's orbit is 93 million miles = 1.49×10^{11} m and there are 24 hr = 8.64×10^4 sec in a day. If the particles are deflected by interplanetary magnetic fields, the above speed represents a lower limit to their actual speed.

13 The solar spectrum is an absorption one, indicating that the surface layers are cooler than those below. An even stronger argument is that only in the sun's interior could conditions of pressure and temperature occur suitable for energy production by nuclear fusion reactions.

CHAPTER 43

1 a. The sun's mass, temperature, spectrum, and position in the Hertzsprung-Russell diagram are those of a typical star.
b. Most stars exhibit dark line spectra, indicative of a cool absorbing gas surrounding a hot, radiating interior.
c. The apparent shift in position of the stars as the earth moves around the sun, called parallax, is exceedingly small, indicative of very great distances to the stars.
d. The spectra of some stars which ap-

pear single in a telescope, photographed on successive days, show small periodic shifts in some lines, periodic doubling of others. These shifts and doubling suggest that two separate stars, revolving about each other, are responsible for the spectra.

e. Stars in the neighborhood of the constellation Cygnus seem to be approaching us and to be radiating out from a point, while those in the opposite part of the sky are apparently receding and coming closer together.

3 The surface temperature of a star determines the radiation it emits per unit area, while its intrinsic brightness is a measure of its total radiation; hence knowing both quantities permits computing the star's surface area and hence its diameter.

5 Stars must be spheres or spheroids because if parts of their surfaces are at different distances from their centers, the resulting pressure differences, due to gravity, would cause the material of the stars to flow until they have spherical or spheroidal (if in rotation) shapes.

7 a. It is large, heavy, hot, and bright, with prominent hydrogen and helium lines in its spectrum.
b. It is small, exceedingly dense, very hot, and dim.
c. It is huge, diffuse, cool, and bright.
d. It is moderately small with moderate temperature, density, and mass, with a spectrum in which lines of metallic elements are prominent.

9 a. It is very hot.
b. Its average density is low.
c. Lines of hydrogen and helium are prominent in its spectrum.
d. Upper end of main sequence.

11 A star must radiate energy because it consists of a large mass of matter in a relatively small volume of space and contains abundant hydrogen, which, owing to conditions created by the gravitational contraction of the stellar material, is converted into helium in nuclear reactions.

13 Heavy stars consume their hydrogen more rapidly than light stars, which ultimately leads to violent explosions which constitute supernovae.

CHAPTER 44

1 a. Stars in a galaxy show a general slow rotation about the galactic center, superimposed upon which is a random motion, stars moving in all directions with various speeds, much like molecules of a gas.
b. The galaxies themselves seem to be drifting apart from one another like the fragments of a bursting rocket, rather than moving in the random manner of molecules of a gas.

3 They are very far from our galaxy and do not partake of the motions of its members.

5 An edgewise view, in order to look for a Doppler shift to the blue in light from the approaching side and a Doppler shift to the red in light from the receding side.

7 Population I stars are found in the disklike configuration we see as the Milky Way, while Population II stars are found in a larger, spherical region surrounding the Milky Way. Population I stars move with relatively slow velocities in orbits about the galactic center in or near the Milky Way, while those of Population II, many of which are concentrated in globular clusters, travel in randomly oriented paths at high speeds about the galactic center. Population II stars are much older than Population I stars, and are relatively cool and faint while the latter are often hot and bright.

ANSWERS TO ODD-NUMBERED EXERCISES

9 Determine the radioactivity of a sample of carbon from the ancient wood and compare it with the radioactivity of a sample of the same mass from a piece of wood of recent origin. The ratio of the two activities may then be converted into an age figure for the ancient wood by taking the half-life of radiocarbon into account.

11 Hydrogen is by far the most abundant element in the universe.

13 The protons pick up electrons and form hydrogen atoms, while most of the neutrons are absorbed by carbon nuclei to form radiocarbon. Some neutrons escape from the earth entirely, and decay into protons and electrons in space.

15 Electrons lose energy by radiation much more readily than do protons in magnetic fields, such as the fields believed responsible for cosmic-ray origin.

17 Radio telescopes do not magnify an image as optical telescopes do, but gather and amplify radio-frequency electromagnetic waves. The larger a radio telescope antenna is, the more sensitive it is and the more accurately it can locate a source of signals in space. The limiting factors in the size of the antenna are mechanical, since it must not sag out of shape when it is turned and winds must not distort it.

CHAPTER 45

1 The spectral lines of spiral nebulae are just what one would expect on the assumption that they are aggregates of stars like our own galaxy, except that the lines are shifted toward the red end of the spectrum. If the shift is due to the Doppler effect, it implies that all of the nebulae are receding from us, or, more reasonably, that the entire visible universe is expanding.

3 All interpret the red shift as indicating the mutual recession of the nebulae, but in the continuous creation theory new nebulae appear spontaneously so that the density of nebulae in space remains constant.

5 Yes; the discovery of nebulae more distant than those known at present.

Acknowledgments

The authors and publisher are deeply indebted to the following for their help in obtaining the illustrations listed by page numbers.

Abbott, Berenice 221
Air Reduction Company 494
Alpha Photo Assoc. 194
American Machine and Foundry Co. 93
American Museum of Natural History 687, 757, 875
American Smelting and Refining Co. 523
Azarraga, Luis 778
Bausch & Lomb, Inc. 247, 343
Beckman Instruments, Inc. 479
Bell Telephone Laboratories 587, 887
Bendix Corp. 243
Brookhaven National Laboratory 99, 211, 246, 302, 311, 322, 327, 330, 334, 339
Brooks, Percy 269
Carnegie Institute of Washington 706, 819
Cooke, Jerry 24
De Beers Consolidated Mines, Ltd. 439
Douglas Aircraft Co., Inc. 782
E. I. du Pont de Nemours & Co. 139, 207, 226, 286, 383, 443, 499, 502, 606
Ealing Corp./Griffin & George, Ltd. 571
Earl, James A./Office of Naval Research and Univ. of Minnesota 449
Eastern Airlines 159
Eisenbeiss, Herman 250
Ellis, Bob/Free Lance Photographers Guild 683
Frankford Arsenal 119
Freeport Sulfur Co. 446
Gardner-Denver Co. 150
General Electric Corp. 230, 264, 323, 401, 442
General Motors Corp. 158
B. F. Goodrich Co. 45, 131
Golden Press 598, 599, 602, 675, 678, 722, 726, 730, 739, 763, 767
W. R. Grace Co. 138, 438, 458, 507
Harper & Row 602, 619
High Voltage Engineering Co. 229
Holt, Rinehart & Winston, Inc. 598, 599
Government of India Tourist Office 742
International Science and Technology Magazine 203, 411, 414
Jensen, Paul 37
Jones & Laughlin Steel Corp. 526
Kimberly-Clark Corp./Vance Johnson & Assoc. 242, 459
Kinne, Russell 646, 658
Lick Observatory 803, 810
Lisle, Capt. Alan G., Jr. 678
Lowell Observatory/E. C. Slipher 786
Luray Caverns 670
McAllister, R. F. 726, 727
McLaughlin, Herb 655
Monsanto Chemical Co. 551, 555
Mount Wilson & Palomar Observatories 783, 785, 790, 791, 818, 820, 823, 847, 851, 854, 855, 860, 861, 863, 866, 870, 871, 884
National Aeronautics & Space Administration 518, 623, 630, 659, 747, 787, 798, 799, 802, 806, 822
National Park Service 675, 766
National Science Foundation 828
Olcott, William 4
Olin Mathieson Chemical Corp. 40
Owens-Illinois, Inc. 583
PPG Industries, Inc. 447
PSSC Physics, 1965, D. C. Heath Co. 29, 169, 172
Pettijohn, Francis J. 668
Photo Researchers, Inc. 16, 36, 77, 282
Princeton University 314
Quest Magazine/© 1961 Putnam Publishing Co. 826, 831
Radio Corp. of America 262
RCA Laboratories 643
RCA Victor Records 179
Reader's Digest, Inc. 586, 590
Reynolds Metals Co. 527
St. Louis Post Dispatch/Black Star 187
Schaefer & Seawell 826, 831, 867
Schaefer, Vincent J. 619
Shanin, R. E. 674
Shaub, Benjamin M. 599
Shell Chemical Co. 243, 326
Shell Oil Co. 539
Smithsonian Institution 761
South Dakota Dept. of Highways 767
Spectra Physics, Inc. 251, 252, 254, 255
Standard Oil Co. of New Jersey 606, 699
Successful Farming Magazine/© Meredith Publishing Co. 603
Tinklenberg, M. 834
Union Carbide Corp./Linde Division 399
Union Pacific Railroad 722
Uniroyal Chemical, Division of Uniroyal Inc. 279
U. S. Coast Guard 663
U. S. Dept. of Agriculture 604, 656, 679
U. S. Dept. of Commerce/E.S.S.I. 639, 662, 667, 682, 686
U. S. Geological Survey/Darton 681, 719, 803
U. S. Navy 643
U. S. Steel Corp. 73, 110, 175, 406
U. S. Weather Bureau 631
University of California at Berkeley/ Rare Book Division, General Library 49
Westinghouse Electric Corp. 638
Wiegel, Robert L. 174
Yerkes Observatory 782, 844

Index

Absolute motion, 4
Absolute temperature, 138, 148
Absolute zero, 148, 148
Absorption spectrum, 343, 821
   origin of, 349
Abyssal plain, 636
Acceleration, 9
   centripetal, 47
   constant, 10
   and distance, 16
   and force, 26
   of gravity, 11
   and speed, 15
Acetic acid, 553
Acetylene molecule, 546
Acetylene torch, 494
Acid, 451, 471, 473
   acetic, 553
   amino, 440, 565
   Brønsted-Lowry, 475
   carbonic, 473
   gram-equivalent of, 480
   hydrochloric, 448, 473
   Lewis, 476
   nitric, 473
   nucleic, 568
   organic, 553
   pH of, 476
   strong, 473
   sulfuric, 446, 473
   weak, 473
      neutralization of, 481
Acidic oxide, 485
Acidic rock, 599
Acoustics, 175
Actinide metals, 371
Action and reaction, 33
Activated molecule, 496
   and reaction rate, 501
Activation energy, 495
Activity, chemical, 363, 383, 515
Adams, W. S. (1876–1956), 841
Air, saturated, 618
Air mass in meteorology, 622
Air resistance, 12
Alchemy, 278
Alcohol, 552
Aldehyde group, 543
Algebra, 893
Alkali, 475
Alkali metals, 364

Alkaline solution, 475
Allowed band in crystal, 411
Alluvial fan, 664
Alpha decay, 329
Alpha helix, 567
Alpha particle, 321
   and Rutherford's experiment, 292
Alternating current, 231
Aluminum, refining of, 530
Amide linkage, 556
Amino acid, 440, 565
Amino group, 566
Ammeter, 228
Ammonia:
   in solution, 475, 483
   synthesis of, 508
Ammonia molecule, 393
Ammonium hydroxide, 474, 483
Ammonium salts, 484
Amorphous solid, 153, 155, 398
Amorphous sulfur, 445
Ampère, André (1775–1836), 209
Ampere, the, 209
Amphoterism, 485
Amplitude of wave, 168
Andes mountains, origin of, 747
Andesite, 597
Andromeda nebula, 861, 867, 870
Angular momentum, 83
   of atom, 352
   conservation of, 83, 85
   quantization of, 352
   of solar system, 777, 891
Annealing, 404
Annihilation of matter, 335
Anode, 200
Antibonding molecular orbital, 391
Anticline, 685
Anticyclone, 621
Antimatter, 335
Antiparticle, 333
Apollo 11, 81, 84
   lunar expedition of, 804
Appalachian mountains, evolution of, 738
Archimedes' principle, 134
Argon, 611
Aristotle (384–322 B.C.), 12, 283, 840
Arrhenius, Svante (1859–1927), 464
Asteroid, 781

Asthenosphere, 736
   convection in, 750
Atmosphere:
   of earth (*see* Earth, atmosphere of)
   of sun, 831
Atmosphere, the (unit of pressure), 132
Atmospheric pressure, 131, 627
Atom, 288
   Bohr model of, 345
   electron shielding in, 376
   energy levels of, 345
   excited states of, 346
   gram, 427
   ground state of, 348
   quantum-mechanical model of, 353
   Rutherford model of, 292
   Thomson model of, 292
Atom group, 422
Atomic bomb, 315
Atomic excitation, 348
Atomic mass unit, 303
Atomic masses of elements, 304
   and nuclear energy, 308
Atomic nucleus, 293
Atomic number, 294
Atomic orbital, 391
Atomic shells and subshells, 371
Atomic spectra, 342
Atomic states, 353
Atomic structure, 292, 371
   and periodic table, 372
Aurora, 830
Avogadro, Amadeo (1776–1856), 289
Avogadro's hypothesis, 289
Avogadro's number, 427, 429
   and electrolysis, 457

Balancing a chemical equation, 424
Balmer series, 344
Barometer, 132
Basalt, 597
Base, 452, 474
   Brønsted-Lowry, 475
   gram-equivalent of, 480
   Lewis, 476
   pH of, 476
   strong, 474
   weak, 474
      neutralization of, 482
Basic oxide, 485

Basic rock, 599
Batholith, 679
Battery, lead-acid, 517
Beats, 176
Beaufort scale of wind speed, 641
Becquerel, Henri (1852–1908), 320
Benzene molecule, 547
Bessel, F. W. (1784–1846), 840
Beta decay, 327
Beta particle, 321
Big bang theory of origin of universe, 885
Binding energy, nuclear, 309
Blast furnace, 529
Blue color of sky, 251
Bode's law, 779
Bohr, Niels (1885–1963), 345
Bohr model of hydrogen atom, 345
Boiling of liquid, 115, 154
Boiling points, table of, 117
Boltzmann's constant, 151
Bond:
  chemical, 380
  covalent, 380, 388
  disulfide, in protein molecule, 567
  and exclusion principle, 420
  hydrogen, 409
    in protein molecule, 567
  ionic, 380, 386
  metallic, 405
  saturation in, 420
  van der Waals, 408
Bonding in solids, 404
Bonding molecular orbital, 391
  hybridization in, 545
Borax, 475, 482
Boyle's law, 136
  origin of, 148
Bragg, W. L. (1890–), 400
Brahe, Tycho (1546–1601), 49, 51
Brecchia, 676
British units of mass and force, 31
Brønsted-Lowry acid and base, 475
Bronze, 528
Brownian movement, 146
Bubble chamber, 97, 300, 334
Buoyancy, 134
Butane, isomers of, 541, 545
Butyric acid, 564

Calcite, 584, 589
Calcium carbonate, 584
Caldera, 803, 812
Caloric, 118

Calorie, 114
Carbide, 442
Carbohydrate, 561
  hydrolysis of, 563
  and photosynthesis, 503
Carbon, 441
  compounds of, 536
Carbon bonds, 544
Carbon cycle, 829
Carbon dioxide, 442
  in atmosphere, 610
  dissolved in oceans, 638
  and photosynthesis, 503
Carbon monoxide, 442
Carbonic acid, 473
Carboniferous period, 759
Carborundum, 405, 444
Carboxyl group, 543, 553
Catalyst, 503
Catastrophism, 716
Cathode, 200
Cathode rays, 200
  behavior in magnetic field, 227
Causality and uncertainty principle, 274
Cavendish, Henry (1731–1810), 57
Cell:
  electrochemical, 516
  fuel, 518
Cellulose, 557, 562
Celsius scale, 112
Cenozoic era, 765
Center of mass, 55
  of earth-moon system, 801
Centrifugal distortion, 694
Centripetal acceleration, 47
Centripetal force, 44, 46
Cepheid variable star, 843, 882
Cerenkov effect, 244, 246
Cesium chloride, crystal structure of, 404
Chadwick, James (1891–), 301
Chain reaction, 312
Chalk, 600
Change of state, 115
Charge:
  electric, 186
    conservation of, 85, 193, 335
  induced, 206
  invariance of, 193
  nature of, 189
  separation of, 205
  of electron, 202
  and electrolysis, 457
Charles' law, 137
  origin of, 148
Chemical activity, 363, 376, 383, 515

Chemical bond, 380
Chemical compound, 280
  carbon, 536
  covalent, 389
  heat of formation of, 492, 495
  ionic, 382
  naming of, 449
  nature of, 288
  organic, 536
  saturated, 547
  solubility of, 461
  stability of, 363
  valences in, 420
Chemical energy, 493
Chemical equation, 424
Chemical equilibrium, 505
Chemical formula, 289
  structural, 538
Chemical reaction, 278
  and activated molecules, 501
  activation energy of, 495
  rate of, 500
  reversible, 504
Chemical symbol, 289, 290
Chemistry:
  inorganic, 535
  organic, 535
Chernozem, 605
Chert, 600
Chitin, 562
Chloride, 447
Chlorine, 446, 464
Chlorophyll, 503
Circuit, electrical, 208
Circuit breaker, 214
Circular motion, 44
Clay, 589
Clay minerals, 588
Cleavage, 585
Climate, 616, 628
Climatic change, 629
Closed shell in atom, 373
Cloud chamber, 300, 872
Clouds, 619
Coal, 497, 729
  origin of, 760
Coal tar, 557
Coke, 498
Color, 249
Combustion, 285, 438
  spontaneous, 525
Comet, 782
Commutator, 228
Component of vector, 38

INDEX

Compound, chemical (*see* Chemical compound)
Compton, Arthur H. (1892–   ), 265
Compton effect, 265
Condensation of vapor, 116
Conductor, electrical, 187, 203
Conglomerate, 600
Conservation of angular momentum, 83, 85
Conservation of charge, 85, 193, 335
Conservation of energy, 80, 119, 287
Conservation of linear momentum, 81, 85
Conservation of mass, 101, 287
Conservation principles and symmetry, 85
Constructive interference, 170
Continental drift, 707
    modern theory of, 746
    Wegener's theory of, 740
Continental rise, 636
Continental shelf, 635
Continental slope, 636
Convection:
    in asthenosphere, 750
    in atmosphere, 625
Copernican system, 50
Copernicus, Nicolaus (1473–1543), 50
Copolymer, 556
Copper, refining of, 528
Core of earth, 700
    composition of, 703
    inner, 702
Coriolis effect, 626
Corona, 832
Cosmic rays, 807, 869
    origin of, 873
    primary, 869
    secondary, 871
    showers of, 872
Coulomb, Charles (1736–1806), 191
Coulomb, the, 191
Coulomb's law, 190
Covalent bond, 380, 388
Covalent compound, 389
    solubility of, 461
Covalent crystal, 405
Crab nebula, 855
Cracking of hydrocarbon, 538
Cross-bedding, 663
Crust of earth, 577, 702
Crystal, 126
    covalent, 405
    defects in, 401
    energy bands in, 411

Crystal:
    ionic, 404
    metallic, 405
    molecular, 408
Crystal form in minerals, 585
Crystal structure, 397, 399
Crystal types, table of, 410
Crystalline solid, 153, 155, 397
Curie, Marie (1867–1934), 320
Current, electric, 207
    alternating, 231
    direct, 232
    in gases, 204
    in liquids, 455
    long-distance transmission of, 234
    mechanisms of flow, 203, 205
Currents, ocean, 642
    rip, 661
    turbidity, 665
Cuvier, Georges (1769–1832), 717
Cyclic theory of universe, 886
Cyclone, 621

Dacron, 556
Dalton, John (1766–1844), 287
Daniell cell, 516
Darwin, Charles (1809–1882), 719
de Broglie, Louis (1892–   ), 268
de Broglie wavelength, 268
Decay of organic matter, 525
Defect, crystal, 401
Definite proportions, law of, 281
Delocalized electrons in benzene molecule, 549
Delta, 664
Denaturation of protein, 568
Density, 132
Deoxyribonucleic acid (DNA), 569
Desert soil, 605
Destructive interference, 170
Deuterium, 306
Devitrification, 581
Diamond, 441
    energy bands in, 412
    structure of, 405
Diastrophism, 673, 681
Diene, 555
Diffraction, 171
    of light, 252
Dike, 678
Dinosaur, 764
Diorite, 597
Dip poles of geomagnetic field, 705
Dipeptide, 567

Dirac, P. A. M. (1902–   ), 355
Direct current, 232
Disaccharide, 562
Dislocation in crystal, 402
Displacement reaction, 514
Dissociation, 462
    energy changes in, 493
Distillation, fractional, 538
Disulfide bond in protein molecules, 567
Doldrums, 621
Domain, magnetic, 220
Doppler effect:
    in sound, 177
    in stellar spectra, 824
Double star, 845
Dynamite, 500
Dynamo effect, 231
Dynamothermal metamorphism, 686

Earth:
    atmosphere of, 610
    composition, 610
        energy, 617
        escape, 611
        general circulation, 621, 624
        nitrogen cycle, 440
        oxygen, 524
        oxygen-carbon dioxide cycle, 504, 610
        regions, 613
    core of, 700, 703
    crust of, 577, 702
    density of, 696
    inner core of, 702
    interior of: composition, 702
        structure, 699
        temperature, 703
    magnetic field of, 704
        origin, 708
        in past, 705, 707, 745, 750
    mantle of, 700
    mass of, 695
    origin of, 889
    shape of, 693
    size of, 692
Earthquake, 681, 697
    epicenter of, 699
    focus of, 697
    magnitude of, 698
Earthquake waves, 698
    shadow zone in, 701
Echo sounding, 744
Eclipse, 797, 834
Einstein, Albert (1879–1955), 91, 261

Elastic limit, 127
Elasticity, 126
Elastomer, 555
Electric charge (see Charge, electric)
Electric circuit, 208
Electric current (see Current, electric)
Electric energy, 215
Electric field, 195
Electric force, 190, 192
Electric induction, 206
Electric motor, 228
Electric power, 213
Electrochemical cell, 516
Electrolysis, 456
 and aluminum refining, 530
 Faraday's laws of, 457
 of water, 436
Electrolyte, 463
 solution of, 465
Electromagnetic, 226
Electromagnetic induction, 231
Electromagnetic interaction, 337
Electromagnetic waves, 238
 spectrum of, 241
 speed of, 242
 types of, 240
Electron, 189, 191, 201
 charge of, 202
 delocalized, in benzene molecule, 549
 and electrolysis, 457
 positive, 333
Electron affinity, 386
Electron configurations of elements, 374
Electron gas in metals, 407
Electron microscope, 269
Electron shielding in atom, 376
Electron spin, 355
Electron transfer:
 in ionic bonding, 381
 in oxidation-reduction reactions, 514
Electron volt, 210
Electronegativity, 390
 and chemical energy, 494
Electroscope, 187
Elementary particles, 332
Elements, 280
 active and inactive, 363
 atomic masses of, 304
 atomic numbers of, 294
 defining property of, 306
 electron configurations of, 374
 families of, 363
 ionization energies of, 385
 isotopes of, 306
 nature of, 288

Elements:
 origin of, 888
 periodic table of, 368
 transition, 371
Ellipse, 52
Emission spectrum, 342, 821
 origin of, 349
Endothermic reaction, 492
Energy, 71
 activation, 495
 chemical, 493
 conservation of, 80, 119, 287
 electric, 215
 internal, 151
 kinetic, 71, 74
 and mass, 100
 nuclear binding, 309
 potential, 75, 77
 quantization of, 121, 348, 352
 rest, 101
 solar, 827
 thermonuclear, 316
 types of, 79
Energy band in crystal, 411
Energy levels:
 atomic, 345
 nuclear, 325
Energy transformations, 76
Engine:
 gasoline, 158
 heat, 156
Enzyme, 563
Epicenter, 699
Epoch, geologic, 730
Equation, 899
 chemical, 424
 ionic, 466
 meaning of, 14
Equilibrium, chemical, 505
Equivalence, principle of, 102
Era, geologic, 730
Eratosthenes (276–194 B.C.), 692
Erosion, 654
 glacial, 659
 stream, 655
 wave, 661
 wind, 660
Escape velocity, 59
 and planetary atmospheres, 612
Ester, 553
Ester group, 564
Ethyl group, 551
Ethylene, 554
 molecular structure of, 546
Evaporation, kinetic theory of, 153

Excited state of atom, 346
Exclusion principle, 356
 and chemical bonding, 380, 387, 420
 and periodic table, 372
Exosphere, 614
Exothermic reaction, 492
Expanding universe, 880
Explosives, 498
Exponents, 896
Extrusive rocks, 598, 676

Fahrenheit scale, 112
Faraday, Michael (1791–1867), 231, 238, 457
Faraday, the, 457
Faraday's laws of electrolysis, 457
Fat, 563
Fatty acid, 563
Fault, 683
 San Andreas, 686
Fault scarp, 684
Feldspar, 588
Fermi, Enrico (1901–1954), 874
Ferric compounds, 522
Ferromagnesian minerals, 588
Ferrous compounds, 522
Fertilizer, 441
Fiber, synthetic, 556
Field:
 electric, 195
 force, 193
 gravitational, 195
 magnetic, 221, 223
Flood plain, 656
 deposition of sediments on, 663
Fluorescence, 201
Flux in iron refining, 528
Focus of earthquake, 697
Folding in earth's crust, 685
Foliation, 601
Foot-pound, 71, 74
Forbidden band in crystal, 411
Force, 25
 and acceleration, 26, 28
 centripetal, 44, 46
 electric, 190
 gravitational, 55
 lines of, 195, 222
 magnetic, 226
 nuclear, 330
 reaction, 34
 tractive, 647
Force field, 193

Formula:
  chemical, 289, 540
  structural, 538
Formula mass, 428
Fossil magnetism, 707
Fossils, 728
Fourth dimension, 94
Fractional distillation, 538
Frame of reference, 4, 89
  inertial, 91
Franklin, Benjamin (1706–1790), 186
Fraunhofer lines, 344
Free fall, 11, 17
Freezing, 116
Freezing points:
  effect of dissolved substances on, 465
  table of, 117
Frequency, 167
  fundamental, 179
  of light waves, 241
  of sound waves, 173
Friction, 30
Friction welding, 119
Front, weather, 624
Fuel, 496
Fuel cell, 518
Fundamental frequency, 179
Fundamental interactions, 336
Fuse, electric, 214
Fusion:
  heat of, 116, 155
  table of, 117

Gabbro, 597
Galactic nebulae, 864
Galactic spectra, red shift in, 880
Galaxy, 860
  magnetic fields in, 874
  spiral, 865
  star clusters in, 863
Galileo Galilei (1564–1642), 12, 776
Galvanometer, 228
Gamma decay, 329
Gamma rays, 321
Gamow, George (1904–1968), 888
Gangue, 527
Gas, 129, 136
  ideal, 140, 432
  inert, 366
  marsh, 537
  natural, 497
  water, 497
Gas volumes in chemical reactions, 431
Gases, kinetic theory of, 147

Gasoline, 497
Gasoline engine, 158
Gay-Lussac, Joseph (1778–1850), 289
Genetic code, 570
Geochronology, 730
Geoid, 694
Geomagnetic field (see Earth, magnetic field of)
Geomagnetic poles, 705
Geosyncline, 738
Gev, 211
Glacial deposits, 664
Glacier, 658
Glass, 581
Globular cluster of stars, 863
Glucose, 562
Glycerin, 552, 563
Glycogen, 562
Gneiss, 602
Gondwanaland, 742
Gradation, 654
Gram, 24
Gram atom, 427
Gram-equivalent, 480
Gram mole, 429
Grand Canyon, 722, 724
Granite, 597, 677
Graphite, 441
Gravitation, 53
  constant of, 55, 58
  law of, 55
  and light, 102, 266
  and relativity, 102
Gravitational force, 55
  and electric force, 192
Gravitational interaction, 337
Gravitational red shift in light, 267
Graviton, 337
Gravity:
  acceleration of, 11
  specific, 517
Ground, electrical, 188
Ground state of atom, 348
Groundwater, 668
Group, chemical, 370
Gulf stream, 644
Gunpowder, 498
Guyot, 749

Haber, Fritz (1868–1934), 508
Hale telescope at Mt. Palomar, 253, 818
Half-life, 324
Half-reaction in electrochemical cell, 516
Halogen derivatives of hydrocarbons, 552

Halogens, 364
  electron affinities of, 386
Hanging valley, 660
Hayford spheroid, 694
Heat, 109, 113, 115, 149
  of formation, 492, 495
  of fusion, 116, 155
  mechanical equivalent of, 119
  nature of, 158
  specific, 114
  and temperature, 111, 113
  of vaporization, 116, 155
Heat death of universe, 160
Heat engine, 156
Heavy water, 306
Heisenberg, Werner (1901–   ), 272, 350
Helium in stars, 853
Hemoglobin, 525
Herschel, William (1738–1822), 790, 817, 868
Hertz, Heinrich (1857–1894), 240
Hertz, the, 167
Hertzsprung-Russell diagram, 849
Heterogeneous substance, 279
Himalayas, evolution of, 742, 748
Hole in electron structure of semiconductor, 413
Homogeneous substance, 278
Hooke's law, 127
Hornfels, 679
Horse latitudes, 621
Horsepower, 71
Hubble, Edwin (1889–1953), 880
Hubble's law, 880
Humidity, 618
  relative, 620
Humus, 603
Hutton, James (1726–1797), 717
Hybridization in molecular bonding, 545
Hydrocarbon derivatives, 549
Hydrocarbons, 536
  cracking of, 538
  paraffin, 536
  polymerization of, 538
Hydrochloric acid, 448, 473
Hydrogen, 362, 435
  in stars, 853
Hydrogen atom, 301
  Bohr model of, 345
Hydrogen bond, 409
  in protein molecules, 567
Hydrogen ion in solution, 472
  concentration of, 476
Hydrogen molecule, 388

Hydrogen-oxygen fuel cell, 518
Hydrogen peroxide, 437
Hydrolysis, 482, 483
   of carbohydrates, 563
   of fats, 564
Hydromagnetism, 708
Hydronium, 472
Hydroxide, 451
Hydroxide group, 474
Hypothesis, 61

Ice:
   on earth's surface, 636
   hydrogen bonding in, 410
   structure of, 410
Ice Age, 769
Icecap, 658
Ideal gas law, 140, 432
Igneous rocks, 596
Indicator of solution pH, 478
Induced charge, 206
Induction:
   electric, 206
   electromagnetic, 231
   magnetic, 224
Inert gases, 366
Inertia, 23
Inertial frame of reference, 91
Inorganic chemistry, 535
Insolation, 617
Instantaneous speed, 8
Insulator:
   electrical, 187, 203
   transparency of, 413
Interactions, fundamental, 336
Interference, 170
   of light waves, 252
   of sound waves, 176
Intermediate boson, 337
Internal energy, 151
Intrusive rocks, 598, 677
Ionic bond, 380, 386
Ionic compound, 382
   solubility of, 461
Ionic crystal, 404
Ionic equation, 466
Ionization, 204
   of solution, 462
Ionization energy, 384
Ionosphere, 615
Ions, 204
   of solution, 462
Iron:
   compounds of, 521

Iron:
   magnetic properties of, 220, 376
   refining of, 528
Iron oxide, reduction of, 521
Island arcs, origin of, 747
Isomer, 541
Isostasy, 736
Isotope, 306

Jet stream, 626
Joule, James Prescott (1818–1889), 119
Joule, the, 69, 74
Jupiter, 788

Kaolin, 589
Kaolinite, 591
Kepler, Johannes (1571–1630), 51, 645, 779
Kepler's laws of planetary motion, 52
Kilocalorie, 113
Kilogram, 24
Kilowatt, 70
Kilowatt-hour, 215
Kinetic energy, 71, 74
Kinetic theory:
   of gases, 147
   of liquids and solids, 152
   of matter, 146
Kundt's tube, 173

Laccolith, 679
Landscape, evolution of, 656
Laser, 251, 254, 255
Latosol soil, 605
Laurasia, 742
Lava, 674
Lavoisier, Antoine (1743–1794), 284, 287, 419
Law, scientific, 61
Lead-acid storage battery, 517
Le Chatelier, H. L. (1850–1936), 506
Le Chatelier's principle, 506
Left-hand rule for magnetic field, 223
Length, relativity of, 96
Lewis acid and base, 476
Life, origin of, 571
Light, 241
   behavior in solids, 413
   diffraction of, 252
   electromagnetic theory of, 241
   frequencies of, 241
   and gravitation, 102
   gravitational red shift in, 267

Light:
   interference of, 252
   quantum theory of, 261
   reflection of, 246
   refraction of, 248
   speed of, 242
      and relativity, 93
   wave versus quantum theories of, 263
   wave nature of, 248
Light rays, 245
Light-year, 840
Limestone, 600
Limewater, 475
Linear accelerator, 327
Linear momentum, 80
   conservation of, 81, 85
Lines of force, 195, 222
Lipid, 563
Liquids, 127
   electric currents in, 455
   kinetic theory of, 152
   nonpolar, 461
   polar, 461
   structure of, 398
Liter, 430
Lithification, 666
Lithosphere, 736
Lithosphere plates, 746
Loess, 661
Longitudinal waves, 165
Lorentz contraction, 96
Lowell, Percival (1855–1916), 786
Lucite, 555
Lyell, Charles (1797–1875), 718

Magma, 674
Magnet, permanent, 220
Magnetic domain, 220
Magnetic field, 221
   and atomic electrons, 352
   of current, 223
   of earth (see Earth, magnetic field of)
Magnetic force, 226
Magnetic induction, 224
Magnetic lines of force, 222
Magnetic pole, 219
Magnetic properties of iron, 220
   origin of, 376
Magnetic storms, 830
Magnetism, nature of, 224
Magnetization of ocean floors, 745, 750
Magnetohydrodynamics, 708
Main sequence of stars, 849
Mammals, 765, 768

INDEX

Man, early history of, 769
Mantle, 700
Marble, 602
Maria of moon, 809
Mariner spacecraft, 787
Mars, 785, 814
    canals of, 787
Marsh gas, 537
Mascon, 810
Mass, 24
    conservation of, 287
    and energy, 100
    formula, 428
    gravitational, 102
    inertial, 102
    relativity of, 25, 100
    and weight, 30
Mass spectrometer, 303
Matter:
    annihilation of, 335
    and energy, 100
Matter waves, 268
    in atom, 347
    probability density of, 271
    wave function of, 270
Mechanical equivalent of heat, 119
Melting, 116, 155
Mendeleev, Dmitri (1834–1907), 367
Mercury (planet), 784
Mercury, compounds of, 521
Meson, 330
Meson theory of nuclear forces, 330
Mesosphere, 614
Mesozoic era, 761
Metal atom, structure of, 376
Metallic bond, 405
Metallic luster, 362
Metallurgy, 526
Metals, 362
    actinide, 371
    alkali, 364
    electron gas in, 407
    order of activity of, 515
    rare-earth, 371
    valences of, 420
Metamorphic rocks, 596, 601
Metamorphism:
    dynamothermal, 686
    thermal, 679
Meteor, 781
Meteorite, 781
Meteoroid, 781
Meteorology, 616
Methane, 537
Methane molecule, 544

Methyl group, 551
Mev, 211
Mica, 588
Michaelson, Albert (1852–1931), 244
Microscope, electron, 269
Mid-Atlantic Ridge, 745, 748
Milky Way, 860, 876
Millibar, 132
Millikan, Robert A. (1868–1953), 202
Minerals, 584
    clay, 588
    ferromagnesian, 588
Moderator in nuclear reactor, 314
Mohorovicic discontinuity, 702
Molarity of solution, 430
Mole, 429
Molecular bond (*see* Bond, chemical)
Molecular crystal, 408
Molecular formula, 540
Molecular mass, 428
Molecular model, 544
Molecular orbital, 391
Molecular spectra, 543
Molecular speeds, 151
Molecule, 146, 288, 379, 381
    activated, 496
    polar, 390, 408
Momentum:
    angular, 83
        conservation of, 83, 85
    linear, 80
        conservation of, 81, 85
Monomer, 554
Monosaccharide, 562
Moon:
    craters of, 811
    density of, 804
    environment of surface, 807
    as falling body, 57
    landscape of, 809
    mass of, 801
    motions of, 796
    origin of, 813
    phases of, 796
    rocks of, 812
Moraine, 664
Motion:
    circular, 44
    first law of, 22
    relative character of, 4, 91
    second law of, 28
    third law of, 34
Motor, electric, 228
Motor effect, 227
Mountain building, 737

Muon (mu meson), 97, 331
Musical sounds, 178
Mutations and DNA, 570

n-type semiconductor, 413
Natural gas, 497
Neap tide, 648
Nebulae, 864
Neoprene, 555
Neptune, 791
    discovery of, 59
Neutral solution, pH of, 477
Neutralization:
    acid-base, 477
    in electrolysis, 456
    energy changes in, 493
Neutrino, 335
Neutron, 301
    decay of, 302
Neutron star, 855
Neutrons in cosmic rays, 871
Newton, Isaac (1642–1727), 43, 53, 695
    law of gravitation of, 55
    laws of motion of, 22, 28, 34
Newton, the, 29
Nitric acid, 473
Nitrocellulose, 500
Nitrogen, 439
    role in explosives, 498
Nitrogen base, 569
Nitrogen cycle, 440
Nitroglycerin, 500, 553
Nonmetal atom, structure of, 376
Nonmetals, 362
    order of activity of, 515
    valences of, 420, 422
Nonpolar liquid, 461
Normality of solution, 481
Nova, 854
Nuclear fission, 311
Nuclear forces, meson theory of, 330
Nuclear fusion, 315
Nuclear interaction:
    strong, 337
    weak, 337
Nuclear reactor, 312
Nucleic acids, 568
Nucleon, 302
Nucleotide, 569
Nucleus, 293
    binding energy of, 309
    energy levels of, 325
    stability of, 308, 325
    symbol for, 307

Nuclide, 307
Nylon, 556

Obsidian, 597
Ocean basins, 635
Ocean currents, 642
Ocean floor:
   magnetization of, 750
   properties of, 744
   spreading of, 746
Ocean waves, 640
Oceans:
   color of, 640
   composition of, 637
   temperatures in, 639
Oersted, Hans Christian (1777–1851), 222
Ohm, Georg Simon (1787–1854), 212
Ohm, the, 212
Ohm's law, 212
Olefin, 554
Olivine, 702
Optical properties of solids, 413
Orbital:
   atomic, 391
   molecular, 391
Orbital quantum number, 352
Ore, 526
Organic acids, 553
Organic chemistry, 535
Orlon, 556
Overtones, 179
Oxidation, 285, 438
   as loss of electrons, 513
Oxidation number, 519
Oxide, 285, 439
   acidic, 485
   basic, 485
Oxygen, 285, 437
   atmospheric, 525
   and photosynthesis, 503
Oxygen-carbon dioxide cycle, 504, 610
Ozone, 614

p-type semiconductor, 413
Pair production, 335
Paleozoic era, 758
Pangaea, 741
Paraffin hydrocarbons, 536
Parallax, stellar, 840
Particle, elementary, 333
Pauli, Wolfgang (1900–1958), 355
Pauling, Linus (1901–   ), 390
Pentane, isomers of, 541

Peptide bond, 566
Period:
   chemical, 370
   geologic, 730
Periodic law, 366
Periodic table of elements, 368
   origin of, 372
Permanent magnet, 220
Peroxides, 451
Petrified wood, 667
Petroleum, 497, 557
   origin of, 761
   refining of, 537
pH, 476
Phases of moon, 796
Phenol, 552
Phlogiston, 283
Phosphate group, 569
Photochemical reaction, 448
Photoelectric effect, 260
Photoelectron, 260
Photon, 261
   and gravity, 266
Photosphere, 825
Photosynthesis, 503
   in ocean, 637
Phytoplankton, 639
Pi (π) molecular orbital, 391
Pion (pi meson), 331
Pitch of sound waves, 168
Planck, Max (1857–1947), 261
Planck's constant, 261
Planetary motion, Kepler's laws of, 52
Planets, 776
   atmospheres of, 611
   properties of, 784
Plankton, 639
Plasma, 129
Plastic deformation of solid, 403
Plastics, 554
Pleistocene epoch, 769
Plexiglas, 555
Pluto, 791
   discovery of, 59
Pluton, 677
Plutonium, 315
Podzol soil, 604
Polar liquid, 461
Polar molecule, 390, 408
Polar wandering, 707
Pole, magnetic, 219
Polonium, 320
Polyethylene, 554
Polymer, 554
Polymerization, 538

Polypeptide, 567
Polysaccharide, 562
Polyvinyl chloride (PVC), 555
Populations, stellar, 863
   Cepheid variables in, 882
Porphyry, 599
Positron, 333
Potential difference, 209
Potential energy, 75
   and reference level, 77
Pound, 31
Power, 69
   electric, 213
   thermonuclear, 316
Powers of ten, 901
Precambrian time, 756
Precipitate, 465
Precipitation (rain and snow), 627
Pressure, 130
   atmospheric, 131, 613, 627
   vapor, 154
Prevailing westerlies, 621
Priestly, Joseph (1733–1804), 285
Primeval fireball, 887
Probability density:
   of atomic electron, 353
   of matter waves, 271
Prominence, solar, 831
Propane, 521
Proportionality, 7, 8
Proportions, law of definite, 281
Propyl group, 551
Protein, 440, 565
   denaturation of, 568
Protein molecule, 567
Protein synthesis and DNA, 570
Proton, 191
   discovery of, 301
Proton-proton cycle, 828
Proton synchrotron, 212
Protoplanets, 890
Protosun, 889
Proxima Centauri, 817, 840
Ptolemaic system, 48
Ptolemy (100–170 A.D.), 47
Pulsar, 856
Pumice, 676
Purine, 569
Pyrimidine, 569
Pythagoras (6th century B.C.), 692

Quantization:
   of angular momentum, 352
   of energy, 121, 348, 352

INDEX

Quantum of light, 261
Quantum mechanics, 350
Quantum numbers of atom, 345, 351
Quantum states of atom, 353
Quantum theory of light, 261
Quark, 337
Quartz, 444, 588
   structure of, 580
Quartzite, 602
Quasar, 884

Radio waves, 240
Radioactive dating, 727
Radioactive decay, 321
Radioactive series, 322
Radioactive tracer, 326
Radioactivity, 320
   in earth's interior, 703
Radiocarbon dating, 873
Radium, 320
Radon, 322
Rain, 621, 627
Rare-earth metals, 371
Rays:
   alpha, 321
   beta, 321
   gamma, 321
   light, 245
Reaction:
   chain, 312
   chemical (*see* Chemical reaction)
   displacement, 514
   endothermic, 492
   exothermic, 492
Reaction force, 34
Reactor, nuclear, 312
Red giant star, 849
Red shift in galactic spectra, 880
Reduction, 285, 438
   as gain of electrons, 513
Reference:
   frame of, 4, 89
   inertial, 91
Reflection, 169
   of light, 246
Refraction, 168
   of light, 248
Relative humidity, 620
Relativity:
   general theory of, 101
   and magnetism, 225
   of mass, 100
   of motion, 4, 91
   of space, 96

Relativity:
   special theory of, 91
   of time, 96
Resistance:
   air, 12
   electrical, 212
Resolution of vector, 38
Resonance, 179
Respiration, 562
Rest energy, 101
Resultant vector, 37
Reverberation, 174
Reversible chemical reaction, 504
Rhyolite, 597
Ribonucleic acid (RNA), 569, 571
Richter scale, 698
Rille, 812
Rip currents, 661
Roche limit, 790
Rock formations, dating of, 721
Rocket propulsion, 83
Rocks:
   classification of, 596
   extrusive, 598, 676
   igneous, 596
   intrusive, 598, 677
   lunar, 812
   metamorphic, 596, 601
   sedimentary, 596, 599, 666
   volcanic, 598, 676
Roentgen, Wilhelm (1845–1922), 264
Rømer, Olaf (1644–1710), 242
Rubber, 555
Ruby, synthetic, 399
Rumford, Count (1753–1814), 118
Rust, 523
Rutherford, Ernest (1871–1937), 292, 301
Rutherford atomic model, 292
Rydberg constant, 344

Salinity of sea water, 637
Salt, 480
San Andreas fault, 686
Sandstone, 600
Saran, 556
Satellite orbits, 58
Saturated air, 618
Saturated organic compound, 547
Saturated solution, 460
Saturation in chemical bonding, 420
Saturn, 789
Scalar quantity, 34
Schist, 602

Schrödinger, Erwin (1887–1961), 350
Scientific method, 60
Scintillation counter, 323
Sea-floor spreading, 746
Sea water:
   color of, 640
   composition of, 637
Sedimentary rocks, 596, 599, 666
Sedimentation, 663
   undersea, 665
Seismograph, 697
Semiconductor, 413
Shadow zone of earthquake waves, 701
Shale, 600
   metamorphism of, 687
Shell, atomic, 371
Shielding, electron, in atom, 376
Sialic rock, 599
Sigma (σ) molecular orbital, 391
Silica, 445
   structure of, 580
Silicates, 444
   chemistry of, 578
   structures of, 580
Siliceous rock, 599
Silicon, 444
   energy bands in, 412
Silicon carbide, 405, 444
Sill, 679
Simatic rock, 599
Simultaneity, 94
Sky, blue color of, 251
Slag, 528
Slate, 602
Slip in structure of solid, 403
Slug, 25, 31
Smith, William (1769–1839), 718
Snow, 621, 627
Soap, 475, 482
Sodium, 448
   energy bands in, 412
Sodium chloride:
   bonding in, 381, 386
   crystal structure of, 404
   electrolysis of, 456
   solution of, 462
Soil, 602
   weathering and, 655
Solar energy, 827
Solar flare, 807
Solar radiation, 617
Solar spectrum, 344
Solar system, 52, 776
   angular momentum of, 891
   models of, 47

Solar system:
  origin of, 889
  other, 891
Solar wind, 832
Solids, 126
  amorphous, 153, 155, 398
  crystalline, 153, 155, 397
  kinetic theory of, 153
  optical properties of, 413
Solubility, 460
  of salts, 480
Solute, 460
Solution, 280
  acid, 473
  alkaline, 475
  basic, 473
  of electrolyte, 465
  molarity of, 430
  normality of, 481
  pH of, 476
  saturated, 460
  supersaturated, 460
  unsaturated, 460
Solvent, 460
Sorting, degree of, 663
Sound waves, 166
  frequency of, 173
  interference of, 176
  loudness of, 177
  musical, 178
  pitch of, 168
  speed of, 171
  wavelength of, 173
Spark chamber, 339
Special relativity, 91
Specific gravity, 517
Specific heat, 114
Spectral classes of stars, 849
Spectral series, 344
Spectrometer, mass, 303
Spectroscope, 249, 342
  in astronomy, 820
Spectrum:
  absorption, 343, 821
  analysis of, 821
  continuous, 249, 342
  electromagnetic, 241
  emission, 342, 821
  of light, 249
  line, 342
  molecular, 543
  origin of, 349
  of sun, 344, 821
Speed, 5
  and acceleration, 15

Speed:
  of chemical reaction, 500
  constant, 6
  instantaneous, 8
  of light, 242
    constancy of, 93
  of sound, 171
  and velocity, 8
Spin, electron, 355
Spiral galaxy, 865
Spontaneous combustion, 525
Spring, 668
Spring tide, 647
Stahl, Georg (1660–1734), 283
Stalactite, 670
Standing waves, 170
Star:
  apparent brightness of, 841
  composition of, 822
  distance of, 840
  double, 845
  intrinsic brightness of, 841
  magnetic field of, 823
  main sequence, 849
  mass of, 845
  motion of, 844
  neutron, 855
  red giant, 849
  size of, 847
  temperature of, 822, 846
  variable, 842
  white dwarf, 267, 849, 851
Star cluster, 863
Starch, 562
State, change of, 115
Steady-state theory of universe, 886
Steam, 115
Steel, 404
Stellar evolution, 849, 852
Stellar populations, 863
Stellar spectra, 846, 848
Stoichiometry, 426
Stratification, 663
Stratosphere, 613
Stream erosion, 655
Structural formula, 538
Sublimation, 117
Subshell, atomic, 371
Sugar, 562
  pentose, 569
Sulfide, 445
Sulfur, 445
Sulfuric acid, 446, 473
Sun, 824
  atmosphere, 831

Sun:
  energy production in, 827
  magnetic fields in, 830
  origin of, 889
  spectrum of, 821
Sunspots, 829
Superconductivity, 187
Supernova, 855
Supersaturated solution, 460
Surface tension, 128
Symbol, chemical, 289
Symmetry and conservation principles, 85
Syncline, 685

Talc, 591
Telescope:
  diffraction as limit on magnification of, 252
  Hale, on Mt. Palomar, 253, 818
Temperature, 110, 149
  absolute, 138
  effect of, on chemical reactions, 500
  and heat, 111, 113
Terminal velocity, 12
Tesla, the, 224
Tethys Sea, 743
Thales of Miletus (c. 600 B.C.), 186
Theory, 61
Thermal metamorphism, 679
Thermocline, 639
Thermocouple, 229
Thermodynamics, laws of, 160
Thermometer, 111
Thermonuclear energy, 316
Thermosphere, 614
Thomson, J. J. (1856–1937), 200, 292
Thomson atomic model, 292
Tidal friction, 797
Tide:
  neap, 648
  origin of, 645
  spring, 647
Till, 664
Timbre, 168
Time dilation, 96
Tin:
  compounds of, 521
  refining of, 528
TNT, 500
Tractive force, 647
Trade winds, 621
Transformer, 232
Transistor, 414

INDEX

947

Transition elements, 371
Transparency, origin of, 413
Transverse waves, 165
Tritium, 306
Troposphere, 613
Tuff, 676
Turbidity current, 665
Twin paradox, 103

Ultrasonic waves, 173
Ultraviolet light, 241
    absorption in atmosphere, 614
Uncertainty principle, 271
    and atomic structure, 350
    and causality, 274
    and nature of moving body, 273
Unconformity, 719, 721
Uniform change, principle of, 719
Universal gas constant, 432
Universe:
    age of, 882
    expansion of, 880
    origin of, 885
Uranium, fission in, 311
Uranium series, 323
Uranus, 59, 790

Valence, 420, 519
    table of, 423
Van der Waals bonds, 408
    in clay, 591
Vapor pressure, 154
Vaporization:
    heat of, 116, 155
    table of, 117
Variable star, 842
    Cepheid, 843, 882
Vector, 35
    resultant, 37
Vector addition, 36
Vector component, 38
Vector quantity, 35
Vector resolution, 38
Vein, 670

Velocity, 8
    escape, 59
    terminal, 12
Venus, 784
"Vital force," 536
Vitreous substance, 398
Volatility of liquid, 153
Volcanic eruption, 675
Volcanic rocks, 598, 676
Volcano, 674
    undersea, 749
Volt, 210
    electron, 210
Volta, Alessandro (1745–1827), 210
Voltmeter, 228
Von Laue, Max (1879–1960), 264
Vulcanism, 673, 674
    problems of, 680

Washing soda, 475, 482
Water:
    changes of state in, 115
    dissociation of, 476
    electrolysis of, 437
    hard, 669
    heavy, 306
    hydrogen bonding in, 409
    sea, 637
    as solvent, 462
Water cycle of earth, 620
Water gas, 497
Water molecule, 392
    polar nature of, 461
    vibrations of, 543
Water table, 668
Water waves, 165
Watt, 70
Wave erosion, 661
Wave function, 270
Wave motion, 164
Wave and quantum theories of light, 263
Waveform, 168
Wavelength, 167
    de Broglie, 268
Waves, 164
    amplitude of, 168

Waves:
    earthquake, 698
    electromagnetic, 238
    light (see Light)
    longitudinal, 165
    matter (see Matter waves)
    ocean, 640
    radio, 240
    sound (see Sound waves)
    standing, 170
    transverse, 165
    ultrasonic, 173
    water, 164
Weather, 616
Weather map, 622
Weather system, 621
Weathering of rocks, 654
Weber per square meter, 224
Wegener, Alfred (1880–1930), 740
Weight, 23
    and mass, 30
Weight density, 133
Werner, Abraham G. (1750–1817), 716
White dwarf star, 267, 849, 851
Wilson, C. T. R. (1869–1959), 300
Wind, 621
    erosion by, 660
    and waves, 641
Wöhler, Friedrich (1800–1882), 536
Wood:
    as fuel, 498
    petrified, 667
Work, 68
Work function of metal, 262
Work hardening, 404

X-ray scattering, 399
X-rays, 264

Yukawa, Hideki (1907–   ), 329

Zeeman effect, 353
    in stellar spectra, 824
Zooplankton, 639

DATE LOANED			
JUN 18 1992			
GAYLORD 3563			PRINTED IN U.S.A.

108104

Q
160
.2
.K7
1971

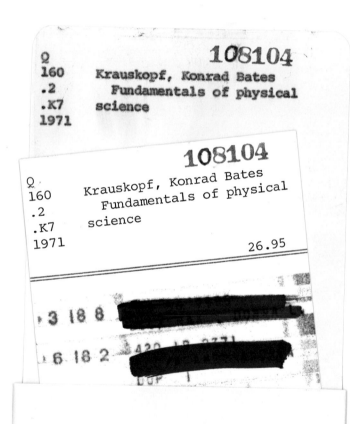

108104

Q
160   Krauskopf, Konrad Bates
.2   Fundamentals of physical
.K7   science
1971

108104

Q
160   Krauskopf, Konrad Bates
.2   Fundamentals of physical
.K7   science
1971   26.95

**LEARNING RESOURCES
BREVARD COMMUNITY COLLEGE
COCOA, FLORIDA**

CONVERSION FACTORS

1 meter (m) = 100 cm = 39.4 in. = 3.28 ft
1 centimeter (cm) = 10 millimeters (mm) = 0.394 in.
1 kilometer (km) = 1,000 m = 0.621 mi
1 foot (ft) = 12 in. = 0.305 m
1 inch (in.) = 0.0833 ft = 2.54 cm
1 mile (mi) = 5,280 ft = 1.61 km
1 day = 86,400 sec = 2.74×10^{-3} year
1 year = 3.16×10^7 sec = 365 days
1 m/sec = 3.28 ft/sec = 2.24 mi/hr = 3.60 km/hr
1 ft/sec = 0.305 m/sec = 0.682 mi/hr = 1.10 km/hr
(Note: 88 ft/sec = 60 mi/hr)
1 mi/hr = 1.47 ft/sec = 0.447 m/sec = 1.61 km/hr
1 kilogram (kg) = 1,000 grams (g) = 0.0685 slug
(Note: 1 kg corresponds to 2.21 lb in the sense that
the weight of 1 kg is 2.21 lb.)
1 slug = 14.6 kg
(Note: 1 slug corresponds to 32.2 lb in the sense that
the weight of 1 slug is 32.2 lb.)
1 mass unit (u) = 1.66×10^{-27} kg
        = 1.49×10^{-10} joule
        = 931 Mev

1 newton (n) = 0.225 lb
1 pound (lb) = 4.45 n
1 joule (j) = 0.738 ft-lb = 2.39×10^{-4} kcal
        = 6.24×10^{18} ev
1 kilocalorie (kcal) = 4,185 j = 3,089 ft-lb
1 foot-pound (ft-lb) = 1.36 j = 3.25×10^{-4} kcal
1 electron volt (ev) = 10^{-6} Mev = 10^{-9} Gev
        = 1.60×10^{-19} J
        = 1.18×10^{-19} ft-lb = 3.83×10^{-23} kcal
1 watt = 1 j/sec = 0.73 ft-lb/sec = 1.34×10^{-3} hp
1 kilowatt (kw) = 1,000 watts = 738 ft-lb/sec = 1.34 hp
1 horsepower (hp) = 550 ft-lb/sec = 746 watts
1 atmosphere of pressure (atm) = 1.013×10^5 n/m²
        = 14.7 lb/in.²

$$°C = \frac{5}{9}(°F - 32°)$$

$$°F = \frac{9}{5}°C + 32°$$

$$°K = °C + 273°$$

PHYSICAL AND CHEMICAL CONSTANTS

Speed of light in vacuum	c	3.00×10^8 m/sec
Charge on electron	e	1.60×10^{-19} coul
Gravitational constant	G	6.67×10^{-11} newton–m²/kg²
Acceleration of gravity at earth's surface	g	9.81 m/sec² 32.2 ft/sec²
Planck's constant	h	6.63×10^{-34} joule–sec
Boltzmann's constant	k	1.38×10^{-23} joule/°K
Electrostatic constant	K	8.99×10^9 newton–m²/coul²
Electron rest mass	m_e	9.11×10^{-31} kg
Neutron rest mass	m_n	1.675×10^{-27} kg
Proton rest mass	m_p	1.673×10^{-27} kg
Avogadro's number	N_0	6.02×10^{23} formula units/mole (or molecules/mole or atoms/gram-atom)
Universal gas constant	R	8.31×10^3 joules/mole–°K
Volume per mole of gas at 0°C and atmospheric pressure		22.4 liters/mole